Modeling of Wind Turbines and Wind Farms

Modeling of Wind Turbines and Wind Farms

Special Issue Editors

Emilio Gomez-Lazaro
Estefania Artigao

MDPI • Basel • Beijing • Wuhan • Barcelona • Belgrade

Special Issue Editors

Emilio Gomez-Lazaro
Universidad de Castilla-La Mancha
Escuela Técnica Superior de Ingenieros
Industriales de Albacete
Spain

Estefania Artigao
Universidad de Castilla-La Mancha
Escuela Técnica Superior de Ingenieros
Industriales de Albacete
Spain

Editorial Office
MDPI
St. Alban-Anlage 66
4052 Basel, Switzerland

This is a reprint of articles from the Special Issue published online in the open access journal *Energies* (ISSN 1996-1073) from 2018 to 2019 (available at: https://www.mdpi.com/journal/energies/special_issues/wind_turbines_wind_farms)

For citation purposes, cite each article independently as indicated on the article page online and as indicated below:

LastName, A.A.; LastName, B.B.; LastName, C.C. Article Title. *Journal Name* **Year**, *Article Number*, Page Range.

ISBN 978-3-03928-756-7 (Hbk)
ISBN 978-3-03928-757-4 (PDF)

Contents

About the Special Issue Editors

Emilio Gomez-Lazaro is a Professor at the University of Castilla-La Mancha (UCLM), Spain (MS.c. and Ph.D. from the Polytechnic University of Valencia, Spain). He has been working at UCLM for more than 15 years, where he holds the position of Director of the Renewable Energy Research Institute. Previously, he was a researcher and a member of the teaching staff at other Spanish universities: The Polytechnic University of Valencia, the Jaume I University in Castellón and the Polytechnic University of Cartagena. As part of his research, he has been involved in more than 100 R&D projects and contracts for public and private institutions, for the most part as the leading researcher. In addition, he has published over 80 papers in ISI indexed peer-reviewed journals and over 200 contributions in books, book chapters, and specialized conferences, and is the co-inventor of 9 national and international patents. Dr. Goméz Lázaro is a member of different IEEE Societies; a member of CIGRE, the Spanish representative on the IEA Wind Annex Task 25 on Design and Operation of Power Systems with Large Amounts of Wind Power, and participates as an expert representing Spain in different international working groups for the International Electrotechnical Commission (IEC). He has been active in the European Wind Energy Platform, TPWIND, as well as in Spanish Wind Energy Platform, REOLTEC. He is a senior member of the IEEE and has worked as an editor and guest editor for various journals related to energy research.

Estefania Artigao (Ph.D.) has been working on the Condition Monitoring of Wind Turbines for over 8 years. She is currently a Research Fellow and an Associate Lecturer at the University of Castilla—La Mancha (UCLM), Spain. She studied Industrial Engineering (BS.c. and MS.c.) at UCLM. During her studies, she was granted an Erasmus Scholarship to study at the University of Modena and Reggio Emilia (Italy) and a Leonardo Scholarship at the University of Birmingham (United Kingdom). After her studies, she moved to Cambridge (UK) where she started her career as a researcher in the Wind Industry. She spent 2 years at TWI Ltd followed by 4 years at Brunel Innovation Centre, both companies based in Cambridge (UK). In July 2015, she moved back to Spain to join the Marie Curie Program under the H2020 AWESOME Project as an Early Stage Researcher where she completed a four-month Industrial Secondment at Ingeteam Service (Albacete, 2016) and a three-month Academic Secondment at the University of Strathclyde (Glasgow, United Kingdom, 2018). In November 2018, she obtained the Cum Laude Ph.D. Degree with European mention, entitled "Current Signature Analysis to Diagnose Doubly-Fed Induction Generators of In-Service Wind Turbines". Her field of research is advanced diagnosis of wind turbines, focusing on electrical analysis.

Preface to "Modeling of Wind Turbines and Wind Farms"

Wind turbine technology has matured over the years and is considered a reliable renewable energy technology. Furthermore, wind energy is a research field characterized by a high degree of interdisciplinarity, given the wide range of technical fields involved, such as aerodynamics, mechanics, meteorology, resource assessment, as well as electrical engineering addressing the generation, transmission, and the integration of wind power plants into the power systems.

The modeling of wind power plants (WPP), wind turbines (WT), and WT components is currently gaining key importance during the design stage, but also over their entire lifetime with respect to operation and maintenance. This book presents solutions to all these challenges, including the development, validation, and implementation of WT components, WTs, and WPP models in applications related to aerodynamics, mechanics, resource assessment, or wind power integration.

Half of the contributions deal with wind integration into power systems, analyzing and forecasting the effects on grid stability and reliability. Different power system agents, such as transmission system operators (TSOs) and distribution system operators (DSOs), are currently engaged in addressing transient analyses, WT frequency provision, WT reactive power capability, or new control strategies to deal with these issues. Assessment and validation of such models is also a major issue due to the importance and difficulty of collecting real data.

The remaining contributions deal with the other technical disciplines mentioned when talking about the interdisciplinarity of wind energy. Works are presented on modeling WT mechanical dynamics, aerodynamics and aeroelasticity, WT blades, WT failures and maintenance, WT lightning protection, wind energy prediction and forecasting, and wind farm (WF) design.

The technical contents will be of great help to researchers as well as practicing engineers in the wind and power industry. The contributions offer a broad view of the relevant, diversified and challenging problems involved in wind turbine technology modeling.

Finally, we would like to thank all the authors for the 25 contributions and the reviewers for the care taken in preparing and assessing the contents. Moreover, we acknowledge the support of the Spanish Ministry of Economy and Competitiveness and the European Union—FEDER Funds, ENE2016-78214-C2-1-R-.

Emilio Gomez-Lazaro, Estefania Artigao
Special Issue Editors

Article

Comparative Analysis of Bearing Current in Wind Turbine Generators

Ruifang Liu [1,*], Xin Ma [1], Xuejiao Ren [1], Junci Cao [1] and Shuangxia Niu [2]

[1] School of Electrical Engineering, Beijing Jiaotong University, Haidian District, Beijing 100044, China; 16126038@bjtu.edu.cn (X.M.); 16121509@bjtu.edu.cn (X.R.); jccao@bjtu.edu.cn (J.C.)

[2] Department of Electrical Engineering, The Hong Kong Polytechnic University, Hong Kong, China; shuangxia.niu@polyu.edu.hk

* Correspondence: rfliu@bjtu.edu.cn; Tel.: +86-10-51684165

Received: 3 April 2018; Accepted: 16 May 2018; Published: 20 May 2018

Abstract: Bearing current problems frequently appear in wind turbine systems, which cause wind turbines the break down and result in very large losses. This paper investigates and compares bearing current problems in three kinds of wind turbine generators, namely doubly-fed induction generator (DFIG), direct-drive permanent magnet synchronous generator (PMSG), and semi-direct-drive PMSG turbines. Common mode voltage (CMV) of converters is introduced firstly. Then stray capacitances of three kinds of generators are calculated and compared through the finite element method. The bearing current equivalent circuits are proposed and simulations of the bearing current are carried out. It is verified that the bearing currents of DFIGs are more serious than the two kinds of PMSG, while common mode current (CMC) of the direct-drive PMSG is much greater than the other two types of wind turbine generators.

Keywords: bearing current; common mode current; doubly fed induction generators; permanent magnet synchronous generators; wind turbine generator

---

1. Introduction

Wind power has become the fastest growing clean energy due to its advantages of being clean, their short construction period, and low operation cost [1,2] among all the renewable energy power generation technologies. At present, megawatt-scale technology has good prospects in wind turbine applications. Doubly fed induction generators (DFIGs) and permanent magnet synchronous generators (PMSGs) are the most widely used in wind power generation. DFIGs adopt rotor converter power supplies, in which the converter only needs 30% of the rated power and the whole cost of the system is greatly reduced. PMSGs require a full power converter, but with no-excitation winding, the operating efficiency is higher. PMSGs can be classified into direct-drive generators and semi-direct-drive generators. Since there is no gearbox in direct-drive PMSGs, the cost for drive parts is saved. PMSGs have low rated speed, multiple poles, and permanent magnets that are usually mounted on the rotor surface. Semi-direct-drive PMSGs need an acceleration gear box. With higher speed and fewer poles, a built-in permanent magnetic structure is usually used in semi-direct-drive PMSGs. Additionally, the size of semi-direct-drive PMSGs is smaller and the installation and transportation of direct-drive PMSGs are more convenient than direct-drive PMSGs.

Whether it is a DFIG or PMSG, the generator power supply current always goes through the converter to the power grid. The switching device of the converter will produce a high-frequency common-mode voltage, which couples with the generator stray capacitance and induces a bearing voltage between the outer raceway and inner raceway of the bearing. If the bearing voltage exceeds the threshold voltage of the bearing lubricating oil film, which is located between the bearing ball and the raceway, the oil film will break down. This would result in the discharge of the bearing current and lead to bearing premature failure. Kevin Alewine, an expert at Shermco Industry in the United States, counted failure types of approximately 1200 wind turbines maintained by the company between 2005 and 2010. The failure rate of the bearings with power from 1 MW to 2 MW is 70% [1], as shown in Figure 1. No matters on land or sea, wind turbines are installed on tall towers, and the maintenance and overhauling caused by bearing failures are complicated and expensive. Therefore, it is of great value to investigate the mechanism of bearing current and investigate mitigation methods.

Aiming to solve the problem of bearing current in a variable frequency AC motor, Chen and Busse [3–5] pointed out that the high frequency common-mode voltage of the inverter and the stray capacitances of motors are the critical reasons of the bearing currents. Muetze [6–9] focused on analyzing the characteristics of Electric discharge machining (EDM) current and circulating bearing current, and put forward the relevant suppression methods [10–13]. With the development of wind power generation technologies, the bearing current problems on wind generators appear. According to [1], the proportion of the bearing electric erosion of the wind turbines is higher than that of the industrial motors, as shown in Figure 2. Zitzelsberger established the DFIG bearing current analysis model [14]. Adabi et al. studied the stray capacitance parameters of the bearing current model, and gave the relevant analytical formulas. At the same time, they analyzed the machine design's influence on the stray capacitance factors [15,16]. The bearing current problem of DFIGs received a significant amount of attention. However, bearing current problems existing in PMSG systems caused by using converters still have not been clearly investigated before.

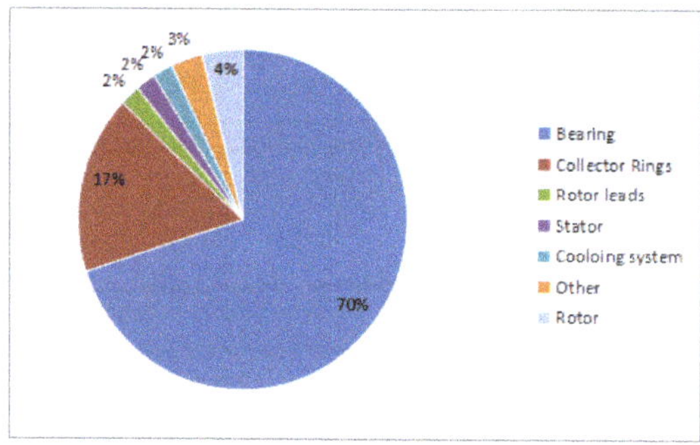

Figure 1. Failure types and occurrences proportion for generators of 1 to 2 MW [1].

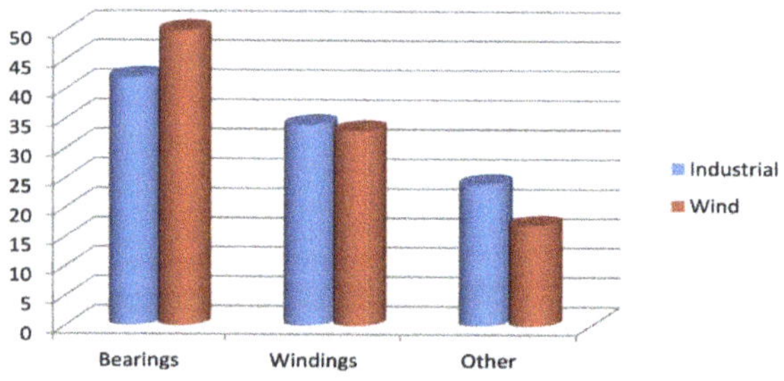

Figure 2. Comparison of wind turbine failures with industrial machine failures. Distribution (%) of the failure types [1].

In this paper, the bearing currents of DFIGs, direct-drive PMSGs, and semi-direct-drive PMSGs are studied and compared. Firstly, the common mode voltage of converter is introduced. Then the calculation method on the stray capacitances of three kinds of generators is put forward and the bearing current equivalent models are built. Finally, the simulation of the converter-wind turbine generator system is carried out and a comparative study of bearing currents of three kinds of generators is carried out.

2. Common Mode Voltage Generation

A two-level voltage converter is often applied to achieve DC/AC conversion in wind generation systems, as shown in Figure 3, where U_d is the DC link voltage. Switches VT1–VT6 are turned on/off by Pulse Width Modulation (PMW) waves, and the Digital Signal Processing (DSP) microcontroller are used to realize the SPWM strategy.

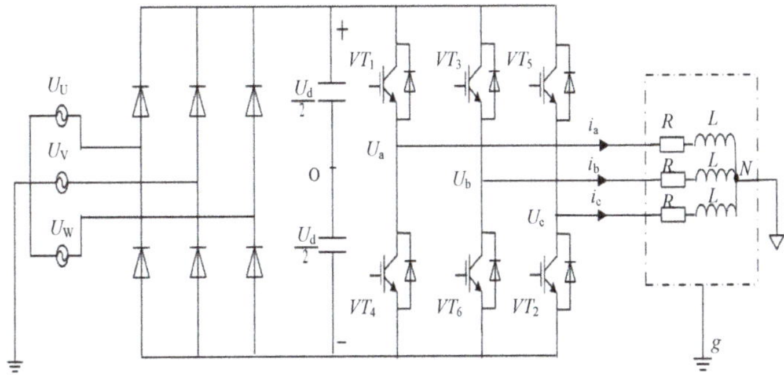

Figure 3. The converter–generator system.

In Figure 3, the voltage equations are:

$$\begin{cases} U_a = Ri_a + L\frac{di_a}{dt} + U_{Ng} \\ U_b = Ri_b + L\frac{di_b}{dt} + U_{Ng} \\ U_c = Ri_c + L\frac{di_c}{dt} + U_{Ng} \end{cases} \tag{1}$$

where U_a, U_b, and U_c are the output three-phase voltage; U_{Ng} is the voltage between the midpoint of the three-phase stator winding and the ground; i_a, i_b, and i_c are the output currents flowing through the generator winding; R and L are the winding resistance and leakage inductance of each phase of the generator.

From Equation (1), the following equations can be derived:

$$U_a + U_b + U_c = R(i_a + i_b + i_c) + L\frac{d(i_a + i_b + i_c)}{dt} + 3U_{Ng} \tag{2}$$

$$U_{Ng} = V_{com} \tag{3}$$

where V_{com} is the common-mode voltage, which is the same as U_{Ng}.

In wind turbines, if the stator three-phase windings are perfectly symmetrical, the sum of i_a, i_b, and i_c is approximately 0, and Equation (2) is rewritten as,

$$V_{com} = \frac{U_a + U_b + U_c}{3} \tag{4}$$

In the system power of a symmetrical three-phase sinusoidal alternating current, the sum of the three-phase AC voltage is zero, which means the common-mode voltage is zero. However, in the inverter power system with SPWM control, the sum of the three-phase voltage is not zero. In order to prevent bridge run-through, the up and down switches of the same legs cannot be turned on at the same time. If '1' indicates the up-side switching on, and '0' indicates the down-side switching on, there are eight switch state combinations in total. If the mid-point of the two electrolytic capacitors in the converter DC link are taken as the system reference point, the common mode voltage and the state of the switch can be expressed in Table 1 or given by Equation (5),

Table 1. Common mode voltages under different switching states.

	S_0	S_1	S_2	S_3	S_4	S_5	S_6	S_7
switching state	000	001	011	010	110	100	101	111
V_{com}	$-\frac{U_d}{2}$	$-\frac{U_d}{6}$	$\frac{U_d}{6}$	$\frac{U_d}{6}$	$\frac{U_d}{6}$	$-\frac{U_d}{6}$	$\frac{U_d}{6}$	$\frac{U_d}{2}$

$$V_{com} = \begin{cases} \pm\frac{U_d}{2} & states\ S_0, S_7 \\ \pm\frac{U_d}{6} & other\ states \end{cases} \tag{5}$$

The common-mode voltage waveform of two-level inverters is shown in Figure 4, which is a staircase wave, and consists of two levels, $\pm U_d/6$ and $\pm U_d/2$.

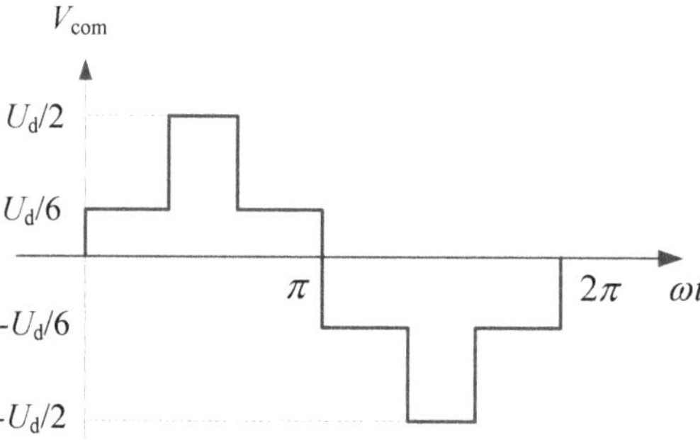

Figure 4. Common mode voltage generated in the converter.

3. Bearing Current Analysis of Doubly-Fed Induction Generator (DFIG)

In DFIGs, the common mode voltage of the converter is applied to the rotor winding. The stray capacitances of the generator constitute the coupling paths of the common mode voltage to the bearing. In order to analyze the bearing currents of DFIG, the stray capacitances of the generator should be acquired. The calculation accuracy of the capacitances would affect the prediction accuracy of bearing currents.

3.1. Stray Capacitances of Doubly-Fed Induction Generator (DFIG)

In DFIGs, there are four parts of the conductor, namely, the stator winding, rotor winding, rotor core, and stator core. The stator core connects to the frame and the rotor core connects to the shaft. Then, the stator core and frame are at the same electric potential, and the rotor core and shaft are also at the same potential. According to the partial capacitance theory of multi-conductor systems, the dielectric materials between the two conductors can induce stray capacitance. Therefore, there are three stray capacitances—C_{rwf}, C_{rwr}, and C_{rf}—in DFIGs, where C_{rwf} is the capacitance between the rotor winding and the frame, C_{rwr} is the capacitance between the rotor winding and the rotor core, and C_{rf} is the capacitance between the rotor core and the frame. The stator winding has no influence on other parts. The specific capacitance distribution in the DFIG is shown in Figure 5.

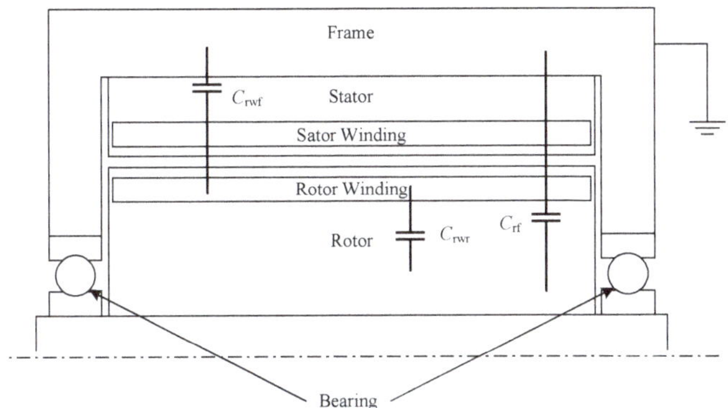

Figure 5. Stray capacitances of the doubly-fed induction generator (DFIG).

Due to the complex structure of the generator, it is not easy to obtain an accurate result from the analytical method, as in [17]. Since the materials and the shapes of different parts are not as simple as the plate capacitor, the assumption used in the analytical method would cause errors. In order to obtain the accurate stray capacitances of the DFIG, the electromagnetic field numerical calculation based on the finite element method (FEM) is adopted in this paper.

Ignoring the generator end effect, a 2D model can be built in ANSYS Maxwell (Version, 16.0,ANSYS, Pittsburgh, PA, USA) based on the generator structure parameters. The electrostatic field solver is adopted. The boundary problem of electric potential φ satisfies following Poisson equation, where x, y are coordinates in space:

$$\frac{\partial^2 \varphi}{\partial x^2} + \frac{\partial^2 \varphi}{\partial y^2} = 0 \tag{6}$$

The stator core is taken as the reference; the rotor winding and rotor core are taken as independent conductors applying different voltages. We set the matrix parameter solving item, then set the proper mesh subdivision of the solution domain and perform the numerical calculation. The electrostatic induction coefficient matrix $[\beta]$ can be obtained, which contains following elements:

$$[\beta] = \begin{bmatrix} \beta_{rw} & \beta_{rw\_r} \\ \beta_{r\_wr} & \beta_r \end{bmatrix} \tag{7}$$

where the subscript rw indicates the rotor winding and r indicates the rotor core. β_{rw} is the induction coefficient between rotor winding and reference conductor, β_r is the induction coefficient between the rotor and reference conductor, β_{rw_r} is the induction coefficient between the rotor winding and rotor. According to the theory of the partial capacitance of multi-conductors, the stray capacitances in the DFIG can be deduced from coefficients of electrostatic induction:

$$\begin{cases} C_{rwf} = \beta_{rw} + \beta_{rw\_r} \\ C_{rwr} = -\beta_{rw\_r} \\ C_{rf} = \beta_r + \beta_{r\_rw} \end{cases} \tag{8}$$

To acquire the actual results, the above values should multiply the generator's effective axial length.

The stray capacitances of a 1.5 MW DFIG are calculated through the above method. The generator model is shown in Figure 6, where the insulation material in the generator is accurately described. The parameter of the DFIG is shown in Table 2.

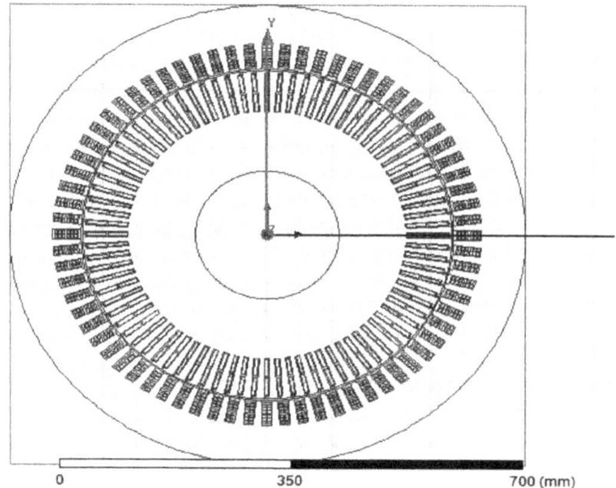

Figure 6. 2D model of a 1.5 MW DFIG.

Table 2. The 1.5 MW doubly-fed induction generator (DFIG) main parameters.

Parameters	Symbol	Value
Stator slot number	N_s	72
Rotor slot number	N_r	84
Length of iron core	L/mm	652
Air gap	δ/mm	2.5
Stator inner radius	R_s/mm	285
Rotor slot wedge thickness	d_{r2}/mm	4.5
Stator slot wedge thickness	d_{s1}/mm	4.5
Rotor slot insulation thickness	d_{rw1}/mm	1.35
Stator slot insulation thickness	d_{sw1}/mm	0.45
Rotor slot width	b_{r1}/mm	3
Stator slot width	b_{s1}/mm	15.2
Relative permittivity of slot wedge	ε_{r1}	3
Relative permittivity of slot insulation	ε_{r2}	3.4

With the finite element numerical calculation, the stray capacitances of the DFIG are obtained as shown in Table 3.

Table 3. Stray capacitances of a 1.5 MW DFIG.

Capacitance/nF	C_{rwf}	C_{rwr}	C_{rf}
DFIG	0.027	152.3	3.3

Among three capacitances, C_{rwr} is much greater than the two other capacitances. This is because C_{rwr} is the capacitance between the rotor core and rotor windings. Comparing the two other capacitances, the insulation distance is shorter, the conductor surface is greater, and the permittivity is larger. All of these factors cause C_{rwr} to be much greater than the two others. C_{rwr} is a critical capacitance in the bearing current problem of DFIGs.

3.2. Bearing Current Model of the DFIG

The rotor windings of the DFIG are connected to the converter. The common mode voltage of the converter exists at the neutral point of the rotor windings and the ground. Coupled by the generator stray capacitances, common mode voltage induces the voltage on the rotor shaft. Common mode current will return to the convertor. Assuming the frame of the generator is well grounded, the equivalent common mode circuit of the DFIG is shown in Figure 7.

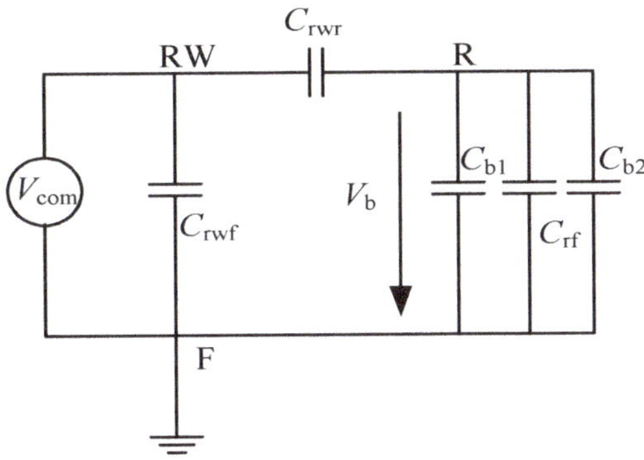

Figure 7. Bearing current model of the DFIG.

In Figure 7, V_{com} is the common-mode voltage of the converter; RW denotes rotor windings; R denotes the rotor core; and F denotes the frame. The bearing inner raceway connects the shaft, namely, the bearing inner raceway is at the same potential as the rotor core. The bearing outer raceway connects the stator end cover, namely, the outer raceway is at the same potential as the stator core and frame. Balls of the bearing separate the inner raceway and outer raceway; lubricating grease exists on the balls and raceways. When the bearing oil film has integrity, the bearing can be taken as a capacitance. C_{b1} and C_{b2} in Figure 7 denote the equivalent capacitance of the drive end and non-driven end bearing, respectively.

Supposing the generator stator frame is well grounded, the high-frequency V_{com} acts on the generator rotor winding. Coupled with stray capacitances, inductive voltage V_b would appear between the rotor and frame. V_b acts simultaneously between the inner and outer bearing raceways. When the electric filed intensity of the oil surpasses the breakdown intensity, it will lead to the oil film breaking down and produce a discharge current.

The bearing voltage ratio (BVR) can be defined as the ratio of the bearing voltage V_b to the rotor winding common-mode voltage V_{com}, which is an indicator of bearing damage. The BVR of the DFIG is shown in Equation (9):

$$\text{BVR} = \frac{V_b}{V_{com}} = \frac{C_{rwr}}{C_{rwr} + C_{rf} + C_{b1} + C_{b2}} \tag{9}$$

The bearing capacitance is much smaller than capacitances C_{rwr} and C_{rf}, and it usually changes with the temperature, speed, and load. If the influence of C_b on the BVR is small, then C_{b1} and C_{b2} can be treated as $C_{b1} = C_{b2} = 0$. Using the capacitance in Table 2, the bearing voltage ratio of this DFIG is 97.9%, which is much greater than the BVR of induction motors supplied by a PWM inverter, which is usually less than 10% [6]. The significant difference of the BVR between the two kinds of machines is because the common mode voltage comes from different sides of the electrical machine. In the DFIG, the rotor winding is connected to the converter, and in induction motors powered by an inverter, the stator winding is connected to the inverter. In the latter condition coupling is on the capacitance of the stator winding to the rotor, which is very small because the two parts are separated by an airgap. Such a high BVR of the DFIG indicates that the bearing voltage of the DFIG is dangerous and harmful to the bearings.

4. Bearing Current Analysis of Permanent Magnet Synchronous Wind Turbine

Permanent magnet synchronous wind generation systems can be divided into direct-drive and semi-direct-drive types. Direct-drive PMSGs have multiple poles and large diameters, in which the permanent magnets are usually surface mounted installations, as shown in Figure 8a. The permanent magnet in the inner rotor PMSG is directly adhered to the outer surface of the rotor core by a specific adhesive. Semi-direct-drive PMSGs are high-speed and with fewer poles. Permanent magnets are usually built-in installations, as shown in Figure 8b. The permanent magnets are placed in pre-opened slots in the designated area of the rotor core.

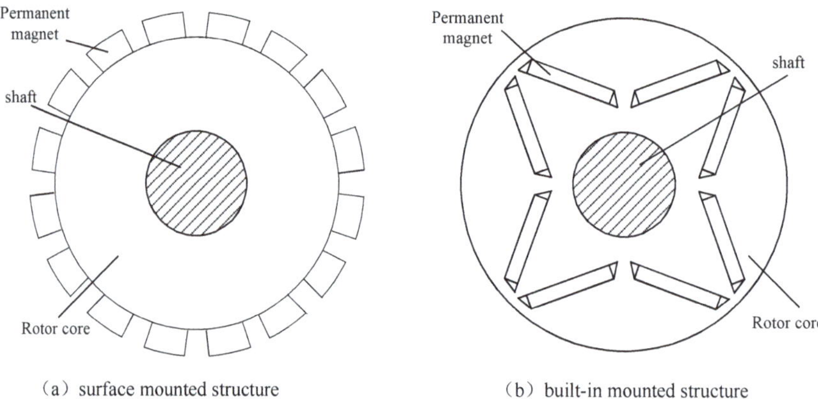

(a) surface mounted structure　　　　　　(b) built-in mounted structure

Figure 8. Permanent magnet installation in a direct-drive permanent magnet synchronous generator (PMSG) (**a**) and semi-direct-drive PMSG (**b**).

4.1. Stray Capacitances of Permanent Magnet Synchronous Generators (PMSGs)

There are four conductor parts in the direct-drive PMSG, which are, respectively, the stator winding, stator core, rotor core, and permanent magnet. The stator windings and the stator core are isolated by the insulating layer. The stator core and the rotor core are separated by air gap. The stator core and the frame are electrically connected. The rotor core and the rotor shaft are connected. The rotor core and the permanent magnet can be regarded as the same potential conductors. Hence, there are three stray capacitances in the generator. Respectively, C_{wf} is the capacitance between the stator windings and the stator, C_{wr} is the capacitance between the stator windings and the rotor and C_{rf} is the capacitance between the rotor core and stator core.

Similar to the direct-drive PMSG, the same four conductor parts also exist in the semi-direct-drive PMSG, which uses a built-in permanent magnet rotor. The rotor core and the permanent magnet can be treated as the same potential conductors, and three stray capacitances C_{wf}, C_{wr}, and C_{rf} exist in the semi-direct-drive PMSGs.

The stray capacitances of one 2 MW direct-drive and one 2 MW semi-direct-drive PMSG are calculated through the electromagnetic field numerical method, respectively. The generator models are shown in Figures 9 and 10. The parameters of the two PMSGs are shown in Table 4.

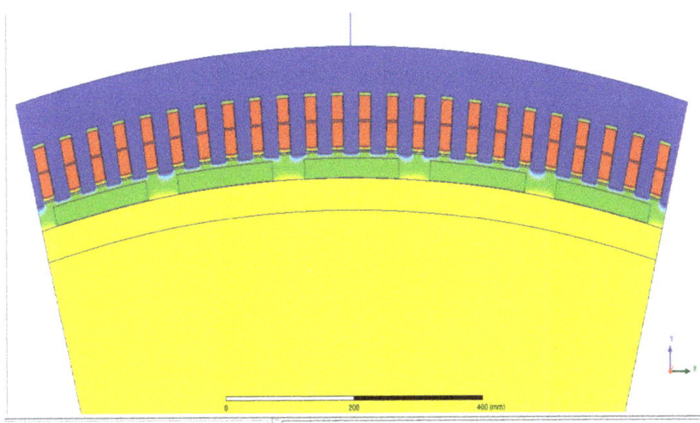

Figure 9. Two megawatt direct-drive permanent magnet synchronous generator (PMSG) model.

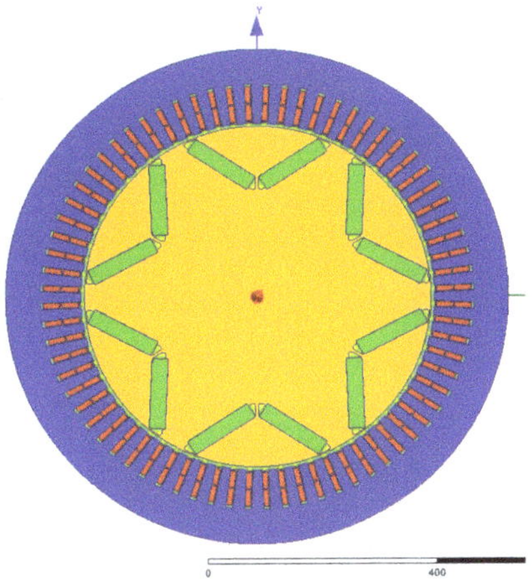

Figure 10. Two megawatt semi-direct-drive PMSG model.

Table 4. Parameters of the 2 MW direct-drive permanent magnet synchronous generator (PMSG) and 2 MW semi-direct-drive PMSG.

Parameters	Symbol	Direct-Drive PMSG	Semi Direct-Drive PMSG
Rated voltage	U_N/V	660	690
Rated frequency	f_N/Hz	8.5	70
Stator slot number	N_s	288	72
Number of pole pairs	P	30	3
Permanent magnet width	W_m/mm	148.4	126
Permanent magnet thickness	h_m/mm	24	28
Iron core length	L_s/mm	1500	680
Air gap	δ/mm	6	6
Rotor outside radius	R_r/mm	1890	306
Stator slot insulation thickness	d_{sw1}/mm	0.5	0.5
Stator slot wedge thickness	d_{s2}/mm	2	2
Stator slot width	b_s/mm	18	14
Relative permittivity of slot wedge	ε_{r1}	3	3
Relative permittivity of slot insulation	ε_{r2}	3.4	3.4

Stray capacitance calculation results of the two generators are shown in Table 5.

Table 5. PMSG stray capacitance calculation results.

Capacitance/nF	C_{wf}	C_{wr}	C_{rf}
Direct-drive PMSG	1607.4	3.32	16.39
Semi Direct-drive PMSG	182.4	0.34	1.42

The difference between these two kinds of generators can be seen from Table 5. The capacitances of the direct-drive PMSG are greater than those of semi-direct-drive PMSG. C_{wf} in the direct-drive PMSG is about nine times of that in the semi-direct-drive PMSG, which is because the slot number of the direct-drive PMSG is much greater than the semi-direct-drive PMSG. Capacitance C_{wf} is proportional to the slot number.

4.2. Bearing Current Model of PMSG

When common-mode voltage of converter exists between stator winding neutral point and ground, the bearing current equivalent model of the PMSG can be obtained, as shown in Figure 11. Both direct-drive and semi-direct-drive PMSGs can adopt this model.

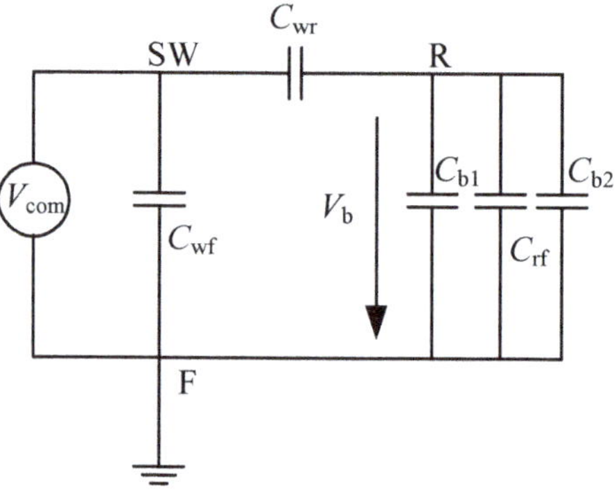

Figure 11. Bearing current equivalent model of the PMSG.

According Figure 11, the BVR of permanent magnet synchronous wind generator is:

$$\text{BVR} = \frac{V_b}{V_{com}} = \frac{C_{wr}}{C_{wr} + C_{rf} + C_{b1} + C_{b2}} \tag{10}$$

Using the capacitance results in Table 4 and neglecting the bearing capacitance, BVR of the direct-drive PMSG and semi-direct-drive PMSG can be acquired. The BVR comparison of three kinds of wind turbine generators is shown in Table 6.

Table 6. BVR comparison of three kinds of wind turbine generators.

	DFIG	Direct-Drive PMSG	Semi Direct-Drive PMSG
BVR	97.9%	16.8%	19.3%

BVRs of two PMSGs are less than 20%, which is much smaller than DFIG. This is because the converter connects to stator windings of PMSGs, and the coupling between the stator and rotor is very weak. Then, the bearing voltage induced by common mode voltage of the converter is lower than that in DFIGs. However, the BVR of the PMSG is greater than that of the induction motor powered by an inverter, which is because the power level and the size of the PMSG is greater than ordinary induction motors. From Table 5 we can see the bearing failure rate of PMSGs will be lower than the DFIG, which is one of the reasons for the wide application of PMSGs in wind power generation systems.

5. Bearing Current Simulations of Wind Turbine Generators

Taking the DFIG system as an example, the converter output is combined with the stray capacitance network of the DFIG to obtain the bearing current simulation model. The structure is shown in Figure 12.

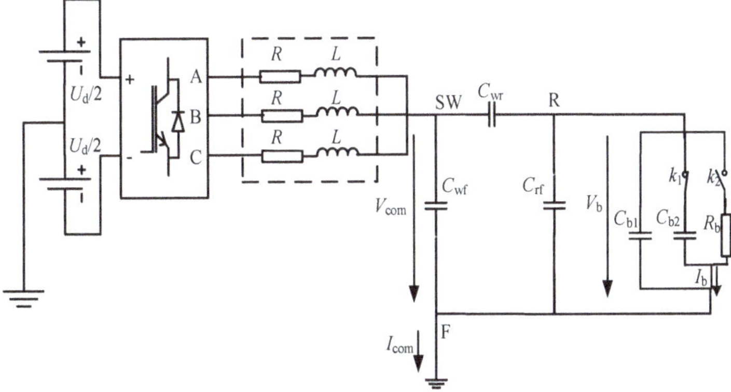

Figure 12. Analysis model of converter—DFIG system.

In Figure 12, the converter is connected to the rotor winding of the DFIG, and a high-frequency common-mode voltage V_{com} exists between the rotor windings and the ground to form a common mode loop through the stray capacitance of the generator. Common mode current I_{com} flows from the stator frame to ground. U_d is the DC bus voltage; resistance and leakage inductance of each phase of the stator windings are represented by R and L. Assuming that the generator frame is well grounded, the ground impedance is ignored. When large breakdown currents flow through the bearing, the bearing is equivalent to a resistance [18]. Equivalent switches k_1 and k_2 are used to simulate the breakdown process of the oil film. Before the oil film breaks down, switch k_1 closes and k_2 opens; the bearing acts as the equivalent of a capacitor. After the film breaks down, the switch k_1 opens and k_2 closes, the bearing acts as the equivalent of a breakdown resistance R_b.

The simulation model is setup with MATLAB/Simulink (R2014a, Mathworks, Natic, MA, USA. The capacitances of the 1.5 MW DFIG in Table 3 are adopted. Assuming $C_{b1} = C_{b2} = 120$ pF, and the bearing equivalent resistance $R_b = 10$ Ω, the threshold voltage of the bearing equivalent capacitor breakdown is ±15 V. The direct current bus voltage U_d is 1100 V, and the carrier frequency f is 5 kHz. The common mode current I_{com} and the bearing current I_b are calculated.

The same simulations are applied to the direct-drive and semi-direct-drive PMSGs. The bearing current (electric discharge machining current) simulation results of the three kinds of generators are shown in Figure 13.

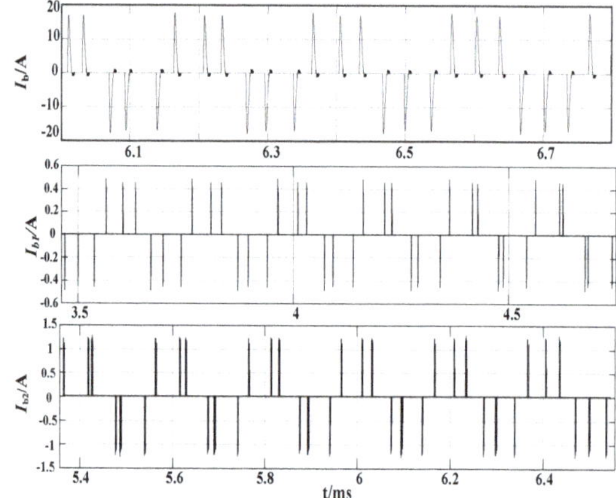

Figure 13. Bearing currents of DFIG, direct-drive PMSG, and semi-direct-drive PMSG from top to bottom.

The comparison of the peak values of common-mode current and bearing current of the three kinds of generator is shown in Table 7.

Table 7. Comparison of common mode current and bearing current in three kinds of wind generators.

Wind Turbine	I_{com}/A	I_b/A
DFIG	20	20
Direct-drive PMSG	150	1.3
Semi-direct-drive PMSG	60	0.5

Table 7 shows that direct-drive and semi direct-drive PMSG bearing currents are much smaller than in the DFIG. However, the common mode current of the direct-drive DMSG is much larger than the DFIG and semi-direct-drive PMSG, which is because the stray capacitance of the PMSG winding and frame is much larger than in the DFIG.

Common mode current through the stray capacitance formed eddy currents in the stator core, which caused generator thermal loss. At the same time, the current flowing through the ground wire into the grid will cause harmonic interference in power grid. Bearing current generated by the breakdown of the bearing will cause the metal to melt near the breakdown point and generate pit points, which leads to premature failure of the bearing. Therefore, it is necessary to take effective measures to suppress common-mode current and bearing current.

The measurement of bearing currents in megawatt wind turbines has not been conducted anywhere in the world. Since the bearing currents cannot be directly measured, the test is a difficult task. This will be our future research content.

6. Conclusions

In this paper, three kinds of widely used wind generators—DFIG, direct-drive PMSG, and semi-direct-drive PMSG—are compared on the bearing current characteristics. The simulation results reveal that the bearing current of DFIG is much larger than the PMSGs. From the perspective of bearing reliability, PMSGs have more advantages than DFIG. From the common mode current point of view, the common-mode current of direct-drive PMSG is much larger than the semi-direct-drive PMSG and DFIG. Bearing current suppression methods for DFIG and common-mode current suppression methods for direct-drive PMSG should be further explored.

Author Contributions: Conceptualization, R.L.; Methodology, R.L.; Software, X.M., X.R.; Formal Analysis, X.M., X.R.; Resources, J.C.; Writing-Original Draft Preparation, X.M., X.R.; Writing-Review & Editing, R.L., S.N. Funding Acquisition, R.L., J.C.

Funding: This research was funded by the National Natural Science Foundation of China grant number [51777008, 51577007].

Acknowledgments: The authors express their gratitude to the National Natural Science Foundation of China.

Conflicts of Interest: The authors declare no conflict of interest.

References

1. Alewine, K.; Chen, W. A review of electrical winding failures in wind turbine generators. *IEEE Trans. Electr. Insul. Mag.* **2012**, *28*, 8–13. [CrossRef]
2. Hoppler, R.; Errath, R.A. Motor Bearings, not just a piece of metal. In Proceedings of the IEEE-IAS/PCA Cement Industry Conference, Charleston, SC, USA, 29 April–3 May 2007.
3. Chen, S.; Lipo, T.; Fitzgerald, D. Source of induction motor bearing currents caused by PWM inverters. *IEEE Trans. Energy Convers.* **1996**, *11*, 25–32. [CrossRef]
4. Busse, D.; Erdman, J.; Kerkman, R.; Schlegel, D.; Skibinski, G. System electrical parameters and their influence effect on bearing currents. *IEEE Trans. Ind. Appl.* **1997**, *33*, 577–584. [CrossRef]
5. Busse, D.; Erdman, J.; Kerkman, R.; Schlegel, D.; Skibinski, G. Bearing Currents and Their Relationship to PWM Drives. *IEEE Trans. Power Electron.* **1997**, *12*, 243–252. [CrossRef]
6. Muetze, A. Bearing Currents in Inverter-Fed AC-Motors. Ph.D. thesis, Darmstadt Univ. of Technology, Shaker Verlag, Aachen, Germany, 2004.
7. Binder, A.; Muetze, A. Scaling Effects of Inverter-Induced Bearing Currents in AC Machines. *IEEE Trans. Ind. Appl.* **2007**, *2*, 769–776.
8. Muetze, A.; Tamminen, J.; Ahola, J. Influence of Motor Operating Parameters on Discharge Bearing Current Activity. *IEEE Trans. Ind. Appl.* **2011**, *47*, 1767–1777. [CrossRef]
9. Muetze, A.; Binder, A. Techniques for Measurement of Parameters Related to Inverter-Induced Bearing Currents. *IEEE Trans. Ind. Appl.* **2007**, *43*, 1274–1283. [CrossRef]
10. Muetze, A. Scaling Issues for Common-Mode Chokes to Mitigate Ground Currents in Inverter-Based Drive Systems. *IEEE Trans. Ind. Appl.* **2009**, *45*, 286–294. [CrossRef]
11. Charles, R.; Muetze, A. Simulation Model of Common-Mode Chokes for High-Power Applications. *IEEE Trans. Ind. Appl.* **2010**, *46*, 884–891.
12. Muetze, A.; Charles, R. Simplified Design of Common-Mode Chokes for Reduction of Motor Ground Currents in Inverter Drives. *IEEE Trans. Ind. Appl.* **2011**, *47*, 2570–2577. [CrossRef]
13. Muetze, A.; Binder, A. Calculation of motor capacitances for prediction of the voltage across the bearings in machines of inverter-based drive systems. *IEEE Trans. Ind. Appl.* **2007**, *43*, 665–672. [CrossRef]
14. Zitzelsberger, J.; Hofmann, W.; Wiese, A.; Stupin, P. Bearing currents in doubly-fed induction generators. In Proceedings of the Power Electronics and Applications Conference, Dresden, Germany, 11–14 Septemper 2005.
15. Adabi, J.; Zare, F.; Ghosh, A.; Lorenz, R.D. Calculations of capacitive couplings in induction generators to analyse shaft voltage. *IET Power Electron.* **2010**, *3*, 379–390. [CrossRef]
16. Adabi, M.E.; Vahed, A. A survey of shaft voltage reduction strategies for induction generators in wind energy applications. *Renewable Energy* **2013**, *50*, 177–187. [CrossRef]
17. Whittle, M.; Trevelyan, J.; Tavner, P.J. Bearing currents in wind turbine generators. *J. Renew. Sustain. Energy* **2013**, *5*, 053128. [CrossRef]
18. Niskanen, V.; Muetze, A.; Ahola, J. Study on bearing impedance properties at several hundred kilohertz for different electric machine operating parameters. *IEEE Trans. Ind. Appl.* **2014**, *5*, 3438–3447. [CrossRef]

Article

Short-Circuit Current Calculation and Harmonic Characteristic Analysis for a Doubly-Fed Induction Generator Wind Turbine under Converter Control

Jing Li *, Tao Zheng and Zengping Wang

State Key Laboratory for Alternate Electrical Power System with Renewable Energy Source, North China Electric Power University, Beijing 102206, China; zhengtao_sf@126.com (T.Z.); wangzp1103@sina.com (Z.W.)
* Correspondence: lijing_ncepu@ncepu.edu.cn; Tel.: +86-158-1146-2585

Received: 30 August 2018; Accepted: 14 September 2018; Published: 17 September 2018

Abstract: An accurate calculation of short-circuit current (SCC) is very important for relay protection setting and optimization design of electrical equipment. The short-circuit current for a doubly-fed induction generator wind turbine (DFIG-WT) under excitation regulation of a converter contains the stator current and grid-side converter (GSC) current. The transient characteristics of GSC current are controlled by double closed-loops of the converter and influenced by fluctuations of direct current (DC) bus voltage, which is characterized as high order, multiple variables, and strong coupling, resulting in great difficulty with analysis. Existing studies are mainly focused on the stator current, neglecting or only considering the steady-state short-circuit current of GSC, resulting in errors in the short-circuit calculation of DFIG-WT. This paper constructs a DFIG-WT total current analytical model involving GSC current. Based on Fourier decomposition of switch functions and the frequency domain analytical method, the fluctuation of DC bus voltage is considered and described in detail. With the proposed DFIG-WT short-circuit current analytical model, the generation mechanism and evolution law of harmonic components are revealed quantitatively, especially the second harmonic component, which has a great influence on transformer protection. The accuracies of the theoretical analysis and mathematical model are verified by comparing calculation results with simulation results and low-voltage ride-through (LVRT) field test data of a real DFIG.

Keywords: doubly-fed generator; converter control; short-circuit current; second harmonic component; low-voltage ride-through (LVRT) field test data

---

1. Introduction

With the worsening global energy crisis and environmental pollution, renewable energy sources have received worldwide attention and undergone rapid development. The doubly-fed induction generator wind turbine (DFIG-WT) is one of the most popular wind turbine generators due to its low manufacturing cost, high efficiency, and high flexibility, and is extensively applied on wind farms [1–3]. In earlier studies, DFIG-WT was regarded as a load or synchronous generator for short-circuit calculation, as the capacity of wind farms is small [4,5]. However, with the increasing capacity of wind power access to the grid, the influence of the short-circuit current (SCC) of DFIG-WT can no longer be ignored. Since accurate SCC calculation is very important to protection settings [6,7], equipment selection, and the optimal design of wind turbine control strategies [8,9], transient characteristics analysis and the SCC calculation model for DFIG-WT have attracted the attention of researchers around the world in recent years.

Grid codes require that wind turbines must remain connected during specific fault conditions and support the grid voltage by providing a reactive current with a magnitude proportional to the voltage deviation. A quick response of reactive current provision is also required, and the response times are explicitly stipulated in the grid codes; for example, it is less than 75 ms in the Chinese criteria [10], and is even more strict in the German criteria, with a value of 20 ms [11]. To comply with the grid code requirements, a crowbar circuit is often utilized by the rotor for protection against excessive current. The short-circuit current characteristics of DFIG-WT with crowbar protection have been extensively discussed [12–14]. However, the crowbar operation is not desired due to the loss of controllability and absorption of reactive power.

Under non-severe fault conditions or for some DFIG-WTs with a higher tolerance for voltage drops, excitation control of converters is retained during a fault. Due to different constrictions compared with traditional generators and inverter interfaced generators, the transient characteristic of DFIG-WT under converter control is determined by both electromagnetic equations of the generator and control strategies of the AC-DC-AC converters, and the short-circuit current of DFIG-WT contains the stator short-circuit current and grid-side converter (GSC) short-circuit current.

In evaluating the transient fault characteristics of DFIG-WT, establishing a mathematical analytical model is effective and helpful in obtaining the physical mechanisms and numerical values of electrical quantities. A simplified stator fault current model was built in [15] by neglecting the dynamic process of stator flux linkage and hypothesizing step mutations of the rotor voltage after a fault, but the model could not fit the actual transient short-circuit current completely. A stator current analytical model was presented in [16] by solving a second-order differential equation related to the rotor current of the time domain. This method produced ideal linearization of input and output characteristics of the converter, and did not consider transient responses in the converter. In [17], a more detailed stator current analytical model was constructed based on transfer functions of the control system. However, the sampling delay of the converter and transfer characteristics of pulse-width modulation (PWM) were not considered in this study, resulting in sudden changes of initial short-circuit current at the time of fault occurrence and inaccuracy of transient current calculation. It is mentioned in [18] that the current reference values of the converter should be limited to prevent overcurrent, and the control limits for the rotor side converter were studied in [19]. However, the influence of current limitation of the converter on SSC calculation was not discussed in the above studies. In addition, the analytical SCC models in these studies did not include GSC current.

The influencing factors of GSC short-circuit current were simulated and analyzed in [20], but no analytical model of GSC-transient SCC was constructed. Based on different control targets, a DFIG-WT steady-state short-circuit current model with a consideration of GSC current was constructed in [21]. By comparing the results calculated with and without GSC steady-state current, that study concluded that GSC current should be considered for accurate fault analysis and protection settings, but it only focused on steady-state current and did not mention the transient characteristics of SCC. Since transient characteristics of GSC fault current are influenced by coupling factors, including control strategies of the two-side convertors, transient fluctuation of DC bus voltage, and the electromagnetic transient response of the generator, the construction of a GSC transient current model is more complex than the stator fault current. Existing studies on transient SCC of DFIG-WT have not fully discussed transient characteristics of GSC current. Moreover, there is a lack of theoretical references on calculation errors of transient short-circuit total current caused by neglecting GSC current.

Moreover, most of the models built in the above studies were validated by simulation. As was mentioned in [22], the wind power industry urgently needs validation in comparison with real measurements to verify the accuracy and corresponding usability of the models. Additionally, the validation of a generic DFIG-WT was presented based on a measurement campaign carried out in a real wind farm, which is of great interest to researchers in the field of wind energy. Field test data of a real DFIG-WT are also presented to verify the short circuit current calculation model in this article.

This paper discusses the following to construct a more accurate transient short-circuit current calculation model for DFIG-WT. First, transient response characteristics of GSC and rotor side convertor (RSC) control systems after a symmetric voltage dip are analyzed based on the transfer functions of the control system. The relation equations among GSC current, DC bus voltage, RSC current, and stator flux linkage are constructed. The coupling mechanisms of the key internal electrical quantities in the converters are thus revealed. Second, the analytical expressions of RSC current, DC bus voltage, and GSC current are deduced based on the above equations. Specifically, a more accurate calculation model of RSC short-circuit current with no sudden changes in the initial time of failure is constructed, considering sampling delay of the control system and small inertial PWM. A detailed calculation model of GSC transient current is established with a consideration of DC bus voltage fluctuation. It was found that there is a high proportion of second harmonic current in the GSC transient fault current, which may result in a false operation of the secondary harmonic restraint relay for transformer protection. Third, the nonlinear characteristics of steady-state fault current of DFIG-WT considering limitations of rotor current are analyzed, and an estimation formula for the maximum steady-state SCC is put forward and verified by simulation. Finally, the accuracy of the theoretical analysis and mathematical models is verified by simulation tests and low-voltage ride-through (LVRT) field test data of a real DFIG. Proportions of GSC current and the second harmonic component in short-circuit total current of DFIG-WT under different fault situations are analyzed.

2. Transient Mathematical Models of DFIG

The electrical parts of DFIG mainly include the induction generator, rotor-side converter, and grid-side converter. RSC and GSC are connected through the DC capacitor [17]. The mathematical models and control strategies of these three parts are briefly introduced in the following section.

2.1. Induction Generator Model

Motor convention is applied on the stator and rotor sides of the induction generator. The magnetic saturation effect is neglected. The mathematical model of the generator in the synchronous reference frame is:

$$\begin{cases} \boldsymbol{u}_s = R_s \boldsymbol{i}_s + j\boldsymbol{\psi}_s + p\boldsymbol{\psi}_s/\omega_1 \\ \boldsymbol{u}_r = R_r \boldsymbol{i}_r + js\boldsymbol{\psi}_r + p\boldsymbol{\psi}_r/\omega_1 \\ \boldsymbol{\psi}_s = L_s \boldsymbol{i}_s + L_m \boldsymbol{i}_r \\ \boldsymbol{\psi}_r = L_r \boldsymbol{i}_r + L_m \boldsymbol{i}_s \end{cases} \tag{1}$$

2.2. RSC Control with Consideration of the Limiting Reference Current

Double closed-loop vector control based on stator voltage orientation was applied on RSC, with the inner loop as the current loop and the outer loop as the power loop. The control mechanism diagram is shown in Appendix A. The reference value of the inner current controller under normal operation control is:

$$\begin{cases} i_{rd0}^* = \min(2L_s P_{s0}^*/3L_m U_{s0}, I_{rd\text{-max}}) \\ i_{rq0}^* = \max(-\dfrac{U_{s0}}{L_m} - \dfrac{2L_s Q_{s0}^*}{3L_m U_{s0}}, -\sqrt{I_{r\text{-max}}^2 - i_{rd,opt}^{*2}}) \end{cases} \tag{2}$$

where $I_{rd\text{-max}}$ is the active current-limiting value and $I_{r\text{-max}}$ is the rotor current-limiting value.

When a three-phase short-circuit fault occurs at the terminal of a wind turbine, the outer power loop of RSC will be open and the reference current of the inner current loop will be given directly in order to quickly respond to the terminal voltage dip and fulfill the grid codes, which require wind turbines to have LVRT capability and provide reactive power to support grid voltage recovery. Considering the current limits of RSC, the reference value of the inner current loop during low-voltage circumstances is:

$$\begin{cases} i_{rd1}^* = \min(2L_s P_{s0}^*/(3L_m U_{s1}), \sqrt{I_{r\text{-max}}^2 - i_{rq,lvrt}^{*2}}, I_{rd\text{-max}}) \\ i_{rq1}^* = \max(-U_{s1}/L_m - K_d(0.9 - U_{s1})L_s/L_m, -I_{r\text{-max}}) \end{cases} \tag{3}$$

where K_d is the reactive current coefficient, usually $K_d \geq 1.5$.

2.3. GSC Control Mechanism

The double closed-loop vector control method based on stator voltage orientation is also applied to GSC control. The inner loop is the current loop, and the outer loop is the DC voltage loop and reactive power loop. The control mechanism diagram is shown in Appendix A.

Viewed from the GSC side, the current from the RSC side can be regarded as an equivalent load. Thus, the DC bus voltage equation is:

$$C\frac{dU_{dc}}{dt} = i_L - i_{g\text{-}dc} = \frac{P_{load} - P_g}{U_{dc}} \tag{4}$$

3. Converter Transient Response Characteristics of DFIG-WT

3.1. GSC Transient Response Characteristics

In GSC control, the d-axis is responsible for maintaining the stability of DC bus voltage and the q-axis is responsible for adjusting the power factor of the wind turbine generator, which is generally operated at a unit power factor. The dq-axis control structure is symmetric, so we just take the d-axis control as an example. According to the control diagram of GSC, the d-axis control framework of the inner current loop and DC voltage loop is shown in Figure 1.

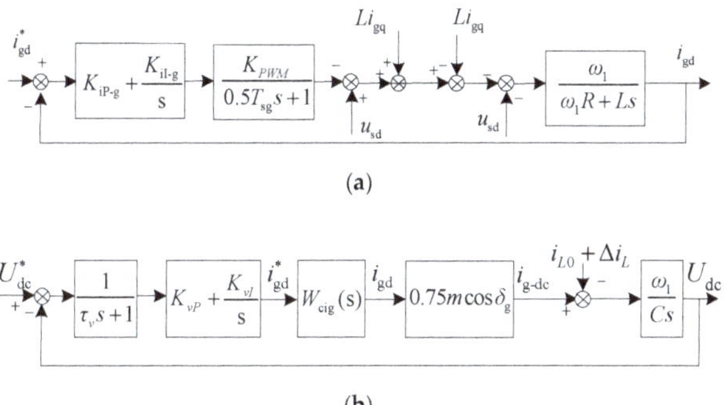

(a)

(b)

Figure 1. Current regulating loop and voltage regulating loop of the grid-side converter (GSC): (a) current regulating loop; (b) voltage regulating loop.

In the figure, $K_{iP\text{-}g}$ and $K_{iI\text{-}g}$ are the proportional gain and integral gain of the inner current controller, respectively; T_{sg} is the switching period of PWM; K_{PWM} is the PWM equivalent gain of the bridge circuit, $K_{PWM} = 0.8165$; τ_v is the period of voltage sampling time; K_{vP} and K_{vI} are the proportional gain and integral gain of the proportional and integral (PI) regulator in the DC voltage loop, respectively; m is the modulation ratio of PWM; i_{L0} is the DC-side pre-fault current of RSC; and Δi_L represents the changes of RSC current on the DC side.

The current loop has to track the reference current quickly, and it is often designed according to a typical first-order system [23,24]. Then, $K_{iP\text{-}g}$ and $K_{iI\text{-}g}$ can be adjusted as:

$$K_{iP\text{-}g} = \frac{R\tau_{ig}}{3T_{sg}K_{PWM}}, \ K_{iI\text{-}g} = \frac{K_{iP\text{-}g}}{\tau_{ig}} = \frac{R}{3T_{sg}K_{PWM}} \tag{5}$$

where $\tau_{ig} = L/(\omega_1 R)$.

According to Equation (5) and Figure 1, the current loop of GSC is equivalent to the first-order inertial element and the transfer function is:

$$W_{ci\text{-}g}(s) = \frac{i_{gd}(s)}{i_{gd}^*(s)} \approx \frac{1}{1 + s/\omega_{ig}} \tag{6}$$

where ω_{ig} is the bandwidth angular velocity of the inner current loop: $\omega_{ig} = 1/(3T_{sg})$.

The outer voltage loop of GSC focuses on disturbance resistance. The PI parameter of the outer voltage loop of GSC is designed according to a typical second-order system:

$$K_{vP} = \frac{2(\omega_{cv}+1)C}{3m\omega_1\omega_{cv}T_{ev}}, K_{vI} = \frac{K_{vP}}{\omega_{cv}T_{ev}} = \frac{2(\omega_{cv}+1)C}{3m\omega_1\omega_{cv}^2 T_{ev}^2} \tag{7}$$

where ω_{cv} is the middle frequency bandwidth of the outer voltage loop and T_{ev} is the equivalent time constant of outer voltage loop: $T_{ev} = \tau_v + 3T_s$.

GSC current is determined by its reference value, which is related to DC bus voltage in the outer voltage loop. According to Figure 1 and Equation (6), the d-axis component of GSC current is:

$$i_{gd}(s) = (K_{vP} + \frac{K_{vI}}{s})\frac{\Delta U_{dc}}{T_{ev}s+1} + i_{gd0} \tag{8}$$

where ΔU_{dc} is the difference between DC bus voltage and its reference value and i_{gd0} is the initial value of the d-axis component of the GSC pre-fault current. GSC active power is about slip times of stator active power, that is, $P_{g0} \approx -sP_{s0}$. Therefore, $i_{gd0} \approx -sP_{s0}/U_{s0}$.

According to Equation (8), in order to calculate GSC current, it is necessary to obtain the DC bus voltage expression first. It can be seen from Figure 1b that the DC bus post-fault voltage is related to $U_{dc}{}^*$, i_{L0}, and Δi_L. DC bus voltage is maintained as constant under the collaborative effect of $U_{dc}{}^*$ and i_{L0} under normal operating conditions. Under fault conditions, the fluctuation of DC bus voltage ΔU_{dc} is mainly influenced by Δi_L. According to Figure 1b, the transfer function from Δi_L to DC bus voltage U_{dc} is:

$$W_{cv}(s) = \frac{U_{dc}(s)}{\Delta i_L(s)} \approx -\frac{1}{C}\frac{s}{(s-\lambda_1)(s-\lambda_1)} \tag{9}$$

where $\lambda_{1,2}$ is the characteristic roots of Equation (9):

$$\lambda_{1,2} = v \pm \gamma = -\frac{1+\omega_{cv}}{4\omega_{cv}T_{ev}} \pm \frac{1}{4\omega_{cv}T_{ev}}\sqrt{(\omega_{cv}+1)(\omega_{cv}-7)} \tag{10}$$

It is worth noting that Δi_L is the DC-side current of RSC. The transfer relationship between DC-side current and AC-side current of RSC is:

$$i_L = S_a i_{ra} + S_b i_{rb} + S_c i_{rc} \tag{11}$$

Since the switching frequency is significantly higher than the grid fundamental frequency and transient analysis mainly focuses on fundamental frequency, the higher harmonic components can be neglected and only low-frequency components of the switch function are considered. Fourier decomposition of the switch function S_{abc} of RSC is carried out as:

$$\begin{cases} S_a \approx 0.5m\cos(\omega_r t - \delta_r) + 0.5 \\ S_b \approx 0.5m\cos(\omega_r t - \delta_r - 120^\circ) + 0.5 \\ S_c \approx 0.5m\cos(\omega_r t - \delta_r + 120^\circ) + 0.5 \end{cases} \tag{12}$$

where δ_r is the fundamental wave initial phase angle of the switch function.

Based on Equations (11) and (12) and the transfer of RSC current from the stationary reference frame to the synchronous reference frame, the DC-side current of RSC can be expressed as:

$$i_L = \begin{bmatrix} S_a & S_b & S_c \end{bmatrix} C_{2s/3s} \begin{bmatrix} i_{rd} \\ i_{rq} \end{bmatrix} = 0.75m(i_{rd}\cos\delta_r - i_{rq}\sin\delta_r) \tag{13}$$

Equation (13) reflects that the DC-side current of RSC can be calculated according to the AC-side current. In fact, the AC-side current of RSC is equal to the rotor current and is related to the transient response of the inner current loop of RSC.

After the occurrence of the three-phase short-circuit fault on the terminal of the wind turbine, GSC is mainly responsible for maintaining the stability of DC bus voltage. Meanwhile, GSC is also able to generate a small amount of reactive power independently to support the grid voltage [25]. To realize the goal of fast regulation, the outer loop of reactive power is open and the reference value of the inner current loop is given directly after the fault. Referring to Equation (6) and considering fast adjustment of the inner current loop, the q-axis component of GSC current can be regarded as equal to the reference value.

3.2. RSC Transient Response Characteristics

According to the above analysis, in order to calculate the DC bus voltage and GSC current, it is necessary to obtain the RSC current, which is determined by the inner current loop of RSC. The dq-axis structure is symmetric. The control framework of the d-axis is shown in Figure 2.

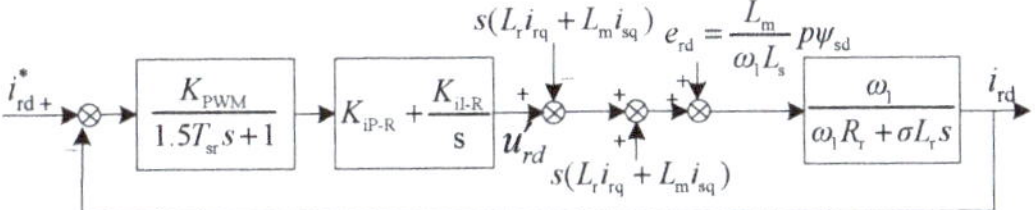

Figure 2. Control framework of inner current loop of RSC.

In Figure 2, $K_{iP\text{-}R}$ and $K_{iI\text{-}R}$ are the proportional gain and integral gain of the PI controller of the inner current controller, respectively; T_{sr} is the switch period of PWM; and e_{rd} is the d-axis component of voltage disturbance.

It can be seen from Figure 2 that changes of reference current $i_r{}^*$ and voltage disturbance e_{rd} will cause transient responses of RSC. When calculating the transient current of RSC, reference [17] neglected the sampling delay of the converter and small inertial characteristics of PWM, causing the calculated rotor short-circuit current to change suddenly at the moment the fault occurred. In the following section, one more accurate expression of rotor transient current is deduced based on the detailed model of RSC, since rotor current is vital to stator current and GSC current.

Similar to the current loop of GSC, the current loop of RSC can be regarded as a first-order inertial element. The closed-loop transfer function of RSC is:

$$W_{ci\text{-}r}(s) = \frac{i_r}{i_r^*} \approx \frac{1}{1 + s/\omega_{ci}} \tag{14}$$

where ω_{ci} is the bandwidth angular frequency of the inner current loop: $\omega_{ci} = 1/(3T_{sr})$.

As stated in Section 3.2, when the terminal voltage drops to lower than 90% U_n, the DFIG-WT will switch to LVRT control and the rotor current reference value will change from a normal operation state (Equation (2)) to an LVRT state (Equation (3)). The rotor transient current caused by sudden changes of the reference value is:

$$\Delta i_{r\text{-}ref}(t) = \Delta i_r^* - \Delta i_r^* e^{-\omega_{ci} t} \tag{15}$$

where $\Delta i_r^* = i_{r1}^* - i_{r0}$, i_{r1}^* is the post-fault reference rotor current and i_{r0} is the pre-fault rotor current.

On the other hand, considering the sampling delay of the converter and small inertia of PWM, the closed-loop transfer function from voltage disturbance to transient current of rotor is:

$$W_{ce}(s) = i_r(s)/e_r(s) = -\frac{2\omega_{ci}^2}{K_{iP}K_{PWM}} \frac{s[s/(2\omega_{ci}) + 1]}{(s + 1/\tau_i)[(s + \omega_{ci})^2 + (\omega_{ci})^2]} \tag{16}$$

According to the conservation principle of flux linkages, the stator flux linkage after the occurrence of the three-phase short-circuit fault is [26]:

$$\boldsymbol{\psi}_s = \boldsymbol{\psi}_{sf} + \boldsymbol{\psi}_{sn} = -j\mathbf{u}_{s1} - j(\mathbf{u}_{s0} - \mathbf{u}_{s1})e^{-t/\tau_s}e^{-j\omega_1 t} \tag{17}$$

where τ_s is the stator attenuation time constant: $\tau_s = (\sigma L_s)/\omega_1 R_s$.

According to Equation (17), the frequency domain expression of voltage disturbance is:

$$e_r(s) = j(\boldsymbol{u}_{s0} - \boldsymbol{u}_{s1})\frac{L_m}{L_s}\frac{1/\tau_s + j\omega_1}{s + 1/\tau_s + j\omega_1} \tag{18}$$

Substituting Equation (18) into Equation (16), the transient current of the rotor caused by voltage disturbance is:

$$\Delta \boldsymbol{i}_{r\text{-}er}(t) = \Delta \boldsymbol{i}_{r1}e^{-j\omega_1 t}e^{-t/\tau_s} + \Delta \boldsymbol{i}_{r2}e^{-t/\tau_i} + [\Delta \boldsymbol{i}_{r3}\cos(\omega_{ci}t) + \Delta \boldsymbol{i}_{r4}\sin(\omega_{ci}t)]e^{-\omega_{ci}t} \tag{19}$$

where $\Delta \boldsymbol{i}_{r1,2,3,4}$ represents coefficients of attenuation components of the rotor current. The detailed coefficient expressions are shown in Appendix B. These coefficients are related to voltage drop amplitude and parameters of the generator and converter, and meet with $\Delta \boldsymbol{i}_{r1} + \Delta \boldsymbol{i}_{r2} + \Delta \boldsymbol{i}_{r3} = 0$, thus assuring $\Delta \boldsymbol{i}_{r\text{-}er}(0) = 0$ at the initial fault stage.

According to Equations (17) and (19), the time domain expression of the rotor current is:

$$\boldsymbol{i}_r = \boldsymbol{i}_{r0} + \Delta \boldsymbol{i}_r = \boldsymbol{i}_{r0}^* + \Delta \boldsymbol{i}_r^* - \Delta \boldsymbol{i}_r^* e^{-\omega_{ci}t} + \Delta \boldsymbol{i}_{r\text{-}er} \tag{20}$$

In this section, the equations between different electrical quantities, such as GSC current, DC bus voltage, and AC-side and DC-side currents of RSC, are deduced based on control mechanisms of GSC and RSC. Moreover, the coupling relationships and variation laws of the DC capacitor and key electrical quantities of GSC and RSC are revealed.

4. Transient Short-Circuit Current Calculation Model of DFIG-WT

4.1. DC Bus Voltage

DC bus voltage has to be calculated first to obtain the GSC current. According to the transfer function in the DC voltage loop and DC-side current of RSC, the expression of DC bus voltage fluctuation ΔU_{dc} can be deduced according to Equations (9), (13), and (19):

$$\Delta U_{dc}(t) = U_{dc1}\cos(\omega_1 t - \beta_1)e^{-t/\tau_s} + U_{dc2}e^{-t/\tau_i} + U_{dc6}e^{\lambda_1 t} + U_{dc7}e^{\lambda_2 t} + [U_{dc3}\cos(\omega_{ci}t + \beta_2) + U_{dc4}\sin(\omega_{ci}t + \beta_2) + U_{dc5}]e^{-\omega_{ci}t} \tag{21}$$

where U_{dc1}, U_{dc2}, U_{dc3}, U_{dc4}, U_{dc5}, U_{dc6}, and U_{dc7} are coefficients of different components. Detailed expressions are shown in Appendix B.

It can be seen from Equation (4) that the fluctuating DC bus voltage is caused by an active power imbalance between the converters at the two sides. The fundamental frequency attenuation component in RSC current may cause a fundamental frequency attenuation component in U_{dc}. This implies that U_{dc1} corresponds to the fundamental frequency attenuation component $\Delta \boldsymbol{i}_{r1}$ in the rotor current. Similarly, U_{dc2}, U_{dc5}, U_{dc6}, and U_{dc7} correspond to $\Delta \boldsymbol{i}_{r2}$, $\Delta \boldsymbol{i}_{r3}$, $\Delta \boldsymbol{i}_{r4}$, and $\Delta \boldsymbol{i}_r^*$, respectively. U_{dc3} and U_{dc4} represent transient response characteristics of the DC voltage loop and their amplitudes are related to all of the transient components in the rotor current.

Equation (21) shows that the DC bus voltage contains complicated frequency components. To elaborate proportions of frequency components in ΔU_{dc} and their relationships with rotor current, a group of data concerning coefficient amplitudes and time constants of the damping components of rotor current and DC bus voltage is given in Table 1. The parameters used in the calculation are from the simulation case. The voltage at the generator terminal dropped to 70% Un. Before the fault, the DFIG-WT was operated at a rated active power and with a unit power factor.

Table 1. Coefficient amplitudes and time constants of attenuation components in i_r and ΔU_{dc}.

i_r Coefficient Amplitude/p.u.		ΔU_{dc} Coefficient Amplitude/p.u.		Attenuation Time Constant	
Δi_{r1}	0.7026	U_{dc1}	0.1351	τ_s	0.0306
Δi_{r2}	0.0316	U_{dc2}	0.0009	τ_i	0.0575
Δi_{r3}	0.2112	U_{dc3}	0.0312	$1/\omega_{ci}$	0.0019
Δi_{r4}	0.2275	U_{dc4}	0.0410	$1/\omega_{ci}$	0.0019
$\Delta i_r{}^*$	0.3332	U_{dc5}	0.0369	$1/\omega_{ci}$	0.0019
		U_{dc6}	0.0505	$-1/\lambda_1$	0.0079
		U_{dc7}	0.0579	$-1/\lambda_2$	0.004

The following conclusions can be drawn from Table 1:

(1) Fundamental frequency attenuation components Δi_{r1} and U_{dc1} account for the highest proportion in rotor current and DC bus voltage, respectively.

(2) DC attenuation components Δi_{r2} and U_{dc2} account for the lowest proportion.

(3) In rotor current, oscillating attenuation components Δi_{r3} and Δi_{r4}, of which both frequencies and time constants are ω_{ci}, will cause the homogeneous components U_{dc3} and U_{dc4} in DC bus voltage. Due to the high switching frequency of RSC, U_{dc3} and U_{dc4} attenuate very quickly. They will attenuate to lower than 10% of their amplitude by about 4 ms.

(4) U_{dc6} and U_{dc7} in DC bus voltage are related to the characteristic roots of the transfer function of the DC voltage loop. According to Equation (10), the characteristic roots are related to the intermediate frequency bandwidth of the DC voltage loop. When $\omega_{cv} < 7$, the characteristic roots of the transfer function are a pair of conjugate complexes. Under this circumstance, U_{dc6} and U_{dc7} are oscillating attenuation components. The oscillation period is the imaginary part of the characteristic roots, and the attenuation time constant is the reciprocal of the real part of the characteristic roots. When $\omega_{cv} \geq 7$, the characteristic roots are two different (or same) real numbers. In this case, U_{dc6} and U_{dc7} are DC attenuation amplitudes, and the attenuation time constant is the reciprocal of the characteristic roots.

4.2. GSC Current

Equation (21) is transferred into the frequency domain and then substituted into Equation (8). The expression of GSC current in the synchronous reference frame is:

$$i_g = -si_{sdf} + ji_{gq}^* + \frac{U_{dc1}K_{vI}}{r}[I_{g1}\cos(\omega t) + I_{g2}\sin(\omega t)]e^{-t/\tau_s} + \frac{U_{dc2}\tau_i}{\tau_i-\tau_v}(K_{vP} - \tau_i K_{vI})e^{-t/\tau_i} + \frac{U_{dc5}}{1-\omega_{ci}\tau_v}(K_{vP} - \frac{1}{\omega_{ci}}K_{vI})e^{-\omega_{ci}t}$$
$$+ \frac{U_{dc6}}{1+\lambda_1\tau_v}(K_{vP} + \frac{1}{\lambda_1}K_{vI})e^{\lambda_1 t} + \frac{U_{dc7}}{1+\lambda_2\tau_v}(K_{vP} + \frac{1}{\lambda_2}K_{vI})e^{\lambda_2 t} + K_{vI}[I_{g3}\sin(\omega_{ci}t) + I_{g4}\cos(\omega_{ci}t)]e^{-\omega_{ci}t} + I_{g5}(K_{vP} - \tau_v K_{vI})e^{-t/\tau_v} \tag{22}$$

where I_{g1}, I_{g2}, I_{g3}, I_{g4}, and I_{g5} are coefficients for different components. Specific expressions are shown in Appendix B.

Comparing Equations (21) and (22), GSC current in the synchronous reference frame contains frequency components corresponding to the transient attenuation components in U_{dc} and additionally generates a DC attenuation component with time constant τ_v due to the inner current loop. According to Table 1, U_{dc2} has a small amplitude, and U_{dc3}, U_{dc4}, and U_{dc5} attenuate quickly. Therefore, it is applicable to neglect the above parameters to obtain a simplified expression of GSC current. Transferred into the stationary reference frame, the a-phase expression of GSC current is:

$$i_{ga} \approx -si_{sdf}\cos(\omega t) - i_{gq}^*\sin(\omega t) + [I_{g5}(K_{vP} - \tau_v K_{vI})e^{-t/\tau_v} + \frac{U_{dc6}}{1+\lambda_1\tau_v}(K_{vP} + \frac{1}{\lambda_1}K_{vI})e^{\lambda_1 t}$$
$$+ \frac{U_{dc7}}{1+\lambda_2\tau_v}(K_{vP} + \frac{1}{\lambda_2}K_{vI})e^{\lambda_2 t}]\cos(\omega t) + \frac{U_{dc1}K_{vI}}{2r}[I_{g1}\cos(2\omega t) + I_{g2}\sin(2\omega t)]e^{-t/\tau_s} + \frac{U_{dc1}K_{vI}}{2r}I_{g1}e^{-t/\tau_s} \tag{23}$$

Equation (23) demonstrates that GSC current in the stationary reference frame contains a steady-state fundamental component, a fundamental component attenuating at different time constants, a DC attenuation component, and a second harmonic frequency attenuation component. Among them, the amplitude of the second harmonic frequency attenuation component is proportional to the fundamental attenuation component in DC bus voltage. According to Table 1, it can be deduced that the second harmonic frequency attenuation component accounts for a large proportion of GSC transient current and takes four to six periods to damp to 0. This second component will influence the second harmonic restraint of transformer protection, which will be further discussed in the simulation section.

4.3. Stator Current and DFIG-WT Total Current

According to Equations (1), (17), and (20), the time domain expression of stator current in the synchronous reference frame is:

$$\mathbf{i}_s = L_{es}\{-\mathbf{i}_r^* - jU_{s1}/L_m - (\Delta\mathbf{i}_{r1} + j\Delta U_s/L_m)e^{-j\omega_1 t}e^{-t/\tau_s} - \Delta\mathbf{i}_{r2}e^{-t/\tau_i} - [\Delta\mathbf{i}_{r3}\cos(\omega_{ci}t) + \Delta\mathbf{i}_{r4}\sin(\omega_{ci}t) + \Delta\mathbf{i}_r^*]e^{-\omega_{ci}t}\} \tag{24}$$

where $L_{es} = L_m/L_s$.

Transferring Equation (24) to the stationary reference frame, the a-phase stator current expression is:

$$\begin{aligned} i_{sa} &= \mathrm{Re}\Big\{-L_{es}\mathbf{i}_r^* e^{j\omega_1 t} - jL_{es}U_{s1}/L_m e^{j\omega_1 t} - L_{es}(\Delta\mathbf{i}_{r1} + j\Delta U_s/L_m)e^{-t/\tau_s} \\ &\quad - L_{es}\Delta\mathbf{i}_{r2}e^{j\omega_1 t}e^{-t/\tau_i} - L_{es}[\Delta\mathbf{i}_{r3}\cos(\omega_{ci}t) + \Delta\mathbf{i}_{r4}\sin(\omega_{ci}t) + \Delta\mathbf{i}_r^*]e^{j\omega_1 t}e^{-\omega_{ci}t}\Big\} \end{aligned} \tag{25}$$

Equation (25) reveals that the a-phase stator short-circuit current contains a steady-state fundamental component, a fundamental attenuation component, a DC attenuation component, and an oscillating attenuation component with ω_{ci} as the period and time constant. It can be concluded from Table 1 that the transient attenuation component of the stator current mainly contains a DC attenuation component, while fundamental attenuation and ω_{ci} relevant components account for a small proportion and attenuate quickly.

With the above analysis and deduction, a short-circuit total current calculation model of DFIG-WT can finally be built as the sum of GSC current and stator current according to Equations (22) and (24):

$$i_T = \mathbf{i}_s + \mathbf{i}_g \tag{26}$$

To demonstrate the transient short-circuit current calculation model of DFIG-WT effectively, a simplified diagram of the relations among key electrical quantities based on the structure of DFIG-WT is shown in Figure 3a, and a detailed flow chart of the entire derivation process is presented in Figure 3b. In the figure, the symbol in bold represents complex vectors in the two phase synchronous rotation coordinate system, and the normal form represents the d (or q)-axis component.

According to Figure 3, the construction process of the transient short circuit current calculation model of DFIG-WT is summarized as follows:

At first, the terminal voltage of DFIG-WT drops to u_{s1}, causing a fundamental frequency attenuation component (ψ_{sn}) that appears in the stator flux, and at the same time, making changes to the reference value of RSC current (i_r^*). The fundamental frequency attenuation component in the stator flux creates a voltage disturbance quantity (e_r) in the current inner loop of RSC, which causes transient fluctuation of the rotor current (Δi_{r-er}), and change of the reference value of RSC current will also cause a transient component in the rotor current. The above two transient components, together with the steady-state component determined by the reference value of the rotor current, constitute the post-fault rotor current. At last, the calculation model of stator short circuit current can be built according to the rotor current and stator flux.

Meanwhile, the rotor current is converted to the DC side of converters as load current (i_L) of the GSC. According to the control diagram of the voltage outer loop of the GSC, changes of load current (Δi_L) will cause a fluctuation component (ΔU_{dc}) in DC bus voltage, causing further fluctuation of the d-axis component of GSC current. The q-axis component of GSC current is generally 0 or very small, which can be considered as equal to its reference value. Finally, the transient short circuit current calculation model of DFIG-WT is the sum of the stator current and GSC current.

According to the above analysis and the flow chart, the generating mechanism and evolution law of transient fluctuations of critical electrical quantities of the DFIG-WT are revealed clearly.

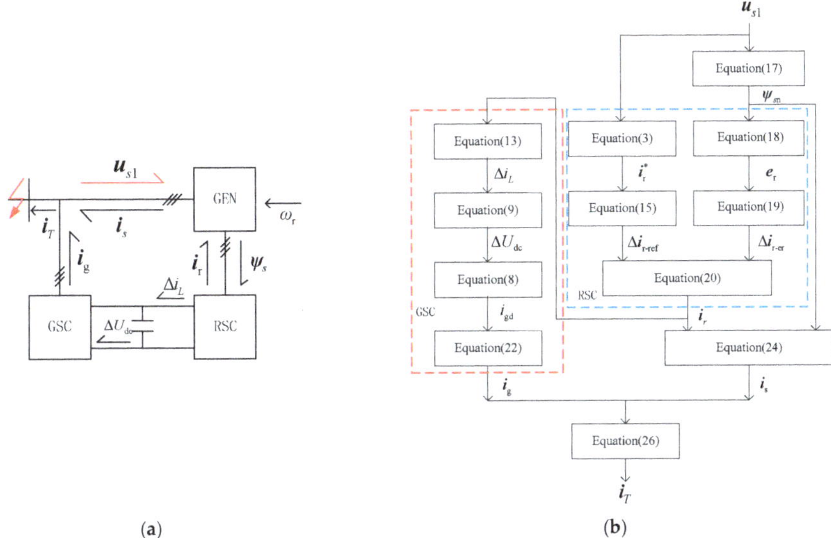

(a)   (b)

Figure 3. Flow chart of the integration process of the transient short-circuit current calculation model for DFIG-WT: (**a**) simplified graphic of the relations among critical quantities; (**b**) detailed flow chart of the integration process.

4.4. Analysis of Steady-State Current of DFIG-WT Considering Current Limits

With a detailed short-circuit current calculation model, quantitative analysis can be carried out to further reveal the fault characteristics of DFIG-WT. Among all the frequency components of the SCC, fundamental components are the most important to protection settings. As shown in Equations (22) and (24), fundamental components of total SCC are mainly composed of steady-state components, while the transient fundamental components attenuate rapidly. Therefore, a detailed analysis of the steady-state current of DFIG-WT was carried out considering the current limit.

According to Equations (22) and (24), the complex of the steady-state component is:

$$i_{Tf} = -\omega_r \frac{L_m}{L_s} i_{rd1}^* - j\left(\frac{U_{s1}}{L_s} + \frac{L_m}{L_s} i_{rq1}^* + i_{gq}^*\right) \tag{27}$$

According to Equation (4) and considering the limit of the converter, the q-axis component of the rotor reference current linearly increases as the voltage drop deepens, while the d-axis component is double-limited by the maximum load current and the maximum rotor current. With parameters from the simulation case, three-dimensional diagrams and contour maps concerning the d-axis rotor reference current and the amplitude of the steady-state short-circuit current of DFIG-WT (I_{Tf}) are shown in Figure 4. The x-axis is the amplitude of post-fault voltage U_{s1} and the y-axis is the pre-fault stator active power P_{s0}.

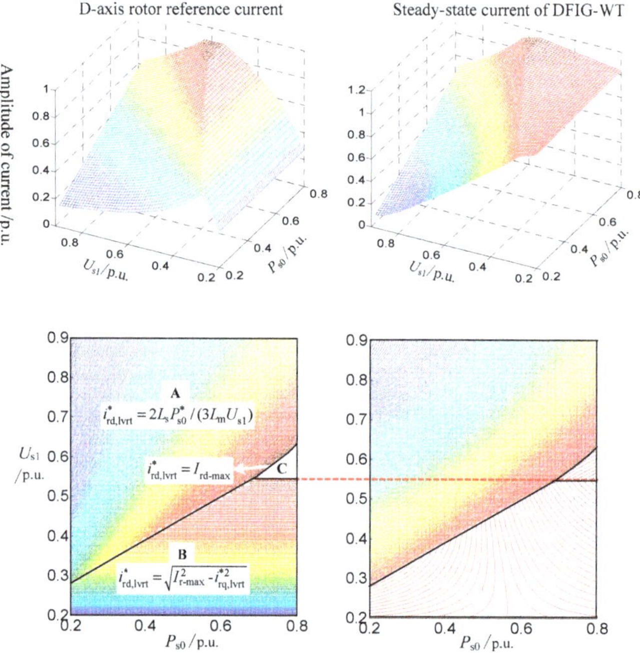

Figure 4. Characteristics of steady-state components of DFIG-WT fault current.

It can be seen from Figure 4 that, due to the influence of current-limiting of the converter, the steady-state short-circuit current of DFIG-WT exhibits nonlinearity when the terminal post-fault voltages and pre-fault power change. It is worth noticing that the deeper the voltage sag is, the larger the steady-state short-circuit current will be. The DFIG steady-state short-circuit current reaches its maximum value when the terminal voltage drops to Us-max and the pre-fault power is full.

According to the contour map of Figure 4, the maximum steady-state short-circuit current of the DFIG appears on the isobaric line between region B and region C of the d-axis rotor reference current. $U_{s\text{-max}}$ can be deduced from the boundary of region B and region C:

$$U_{s\text{-max}} = \frac{0.9K_d - \sqrt{I_{r\text{-max}}^2 - I_{rd\text{-max}}^2}}{K_d - 1/L_m} \tag{28}$$

As can be seen from the above equation, $U_{s\text{-max}}$ only relates to the RSC reference current limit and reactive current coefficient.

Combining Equations (27) and (28) and ignoring the term $1/L_m$ in the denominator of Equation (28), the estimation formula for the maximum steady-state short-circuit current of DFIG $I_{T\text{-max}}$ under different voltage drops and pre-fault conditions is simplified and obtained as:

$$I_{T\text{-max}} \approx \sqrt{I_{r\text{-max}}^2 + \left[\left(\omega_{r\text{max}} \frac{L_m}{L_s}\right)^2 - 1\right] I_{rd\text{-max}}^2} \tag{29}$$

where $\omega_{r\text{max}}$ is the maximum rotor speed frequency in per-unit value, usually about 1.2–1.3.

Parameters in the simulation case are used to verify the accuracy of Equation (29). When the DFIG steady-state short-circuit current reaches its maximum, $U_{s\text{-max}}$ is about 0.55 p.u. and calculated $I_{T\text{-max}}$

is 1.2263 p.u. Compared with the simulation result, the error of the maximum steady-state short-circuit current estimation formula is:

$$\varepsilon = \frac{1.2263 - 1.1821}{1.1821} \times 100\% \approx 3.74\% \tag{30}$$

The estimated value is slightly larger than the actual value, and the error comes from ignoring the term $1/L_m$, which results in the $U_{s\text{-max}}$ calculated value decreasing and the short-circuit current reactive current increasing, leading to an increase in the $I_{T\text{-max}}$ calculated value. However, due to the excitation inductance, $1/L_m$ is generally large, and the error caused by ignoring this item is small.

According to Equation (30), because of convertor current limiting, the steady-state short-circuit current provided by the DFIG-WT is relatively small and has nonlinear relationships with terminal voltage and pre-fault power. This is quite different from the synchronous generators, whose short-circuit current calculation model is a constant internal voltage behind a linear transient impedance, which could bring new problems to the traditional relay setting calculation.

5. Simulation Analysis and Verification

To verify the accuracy of the constructed short-circuit current calculation model and the transient response characteristics of converters during a fault, a simulation system of DFIG-WT with an LVRT control strategy was built in the MATLAB/Simulink simulation platform based on the demo for detailed DFIG-WT. The simulation system is shown in Figure 5. Major parameters of the system are listed in Table 2.

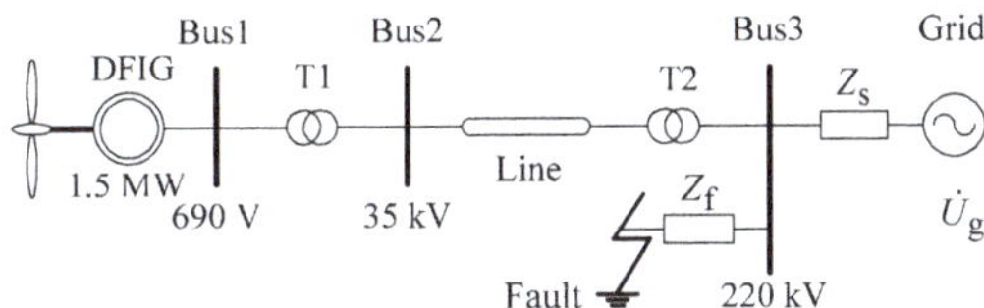

Figure 5. Simulation system.

Table 2. Parameters of the DFIG-WT simulation model.

Parameter	Value	Parameter	Value
Rated capacity	1.5 MW	Switching frequency of RSC	1.6 kHz
Rated voltage of stator	690 V	Proportionality coefficient of inner current loop	0.578
Stator resistance	0.023 p.u.	Integral coefficient of inner current loop	10.58
Rotor resistance	0.016 p.u.	Switching frequency of GSC	2.7 kHz
Stator inductance	3.08 p.u.	Proportionality coefficient of DC voltage loop	6.17
Rotor inductance	3.06 p.u.	Integral coefficient of DC voltage loop	400
Rated DC bus voltage	1150 V	Reactive current coefficient K_d	1.5
Active current-limiting of rotor	0.9 p.u.	Maximum current-limiting value of rotor	1.15 p.u.
DC bus capacitor	0.0032 p.u.	Modulation coefficient of PWM	0.95

5.1. Contrast Verification of Key Electrical Quantities of Generator and Converter

In this simulation case, DFIG-WT operates at a supersynchronous state with a unit power factor when a three-phase-to-ground fault through transition resistance Z_f occurs on Bus 3 at 0.113 s, making the voltage at the generator terminal drop to 0.65 p.u. Before the fault, the stator active power is about 0.82 p.u. and GSC active power is about 0.18 p.u. The slip ratio is about -0.21.

5.1.1. Verification of DC Bus Voltage

The postfault DC bus voltage was calculated according to Equation (21) and compared with the simulation waveform under the fault condition. According to Equation (4) and the analysis in Section 3.1, when the voltage at the generator terminal drops suddenly, there will be transient fluctuation in U_{dc}, which is caused by the power imbalance of the converters at the two sides. As shown in Figure 6a, the calculated waveform coincides with the simulated waveform, verifying the accuracy of Equation (21).

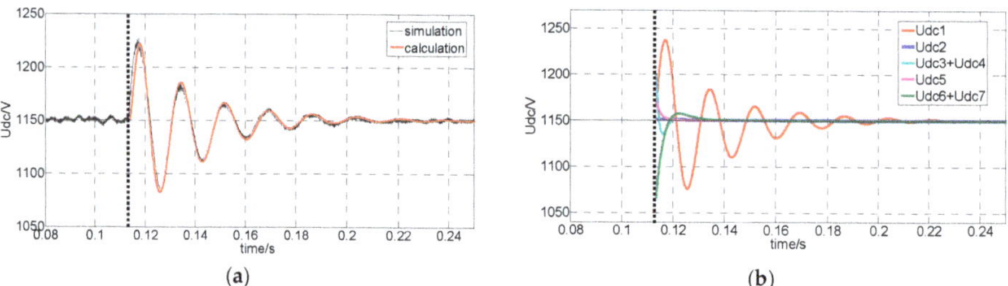

Figure 6. Comparison of DC bus voltage and decomposition of frequency components: (**a**) comparison between simulated and calculated waveform; (**b**) decomposition of frequency components.

The decomposition of transient attenuation components in U_{dc}, which are calculated from Equation (21), is shown in Figure 6b. Different frequency components are represented by different colors and their relevant coefficients. As shown in Figure 6b, the fundamental component takes the dominant role among all attenuation components of U_{dc}. The remaining components attenuate to 0 in less than one period and their amplitudes are smaller than the fundamental component.

5.1.2. Verification of GSC Current

A-phase GSC current is calculated according to Equation (23) and is compared with the simulated waveform, as shown in Figure 7a. Under the simulation condition, the peak value of the GSC A-phase short-circuit current is about 0.39 p.u., and generally damps to the steady-state value of 0.16 p.u. in about four periods. The theoretical calculation accurately coincides with simulation results.

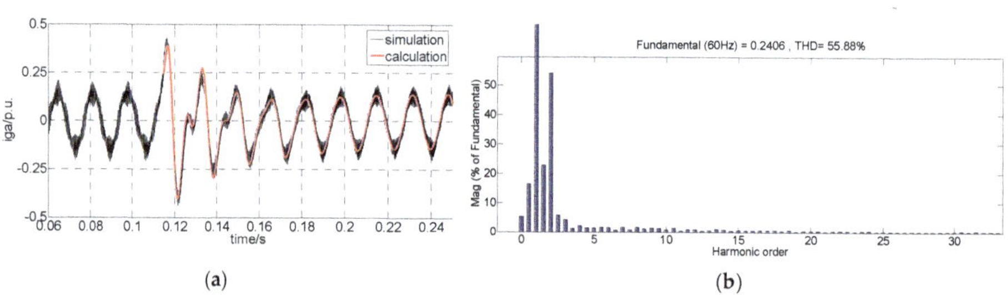

Figure 7. GSC short-circuit current and its frequency spectral analysis: (**a**) comparison between simulated and calculated waveform; (**b**) frequency spectral analysis.

It can be observed from Figure 7a that there are evident harmonic components in the first three cycles. The frequency spectral analysis result of the first cycle of simulated A-phase GSC current is shown in Figure 7b. The total harmonic distortion (THD) of the current of the selected window is about 55.88%, and the second harmonic content is about 54.2%. According to the theoretical analysis in Section 4, the fundamental attenuation component, which accounts for a high proportion in DC bus voltage, will induce the second harmonic frequency component in the GSC current. The simulation results coincide with the theoretical analysis.

The characteristics of the second harmonic component in the GSC current are studied further under supersynchronous and subsynchronous conditions. The second harmonic frequency component in GSC is positively related to the voltage dip level according to Equation (23), meaning that a lower residual voltage will cause a higher proportion of the second harmonic component in the short-circuit current. When the terminal voltage drops to 40% U_n, the proportions of the second harmonic component in the short-circuit current of DFIG-WT and GSC are as shown in Figure 8. Moreover, the proportion gets higher under subsynchronous conditions, with up to 68.4% in GSC current and 26.6% in DFIG-WT current.

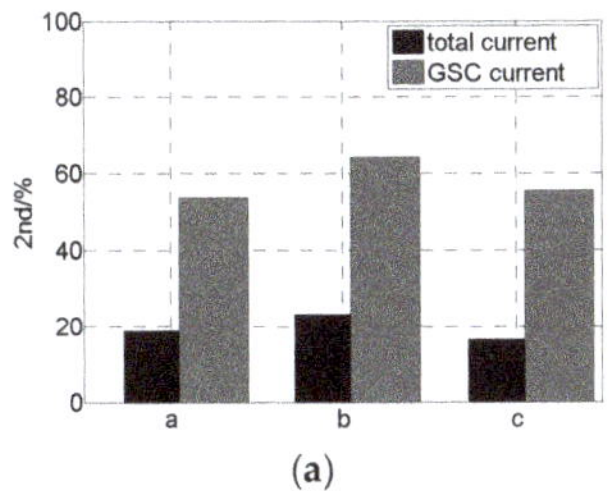
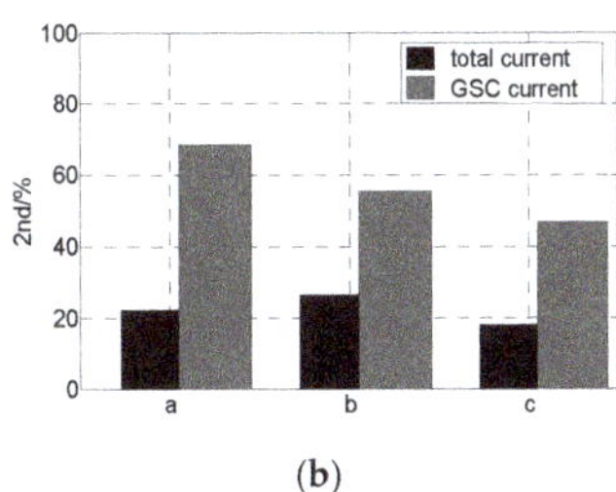

(a)   (b)

Figure 8. Proportions of the second harmonic component in short-circuit current of DFIG-WT and GSC: (**a**) supersynchronous, $s = -0.2$; (**b**) subsynchronous, $s = 0.2$.

Since the method of second harmonic restraint is widely used in transformer protection to overcome maloperation caused by a magnetizing inrush current, and the threshold for second harmonic content is usually set at 15–20%, the high proportion of the second harmonic component generated by DFIG-WT may cause the transformer protection failure to operate.

5.1.3. Verification of Rotor Current

A-phase rotor current was calculated according to Equation (20) and transferred to the rotor coordinate system. A comparison between the calculated waveform and simulated waveform is shown in Figure 9a. According to Equation (19) and Table 1, the fundamental component Δi_{r1} is the key attenuation component of rotor short-circuit current in the synchronous reference frame, while in the rotor coordinate system, its frequency is $(1 - s)$ times the rotor fundamental frequency. In this simulation case, with a slip ratio of -0.21, the frequency of Δi_{r1} is about 72 Hz and it attenuates to 0 by about 0.064 s.

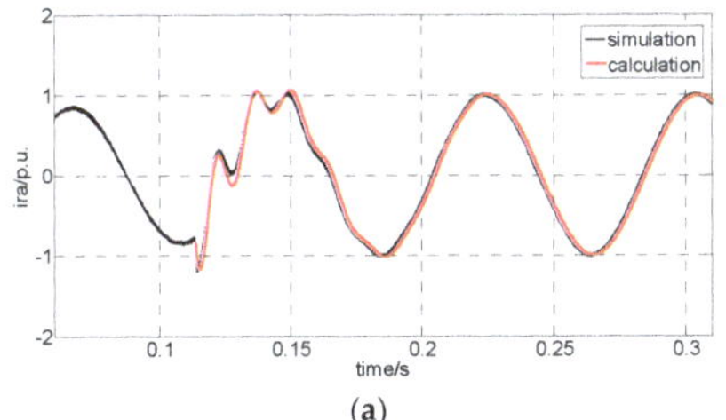
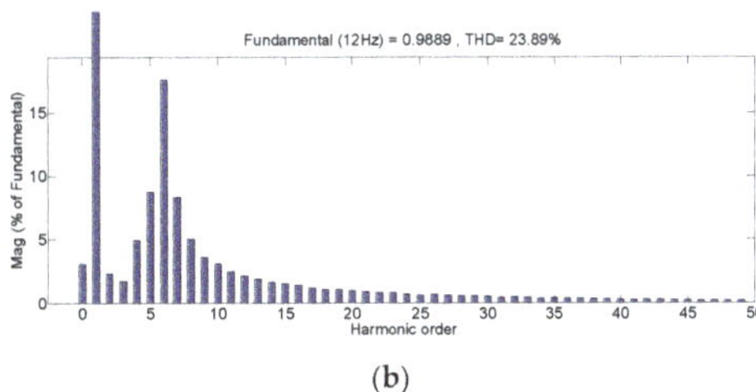

(a)   (b)

Figure 9. Rotor short-circuit current and its frequency spectral analysis: (**a**) comparison between simulated and calculated waveform; (**b**) frequency spectral analysis.

Frequency spectral analysis results of the first cycle of simulated rotor current are shown in Figure 9b. It can be seen from Figure 9b that among different harmonic components, the sixth harmonic accounts for the largest proportion, which is 72 Hz, with a fundamental of 12 Hz. Its proportion relative to the fundamental wave is 17.6%. Besides, as discussed in Section 3.2, there is no sudden change of rotor current at the moment the fault happens because the calculation considers the sampling delay and PWM small inertial characteristics. Theoretical analysis coincides with simulation results, which verifies the accuracy of the theory.

5.1.4. Verification of Stator Current

A comparison between the A-phase stator current calculated by Equation (25) and the simulated waveform is shown in Figure 10. When the voltage at the generator terminal drops to 0.65 p.u., the peak value of A-phase stator short-circuit current is about 1.37 p.u. and the steady-state value is about 0.81 p.u. According to Equation (25) and as shown in Figure 10b, DC is the main transient attenuation component of the stator short-circuit current.

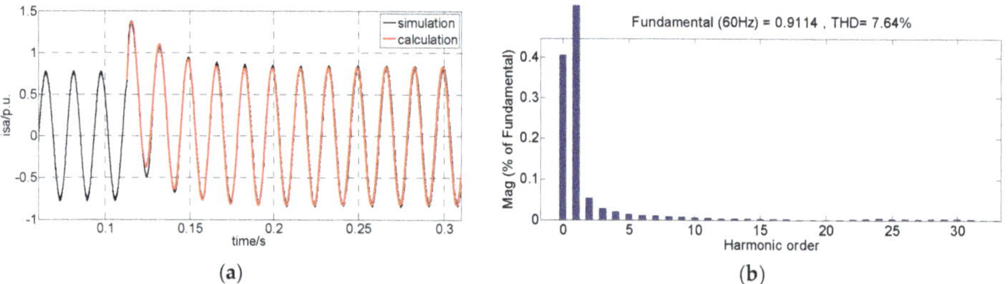

Figure 10. Stator short-circuit current and its frequency spectral analysis: (**a**) comparison between simulated and calculated waveform; (**b**) frequency spectral analysis.

Compared with Figure 7, the short-circuit current of GSC is about 20% of the stator short-circuit current in a steady state, while the peak value of GSC SCC is about 28.5% of the stator SCC. In [15–17], the GSC current is neglected and the stator current is used to replace the short-circuit total current of DFIG-WT, which will surely result in a large calculation error.

5.2. Contrast Verification of DFIG-WT Transient Characteristics under Different Situations

For a more comprehensive analysis and verification of the theoretical analysis, 15 groups of experiments were carried out under different operating conditions and voltage drops. A comparison between simulated results and calculated results of key characteristic parameters of the short-circuit current is shown in Figure 11, with solid lines representing simulation results and dotted lines representing calculated results.

In the experiments, three typical working conditions of DFIG-WT were chosen: supersynchronous ($s = -0.2$), synchronous ($s = 0.01$), and subsynchronous ($s = 0.2$). The voltage at the generator terminal dropped to 0.35 p.u., 0.47 p.u., 0.6 p.u., 0.75 p.u., and 0.9 p.u., respectively. Key characteristic parameters include peak values of DFIG-SCC and stator SCC, and steady-state values of DFIG-SCC and GSC-SCC.

The diagrams in the first row of Figure 11 show variations of the four key characteristic parameters with different residual voltages and operating conditions. Under the supersynchronous condition, the peak value of total current is higher than that of stator current. However, the opposite phenomenon is observed under the subsynchronous condition, while under the synchronous condition, these two parameters are almost equal to each other. This is because GSC current has the same direction as stator current under the supersynchronous condition, the opposite direction under the subsynchronous condition, and almost 0 under the synchronous condition.

The diagrams in the second row of Figure 11 show steady-state value proportions of GSC-SCC of DFIG-SCC. Under the subsynchronous condition, the proportion of GSC current is higher than that in other conditions, reaching 35% at the most.

Among the comparisons in all cases, the maximum calculation error is about 6.8%, as shown in Figure 11, which occurred at the peak value of DFIG-SCC when $s = -0.2$ and the voltage dropped to 0.35 p.u.

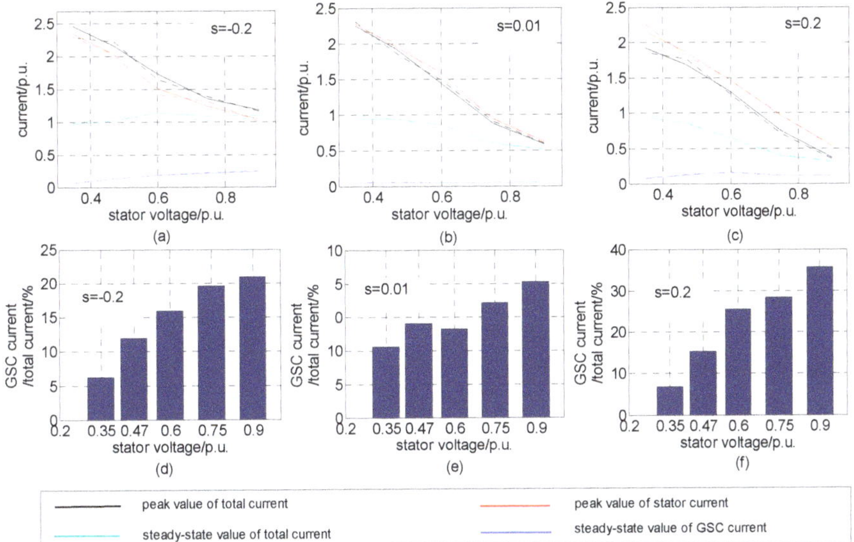

Figure 11. Contrast verification of key characteristic parameters of short-circuit current under different conditions, as well as proportions of steady-state GSC-SCC.

6. Verification with LVRT Test Data

According to the grid criteria for wind power [27], the LVRT test must be conducted with all kinds of wind turbines before they have access to the grid, which provides reference data for studying the fault characteristics of wind turbine generators. The LVRT test schematic diagram is shown in Figure 12a. A movable vehicle-mounted container structure is adopted by LVRT test devices, including a voltage sag generator and remote console cabinet, as shown in Figure 12b. The terminal voltage of tested wind turbines is remotely controlled by adjusting the voltage division ratio of the current, limiting reactance, and short-circuit reactance. The three-phase voltage and current at the terminal of tested wind turbines are saved by a Dewetron DEWE-5000 high-precision recorder (DEWETRON, Grambach, Austria).

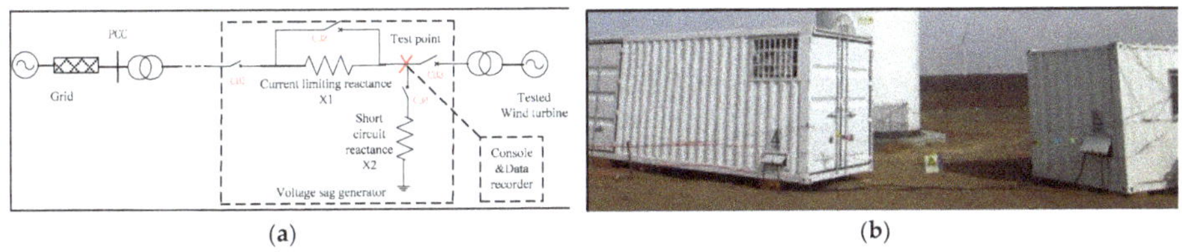

Figure 12. (a) LVRT test schematic diagram, (b) picture of the LVRT field test devices.

In the following section, a group of field test data are analyzed and compared with the proposed analytic expressions. Parameters of the tested DFIG-WT are listed in Table 3. In the test, the terminal voltage dropped to about 0.23 p.u. under subsynchronous ($P = 0.28$ p.u., $s = 0.2$) and supersynchronous ($P = 0.97$ p.u., $s = -0.2$) conditions.

Table 3. Parameters of the tested DFIG-WT.

Parameter	Value	Parameter	Value
Rated capacity	1.5 MW	Leakage inductance of rotor	0.0162 Ω
Rated voltage of stator	690 V	Exciting inductance	1.123 Ω
Rated voltage of rotor	1800 V	Switching frequency of converter	2.5 kHz
Stator resistance	0.0023 Ω	Current-limiting value of rotor	1.5 p.u.
Rotor resistance	0.0024 Ω	Reactive current coefficient	1.8
Leakage inductance of stator	0.0184 Ω	Rated voltage of DC bus	1150 V
Proportionality coefficient of inner current loop	0.6	Proportionality coefficient of DC voltage loop	10
Integral coefficient of inner current loop	15	Integral coefficient of DC voltage loop	500
DC bus capacitor	0.001 F	Modulation coefficient of PWM	0.95

Active and reactive power generated by wind turbines under two test conditions are shown in Figure 13. As the grid codes require, wind power generators should output reactive power during the LVRT process to support grid voltage recovery. It can be concluded from Equation (3) that active power during the LVRT process under different operating conditions should be the same with the same voltage dip level. As shown in Figure 13, the reactive power is about 0.3 p.u. under both test conditions during the LVRT process as the residual voltages are the same. Reactive power output is the primary concern for DFIG-WT control, limiting the active power output capacity. The active power in the supersynchronous condition (about 0.21 p.u.) is slightly larger than that in the subsynchronous condition (about 0.14 p.u.). This is because GSC outputs active power under the supersynchronous condition and absorbs active power under the subsynchronous condition.

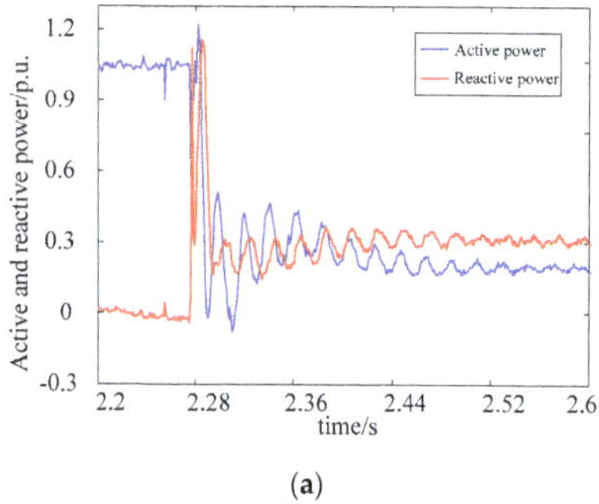

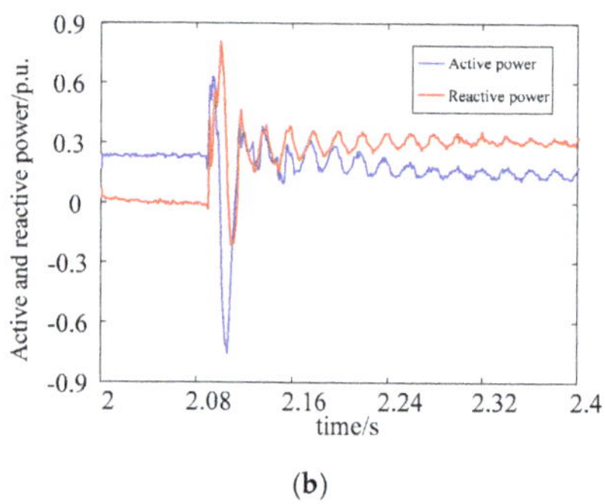

(a)   (b)

Figure 13. Active and reactive power under different conditions: (**a**) supersynchronous condition; (**b**) subsynchronous condition.

Parameters of the tested DFIG-WT are substituted into Equations (21), (24), and (26) to obtain the calculated waveforms of the DFIG-SCC under different conditions. A comparison between the actual waveforms and calculated waveforms is shown in Figure 14. Transient characteristics and the attenuation law of the calculated waveform and simulated waveform are consistent. However, the first three circles do not match well. This is because the controller may not be able to perform as ideally as the simulated model under severe fault conditions. Moreover, the steady-state short-circuit current agrees precisely.

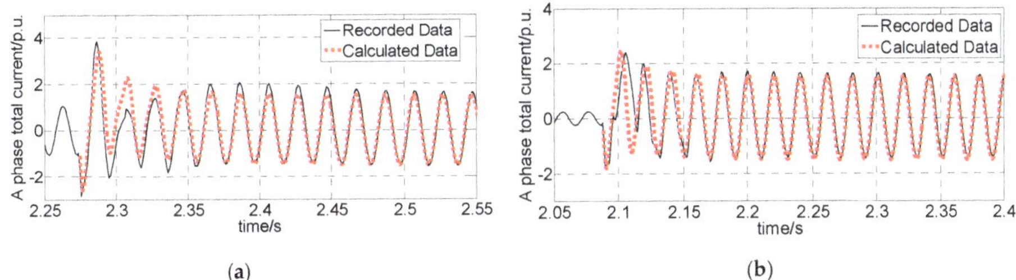

Figure 14. Comparison between recorded data and calculated data under two operating conditions: (**a**) supersynchronous condition; (**b**) subsynchronous condition.

According to the test data, the steady-state value of the short-circuit current is 1.642 p.u. under the supersynchronous condition and 1.507 p.u. under the subsynchronous condition. Parameters of DFIG-WT are transferred into per-unit values and substituted into Equation (3). The reference values of rotor current are gained as:

$$\begin{cases} i^*_{rq,lvrt} = -[1.8 \times (0.9 - 0.23)] \times \frac{3.5961}{3.5381} - \frac{0.23}{3.5961} = -1.2908 \\ i^*_{rd,lvrt} = \sqrt{1.5^2 - i^{*2}_{rq,lvrt}} = 0.7641 \end{cases} \tag{31}$$

The reference values of rotor current are substituted into Equation (26). The steady-state short-circuit current under the two conditions is calculated as:

$$\begin{cases} I_{Tf\text{-}full} &= \left| -1.2 \times \frac{3.5381}{3.5961} \times i^*_{rd,lvrt} - j(\frac{0.23}{3.5961} - i^*_{rq,lvrt}) \right| \\ &= |-0.9021 - j1.3339| = 1.6103 \\ I_{Tf\text{-}low} &= \left| -0.8 \times \frac{3.5381}{3.5961} \times i^*_{rd,lvrt} - j(\frac{0.23}{3.5961} - i^*_{rq,lvrt}) \right| \\ &= |-0.6041 - j1.3339| = 1.4632 \end{cases} \tag{32}$$

where $I_{Tf-full}$ and I_{Tf-low} are amplitudes of short-circuit currents of DFIG-WT under full and sub synchronous conditions, respectively. Their relative errors with test data are 1.9% and 2.9%, respectively, verifying the accuracy of the proposed analytical expression.

The second harmonic contents of the first cycle in the three-phase short-circuit current under two conditions are shown in Table 4. According to Table 4, the following conclusions can be drawn as follows:

(1) According to theoretical derivation and Equation (23), the deeper the voltage drops, the higher the second harmonic content will be. It can be seen from Table 4 that when the terminal voltage drops to 0.23 p.u., the second harmonic content in the three-phase short circuit current can reach up to 51.74%, far exceeding the transformer's second harmonic setting value (15–20%).

(2) Under the same voltage drop level, the second harmonic content is higher under the sub-synchronous condition than the super-synchronous condition. This is because the secondary harmonics are mainly generated by GSC, and the proportion of GSC current to the total current is higher under the sub-synchronous condition, which is consistent with the result in Figure 8.

(3) Then second harmonic content in the three-phase current is different, and this phenomenon can be explained by the mechanism of second harmonic generation. According to the theoretical derivation and the flow chart in Figure 4, the second harmonic component in three phase short circuit currents is originally caused by the DC component in stator flux, which is generated due to the conservation law of flux linkage and its amplitude is determined by the instantaneous value of stator flux at the time the fault occurs. Therefore, the DC attenuation components in the three phases are not equal, so that the second harmonic content in three phases is also different.

(4) The phase with the highest second harmonic content is related to the time of failure due to the same reason at the point (3). In Table 4, the highest second harmonic content is in phase C when the fault occurred at 2.27 s under the super synchronous condition, and it is phase A when the fault occurred at 2.08 s under the sub synchronous condition. As shown in Figure 8, in the simulation study, the fault occurs at the same time in both cases, and the phase with the highest second harmonic content is phase B.

Table 4. Second harmonic contents of three-phase short-circuit current under two conditions.

Operating Condition	A-Phase	B-Phase	C-Phase
Super-synchronous	32.68%	25.16%	43.40%
Subs-synchronous	49.45%	34.58%	51.74%

7. Conclusions

In this paper, transient characteristics of DFIG-WT short-circuit current under converter control are analyzed and a detailed analytical calculation model of short-circuit total current of DFIG-WT, including GSC current and stator current, is deduced. Based on the established model, the characteristics of steady-state fault current of DFIG-WT are further studied. The accuracy of the theoretical analysis and mathematical deduction is verified by comparing simulation test and LVRT field test data. The main work of this paper can be summarized and conclusions can be drawn as follows:

(1) Equations for GSC current, DC bus voltage, and rotor current are constructed. The disturbance evolution mechanism in converters and the coupling relationships of the electrical quantities are thus revealed.

(2) The transient analytical models of GSC current, DC bus voltage, stator current, and rotor current are constructed. With the transient analytical model, short-circuit current frequency components, key influencing factors, and attenuation characteristics are quantized.

(3) The amplitude of steady-state short-circuit current of DFIG-WT varies with the post-fault terminal voltage and pre-fault power nonlinearly, considering the multi-limitation of rotor current. Estimation formulas for the maximum steady-state SCC of DFIG-WT and the corresponding voltage are put forward and verified by simulation.

(4) It is proved by theoretical analysis and simulation that the proportion of short-circuit current of GSC is related to the pre-fault operation state. The proportion of GSC current is higher under subsynchronous conditions. The accurate calculation of the total short-circuit current of DFIG-WT should take the influence of GSC current into account, or it will generate up to 30% error.

(5) Reasons for high second harmonic contents contained in the transient short-circuit current of GSC are disclosed. It is concluded that the second harmonic content is positively related to the voltage dip level, and the second harmonic component might have an adverse impact on transformer differential current protection.

The results and conclusions in this paper could provide theoretical references for short-circuit current calculation of power systems with DFIG-WTs connected to the grid, as well as optimizing settings and redesigns for relevant protection, such as transformer protection with second harmonic restraint. Moreover, from the aspect of control strategies of convertors, if the fluctuation of DC bus voltage could be suppressed, the second harmonic content contained in the short-circuit current should also be decreased.

Author Contributions: Conceptualization, J.L. and T.Z.; methodology, J.L.; software, J.L.; validation, J.L. and T.Z.; formal analysis, J.L.; investigation, J.L.; resources, Z.W.; data curation, J.L.; writing—original draft preparation, J.L.; writing—review and editing, J.L., T.Z. and Z.W.; visualization, J.L.; supervision, T.Z. and Z.W.; project administration, T.Z. and Z.W.; funding acquisition, Z.W.

Funding: This work was funded by the National Key Research and Development Plan of China (2016YFB0900640) and Research Project of State Grid Corporation of China (5211TZ16000F).

Acknowledgments: We acknowledge the North China Electric Power Research Institute (NCEPRI), which provided technical support and LVRT field test data for this research.

Conflicts of Interest: The authors declare no conflict of interest.

Nomenclature

u_s, u_r	Stator and rotor voltage vectors	L_s, L_r	Stator and rotor self-inductances
i_s, i_r	Stator and rotor current vectors	R_s, R_r	Stator and rotor resistances
ψ_s, ψ_r	Stator and rotor flux vectors	L_m	Mutual inductance
i_g	GSC current vector	C	Capacitance of the capacitor
i_T	Total current vector of DFIG-WT	L, R	Filter inductance and resistance
U_{dc}	DC bus voltage	P_g	Active power of the GSC
$i_L, i_{g\text{-}dc}$	DC-side currents of RSC and GSC	$i_{ra,b,c}$	AC-side three-phase rotor currents
P_{load}	DC-side active power of the RSC	$*$	Superscript donating reference value
ω_1	Synchronous angular frequency	s	Slip of DFIG
U_{s0}, U_{s1}	Amplitudes of pre-fault and post-fault stator voltages	P^*_{s0}, Q^*_{s0}	Reference values of pre-fault stator active and reactive powers
p	Differential operator	S_{abc}	Switch function of the converter
d, q	Subscripts donating d-axis component and q-axis component	f, n	Subscripts donating forced component and natural component

Note: a symbol in bold form denotes a complex vector.

Appendix A

Control diagrams of the RSC and GSC

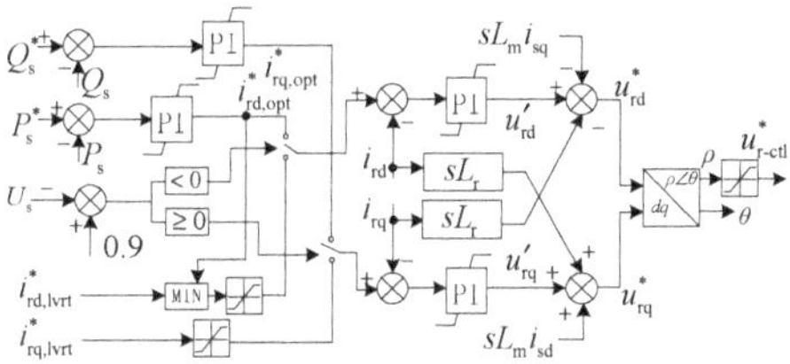

Figure A1. Control diagram of the RSC.

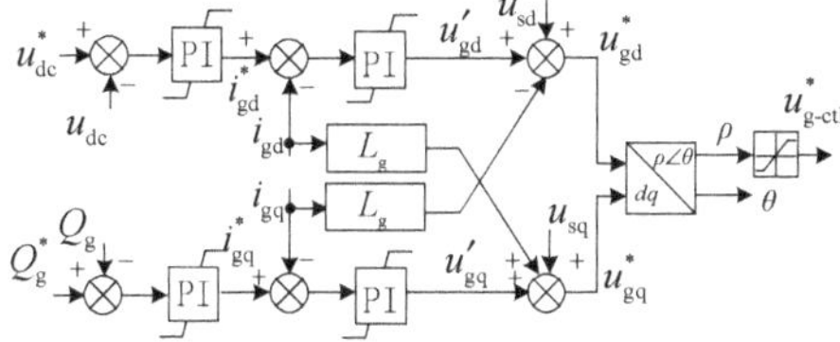

Figure A2. Control diagram of the GSC.

Appendix B

1. Coefficients of rotor current in Equation (19):

$$\Delta i_{r1} = k_0 \frac{2\omega_{ci}k_4 + k_2}{k_1k_4 - k_2k_3}\left(\frac{1}{\tau_s} + j\omega_1\right)$$

$$\Delta i_{r2} = -k_0 \frac{k_1 + 2\omega_{ci}k_3}{k_1k_4 - k_2k_3}\frac{1}{\tau_i}$$

$$\Delta i_{r3} = k_0[2\omega_{ci}(\frac{k_1 + 2\omega_{ci}k_3}{k_1k_4 - k_2k_3} - \frac{2\omega_{ci}k_4 + k_2}{k_1k_4 - k_2k_3}) - k_5]$$

$$\Delta i_{r4} = k_0 k_5$$

where

$$k_0 = -j(\frac{1}{\omega_1 \tau_s} + j)\frac{L_m}{sL_s}\frac{\Delta U_s}{3T_s K_{p\_i}K_{pwm}}$$

$$k_1 = \frac{1}{\tau_i}[(\omega_{ci} - \frac{1}{\tau_s} - j\omega_1)^2 + \omega_{ci}^2]$$

$$k_2 = (\frac{1}{\tau_s} + j\omega_1)[(\frac{1}{\tau_i} - \omega_{ci})^2 + \omega_{ci}^2]$$

$$k_3 = -\omega_{ci}^2 - [\omega_{ci} - (\frac{1}{\tau_s} + j\omega_1)]^2$$

$$k_4 = -\omega_{ci}^2 - (\omega_{ci} - \frac{1}{\tau_i})^2$$

$$k_5 = \frac{(2\omega_{ci} - \frac{1}{\tau_i})(k_1 + 2\omega_{ci}k_3) - [2\omega_{ci} - (\frac{1}{\tau_s} + j\omega_1)](k_2 + 2\omega_{ci}k_4)}{k_1k_4 - k_2k_3}$$

2. Coefficients of DC bus voltage in Equation (21):

$$U_{dc1} = -0.75m|\Delta i_{r1}|\frac{\omega_1 a}{Cn}$$

$$U_{dc2} = 0.75m\frac{\omega_1}{C}|\Delta i_{r2}|\cos(\alpha_2 + \delta)\frac{1/\tau_i}{(R + 1/\tau_i)^2 - M^2}$$

$$U_{dc3} = -0.75m\frac{\omega_1}{C}|\Delta i_{r3}|\cos(\alpha_3 + \delta)\frac{\sqrt{2}\omega_{ci}}{x}$$

$$U_{dc4} = -0.75m\frac{\omega_1}{C}|\Delta i_{r4}|\cos(\alpha_4 + \delta)\frac{\sqrt{2}\omega_{ci}}{x}$$

$$U_{dc5} = -0.75m\frac{\omega_1}{C}|\Delta i_r^*|\cos(\alpha_5 + \delta)[\frac{\omega_{ci}}{(R + \omega_{ci})^2 - M^2}]$$

$$U_{dc6} = \frac{1}{2M}[U_{dc1}\frac{n}{a}\frac{\lambda_1\cos(\alpha_1 + \delta - \theta_4)}{\sqrt{(\lambda_1 + 1/\tau_s)^2 + \omega^2}} + U_{dc2}\frac{(R + 1/\tau_i)^2 - M^2}{1/\tau_i}\frac{\lambda_1}{\lambda_1 + 1/\tau_i}$$
$$+ U_{dc3}\frac{x}{\sqrt{2}\omega_{ci}}\frac{\lambda_1 + \omega_{ci}}{(\lambda_1 + \omega_{ci})^2 + (\omega_{ci})^2} + U_{dc4}\frac{x}{\sqrt{2}}\frac{\lambda_1}{(\lambda_1 + \omega_{ci})^2 + (\omega_{ci})^2} + U_{dc5}\frac{(R + \omega_{ci})^2 - M^2}{\lambda_1 + \omega_{ci}}]$$

$$U_{dc7} = -\frac{1}{2M}[U_{dc1}\frac{n}{a}\frac{\lambda_2\cos(\alpha_1 + \delta - \theta_5)}{\sqrt{(\lambda_2 + 1/\tau_s)^2 + \omega^2}} + U_{dc2}\frac{(R + 1/\tau_i)^2 - M^2}{1/\tau_i}\frac{\lambda_2}{\lambda_2 + 1/\tau_i}$$
$$+ U_{dc3}\frac{x}{\sqrt{2}\omega_{ci}}\frac{\lambda_2 + \omega_{ci}}{(\lambda_2 + \omega_{ci})^2 + (\omega_{ci})^2} + U_{dc4}\frac{x}{\sqrt{2}}\frac{\lambda_2}{(\lambda_2 + \omega_{ci})^2 + (\omega_{ci})^2} + U_{dc5}\frac{(R + \omega_{ci})^2 - M^2}{\lambda_2 + \omega_{ci}}]$$

where $\lambda_{1,2} = R \pm M$,

$$\alpha_1 = a\tan\frac{\mathrm{Re}[\Delta i_{r1}]}{\mathrm{Im}[\Delta i_{r1}]}, \quad \alpha_2 = a\tan\frac{\mathrm{Re}[\Delta i_{r2}]}{\mathrm{Im}[\Delta i_{r2}]}, \quad \alpha_3 = a\tan\frac{\mathrm{Re}[\Delta i_{r3}]}{\mathrm{Im}[\Delta i_{r3}]}, \quad \alpha_4 = a\tan\frac{\mathrm{Re}[\Delta i_{r4}]}{\mathrm{Im}[\Delta i_{r4}]}, \quad \alpha_5 = a\tan\frac{\mathrm{Re}[\Delta i_r^*]}{\mathrm{Im}[\Delta i_r^*]}$$

$$ae^{j\theta_1} = 1/\tau_s + j\omega$$

$$ne^{j\theta_2} = (R + 1/\tau_s)^2 - \omega^2 + M^2 + 2j\omega(R + 1/\tau_s)$$

$$xe^{j\theta_3} = (R + 1/3T_s + j/3T_s)^2 + M^2$$

$$\theta_4 = a\tan\frac{\omega_1}{\lambda_1 + 1/\tau_s}$$

$$\theta_5 = a\tan\frac{\omega_1}{\lambda_2 + 1/\tau_s}$$

$$\beta_1 = \alpha_1 + \delta + \theta_1 - \theta_2$$

$$\beta_2 = \theta_3 - \frac{\pi}{4}$$

3. Coefficient of GSC current in Equation (22):

$$I_{g1} = \left(\omega_{cv} - \frac{1}{a^2 \tau_v \tau_s}\right)\cos\beta_1 - \frac{\omega}{a^2 \tau_v}\sin\beta_1$$

$$I_{g2} = \left(\omega_{cv} - \frac{1}{a^2 \tau_v \tau_s}\right)\sin\beta_1 + \frac{\omega}{a^2 \tau_v}\cos\beta_1$$

$$I_{g3} = U_{dc3}\left(\cos\beta_2 - \frac{\cos\beta_2 - \sin\beta_2}{2\omega_{ci}^2 \tau_v}\right) - U_{dc4}\left(\sin\beta_2 - \frac{\sin\beta_2 - \cos\beta_2}{2\omega_{ci}^2 \tau_v}\right)$$

$$I_{g4} = U_{dc3}\left(\sin\beta_2 - \frac{\sin\beta_2 + \cos\beta_2}{2\omega_{ci}^2 \tau_v}\right) - U_{dc4}\left(\cos\beta_2 - \frac{\cos\beta_2 + \sin\beta_2}{2\omega_{ci}^2 \tau_v}\right)$$

$$\begin{aligned}
I_{g5} = & -\frac{U_{dc1}}{\tau_v}\left[\frac{\cos\beta_1(1/\tau_v - 1/\tau_s) - \omega\sin\beta_1}{(1/\tau_v - 1/\tau_s)^2 + \omega^2}\right] + \frac{U_{dc2}\tau_i}{\tau_v - \tau_i} - \frac{U_{dc3}}{\tau_v}\left[\frac{(1/\tau_v - \omega_{ci})\cos\beta_2 + \omega_{ci}\sin\beta_2}{(1/\tau_v - \omega_{ci})^2 + \omega_{ci}^2}\right] \\
& -\frac{U_{dc4}}{\tau_v}\left[\frac{(1/\tau_v - \omega_{ci})\sin\beta_2 - \omega_{ci}\cos\beta_2}{(1/\tau_v - \omega_{ci})^2 + \omega_{ci}^2}\right] + \frac{U_{dc5}}{\omega_{ci}\tau_v - 1} + \frac{U_{dc6}}{-\lambda_1 \tau_v - 1} + \frac{U_{dc7}}{-\lambda_2 \tau_v - 1}
\end{aligned}$$

References

1. Mai, T.; Wiser, R.; Sandor, D.; Brinkman, G.; Heath, G.; Denholm, P.; Hostick, D.J.; Darghouth, N.; Schlosser, A.; Strzepek, K. *Renewable Electricity Futures Study Volume 1: Exploration of High-Penetration Renewable Electricity Futures*; National Renewable Energy Laboratory: Golden, CO, USA, 2017.
2. Li, H.; Eseye, A.T.; Zhang, J.; Zheng, D. Optimal energy management for industrial microgrids with high-penetration renewables. *Prot. Control Mod. Power Syst.* **2017**, *2*, 12. [CrossRef]
3. Tazil, M.; Kumar, V.; Bansal, R.C.; Kong, S.; Dong, Z.Y.; Freitas, W.; Mathur, H.D. Three-phase doubly fed induction generators: An overview. *IET Electr. Power Appl.* **2010**, *4*, 75–89. [CrossRef]
4. Strezoski, V.L.; Prica, D.M. Short-circuit analysis in large-scale distribution systems with high penetration of distributed generators. *IEEE/CAA J. Autom. Sin.* **2017**, *4*, 243–251. [CrossRef]
5. Wang, X.; Pan, X.; Wan, J.; Qi, J.; Zhang, H.; Li, S. A wind farm short-circuit current calculation practical model applied to high voltage power grid simulation. In Proceedings of the 2014 International Conference on Mechatronics, Electronic, Industrial and Control Engineering, Shenyang, China, 17–19 November 2014. [CrossRef]
6. Shen, S.; Lin, D.; Wang, H.; Hu, P.; Jiang, K.; Lin, D.; He, B. An adaptive protection scheme for distribution systems with DGs based on optimized Thevenin equivalent parameters estimation. *IEEE Trans. Power Deliv.* **2017**, *32*, 411–419. [CrossRef]
7. Kauffmann, T.; Karaagac, U.; Kocar, I.; Jensen, S.; Mahseredjian, J.; Farantatos, E. An accurate type III wind turbine generator short circuit model for protection applications. *IEEE Trans. Power Deliv.* **2017**, *32*, 2370–2379. [CrossRef]
8. Zhou, N.; Wu, J.; Wang, Q. Three-phase short-circuit current calculation of power systems with high penetration of VSC-based renewable energy. *Energies* **2018**, *11*, 537. [CrossRef]
9. Dosoglu, M.K.; Guvenc, U.; Sonmez, Y.; Yilmaz, C. Enhancement of demagnetization control for low-voltage ride-through capability in DFIG-based wind farm. *Electr. Eng.* **2018**, *100*, 491–498. [CrossRef]
10. General Administration of Quality Supervision, Inspection and Quarantine of the People's Republic of China and the Standardization Administration of the People's Republic of China. *Technical Rule for Connecting Wind Farm to Power System*; Xinhua Press: Beijing, China, 2012.
11. VDE Verlag GmbH. *Technical Requirements for the Connection to and Parallel Operation with Low-Voltage Distribution Networks*; VDE Verlag GmbH: Berlin, Germany, 2017.
12. Swain, S.; Ray, K.P. Short circuit fault analysis in a grid connected DFIG based wind energy system with active crowbar protection circuit for ride-through capability and power quality improvement. *Electr. Power Energy Syst.* **2017**, *84*, 64–75. [CrossRef]
13. Zhu, Q.; Ding, M.; Han, P. Equivalent modeling of DFIG-based wind power plant considering crowbar protection. *Math. Probl. Eng.* **2016**, *2016*, 8426492. [CrossRef]
14. Yuan, Y.H.; Wu, F. Short-circuit current analysis for DFIG wind farm considering the action of a crowbar. *Energies* **2018**, *11*, 425. [CrossRef]

15. Yin, J.; Bi, T.; Xue, A.; Yang, Q. Study on short circuit current and fault analysis method of double fed induction generator with low voltage ride-through control strategy. *Trans. China Electr. Tech. Soc.* **2015**, *30*, 118–125.
16. Ouyang, J.; Zheng, D.; Xiong, X.; Xiao, C.; Yu, R. Short-circuit current of doubly fed induction generator under partial and asymmetrical voltage drop. *Renew. Energy* **2016**, *88*, 1–11. [CrossRef]
17. Kong, X.; Zhang, Z.; Yin, X.; Wen, M. Study of fault current characteristics of the DFIG considering dynamic response of the RSC. *IEEE Trans. Energy Convers.* **2014**, *29*, 278–287.
18. Abad, G.; Lopez, J.; Rodrlguez, A.M.; Marroyo, L.; Iwanski, G. *Doubly Fed Induction Machine: Modeling And Control For Wind Energy Generation*; Wiley-IEEE Press: Hoboken, NJ, USA, 2014; pp. 246–247.
19. Xiao, S.; Geng, H.; Zhou, H.; Yang, G. Analysis of the control limit for rotor-side converter of doubly fed induction generator-based wind energy conversion system under various voltage dips. *IET Renew. Power Gener.* **2013**, *7*, 71–81. [CrossRef]
20. Hu, J.; He, Y. DFIG wind generation systems operating with limited converter rating considered under unbalanced network conditions—Analysis and control design. *Renew. Energy* **2011**, *36*, 829–847. [CrossRef]
21. Chen, P.; Zhang, Z.; Yin, X.; Xiao, F.; Yang, Z.; Wang, L. Steady fault current calculation model of doubly-fed induction generator considering grid-side converter current and different control strategies. *Autom. Electr. Power Syst.* **2016**, *40*, 8–16.
22. Honrubia-Escribano, A.; Jiménez-Buendía, F.; Gómez-Lázaro, E.; Fortmann, J. Field validation of a standard type 3 wind turbine model for power system stability, according to the requirements imposed by IEC 61400-27-1. *IEEE Trans. Energy Convers.* **2018**, *33*, 137–145. [CrossRef]
23. Zhang, C.W.; Zhang, X. *PWM Rectifier and Control*; Machine Press: Beijing, China, 2003; pp. 114–117.
24. Xiao, F.; Zhang, Z.; Yin, X. Fault current characteristics of the DFIG under asymmetrical fault conditions. *Energies* **2015**, *8*, 10971–10992. [CrossRef]
25. Liu, S.M.; Bi, T.S.; Jia, K.; Yang, Q.X. Coordinated fault-ride-through strategy for doubly-fed induction generators with enhanced reactive and active power support. *IET Renew. Power Gener.* **2016**, *10*, 203–211. [CrossRef]
26. Lopez, J.; Sanchis, P.; Roboam, X.; Marroyo, L. Dynamic behavior of the doubly fed induction generator during three-phase voltage dips. *IEEE Trans. Energy Convers.* **2007**, *22*, 709–717. [CrossRef]
27. National Energy Administration. *Test Procedures of WIND Turbine Low Voltage Ride Through Ability*; Xinhua Press: Beijing, China, 2015.

Article

Numerical Investigation of Terrain-Induced Turbulence in Complex Terrain by Large-Eddy Simulation (LES) Technique

Takanori Uchida

Research Institute for Applied Mechanics (RIAM), Kyushu University, 6-1 Kasuga-kouen, Kasuga, Fukuoka 816-8580, Japan; takanori@riam.kyushu-u.ac.jp

Received: 4 July 2018; Accepted: 18 September 2018; Published: 2 October 2018

Abstract: In the present study, field observation wind data from the time of the wind turbine blade damage accident on Shiratakiyama Wind Farm were analyzed in detail. In parallel, high-resolution large-eddy simulation (LES) turbulence simulations were performed in order to examine the model's ability to numerically reproduce terrain-induced turbulence (turbulence intensity) under strong wind conditions (8.0–9.0 m/s at wind turbine hub height). Since the wind velocity and time acquired from the numerical simulation are dimensionless, they are converted to full scale. As a consequence, both the standard deviation of the horizontal wind speed (m/s) and turbulence intensity evaluated from the field observation and simulated wind data are successfully in close agreement. To investigate the cause of the wind turbine blade damage accident on Shiratakiyama Wind Farm, a power spectral analysis was performed on the fluctuating components of the observed time series data of wind speed (1 s average values) for a 10 min period (total of 600 data) by using a fast Fourier transform (FFT). It was suggested that the terrain-induced turbulence which caused the wind turbine blade damage accident on Shiratakiyama Wind Farm was attributable to rapid wind speed and direction fluctuations which were caused by vortex shedding from Tenjogadake (elevation: 691.1 m) located upstream of the wind farm.

Keywords: complex terrain; terrain-induced turbulence; turbulence intensity; LES; vortex shedding

---

1. Introduction

Recently, the number of accidents which have involved wind turbines constructed on complex terrain in the mountains is increasing rapidly. Recent studies by the author of the present study and others indicate that these investigated accidents are strongly associated with terrain-induced turbulence [1,2]. In these studies, terrain-induced turbulence is defined as "temporal and spatial fluctuations of airflow that are mechanically generated due to terrain irregularities." Furthermore, the author of the present study classified terrain-induced turbulence roughly into two kinds.

The first kind is "extraordinary" terrain-induced turbulence, which is generated with the passage of a typhoon or a rarely occurring meteorological phenomena. That is, this kind of turbulence is terrain-induced turbulence that is commonly generated in wind directions that are different from the prevailing wind direction, and that occur infrequently throughout a year. It has been reported that this kind of terrain-induced turbulence caused a serious accident that involved cracks on the wind turbine blades and other damage [1].

The other kind of terrain-induced turbulence is "ordinary" terrain-induced turbulence, which is generated under the prevailing wind direction. This turbulence has caused reduced power output from wind turbines, and damage to the interior and exterior of wind turbines (e.g., the breakdown of yaw motors and yaw gears), which have become evident issues [2].

Uchida [1] investigated "extraordinary" terrain-induced turbulence. When Typhoon No. 0918 (Melor) passed the southern part of the main island of Japan between 7 and 8 October 2009, strong winds with extremely strong turbulence fluctuations were observed over the ridge of Mt. Shirataki and the surrounding ridges in Houhoku-cho, Shimonoseki City, Yamaguchi Prefecture, Japan. This strong wind caused damage to a turbine blade on Shiratakiyama Wind Farm owned by Kinden Corporation (Figure 1). Uchida [1] studied the cause of this accident by reconstructing the atmospheric phenomena occurring on spatial scales between a few meters and a few hundred kilometers using a computer simulation. In this study, the airflow field from the time of the accident was initially reconstructed in detail using a combination of a mesoscale meteorological model and RIAM-COMPACT, which is based on a large-eddy simulation (LES) turbulence model. Subsequently, the airflow fields in the vicinity of wind turbine blades were reconstructed separately using a Reynolds-averaged Navier-Stokes equations (RANS) turbulence model in order to evaluate the wind pressure on the wind turbine blades. For this analysis, the time-averaged flow field data from the LES simulation were used for the boundary conditions. Finally, stress exerted on the blades was calculated using a finite element method (FEM) with the RANS analysis results as the boundary conditions.

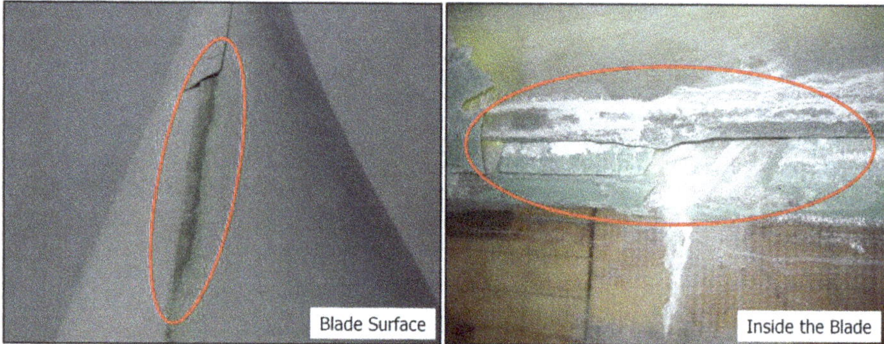

Figure 1. Photos of the blade damage.

The results of the above analyses revealed that large values of stress occurred at the junction between the dorsal and ventral sides of the leading edge (LE) (the LE dorsal–ventral junction, hereafter), and that the locations at which these values occurred matched the locations of the actual damage. The static strength of the adhesive bond used for the LE dorsal-ventral junction is quite high. However, adhesive bonds will likely break when repeated loading, as the one discussed above, is applied to LE dorsal–ventral junctions as a result of low-cycle fatigue; that is, fatigue associated with approximately 10^4 or fewer cycles for the number of cycles-to-failure. During the passage of Typhoon 0918, a large number of wind direction deviations were observed under high wind conditions. Therefore, it was surmised that damage such as cracks formed at the LE dorsal–ventral junction as a result of low-cycle fatigue, and this damage propagated further during subsequent wind turbine operation.

In contrast, Uchida [2] probed into "ordinary" terrain-induced turbulence by studying Taikoyama Wind Farm (located on Mt. Taikoyama, Nomura, Ine Town, Yosa District, Kyoto Prefecture, Japan). At this wind farm, at around 19:30 on 12 March 2013, a major accident occurred in which the nacelle and blades (hub height: approx. 50.0 m; weight: approx. 45.0 t) of Wind Turbine No. 3 fell to the ground as a result of a rupture of the tower in the vicinity of the tower top flange. Since the wind speed at the time of the accident was about 15.0 m/s, and this was within the design limit, investigations were conducted based on a standpoint that metal fatigue was the main cause of the accident. As a result of detailed investigations of the accident, the following was revealed: tensile stress on the welded joint between the tower shell and the tower top flange increased significantly because of the breakage of tower top bolts, which led to the formation of fatigue cracks on the internal wall of the tower shell in the vicinity of the weld toe. The formation of these fatigue cracks, in turn, caused the nacelle and blades to fall to the ground.

Detailed numerical wind simulation results for the prevailing wind direction showed that many wind velocity shears that deviated from those predicted by a power law occurred at all of the wind turbine sites, including Wind Turbine No. 3, the nacelle and blades of which fell to the ground. The simulation results also revealed that large wind direction deviations in the yaw direction of the wind turbines occurred frequently. Furthermore, the streamwise component of the wind velocity was relatively large, and the standard deviation of the vertical component of the wind velocity was large. The values of the standard deviation of the spanwise component of the wind velocity were approximately the same as those of the streamwise component of the wind velocity. From these findings, the following was conjectured. The exciting force on Wind Turbine No. 3 increased due to the effect of terrain-induced turbulence. As a result of the increased exciting force, additional load was imposed in the vicinity of the tower top flange, and thus increased metal fatigue in multiple bolts.

In light of the recent increase in wind turbine accidents such as the ones described above, laws and regulations about wind power generation in Japan have been reviewed and amended. Specifically, the Nuclear and Industrial Safety Agency of the Ministry of Economy, Trade, and Industry amended a part of the Ministerial Ordinance for Establishing Technical Standards for Wind Power Generation Facilities Based on the Electricity Business Act, Interpretations of Technical Standards, and the Rules for the Electricity Business Act. Regarding the "wind pressure" that is stipulated in Article No. 4 of the Ministerial Ordinance for Establishing Technical Standards for Wind Power Generation Facilities Based on the Electricity Business Act and that is relevant for examining safety in terms of wind turbine structures, it was specified by the Nuclear and Industrial Safety Agency that "wind pressure" is to be calculated by taking into account wind conditions at a wind turbine site that include extreme values of wind speed and turbulence fluctuations in the streamwise, spanwise, and vertical directions at the hub height of a wind turbine (27 June 2014).

As a result of this clarification on the use of wind conditions (turbulence) at a wind power generation facility, there is no doubt that the prediction and evaluation of turbulence intensity over complex terrain by computational fluid dynamics (CFD) such as LES and RANS will become increasingly important [3–13]. Generally, in RANS models, such as k–ε models, the standard deviation of the streamwise wind velocity which is attributable to terrain and/or surface roughness, σ_u^{Surf}, is calculated using the values of turbulence kinetic energy, k; then, the standard deviation of the streamwise wind velocity, σ_u, is calculated by taking into account the background atmospheric turbulence intensity, I_a [14].

$$\sigma_u^{Surf} = 0.93\sqrt{k} \tag{1}$$

$$\sigma_u = U \times \sqrt{\left(\frac{\sigma_u^{Surf}}{U}\right)^2 + I_a^2} \tag{2}$$

where the background atmospheric turbulence intensity, I_a, is taken as 0.1. The standard deviations of the spanwise and vertical wind velocities are calculated automatically using the ratios of these standard deviations to σ_u based on, for instance, an International Electrotechnical Commission (IEC) standard. However, this approach does not make it possible to evaluate the standard deviations of the three wind velocity components that result from turbulence structures that develop over complex terrain, and thus which deviate from those over flat, homogeneous terrain. In contrast, LES, which allows for unsteady simulations, does not involve the above-mentioned assumptions and makes it possible to directly evaluate the standard deviations of the three wind velocity components using time series data of the three wind velocity components as is the case for evaluations of the standard deviations with field wind observation data. Therefore, LES is a highly effective means to predict and evaluate turbulence intensity over complex terrain.

In the present study, field observation wind data from the time of the wind turbine blade damage accident at Shiratakiyama Wind Farm, which was investigated in Uchida [1], are analyzed in detail.

In addition, a high-resolution LES turbulence simulation is performed by refocusing attention on the connection between the blade damage and the terrain-induced turbulence that caused the blade damage. Based on the results from this simulation, the numerical reproducibility of terrain-induced turbulence by the LES model is examined.

2. Shiratakiyama Wind Farm and Airflow Characteristics from the Time of the Blade Damage Accident

The Shiratakiyama Wind Farm owned by Kinden Corporation is located on a ridge near Mt. Shirataki in Houhoku-cho, Shimonoseki City, Yamaguchi Prefecture, Japan. On this wind farm, twenty 2500 kW wind turbines manufactured by General Electric Company (wind turbine hub height: 85.0 m; swept-area diameter: 88.0 m; height of the upper end of the swept area above the ground surface: 129.0 m; height of the lower end of the swept area above the ground surface: 41.0 m) are deployed (see Figures 2 and 3).

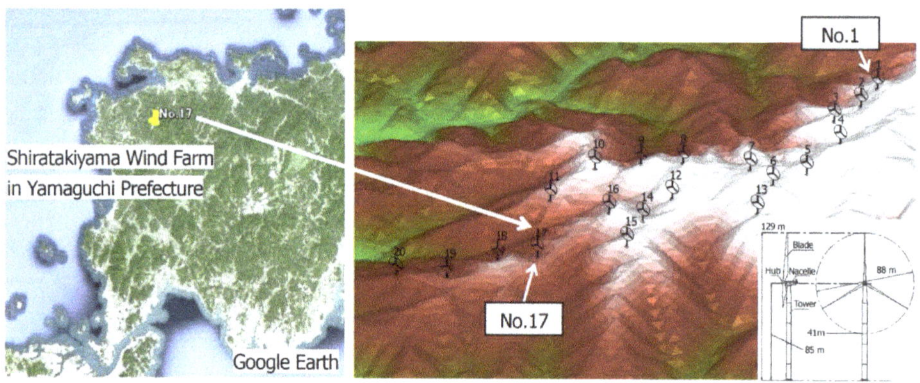

Figure 2. Location of Shiratakiyama Wind Farm in Yamaguchi Prefecture, Japan.

Figure 3. Photo of the Shiratakiyama Wind Farm taken from Wind Turbine No. 1.

Figure 4 shows the trajectory of Typhoon No. 0918 (Melor), pressure charts from 9:00, 7 October 2009 and 9:00, 8 October 2009, and the number of wind turbine incidents on these two days. Generally, strong winds blow to the east of the direction of motion of a typhoon. However, on 7 October 2009, on which the blade damage accident occurred, the center of the typhoon was located to the south of the main island of Japan, as shown by the trajectory of the typhoon. Therefore, the position of Shiratakiyama Wind Farm on this day was in the range between north and west of the center of the typhoon. At that time, a steep horizontal pressure gradient (i.e., a zone with densely packed isobars extending from east to west) was present between the typhoon and the high pressure located to the north of the typhoon. As a result, the wind was flowing from the east around the typhoon to the north of the typhoon, which caused a strong north-easterly wind to flow into the site of the blade damage accident, causing the accident.

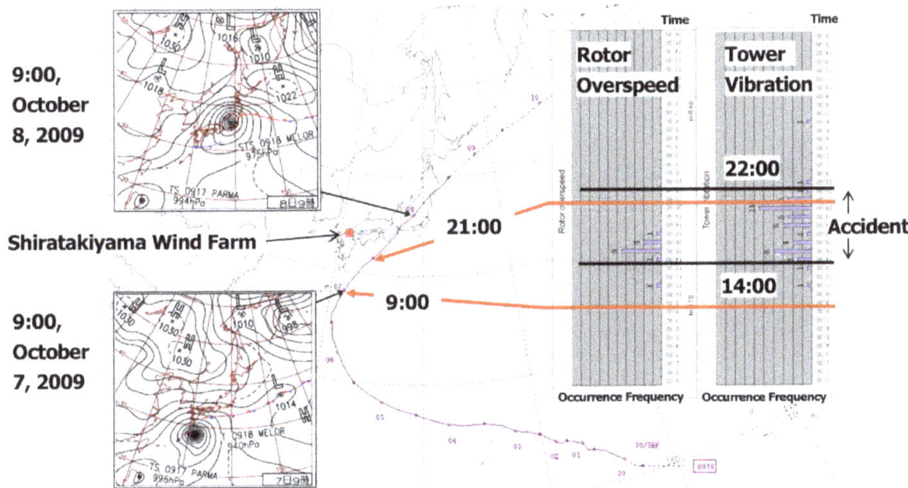

Figure 4. Tracking of Typhoon No. 0918 (Melor) and pressure charts.

As indicated on Figure 4, the time period in which the wind turbine blade damage accident occurred was between 14:00 and 22:00 on 7 October 2009. In this time period, alarms for tower vibration, excessive rotation, and abnormalities in wind direction deviations were frequently set off for the wind turbines in operation. Furthermore, wind speed and wind direction deviations were subject to large fluctuations, which caused severe fluctuations in the rotor speeds and pitch angles of the wind turbine blades.

Figure 5 shows a photo of wind vanes and anemometers mounted on the top of the nacelle of a wind turbine No. 1. The field observation wind data used in the present study are those acquired from these wind vanes and anemometers.

Figure 6 shows time series data from one of the wind vanes and one of the anemometers mounted on the nacelle of Wind Turbine No. 17 (50.0 m above the ground surface). In Figure 6a–d, the horizontal axes indicate the time (Midnight, 7 October 2009 to noon, 8 October 2009). These figures also include indications of (1) the time period in which the wind turbine blade damage accident occurred (14:00–22:00, 7 October 2009), and (2) the start time of the tower vibration (15:50). On 7 October 2009, the wind speed started increasing gradually in the area of Shiratakiyama Wind Farm in the afternoon. As described earlier, at around 14:00, 7 October 2009, the time at which the wind turbine blade damage accident began, a north-easterly wind was flowing into the farm near the ground surface. As shown in Figure 6c, at the time at which the tower vibrations started (15:50), the turbulence intensity reached 0.50.

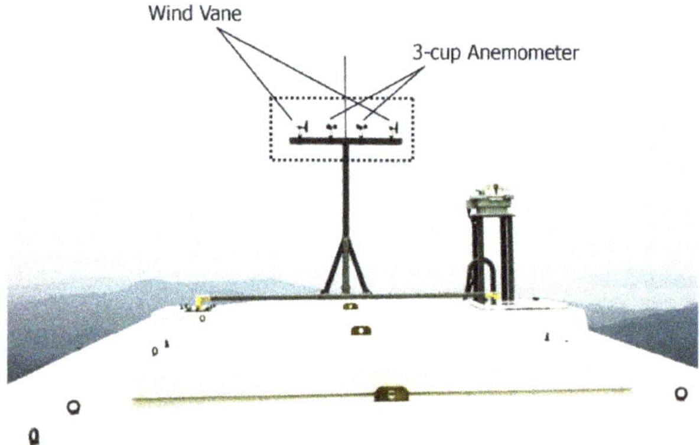

Figure 5. Wind vanes and 3-cup anemometers (dotted line) mounted on a nacelle of Wind Turbine No. 1.

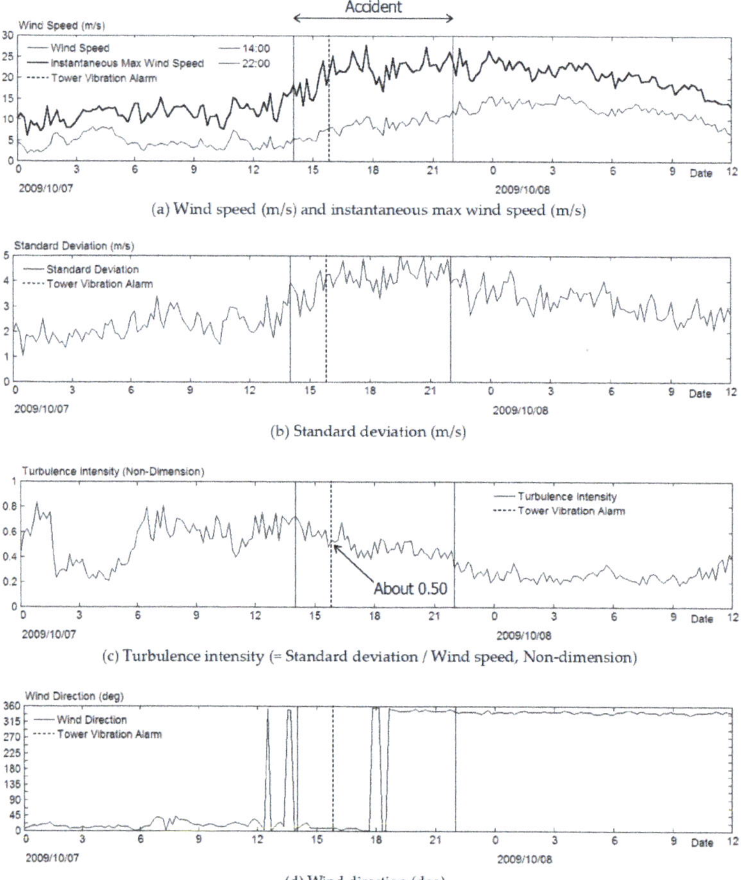

(a) Wind speed (m/s) and instantaneous max wind speed (m/s)

(b) Standard deviation (m/s)

(c) Turbulence intensity (= Standard deviation / Wind speed, Non-dimension)

(d) Wind direction (deg)

Figure 6. Time series of 10 min average data from the hub height (50.0 m) of Wind Turbine No. 17. Midnight, 7 October 2009 to noon, 8 October 2009.

Figure 7 shows the time series of wind speed (1 s average values) for a 10 min period which includes the start time of the tower vibrations (15:50). It can be seen from this figure that the observed wind speed fluctuated significantly with periodicity. Concurrently with the significant wind speed fluctuations, very frequent wind direction deviations were also observed (not shown). The most frequent wind direction at Shiratakiyama Wind Farm is northerly, and it is rare that strong easterly or north-easterly wind flows into this wind farm, as in the case under investigation in the present study. As discussed in Uchida [1], when a strong northeasterly wind flows into Shiratakiyama Wind Farm, airflows characterized by rapid fluctuations in wind direction and speed (terrain-induced turbulence), which originate from a relatively large terrain feature located upwind of the wind farm, flow into the wind turbines.

Under strong wind conditions such as those described above, the pitch control of the wind turbine blades enters a state of not being able to respond appropriately to rapid wind velocity fluctuations. As a result, large amounts of wind pressure (wind load or stress) that exceed the wind pressure presumed at the time of design may be exerted on the LE dorsal–ventral junction. It was conjectured in Uchida [1] that, in the wind turbine accident at Shiratakiyama Wind Farm, repeated exertion of such wind pressure on the LE dorsal–ventral junction caused damage (cracks) at this junction due to low-cycle fatigue, and that this damage propagated further during subsequent wind turbine operation.

In the present study, a new high-resolution LES turbulence simulation was performed to investigate the numerical reproducibility of the terrain-induced turbulence that caused the wind turbine blade damage accident.

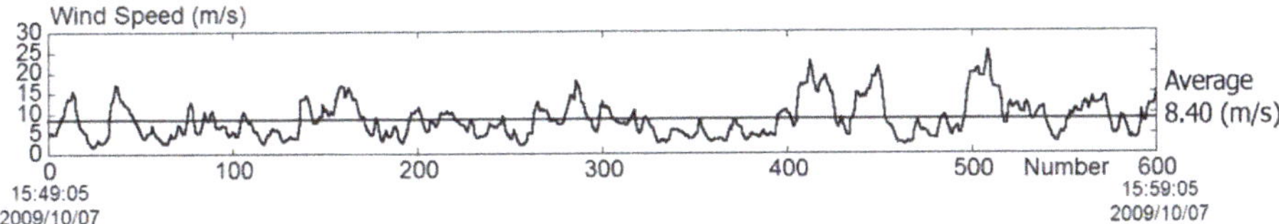

Figure 7. Tine series of 1 s average wind-speed data from the hub height (50.0 m) of Wind Turbine No. 17.

3. Overview of the Numerical Simulation Method

The present study employed the wind farm design tool called RIAM-COMPACT, which is based on a large-eddy simulation (LES) turbulence model [1,2,15–25]. This software simulates the local airflow with the use of a collocated grid arrangement in a general curvilinear coordinate system. For the governing equations of the flow, a filtered continuity equation for incompressible fluid (Equation (3)) and a filtered Navier-Stokes equation (Equation (4)) are used. In the present study, the LES was assumed to reproduce wind tunnel testing. Therefore, the effects of atmospheric stability associated with vertical thermal stratification of the atmosphere were neglected. In addition, as in Uchida [1,2], the effects of the surface roughness were taken into consideration by reconstructing surface irregularities in high resolution. The comparison between the RANS results and the present LES results are summarized in the latest article [15], and the prediction accuracy of the present LES approach by comparison with wind tunnel experiments are discussed in the article [25].

For the computational algorithm, a method similar to a fractional step (FS) method [26] was used, and a time marching method based on the Euler explicit method was adopted. The Poisson's equation for pressure was solved by the successive over-relaxation (SOR) method. For the discretization of all the spatial terms, except for the convective term in Equation (4), a second-order central difference scheme was applied. For the convective term, a third-order upwind difference scheme was applied. The interpolation technique by Kajishima et al. [27] was used for the fourth-order central differencing that appeared in the discretized form of the convective term. For the weighting of the numerical diffusion term in the convective term discretized by third-order upwind differencing, $\alpha = 0.5$ was used, as opposed to $\alpha = 3.0$ from the Kawamura–Kuwahara scheme [28], in order to minimize the influence of numerical diffusion. For LES subgrid-scale modeling, the standard Smagorinsky model [29] was adopted with a model coefficient of 0.1 in conjunction with a wall-damping function:

$$\frac{\partial \overline{u}_i}{\partial x_i} = 0 \tag{3}$$

$$\frac{\partial \overline{u}_i}{\partial t} + \overline{u}_j \frac{\partial \overline{u}_i}{\partial x_j} = -\frac{\partial \overline{p}}{\partial x_i} + \frac{1}{Re} \frac{\partial^2 \overline{u}_i}{\partial x_j \partial x_j} - \frac{\partial \tau_{ij}}{\partial x_j} \tag{4}$$

$$\tau_{ij} \approx \overline{u_i' u_j'} \approx \frac{1}{3} \overline{u_k' u_k'} \delta_{ij} - 2\nu_{SGS} \overline{S}_{ij} \tag{5}$$

$$\nu_{SGS} = \left(C_s f_s \Delta \right)^2 \left| \overline{S} \right| \tag{6}$$

$$\left| \overline{S} \right| = \left(2\overline{S}_{ij} \overline{S}_{ij} \right)^{1/2} \tag{7}$$

$$\overline{S}_{ij} = \frac{1}{2} \left(\frac{\partial \overline{u}_i}{\partial x_j} + \frac{\partial \overline{u}_j}{\partial x_i} \right) \tag{8}$$

$$f_s = 1 - exp\left(-z^+ / 25 \right) \tag{9}$$

$$\Delta = \left(h_x h_y h_z \right)^{1/3} \tag{10}$$

4. Overview of the Numerical Simulation Set-Up

The computational domain used in the present study extended over the space of 21.0 km (x) × 10.0 km (y) × 3.4 km (z), where x, y, and z are the streamwise, spanwise, and vertical directions, respectively. The maximum terrain elevation within the computational domain was 686.0 m, and the minimum terrain elevation was the sea surface (0.0 m). Terrain elevation data with 10.0 m and 50.0 m spatial resolutions from the Geospatial Information Authority of Japan (GSI) were used. The total number of computational grid points, 421 (x) × 201 (y) × 51 (z), was approximately 4.3 million, and the grid spacing was uniform (50.0 m) in both the x- and y-directions. Figure 8 shows a comparison of terrain configurations constructed from two datasets, i.e., 10.0 m and 50.0 m resolutions, for the area of Shiratakiyama Wind Farm. The comparison revealed no significant difference, and the accuracy of the reconstructed terrain was approximately the same between the two datasets. The grid spacing was non-uniform in the z-direction so that the density of grid points increased smoothly toward the ground surface. The minimum vertical grid spacing was 2.0 m. An earlier study of the wind blade damage accident [1] conjectured that the accident occurred with a northeasterly wind. Therefore, the simulations in the present study were also performed for northeasterly wind conditions. Regarding the boundary conditions, the wind velocity profile applied at the inflow boundary was based on a commonly used empirical power law (see Figure 9). A power law index was set to 5. At the side and upper boundaries, free-slip conditions were applied, and convective outflow conditions were applied at the outflow boundary. On the ground surface, a non-slip boundary condition was imposed. The non-dimensional parameter Re in Equation (4) is the Reynolds number ($=Uh/\nu$). For the simulations, $Re = 10^4$ was used. Figure 10 illustrates the characteristic scales used in the present simulations: h is the difference between the minimum and maximum terrain elevations within the computational domain ($=686.0$ m), U is the wind velocity at the inflow boundary at the height of the maximum terrain elevation within the computational domain, and ν is the kinematic viscosity of air. The time increment is set to $\Delta t = 2 \times 10^{-3} \, h/U$.

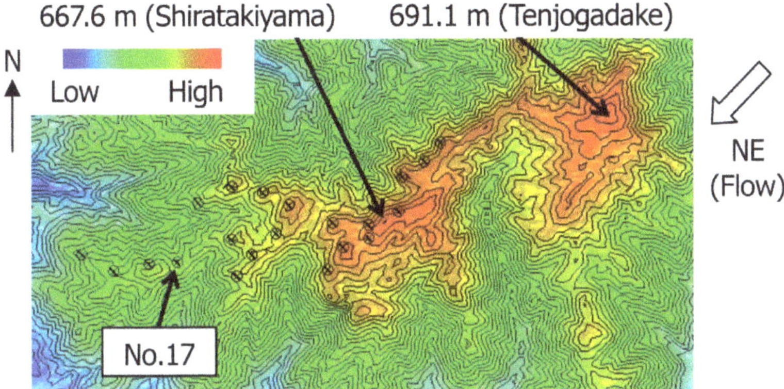

Figure 8. Comparison between 10.0 m resolution terrain elevation data (contour line) and 50.0 m resolution terrain elevation data (shading).

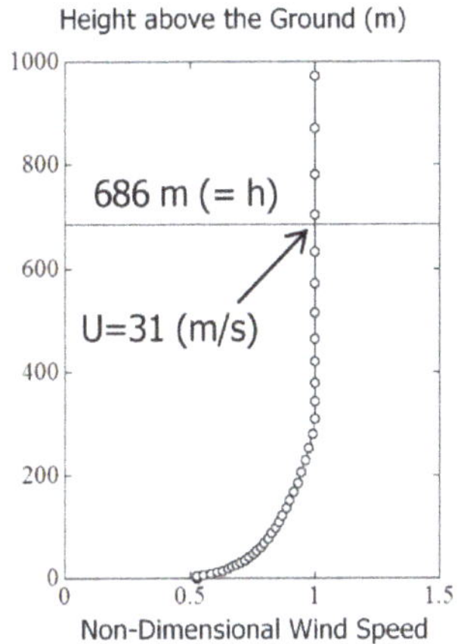

Figure 9. Inflow wind velocity profile.

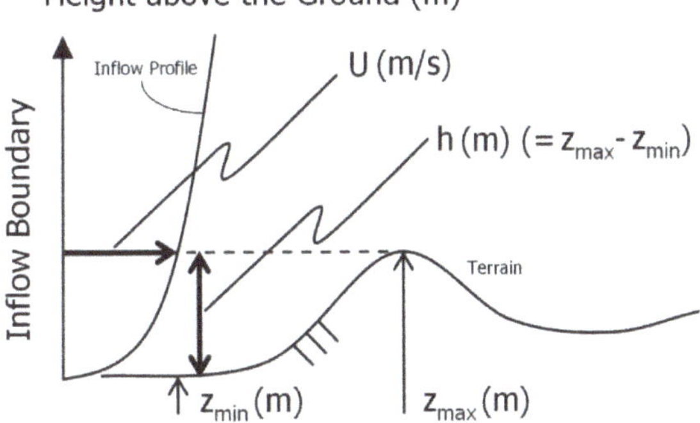

Figure 10. Characteristic scales U and h.

5. Results and Discussions Based on the Non-Dimensional Simulation Outputs

First, a comparison was made on the results from the simulations which used terrain elevation data with 10.0 m and 50.0 m resolutions from the Geospatial Information Authority of Japan (GSI). Figure 11 shows the vertical profiles of the mean streamwise wind velocity from the two simulations. Specifically, the plotted values were acquired by time-averaging (frame-averaging) the streamwise wind velocity at Wind Turbine No. 17 over the non-dimensional time period of $t = 200.0$–400.0. These values correspond to the output from a RANS model. The variable z^* on the vertical axis represents the height above the terrain surface (m), and the horizontal axis represents the averaged streamwise wind velocity normalized by the inflow wind velocity U (m/s). As can be presumed from the fact that no significant differences existed between the two terrain datasets (see Figure 8), the tendencies of the results from the two simulations, including those plotted in Figure 11, were nearly the same. In light of this finding, further discussions are presented based on the results from the simulation, which used the 10.0 m resolution terrain elevation data.

Figure 12 shows the vertical profiles of the standard deviations of the three wind velocity components at Wind Turbine No. 17. The standard deviations were calculated with respect to the time-averaged (frame-averaged) wind velocity components from the non-dimensional time period of $t = 200.0$–400.0. In the present study, the fluctuating wind velocity components (gusts) that are present in the observed inflow wind were not included in the inflow wind in the simulation; thus, only the airflow fluctuations of terrain-induced turbulence that were generated due to the terrain irregularities were evaluated. The left panel of Figure 12 shows the entire range, and the right panel of Figure 12 shows an enlarged view for the range of $z^* = 0.0$–200.0 m. The left panel (entire range) indicates that the maximum value of the standard deviation of the x-component of the wind velocity occurred in the vicinity of $z^* = 300.0$ m. The right panel (enlarged view) shows that the values of the standard deviations of the three wind velocity components were relatively large within the swept area of the wind turbine. In particular, it should be especially mentioned that the values for the y-component exceeded those for the x-component at the wind turbine hub height (=85.0 m) and below. This finding suggests the formation of an anisotropic turbulent flow field, in which the turbulent eddies are distorted significantly away from isotopic forms.

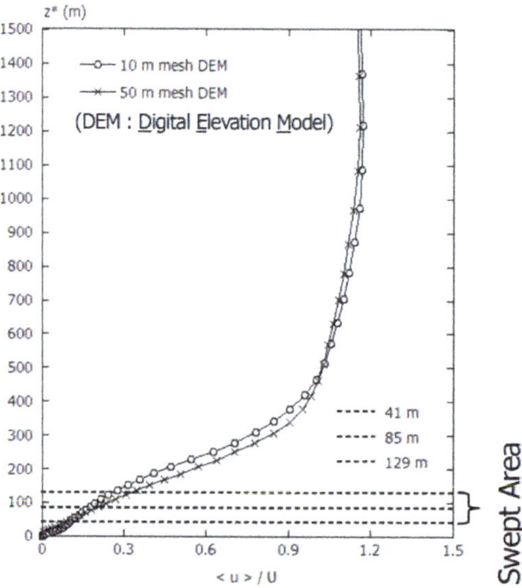

Figure 11. Profiles of mean streamwise wind velocity at Wind Turbine No. 17.

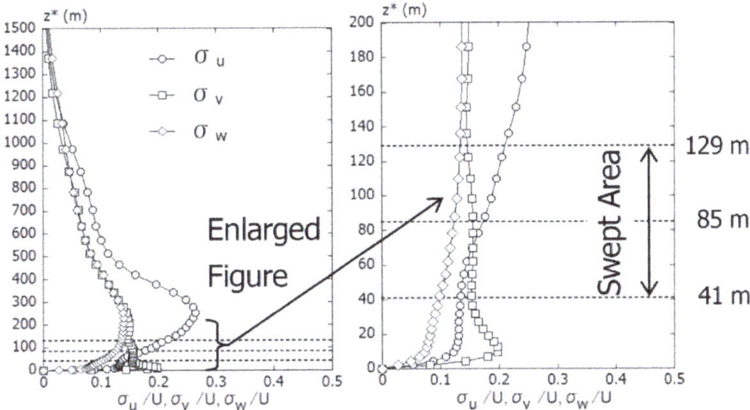

Figure 12. Profiles of the standard deviations of the three wind velocity components at Wind Turbine No. 17, resolution of terrain elevation data = 10.0 m.

6. Results and Discussions Based on the Rescaled-Simulation Outputs

Based on the rescaled-simulation outputs, we tried to examine the model's ability to numerically reproduce the terrain-induced turbulence (turbulence intensity) under strong wind conditions (8.0–9.0 m/s at wind turbine hub height). Since the wind velocity and time (t = 200.0–300.0, time step interval: 0.002, number of data values: approximately 50,000) acquired from the numerical simulation are dimensionless, they were converted to full scale with the procedure described below.

For the field observation wind dataset, scalar horizontal wind speed, U_{scalar}, was measured and recorded. This scalar horizontal wind speed, U_{scalar}, can be related to the horizontal wind speed acquired from the simulation as $U_{scalar} = (u^2 + v^2)^{1/2}$, where u and v are the streamwise and spanwise wind velocities acquired from the simulation. As shown in Figure 7, the mean horizontal wind speed acquired from the field wind observations was 8.40 m/s. On the other hand, the non-dimensional mean horizontal wind speed from the numerical simulation was 0.27. Therefore, all of the wind velocity data values from the numerical simulation were multiplied by (8.40/0.27) to ensure that the mean horizontal wind speed from the numerical simulation would be equal to 8.40. Accordingly, the inflow wind velocity was set to approximately 31.0 m/s for the numerical simulation (see Figure 9). This value was in close agreement with the value of the wind velocity obtained from the mesoscale meteorological model in Uchida [1]. The non-dimensional time on the horizontal axis t (=200.0–300.0) was converted to full scale (s) based on $T = t\,(h/U)$, where h = 686.0 m and U = 31.0 m/s for this case. As a result, the non-dimensional time interval of 100 was converted to approximately 2209 s (approximately 37 min) in full scale (time step interval: 0.044 s).

Figure 13 shows the temporal change (time series) of the horizontal wind speed from the hub height (85.0 m above the terrain surface) of Wind Turbine No. 17. In this figure, both the wind speed (m/s) on the vertical axis and time (s) on the horizontal axis represent those in full scale. Figure 13 also shows the field observation wind data, which are compared against the simulated wind data. These field observation wind data are the time series (1 s average values) from the 10 min period from Figure 7, which includes the time at which tower vibrations started (15:50). In Figure 13, the red solid line indicates field wind observation data (scalar horizontal wind speed) for 600 s (10 min), and the blue solid line indicates the horizontal wind speed calculated from the numerical simulation for approximately 2209 s (approximately 37 min). An examination of the numerical simulation results reveals the following. Rapidly fluctuating high-frequency components that were present in the field wind observation data were not fully reproduced in the simulation data. However, wind velocity fluctuations with very short periodic cycles which were attributable to the generation of terrain-induced turbulence, were reproduced for the most part. As a consequence, both the standard deviation of the horizontal wind speed (m/s) and turbulence intensity evaluated from the field observation and simulated wind data were successfully in close agreement (see numerical value shown in Figure 13). In order to obtain the time series with the above-mentioned periodicity, the settings of the parameters that are discussed below (i.e., horizontal grid resolution and time increment) were particularly important.

When strong north-easterly winds flow into Shiratakiyama Wind Farm, airflow characterized by rapid fluctuations in wind direction and speed (terrain-induced turbulence), which originate from a relatively large terrain feature located upwind of the wind farm, flow into the wind turbines. Therefore, it is necessary to accurately reproduce (1) a relatively large-scale terrain feature that is present upwind of the wind farm, and (2) the three-dimensional structure of the terrain-induced turbulence that is generated due to the terrain features, by sufficiently resolving the turbulence both temporarily and spatially. To achieve this goal, a grid resolution of approximately 50.0 m is required for the horizontal cross sections. Furthermore, in order to capture the unsteady fluid properties of terrain-induced turbulence, a sufficiently small time increment ($\Delta t = 2 \times 10^{-3}\, h/U$ in the present study) was required.

Finally, to investigate the cause of the wind turbine blade damage accident on Shiratakiyama Wind Farm, a power spectral analysis was performed on the fluctuating components of the observed time series data of wind speed (1 s average values) for a 10 min period (total of 600 data), indicated by a red line in Figure 13 by using a fast Fourier transform (FFT). The obtained results of this spectral analysis are shown in Figure 14. Here, the observed time series data (red line) shown in Figure 13 is the same as the data shown in Figure 7. The observed time series data of wind speed (1 s average values) for a 10 min period (total of 600 data), indicated by a red line shown in Figure 13, was divided into a 256 dataset. Then, spectral analysis was performed on the divided time series dataset, and a standard triangular window was also applied. In Figure 14, the vertical axis shows the power spectra, which is non-dimensionalized by the frequency f (Hz) and standard deviation σ (m/s), and the horizontal axis shows the frequency f (Hz). By using the dominant frequency from Figure 14, $f = 0.04$ Hz, the elevation of Tenjogadake, $h = 691.1$ m (see Figure 8), and the choice of the inflow wind velocity $U = 31.0$ m/s together yields 0.89 for the non-dimensional frequency, Strouhal number (St) ($=f\,h/U$). This value nearly agrees with the value of the vortex shedding frequency, $St = 0.87$, which was obtained from wind tunnel experiments with simple topography (a two-dimensional ridge and a three-dimensional isolated hill) [24]. Thus, it is likely that the terrain-induced turbulence that caused the wind turbine blade damage accident on Shiratakiyama Wind Farm was attributable to rapid wind speed and direction fluctuations, which were caused by vortex shedding from Tenjogadake (elevation: 691.1 m) located upstream of the wind farm.

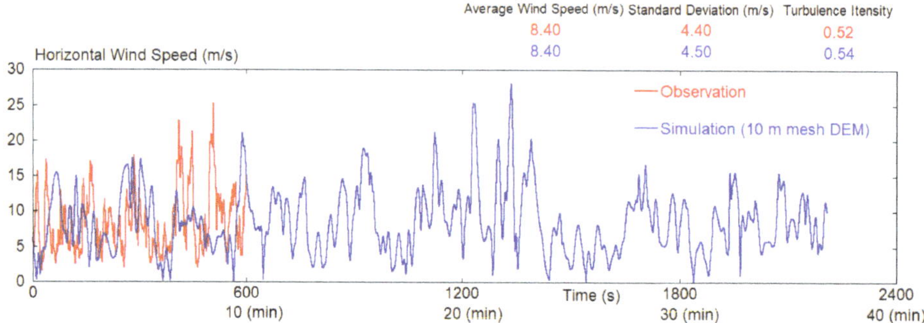

Figure 13. Comparison of horizontal wind speed between the observed data (red line) and the simulated data converted to full scale (blue line).

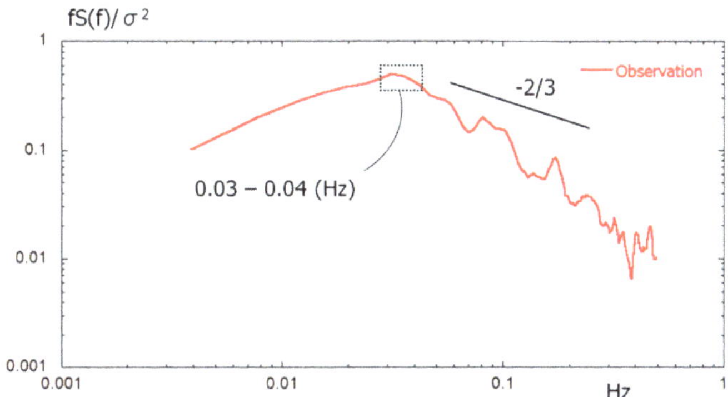

Figure 14. Power spectra of the observed data (red line) from Figure 13.

7. Conclusions

In the present study, field observation wind data from the time of the wind turbine blade damage accident on Shiratakiyama Wind Farm, which was studied by Uchida [1], were analyzed in detail. In parallel, high-resolution LES turbulence simulations were performed in order to examine the model's ability to numerically reproduce terrain-induced turbulence.

First, a comparison was made between terrain elevation data with 10.0 m and 50.0 m spatial resolutions from the Geospatial Information Authority of Japan (GSI). When a uniform grid spacing of 50.0 m was used for the horizontal grid for the airflow calculation, the accuracy of the simulation results was approximately the same between the simulations which used the two datasets. Furthermore, in order to capture unsteady fluid properties of terrain-induced turbulence, a sufficiently small time increment ($\Delta t = 2 \times 10^{-3} \, h/U$) is required. Here, h is the difference between the minimum and maximum terrain elevations within the computational domain; U is the wind velocity at the inflow boundary at the height of the maximum terrain elevation within the computational domain.

Secondly, vertical profiles of the standard deviations of the three wind velocity components at Wind Turbine No. 17 were examined based on the non-dimensional simulation outputs. This examination revealed that the value of the standard deviation for the y-component of wind velocity exceeded that for the x-component of wind velocity at the wind turbine hub height (=85.0 m) and below.

Thirdly, based on the rescaled-simulation outputs, we try to examine the model's ability to numerically reproduce terrain-induced turbulence (turbulence intensity) under strong wind conditions (8.0–9.0 m/s at wind turbine hub height). Since the wind velocity and time acquired from the numerical simulation are dimensionless, they are converted to full scale. As a consequence, both the standard deviation of the horizontal wind speed (m/s) and turbulence intensity evaluated from the field observation and simulated wind data are successfully in close agreement.

Finally, to investigate the cause of the wind turbine blade damage accident on Shiratakiyama Wind Farm, a power spectral analysis was performed on the fluctuating components of the observed time series data of wind speed (1 s average values) for a 10 min period (total of 600 data) by using a fast Fourier transform (FFT). It was suggested that the terrain-induced turbulence that caused the wind turbine blade damage accident on Shiratakiyama Wind Farm was attributable to rapid wind speed and direction fluctuations that were caused by vortex shedding from Tenjogadake (elevation: 691.1m) located upstream of the wind farm.

Funding: This work was supported by JSPS KAKENHI Grant Number 17H02053.

Acknowledgments: In carrying out the present study, the author received valuable input from Takashi Maruyama of the Disaster Prevention Research Institute, Kyoto University, Japan. In addition, the field observation wind data from Shiratakiyama Wind Farm which were analyzed in the present study were provided by Kinden Corporation. The author wishes to express his gratitude to those who have contributed to the present study.

Conflicts of Interest: The authors declare no conflict of interest.

References

1. Uchida, T. Computational Investigation of the Causes of Wind Turbine Blade Damage at Japan's Wind Farm in Complex Terrain. *J. Flow Control Meas. Vis.* **2018**, *6*, 152–167. [CrossRef]
2. Uchida, T. Computational Fluid Dynamics (CFD) Investigation of Wind Turbine Nacelle Separation Accident over Complex Terrain in Japan. *Energies* **2018**, *11*, 1485. [CrossRef]
3. Palma, J.M.L.M.; Castro, F.A.; Ribeiro, L.F.; Rodrigues, A.H.; Pinto, A.P. Linear and nonlinear models in wind resource assessment and wind turbine micro-siting in complex terrain. *J. Wind Eng. Ind. Aerodyn.* **2008**, *96*, 2308–2326. [CrossRef]
4. Berg, J.; Mann, J.; Bechmann, A.; Courtney, M.S.; Jørgensen, H.E. The Bolund experiment, part I: Flow over a steep, three-dimensional hill. *Bound.-Lay. Meteorol.* **2011**, *141*, 219–243. [CrossRef]
5. Bechmann, A.; Sørensen, N.N.; Berg, J.; Mann, J.; Réthoré, P.-E. The Bolund experiment, part II: Blind comparison of microscale flow models. *Bound.-Lay. Meteorol.* **2011**, *141*, 245–271. [CrossRef]
6. Prospathopoulos, J.M.; Politis, E.S.; Chaviaropoulos, P.K. Application of a 3D RANS solver on the complex hill of Bolund and assessment of the wind flow predictions. *J. Wind Eng. Ind. Aerodyn.* **2012**, *107*, 149–159. [CrossRef]
7. Diebold, M.; Higgins, C.; Fang, J.; Bechmann, A.; Parlange, M.B. Flow over hills: A large-eddy simulation of the Bolund case. *Bound.-Lay. Meteorol.* **2013**, *148*, 177–194. [CrossRef]
8. Porté-Agel, F.; Wu, Y.-T.; Chen, C.-H. A Numerical Study of the Effects of Wind Direction on Turbine Wakes and Power Losses in a Large Wind Farm. *Energies* **2013**, *6*, 5297–5313. [CrossRef]
9. Yeow, T.S.; Cuerva, A.; Conan, B.; Pérez-Álvarez, J. Wind tunnel analysis of the detachment bubble on Bolund Island. *J. Phys. Conf. Ser.* **2014**, *555*, 012021. [CrossRef]
10. Yeow, T.S.; Cuerva-Tejero, A.; Pérez-Álvarez, J. Reproducing the Bolund experiment in wind tunnel. *Wind Energy* **2015**, *18*, 153–169.
11. Chaudhari, A.; Hellsten, A.; Hämäläinen, J. Full-Scale Experimental Validation of Large-Eddy Simulation of Wind Flows over Complex Terrain: The Bolund Hill. *Adv. Meteorol.* **2016**. [CrossRef]
12. Conan, B.; Chaudhari, A.; Aubrun, S.; van Beeck, J.; Hämäläinen, J.; Hellsten, A. Experimental and numerical modelling of flow over complex terrain: The Bolund hill. *Bound.-Lay. Meteorol.* **2016**, *158*, 183–208. [CrossRef]
13. Sessarego, M.; Shen, W.Z.; van der Laan, M.P.; Hansen, K.S.; Zhu, W.J. CFD Simulations of Flows in a Wind Farm in Complex Terrain and Comparisons to Measurements. *Appl. Sci.* **2018**, *8*, 788. [CrossRef]
14. Japan Society of Civil Engineers. *Wind Power Generation Equipment Support Structure Design Guideline*; Japan Society of Civil Engineers: Tokyo, Japan, 2010; ISBN 978-4-8106-0705-5. (In Japanese)
15. Uchida, T.; Li, G. Comparison of RANS and LES in the Prediction of Airflow Field over Steep Complex Terrain. *Open J. Fluid Dyn.* **2018**, *8*, 286–307. [CrossRef]
16. Uchida, T. Design Wind Speed Evaluation Technique in Wind Turbine Installation Point by Using the Meteorological and CFD Models. *J. Flow Control Meas. Vis.* **2018**, *6*, 168–184. [CrossRef]
17. Uchida, T. LES Investigation of Terrain-Induced Turbulence in Complex Terrain and Economic Effects of Wind Turbine Control. *Energies* **2018**, *11*, 1530. [CrossRef]
18. Uchida, T. Computational Fluid Dynamics Approach to Predict the Actual Wind Speed over Complex Terrain. *Energies* **2018**, *11*, 1694. [CrossRef]
19. Uchida, T. Large-Eddy Simulation and Wind Tunnel Experiment of Airflow over Bolund Hill. *Open J. Fluid Dyn.* **2018**, *8*, 30–43. [CrossRef]
20. Uchida, T. High-Resolution LES of Terrain-Induced Turbulence around Wind Turbine Generators by Using Turbulent Inflow Boundary Conditions. *Open J. Fluid Dyn.* **2017**, *7*, 511–524. [CrossRef]
21. Uchida, T. High-Resolution Micro-Siting Technique for Large Scale Wind Farm Outside of Japan Using LES Turbulence Model. *Energy Power Eng.* **2017**, *9*, 802–813. [CrossRef]
22. Uchida, T. CFD Prediction of the Airflow at a Large-Scale Wind Farm above a Steep, Three-Dimensional Escarpment. *Energy Power Eng.* **2017**, *9*, 829–842. [CrossRef]

23. Uchida, T.; Ohya, Y. Latest Developments in Numerical Wind Synopsis Prediction Using the RIAM-COMPACT CFD Model-Design Wind Speed Evaluation and Wind Risk (Terrain-Induced Turbulence) Diagnostics in Japan. *Energies* **2011**, *4*, 458–474. [CrossRef]
24. Uchida, T.; Maruyama, T.; Ohya, Y. New Evaluation Technique for WTG Design Wind Speed using a CFD-model-based Unsteady Flow Simulation with Wind Direction Changes. *Model. Simul. Eng.* **2011**. [CrossRef]
25. Uchida, T.; Ohya, Y. Micro-siting Technique for Wind Turbine Generators by Using Large-Eddy Simulation. *J. Wind Eng. Ind. Aerodyn.* **2008**, *96*, 2121–2138. [CrossRef]
26. Kim, J.; Moin, P. Application of a fractional-step method to incompressible Navier-Stokes equations. *J. Comput. Phys.* **1985**, *59*, 308–323. [CrossRef]
27. Kajishima, T. Upstream-shifted interpolation method for numerical simulation of incompressible flows. *Bull. Jpn. Soc. Mech. Eng. B* **1994**, *60*, 3319–3326. (In Japanese) [CrossRef]
28. Kawamura, T.; Takami, H.; Kuwahara, K. Computation of high Reynolds number flow around a circular cylinder with surface roughness. *Fluid Dyn. Res.* **1986**, *1*, 145–162. [CrossRef]
29. Smagorinsky, J. General circulation experiments with the primitive equations, Part 1, Basic experiments. *Mon. Weather Rev.* **1963**, *91*, 99–164. [CrossRef]

Article

Fast Power Reserve Emulation Strategy for VSWT Supporting Frequency Control in Multi-Area Power Systems

Ana Fernández-Guillamón [1,*], Antonio Vigueras-Rodríguez [2], Emilio Gómez-Lázaro [3] and Ángel Molina-García [1]

[1] Department of Electrical Engineering, Universidad Politécnica de Cartagena, 30202 Cartagena, Spain; angel.molina@upct.es
[2] Department of Civil Engineering, Universidad Politécnica de Cartagena, 30203 Cartagena, Spain; avigueras.rodriguez@upct.es
[3] Renewable Energy Research Institute and DIEEAC-EDII-AB, Universidad de Castilla-La Mancha, 02071 Albacete, Spain; emilio.gomez@uclm.es
* Correspondence: ana.fernandez@upct.es; Tel.: +34-968-325357

Received: 2 October 2018; Accepted: 11 October 2018; Published: 16 October 2018

Abstract: The integration of renewables into power systems involves significant targets and new scenarios with an important role for these alternative resources, mainly wind and PV power plants. Among the different objectives, frequency control strategies and new reserve analysis are currently considered as a major concern in power system stability and reliability studies. This paper aims to provide an analysis of multi-area power systems submitted to power imbalances, considering a high wind power penetration in line with certain European energy road-maps. Frequency control strategies applied to wind power plants from different areas are studied and compared for simulation purposes, including conventional generation units. Different parameters, such as nadir values, stabilization time intervals and tie-line active power exchanges are also analyzed. Detailed generation unit models are included in the paper. The results provide relevant information on the influence of multi-area scenarios on the global frequency response, including participation of wind power plants in system frequency control.

Keywords: frequency control; wind power integration; power system stability

---

1. Introduction

Traditionally, synchronous generators have provided frequency control reserves, which are released under power imbalance conditions to recover grid frequency [1]. In fact, any generation-demand imbalance leads the grid frequency to deviate from its nominal value, which can cause serious scale stability problems [2]. With the significant penetration of renewables, mainly wind power plants, a proportional capacity of the system reserves must be provided by these new resources [3]. In this way, reference [4] considers that wind power plant participation in grid frequency control is imminent. However, wind turbines usually include back-to-back converters, and they are electrically decoupled from the grid through power electronic converters [5]. Consequently, with the significant integration of wind power into power systems, grid frequency tends to degrade progressively due to the reduction of the grid inertial responses [6]. Therefore, this new scenario presents a preliminary reduction of reserves from conventional generation units, mainly in weak and/or isolated power systems with high renewable resource penetration [7,8]. Moreover, these problems would be exacerbated in micro-grids, with a high share of power-electronically interfaced and thus a low grid inertia [9,10]. Under this framework, frequency control strategies must be included in wind power plants to provide additional active power under disturbances [11]. These new strategies would allow us to integrate Variable Speed Wind Turbines (VSWTs) into these services, replacing conventional power plants by renewables [12] and maintaining a reliable power system operation [13]. Most of the proposed strategies for VSWTs are based on 'hidden inertia emulation', enhancing their inertia response [14–16]. According to the specific literature, '*Fast power reserve emulation*' has been proposed as a suitable solution. It is based on supplying the kinetic energy stored in the rotating masses to the grid as an additional active power, being subsequently recovered through an under-production period (recovery) [17–19]. Different studies can be found to discuss the definition of overproduction period and the transition from overproduction to recovery period [20–25]. These studies are mainly focused on analyzing the inertia reduction problem on isolated power systems [20,21,23–27]. However, there is a lack of contributions focused on large interconnected power systems with high wind power penetration [28]. These new scenarios are in line with current wind generation units, covering more than 20% in different power systems. Moreover, renewables have accounted for more than 50% at different times in some European countries such as Spain, Portugal, Ireland, Germany or Denmark [29].

In general, synchronous generators inherently release or absorb kinetic energy as an inertial response to imbalance situations [24]. However, to recover the grid frequency at the nominal value, an additional control system is needed as well [30]. Automatic Generation Control (AGC) is thus considered as one of the most important ancillary services in power systems. AGC is used to match the total generation with the total demand, including power system losses [31]. Over the last decade, different authors have proposed several control strategies and optimization techniques. A modified AGC for an interconnected power system in a deregulated environment is described in [32]. A similar contribution can be found in [33], where an energy storage system is added to a multi-area power system, and the I controller gains are optimized by using the Opposition-based Harmony Search algorithm. A teaching-learning process based on an optimization algorithm to tune both I and PID controller parameters in single and multi-area power systems is described in [34]. In [35], a hybrid fuzzy PI controller is proposed for AGC of multi-area systems, yielding significant improvements compared to previous approaches. In [36], the gray wolf optimization method is proposed to tune the controller gains of an interconnected power system. This solution presented a more suitable tuning capability than other population-based optimization techniques. An optics inspired optimization algorithm is proposed in [37] and compared to other optimization algorithms, reaching a better performance for maximum overshoot and settling time values. However, in these contributions, only thermal, gas and hydro-power plants are considered from the supply side [32–36]. Therefore, multi-area power system modeling by including wind power plants are required to simulate frequency excursions under power imbalance conditions. Consequently, and by considering previous contributions, this paper analyzes different power imbalance situations and the corresponding frequency deviations in a multi-area interconnected power system with high wind power penetration. The main contributions of the paper are summarized as follows:

- Different multi-area power systems are analyzed with significant wind power integration, in line with current shares of renewables accounting for between 25% and 40%. Most previous studies on multi-area power systems only consider conventional generating units, such as thermal, gas and hydro-power [38–41].
- Wind power plants include a fast power reserve emulation control strategy in order to provide frequency response under power imbalances. Indeed, there is a lack of contributions describing frequency control response in wind power plants without energy storage solutions under multi-area power systems [42–45].
- The total power exchanged between areas is in line with the recent EU-wide targets, assuming a power interconnection share of 10% [46].
- The impact of wind power plants located in different areas on the frequency evolution is included in our model and dicussed in detail.

The rest of the paper is organized as follows: Section 2 presents the frequency control strategy for VSWTs. The implemented multi-area interconnected power system is described in Section 3. The results are provided and widely discussed in Section 4. Finally, the conclusions are presented in Section 5.

2. Improving Frequency Control Strategy of Wind Turbines

According to the specific literature, different methods for VSWTs have been proposed to provide frequency control. Figure 1 summarizes the corresponding solutions to be implemented in wind power plants: (i) de-loading, (ii) droop control and (iii) inertial response [47]. With regard to de-loading control methods, they are based on operating VSWTs below their optimal generation point. A certain amount of active power reserve is thus available to supply additional generation under a contingency [48]. It can be implemented by regulating the pitch angle from β_{min} to a maximum value or by increasing the rotational speed above the Maximum Power Point Tracking (MPPT) speed (over-speeding) [49]. An extension of de-loading strategy applied to Photovoltaic system (PV) taking into account a percentage of the PV power production for back-up reserve can be found in [50]. Secondly, droop control solutions have a significant influence on the frequency minimum value (nadir) and the frequency recovery [51]. The controller is based on considering the torque/power-set point as a function of the frequency excursion (Δf) and the rate of change of frequency (ROCOF) [52–56]. Finally, 'hidden inertia' controllers introduce a supplementary loop into the active power control. This additional loop control is only added under frequency deviations. Both blades and rotor inertia are then used to provide primary frequency response. Different approaches can be found in the specific literature. One solution is based on emulating similar inertia response to conventional generation units, shifting the torque/power reference proportionally to the ROCOF [51,57–60]. Another study uses the fast power reserve emulation. Constant overproduction power is released from the kinetic energy stored in the rotating mass of the wind turbine, with the rotational speed being recovered later through an underproduction period [17,20,21,25,47,61].

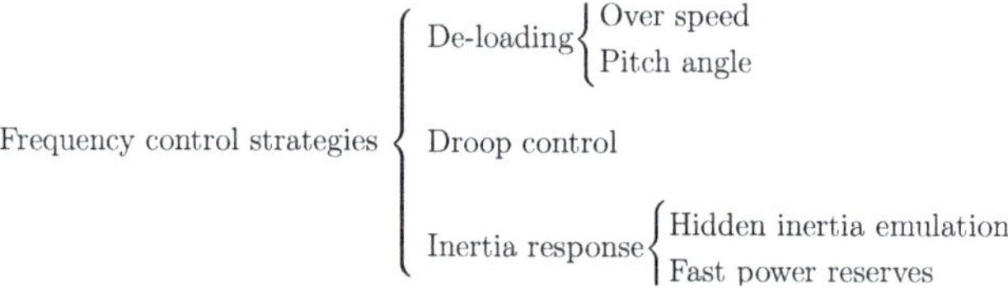

Figure 1. Wind power plant frequency control: general overview [28,47].

In line with previous contributions, the frequency control strategy for VSWTs implemented in this work is based on the fast power reserve emulation technique developed by the authors in [25]. This approach improves an initial proposal described in [61], by minimizing frequency oscillations and smoothing the wind power plant frequency response. Three operation modes are considered: normal operation mode, overproduction mode and recovery mode, see Figure 2. Different active power (P_{cmd}) values are determined aiming to restore the grid frequency under power imbalance conditions. Figure 2b depicts the VSWTs active power variations (ΔP_{WF}) submitted to an under-frequency excursion, being $\Delta P_{WF} = P_{cmd} - P_{MPPT}(\Omega_{MPPT})$.

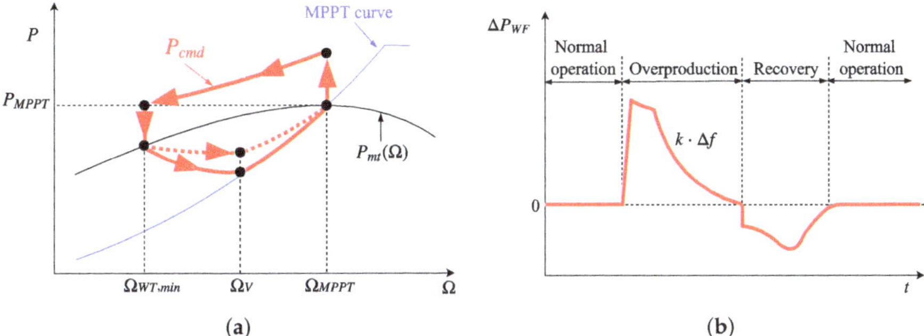

Figure 2. Wind frequency control strategy and VSWTs' active power variation (ΔP_{WF}) [25]; (a) frequency control strategy used for VSWTs; (b) ΔP_{WF} with frequency control strategy.

1. *Normal operation mode.* The VSWTs operate at a certain active power value (P_{cmd}), according to the available mechanical power for a specific wind speed, $P_{mt}(\Omega_{WT})$. It matches the maximum available active power for this current wind speed $P_{MPPT}(V_W)$; see Figure 2a,

$$P_{cmd} = P_{mt}(\Omega_{WT}) = P_{MPPT}(V_W). \tag{1}$$

Under power imbalance conditions, and assuming an under-frequency deviation, the frequency controller strategy changes to the overproduction mode and, subsequently, $\Delta f < -\Delta f_{lim} \to$ Overproduction.

2. *Overproduction mode.* The active power supplied by the VSWTs involves (i) mechanical power P_{mt} available from the $P_{mt}(\Omega_{WT})$ curve and (ii) additional active power ΔP_{OP} provided by the kinetic energy stored in the rotational masses,

$$P_{cmd} = P_{mt}(\Omega_{WT}) + \Delta P_{OP}(\Delta f). \tag{2}$$

ΔP_{OP} is estimated proportionally to the evolution of frequency excursion in order to emulate primary frequency control of conventional generation units [26,62]. Most previous approaches assume ΔP_{OP} as a constant value independent of the frequency excursion [22,23,61]. Moreover, the mechanical power P_{mt} is also considered as constant by most authors, even when rotational speed decreased [20–24,61]. This overproduction strategy remains active until one of the following conditions is met: the frequency excursion disappears, the rotational speed reaches a minimum allowed value, or the commanded power is lower than the maximum available active power,

$$\left.\begin{array}{ll} \Delta f & > -\Delta f_{lim} \\ \Omega_{WT} & < \Omega_{WT,min} \\ P_{cmd} & < P_{MPPT}(\Omega_{MPPT}) \end{array}\right\} \to \text{Recovery.} \tag{3}$$

3. *Recovery mode.* With the aim of minimizing frequency oscillations, wind power plants have to move from overproduction mode to recovery mode as smoothly as possible, avoiding abrupt power changes and, subsequently, undesirable secondary frequency shifts [20,22,24,61]. With this aim, the authors' solution described in [25] follows the mechanical power curve $P_{mt}(\Omega_{WT})$ according to the wind speed instead of the maximum power curve $P_{MPPT}(\Omega_{WT})$ [22]. The power provided by the VSWTs in this mode is based on two periods according to [25]: (i) a parabolic trajectory and (ii) following the P_{MPPT} curve proportional to the difference between $P_{mt}(\Omega_{WT})$ and $P_{MPPT}(\Omega_{WT})$. The normal operation mode then can be recovered when either Ω_{MPPT} or $P_{MPPT}(\Omega_{MPPT})$ are respectively reached by the wind turbine.

This strategy was evaluated in [25] and compared to [61] for single-are power system modeling, providing an improved frequency response under power imbalance conditions. This approach is considered in the present paper and extended to a multi-area power system with significant wind power integration into different areas.

3. Power System Modeling

3.1. General Overview

Traditional power system modeling for frequency deviation analysis under imbalance conditions is usually based on the following expression [63],

$$\Delta f = \frac{1}{2\,H_{eq}\,s + D_{eq}} \cdot (\Delta P_g - \Delta P_L),$$

(4)

where Δf is the frequency variation from nominal system frequency, H_{eq} is the equivalent inertia constant of the system, D_{eq} is the equivalent damping factor of the loads, and $\Delta P_g - \Delta P_L$ is the power imbalance. H_{eq} is estimated from Equation (5), H_m is the inertia constant of m-power plant, $S_{B,m}$ is the rated power of the m-generating unit, CG is the total number of conventional synchronous generators and S_B is the base power system:

$$H_{eq} = \frac{\sum_{m=1}^{CG} H_m \cdot S_{B,m}}{S_B}.$$

(5)

Transmission level voltage is usually considered for multi-area interconnection purposes through tie-lines. Frequency and tie-line power exchange can vary according to variations in power load demand [64–68]. The total tie-line power exchange between two areas is determined by

$$\Delta P_{tie_{i,j}} = \frac{2 \cdot \pi \cdot T_{i,j}}{s} \cdot (\Delta f_i - \Delta f_j),$$

(6)

where $T_{i,j}$ is the synchronizing moment coefficient of the tie-line between i and j areas.

When a frequency deviation is detected, the balance between an interconnected power system is determined by generating the Area Control Error signal (ACE), expressed as a linear combination of the tie-line power exchange and the frequency deviation [69]

$$ACE_i = B_i \cdot \Delta f_i + \sum_{\substack{j=1 \\ j \neq i}}^{N} \Delta P_{tie_{i,j}},$$

(7)

where i, j refers to i and j areas, respectively, B is the bias-factor, ΔP_{tie} is the variation in the exchanged tie-line power and N is the total number of interconnected areas. Figure 3 schematically shows these power exchanges for a three-area power system example. Recent contributions focused on a new control logic of the Balancing Authority Area Control Error Limit (BAAL) Standard adopted in the North American power grid can be found in [70].

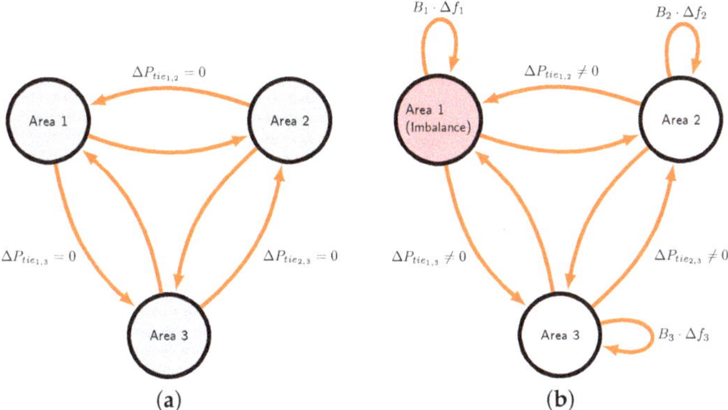

Figure 3. Multi-area power system. (**a**) balanced situation; (**b**) imbalanced situation in Area 1.

3.2. Supply-Side Modeling

From the supply-side, the power systems considered for simulation purposes involve conventional generating units (such as non-reheat thermal and hydro-power) and renewable energy sources (wind and PV power plants). One equivalent generator is used for each type of production to model the supply-side. This assumption is in line with previous contributions focused on frequency strategy control analysis.

The conventional generating unit models considered for simulations can be seen in Figure 4. Taking into account the specific literature, they are modeled according to the simplified governor-based models widely used and proposed in [62]. Parameters are provided in Tables 1 and 2, respectively. The different transfer functions of governor and turbine are indicated in Figure 4.

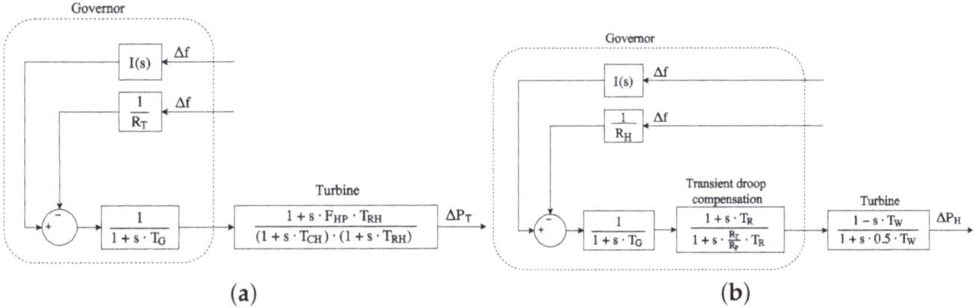

Figure 4. Conventional generation modeling. (**a**) thermal plant model; (**b**) hydro-power plant model.

Table 1. Thermal power plant parameters [62].

Parameter	Name	Value ($\mathrm{pu_{thermal}}$)
T_G	Speed relay pilot valve	0.20
F_{HP}	Fraction of power generated by high pressure section	0.30
T_{RH}	Time constant of reheater	7.00
T_{CH}	Time constant of main inlet volumes and steam chest	0.30
R_T	Speed droop	0.05
$I(s)$	Integral controller	1.00
$H_{thermal}$	Inertia constant	5.00 s

Table 2. Hydro-power plant parameters [62].

Parameter	Name	Value (pu$_{hydro}$)
T_G	Speed relay pilot valve	0.20
T_R	Reset time	5.00
R_T	Temporary droop	0.38
R_P	Permanent droop	0.05
T_W	Water starting time	1.00
R_H	Speed droop	0.05
$I(s)$	Integral controller	1.00
H_{hydro}	Inertia constant	3.00 s

Wind power plants are able to provide frequency response according to the strategy discussed in Section 2. An aggregated model for wind power plants is considered for the simulation purposes. They are represented by one equivalent generator, which is generally accepted in the specific literature for frequency response simulations (Figure 5). The equivalent wind turbine has n-times the size of each individual wind turbine, with n being the number of wind turbines [71,72]. The equivalent wind turbine model is based on [73,74], which have been widely used in recent publications [22,23,25,75–77]. Parameters are shown in Table 3. The remaining renewable generation is modeled through an equivalent PV power plant connected to the grid. It represents a renewable non-dispatchable energy source, following recent contributions [78]. Due to the short period of simulated time (under 5 min), a constant active power provided by this non-dispatchable resource is considered for our analysis.

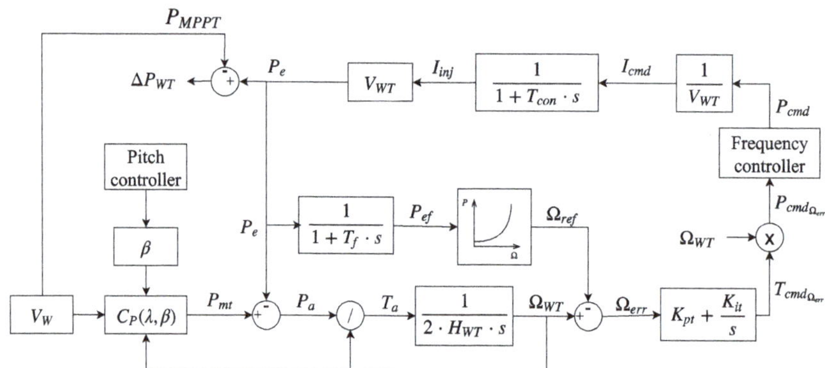

Figure 5. Aggregated wind power plant model with frequency controller.

Table 3. Wind turbine parameters [74].

Parameter	Name	Value
V_w	Wind speed	10 m/s
S_n	Rated power	3.6 MW
H_{WT}	Inertia constant	5.19 s
Ω_0	Base rotational speed	1.335 rad/s
T_f	Time delay to measure electric power	5 s
T_{con}	Time delay to generate the injected current I_{inj}	0.020 s
V_{WT}	Wind turbine voltage	1 pu$_{WT}$
K_{pt}	Proportional constant of speed controller	3 pu$_{WT}$
K_{it}	Integral constant of speed controller	0.6 pu$_{WT}$

3.3. Area Descriptions

Figure 6 summarizes the percentages for the different generating units of each area. Previous studies address the problem of multi-area power systems considering only conventional power plants (mainly thermal, hydro-power and gas) and assuming two or three areas [32–36,38–41]. In this work, two different interconnected multi-source power systems are analyzed: (i) a two-area power system (considering only Areas 1 and 2) and (ii) a three-area power system. Both systems allow us to study in detail the relationships between the number of areas and the exchanged power between them when a significant number of renewable energies are considered from the supply-side. A base power of 2000 MW per area is assumed that corresponds to the capacity of each area. In Europe, it is expected that wind and PV will cover up to 30% and 18% of the demand respectively by 2030 [79,80]. Therefore, the integration of these sources in the areas considered in this paper are in line with current European road-maps, having a RES/non-dispatchable integration lying between 25% to 50%. In addition, $\Delta P_{tie_{i,j}}$ is limited to a maximum value of 10%. This limit agrees with recent EU-wide targets, which expect to have an interconnection power of 10% in the year 2020 [46]. Most contributions found in the literature review either do not limit the maximum tie-line power, or it is not indicated [81–84].

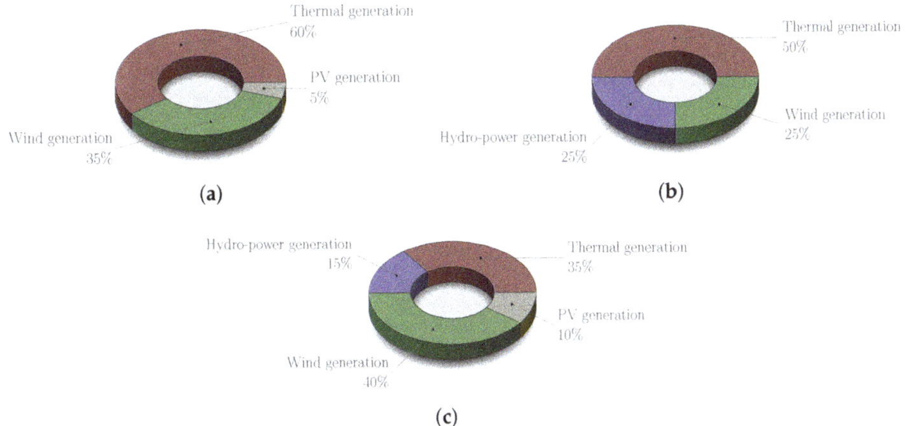

Figure 6. Generation contribution per area. (**a**) Area 1; (**b**) Area 2; (**c**) Area 3.

$T_{i,j}$ and B values are provided in Table 4 for the two-interconnected areas [30,35] and in Table 5 for three-interconnected areas [30]. The equivalent inertia H_{eq} of each area is calculated according to Equation (5), and taking into account the inertia constants of thermal and hydro-power plants indicated in Section 3.2. With regard to the damping factor, the impact of an inaccurate value is relatively small if the power system is stable [85]. Moreover, it is expected to decrease accordingly to the use of variable frequency drives [86]. Table 6 summarizes different values proposed for the damping factor in the literature over recent decades. A value of $D_{eq} = 1$ is considered for simulation purposes, which is in line with recent contributions and is lower than values corresponding to previous works. A general overview of a two-area power system can be seen in Figure 7.

Table 4. Interconnected two-area power system parameters [30,35].

Parameter	Name	Value
B_1	Bias factor of Area 1	0.425
B_2	Bias factor of Area 2	0.425
$T_{1,2}$	Synchronizing moment coefficient between Areas 1 and 2	0.545
$H_{eq,1}$	Equivalent inertia constant of Area 1	2.997 s
$H_{eq,2}$	Equivalent inertia constant of Area 2	3.324 s

Table 5. Interconnected three-area power system parameters [30].

Parameter	Name	Value
B_1	Bias factor of Area 1	0.3483
B_2	Bias factor of Area 2	0.3827
B_3	Bias factor of Area 3	0.3629
$T_{1,2}$	Synchronizing moment coefficient between Areas 1 and 2	0.2
$T_{2,3}$	Synchronizing moment coefficient between Areas 2 and 3	0.12
$T_{3,1}$	Synchronizing moment coefficient between Areas 3 and 1	0.25
$H_{eq,1}$	Equivalent inertia constant of Area 1	2.997 s
$H_{eq,2}$	Equivalent inertia constant of Area 2	3.324 s
$H_{eq,3}$	Equivalent inertia constant of Area 3	2.246 s

Table 6. Damping factor values.

Ref.	Value (pu/Hz)	Analysis	Year
[62]	1–2	Power system stability	1994
[87]	0.83	Two areas with non-reheat thermal units	2011
[88]	1.66	Two areas with thermal units	2011
[89]	1–1.8	Three areas with non-reheat thermal units	2012
[90]	2	One area with nuclear, thermal, wind and PV	2012
[91]	0.5–0.9	Three areas with nonlinear thermal units	2013
[92]	0.83	Two areas non-reheat thermal units	2013
[93]	0.83	Two areas with thermal units	2013
[67]	0.83	Two areas with reheat units	2015
[94]	0.8	IEEE 9 bus system with hydro-power, gas and wind turbines	2016
[95]	1–1.8	One and three areas with non-reheat thermal units	2017
[96]	1–1.8	Three areas with non-reheat thermal units	2018
[97]	1	Two areas with non-reheat thermal units	2018

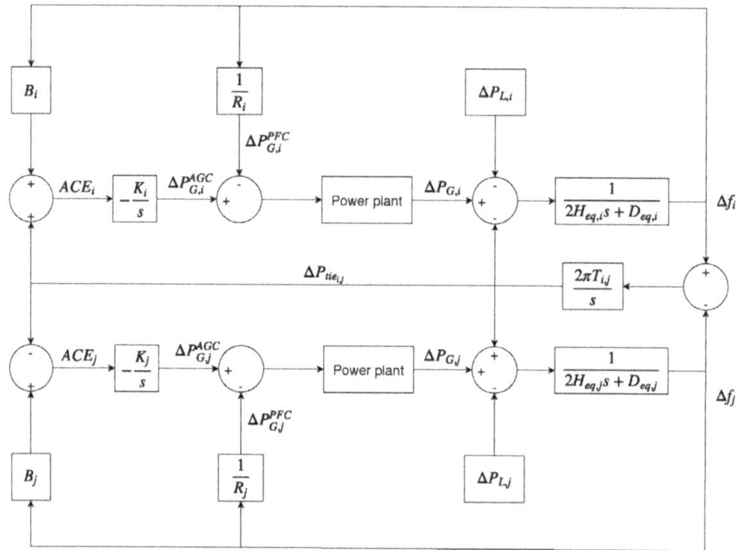

Figure 7. Two-area power system modeling for frequency control.

4. Results

As was discussed in Section 3, and with the aim of evaluating frequency oscillations and power system performances under imbalance conditions with different number of areas, two different multi-area power systems were simulated: (i) a two-area power system and (ii) a three-area power system. Both power systems were implemented in Matlab/Simulink© (2016, MathWorks, Natick, MA, USA). Source codes are available under request.

4.1. Two-Area Interconnected Power System

Firstly, and in order to evaluate the sensitivity of frequency excursions in a multi-area power system, two different imbalance conditions were simulated. In both cases, one area is submitted to imbalances while the other area maintains a balanced condition. A 5% increase in demand of the base power is assumed in all simulations as imbalance power ($\Delta P_{L,1} = \Delta P_{L,2} = 100$ MW). Under these scenarios, with an active-power deficit, different frequency control strategies are addressed by the simulations depending on the generation units involved in the frequency response: *case* (1) whole conventional generation units of the multi-area power system; *case* (2) whole conventional generation units and only wind power plants within the area submitted to imbalances; and *case* (3) whole conventional generation units and wind power plants.

Figure 8a,b shows the frequency oscillations in both areas when a power imbalance is applied to Area 1 ($\Delta P_{L,1}$). As can be seen, the maximum nadir is achieved in both areas when *case* (1) is conducted. The nadir values are improved when wind power plants are considered for frequency control: *cases* (2) and (3). Indeed, *case* (2) offers a smoother and less oscillatory response than *case* (3), yielding a stabilization time interval very similar to *case* (1). Moreover, *case* (3) causes three different well-identified frequency shifts: the first one is due to the power imbalance; the second one occurs due to the lack of coordination between power plants as well as the different time response of the supply-side operation units (see Figure 9); and the last one depends on the transition from overproduction mode to recovery mode of the wind power plant located in Area 2 (see Figure 2b and the active power decrease in WPP2 Figure 9).

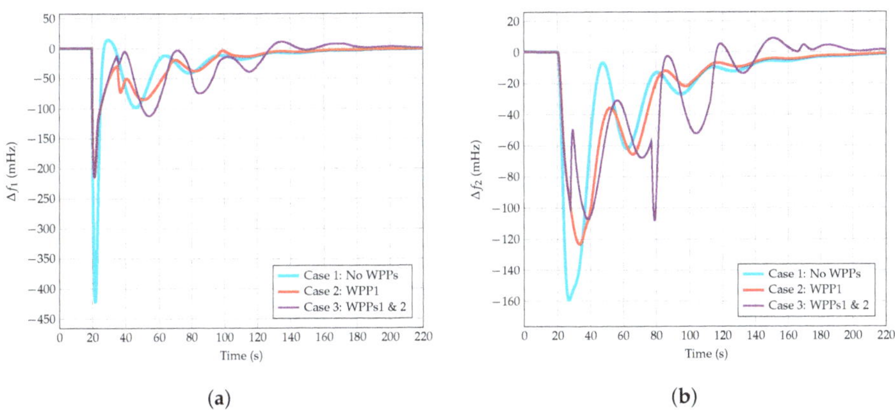

Figure 8. Area 1 under power imbalance ($\Delta P_{L,1}$); (**a**) frequency oscillations in Area 1; (**b**) frequency oscillations in Area 2.

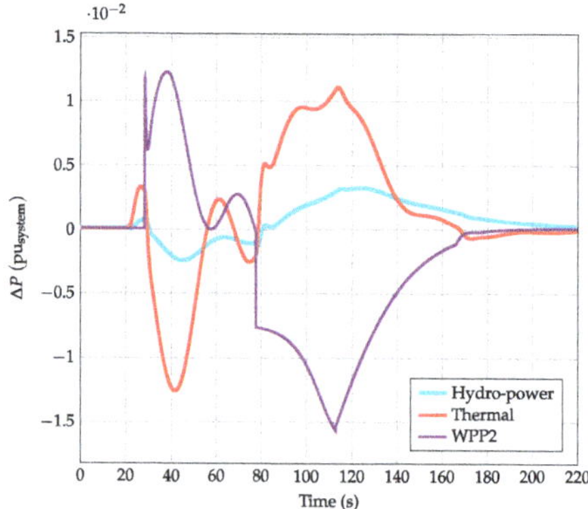

Figure 9. Area 1 under power imbalance ($\Delta P_{L,1}$): generation deviations in Area 2.

A similar study can be carried out by considering power imbalance conditions in Area 2 ($\Delta P_{L,2}$). Figure 10 compares the results in terms of nadir for both scenarios ($\Delta P_{L,1}$ and $\Delta P_{L,2}$) and considering the different frequency control strategies. As can be seen, minor differences are found in both analyses. In addition, Figure 11 compares the tie-line power evolution under both imbalance conditions, $\Delta P_{L,1}$ and $\Delta P_{L,2}$ accordingly, and peak-to-peak tie-line power exchange. Subsequently, and according to the generation mix considered in each area, frequency oscillations and active tie-line power results present similar values regardless of the area submitted to imbalances. Based on these results, and taking into account the different frequency control strategies implemented and simulated, lower frequency oscillations are obtained when only wind power plants within the area submitted to imbalance conditions are considered. Therefore, the contribution of wind power plants from other areas under frequency excursions would provide additional oscillation responses.

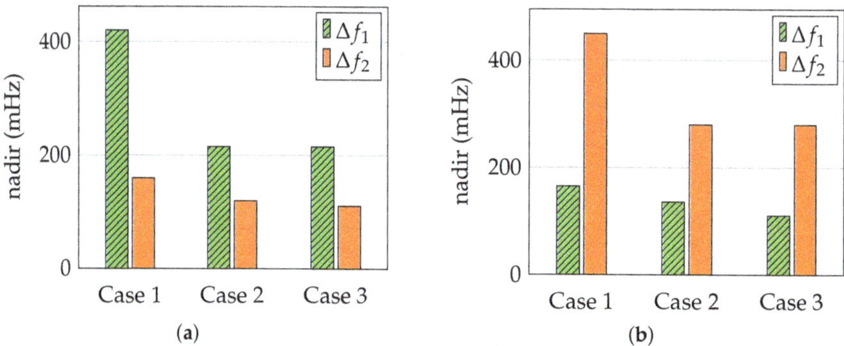

Figure 10. Nadir: Comparison of $\Delta P_{L,1}$ and $\Delta P_{L,2}$ scenarios. (**a**) Area 1 submitted to imbalance ($\Delta P_{L,1}$); (**b**) Area 2 submitted to imbalance ($\Delta P_{L,2}$).

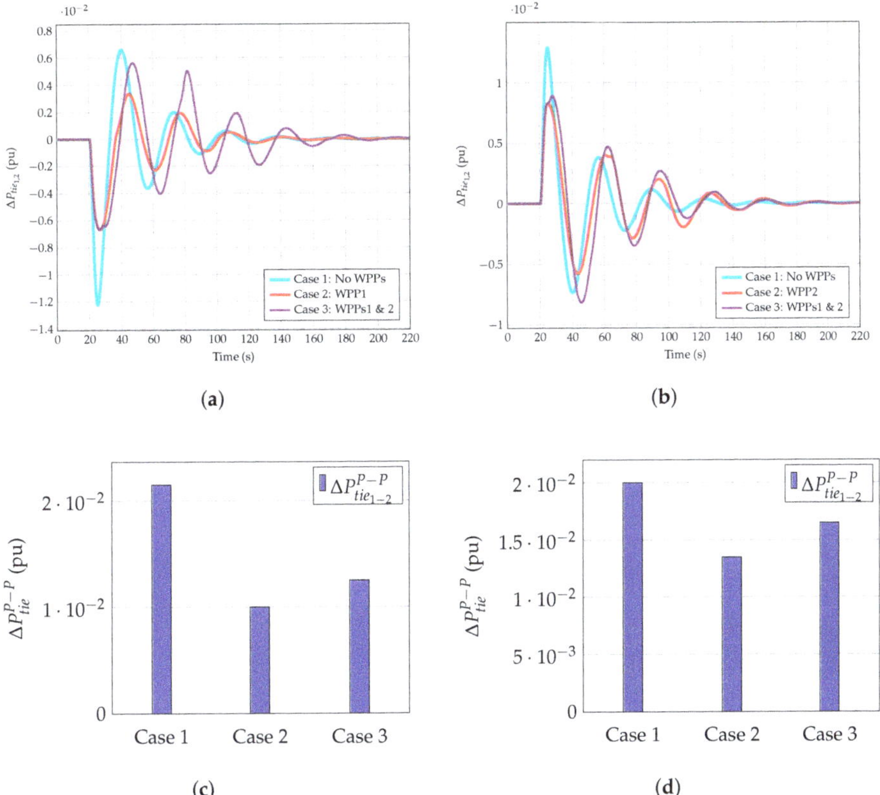

Figure 11. Active tie-line power evolution: comparison of $\Delta P_{L,1}$ and $\Delta P_{L,2}$ scenarios. (**a**) Area 1 submitted to power imbalance ($\Delta P_{L,1}$); (**b**) Area 2 submitted to power imbalance ($\Delta P_{L,2}$); (**c**) Area 1 submitted to imbalance ($\Delta P_{L,1}$); (**d**) Area 2 submitted to imbalance ($\Delta P_{L,2}$).

4.2. Three-Area Interconnected Power System

Considering the preliminary conclusion given in Section 4.1, where similar results are obtained independently of the area submitted to imbalances, the authors reduce the number of simulations in this three-area interconnected power system, assuming only that one area is submitted to imbalance conditions. Different frequency control strategies are then simulated by including an active-power deficit applied to Area 1, $\Delta P_{L,1}$. It is also defined as a step of 5% with respect to the base power ($\Delta P_{L,1} = 100$ MW).

Figure 12 depicts the frequency deviation of each area and the nadir comparison according to the different frequency control strategies. As can be seen, the results are in line with those obtained previously, when a two-area power system was considered. Therefore, the maximum nadir values are obtained in all areas when wind power plants are not included for frequency control. When wind power plants provide frequency response, the nadir values of all areas are considerably improved. Regarding *case* (2) and *case* (3), nadir values give similar results. However, larger frequency oscillations are identified when *case* (3) is conducted, especially in Area 2 and area 3. This behavior is a consequence of the wind power variations due to the different operation modes of each frequency controller, increasing the tie-line power exchanged between these two areas. Stabilization time presents similar values ($t_{stab} \simeq 100$ s) in all cases and areas.

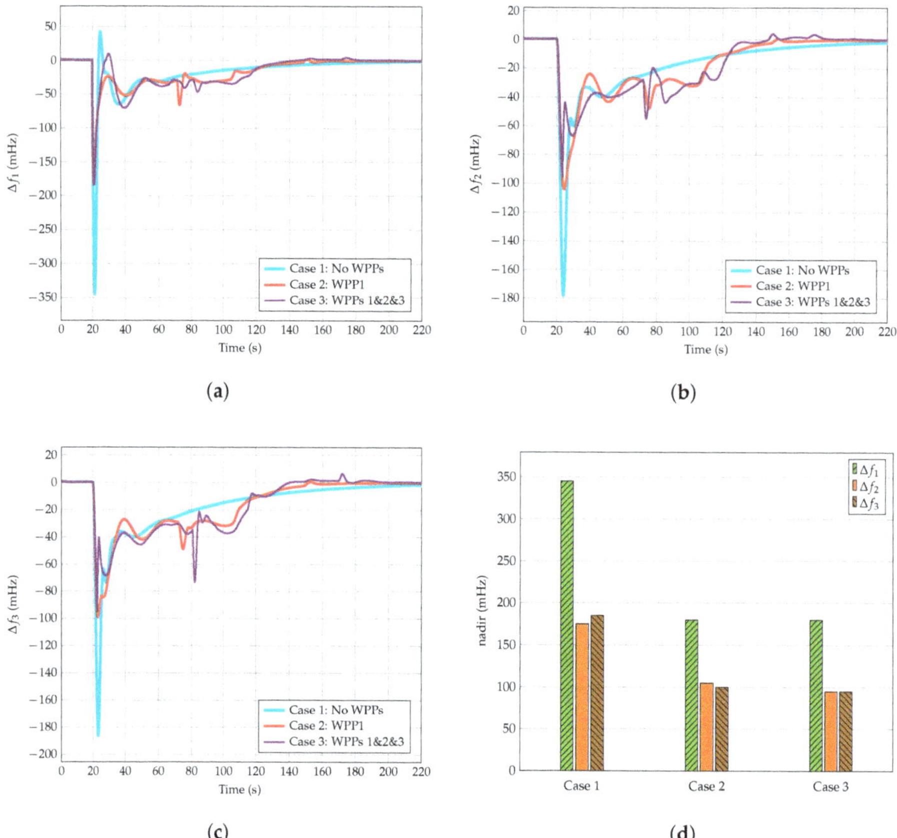

Figure 12. Frequency oscillations: Area 1 under power imbalance ($\Delta P_{L,1}$); (**a**) frequency oscillations in Area 1 (Δf_1); (**b**) frequency oscillations in Area 2 (Δf_2); (**c**) frequency oscillations in Area 3 (Δf_3); (**d**) nadir values: case comparison.

Figure 13 shows and compares the tie-line power variation and its peak-to-peak value. As can be seen, tie-line power exchange does not overcome the maximum restriction of 10% under any circumstances. Power exchanged between areas 2–3 is practically negligible regardless of the frequency control strategy, as the frequency deviations in these areas are a consequence of imbalances subsequently induced by Area 1 (see Section 3.1). As was previously mentioned, with the use of the wind power plants in all the areas $\Delta P_{tie_{2,3}}$ increases due to the wind power plants variations. Actually, $\Delta P_{tie_{2,3}}$ *case* (3) doubles the value of *case* (2), subsequently producing more oscillations in frequency deviations in those areas, as depicted in Figure 12. Therefore, *case* (2) is suggested by the authors under imbalance conditions to reduce frequency oscillations and power flow between areas.

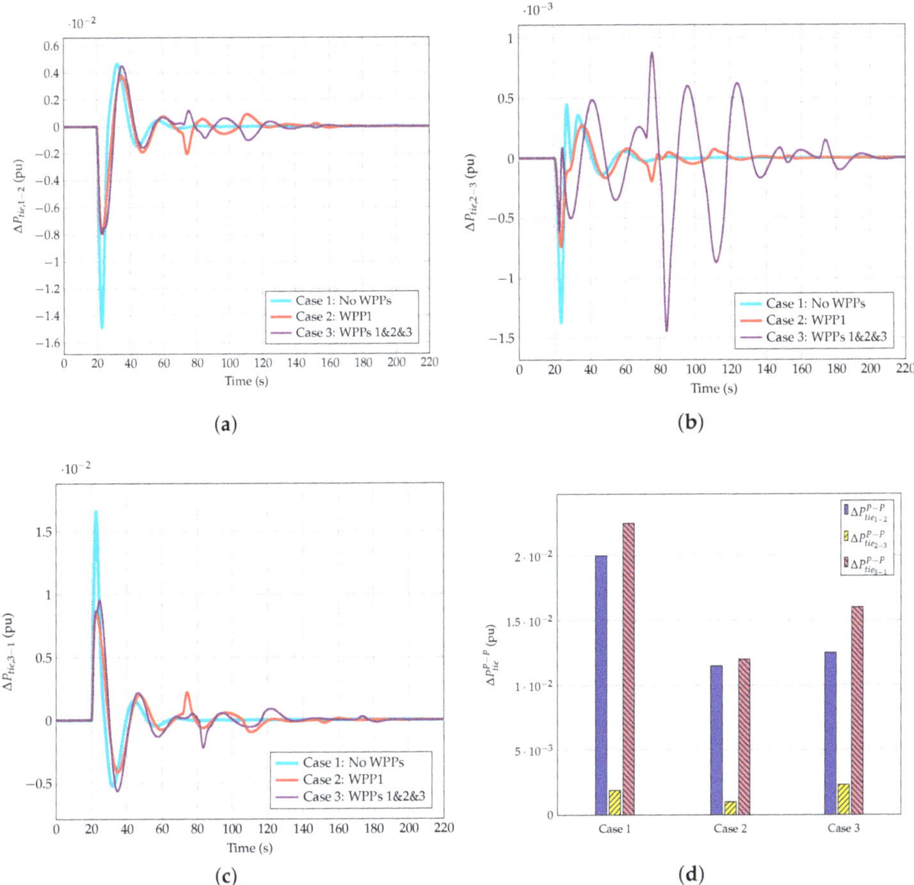

Figure 13. Area 1 under power imbalance ($\Delta P_{L,1}$): tie-line power comparison; (**a**) tie-line power variation between Areas 1 and 2; (**b**) tie-line power variation between Areas 2 and 3; (**c**) tie-line power variation between Areas 3 and 1; (**d**) comparison among peak-to-peak tie-line power variation exchange.

5. Conclusions

Multi-areas interconnected power systems are analyzed under power imbalance conditions and with high wind energy integration. From the supply-side, conventional and renewable resources are considered, including thermal, hydro-power, wind and PV power plants. Wind power integration accounts for between 25% and 40%, corresponding to current percentages in some European countries. Tie-line power is limited to a maximum value of 10%, in line with recent EU directives. Different cases are compared and analyzed, depending on frequency control strategies applied by wind power plants. According to the results, frequency responses are improved by including wind power plants in frequency control, in comparison with simulations where this task is only performed by conventional generation units. Of the different cases, the nadir reductions are maximized when only wind power plants within the area submitted to imbalances are considered. In this case, the nadir is reduced between 40% and 50% in the area submitted to imbalanced areas in comparison to conventional generational unit scenarios. Moreover, these nadir values are also reduced in the other areas between 20% and 30%. When wind power responses of all areas are considered, higher frequency oscillations and lower nadir reductions can be reached in comparison with only conventional generation unit scenarios. Stabilization time remains almost constant under different situations, and very similar to simulations where only conventional units respond under frequency excursions. Subsequently, the authors suggest including only wind power plant frequency response within the area submitted to imbalances, avoiding additional frequency oscillations coming from wind power plants located in the other areas.

Author Contributions: Data curation, A.F.-G.; formal analysis, A.F.-G.; methodology, A.F.-G., A.V.-R. and A.M.-G.; software, A.F.-G.; supervision, A.V.-R. and E.G.-L.; visualization, A.M.-G.; writing—original draft, A.M.-G.; writing—review & editing, E.G.-L.

Funding: This work was supported by 'Ministerio de Educación, Cultura y Deporte' of Spain (ref. FPU16/04282). The authors are grateful for the financial support from the Spanish Ministry of the Economy and Competitiveness and the European Union —ENE2016-78214-C2-2-R.

Conflicts of Interest: The authors declare no conflict of interest.

Abbreviations

The following abbreviations are used in this manuscript:

ACE	Area Control Error
AGC	Automatic Generation Control
CG	Total number of conventional synchronous generators
$ROCOF$	Rate of Change of Frequency
$VSWTs$	Variable Speed Wind Turbines
WPP	Wind Power Plant
$nadir$	Minimum value of the frequency excursion
n	Number of VSWT in the wind power plant
t_{stab}	Stabilization time
B	Bias factor
D_{eq}	Equivalent damping factor of the power system
H_{eq}	Equivalent inertia constant of the power system
H_m	Inertia constant of generating unit m
N	Number of interconnected areas
P_{cmd}	Commanded power of the VSWT
P_{MPPT}	Maximum power point tracking of the VSWT
P_{mt}	Mechanical power of the VSWT
S_B	Rated power of the power system
$S_{B,m}$	Rated power of generating unit m
S_n	Rated power of a VSWT
$T_{i,j}$	synchronizing moment coefficient of a tie-line between areas i and j
V_W	Wind speed
β	Pitch angle
Δf	Frequency excursion
Δf_{lim}	Value at which frequency controller of the VSWT activates
ΔP_g	Variation of active power of the power system
ΔP_L	Variation of power demand
ΔP_{OP}	Additional active power in overproduction operation mode
$\Delta P_{tie_{i,j}}$	Tie-line power changed between areas i and j
$\Delta P_{tie_{i,j}}^{P-P}$	Peak-to-peak tie-line power changed between areas i and j
ΔP_{WF}	Variation of active power of the wind power plant
Ω_{WT}	Rotational speed of the VSWT
$\Omega_{WT,min}$	Minimum rotational speed of the VSWT
Ω_{MPPT}	Rotational speed at maximum power point tracking

References

1. Li, D.Y.; Li, P.; Cai, W.C.; Song, Y.D.; Chen, H.J. Adaptive Fault Tolerant Control of Wind Turbines with Guaranteed Transient Performance Considering Active Power Control of Wind Farms. *IEEE Trans. Ind. Electron.* **2017**, *65*, 3275–3285. [CrossRef]
2. Rasolomampionona, D. A modified power system model for AGC analysis. In Proceedings of the 2009 IEEE Bucharest PowerTech, Bucharest, Romania, 28 June–2 July 2009; pp. 1–6.
3. Du, P.; Matevosyan, J. Forecast system inertia condition and its impact to integrate more renewables. *IEEE Trans. Smart Grid* **2018**, *9*, 1531–1533. [CrossRef]
4. Toulabi, M.; Bahrami, S.; Ranjbar, A.M. Application of Edge theorem for robust stability analysis of a power system with participating wind power plants in automatic generation control task. *IET Renew. Power Gener.* **2017**, *11*, 1049–1057. [CrossRef]
5. Toulabi, M.; Bahrami, S.; Ranjbar, A.M. An Input-to-State Stability Approach to Inertial Frequency Response Analysis of Doubly-Fed Induction Generator-Based Wind Turbines. *IEEE Trans. Energy Convers.* **2017**, *32*, 1418–1431. [CrossRef]
6. Ochoa, D.; Martinez, S. Frequency dependent strategy for mitigating wind power fluctuations of a doubly-fed induction generator wind turbine based on virtual inertia control and blade pitch angle regulation. *Renew. Energy* **2018**, *128*, 108–124. [CrossRef]
7. Ochoa, D.; Martinez, S. Proposals for Enhancing Frequency Control in Weak and Isolated Power Systems: Application to the Wind-Diesel Power System of San Cristobal Island-Ecuador. *Energies* **2018**, *11*, 910.10.3390/en11040910. [CrossRef]
8. Bao, Y.; Xu, J.; Liao, S.; Sun, Y.; Li, X.; Jiang, Y.; Ke, D.; Yang, J.; Peng, X. Field Verification of Frequency Control by Energy-Intensive Loads for Isolated Power Systems with High Penetration of Wind Power. *IEEE Trans. Power Syst.* **2018**, in press. [CrossRef]
9. Fini, M.H.; Golshan, M.E.H. Determining optimal virtual inertia and frequency control parameters to preserve the frequency stability in islanded microgrids with high penetration of renewables. *Electr. Power Syst. Res.* **2018**, *154*, 13–22. [CrossRef]
10. Li, H.; Wang, X.; Xiao, J. Differential Evolution-Based Load Frequency Robust Control for Micro-Grids with Energy Storage Systems. *Energies* **2018**, *11*, 686. [CrossRef]
11. Huang, L.; Xin, H.; Zhang, L.; Wang, Z.; Wu, K.; Wang, H. Synchronization and Frequency Regulation of DFIG-Based Wind Turbine Generators With Synchronized Control. *IEEE Trans. Energy Convers.* **2017**, *32*, 1251–1262. [CrossRef]
12. Luo, X.; Wang, J.; Dooner, M.; Clarke, J. Overview of current development in electrical energy storage technologies and the application potential in power system operation. *Appl. Energy* **2015**, *137*, 511–536. [CrossRef]
13. Xue, Y.; Tai, N. Review of contribution to frequency control through variable speed wind turbine. *Renew. Energy* **2011**, *36*, 1671–1677.
14. Sun, D.; Sun, L.; Wu, F.; Zu, G. Frequency Inertia Response Control of SCESS-DFIG under Fluctuating Wind Speeds Based on Extended State Observers. *Energies* **2018**, *11*, 830. [CrossRef]
15. Tavakoli, M.; Pouresmaeil, E.; Adabi, J.; Godina, R.; Catalao, J.P. Load-frequency control in a multi-source power system connected to wind farms through multi terminal HVDC systems. *Comput. Oper. Res.* **2018**, *96*, 305–315. [CrossRef]
16. Saeed Uz Zaman, M.; Bukhari, S.B.A.; Hazazi, K.M.; Haider, Z.M.; Haider, R.; Kim, C.H. Frequency Response Analysis of a Single-Area Power System with a Modified LFC Model Considering Demand Response and Virtual Inertia. *Energies* **2018**, *11*, 787. [CrossRef]
17. Sun, Y.; Zhang, Z.; Li, G.; Lin, J. Review on frequency control of power systems with wind power penetration. In Proceedings of the 2010 IEEE International Conference on Power System Technology (POWERCON), Hangzhou, China, 24–28 October 2010; pp. 1–8.
18. Gonzalez-Longatt, F.M.; Bonfiglio, A.; Procopio, R.; Verduci, B. Evaluation of inertial response controllers for full-rated power converter wind turbine (Type 4). In Proceedings of the 2016 IEEE Power and Energy Society General Meeting (PESGM), Boston, MA, USA, 17–21 July 2016; pp. 1–5.
19. Wu, Z.; Gao, W.; Gao, T.; Yan, W.; Zhang, H.; Yan, S.; Wang, X. State-of-the-art review on frequency response of wind power plants in power systems. *J. Modern Power Syst. Clean Energy* **2018**, *6*, 1–16. [CrossRef]

20. Keung, P.K.; Li, P.; Banakar, H.; Ooi, B.T. Kinetic energy of wind-turbine generators for system frequency support. *IEEE Trans. Power Syst.* **2009**, *24*, 279–287. [CrossRef]

21. El Itani, S.; Annakkage, U.D.; Joos, G. Short-term frequency support utilizing inertial response of DFIG wind turbines. In Proceedings of the 2011 IEEE Power and Energy Society General Meeting, Detroit, MI, USA, 24–28 July 2011; pp. 1–8.

22. Hansen, A.D.; Altin, M.; Margaris, I.D.; Iov, F.; Tarnowski, G.C. Analysis of the short-term overproduction capability of variable speed wind turbines. *Renew. Energy* **2014**, *68*, 326–336. [CrossRef]

23. Hafiz, F.; Abdennour, A. Optimal use of kinetic energy for the inertial support from variable speed wind turbines. *Renew. Energy* **2015**, *80*, 629–643. [CrossRef]

24. Kang, M.; Kim, K.; Muljadi, E.; Park, J.W.; Kang, Y.C. Frequency control support of a doubly-fed induction generator based on the torque limit. *IEEE Trans. Power Syst.* **2016**, *31*, 4575–4583. [CrossRef]

25. Fernández-Guillamón, A.; Villena-Lapaz, J.; Vigueras-Rodríguez, A.; García-Sánchez, T.; Molina-García, Á. An Adaptive Frequency Strategy for Variable Speed Wind Turbines: Application to High Wind Integration into Power Systems. *Energies* **2018**, *11*, 1436. [CrossRef]

26. Margaris, I.D.; Papathanassiou, S.A.; Hatziargyriou, N.D.; Hansen, A.D.; Sorensen, P. Frequency control in autonomous power systems with high wind power penetration. *IEEE Trans. Sustain. Energy* **2012**, *3*, 189–199. [CrossRef]

27. Vidyanandan, K.; Senroy, N. Primary frequency regulation by deloaded wind turbines using variable droop. *IEEE Trans. Power Syst.* **2013**, *28*, 837–846. [CrossRef]

28. Alsharafi, A.S.; Besheer, A.H.; Emara, H.M. Primary Frequency Response Enhancement for Future Low Inertia Power Systems Using Hybrid Control Technique. *Energies* **2018**, *11*, 699. [CrossRef]

29. Tielens, P.; Van Hertem, D. Receding horizon control of wind power to provide frequency regulation. *IEEE Trans. Power Syst.* **2017**, *32*, 2663–2672. [CrossRef]

30. Sahu, R.K.; Gorripotu, T.S.; Panda, S. Automatic generation control of multi-area power systems with diverse energy sources using teaching learning based optimization algorithm. *Eng. Sci. Technol. Int. J.* **2016**, *19*, 113–134. [CrossRef]

31. Shayeghi, H.; Shayanfar, H.; Jalili, A. Load frequency control strategies: A state-of-the-art survey for the researcher. *Energy Convers. Manag.* **2009**, *50*, 344–353. [CrossRef]

32. Parmar, K.S.; Majhi, S.; Kothari, D. LFC of an interconnected power system with multi-source power generation in deregulated power environment. *Int. J. Electr. Power Energy Syst.* **2014**, *57*, 277–286. [CrossRef]

33. Shankar, R.; Chatterjee, K.; Bhushan, R. Impact of energy storage system on load frequency control for diverse sources of interconnected power system in deregulated power environment. *Int. J. Electr. Power Energy Syst.* **2016**, *79*, 11–26. [CrossRef]

34. Barisal, A. Comparative performance analysis of teaching learning based optimization for automatic load frequency control of multi-source power systems. *Int. J. Electr. Power Energy Syst.* **2015**, *66*, 67–77. [CrossRef]

35. Sahu, R.K.; Panda, S.; Sekhar, G.C. A novel hybrid PSO-PS optimized fuzzy PI controller for AGC in multi area interconnected power systems. *Int. J. Electr. Power Energy Syst.* **2015**, *64*, 880–893. [CrossRef]

36. Guha, D.; Roy, P.K.; Banerjee, S. Load frequency control of interconnected power system using grey wolf optimization. *Swarm Evol. Comput.* **2016**, *27*, 97–115. [CrossRef]

37. Ozdemir, M.T.; Ozturk, D. Comparative Performance Analysis of Optimal PID Parameters Tuning Based on the Optics Inspired Optimization Methods for Automatic Generation Control. *Energies* **2017**, *10*, 2134. [CrossRef]

38. Mohanty, B.; Panda, S.; Hota, P. Controller parameters tuning of differential evolution algorithm and its application to load frequency control of multi-source power system. *Int. J. Electr. Power Energy Syst.* **2014**, *54*, 77–85. [CrossRef]

39. Zare, K.; Hagh, M.T.; Morsali, J. Effective oscillation damping of an interconnected multi-source power system with automatic generation control and TCSC. *Int. J. Electr. Power Energy Syst.* **2015**, *65*, 220–230. [CrossRef]

40. Gorripotu, T.S.; Sahu, R.K.; Panda, S. AGC of a multi-area power system under deregulated environment using redox flow batteries and interline power flow controller. *Eng. Sci. Technol. Int. J.* **2015**, *18*, 555–578. [CrossRef]

41. Hota, P.; Mohanty, B. Automatic generation control of multi source power generation under deregulated environment. *Int. J. Electr. Power Energy Syst.* **2016**, *75*, 205–214. [CrossRef]

42. Ketabi, A.; Fini, M.H. An underfrequency load shedding scheme for hybrid and multiarea power systems. *IEEE Trans. Smart Grid* **2015**, *6*, 82–91. [CrossRef]

43. Mi, Y.; Fu, Y.; Li, D.; Wang, C.; Loh, P.C.; Wang, P. The sliding mode load frequency control for hybrid power system based on disturbance observer. *Int. J. Electr. Power Energy Syst.* **2016**, *74*, 446–452. [CrossRef]

44. Kerdphol, T.; Rahman, F.S.; Mitani, Y. Virtual Inertia Control Application to Enhance Frequency Stability of Interconnected Power Systems with High Renewable Energy Penetration. *Energies* **2018**, *11*, 981. [CrossRef]

45. Babahajiani, P.; Shafiee, Q.; Bevrani, H. Intelligent demand response contribution in frequency control of multi-area power systems. *IEEE Trans. Smart Grid* **2018**, *9*, 1282–1291. [CrossRef]

46. 2030 Energy Strategy. Available online: https://ec.europa.eu/energy/en/topics/energy-strategy-and-energy-union/2030-energy-strategy (accessed on 15 August 2018).

47. Dreidy, M.; Mokhlis, H.; Mekhilef, S. Inertia response and frequency control techniques for renewable energy sources: A review. *Renew. Sustain. Energy Rev.* **2017**, *69*, 144–155. [CrossRef]

48. Wang, H.; Yang, J.; Chen, Z.; Ge, W.; Hu, S.; Ma, Y.; Li, Y.; Zhang, G.; Yang, L. Gain Scheduled Torque Compensation of PMSG-Based Wind Turbine for Frequency Regulation in an Isolated Grid. *Energies* **2018**, *11*, 1623. [CrossRef]

49. Jallad, J.; Mekhilef, S.; Mokhlis, H. Frequency Regulation Strategies in Grid Integrated Offshore Wind Turbines via VSC-HVDC Technology: A Review. *Energies* **2017**, *10*, 1244. [CrossRef]

50. Tavakkoli, M.; Adabi, J.; Zabihi, S.; Godina, R.; Pouresmaeil, E. Reserve Allocation of Photovoltaic Systems to Improve Frequency Stability in Hybrid Power Systems. *Energies* **2018**, *11*, 583. [CrossRef]

51. Kayikçi, M.; Milanovic, J.V. Dynamic contribution of DFIG-based wind plants to system frequency disturbances. *IEEE Trans. Power Syst.* **2009**, *24*, 859–867. [CrossRef]

52. Morren, J.; de Haan, S.W.H.; Kling, W.L.; Ferreira, J.A. Wind turbines emulating inertia and supporting primary frequency control. *IEEE Trans. Power Syst.* **2006**, *21*, 433–434. [CrossRef]

53. Ramtharan, G.; Jenkins, N.; Ekanayake, J. Frequency support from doubly fed induction generator wind turbines. *IET Renew. Power Gener.* **2007**, *1*, 3–9. [CrossRef]

54. Chowdhury, B.H.; Ma, H.T. Frequency regulation with wind power plants. In Proceedings of the 2008 IEEE Power and Energy Society General Meeting-Conversion and Delivery of Electrical Energy in the 21st Century, 20–24 July 2008; pp. 1–5.

55. Conroy, J.F.; Watson, R. Frequency response capability of full converter wind turbine generators in comparison to conventional generation. *IEEE Trans. Power Syst.* **2008**, *23*, 649–656. [CrossRef]

56. Mauricio, J.M.; Marano, A.; Gómez-Expósito, A.; Ramos, J.L.M. Frequency regulation contribution through variable-speed wind energy conversion systems. *IEEE Trans. Power Syst.* **2009**, *24*, 173–180. [CrossRef]

57. Bassi, F.; Caciolli, L.; Giannuzzi, G.; Corsi, N.; Giorgi, A. Use of hidden inertia from wind generation for frequency support in power grids. In Proceedings of the 2016 IEEE AEIT International Annual Conference (AEIT), Capri, Italy, 5–7 October 2016; pp. 1–6.

58. Gonzalez-Longatt, F.M. Impact of emulated inertia from wind power on under-frequency protection schemes of future power systems. *J. Modern Power Syst. Clean Energy* **2016**, *4*, 211–218. [CrossRef]

59. Hu, J.; Sun, L.; Yuan, X.; Wang, S.; Chi, Y. Modeling of type 3 wind turbines with df/dt inertia control for system frequency response study. *IEEE Trans. Power Syst.* **2017**, *32*, 2799–2809. [CrossRef]

60. Bonfiglio, A.; Invernizzi, M.; Labella, A.; Procopio, R. Design and Implementation of a Variable Synthetic Inertia Controller for Wind Generating Units. *IEEE Trans. Power Syst.* **2018**, in press.

61. Tarnowski, G.C.; Kjar, P.C.; Sorensen, P.E.; Ostergaard, J. Variable speed wind turbines capability for temporary over-production. In Proceedings of the 2009 IEEE Power & Energy Society General Meeting (PES'09), Calgary, AB, Canada, 26–30 July 2009; pp. 1–7.

62. Kundur, P.; Balu, N.J.; Lauby, M.G. *Power System Stability and Control*; McGraw-Hill: New York, NY, USA, 1994; Volume 7.

63. Tan, W. Load frequency control: Problems and solutions. In Proceedings of the 2011 IEEE 30th Chinese Control Conference (CCC), Yantai, China, 22–24 July 2011; pp. 6281–6286.

64. Ma, M.; Chen, H.; Liu, X.; Allgöwer, F. Distributed model predictive load frequency control of multi-area interconnected power system. *Int. J. Electr. Power Energy Syst.* **2014**, *62*, 289–298. [CrossRef]

65. Yousef, H.A.; Khalfan, A.K.; Albadi, M.H.; Hosseinzadeh, N. Load frequency control of a multi-area power system: An adaptive fuzzy logic approach. *IEEE Trans. Power Syst.* **2014**, *29*, 1822–1830. [CrossRef]

66. Pan, I.; Das, S. Fractional-order load-frequency control of interconnected power systems using chaotic multi-objective optimization. *Appl. Soft Comput.* **2015**, *29*, 328–344. [CrossRef]

67. Sathya, M.; Ansari, M.M.T. Load frequency control using Bat inspired algorithm based dual mode gain scheduling of PI controllers for interconnected power system. *Int. J. Electr. Power Energy Syst.* **2015**, *64*, 365–374. [CrossRef]

68. Abdelaziz, A.; Ali, E. Cuckoo search algorithm based load frequency controller design for nonlinear interconnected power system. *Int. J. Electr. Power Energy Syst.* **2015**, *73*, 632–643. [CrossRef]

69. Padhan, D.G.; Majhi, S. A new control scheme for PID load frequency controller of single-area and multi-area power systems. *ISA Trans.* **2013**, *52*, 242–251. [CrossRef] [PubMed]

70. Chang, Y.; Liu, R.; Ba, Y.; Li, W. A New Control Logic for a Wind-Area on the Balancing Authority Area Control Error Limit Standard for Load Frequency Control. *Energies* **2018**, *11*, 121. [CrossRef]

71. Pyller, M.; Achilles, S. Aggregated wind park models for analyzing power system dynamics. In Proceedings of the 4th International Workshop Large-Scale Integration of Wind Power and Transmission Networks, Billund, Denmark, 20–21 October 2003.

72. Mokhtari, M.; Aminifar, F. Toward wide-area oscillation control through doubly-fed induction generator wind farms. *IEEE Trans. Power Syst.* **2014**, *29*, 2985–2992. [CrossRef]

73. Miller, N.W.; Sanchez-Gasca, J.J.; Price, W.W.; Delmerico, R.W. Dynamic modeling of GE 1.5 and 3.6 MW wind turbine-generators for stability simulations. In Proceedings of the 2003 IEEE Power Engineering Society General Meeting, Toronto, ON, Canada, 13–17 July 2003; Volume 3, pp. 1977–1983.

74. Ullah, N.R.; Thiringer, T.; Karlsson, D. Temporary primary frequency control support by variable speed wind turbines—Potential and applications. *IEEE Trans. Power Syst.* **2008**, *23*, 601–612. [CrossRef]

75. Baccino, F.; Conte, F.; Grillo, S.; Massucco, S.; Silvestro, F. An optimal model-based control technique to improve wind farm participation to frequency regulation. *IEEE Trans. Sustain. Energy* **2015**, *6*, 993–1003. [CrossRef]

76. Zhao, J.; Lyu, X.; Fu, Y.; Hu, X.; Li, F. Coordinated microgrid frequency regulation based on DFIG variable coefficient using virtual inertia and primary frequency control. *IEEE Trans. Energy Convers.* **2016**, *31*, 833–845. [CrossRef]

77. Martínez-Lucas, G.; Sarasúa, J.I.; Sánchez-Fernández, J.Á. Frequency Regulation of a Hybrid Wind–Hydro Power Plant in an Isolated Power System. *Energies* **2018**, *11*, 239. [CrossRef]

78. Perera, A.T.D.; Nik, V.M.; Mauree, D.; Scartezzini, J.L. Electrical hubs: An effective way to integrate non-dispatchable renewable energy sources with minimum impact to the grid. *Appl. Energy* **2017**, *190*, 232–248. [CrossRef]

79. Wind energy in Europe: Scenarios for 2030. *Wind Eur.* **2017**. Available online: https://windeurope.org/wp-content/uploads/files/about-wind/reports/Wind-energy-in-Europe-Scenarios-for-2030.pdf (accessed on 10 August 2018).

80. Technology Roadmap: Solar Photovoltaic Energy. *Int. Energy Agency* **2014**. Available online: https://www.iea.org/publications/freepublications/publication/TechnologyRoadmapSolarPhotovoltaicEnergy_2014edition.pdf (accessed on 10 August 2018).

81. Huang, C.; Yue, D.; Xie, X.; Xie, J. Anti-windup load frequency controller design for multi-area power system with generation rate constraint. *Energies* **2016**, *9*, 330. [CrossRef]

82. Zeng, G.Q.; Xie, X.Q.; Chen, M.R. An Adaptive Model Predictive Load Frequency Control Method for Multi-Area Interconnected Power Systems with Photovoltaic Generations. *Energies* **2017**, *10*, 1840. [CrossRef]

83. Yang, L.; Liu, T.; Hill, D.J. Decentralized periodic event-triggered frequency regulation for multi-area power systems. In Proceedings of the 2018 IEEE Power Systems Computation Conference (PSCC), Dublin, Ireland, 11–15 June 2017; pp. 1–7.

84. Lu, K.; Zhou, W.; Zeng, G.; Zheng, Y. Constrained population extremal optimization-based robust load frequency control of multi-area interconnected power system. *Int. J. Electr. Power Energy Syst.* **2019**, *105*, 249–271. [CrossRef]

85. Huang, H.; Li, F. Sensitivity analysis of load-damping characteristic in power system frequency regulation. *IEEE Trans. Power Syst.* **2013**, *28*, 1324–1335. [CrossRef]

86. Tielens, P.; Van Hertem, D. The relevance of inertia in power systems. *Renew. Sustain. Energy Rev.* **2016**, *55*, 999–1009. [CrossRef]

87. Ali, E.; Abd-Elazim, S. Bacteria foraging optimization algorithm based load frequency controller for interconnected power system. *Int. J. Electr. Power Energy Syst.* **2011**, *33*, 633–638. [CrossRef]

88. Sudha, K.; Santhi, R.V. Robust decentralized load frequency control of interconnected power system with generation rate constraint using type-2 fuzzy approach. *Int. J. Electr. Power Energy Syst.* **2011**, *33*, 699–707. [CrossRef]

89. Jiang, L.; Yao, W.; Wu, Q.; Wen, J.; Cheng, S.; others. Delay-dependent stability for load frequency control with constant and time-varying delays. *IEEE Trans. Power Syst.* **2012**, *27*, 932. [CrossRef]

90. Masuta, T.; Yokoyama, A. Supplementary load frequency control by use of a number of both electric vehicles and heat pump water heaters. *IEEE Trans. Smart Grid* **2012**, *3*, 1253–1262. [CrossRef]

91. Shabani, H.; Vahidi, B.; Ebrahimpour, M. A robust PID controller based on imperialist competitive algorithm for load-frequency control of power systems. *ISA Trans.* **2013**, *52*, 88–95. [CrossRef] [PubMed]

92. Rout, U.K.; Sahu, R.K.; Panda, S. Design and analysis of differential evolution algorithm based automatic generation control for interconnected power system. *Ain Shams Eng. J.* **2013**, *4*, 409–421. [CrossRef]

93. Sahu, R.K.; Panda, S.; Rout, U.K. DE optimized parallel 2-DOF PID controller for load frequency control of power system with governor dead-band nonlinearity. *Int. J. Electr. Power Energy Syst.* **2013**, *49*, 19–33. [CrossRef]

94. Ruiz, S.; Patiño, J.; Espinosa, J. Load Frequency Control of a Multi-area Power System Incorporating Variable-speed Wind Turbines. **2016**, in press.

95. Dong, L.; Tang, Y.; He, H.; Sun, C. An event-triggered approach for load frequency control with supplementary ADP. *IEEE Trans. Power Syst.* **2017**, *32*, 581–589. [CrossRef]

96. Peng, C.; Zhang, J.; Yan, H. Adaptive Event-Triggering H_∞ Load Frequency Control for Network-Based Power Systems. *IEEE Trans. Ind. Electr.* **2018**, *65*, 1685–1694. [CrossRef]

97. Wu, Y.; Wei, Z.; Weng, J.; Li, X.; Deng, R.H. Resonance attacks on load frequency control of smart grids. *IEEE Trans. Smart Grid* **2018**, *9*, 4490–4502. [CrossRef]

Article

On Modelling Wind-Farm Wake Turbulence Autospectra and Coherence from a Database

Kyle A. Schau [1], Gopal Gaonkar [2] and Vaishakh Krishnan [2],*

[1]   School of Aerospace Engineering, Georgia Institute of Technology, Atlanta, GA 30332, USA;
      kschau@gatech.edu
[2]   Department of Ocean and Mechanical Engineering, Florida Atlantic University, Boca Raton, FL 33431, USA;
      gaonkar@fau.edu
*   Correspondence: krishnanv2016@fau.edu

Received: 30 November 2018; Accepted: 29 December 2018; Published: 30 December 2018

Abstract: This study addresses the feasibility of modeling wind-farm wake-turbulence autospectra and coherences from a database: flow velocity points from experimental and computational fluid dynamics (CFD) investigations. Specifically, it first applies an earlier-exercised framework to construct the autospectral models from a database and then it adopts a recently proposed framework to construct the coherence models from a database. While this proposed framework has not been tested against a database, the methodology has been completely formulated with a theoretical basis. These models of autospectrum and coherence are interpretive, and in closed form. Both frameworks basically involve the perturbation series expansion of the autospectra and coherences. The framework for modeling autospectra is tested against a demanding database of wake turbulence inside a wind farm over a complex terrain from a full-scale test. The suitability of these autospectral models for simulation through white-noise driven filters is also demonstrated. Finally, coherence models are generated for assumed values of the perturbation series constants, and these coherence models are used to demonstrate how the coherence models of homogeneous isotropic turbulence deviate from the coherence models of non-homogeneous non-isotropic turbulence such as wind-farm wake turbulence. This feasibility of extracting both the one-point statistics of autospectral models and the two-point statistics of coherence models from a database represents a research avenue that is new and promising in the treatment of wind-farm wake turbulence. This paper also demonstrates the feasibility of fruitfully exploiting the wake treatment methods developed in other fields.

Keywords: turbulence; statistical modelling

---

1. Introduction

During the past thirty years, wake turbulence and its effects on wind turbines and wind farms have been extensively investigated, primarily analytically and to some extent experimentally. The extensive literature up to 2010 has been well covered in the widely used text of Manwell et al. [1]. As for the extensive analytical investigations since, suffice it to mention, as representative samples, Keck et al. [2] for wake-turbulence modeling from the low-fidelity CFD treatment of the Navier-Stokes (NS) equations, and Carrion et al. [3] for wake-turbulence modeling from the high-fidelity treatment of NS equations. The work of Carrion et al. [3] for example, includes a concise account of the state of the art of modeling wind-farm wake turbulence.

An overview of these investigations [1–3] is included here; although extremely brief, this should help appreciate how the present work serves as a desirable adjunct of experimental and high-fidelity CFD based investigations, and why it represents a new and promising avenue of wake-turbulence modeling. Wake-turbulence modeling falls into three categories:

1. Semi-empirical models: These models are based on conservation of momentum and such simplified assumptions; and they typically contain an empirical constant. For example, the widely used model due to Katic et al. [1] belongs to this category; therein, the empirical constant is referred as wake decay constant. A recent work of Ge M. et al. [4] merits mention: it gives a thorough account of the key features of this class of models and their continual evolution as well as their utility base in the treatment of wind energy applications such as design of wind-farm layout.

2. Low-fidelity CFD solutions: Several numerical schemes have been continually proposed (e.g., dynamic wake meandering model [2]); these are based on a wide range of physics-based approximations to NS equations. As more and more experimental databases become available, they are being continually updated and they merit further validation.

3. High-fidelity CFD solutions: Despite the severe CPU-hour constraint, these high-fidelity approaches are indispensable in generating a database that serves as a reference point and in supplementing databases from experiments.

While major strides have been made in generating databases and in providing a much-improved understanding of wake turbulence, the current capability for modeling the one-point statistics of autospectrum, much more so, for modeling the two-point statistics of coherence, merits significant improvement. In fact, empirical exponential coherence functions are still being used [1]. Accordingly, the present study explores the feasibility of extracting the one-point statistics of autospectrum and the two-point statistics of coherence from a database.

The autospectral model extraction from a database of the present study is based on the framework due to Schau, Gaonkar and Polsky [5]. This framework guarantees that the extracted model and the autospectral data points have the same mean square value (a measure of turbulence energy), time scale and the Kolmogorov $-5/3$ spectral decay. For completeness, an earlier study by Gaonkar [6] should be mentioned as well; therein, the autospectral model extraction from a database is based on the framework of reference [7], in which the $-5/3$ spectral law is bypassed. Now it is expedient to address the development of a framework for extracting the two-point statistics models from a database. This can be approached either through cross-spectrum, which is a complex quantity involving the magnitude and phase or through coherence, which, as a spectral correlation coefficient, is a real quantity. The first approach generally leads to modeling the magnitude and not the phase, as was the case in Ref. [7]. The second approach through coherence is relatively more convenient and provides a means of capturing the two-point statistics from a database completely. The recent study due to Krishnan and Gaonkar [8] follows this second approach; although not tested against a database, the framework is formulated with a mathematical basis and the present study adopts this framework [8].

By design, these autospectral and coherence models are in closed form and they have a simple analytical structure to facilitate interrogation and interpretation of voluminous data points on autospectra and coherences. And they lend themselves well to routine use as a predictive tool. Compared to a description through such voluminous, numerically generated, autospectral and coherence data points, they describe wake turbulence analytically with better transparency and bring better understanding. Thus, these interpretive models broaden the scope and utility base of the database that invariably involves enormous resources. While the extracted models are database-specific (thus they are not predictive by themselves), the framework can be applied to any database and the model extraction is a routine exercise.

In the treatment of coherence for homogeneous isotropic turbulence for which the frozen turbulence hypothesis is applicable (HIT), the present study is motivated by and built on the earlier studies of Burton et al. [9], Houbolt and Sen [10], Frost et al. [11] and Irwin [12]. This treatment of coherence for HIT is found to show differences among these studies [9–12], and the present study, after an in-depth examination, follows Frost et al. [10] for cross-spectra and Irwin [12] for coherences. Given this background, the present study seems to provide a unified account of coherence for HIT in the treatment of wake turbulence.

To sum up: These interpretive models complement the experimental and CFD-based investigations as surrogate analytical models for both the one-point statistics of autospectrum and the two-point statistics of coherence. Moreover, this paper also demonstrates the feasibility of fruitfully exploiting the methodologies from other fields to the treatment of wind-turbine wake turbulence. And these methodologies offer promise towards providing a foothold on a formidably complex flow field inside a windfarm for engineering analysis.

Basic of Modeling

A comparison of the measured autospectra of ambient atmospheric boundary layer turbulence (ABL) and wake turbulence shows that the ABL autospectrum has gone through changes in energy distribution with respect to frequency. Figure 1 [13] should help bring a better understanding of this comparison; specifically, it shows measured dimensionless longitudinal autospectrum $f\check{S}_{uu}/\sigma_u^2$ versus dimensionless frequency fz/U, where z is the mast height and U is the mean wind speed. These autospectra were experimentally generated at the same location in a wind farm over a complex terrain. Figure 1a refers to ABL with a turbulence intensity of 0.103, when the turbines were under stand-still conditions. Furthermore, Figure 1b refers to wake turbulence with a turbulence intensity of 0.204, when the turbines were fully operational. This change in the shape of the wake turbulence autospectrum cannot be realized through a linear superposition of a series of independently occurring changes at different frequencies on the ABL autospectrum. Thus, the autospectral morphing must be due to a nonlinear transformation of ABL. Stated otherwise, wind-farm wake turbulence could be idealized as nonlinearly transformed ABL and in turn, an earlier-developed mathematical framework for autospectral modeling of airwake-downwash turbulence could be adapted to modeling wind-farm wake turbulence as well [5,14]. (Airwake-downwash turbulence refers to the coupled flow-field of ship's airwake shed from the superstructure and the helicopter downwash. Therein [5,14], the mathematical framework "posits" that airwake-downwash turbulence is nonlinearly transformed ABL. Regarding the coherence, the framework of Ref. 8 is adopted with the same justification that is used for the autospectrum.)

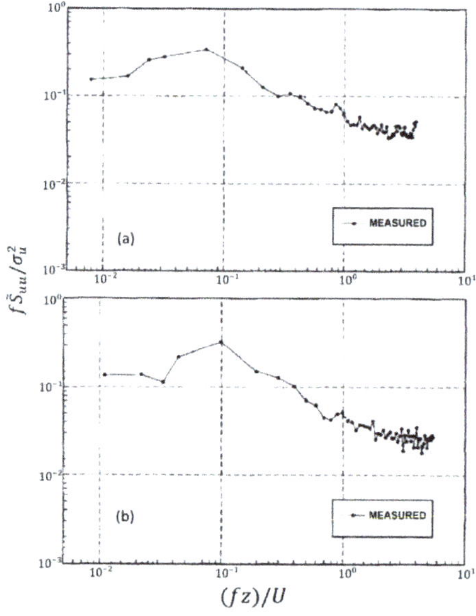

Figure 1. (**a**) Measured longitudinal velocity autospectrum of ambient atmospheric boundary layer turbulence, and (**b**) Measured longitudinal velocity autospectrum of wake turbulence.

2. Methodology of AutoSpectra

The Lateral component $v(t)$ is selected for providing details of the autospectral model extraction methodology [5]. The methodology remains the same for the vertical component with minor changes in the parameters used in the constraint equations. However, for the Longitudinal component, additional changes in the constraint equations are also required [14]. One-sided autospectrum is used throughout.

Statistical independence of velocity components is assumed [9]. The mean square value, time scale and autospectral asymptotic limit law are different for each component of non-homogeneous turbulence according to Kolmogorov's $-5/3$. The framework combines four elements: (1) A mathematical formulation based on a perturbation series expansion of the autocorrelation/autospectrum functions; (2) Extraction of time scale and autospectral asymptotic limit from the database; (3) Development of constraint equations in closed form to ensure that the developed model satisfies the requirements related to normalization, time scale, and autospectral asymptotic limit; and (4) Evaluation of the constants in the series expansion subject to satisfying the constraint equations and fitting a curve on a set of selected autospectral data points in a least squares sense.

2.1. Lateral Wake Turbulence

The perturbation series for the autocorrelation of lateral wake turbulence velocity $v(t)$ can be expressed as in Equation (1).

$$\tilde{R}_{vv}(\tau) = \beta_{1v} R_{vv}(\tau) + \beta_{2v} R_{vv}^2(\tau) + \beta_{3v} R_{vv}^3(\tau) + \ldots + \beta_{nv} R_{vv}^n(\tau) \tag{1}$$

The calculated autocorrelation as well as series expansion autocorrelation follow the properties of normalized autocorrelations, that is, $\tilde{R}_{vv}(0) = R_{vv}(0) = 1$. The Fourier transform of Equation (1) gives the series expansion for the autospectrum $\tilde{S}_{vv}(f)$:

$$\tilde{S}_{vv_{vv}}(f) = \beta_{1v} S_{vv1}(f) + \beta_{2v} S_{vv2}(f) + \beta_{3v} S_{vv3}(f) + \ldots + \beta_{nv} S_{vvn}(f) \tag{2}$$

where $S_{vvn}(f)$ is the Fourier transform of $R_{vv}^n(\tau)$.

$$S_{vvn}(f) = 4 \int_0^\infty R_{vv}^n(\tau) \cos(2\pi f \tau) d\tau \tag{3}$$

The autospectrum is typically normalized with respect to dimensional time scale T_v, which is traditionally defined as $T_v = \int_0^\infty \tilde{R}_{vv}(\tau) d\tau$. With σ_v^2, the mean square value, Equations (4) and (5) typify the normalization:

$$\tilde{S}_{vv}(0) = 4\sigma_v^2 \int_0^\infty \tilde{R}_{vv}(\tau) d\tau = 4\sigma_v^2 T_v \tag{4}$$

$$\frac{1}{\sigma_v^2} \int_0^\infty \tilde{S}_{vv}(f) df = \tilde{R}_{vv}(0) = 1 \tag{5}$$

According to Kolmogorov's spectral law, the autospectrum model should decay as given in Equation (6), where A_v is a scaling parameter determined from the data, and f_{hf} represents high frequencies.

$$\frac{f S_{vv}\left(f_{hf}\right)}{\sigma_v^2} = A_v (f T_v)^{-2/3} \tag{6}$$

The basis function $R_{vv}(\tau)$ on the right-hand side of Equation (1) is the von Karman lateral correlation function as given in Equation (7).

$$R_{vv}(x) = \frac{2^{2/3}}{\Gamma(1/3)} \left(\frac{\alpha_v \tau}{T_v}\right)^{1/3} \left[K_{1/3}\left(\frac{\alpha_v \tau}{T_v}\right) - \frac{1}{2}\left(\frac{\alpha_v \tau}{T_v}\right) K_{2/3}\left(\frac{\alpha_v \tau}{T_v}\right)\right] \tag{7}$$

The scaling parameter α_v in Equation (7) ensures that the relation in Equation (4) is satisfied.

2.2. Constraint Equations

The autospectrum model is constrained by Equations (4)–(6). The extracted model as given in Equation (1) is substituted to obtain the constraint Equations (8)–(10) for the expansion series co-efficients [5].

Because $\tilde{R}_{vv}(0) = R_{vv}(0) = 1$, Equation (1) gives:

$$1 = \beta_{1v} + \beta_{2v} + \cdots + \beta_{nv} \tag{8}$$

Satisfying Equation (4), and integrating both sides by Equation (1) gives [5,14]

$$\alpha_v = 0.373417\beta_{1v} + 0.199591\beta_{2v} + 0.12236\beta_{3v} + \cdots \tag{9}$$

Similarly, satisfying Equation (6) leads to [5,14]

$$\frac{A_v}{\alpha_v^{2/3}} = 0.186176(\beta_{1v} + 2\beta_{2v} + 3\beta_{3v} + \cdots) \tag{10}$$

3. Database

The database is generated from a full-scale experimental study of wake turbulence by Mofiadakis et al. [13]. Specifically, data is collected at several points along a complex windy terrain at an altitude of 320–330 m with seven Vestas (27–225 kW) installed in a row. The thirty cases of autospectra that were generated from the database represent free stream to fully wake affected conditions. The autospectral decay was found to be in the range of -1.36 to -1.75 for free stream conditions (all wind turbines at stand-still condition). Moreover, for notational simplicity, β_{iu}, β_{iv} and β_{iw} are simply referred to as β coefficients in this section.

The database typically comprises the temporal flow velocity points. In this case, however, the database [13] has already been transformed from the temporal to the frequency domain and normalized to dimensionless form $fS_{ii}(f)/\sigma_i^2$. Moreover, the temporal autocorrelation is not available, and the dimensionless autospectra are presented on a log-log scale and in turn $S_{ii}(f=0)$ is not available. This limitation is overcome by assuming that $S_{ii}(0) = S_{ii}(f)$ as f approaches zero. It is emphasized that the framework is designed to develop autospectral models from a database; thus the lack of a temporal database ceases to be a major issue. Having extracted σ_i^2 from Equation (5), key information to be extracted from the database is T_i, the time scale. As for T_i, it is calculated using $S_{ii}(f)$ at the lowest frequency in a log-log plot; see Equation (4). Finally, the autospectral decay constant A_i is graphically evaluated from Equation (6).

Having thus generated time scale T_i and autospectral asymptotic limit A_i, the numerical scheme now focuses on computing the series expansion β coefficients. It is emphasized that these β coefficients determine the scaling parameter α_i; see Equation (9). As an iterative procedure, the scheme involves selecting the β coefficients, beginning with the von Karman model (e.g., $\beta_{1v} = 1$ for the lateral component) and strictly enforcing the constraint as typified by Equation (8). For completeness, the gist of the iterative procedure is included; for details see [14].

The numerical scheme minimizes the sum of two errors in a least squares sense: model's deviation from the autospectral data and the A_i constraint error. That is, a selected set of β coefficients gives a model with a value of A_i; stated otherwise, these β coefficients carry a least squares error with respect to the measured autospectral data and an error with respect to the graphically measured A_i value. The resulting A_i error is expected to be within acceptable limits for wind turbine applications (<<10%). This error is perhaps acceptable, after all, A_i is not rigorously defined with respect to the data sets, nor is there a standard method of determining when a computed autospectrum has reached its point of asymptotic decay. This lack of precision also means A_i will vary somewhat from user to user. The A_i constraint typified by Equation (10) also merits one final comment. For some isolated cases of data sets, the high frequency limit does not exhibit accurately the $-5/3$ spectral decay [13]. For these cases, it is sensible to exclude this constraint. The next section elaborates this scenario.

Computationally, the above numerical scheme is found to be inexpensive. For example, a gradient based search algorithm (MATLAB SQP) starting with the von Karman model does not require large iteration counts for a converged model. The reason is that the computational cost, now, depends only on the length of the discrete data array, as an error between the model and this data array, and this error has to be calculated at each iteration of the search algorithm. Given the thoughtful selection of search parameters such as step size in β-space and convergence criteria, this numerical scheme should prove inexpensive computationally.

4. Result of Autospectra

For illustration, just one example of the vertical component $w(t)$ is selected. Modeling is presented based on both first-order (two-term series) and second-order (three-term series) correction. And in each case, modeling covers two approaches. In the first approach, the Kolmogorov $-5/3$ law is not enforced; that is, by satisfying only the first two constraints, typified by Equations (8) and (9). In the second approach, all three constraints are satisfied; that is, in addition to satisfying these two constraints, the model also satisfies the Kolmogorov $-5/3$ law (see Equation (10)) by a minimized error. This enforcement is identified in the respective figures by "*A error* $= xx\%$", where $xx\%$ indicates the percentage error in satisfying the $-5/3$ law. Typically, an "*A error*" of less than 10% is considered satisfactory. Throughout, the dimensionless autospectrum $f\tilde{S}_{ii}(f)/\sigma_i^2$ is presented against frequency f (Hz). Furthermore, in each figure, the corresponding von Karman model (e.g., $\beta_{1v} = 1$ for the lateral) is also included; this helps assess how far the developed model is an improvement over the von Karman, a widely used model for the free-stream case [13]. For additional results, see Schau [14].

For the vertical component in Figure 2, the corresponding first-order-correction (a two-term series) models represent appreciable improvement over the von Karman, particularly for $f > 10^{-1}$ Hz. Overall, modeling still merits further improvements for $f > 10^{-1}$ Hz. For the vertical component (Figure 2b), the enforcement of the Kolmogorov $-5/3$ law in a least squares sense involves "*A error* $= 23.56\%$", well above the stipulated "*A error*" of 10%. These two features, the feasibility of improving the correlation and reducing the "*A error*", is explored in the next figure based on the second-order-correction (a three-term series). The results of Figure 3 are extremely instructive in two respects. First, a comparison of the respective figures (Figure 2a compared to Figure 3a, and Figure 2b compared to Figure 3b) shows that the three-term series model improves the correlation throughout, particularly for $f > 10^{-1}$ Hz. Second, the "*A error*", which is 23.56% for the two-term series model comes down to 7.64%. Thus, this comparison shows that the three-term series model is a noteworthy improvement over the two-term series model, without or with the enforcement of the $-5/3$ law. To sum up: modeling based on first-order correction (a two-term series) is generally adequate, and further improvement in correlation and further reduction in "*A error* $= x$" can be achieved through modeling based on second-order correction (a three-term series).

Figure 4 shows how the autospectrum from the white-noise-driven filter for the developed vertical model compares with the one from the database and the developed model itself (specifically, Figure 4 refers to Figure 3a). As seen from this Figure 4, the developed model and simulation are almost indistinguishable. (The filter represents a single-input, single-output system driven by white noise; the design is routine and thus the details are omitted [14]).

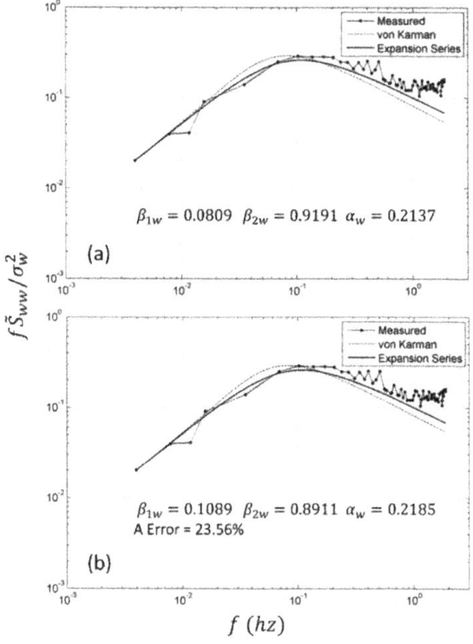

Figure 2. Two-term series modeling for the vertical component (**a**) without the 'A' constraint, and (**b**) with the 'A' constraint.

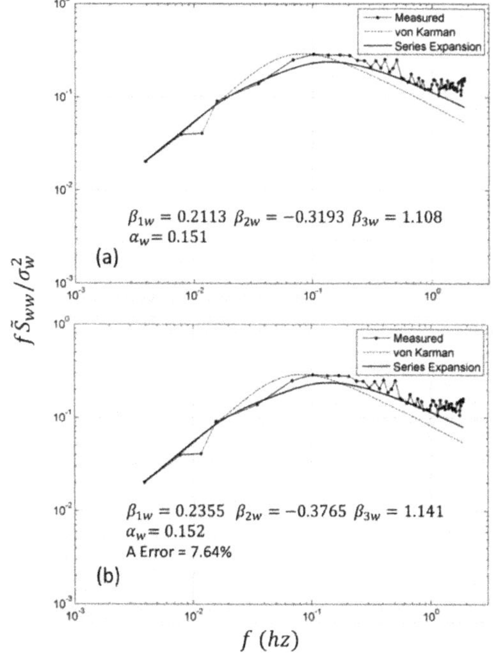

Figure 3. Three-term series modeling for the vertical component (**a**) without the 'A' constraint, and (**b**) with the 'A' constraint.

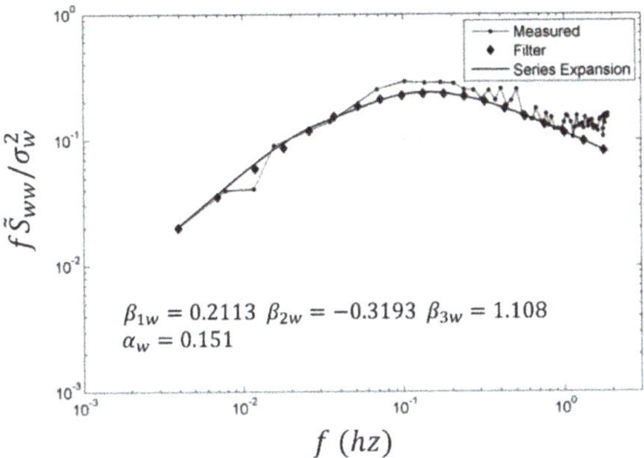

Figure 4. Autospectra from white-noise-driven filter simulation, measurements and a three-term series.

5. Methodology of Coherence

As done for autospectra, for coherences also, a mathematical framework is developed for extracting interpretive coherence models from a database of flow velocity points from experimental and CFD investigations. Here as well, each velocity component is considered statistically independent of the other two. For each velocity component, the framework begins with a perturbation series expansion of the coherence; therein, the basis function or the first term of the series is represented by the corresponding coherence for HIT. The perturbation coefficients are evaluated by satisfying the theoretical constraints and fitting a curve on a set of numerically generated coherence points from a database.

In the literature, the development of the cross-spectra and coherences for the longitudinal, vertical and lateral components is scattered and piecemeal; what is more, the expressions for these cross-spectra and coherences show difference among these studies. Accordingly, this section first presents the cross-spectrum, after all, coherence is cross-spectrum that is normalized by the corresponding autospectrum (details to follow). Then it presents the coherences and finally a perturbation theory scheme for the wind-farm wake-turbulence coherence.

5.1. Construction of the Vertical Cross Spectrum

For illustration, vertical turbulence $w(t)$ is considered under headwind conditions. Given V, the mean wind velocity, τ, the elapsed time $(t_2 - t_1)$ and the correlation distance $x = V\tau$, the von Karman correlation function $R_{ww}(x)$ for vertical turbulence $w(t)$ is given by Equation (11) [15]:

$$R_{ww}(x) = \sigma_w^2 \frac{2^{2/3}}{\Gamma(1/3)} \left[(u)^{1/3} K_{1/3}(u) - \frac{1}{2}(u)^{4/3} K_{2/3}(u) \right] \tag{11}$$

where $u = x/1.339L$, L is the scale length and K_n is the modified Bessel function of the second kind. Now consider the cross-correlation between vertical turbulence $w_1(t)$ at Point 1 and $w_2(t)$ at Point 2, where these two points are separated by the across-wind distance S, as typified by Figure 5. For this scenario, Figure 5 shows that the correlation distance changes to the expression given in Equation (12a):

$$u = \frac{\sigma}{1.339L} \sqrt{1 + \left(\frac{V\tau}{S} \right)^2} \tag{12a}$$

where $\sigma = S/L$. Now, the cross-correlation $R_{w_1 w_2}(x)$ can be expressed as in Equation (12b) [10]:

$$R_{w_1 w_2}(x) = \sigma_w^2 \frac{2^{2/3}}{\Gamma(1/3)}\left[(u)^{1/3}K_{1/3}(u) - \frac{1}{2}(u)^{4/3}K_{2/3}(u)\right] \tag{12b}$$

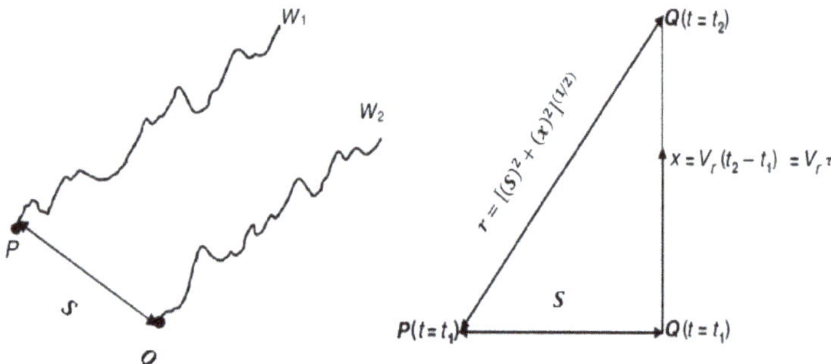

Figure 5. Correlation Distance $P(t_1)\cdot Q(t_2)$ for negligible Stream-wise Separation and across-wind separation 'S'.

The Fourier transform of $R_{w_1 w_2}(\tau)$ is the cross-spectrum $S_{w_1 w_2}(\nu)$ [10]:

$$S_{w_1 w_2}(\nu) = \sigma_w^2 \frac{2^{\frac{2}{3}}}{\Gamma\left(\frac{1}{3}\right)} \cdot \frac{1}{\sqrt{2\pi}}\left(\frac{1}{1.339}\right)^{-\frac{8}{3}}\left[\frac{8}{3}1.339^2\left(\frac{\sigma^{\frac{5}{3}}}{z^{\frac{5}{6}}}\right)K_{\frac{5}{6}}(z) - \left(\frac{\sigma^{\frac{11}{3}}}{z^{\frac{11}{6}}}\right)K_{\frac{11}{6}}(z)\right] \tag{12c}$$

where,

$$z = \frac{\sigma}{1.339}\sqrt{1 + (1.339\nu)^2} \tag{12d}$$

In Equation (12d), ν represents the dimensionless frequency $\nu = \omega L/V$ and $\sigma = S/L$, the dimensionless distance.

5.2. Coherence for HIT

Coherence is also referred to as spectral correlation coefficient in that it quantifies the normalized cross-correlation between the turbulence velocities at two points as a function of frequency. For illustration, consider the vertical turbulence velocities at two points which are separated by a distance S, as typified by Figure 5. By definition, coherence is given by:

$$C_{w_1 w_2}(\sigma, \nu) = \frac{|S_{w_1 w_2}(\nu)|}{\sqrt{S_{w_1 w_1}(\nu)S_{w_2 w_2}(\nu)}} \tag{13}$$

where $S_{w_1 w_2}(\nu)$ is the cross-spectrum between vertical turbulence $w_1(t)$ at Point 1 and $w_2(t)$ at Point 2, and similarly $S_{w_1 w_1}(\nu)$ and $S_{w_2 w_2}(\nu)$ are the corresponding autospectra of $w_1(t)$ and $w_2(t)$. For HIT, cross-spectrum is real and $S_{w_1 w_1}(\nu) \approx S_{w_2 w_2}(\nu)$. Therefore, coherence from Equation (13) simplifies to Equation (14).

$$C_{w_1 w_2}(\sigma, \nu) = \frac{|S_{w_1 w_2}(\nu)|}{S_{w_1 w_1}(\nu)} \tag{14}$$

where $S_{w_1 w_2}(\nu)$ is given by Equation (12c) and $S_{w_1 w_1}(\nu)$ is the von Karman vertical spectrum as given in Equation (15) [15].

$$S_{w_1 w_1}(\nu) = \frac{\sigma_w^2}{\pi}\left[\frac{1 + \frac{8}{3}(1.339)^2}{(1 + 1.339^2)^{11/6}}\right] \tag{15}$$

As seen from Equation (14), $C_{w_1 w_2}(\sigma, v)$ is a ratio of the cross-spectrum from Equation (12c) and the autospectrum from Equation (15). It is expedient to reiterate that this autospectrum is due to von Karman [15] and that the cross-spectrum is due to Houbolt and Sen [10], as an extended version of the von Karman spectral equations that accounts for the cross-correlation between vertical turbulence velocities at two points; also see Figure 5. After some algebra, Equation (14) simplifies to Equation (16) [8].

$$C_{w_1 w_2}(\sigma, v) = \frac{0.597}{23869(z/\sigma)^2 - 1}\left[4.781(z/\sigma)^2 z^{5/6} K_{5/6}(z) - \frac{1}{2}z^{11/6}K_{11/6}(z)\right] \tag{16}$$

As for the longitudinal and lateral velocity components, the cross-spectra are given by Equations (17a) and (18a) and the coherences are given by Equations (17b) and (18b). In the literature (e.g., [9–12]), the expressions for cross-spectra and coherences from one set of study do not completely agree from another set.

$$S_{u_1 u_2}(v) = 0.1946\sigma_u^2 \frac{z^{5/6}}{\left(1 + (1.339v)^2\right)^{11/6}}\left[K_{5/6}(z) - \frac{z}{2}K_{1/6}(z)\right] \tag{17a}$$

$$C_{u_1 u_2}(\sigma, v) = 0.9944 z^{5/6}\left[K_{5/6}(z) - \frac{z}{2}K_{1/6}(z)\right] \tag{17b}$$

$$S_{v_1 v_2}(v) = 0.0727\sigma_v^2\left(\frac{\sigma^{5/3}}{z^{5/6}}\right)\left[\frac{8}{3}K_{5/6}(z) - \frac{\sigma^2}{1.339^2 z}K_{11/6}(z) + \frac{z}{2}K_{1/6}(z)\right] \tag{18a}$$

$$C_{v_1 v_2}(\sigma, v) = \frac{0.597}{2.8687(z/\sigma)^2 - 1}\left[4.781(z/\sigma)^2 z^{5/6} K_{5/6}(z) - \frac{1}{2}z^{11/6}K_{11/6}(z)\right] \tag{18b}$$

Given this background, it is emphasized that in the present study, the expressions of cross-spectra as typified by Equations (12c), (17a) and (18a) for the vertical, longitudinal and lateral components agree with those of Frost et al. [11]. As for coherence, the corresponding expressions given by Equations (16), (17b) and (18b) agree with those of Irwin [12]. Figures 6–8, respectively, show vertical, longitudinal and lateral coherence between Points 1 and 2 as a function of dimensionless frequency $v = \omega L/V$ for $\sigma = S/L = 0, 0.1, 0.2, \ldots 1$. For $\sigma = 0$, Point 2 merges into Point 1 and in turn the cross-spectra become the respective autospectra and thus the coherence represents the perfect coherence. For example, as seen from Figure 6 for the vertical coherence, $C_{w_1 w_2}(\sigma, v) \rightarrow 1$. Similarly, as seen from Figures 7 and 8, $C_{u_1 u_2}(\sigma, v) \rightarrow 1$ and $C_{v_1 v_2}(\sigma, v) \rightarrow 1$. Exactly the opposite happens with increasing $\sigma = S/L$. That is, with increasing σ, the correlation between these two points decreases and so does the corresponding coherence. For example, as seen from Figures 6–8, $C_{w_1 w_2}(\sigma, v) \rightarrow 0$, $C_{u_1 u_2}(\sigma, v) \rightarrow 0$ and $C_{v_1 v_2}(\sigma, v) \rightarrow 0$ as $\sigma \rightarrow \infty$. Moreover, as seen from these figures, the coherence decreases rapidly for $v > 1$ or so.

The longitudinal cross-spectrum $S_{u_1 u_2}(v)$ and coherence $C_{u_1 u_2}(\sigma, v)$ are typified by Equations (17a) and (17b), respectively, and Figure 7 shows coherence $C_{u_1 u_2}(\sigma, v)$ as a function of dimensionless frequency $v = \omega L/V$; all of this merits revisiting. The reason is that $S_{u_1 u_2}(v)$ and in turn the corresponding coherence can become negative at high frequencies. As seen from Equations (17a) and (17b), respectively, $S_{u_1 u_2}(v)$ and $C_{u_1 u_2}(\sigma, v)$ can take on negative values for $K_{5/6}(z) \leq z/2K_{1/6}(z)$. See Figure 9, which is a recasting of Figure 7 for a much expanded vertical scaling (1 to 10^{-6} in Figure 9 in comparison to 1 to 10^{-2} in Figure 7). The crosses (*) in Figure 9 indicate the termination of the curve to avoid generating negative coherence values. Given the state of the art and one's initiation into cross-spectra and coherence, it is difficult to come up with a basis for these negative values of cross-spectrum and coherence for HIT; a resolution of this difficulty would require further research [11].

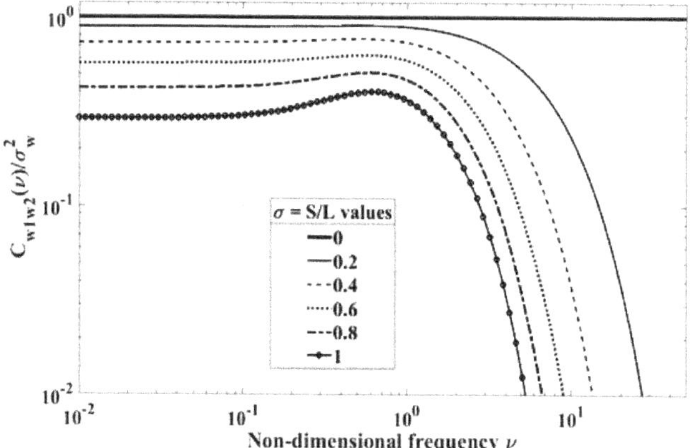

Figure 6. Vertical Coherence $C_{w_1 w_2}(\sigma, \nu)$.

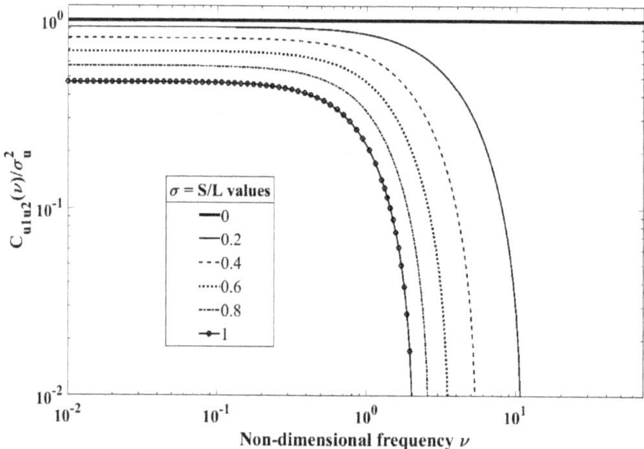

Figure 7. Longitudinal Coherence $C_{u_1 u_2}(\sigma, \nu)$.

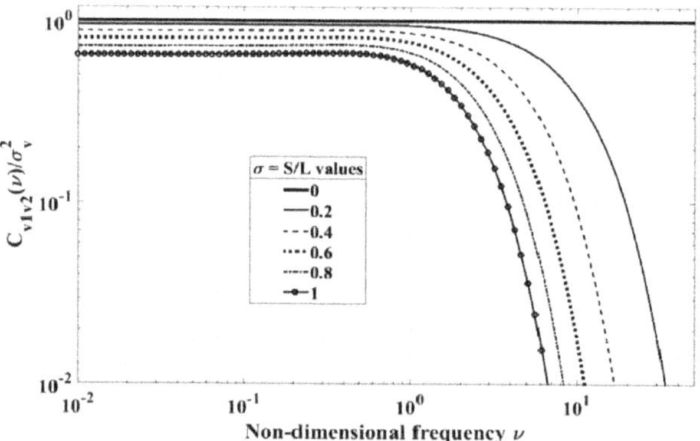

Figure 8. Lateral Coherence $C_{v_1 v_2}(\sigma, \nu)$.

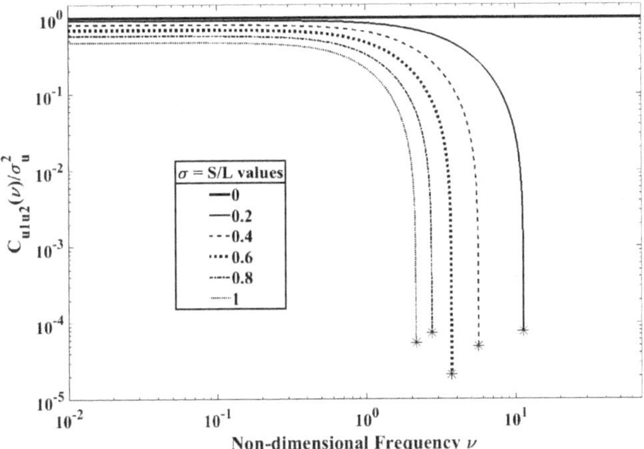

Figure 9. Longitudinal Coherence with an expanded scale.

5.3. Coherence Modeling for Wind-Farm Wake Turbulence

Wind farm wake turbulence deviates from HIT. Accordingly, the framework for coherence modeling from a database accounts for this deviation based on perturbation theory. Here as well, the framework assumes the same topology that was assumed in the development of the basis functions; for illustration vertical coherence $C_{w_1 w_2}(\sigma, \nu)$ is selected.

Let $C_{w_1 w_2}(\sigma, \nu)$ represent the vertical coherence of wake turbulence. The framework begins with a perturbation series expansion of $\tilde{C}_{w_1 w_2}(\sigma, \nu)$ (essentially the same procedure applies to the other two components):

$$\tilde{C}_{w_1 w_2}(\sigma, \nu) = C_{1w} C_{w_1 w_2}(\sigma, \nu) + C_{2w} C^2_{w_1 w_2}(\sigma, \nu) + \cdots + C_{nw} C^n_{w_1 w_2}(\sigma, \nu) \tag{19}$$

The basis function or the first term of the series is given by Equation (16). Since $C_{w_1 w_2}(\sigma, \nu) = 1$ for $\sigma = 0$, Equation (17) is subject to the constraint:

$$C_{1w} + C_{2w} + \cdots + C_{nw} = 1 \tag{20}$$

The second condition that $\tilde{C}_{w_1 w_2}(\sigma, \nu) = 0$ for $\sigma = \infty$ is automatically satisfied since $C_{w_1 w_2}(\sigma, \nu) = 0$ for $\sigma = \infty$. The coefficients in the series C_{iw} are evaluated by satisfying the theoretical constraint of Equation (20) and fitting a curve on a set of selected numerically generated coherence points in a least squares sense.

For illustrations, longitudinal coherence of wake turbulence $\tilde{C}_{u_1 u_2}(\sigma, \nu)$ is selected with a two-term perturbation series (also see Equation (19)):

$$\tilde{C}_{u_1 u_2}(\sigma, \nu) = C_{1u} C_{u_1 u_2}(\sigma, \nu) + C_{2u} C^2_{u_1 u_2}(\sigma, \nu) \tag{21}$$

Specifically, consider $C_{1u} = 0.7$ and $C_{2u} = 0.3$ (also see constraint Equation (20)). Descriptively stated, this case represents wake turbulence, which deviates weakly from HIT. It is plausible that this case belongs to wake turbulence at locations that are downwind of the first two rows or so. Therein, wake turbulence is expected to deviate only weakly from HIT as depicted in Figure 10.

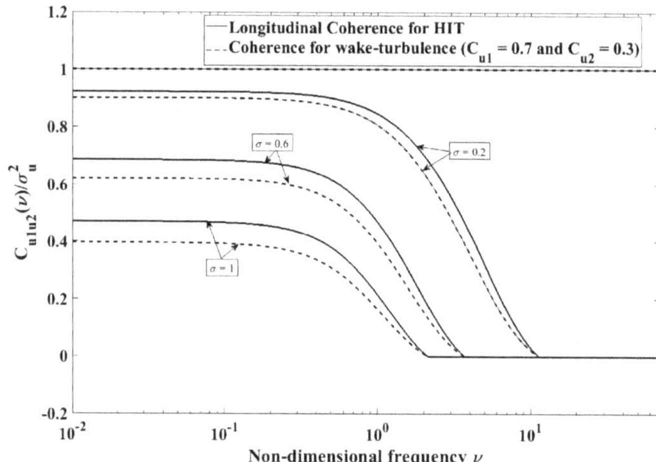

Figure 10. Longitudinal Coherence for HIT and for wake turbulence weakly deviating from HIT.

6. Conclusions

This study has shown that an earlier-exercised mathematical framework lends itself well to extracting interpretive autospectral models of wind farm wake turbulence from a database. While these models are database specific, the framework can be applied to any database and the model construction is straightforward. As to the two-point statistics of wake-turbulence, this study first presents a unified account of cross-spectrum and coherence for HIT; this account is of considerable utility in that, in the literature, the expressions of cross-spectrum and coherence show differences from one study to the other. Given these expressions of coherences, this study then builds a framework for extracting wake turbulence coherence models from a database. The frameworks for autospectra and coherences do not follow classical perturbation theory approach of a solution for a linearized problem along with successively added corrections. Both frameworks represent a practical combination of a series expansion, exploitation of a database, and theoretical constraints in closed form.

This study also leads to following specific findings:

1. Generally, no more than a three-term series (second-order correction) is necessary to develop an autospectral model; in most cases, a two-term series (first-order correction) is found to be adequate for wind engineering applications.
2. The addition of a third term to the series has significant power in reducing the "*A error*" between the model and the data. Recall that the "*A error*" refers to minimizing the sum of the errors in a least squares sense: the model deviation from the autospectral data points and from the measured high frequency autospectral decay level (when applicable).
3. These developed models lend themselves well to design of filters driven by white noise; that is, the filter design is as routine as the currently used procedure for the von Kármán models.
4. While this framework to constructing the coherence models from a database has not been tested against a database, it has been formulated from first principles and with a theoretical basis.
5. This study has shown the feasibility of constructing both the one-point statistics of autospectral models and the two-point statistics of coherence models from a database. These models could serve as surrogate analytical models in the experimental and CFD investigations; thus, this feasibility offers promise in providing an improved understanding of wake turbulence.
6. The two frameworks for the autospectrum and coherence increase the utility base of the database, involving enormous resources. Given the simple analytical structure of these models, they bring better understanding and transparency to a dataset.

Author Contributions: All authors contributed equally and effectively to the paper.

Funding: This research received no external funding.

Conflicts of Interest: The authors declare no conflict of interest.

Nomenclature

A_w	Autospectral high-frequency asymptotic limit of $f\widetilde{S}_{w_1 w_2}(f)/\sigma_w^2$ (similarly for A_v and A_u)
ABL	Atmospheric Boundary Layer turbulence
$C_{w_1 w_2}(\sigma, \nu)$	Vertical coherence between $w_1(t)$ and $w_2(t)$ for HIT
$\widetilde{C}_{w_1 w_2}(\sigma, \nu)$	Modeled series expansion of coherence for vertical wake turbulence velocity w (likewise, for u and v)
C_{iw}	Constants in the perturbation series expansion of $\widetilde{C}_{w_1 w_2}(\sigma, \nu)$
f	Frequency (Hz)
HIT	Homogeneous Isotropic Turbulence for which the frozen turbulence hypothesis is applicable
L	Turbulence length scale
$R_{ww}(\tau)$	von Karman vertical autocorrelation function (likewise, for u and v) for HIT; also, the first term in the perturbation series expansion of $\widetilde{R}_{ww}$
$\widetilde{R}_{ww}(\tau)$	Modeled series expansion of autocorrelation for vertical wake turbulence velocity w (similarly, for u and v)
S	Separation between Point 1 and Point 2
$S_{ww}(\nu)$	Autospectrum of vertical turbulence velocity w (likewise, for u and v) for HIT
$\widetilde{S}_{ww}(\nu)$	Modeled series expansion of autospectrum for vertical wake turbulence velocity w (similarly, for u and v)
T_u, T_v, T_w	Times scales of longitudinal, lateral and vertical turbulence
T	Time(s)
u, v, w	Longitudinal, lateral and vertical turbulence (u is also used as a general variable)
V	Mean Velocity
z	Non-dimensional general variable
α_u, α_v & α_w	Time-scale preservation parameters for autocorrelations (or autospectra) of u, v and w
β_{iw}	Constants in the perturbation series expansion of $\widetilde{R}_{ww}$
σ_u^2, σ_v^2 & σ_w^2	Variance of turbulence components u, v and w
ω	Angular frequency (rad/s)
ν	Non-dimensional frequency

References

1. Manwell, J.F.; McGowan, J.G.; Rogers, A.L. Wind Turbine Siting, System Design, and Integration. In *Wind Energy Explained: Theory, Design and Applications*, 2nd ed.; John Wiley & Sons Publications: West Sussex, UK, 2015; pp. 430–432. ISBN 978-0-470-01500-1.
2. Keck, R.; Veldkamp, D.; Madsen, H.A.; Larsen, G. Implementation of a Mixing Length Turbulence Formulation into the Dynamic Wake Meandering Model. *J. Sol. Energy Eng.* **2012**, *134*, 021012. [CrossRef]
3. Carrion, M.; Woodgate, M.; Steijl, R.; Barakos, G.; Munduate, X. Understanding Wind Turbine Wake Breakdown Using Computational Fluid Dynamics. *AIAA J.* **2015**, *53*, 588–602. [CrossRef]
4. Ge, M.; Wu, Y.; Liu, Y.; Li, Q. A Two-Dimensional Model Based on the Expansion of Physical Wake Boundary for Wind-Turbine Wakes. *Appl. Energy* **2019**, *233*, 975–984. [CrossRef]
5. Schau, K.A.; Gaonkar, G.; Polsky, S. Rotorcraft Downwash Impact on Ship airwake: Statistics Modeling and Simulation. *Aeronaut. J.* **2016**, *120*, 1025–1048. [CrossRef]
6. Gaonkar, G. Extracting Stochastic Models of Airwake-Downwash Turbulence from a Database for Simulation. *J. Aircr.* **2013**, *50*, 1309–1311. [CrossRef]
7. Gaonkar, G.; Mohan, R. Extracting Stochastic Models of Airwake-Downwash Turbulence from a Database for Engineering Analysis and Simulation. *J. AHS* **2013**, *50*, 1309–1312.

8. Krishnan, V.; Gaonkar, G. A Framework for Modelling Two-Point Statistics of Coherence from a Database for Airwake with Helicopter Downwash. In Proceedings of the AHS 75th Annual Forum, Philadelphia, PA, USA, 13–15 May 2019; (Abstract accepted, paper to be included in Proceedings).

9. Burton, T.; Jenkins, N.; Sharpe, D.; Bossanyi, E. The Wind Resource. In *Wind Energy Handbook*, 2nd ed.; John Wiley & Sons Publications: West Sussex, UK, 2011; pp. 20–27. ISBN 978-0-470-69975-1.

10. Houbolt, J.; Sen, A. *Cross-Spectral Functions Based on Von Karman's Spectral Equations*; Technical Report NASA-CR-2011; NASA: Washington, DC, USA, 1972.

11. Frost, W.; Chang, H.P.; Ringnes, E.A. *Analysis and Assessments of Span Wise Gust Gradient Data from NASA B-57B Aircrafts*; NASA CR 178288; NASA: Washington, DC, USA, 1987.

12. Irwin, H.P.A.H. Cross-spectra of Turbulence Velocity in Isotropic Turbulence. *Bound. Lay. Meteorol.* **1979**, *16*, 234–243. [CrossRef]

13. Morfiadakis, E.E.; Glinou, G.G.; Koulouvari, M.J. The Suitability of the von Karman Spectrum for the Structure of Turbulence in a Complex Terrain Wind Farm. *J. Wind Eng. Ind. Aerodyn.* **1996**, *62*, 237–257. [CrossRef]

14. Schau, K. Developing Interpretive Turbulence Models from a Database with Applications to Wind Farms and Shipboard Operations. Master's Thesis, Florida Atlantic University, Boca Raton, FL, USA, October 2013.

15. Gaonkar, G. Review of turbulence modeling and related applications to some problems of helicopter flight dynamics. *J. AHS* **2012**, *53*, 87–107. [CrossRef]

Article

Dynamic Modeling and Robust Controllers Design for Doubly Fed Induction Generator-Based Wind Turbines under Unbalanced Grid Fault Conditions

Imran Khan [1], Kamran Zeb [1,2], Waqar Ud Din [1], Saif Ul Islam [1], Muhammad Ishfaq [1], Sadam Hussain [1] and Hee-Je Kim [1,*]

[1] School of Electrical Engineering, Pusan National University, San 30, ChangJeon 2 Dong, Pusandaehak-ro 63 beon-gil 2, Geumjeong-gu, Busan 46241, Korea; imrankhan@pusan.ac.kr (I.K.); kami_zeb@yahoo.com (K.Z.); waqudn@pusan.ac.kr (W.U.D.); shaheen_575@yahoo.com (S.U.I.); engrishfaq1994@gmail.com (M.I.); sadamengr15@gmail.com (S.H.)

[2] School of Electrical Engineering and Computer Science, National University of Sciences and Technology, Islamabad 44000, Pakistan

* Correspondence: heeje@pusan.ac.kr; Tel.: +82-10-3462-1990

Received: 16 December 2018; Accepted: 30 January 2019; Published: 31 January 2019

Abstract: High penetration of large capacity wind turbines into power grid has led to serious concern about its influence on the dynamic behaviors of the power system. Unbalanced grid voltage causing DC-voltage fluctuations and DC-link capacitor large harmonic current which results in degrading reliability and lifespan of capacitor used in voltage source converter. Furthermore, due to magnetic saturation in the generator and non-linear loads distorted active and reactive power delivered to the grid, violating grid code. This paper provides a detailed investigation of dynamic behavior and transient characteristics of Doubly Fed Induction Generator (DFIG) during grid faults and voltage sags. It also presents novel grid side controllers, Adaptive Proportional Integral Controller (API) and Proportional Resonant with Resonant Harmonic Compensator (PR+RHC) which eliminate the negative impact of unbalanced grid voltage on the DC-capacitor as well as achieving harmonic filtering by compensating harmonics which improve power quality. Proposed algorithm focuses on mitigation of harmonic currents and voltage fluctuation in DC-capacitor making capacitor more reliable under transient grid conditions as well as distorted active and reactive power delivered to the electric grid. MATLAB/Simulink simulation of 2 MW DFIG model with 1150 V DC-linked voltage has been considered for validating the effectiveness of proposed control algorithms. The proposed controllers performance authenticates robust, ripples free, and fault-tolerant capability. In addition, performance indices and Total Harmonic Distortions (THD) are also calculated to verify the robustness of the designed controller.

Keywords: Wind Turbine (WT); Doubly Fed Induction Generator (DFIG); unbalanced grid voltage; DC-linked voltage control; Proportional Resonant with Resonant Harmonic Compensator (PR+HC) controller; Adaptive Proportional Integral (API) control; power control

---

1. Introduction

Extinction and environmental concerns regarding the use of fossil fuels for power generation have shifted the attention of scientists towards Renewable Energy (RE). Among all RE resources, wind power generation has recorded significant growth in the last decade. With energy saving ambitions, by 2030 wind power will be able to supply 29.1% of the electricity needed worldwide and 34.5% by 2050 [1,2]. Energy quality is a significant feature in grid-connected converters, and wind power generators have a high influence on the stability and security of the power grid. To meet the required results, WT systems must be continuously developed and their performance improved. In recent years, DFIG based WT have become a well-known and widely installed due to their high efficiency, variable speed operation ($\pm$33% around the synchronous speed), four quadrant active and reactive power capability, less power losses, small converter rating (around 30% of generator rating), reduced mechanical stress and hence minimized pulsating power and torque [3–6].

Since the DFIG stator and the grid are connected directly, during unbalanced grid voltage conditions a negative sequence is added to stator flux, resulting in a flow of large negative sequential currents in the rotor and stator causing second-order harmonic fluctuating power and electromagnetic torque [7,8]. From both the Rotor Side Converter (RSC) and Grid Side Converter (GSC), active power fluctuations flow through DC-linked capacitors as shown in Figure 1. resulting in voltage ripples in the DC-link capacitor as well as significant second-order harmonic currents in the DC-capacitor [9], which affect the DC-capacitor causing high power losses and increased operational temperature which may evaporate the electrolyte faster making their lifespan shorter. In addition, fluctuations in torque can cause wear and tear of mechanical parts such as the shaft and gear box [10]. Further, a comparison of the high and low frequency ripple currents shows that ripple currents with low frequency are more detrimental [11,12]. Hence, voltage ripples and converter DC-linked capacitor with large low frequency currents under unbalanced conditions are the most serious issues of DFIG [8,9]. Under the unbalanced condition the DC-voltage control in GSC differs slightly from the GSC for the DFIG, because the DC-voltage ripples are not only caused by the unbalanced grid voltage, but also by RSC fluctuating active power. These two disturbances i.e., active power fluctuation of RSC and unbalanced grid voltage, should be rejected by GSC to ensure a constant DC-voltage.

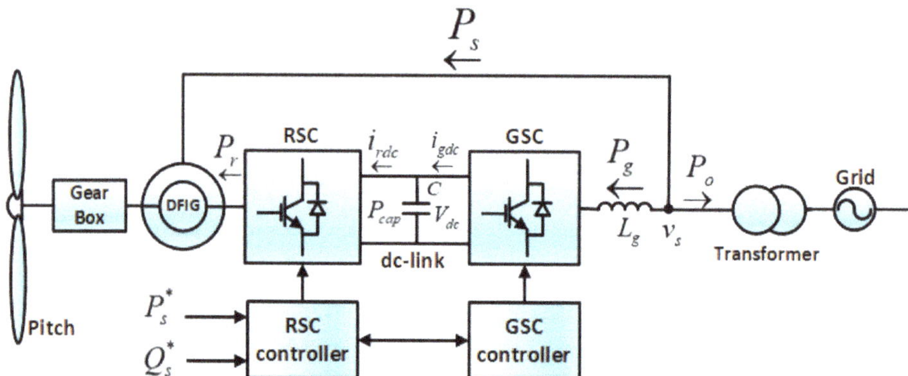

Figure 1. Active power flow in a DFIG wind turbine.

Numerous control strategies have been presented to decrease the voltage ripple for GSC controllers under unbalanced voltage conditions. To regulate negative sequence current and positive currents at the same time dual current control methods were designed [9,13–15]. Grid voltage and the desired power ensure the calculation of negative and positive reference currents. By setting of the references multiple control targets are available, like constant DC voltage, constant electromagnetic power, constant stator power and balanced stator currents [14,15]. The GSC fluctuating active power output must be equal to that of RSC under unbalanced conditions. Then the GSC reference current depends on the RSC fluctuating active power [9,14]. Consequently, implementation of dual current control method is not applicable in modular structural wind power converters. Another method to reduce voltage ripples during unbalance grid voltage conditions is feed-forward control which comprising RSC DC-current feed-forward control [16–19] and grid voltage feedforward control [20,21]. Feed-forward control for RSC DC-current reduces the impact of fluctuating RSC active power while feed-forward control for grid voltage reduces the impact on DC-capacitor due to unbalanced grid voltages.

The feed-forward technique control performance may be degraded by the control delay, which results in an addition of high-frequency noise to the feed-forward term. Moreover, additional hardware of the load current detection may require detecting the DC current of the RSC [17,18]. An alternate approach is used to get rid of additional detection circuits, whereby the RSC real-time active power is calculated by GSC based on rotor voltage reference and rotor current [16,19] which require integration of both the RSC controller and GSC controller into a single controller. This integration results in loss of the modular structure of DFIG converters. For high maintenance and reliability, DFIG converter exhibits modularity which is not achieved in this technique Automatic generation control employed with inertia support for load frequency control was analyzed in an interconnected multigeneration wind power system [22]. For mitigation of subsynchronous resonance, a non-linear damping controller was designed using a partial feed-back linearization technique in series compensated DFIG-based wind farms [23]. To mitigate subsynchronous resonance (SSR) oscillations, doubly fed induction generator (DFIG) supplemental control is used [24], in which a supplemental signal is introduced into the control loop of the DFIG voltage source converter. Furthermore, two-degree-of -freedom along with a damping control loop is used [25] to mitigate SSR which is caused by induction generator effects and thus enhance the system stability. In [26] two SSR oscillation mitigating strategies were compared, which generate supplementary damping control signal; integrated on the rotor side converter and grid side converter. A hybrid scheme for enhancing fault ride through capability of DFIG under symmetric and asymmetric faults was presented [27], comprising an energy storage system, break chopper and switch type fault current limiter.

The main contributions of this paper may be summarized as follows:

(1)   A simplified and comprehensive study about dynamics characteristics and modelling of DFIG based grid connected wind turbine system is presented.

(2)   Active and reactive power stability and elimination of voltage fluctuation and harmonic current of DC-capacitor using API and PR+RHC as a grid side control algorithm are discussed.

(3)   A comprehensive performance analysis under normal condition and various faults, i.e.: Under Voltage, Over Voltage, Single Phase, and Double Phase faults conditions to validate the active power, reactive power, and DC-link voltage performance of the proposed API and PR+RHC controllers is performed.

(4)   A comparative assessment of designed controllers such as API and PR+RHC with a conventionally tuned PI controller is also carried out.

(5)   A FFT analysis of a PI controller, and the proposed API and PR+RHC controller by calculating the total harmonics distortion of grid current to validate the robustness of proposed PR controller is presented.

(6)   The performance of various controllers (PI, API & PR+RHC) was evaluated by calculating three control parameters i-e. Integral Absolute Error (IAE), Integral Square Error (ISE) and Integral Time-weighted Absolute Error (ITAE) which precisely compare their performances.

The remaining paper is organized as follows: in Section 2, detailed modeling of DFIG is discussed. The proposed WTs model is explained in Section 3. The proposed API and PR+RHC controllers are designed in Section 4. Results and discussion are presented in Section 5. The paper is concluded in Section 6.

2. Modeling of DFIG

The configuration of a DFIG-based wind turbine is illustrated in Figure 1. The stator and grid voltage are directly linked to each other while the rotor and back-to-back converter are interfaced, comprising a GSC common DC-link and a RSC [28]. The generator output power is controlled by the RSC while GSC ensures the stability of the DC-link voltage irrespective of the direction and magnitude of the rotor power [29]. At the wind turbine the terminal grid active power P_O is equal to the sum of the stator active power P_s and the grid active power P_g. The current and power reference directions are shown in Figure 1. The equivalent circuit of DFIG is shown in a *dq*-synchronous reference frame in Figure 2.

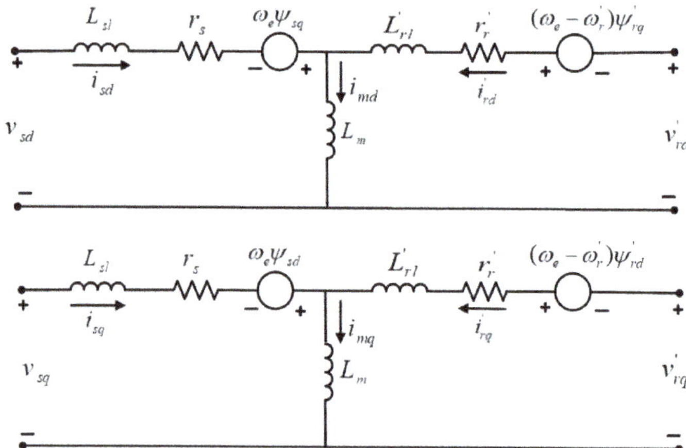

Figure 2. Equivalent circuit of the DFIG in the *dq*-synchronous reference frame.

The DFIG mathematical model is analyzed in the *dq* reference frame and is defined by Equations (1) to (6) [30,31]:

$$\left. \begin{aligned} v_{sd} &= r_s i_{sd} + \frac{d\psi_{sd}}{dt} - \omega_e \psi_{sq} \\ v_{sq} &= r_s i_{sq} + \frac{d\psi_{sq}}{dt} + \omega_e \psi_{sd} \end{aligned} \right\} \tag{1}$$

$$\left. \begin{aligned} v'_{rd} &= r'_r i'_{rd} + \frac{d\psi'_{rd}}{dt} - \omega_{sl} \psi'_{rq} \\ v'_{rq} &= r'_r i'_{rq} + \frac{d\psi'_{rq}}{dt} - \omega_{sl} \psi'_{rd} \end{aligned} \right\} \tag{2}$$

$$\omega_{sl} = \omega_e - \omega'_r \tag{3}$$

$$\left. \begin{aligned} \psi_{sd} &= L_s i_{sd} + L_m i'_{rd} \\ \psi_{sq} &= L_s i_{sq} + L_m i'_{rq} \end{aligned} \right\} \tag{4}$$

$$\left. \begin{aligned} \psi'_{rd} &= L'_r i'_{rd} + L_m i_{sd} \\ \psi'_{rq} &= L'_r i'_{rq} + L_m i_{sq} \end{aligned} \right\} \tag{5}$$

$$\left. \begin{aligned} L_s &= L_{sl} + L_m \\ L'_r &= L'_{rl} + L_m \end{aligned} \right\} \tag{6}$$

where V_{sd}, V_{sq} and V'_{rd}, V'_{rq} are the stator and rotor voltages in the *dq* reference frame, r_s and r'_r are the stator and rotor per phase electrical resistances, i_{sd}, i_{sq} and i'_{rd}, i'_{rq} are stator and rotor currents in the *d-q* reference frame, ψ_{sd}, ψ_{sq} and ψ'_{rd}, ψ'_{rq} are stator and rotor fluxes in the *dq* reference frame, L_s, L'_r and L_m are stator, rotor and magnetizing per phase inductances, L_{sl} and L'_{rl} are stator and rotor leakage inductance, ω_e and ω'_r are the synchronous and rotor speeds.

The magnetic flux in the stator in *d* and *q* axis is determined by Equation (7) and it is assumed that all magnetic fluxes are aligned with the *d* axis:

$$\left. \begin{aligned} \psi_{sq} &= 0 \quad and \quad \frac{d\psi_{sq}}{dt} = 0 \\ \psi_s &= \psi_{sd} = L_m i_{ms} \quad and \quad \frac{d\psi_{sq}}{dt} = 0 \end{aligned} \right\} \tag{7}$$

The DFIG stator active and reactive power are computed for rotor side after simplification as:

$$P_s = -\frac{3}{2}\frac{L_m}{L_s}v_s i'_{rq}$$ (8)

$$Q_s = \frac{3}{2}\frac{L_m}{L_s}v_s\left(\frac{v_s}{(\omega_e L)_m} - i'_{rd}\right)$$ (9)

From Equations (8) and (9), one observes that the active and reactive powers can be controlled by the quadrature components of rotor current, considering the constant voltage. The converter controls the active and reactive powers of the DFIG stator, where $1 - L_m^2/L_s L'_r$ and i_{ms} is the magnetizing current.

The GSC block diagram uses current loops to i_d and i_q, having i_d^* as reference from the DC-link. Since $i_q^* = 0$, the converter operates at a unity power factor. The reference signal generator produces the current reference (i_d^*, i_q^*), from Equations (10) and (11):

$$P_{ref} = \frac{3}{2}[v_d i_d^*]$$ (10)

$$Q_{ref} = \frac{3}{2}[v_q i_d^*]$$ (11)

3. Proposed Model

An overview of the control structure of a wind turbine system (WTS) [4,32,33] is shown in Figure 3. For maximum power extraction, the generator is controlled by a power converter, thereafter electrical parameters are generated based on generator and control algorithm while the generator torque ω_m is obtained from the turbine model [30].

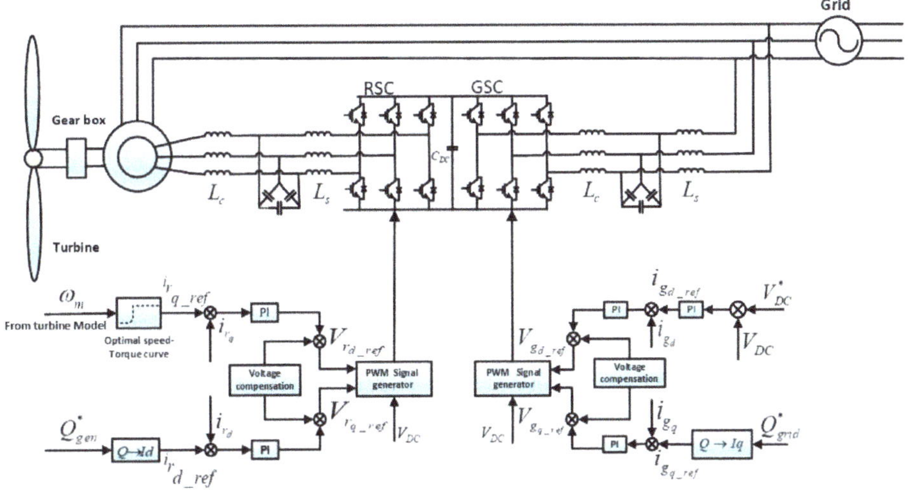

Figure 3. Control schematics for a DFIG wind turbine.

The electric and control models are classified into grid side and generator side as shown in Figure 3. The generator side control deals with two parameters, generator current and the duty cycle. DC-linked voltage alone with these two parameters is used to model generator side converters using the following Equations (12) and (13):

$$V_{s_{dq}} = D_{dq} \times V_{DC}$$ (12)

$$I_{dc} = D_d \times I_{s_d} \times D_q \times I_{s_q} \tag{13}$$

where D is the duty ratio, V_{DC} is the DC-link voltage, I_{DC} is the current flow into DC link, I_s is the stator current V_s is the stator voltage.

Based on the vector control of generator the control algorithm implemented here is for maximum power extraction. The control structure works in the following sequence: first in the reference current generation phase, the rotor's rotational speed is measured which is used to generate the reference torque from the maximum power/torque curve based on the turbine design and characteristic. Using this reference torque, a reference current signal is generated for the generator-side converter in the *dq* frame. In the current control loop phase, an error signal is generated by comparing the generated reference current and the measured current in the *dq* reference frame, which then generate a voltage reference for the converter by feeding through Proportional Integral (PI) controllers. In the modulation phase, the resulting reference voltages should be converted into a duty ratio for the generator side converter, and finally this will result in a PWM switching signal for the converter as shown in Figure 4.

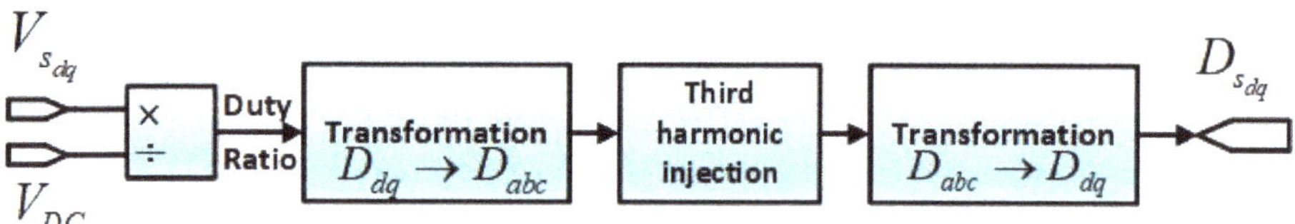

Figure 4. Modulation of generator-side converter in proposed model.

The converter model on the grid-side is elaborated by three differential Equations (14)–(16), which use the voltage of the grid and the resistance and inductance of the grid-side filter as input:

$$L_f \frac{di_{g_d}}{dt} + R_f i_{g_d} = \omega L_f i_{g_q} + V_{conv_d} - V_{grid_d} \tag{14}$$

$$L_f \frac{di_{g_q}}{dt} + R_f i_{g_q} = -\omega L_f i_{g_d} + V_{conv_q} - V_{grid_q} \tag{15}$$

$$C_{DC} \frac{dV_{DC}}{dt} = i_{DC} - k\left(i_{g_d} D_d + i_{g_q} D_q\right) \tag{16}$$

where the k value is dependent on the transformation technique used to convert *abc* values to *dq* values. The k value must be 1 is when using a normalized Clarke transformation and in case of a non-normalized transformation $k = 3/2$. Further, V_{DC} is the DC-link voltage, i_g is the grid current, R_f is the filter resister, D is the duty cycle, C_{DC} is the DC-linked capacitor, L_f is inductance of filter and V_{grid} is the voltage of grid.

In the *dq* reference frame the grid-side converter is controlled with the grid voltage. The reactive power which is transferred to the grid is controlled by i_{g_q}. Similarly, by maintaining the DC-linked voltage real power transferred to the grid is regulated by i_{g_d} current. Both the generator-side as well as the grid-side controller have the same limiting algorithms and modulation techniques.

4. Controller Design

4.1. API Controller

Control of traditional processes always depends on creating a mathematical model of the required system. An expert system was established to mimic the behavior of a skilled human operator for those processes too complex to be mathematically modeled in real time. Fuzzy logic controller (FLC) engines use as expert system paradigm for automatic process control. In addition, intuition and heuristics knowledge are also included into the system. This feature ranked FLC high in application where the existing models are ill defined, complex and not adequately reliable. FLC can mainly be classified into four main parts: fuzzifier, rules, inference engine and de-fuzzifier [34] as illustrated in Figure 5:

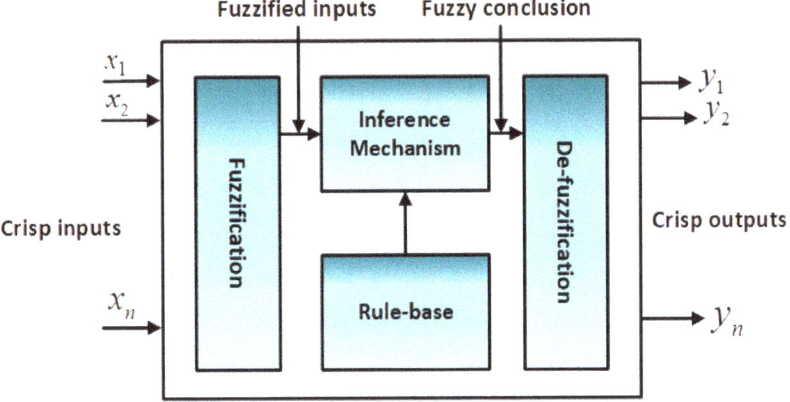

Figure 5. Fuzzy controller architecture.

4.2. Fuzzy PI Controller

The PI controller comprising constant integral and proportional gain k_i and k_p, respectively. Control scheme performance is enhanced by adaption of gain with respect to error. This distinguish feature of adaption can be achieved by applying fuzzy rules as illustrated in Table 1:

Table 1. Fuzzy rules.

Absolute Error $\|e(t)\|$	Proportional Gain (k_p)	Integral Gain (k_i)
Zero	large	small
Small	large	zero
Large	large	large

Gaussian Member function (GMF) is applied here in the rules that needs two parameters i.e., center c_i and σ_i standard variance or deviation as:

$$\mu(x) = exp\left(-\frac{1}{2}\left(\frac{x_{i-c_i}}{\sigma_i}\right)^2\right) \tag{17}$$

Mathematical description of PI controller is illustrated as:

$$v_{dc}^*/i_{sd}^*/i_{sq}^*(PI) = k_p e(t) + k_i \int e(t)dt \tag{18}$$

where $v_{dc}^*/i_{sd}^*/i_{sq}^*$ is output of the controller, k_i and k_p is integral and proportional gain respectively and $e(t)$ is input of controller, furthermore PI controller gains are constant in the preceding equation that requires adaptation with respect to electrical fault perturbation, parameter uncertainties, load variation and load disturbances.

$$v_{dc}^*/i_{sd}^*/i_{sq}^*(Fuzzy) = F_1 k_1 e(t) + F_2 k_2 \int e(t)dt \tag{19}$$

where k_p and k_i results in fuzzy controller's output F_1 and F_2 respectively, and k_1 and k_2 are learning rates constant for k_p and k_i respectively as mentioned in Figure 6.

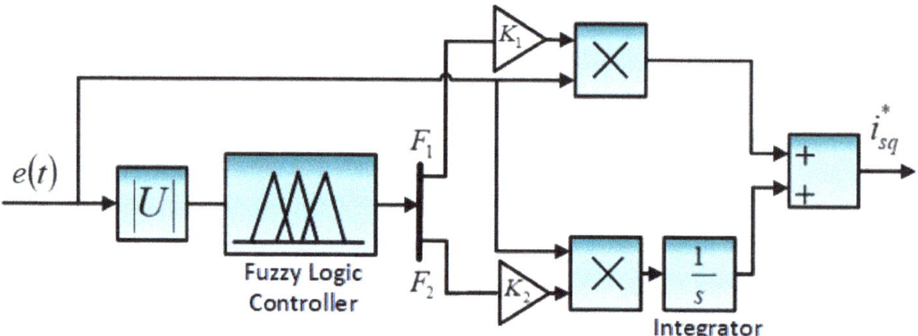

Figure 6. Adaptive PI controller.

A comparison of FLC-based adaptive PI control with PI conventionally tuned control as benchmark is provided in [35]. The gain for integral and proportional constant are calculated for the operating conditions by linearizing the system for numerous control loops.

4.3. Proportional Resonant Controller with Hormonic Compensator (PR+HC)

A PR controller has distinguished integration features. Due to the action of integration of frequencies near and around the resonance frequency; phase shift and static error do not occur in a PR controller. Although high order filters are used to obtain optimized current waves at the grid side during unbalanced grid conditions, in practical applications the current wave is not exactly the normal one, but has time varying elements of grid voltage with small deviations which result in poor THD of the feed-in current, but it is demanded in most grid standards [36,37] that the grid connected devices should be operated within certain frequencies range. To meet grid standards by improving the current quality a harmonic compensator is employed along with the PR controller as shown in Figure 7.

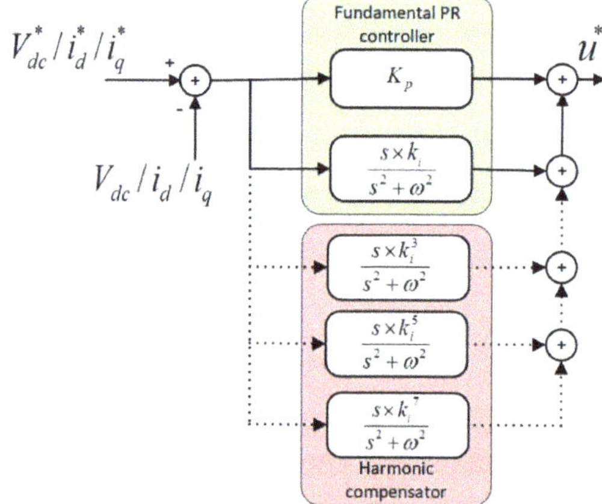

Figure 7. Combined structure of PR with harmonic compensator.

The PR controller consists of two parts i.e., proportional and resonant part, expressed by Equation (20) below:

$$G_{PR}(s) = K_p + K_i \left(\frac{s}{s^2 + \omega^2} \right) \tag{20}$$

Here, ω is a resonant frequency. Due to the high gain at narrow band at the resonant frequency, PR can eliminate steady-state error. K_i is the time constant integral which is related to band width, and

K_p is proportional gain determines the phase of band width and gain of margin [38]. The harmonic compensator is parallelized with the PR controller for the sake of quality of grid current [39]. Harmonic compensators can be mathematically expressed as:

$$G_{HC}(s) = \sum_{h=3,5,7,\dots} G_{HC}^h(s) \tag{21}$$

Here, $G_{HC}^h(s)$ is resonant controller with h^{th} order, where "h" is harmonic order. However, particularly

$$G_{HC}^h(s) = \frac{k_i^h\,s}{s^2 + (h\omega)^2} \tag{22}$$

where, k_i^h is the gain of particular order resonant controller.

5. Results and Discussion

To verify the proposed control strategies, a MATLAB/Simulink-based simulation have been carried out. The nominal parameters of the 2 MW system are listed in Table A1 (Appendix A). Control strategies (PI, API and PR+RHC) were simulated and compared under different conditions, i.e., rated, single-phase fault, two-phase fault, under-voltage, and over-voltage fault. The faults are applied for 200 ms which occurs from 1 s and cleared at 1.2 s, whereas the grid-side voltage was dropped and raised to 50% of its normal values in the under- and over-voltage cases, respectively. The performance of PI controller and proposed PR control strategy is evaluated by considering the following parameters: DC-linked voltage V_{dc}, stator voltage V_s, active current component I_d, reactive current component I_q, grid current Ig, rotor current I_r, rotor real power P_r, rotor voltage V_r, electro-magnetic torque T_{em}, stator real power P_s, stator reactive power P_{s_react}. Finally, THD and control performance measures are calculated to examine the controller's performance.

5.1. Rated Voltage

Conventional (PI) and Proposed (API & PR+RHC) control strategies are analyzed considering rated voltages. Figure 8a illustrates the DC-linked voltage responses of all control strategies; the PR+RHC and API controller responses are robust, faster and stabilize quickly, whereas the PI controller takes 1.3 s to attains stability. The API controller updates its parameters adoptively to minimize errors abruptly. The PR+RHC, due to the harmonic compensation, effectively tracks the reference, compared to PI. Figure 8b shows the rated stator voltage waveform for all control schemes. Figure 8c–e shows I_d for PI, API and PR+RHC control schemes, where both the designed controllers currents are efficiently tracking the reference currents. They have stable, robust, and chatter-free responses. The API and PR+RHC strategy responses for the rotor current are stable and less oscillatory with respect to the PI response as presented in Figure 8f. I_q is depicted in Figure 8g and the I_g response is illustrated for all controllers in Figure 8h. The API and PR+RHC response is faster and globally convergent. In case of P_s and P_r the API and (PR+RHC) controller responses are stable and robust, which reduces the acoustic noise, reduces stress on both drive trains and mechanical components which is a desired requirement as shown in Figure 8i,j. The T_{em} response is observed in Figure 8k, which shows minimum oscillation or almost stable responses for the API and PR+RHC control schemes, something that could be harmful from a mechanical view point. Figure 8l describes the P_{s_react} response which is quite stable and ripple less, which is desired in proposed control strategies. The rotor voltage response shows that API and PR+RHC strategies' responses are stable and less oscillatory with respect to the PI response as shown in Figure 8m. The performance indices of all the control schemes are evaluated in Tables 2–4 for V_{dc}, I_d, I_q, respectively. Three control measuring parameters, i.e., Integral Absolute Error (IAE), Integral Square Error (ISE) and Integral Time-weighted Absolute Error (ITAE) are calculated for all controllers which precisely compare their performances. The performance of a controller is based on its minimum value, where the smaller the value of parameters, the better the controller performance. In all three parameters API and PR+RHC controllers' values are the minimum compared with the PI controller, which proves the robust performance of the proposed controllers. Finally, the control schemes (PI, API & PR+RHC) are further investigated using FFT analysis of the grid current, which shows that the proposed API and PR+RHC strategies' grid currents are more robust and less harmonic with THD 0.02% and 0.06% respectively, as compared to 0.07% THD of the PI controller as shown in Figure 8n–p.

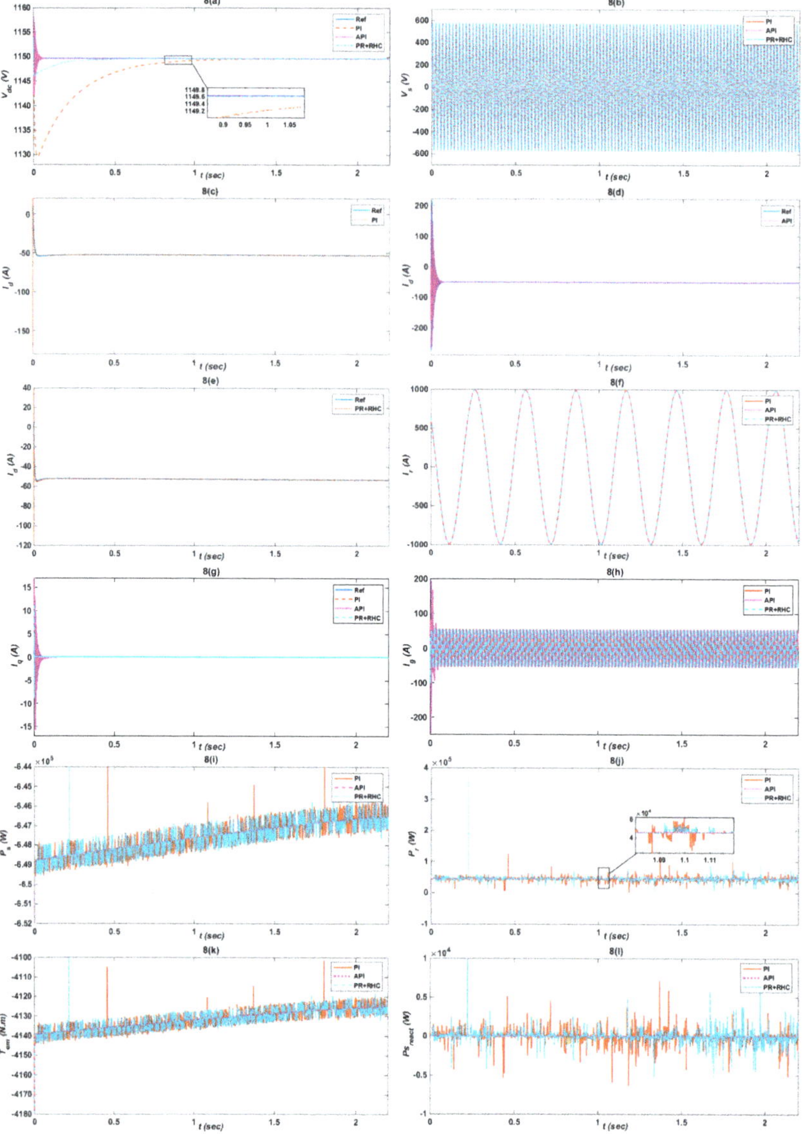

Figure 8. *Cont.*

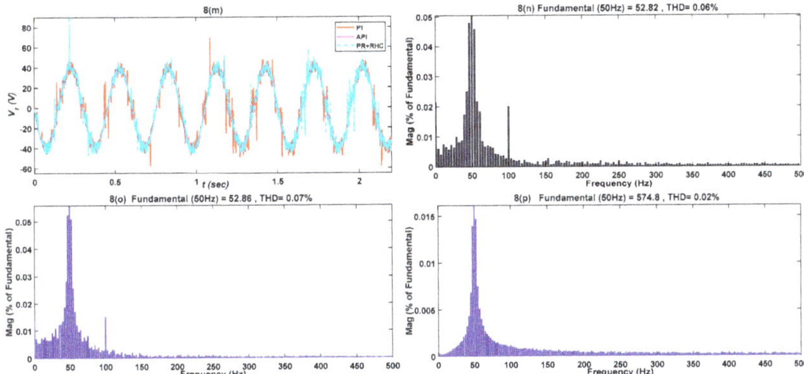

Figure 8. Comparison of PI and Proposed API and PR+RHC controllers responses under rated voltage, considering: (**a**) Dc-link voltage V_{dc}; (**b**) Stator voltage V_s; (**c–e**) Active component of current I_d; (**f**) Rotor current I_r; (**g**) Reactive component I_q; (**h**) Grid current I_g; (**i**) Stator active power P_s; (**j**) Rotor active power P_r; (**k**) Electromagnetic torque T_{em}; (**l**) Stator reactive power Ps_{react}; (**m**) Rotor voltage V_r; (**n**) PR+RHC controller THD; (**o**) PI controller THD; (**p**) API controller THD.

Table 2. Performance evaluation of designed control strategies for V_{dc}.

Control Strategies	Performance Index		
	IAE	ISE	ITAE
PI	5.473	55.95	1.991
API	0.1145	0.659	0.0810
PR+RHC	0.46	1.37	0.0325

Notes: IAE: Integral Absolute Error, ISE: Integral Square Error, ITAE: Integral of Time-Weighted Absolute Error.

Table 3. Performance evaluation of designed control strategies for I_d.

Control Strategies	Performance Index		
	IAE	ISE	ITAE
PI	2.50	20.78	0.0749
API	0.96	6.34	0.0224
PR+RHC	0.0117	5.68	0.0094

Notes: IAE: Integral Absolute Error, ISE: Integral Square Error, ITAE: Integral of Time-Weighted Absolute Error.

Table 4. Performance evaluation of designed control strategies for I_q.

Control Strategies	Performance Index		
	IAE	ISE	ITAE
PI	0.18	1.208	0.0066
API	0.01	0.062	0.0016
PR+RHC	0.017	0.069	0.004

Notes: IAE: Integral Absolute Error, ISE: Integral Square Error, ITAE: Integral of Time-Weighted Absolute Error.

5.2. Under-Voltage

The grid voltage is dropped to 50% of its rated value for 200 ms from 1 s to 1.2 s during the under-voltage case, as illustrated in Figure 9b. The proposed controller V_{dc} response, shown in Figure 9a, is less oscillatory, fast, and robust for the API and PR+RHC algorithms, as compared to PI's response which is unstable and out of limits. Figure 9c–e clearly shows that I_d completely traces the reference value which indicates the robustness of the proposed (API & PR+RHC) strategies. The API controller updates its parameters using fuzzy rules to track the reference abruptly and the PR+RHC, due to its harmonic compensation, effectively minimizes the error, in comparison to the PI controller. The proposed controller responses in the case of I_r is shown in Figure 9f. Figure 9g depicts I_q having smooth response for the proposed controllers which gain stability soon after voltage the reaches a normal value. Figure 9h illustrates the I_g response for the API & (PR+RHC) controllers with respect to the PI controller which ensures grid stability. The P_r and P_s responses are described in Figure 9i,j which show that the API & (PR+RHC) controller responses are less oscillatory, and more stable as compared to the PI controller which reduces mechanical stress y as well as stress on drives.

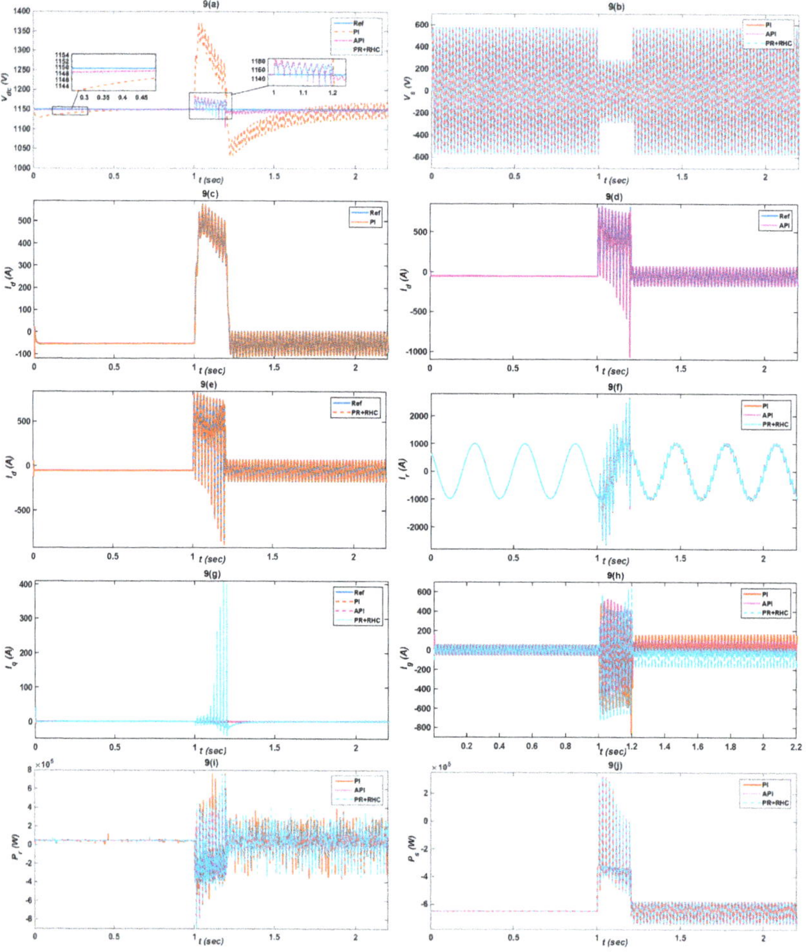

Figure 9. *Cont.*

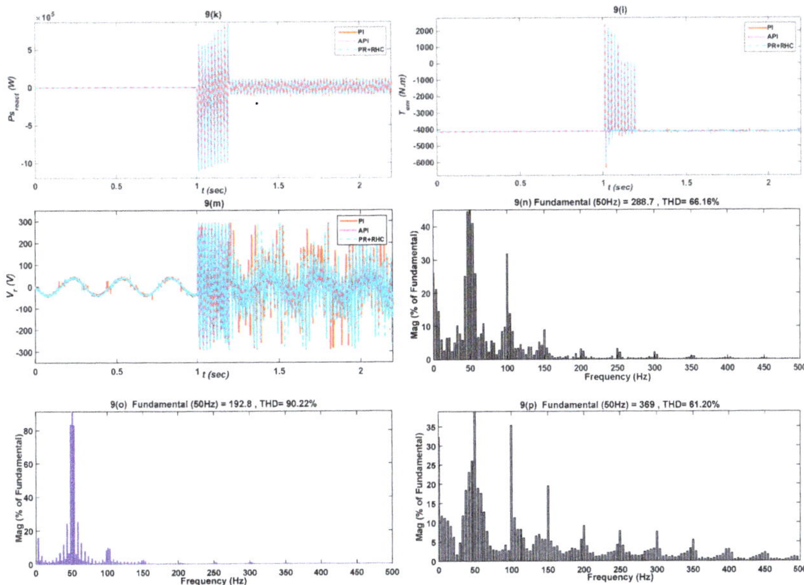

Figure 9. Comparison of PI and Proposed API and PR+RHC controller responses under undervoltage fault considering: (**a**) Dc-link voltage V_{dc}; (**b**) Stator voltage V_s; (**c–e**) Active component of current I_d; (**f**) Rotor current I_r; (**g**) Reactive component I_q; (**h**) Grid current I_g; (**i**) Rotor active power P_r; (**j**) Stator active power P_s; (**k**) Electromagnetic torque T_{em}; (**l**) Stator reactive power Ps_{react}; (**m**)Rotor voltage V_r; (**n**) PR+RHC controller THD; (**o**) PI controller THD; (**p**) API controller THD.

The Ps_{react}, T_{em} and V_r responses for both the proposed and conventional strategy are shown in in Figure 9k–m. Finally, the robustness of the proposed controllers over the PI conventional controller was proved by harmonic spectrum analysis of I_g, The THD value for the PI controller was 90.22% which is reduced to 61.20% and 66.16% in the case of the API and PR+RHC, respectively, and demonstrated in Figure 9n–p. The performance indices of all the control schemes are evaluated in Tables 5–7 for V_{dc}, I_d, and I_q, respectively. In the case of the API & PR+RHC controllers, all three parameter values are the minimum compared with the PI controller, which proves the better performance of the proposed controllers in under-voltage conditions.

Table 5. Performance evaluation of the designed control strategies for V_{dc}.

Control Strategies	Performance Index		
	IAE	ISE	ITAE
PI	99.7	6240	211.8
API	16.2	120.8	37.46
PR+RHC	13.36	98.36	29.90

Table 6. Performance evaluation of the designed control strategies for I_d.

Control Strategies	Performance Index		
	IAE	ISE	ITAE
PI	18.74	583.4	53.43
API	3.015	142.1	5.431
PR+RHC	4.59	154.26	6.55

Table 7. Performance evaluation of the designed control strategies for I_q.

Control Strategies	Performance Index		
	IAE	ISE	ITAE
PI	1.601	0.8957	4.672
API	0.0019	0.0021	0.0024
PR+RHC	0.026	0.102	0.069

5.3. Over-Voltage

In over-voltage conditions the grid voltage is increased 50% of its rated value for 200 ms from 1 s to 1.2 s as shown in Figure 10b. The V_{dc} of the proposed controllers is robust, faster, and stable soon after the grid voltage recovers as shown in Figure 10a. I_d for the PI, API and PR+RHC control controllers are clearly depicted in Figure 10c–e which prove that the proposed controllers are exactly following the reference value. Due to adaptiveness of the API and harmonic compensation of PR+RHC, both controllers are less sensitive to faults and the response is faster. I_r are also depicted in Figure 10f for all controllers. In case, the I_g responses in the API and PR+RHC controllers are fast and attain stability quickly after 1.2 s as shown in Figure 10g. Similarly I_q, the API and PR+RHC controller responses are fast and achieve stability soon after 1.2 s, while the PI controller responds after 1.5 s as elaborated in Figure 10h. The proposed controllers' responses in the case of P_r and P_s is less oscillatory and stable, which ensures stable performance is shown in Figure 10i,j. The proposed controllers' performances in the case of Ps_{react}, T_{em} and V_r are also dominant and less harmonic as shown in Figure 10k–m. Finally, THD of I_g is calculated, which is 1046.10% using the PI controller while it reduces to 446.52% and 684.51% in the case of the API and PR+RHC controllers which makes the proposed controllers more reliable and efficient in over-voltage conditions as shown in Figure 10n–p. The performance indices of all the control schemes are evaluated in Tables 8–10 for V_{dc}, I_d, and I_q, respectively. In the case of the API and PR+RHC controllers, all three parameters values are minimum compared with the PI controller, which validates the better performance of the proposed controllers.

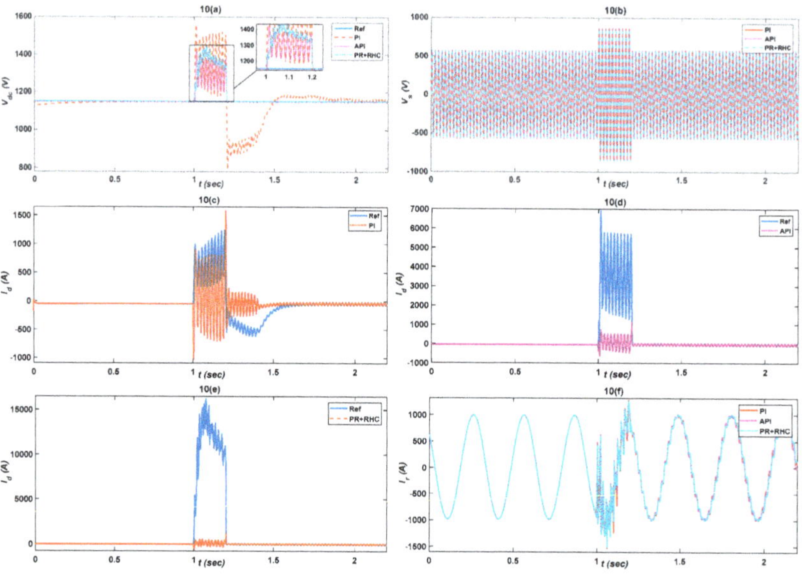

Figure 10. *Cont.*

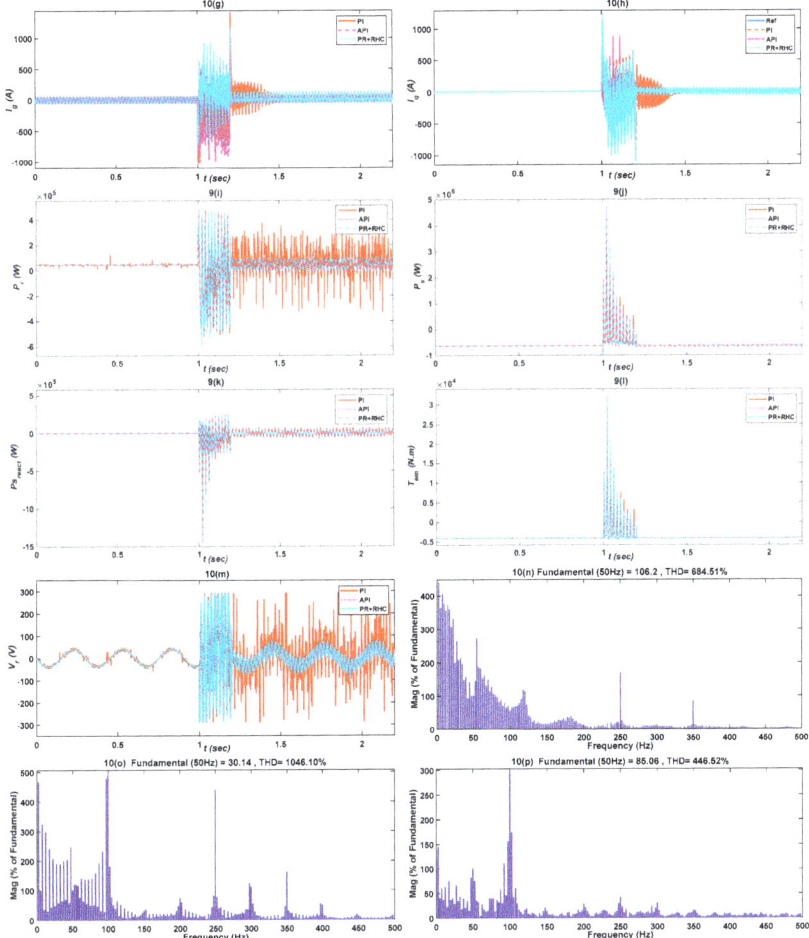

Figure 10. Comparison of PI and Proposed API and PR+RHC controller responses under overvoltage fault, considering: (**a**) Dc-link voltage V_{dc}; (**b**) Stator voltage V_s; (**c–e**) Active component of current I_d; (**f**) Rotor current I_r; (**g**) Reactive current component I_q; (**h**) Grid current I_g; (**i**) Rotor active power P_r; (**j**) Stator active power P_s; (**k**) Electromagnetic torque T_{em}; (**l**) Stator reactive power Ps_{react}; (**m**) Rotor voltage V_r; (**n**) PR+RHC controller THD; (**o**) PI controller THD; (**p**) API controller.

Table 8. Performance evaluation of the designed control strategies for V_{dc}.

Control Strategies	Performance Index		
	IAE	ISE	ITAE
PI	55.17	23.86	70.02
API	10.49	4.09	15.63
PR+RHC	14.49	6.89	9.02

Table 9. Performance evaluation of the designed control strategies for I_d.

Control Strategies	Performance Index		
	IAE	ISE	ITAE
PI	5.6	35.26	7.65
API	2.6	16.32	3.27
PR+RHC	3.1	15.36	4.09

Table 10. Performance evaluation of the designed control strategies for I_q.

Control Strategies	Performance Index		
	IAE	ISE	ITAE
PI	73.01	63.86	88.97
API	36.73	20.71	55.23
PR	35.29	19.06	49.74

5.4. Single Phase Fault

A single-phase fault is applied to evaluate the performance of the proposed controllers. The fault is applied for 200 ms from 1 s to 1.2 s as depicted in Figure 11b. The V_{dc} responses of the API and PR+RHC controllers are robust and attain stability soon after the fault is cleared, while the PI controller response is oscillatory and delayed in accomplishing stability after the fault is cleared as illustrated in Figure 11a. The I_d responses for the PI, API and PR+RHC controllers are shown in Figure 11c–e. The API controller updates its parameters using fuzzy rules to track the reference abruptly and the PR+RHC controller, due to its harmonic compensation, effectively minimizes the error, in comparison to the PI controller. I_r values for the conventional and proposed controllers are illustrated in Figure 11f. The I_q and I_g responses of the proposed controllers are more stable and less oscillatory as shown in Figure 11g,h. The responses of P_s and P_r powers, T_{em}, P_{Sreact}, and V_r are shown in Figure 11i–m. Analyzing the controllers on the basis of the grid current I_g THD values, it clearly shows that the proposed API controller with 55.43% THD and PR+RHC with 60.91% THD show less harmonics with respect to the 76.35% THD of the PI controller with increased harmonics which shows that the proposed controllers' responses in case of a single-phase fault are robust and stable as compared to the PI controller as shown in Figure 11n–p. The performance indices of all the control schemes are evaluated in Tables 11–13 for V_{dc}, I_d, and I_q, respectively. In the case of proposed API and PR+RHC controllers, all three parameter values are minimum compared with the PI controller, which guarantees the better performance of the proposed controllers under single-phase fault conditions.

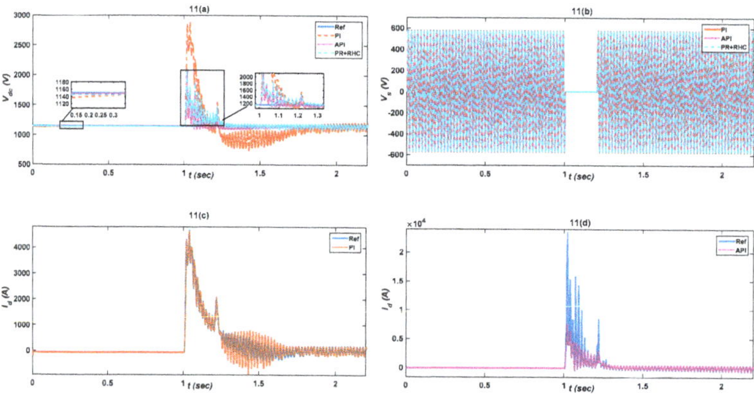

Figure 11. *Cont.*

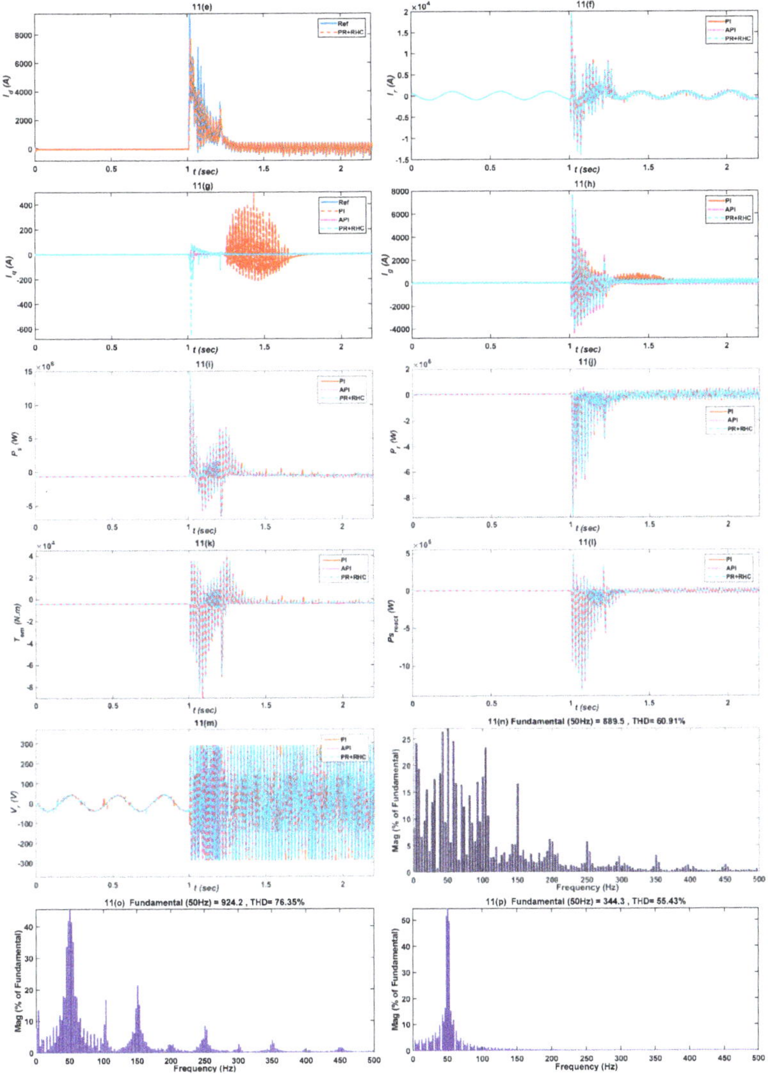

Figure 11. Comparison of PI and Proposed API and PR+RHC controller responses under Single-phase fault, considering: (**a**) Dc-link voltage V_{dc}, (**b**), Stator voltage V_s, (**c–e**) Active component of current I_d, (**f**) Rotor current I_r, (**g**) Reactive component I_q, (**h**) Grid current I_g, (**i**) Stator active power P_s, (**j**) Rotor active power P_r, (**k**) Electromagnetic torque T_{em}, (**l**) Stator reactive power Ps_{react}, (**m**) Rotor voltage V_r, (**n**) PR+RHC controller THD, (**o**) PI controller THD, (**p**) API controller.

Table 11. Performance evaluation of the designed control strategies for V_{dc}.

Control Strategies	Performance Index		
	IAE	ISE	ITAE
PI	366.1	63.23	754.1
API	80.64	32.36	170.7
PR+RHC	84.64	39.36	111.7

Table 12. Performance evaluation of the designed control strategies for I_d.

Control Strategies	Performance Index		
	IAE	ISE	ITAE
PI	190.4	5.323	35.5
API	0.20	1.916	0.25
PR+RHC	1.06	3.09	2.36

Table 13. Performance evaluation of the designed control strategies for I_q.

Control Strategies	Performance Index		
	IAE	ISE	ITAE
PI	45.59	456	73.66
API	11.58	154	15.51
PR+RHC	15..69	93	29.6

5.5. Two-Phase Faults

A two-phase fault is applied to evaluate the performance of the control strategies. The fault is applied for 200 ms from 1 s and cleared at 1.2 s, as shown in Figure 12b. The V_{dc} responses of the API and PR+RHC controllers are more stable, quickly tracking the reference value after the fault is cleared, as compared to the unstable response of the PI controller as presented in Figure 12a. A comparison of the I_d of all controllers (Figure 12c–e) indicates that the API and PR+RHC controllers clearly track the reference value while PI goes unstable as it proceeds after 1.2 s. The API controller employs fuzzy rules adoptively with robust response and the PR+RHC due to its harmonic compensation effectively minimizes the error, in comparison to the PI controller. Figure 12f describes the I_r responses for all the controllers. Similarly, the I_q and I_g responses are more stable and robust in the API and PR+RHC controllers' case as elaborated in Figure 12g,h. The responses of other parameters of WTs i.e., P_s, P_r, T_{em}, Ps_{react} and V_r are shown in Figure 12i–m. The grid current I_g THDs of all controllers are presented in Figure 12n–p.

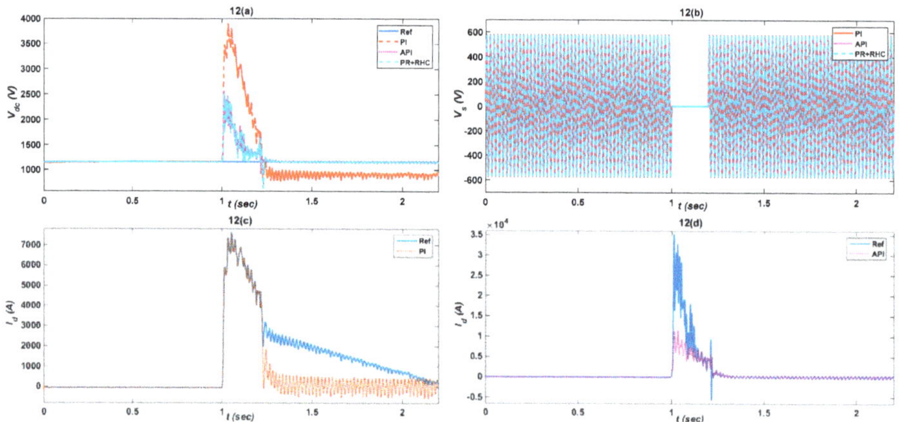

Figure 12. *Cont.*

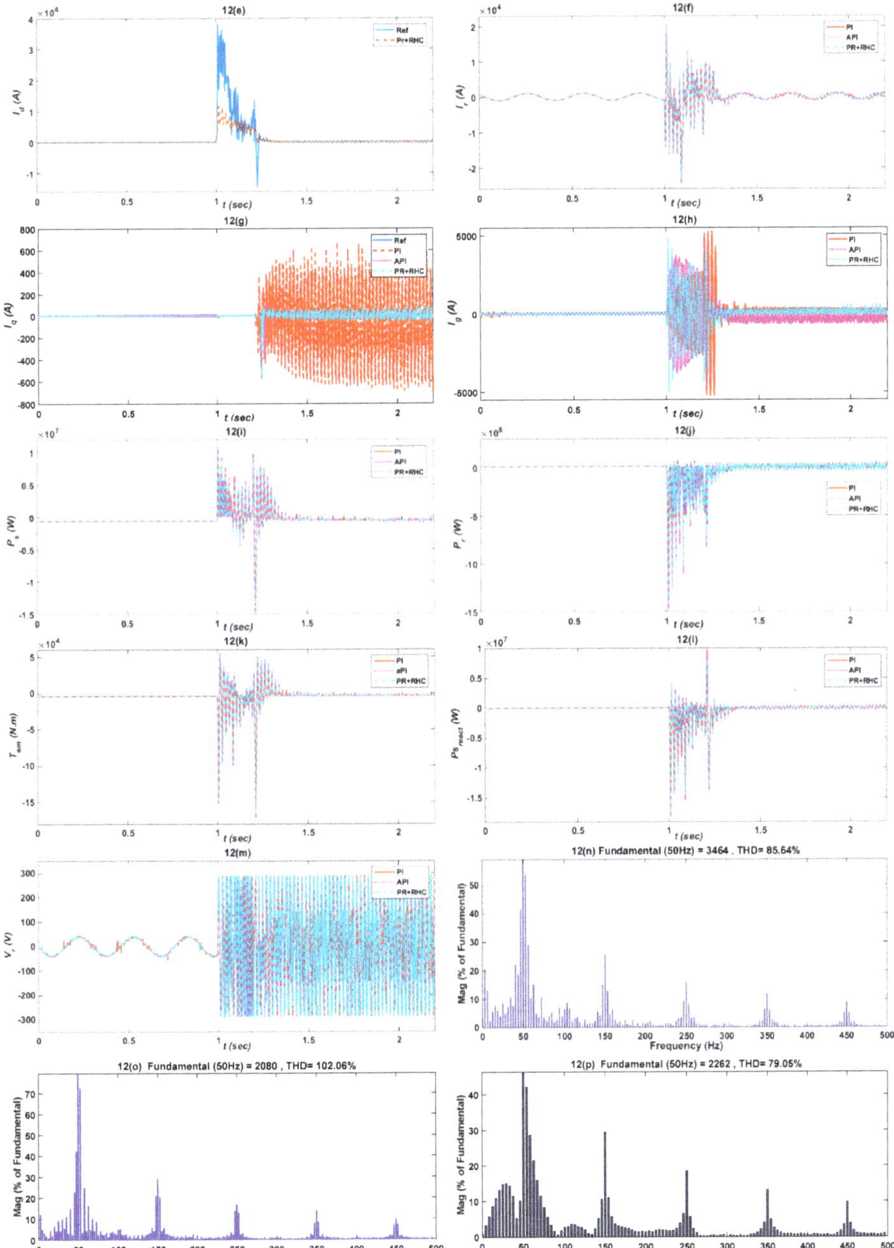

Figure 12. Comparison of PI and Proposed API and PR+RHC controller responses under two-phase fault, considering: (**a**) Dc-link voltage V_{dc}, (**b**) Stator voltage V_s, (**c–e**) Active component of current I_d, (**f**) Rotor current I_r, (**g**) Reactive component I_q, (**h**) Grid current I_g, (**i**) Rotor active power P_r, (**j**) Stator active power P_s, (**k**) Electromagnetic torque T_{em}, (**l**) Stator reactive power Ps_{react}, (**m**) Rotor voltage V_r, (**n**) PR+RHC controller THD, (**o**) PI controller THD and (**p**)API controller.

The API and PR+RHC controllers have THDs of 79.03% and 85.64% while the PI controller has 102.06% THD which demonstrates the effectiveness and dominance of the proposed (API & PR+RHC) controllers over PI. The performance indices of all the control schemes (PI, API & PR+RHC) are evaluated in Tables 14–16 for V_{dc}, I_d, and I_q, respectively. In the case of the proposed (API & PR+RHC) controllers, all three parameter values are minimum compared with the PI controller, which authenticates the better performance of the proposed controllers under two-phase fault conditions.

Table 14. Performance evaluation of the designed control strategies for V_{dc}.

Control Strategies	Performance Index		
	IAE	ISE	ITAE
PI	96.39	12.36	96.4
API	24.4	2.36	35.32
PR+RHC	29.31	3.59	39.85

Table 15. Performance evaluation of the designed control strategies for I_d.

Control Strategies	Performance Index		
	IAE	ISE	ITAE
PI	65.75	37.77	51.23
API	9.32	13.26	17.34
PR+RHC	14.60	19.32	24.09

Table 16. Performance evaluation of the designed control strategies for I_q.

Control Strategies	Performance Index		
	IAE	ISE	ITAE
PI	59.32	16.96	86.36
API	15.30	6.32	19.32
PR+RHC	20.96	9.96	24.49

6. Conclusions

Dynamic behaviors and critical issues like the stability of DC-link capacitor voltage and grid injected active and reactive power in DFIG-based WTs under voltage sags and grid faults were investigated and robust and novel Adaptive Proportional Integral (API) and Proportional Resonant with Resonant Harmonic Compensator (PR+RHC) controllers were proposed. The proposed DC-voltage control method is implemented independent of rotor side control which mitigates voltage harmonics in DC-capacitors and stabilizes active and reactive power which results in enhanced reliability of DC-link capacitor, WT stability, and makes control systems adoptable for large scale DFIG converters.

The performance of the PI control scheme shows sensitivity, large oscillations, and slow convergence to normal and abnormal conditions as verified from our simulation results, Total Harmonic Distortion (THD) analysis, and performance indices tables (Integral Absolute Error (IAE), Integral Square Error (ISE) and Integral Time-weighted Absolute Error (ITAE). However, comparatively the proposed controllers, i.e., API and PR+RHC, provide a better dynamic response, less sensitivity, fast convergence, less oscillation, robust, ripple-free and fault tolerant performance under normal and abnormal conditions.

Author Contributions: I.K., K.Z. and W.U.D. propose the main idea of the paper. I.K. implements the mathematical derivations, simulation verifications and analyses. The paper is written by I.K., and is revised by K.Z., W.U.D., S.U.I., M.I., S.H. and H.-J.K. All the authors were involved in preparing the final version of this manuscript. Besides, this whole work was supervised by H.-J.K.

Acknowledgments: This Research was supported by BK21PLUS, Creative Human Resource Development Program for IT Convergence.

Conflicts of Interest: The authors declare no conflict of interest.

Appendix A

Table A1. Model nominal parameters.

Generator Parameters	Values	Back-to-Back Converter	Data
Rated grid Power	2MW	Parallel converters	2
Polar pairs	2	Rated active power	400 kW
Gear ratio	95	DC-link voltage	1150 V
Rated shaft speed	1800 rpm	Switching frequency	2 kHz
Stator leakage inductance	0.038 mH	*Grid-side converter*	
Magnetizing inductance	2.91 mH	Rated output voltage	704 V
Rotor Leakage inductance	0.034 mH	Filter inductance	0.5 mH
Stator/rotor turns ratio	0.369	*Generator-side converter*	
		Rated output voltage	560 V

Table A2. Control schemes constants.

Control Schemes	Parameters	V_{dc}	I_d	I_q
PI	k_p	2.5	1.09	1.09
	k_i	10	17.25	17.25
API	k_p	25	250	250
	k_h	27500	200	200
PR+RHC	k_p	0.001	28	28
	k_i	0.01	1.5	1.5
	k_i^3 3rd	2	1.2	1.2
	k_i^5 5th	8	10	10
	k_i^7 7th	10	90	90

References

1. International Energy Agency. Wind Energy Technology Roadmap. 2009. Available online: http://www.iea.org/ (accessed on 22 November 2018).
2. Kaldellis, J.K.; Zafirakis, D. The wind energy revolution: a short review of a long history. *Renew. Energy* **2011**, *36*, 1887–1901. [CrossRef]
3. Abdeddaim, S.; Betka, A. Optimal tracking and robust power control of DFIG wind turbine. *Electr. Power Energy Syst.* **2013**, *49*, 234–242. [CrossRef]
4. Pena, R.; Clare, J.C.; Asher, G.M. Doubly fed induction generator using back-to-back PWM converters and its application to variable-speed wind-energy generation. *IEE Proc. Electr. Power Appl.* **1996**, *143*, 231–241. [CrossRef]
5. Ekanayake, J.B.; Holdsworth, L.; Wu, X.; Jenkins, N. Dynamic modeling of doubly fed induction generator wind turbines. *IEEE Trans. Power Syst.* **2003**, *18*, 803–809. [CrossRef]
6. Hu, J.; He, Y. DFIG wind generation systems operating with limited converter rating considered under unbalanced network conditions-analysis and control design. *Renew. Energy* **2011**, *36*, 829–847. [CrossRef]
7. Muljadi, E.; Yildirim, D.; Batan, T.; Butterfield, C.P. Understanding the unbalanced-voltage problem in wind turbine generation. In Proceedings of the 1999 IEEE Industry Applications Conference: Thirty-Forth IAS Annual Meeting, Phoenix, AZ, USA, 3–7 October 1999; pp. 1359–1365. [CrossRef]
8. Abad, G.; Lopez, J.; Rodriguez, M.; Marroyo, L.; Iwanski, G. *Doubly Fed Induction Machine: Modeling and Control for Wind Energy Generation Applications*; Wiley-IEEE Press: Hoboken, NJ, USA, 2011.
9. Yi, Z.; Bauer, P.; Ferreira, J.A.; Pierik, J. Operation of grid-connected DFIG under unbalanced grid voltage condition. *IEEE Trans. Energy Convers.* **2009**, *24*, 240–246.

10. Tazil, M.; Kumar, V.; Bansal, R.C.; Kong, S.; Dong, Z.Y.; Freitas, W.; Mathur, H.D. Three-phase doubly fed induction generators: An overview. *IET Electr. Power Appl.* **2010**, *4*, 75–89. [CrossRef]

11. EPCOS. Aluminum Electrolytic Capacitors General Technical Information. 2010. Available online: http://www.epcos.com (accessed on 24 November 2018).

12. Hitachi AIC. Aluminum Electrolytic Capacitors. 2011. Available online: http://www.hitachi-aic.com (accessed on 1 December 2018).

13. Song, H.; Nam, K. Dual current control scheme for PWM converter under unbalanced input voltage conditions. *IEEE Trans. Ind. Electron.* **1999**, *46*, 953–959. [CrossRef]

14. Xu, L. Coordinated control of DFIG's rotor and grid side converters during network unbalance. *IEEE Trans. Power Electron.* **2008**, *23*, 1041–1049.

15. Xu, L.; Wang, Y. Dynamic modeling and control of DFIG-based wind turbines under unbalanced network conditions. *IEEE Trans. Power Syst.* **2007**, *22*, 314–323. [CrossRef]

16. Pena, R.S.; Cardenas, R.J.; Clare, J.C.; Asher, G.M. Control strategies for voltage control of a boost type PWM converter. In Proceedings of the 2001 IEEE 32nd Annual Power Electronics Specialists Conference, Vancouver, BC, Canada, 17–21 June 2001; pp. 730–735. [CrossRef]

17. Malesani, L.; Rossetto, L.; Tenti, P.; Tomasin, P. AC/DC/AC PWM converter with reduced energy storage in the dc link. *IEEE Trans. Ind. Appl.* **1995**, *31*, 287–292. [CrossRef]

18. Liu, C.; Huang, X.; Chen, M.; Xu, D. Flexible control of dc-link voltage for doubly fed induction generator during grid voltage swell. In Proceedings of the 2010 IEEE Energy Conversion Congress and Exposition, Atlanta, GA, USA, 12–16 September 2010; pp. 3091–3095. [CrossRef]

19. Yao, J.; Li, H.; Liao, Y.; Chen, Z. An improved control strategy of limiting the dc-link voltage fluctuation for a doubly fed induction wind generator. *IEEE Trans. Power Electron.* **2008**, *23*, 1205–1213.

20. Kim, H.S.; Mok, H.S.; Choe, G.H.; Hyun, D.S.; Choe, S.Y. Design of current controller for 3-phase PWM converter with unbalanced input voltage. In Proceedings of the 29th Annual IEEE Power Electronics Specialists Conference, Fukuoka, Japan, 18–21 May 1998; pp. 503–509. [CrossRef]

21. Vincenti, D.; Jin, H. A three-phase regulated PWM rectifier with online feedforward input unbalance correction. *IEEE Trans. Ind. Electron.* **1994**, *41*, 526–532. [CrossRef]

22. Aziz, A.; Shafiullah, G.M.; Stojcevski, A.; Mto, A. Participation of DFIG based wind energy system in load frequency control of interconnected multi generation power system. In Proceedings of the 2014 Australasian Universities Power Engineering Conference, Perth, WA, Australia, 28 September–1 October 2014; pp. 1–6. [CrossRef]

23. Chowdhury, M.A.; Shafiullah, G.M. SSR mitigation of series-compensated DFIG wind farms by a nonlinear damping controller using partial feedback linearization. *IEEE Trans. Power Syst.* **2018**, *33*, 2528–2538. [CrossRef]

24. Faried, S.O.; Unal, I.; Rai, D.; Mahseredjian, J. Utilizing DFIG-based wind farms for damping subsynchronous resonance in nearby turbine-generators. *IEEE Trans. Power Syst.* **2013**, *28*, 452–459. [CrossRef]

25. Huang, P.-H.; EI Moursi, M.S.; Xiao, W. Subsynchronous resonance mitigation for series-compensated DFIG-based wind farm by using two-degree-of-freedom control strategy. *IEEE Trans. Power Syst.* **2015**, *30*, 1442–1454. [CrossRef]

26. Leon, A.E.; Solsona, J.A. Sub-synchronous interaction damping control for DFIG wind turbines. *IEEE Trans. Power Syst.* **2015**, *30*, 419–428. [CrossRef]

27. Uddin, W.; Zeb, K.; Tanoli, A.; Haider, A. Hardware-based hybrid scheme to improve the fault ride through capability of doubly fed induction generator under symmetrical and asymmetrical faults. *IET Gener. Trans. Distrib.* **2017**, 200–206. [CrossRef]

28. Muller, S.; Deicke, M.; De Doncker, R.W. Doubly fed induction generator systems for wind turbines. *IEEE Ind. Appl. Mag.* **2002**, *8*, 26–33. [CrossRef]

29. Zeb, K.; Khan, I.; Uddin, W. A Review on Recent Advances and Future Trends of Transformerless Inverter Structures for Single-Phase Grid-Connected Photovoltaic Systems. *Energies* **2018**, *11*, 1968. [CrossRef]

30. Blaabjerg, F.; Lonel, D.M. Power Electronics and Control for Large Wind Turbines and Wind Farms. In *Renewable Energy Devices and System with Simulation in MATLAB and ANSYS*, 1st ed.; CRC Press Taylor & Francis Group: Boca Raton, FL, USA, 2017; pp. 194–195. ISBN 9781498765831.

31. Erdogan, N.; Henao, H.; Grisel, R. An improved methodology for dynamic modelling and simulation of electromechanically coupled drive systems: An experimental validation. *Sadhana* **2015**, *40*, 2021–2043. [CrossRef]

32. Errami, Y.; Maaroufi, M.; Ouassaid, M. Modelling and control strategy of PMSG based variable speed wind energy conversion system. In Proceedings of the 2011 International Conference on Multimedia Computing and Systems, Ouarzazate, Morocco, 7–9 April 2011; pp. 1–6. [CrossRef]

33. Kundur, P. *Power Systems Stability and Control*, 1st ed.; McGraw-Hill Education: New York, NY, USA, 1994.

34. Bose, B.K. *Modern Power Electronics and AC Drives*; Prentice-Hill Inc.: Upper Saddle River, NJ, USA, 2008.

35. Mohan, N. *Advance Electric Drives Analysis, Control and Modeling using Simulink*, 1st ed.; Wiley: Hoboken, NJ, USA, 2001; ISBN 978-1-118-91113-6.

36. Boemer, J.C.; Burges, K.; Zolotarev, P.; Lehner, J.; Wajant, P.; Fürst, M.; Brohm, R.; Kumm, T. Overview of German grid issues and retrofit of photovoltaic power plants in German for the prevention of frequency stability problems in abnormal system conditions of the ENTSO-E region continental Europe. In Proceedings of the 1st Solar International Workshop on Integration of Solar Power Systems, Aarhus, Denmark, 24 October 2011.

37. Ackermann, T.; Ellis, A.; Fortmann, J.; Matevosyan, J.; Muljadi, E.; Piwko, R.; Pourbeik, P.; Quitmann, E.; Sorensen, P.; Urdal, H.; et al. Code shift: Grid specifications and dynamic wind turbine models. *IEEE Power Energy Mag.* **2013**, *11*, 72–82. [CrossRef]

38. Yuan, X.; Merk, W.; Stemmler, H.; Allmeling, J. Stationary-Frame Generalized Integrators for Current Control of Active Power Filters with Zero Steady-State Error for Current Harmonics of Concern Under Unbalanced and Distorted Operating Conditions. *IEEE Trans. Ind. App.* **2002**, *38*, 523–532. [CrossRef]

39. Zeb, K.; Uddin, W.; Khan, M.A. A comprehensive review on inverter topologies and control strategies for Grid connected photovoltaic system. *Renew. Sustain. Energy Rev.* **2018**, *94*, 1120–1141. [CrossRef]

Article

Development of a Computational System to Improve Wind Farm Layout, Part I: Model Validation and Near Wake Analysis

Rafael V. Rodrigues * and Corinne Lengsfeld

Department of Mechanical and Materials Engineering, University of Denver, Denver, CO 80210, USA;
corinne.lengsfeld@du.edu
* Correspondence: Rafael.rodrigues@du.edu; Tel.: +1-720-810-8349

Received: 23 January 2019; Accepted: 6 March 2019; Published: 12 March 2019

Abstract: The first part of this work describes the validation of a wind turbine farm Computational Fluid Dynamics (CFD) simulation using literature velocity wake data from the MEXICO (Model Experiments in Controlled Conditions) experiment. The work is intended to establish a computational framework from which to investigate wind farm layout, seeking to validate the simulation and identify parameters influencing the wake. A CFD model was designed to mimic the MEXICO rotor experimental conditions and simulate new operating conditions with regards to tip speed ratio and pitch angle. The validation showed that the computational results qualitatively agree with the experimental data. Considering the designed tip speed ratio (TSR) of 6.6, the deficit of velocity in the wake remains at rate of approximately 15% of the free-stream velocity per rotor diameter regardless of the free-stream velocity applied. Moreover, analysis of a radial traverse right behind the rotor showed an increase of 20% in the velocity deficit as the TSR varied from TSR = 6 to TSR = 10, corresponding to an increase ratio of approximately 5% m·s^{-1} per dimensionless unit of TSR. We conclude that the near wake characteristics of a wind turbine are strongly influenced by the TSR and the pitch angle.

Keywords: wind turbine near wake; wind turbine wakes; wake aerodynamics; computational fluid dynamics; rotor aerodynamics; wind turbine validation; MEXICO experiment

---

1. Introduction

Limited carbon resources and environmental concerns are some of the reasons leading the energy industry to exploit alternative energy sources. Wind energy systems have been developed and applied for sites with suitable conditions, the first modern commercial-scale wind turbines were placed in United States approximately 40 years ago. Nowadays, the most common and profitable applications for wind energy systems are the large wind farms. Commercial-scale wind generators for wind farms are within 3 MW and 5 MW, and all have a predominantly horizontal axis and are three bladed. One problem of these large wind farms is the row arrangement of the generators. The towers are usually placed in rows, requiring large areas of land for rotors up to 100 m in diameter. Previous research has suggested safe distances to avoid the wind turbines blade/components damage and output power waste. However, the optimum spacing between turbines in a wind farm is still a challenging and open question in wind energy research.

Several efforts using different methodologies have been done to achieve layout optimization, focusing on finding optimal spacing between turbines in a wind farm. Park and Law [1] applied sequential convex programming to maximize wind farm output power by optimizing the placement of wind turbines of the Horns wind farm in Denmark. They found that the optimal spacing between wind turbines is dependent on the wind direction. Scattering the turbines helped to avoid wake chain effects, so that downstream rotors were not significantly affected. Moreover, the same study considered wind statistical data to optimize the wind farm power production over a long period, resulting in a 7.3% power increase. Son et al. [2] found that the total wind farm output power is strongly related to the distance between the first and second wind turbine rows. When the referred distance became larger, the output power considerably dropped in comparison to smaller distances. This means that the increase of the spacing between the first and second rows is ineffective in improving output power. On the other hand, decreased distances made the second wind turbine row much less efficient. They discovered the importance of keeping turbines as close as possible, but with enough space so that the second row can have guaranteed output power. Longer distances did not contribute to increase the total output power. Further, increasing the space between the fourth and the fifth rows has a better contribution than increasing the space between the first and second rows. Wu and Porté-Agel [3] investigated two layout configurations in the same area with 30 turbines either arranged in aligned or staggered conditions. In comparison to the aligned configuration, the staggered one allows better wake recovery. This exposes the downstream turbines to higher local wind speeds (consequently higher performance) and lower turbulence intensity. Stevens et al. [4] found that the distance of 10 diameters (or higher) would minimize the cost per unit of energy production, and the same is true for a distance of 15 diameters if the objective function was evaluated using dimensionless parameters. Those value are significantly higher than applied values in wind farms (6–10 turbine diameters). Meyers and Meneveau [5] found that the current wind farms layout solutions in literature have characteristics with considerably lower spacing than computationally optimized layout solutions.

Moreover, other efforts have attempted to achieve wind farm optimization using control strategies to mitigate wake effects, applying sub-optimal operating conditions. This means that each rotor will not necessarily deliver the best aerodynamic performance, but the goal is to find the best solution that avoids wake interaction effects, increasing the total wind farm output power. Park and Law et al. [6] studied control strategies for wake effects mitigation, showing that control techniques can be applied for each individual rotor to improve overall wind farm efficiency. González et al. [7] proposed the individual selection of an operating point on each wind turbine in order to maximize the overall wind farm output power. This is performed by studying the optimal pitch angle and Tip Speed Ratio (TSR) of each rotor in regards to the total wind farm output power. Additionally, the methodology also allows decreased turbulence intensity levels in the produced wakes. The results showed increased power production when the wind speed is lower than the rated wind speed, and for non-prevailing wind directions. Lee et al. [8] found an increase of 4.5% in the total output power by applying pitch angle control for the Horns Rev wind farm. Kazda et al. [9] applied weakened wake conditions for upstream turbines by using sub-optimal operations through control strategies. They found that a 12.5% reduction for the upstream turbines resulted in a 2.5% increase in the sum of the upstream and downstream turbines. This could be achieved by either a change of 3.5° in the pitch angle or by a 24% reduction in TSR compared to optimum TSR. For the case of two upstream turbines operating at 87.5% of optimal conditions, the sum of total power of the upstream and downstream turbines increased by 9.7%. Gil et al. [10] applied control strategies, achieving from 1.86% up to 6.24% in energy captured by using sub-optimal operating points. Chowdhurry et al. [11] found that using variable rotor diameters improved efficiency, achieving 30% increase in the total power generation.

All these efforts in the literature described above provided relevant contributions to wind farm optimization and turbine spacing research. However, they did not consider a rigorous evaluation of three-dimensional wake effects, which this study will achieve. Most of the computational studies from literature proposed design optimization frameworks in which the wind turbine models were not based on data validated against experimental measurements of the wake flow field. In the context of science applied to wind farm optimization, this work proposes a numerical Computational Fluid Dynamics (CFD) model to rigorously analyze wind turbine wake flow field, characterizing wake flow characteristics for the most relevant wind farm parameters: velocity flow field and turbulence intensity. The current study will do a full computational analysis of the near-wake aerodynamic behavior, considering configurations not analyzed before in literature: several different loading, free-stream velocity and pitch angle conditions. The goal is to achieve a validated model by comparing computational and experimental data from existing literature. Engineering tools such as CFD or wake analytical methods have been improved to accurately characterize wake characteristics, but there are few experiments to effectively validate wind turbine wake flow. The present study provides an advance in relation to previous models as it propose a physical model in which the near wake is validated against a critical wind farm design parameter: the velocity flow field. With a reliable CFD model validated in terms of flow field, this model could be applied to improve wind farm layout design by including both near and far wake regions. Literature shows a variety of techniques and different goals in regards to wind turbine CFD. The next section shows a description of the main experimental approaches found in literature, which will be useful to provide data to develop and validate the numerical model in this study.

1.1. Brief Description of Wind Tunnel Experiments

A full review of low-speed wind tunnel studies and scaled turbines is provided by Crespo et al. [12]; additionally, other recent relevant studies can be found in the literature [13–22]. Most of these studies are meant to validate wind turbine simulations, and some of them are described below to provide an overview of low-speed wind tunnel experiments. The objective of this literature review is to show the way that experimental data can be used in order to validate wind turbine simulations.

Wind turbine experiments conducted by the Norwegian University of Science and Technology validated the numerical results against wind tunnel measurements in terms of mean velocity, turbulence intensity and the power and thrust coefficients. This research center has low-speed wind tunnel facilities, with dimensions of 2.71 m wide, 1.8 m high and 11.1 m long. An experimental study was performed using two aligned prototype rotors of 0.944 m and 0.894 m, and the blade consists of 14% S826 NREL (National Renewable Energy Laboratory) profile for the two rotors [13]. The velocity profile was characterized using Pitot-Static tubes, and the thrust force was determined using a Six-Component Balance Force.

A qualitative study of the rotor wake behavior by Chamorro and Porté-Angel [14] analyzed a 150 mm diameter three-bladed wind turbine prototype, which was tested using a wind tunnel with 37.5 m length driven by a 200 hp fan. The experimental data was used to produce a qualitative study of the wake behavior, since the Reynolds number is different compared to full-scale wind turbines. A particularly interesting aspect that distinguishes this study from the others is that the authors were able to characterize the surface roughness by placing straight chains of approximately 5 mm height covering a 10 m section of the tunnel. These chains were aligned perpendicular to the flow direction and separated from each other by 0.20 m. The mean wind velocity in the tunnel was measured using Pitot static tubes, and constant TSR values ($\lambda = 4.2$ for smooth surfaces and $\lambda = 4.4$ rough surfaces) were maintained in order to reflect the typical operational conditions of full-scale field turbines (typically $3 < \lambda < 6$).

In another experiment, a virtual wind-tunnel model (24.4 m $\times$ 36.6 m) with the same dimension of the NASA (National Aeronautics and Space Administration) wind tunnel was analyzed using the ANSYS Fluent package [15]. The model validation was performed comparing the pressure coefficient at different span-wise sections along the turbine blade. In addition, the wind turbine output power was compared to published experimental results for the NREL phase VI rotor tested in the NASA wind tunnel. Several other studies in literature utilized data from the NREL/NASA framework to develop CFD studies using pressure coefficient values on the blades and aerodynamic torque data for comparison and validation. Zhou et al. [23] performed LES (Large Eddy Simulation) of the NREL phase IV to evaluate the effect of different inflow conditions on the aerodynamic loading and near wake characteristics. Hsu et al. [24] implemented a finite-element (Lagrangian-Eulerian) model of the NREL Phase IV using a non-structured rotating mesh refined close to the rotor disc. Wake characterization was not the focus of the study, what explains the wake made out of coarse non-structured cells with no refinement. Gundling et al. [25] evaluated low and high fidelity models using the NREL Phase

VI for predicting wind turbine performance, aeroelastic behavior and wakes: 1) The Blade Element method with a free-vortex wake; 2) The actuator disc method; 3) The full-rotor method. Mo et al. [26] did a study in more depth to understand wake aerodynamics performing a LES of the NREL Phase VI using dynamic Smagorinsky-model, additionally verification of the average Turbulence Intensity was performed against an analytical model. They found that the downstream distance where instability and vortex breakdowns occur is dependent on wind free-stream inlet conditions (7 m·s^{-1} happens at four rotor diameters, while 15.1 m·s^{-1} between 11 and 13 diameters), and a decrease of the turbulence intensity happened after instability and vortex breakdowns. Choudhry et al. [27] performed a very similar CFD study of the NREL phase VI using computational methods similar to the ones found in the study conducted by Mo et al. [26], finding that regions of velocity deficit and high turbulence intensity are within the high vorticity region.

Sturge et al. [16] utilized an open-circuit suction tunnel, driven by an eight-blade axial fan positioned at the outlet. In this experiment, the wind speed is controlled by using a variable frequency drive. The air flow passes through a honeycomb mesh with cells 0.01 m wide and 0.1 m long. The dimensions vary along the tunnel, with a 6.25:1 contraction section and 1.2 m high $\times$ 1.2 m wide $\times$ 3 m long test section. Afterwards, analysis of static pressure along the blade showed a large reduction in the suction peak along the leading edge, which reduced the lift generated by the rotor and consequently the torque production.

The wake flow of a 5 $\times$ 5 array of 50 mm micro-wind turbines was studied and analyzed by Houssain et al. [18] using a wind tunnel. These 1/10 scaled prototypes were placed in a 3 m $\times$ 1.8 m wind tunnel, allowing the velocity profile and turbulence intensity (velocity fluctuations) behind the array to be measured at different downstream locations. The wake flow was characterized by using hot-wire anemometer, ultrasonic anemometer measurements, and Particle Image Velocimetry (PIV). The full-scale rotor of 500 mm diameter was analyzed as well. The results for velocity deficit and the turbulence intensity were similar for both rotors.

In this sense, the MEXICO (Model Experiments in Controlled Conditions) experiment [28] was one of the most comprehensive collaborative efforts by the International Energy Agency (IEA), who created the task 29 to gain understanding about wind turbine aerodynamics, as well as to improve aerodynamic models used for wind turbine design. A series of tests for a small wind turbine prototype were performed using the DNW German Dutch open section wind tunnel. Although the rotor wake measurements comprised only the near wake region right behind to the wind turbine (up to 1.33D downstream of the rotor), the experiment is a very rich source of data useful to validate wind turbine CFD wake models.

This present work covers the gap of characterizing the wind turbine wake flow field based on experimental data from existing literature, which describes the validation of a wind turbine CFD simulation using velocity wake data from the MEXICO experiment. The goal is to extend the understanding of the wake flow field beyond the distances analyzed in these experiments, and also analyzing the influence of variable operating conditions on near wake aerodynamic behavior. In order to do so, variable operating conditions with regards to the TSR and the Pitch Angle (θ) were simulated to understand how these specific design parameters affect the flow field. The second part of this work will extend the analysis beyond the near wake, characterizing the far wake aerodynamic behavior according to the same TSR and Pitch Angle (θ) conditions.

1.2. Detailed Overview of the MEXICO Experiment

The experiments described in the previous section only performed rotor measurements. However, computational models based on CFD assumptions also need flow field measurements to be successfully validated. The most comprehensive experimental flow field measurement study was the MEXICO Experiment [28], which used a rotor prototype of 4.5 m diameter and the largest wind tunnel existent in the European continent. PIV techniques were employed to collect flow field measurements around the rotor plane (Figure 1). Several recent studies utilized data from the MEXICO experiment to validate their CFD models [29–45] with different research goals as detailed below.

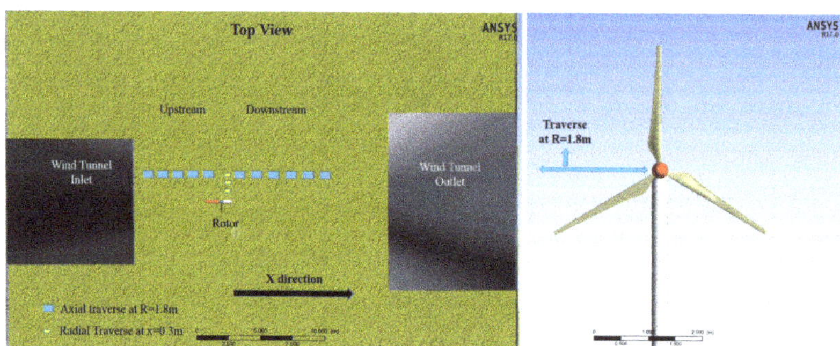

Figure 1. Sketch showing an overview of the MEXICO (Model Experiments in Controlled Conditions) Experiment (Top View).

In regards to Lifting Line codes, Yang et al. [29] showed the necessity for developing new techniques to account for 3D rotational effects on predicting loading for rotors. They created a new technique to determine the angle of attack on rotating blades using data from the MEXICO experiment, a Blade Element Momentum (BEM) code relying on 2D airfoil data was found to over-predict the loading of the rotor; this discrepancy was attributed to the 3D effects originated from the rotor geometry. Xudong et al. [30] developed an aerodynamic/aero-elastic design tool to optimize wind turbine blades and validated the results using MEXICO data for turbine loading.

Regarding the first round of PIV wake measurements (axial flow), Bechmann et al. [31] performed a CFD simulation of the MEXICO rotor using RANS (Reynolds Averaged Navier Stokes) equations further downstream up to 2.5 diameters behind the rotor. All the simulations were done fully turbulent, but there might be laminar flow at the leading edge of the blades; further work is needed to demonstrate the length of accuracy of laminar turbulent-transition models. Micallef et al. [32] characterized the radial velocities in the near wake close to the MEXICO rotor using a potential-flow panel model to characterize the wake radial induction. Tip vortex characterization performed by tracking its location showed that the radial flow velocity in the rotor plane is not fully dominated by the blade vorticity. Carrión et al. [33] assumed periodic boundary conditions to model only one of the MEXICO rotor blades under axial flow conditions, finding good agreement for the wake flow field by using a compressible multi-block solver without needing to switch between compressible and incompressible flow. Herraez et al. [34] validated a CFD model in OpenFoam using the Spalart-Allmaras turbulence model, showing comparisons for pressure distributions from several blade sections, and PIV near wake measurements. Shen et al. [35] performed CFD simulations of the MEXICO rotor including the geometry of the wind tunnel, and regarding tunnel wall effects this study found that tunnel effects are not significantly influenced by the fluid flow. Garcia et al. [36] developed a hybrid filament-mesh vortex method to improve computational efficiency, using the MEXICO experimental dataset for near wake validation. Nilsson et al. [37] described vortex structures in the near wake of the MEXICO rotor using the actuator line method. The trajectory of the tip vortices and wake expansion were described according to the TSR, implementing a RANS LES model. Wimshurst and Willden [38] simulated the near wake flow field of the MEXICO rotor using multiple reference frame approach. The actuator line method using 2D aerodynamic data was compared to a 3D polar actuator line model. Zhong et al. [39] developed a numerical tool combining Lagrangian dynamic large-eddy and actuator line models using PIV wake data for validation, finding that the tip vortices contribute to a maximum velocity deficit peak and turbulence intensity peak near the blade tip. Guntur and Sørensen [40] developed a full rotor CFD model of the MEXICO rotor focusing on the flow at the inboard part of the blades, analyzing the boundary layer separation at this region to understand differences in behavior between 3D flow and 2D flow. This latter study showed that the fluid flow separation starts at a higher angle of attack for the 3D case.

In regards to the second round of measurements (yawed flow), Sørensen et al. [41] did the first attempts to validate the near wake flow field in yawed flow. Tsalicoglou et al. [42] performed RANS computations of the MEXICO rotor wake for yawed and uniform flow cases, showing that the velocity deficit in the near wake (up to two diameters downstream) does not follow a Gaussian distribution. Additionally, the interaction with structures of the wind turbine (nacelle and tower) is more significant for yawed flows. The effects on the wake caused by the tower and the blade could still be observed at the end of the near wake. Grasso and Garrel [43] showed that the lifting line code coupled with the free wake method can accurately represent the near wake at uniform or yawed conditions. Shen et al. [44] developed an actuator line/Navier-Stokes model using the MEXICO rotor experimental dataset under yawed flow for flow field validation, considering both loading and velocity flow field for the simulation.

2. Computational Methods

2.1. Rotor Blade Geometry

The MEXICO experiment performed several different flow field measurements to characterize the three-dimensional velocity flow field in the near wake. Experimental measurements such as traverse and longitudinal wind velocity, both upwind and downwind of the rotor, were performed at a few specific locations. Here, we validated the computational model by plotting the velocity in the wake region of the blade and directly comparing the simulation results with experimental data from the MEXICO rotor. Because our hope is to implement a rapid computational simulation, the objective is to obtain agreement between experimental and computational velocities within 5%. The rotor simulated in this work was the MEXICO Rotor (Figure 2); the three-bladed model has three types of airfoil: DU91-W2-250 (20% to 45%), Riso-A1-21 (54% to 65%), and NACA 64-418 (75% until the blade tip). The blade is also twisted, and a pitch angle of $-2.3°$ was applied for the measurements. The blade geometry can be found in the final report of this experiment [28]. Since some of the airfoil data are not publicly available, a reverse engineering process was performed to find the airfoil coordinates.

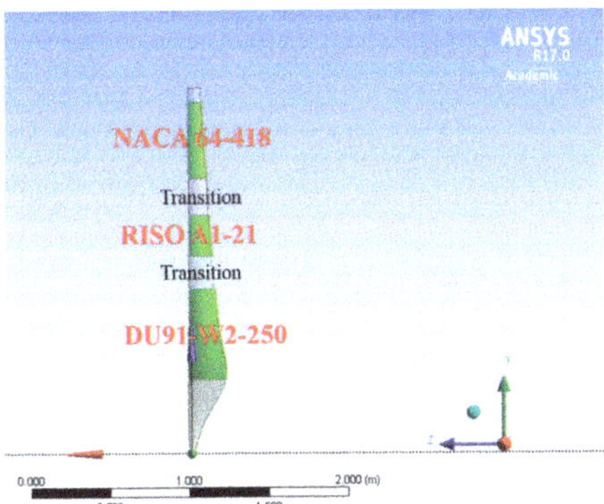

Figure 2. MEXICO rotor geometry, a three-bladed rotor with 4.5 m diameter. Reference for the blade geometry data: Scheppers et al. [28].

2.2. Layout and Boundary Conditions

We broke down the computational domain (Figure 3) into smaller parts for two reasons. First, local mesh sizing: the meaningful region can be refined to correctly characterize the flow field. Second, pressure-far-field boundary conditions for the lateral and superior boundaries require a larger domain to keep straight streamlines at the boundaries to achieve numerical convergence. The dimensions of the square part containing the wind tunnel and the rotor extends from −2.5D to 2.5D, while the exterior part corresponding to the surroundings extends from −10D to 10D.

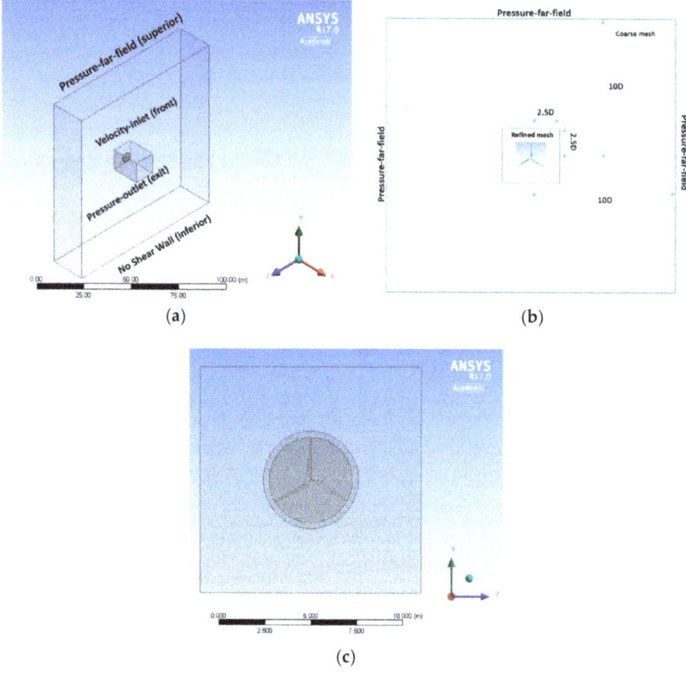

Figure 3. Layout of the computational domain and boundary conditions. (**a**) Perspective view and boundary conditions; (**b**) Information about the physical domain; (**c**) Front view of the central disc.

2.3. Computational Fluid Dynamics Modeling

CFD assumptions are based on the Finite Volume Method (FVM) for representing and evaluating partial differential equations in the form of algebraic equations. The domain of interest is divided into small cells, reducing the Navier-Stokes equations to algebraic or simple differential equations. Integration of the volume is conducted to obtain surface fluxes because the flux entering a given volume is identical to that leaving the adjacent volume. The CFD solver implemented in this work was ANSYS Fluent 17, housed in two computers, each with 64 GB RAM/8 processes. ANSYS Fluent solves the equations of fluid flow and heat transfer by default using a stationary (or inertial) reference frame. However, a Moving (or non-inertial) Reference Frame (MRF) can bring advantages in solving the equations for some problems involving moving parts, such as rotating blades. In those problems, the flow around the moving parts is the variable of interest to be determined. In the case of this work, the region behind of the wind turbine corresponding to the wake flow field is the region of interest. The MRF technique models the flow around the moving part as a steady-stead problem with respect to the moving frame, allowing to activate reference frames in selected cell zones. The ANSYS Fluent MRF modeling modify the equations of motion to incorporate additional acceleration terms that occur due to the transformation from the stationary to the moving reference. The main reason for employing a MRF is to solve a problem that is unsteady in the stationary (inertial) frame but steady with respect to the moving frame. In this work, the simulation was performed using a steady state MRFapproach, and setting the rotational speed to match experimental conditions. The turbulence model selected was the k–ω SST, which is suitable for swirl flow, and it was used in the literature studies as their main turbulence modelling technique. Because there is no public information from the reports of the MEXICO experiment regarding the inlet inflow conditions, default values of 5% and 10% were assumed for the inflow turbulence intensity (TI) and the viscosity ratio (VR) at the inlet, respectively. The Reynolds number based on the average chord length is approximately 1.5×10^5. Pressure-far-field boundaries are applied for the lateral and superior boundaries, pressure-outlet for the exit, velocity-inlet for the front boundary, and a special type of wall with no shear (named No Shear Wall inferior) for the inferior boundary, which represents the bottom of the physical domain (Figure 3). Different operating conditions were tested in this experiment, and some of them were mimicked in this computational study for the validation: ω = 424.5 rpm, U = 15 m·s^{-1} (which results in a TSR = λ = 6.6), and U = 10 m·s^{-1} (TSR = 10). Additionally, several other operating conditions regarding Free-Stream Velocity, TSR and Pitch Angle were simulated to characterize the wake aerodynamic behavior.

The physical domain was meshed using unstructured elements (Figure 4), which are suitable for CFD applications because of its good convergence rate. The mesh sensitivity study showed a total of approximately 10 million cell elements to be sufficient to accurately validate the model and describe the near wake (Appendix A). The meshing process consisted of a sphere of influence with 0.1m cell elements in a radial distance of 6 m surrounding the rotor, and a square part extending from $-0.5D$ to $3D$ with 0.25 m cell elements. The blade surface mesh was dimensioned using local edge sizing to reduce the skewness of the cells, resulting in 175 nodes spanwise and 75 nodes chordwise at the blade tip. Additionally, 10 inflation layers with a ratio of 1.1 were built to ensure y+ < 1 next to the blade surface. The physical domain needs to be large enough to result in a good simulation convergence, since pressure-far-field boundaries (lateral boundaries) require straight streamlines to avoid divergence for the residuals. However, the mesh at the exterior part surrounding the wind turbine and the rotor domain is coarse, since this region is not meaningful for the CFD analysis.

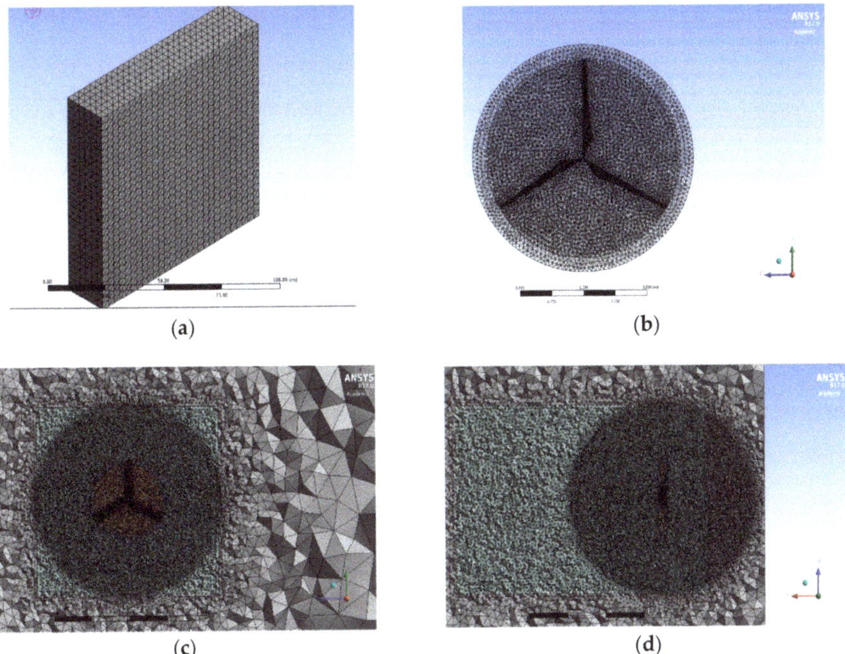

(a)

(b)

(c)

(d)

Figure 4. *Cont.*

Figure 4. Mesh of the Computational Domain. (**a**) Computational domain; (**b**) Rotational disc; (**c**) Front view: sectional plane showing details of the rotational central disc and the sphere of influence; (**d**) Lateral view: sectional plane showing details of the sphere of influence and the rectangular near wake region; (**e**) Sectional plane showing the mesh close to the blade surface; (**f**) Sectional plane showing inflation layers close to the blade surface.

2.4. TSR (λ) Effect on the Near Wake

A very important design parameter for wind farms is the TSR, which is defined as the ratio between the blade tip speed velocity and the free-stream velocity (Equation (1)). The TSR and other parameters such as free-stream velocity are critical to determine the wake behavior:

$$\lambda = \frac{\omega \cdot R}{U_{freestream}} \tag{1}$$

where ω is the rotor rotational speed, R is the blade radius and U is the free stream velocity.

Another important design parameter is the Turbulence Intensity (TI). This parameter can be calculated using the Equation (2):

$$TI = \frac{\sigma_U}{U_{freestream}} \tag{2}$$

where σ_U is the velocity standard deviation.

2.5. Wake Validation

The flow field at the wake of the rotor is validated by comparison between experimental [28] and computational data from the CFD simulation. The axial and radial traverses at the wake described in the section 1.2 (Figure 1) are considered for the validation. The MEXICO experimental dataset is an extensive one, with different turbine configurations (axial, yawed, azimuth angle) under multiple operating conditions (velocity, TSR, stand-still). In this work, specific conditions were selected for the validation: axial flow for U = 10 m·s⁻¹ and U = 15 m·s⁻¹. This decision is related to the complexity of the dataset, which would make it challenging to mimic using a steady-state CFD model. Rigorously, a CFD model would not only need to replicate velocity flow field but also tip vortex tracking agreement. There is no such a work in literature, the vast majority of the studies selected specific operational conditions and turbine configurations in order to narrow the scope of the research. Even though a wind speed of 24 m·s⁻¹ is available in the MEXICO experimental dataset, different near wall resolutions may have a considerable effect on the prediction the wake flow field. Particularly, the turbulence profile in the wake could be significantly affected by different orders of near wall resolution. As pointed out by Shen et al. [35], higher free stream wind speeds would require a more refined mesh for a more accurate prediction of the wake flow field.

2.6. Near Wake Analysis

Besides implementing and validating the CFD model of the MEXICO rotor, simulations were carried out considering variable operating conditions other than the ones analyzed in the original experiment. As a result of such simulations, a detailed study on near wake aerodynamics was developed to assess numerical sensitivity of the wake in regards to TSR, Pitch Angle and Free-Stream Velocity.

3. Results

3.1. Validation Dataset

As part of the model validation process, the loading on the blades from our CFD results were computed considering the total pressure on each blade (static and dynamic pressure), which produces an axial net force on the blade surfaces. The validation presented in Figure 5 is a result of the axial induction exerted through the rotating frame, which induces an axial net force on the blade and consequently a velocity deficit in the wake. The tangential loads presented represent loading in the chord wise direction. The comparison between computational and experimental data shows acceptable agreement for the two velocity values (Figure 5a,b). Figures 6 and 7 present graphs of the pressure coefficient on the blade surface, showing acceptable agreement between CFD results and experimental data from the MEXICO experiment. The pressure coefficients presented in Figures 6 and 7 are normalized, setting up a value of one at the stagnation point. The radial locations at r/R = 0.25 and r/R = 0.35 show less accuracy than the others because of the airfoil extrapolated data for this blade location. Figures 8 and 9 show the validation based on data for the wake velocity flow field. Figure 8 shows the validation of the axial traverse (R=1.8m) considering the free stream velocity = 15 m·s^{-1} and 10 m·s^{-1} at one radial and one axial downstream position: R = 1.8 m (axial) and x=0.3 m (radial). Figure 9 shows the validation for the radial traverse at 0.3 m downstream of the rotor, while considering free stream velocity of 15 m·s^{-1} and 10 m·s^{-1}. The computational results match the experimental data very well for the axial traverse at R = 1.8 m (Figure 8), and almost entirely match the radial traverse at x = 0.3 m (Figure 9). This demonstrates that this CFD model can accurately reflect the real rotor behavior. The computational results qualitatively agree with the experimental results; however, there are minor numerical discrepancies. Even though the velocity values do not completely overlap, the shape of the computational curve is very similar to the shape of the curve obtained with the experimental procedure.

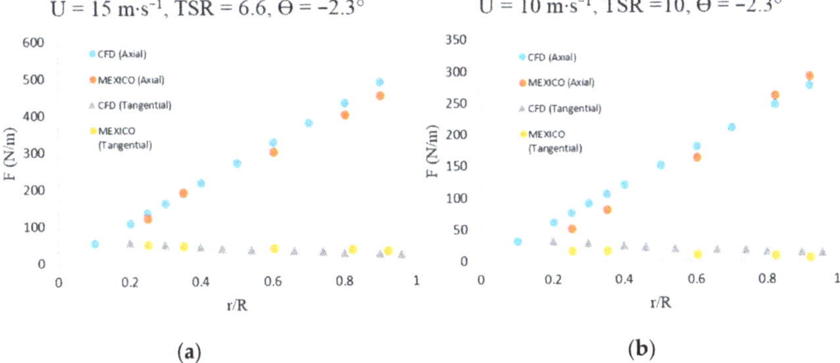

Figure 5. Axial and tangential forces on the rotor, showing comparison between computational and experimental results for (**a**) 15 m·s^{-1} and TSR = 6.6; (**b**) 10 m·s^{-1} and TSR = 10.

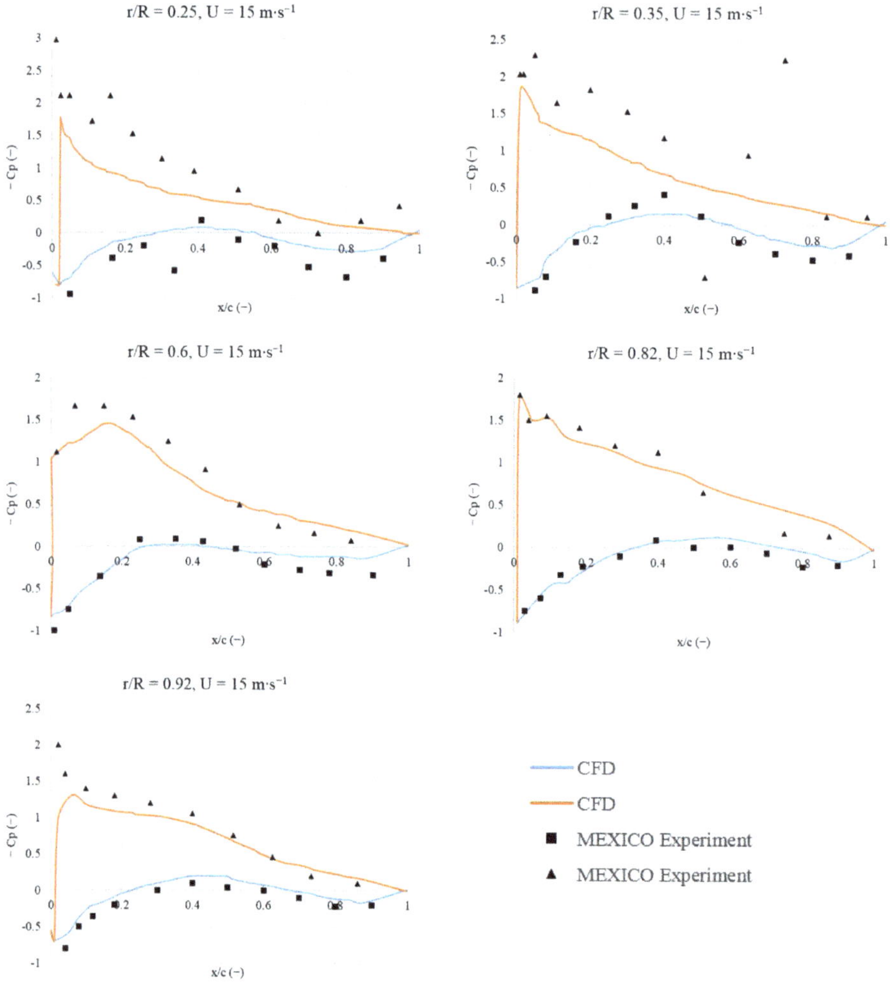

Figure 6. Normalized pressure coefficient on the blades for U = 15 m·s⁻¹, TSR = 6.6 and θ = −2.3°.

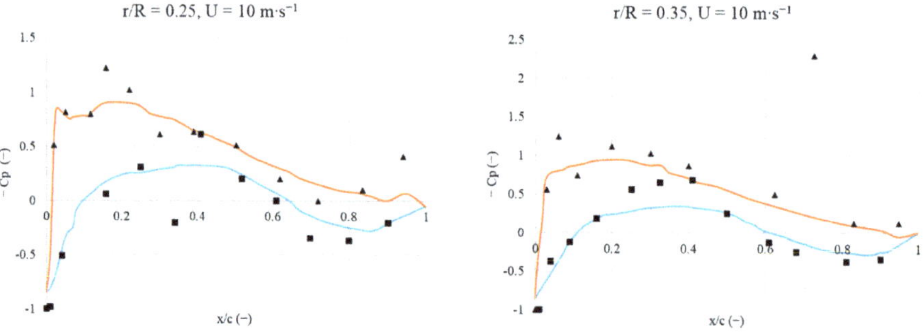

Figure 7. *Cont.*

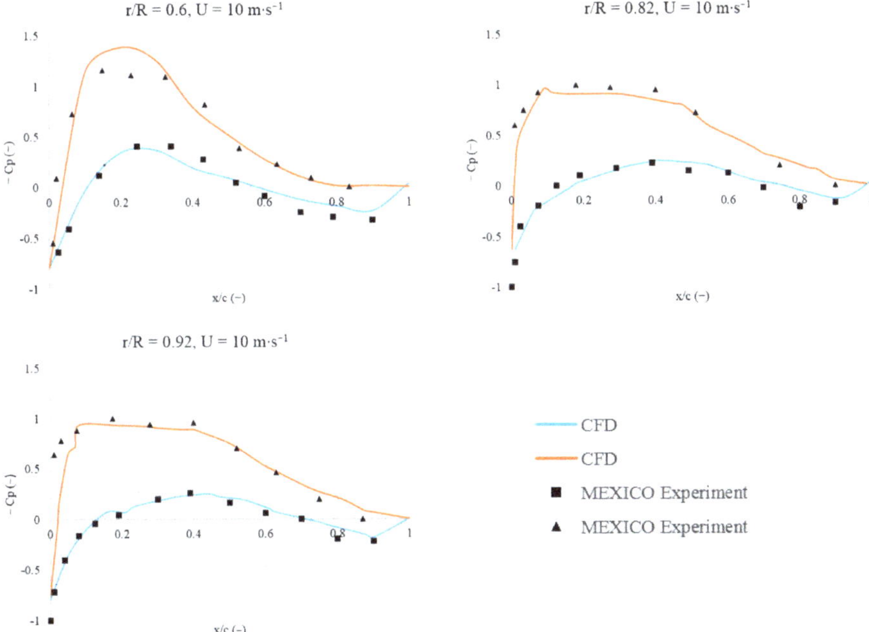

Figure 7. Normalized pressure coefficient on the blades for $U = 10$ m·s^{-1}, TSR = 10 and $\theta = -2.3°$.

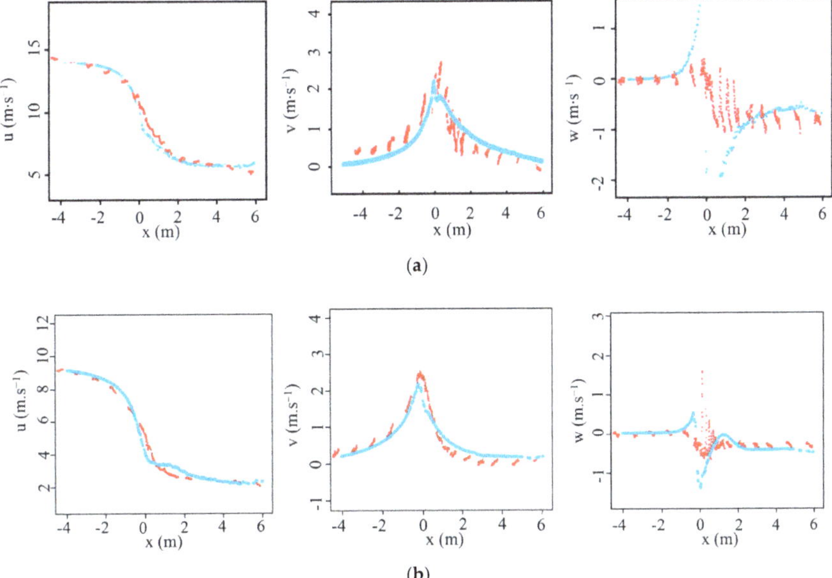

Figure 8. Validation dataset for an axial traverse at R = 1.8 m, showing comparison between computational and experimental data for (**a**) Free-Stream Velocity $U = 15$ m·s^{-1}; (**b**) Free-Stream Velocity $U = 10$ m·s^{-1}. The blue lines represent the computational data and the red lines represent experimental data.

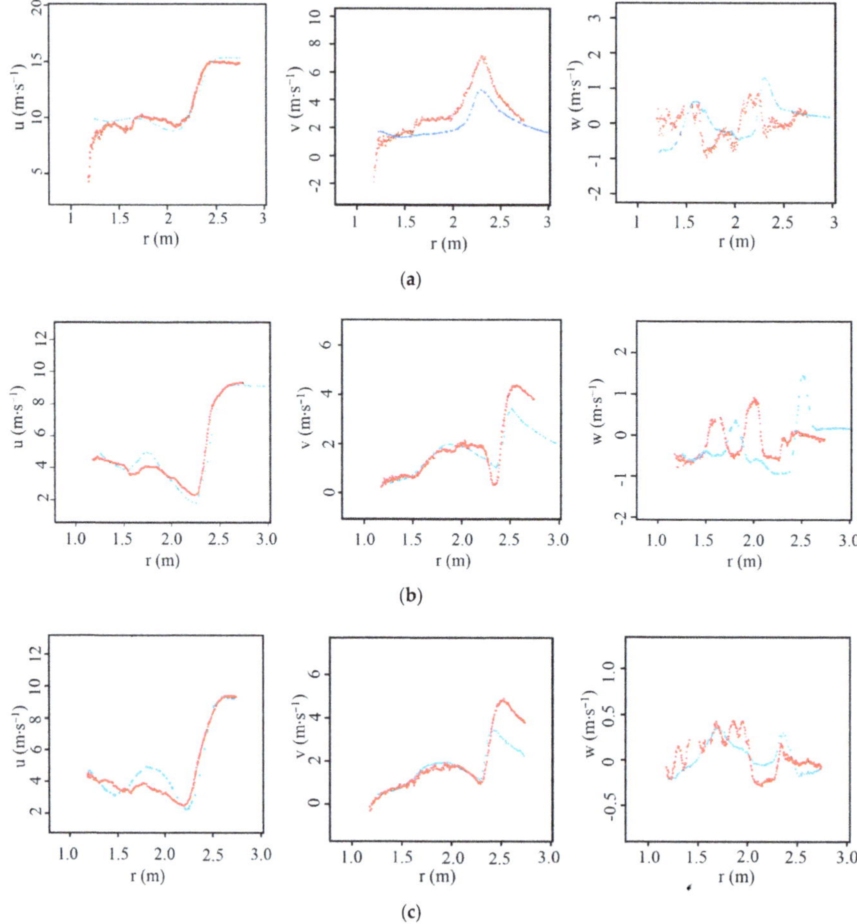

Figure 9. Validation of the radial traverse at x = 0.30 m, showing comparison between computational and experimental data for (**a**) U = 15 m·s^{-1}; (**b**) Free-Stream Velocity U = 10 m·s^{-1}; (**c**) Free-Stream Velocity U = 10 m·s^{-1} for Azimuth = 100°. The blue lines represent the computational data and the red lines represent MEXICO rotor experimental data.

A possible explanation for the minor discrepancies comes from the MRF approach utilized in the numerical method applied here, which assumes steady state behavior. This means that the Navier-Stokes equations are averaged by the Reynolds number. In spite of that, the simulation is suitable to determine how design parameters (such as TSR, velocity and pitch angle) affect the wake aerodynamic behavior.

3.2. TSR (λ) Effect on the Near Wake

3.2.1. Velocity Profile at the Near Wake

The near wake aerodynamic behavior is dependent on the rotor loading, which is dependent on the TSR. The rotor loading increases as the TSR increases, leading to an increase of the velocity deficit at the wake. Figure 10 shows the streamwise velocity-deficit evolution at five downstream positions in intervals of 0.5D, under different loading (or TSR) and upstream velocity conditions. The x-axis shows a radial traverse downstream of the rotor, while the y-axis shows the velocity at the wake. First of all, the axial induction increases as the rotor loading/TSR increases. As a consequence, the velocity deficit in the near wake increases as the rotor loading (or TSR) increases. A TSR = 6.6 results in a higher rotor loading and more produced power compared to a TSR = 4, thus extracting more energy from the incident wind. The shape of the curves with the same TSR is very similar, regardless of the incident upstream velocity. For a TSR = 6.6 and U = 10 m·s^{-1} (Figure 10), the velocity increases from approximately 4 m·s^{-1} at 1D downstream of the rotor to 7 m·s^{-1} at 3D downstream of the rotor, showing an increased rate of 1.5 m·s^{-1} for each diameter or 15% of the free-stream velocity per rotor diameter at the wake. From the perspective of the same analysis, but considering the case of TSR = 6.6 and U = 15 m·s^{-1}, the velocity increases from approximately 6 m·s^{-1} at 1D downstream of the rotor to approximately 11 m·s^{-1} at 3D downstream of the rotor. This corresponds to an increased ratio of 2.5 m·s^{-1} for each rotor diameter or approximately 15% of the free-stream velocity per rotor diameter at the wake. Moreover, the radial traverse right behind the rotor in Figure 11 shows an increase of 20% in the velocity deficit as the TSR varies from 6 to 10, corresponding to an increased ratio of approximately 5% m·s^{-1} per dimensionless unit of TSR.

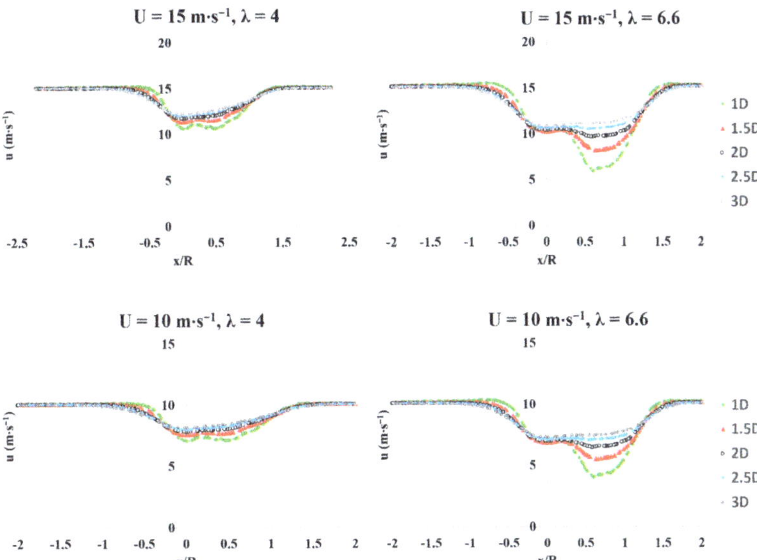

Figure 10. Wake development for two different velocity and TSR (λ) values.

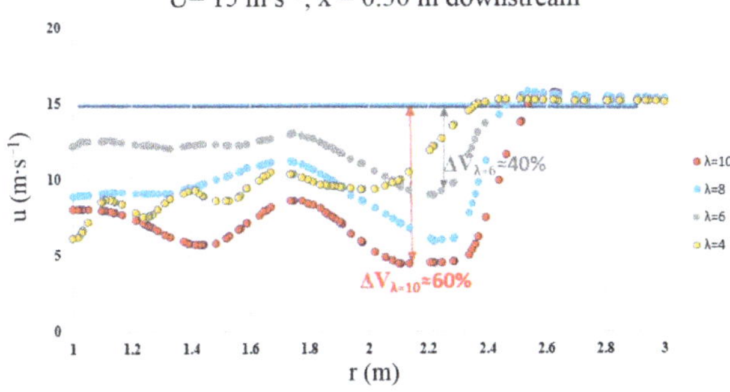

Figure 11. Axial Velocity profile for a radial traverse, and several TSR values.

3.2.2. Turbulence Intensity Profile at the Near Wake

Figure 12a shows a plot of the TI profile in the y-axis as a function of the radial position in the x-axis, for three free-stream velocity values. The first thing to notice is that the TI profile is relatively more symmetric in comparison to the velocity profile, especially for the downstream positions corresponding to 2D and 3D. Moreover, the TI reaches a maximum peak at a location right behind the rotor in the wake at 1D, decreasing through the wake for the subsequent radial positions of 2D and 3D (Figure 12a). This trend is observed for all the three free-stream velocities analyzed in this work. Additionally, when comparing the TI profile between 1D and 2D/3D it is also possible to see the wake expansion effects as the fluid flow develops in the wake: the shape of the curves is slightly tighter for 1D than for 2D or 3D. Furthermore, the TI peak increases as the free-stream velocity increases. When considering a downstream position of 1D (Figure 12a): the TI reaches a maximum value of 0.35 for U = 10 m·s^{-1}, while TI reaches a maximum peak of 0.65 for U = 15 m·s^{-1}, and finally TI reaches 0.90 maximum peak for U = 24 m·s^{-1}. This shows that there is a dependence of the TI behavior according to the free-stream velocity, and the same trend can be extended to the downstream positions of 2D and 3D. Figure 12b shows plots for the TKE as a function of the velocity and downstream distances (in rotor diameters) in the near wake. The TKE has some components: the advection by the mean flow, the transport by the vorticity, the TKE production, and the TKE dissipation. The TKE presents a similar trend observed in the TI, where the near wake immediately next to the rotor at 1D presents the TKE peak for all the velocities.

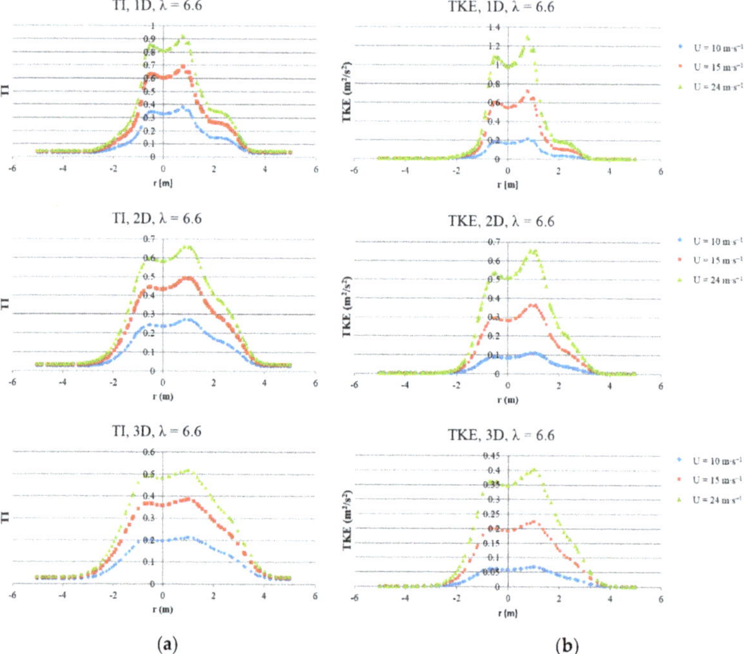

Figure 12. (**a**) Turbulence Intensity (TI) as a function of Velocity and downstream distances in the near wake; (**b**) Turbulence Kinetic Energy (TKE) as a function of Velocity and downstream distances in the near wake.

3.2.3. Pitch Angle (θ) Effect on the Near Wake

The Pitch Angle (θ) influences the near wake development in regards to the velocity deficit (Figure 13). The rotor design process aims to deliver the best aerodynamic performance according to the blade geometry (chord length, airfoil, rotor diameter), and a specific set of operating conditions. It is important to point out that the designed pitch angle for the MEXICO rotor blade is $\theta = -2.3°$, corresponding to a TSR of $\lambda = 6.6$ for $U = 15$ m·s^{-1} and $\omega = 424.5$ rpm. The pitch angle θ can significantly influence the near wake aerodynamic behavior. However, the far wake will not be significantly affected if the pitch angle is close to the designed condition. As can be seen by the axial velocity behavior (Figure 13), the velocity deficit is greater for negative pitch angle values than for positive values. This happens because in the case of the MEXICO rotor, negative pitch angle values are closer to the designed condition, thus extracting more energy from the incident wind. Consequently, the axial induction is greater for those pitch angle values close to the designed condition. Additionally, the velocity deficit increases as the pitch angle becomes more negative. This can be verified in Figure 13a,b, where a pitch angle of $-1°$ resulted in a smaller velocity deficit in comparison to a pitch angle of $-2.3°$ or $-3°$.

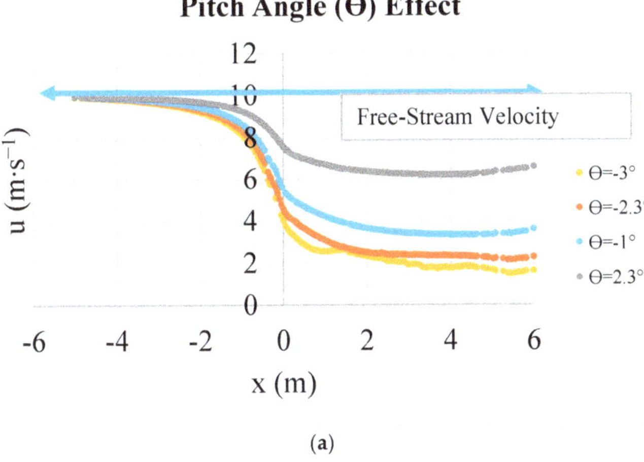

(a)

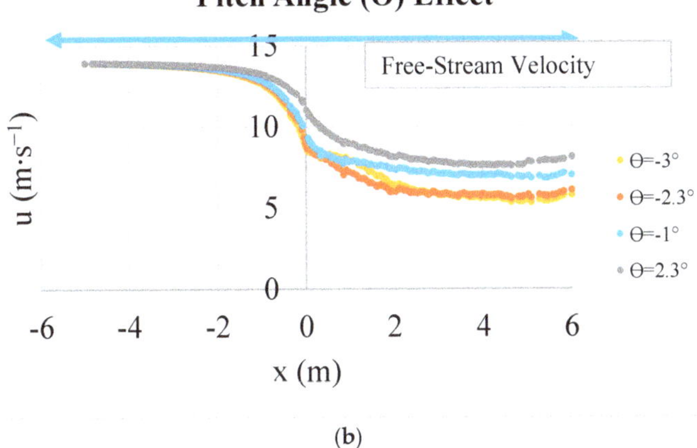

(b)

Figure 13. Influence of the pitch angle (θ) in the wake for: (**a**) $U = 10$ m·s^{-1}; (**b**) $U = 15$ m·s^{-1}. The designed pitch angle is $\theta = -2.3°$.

4. Discussion

Discrepancies between experimental and computational data were also verified in other studies. First of all, the type of experiment apparently plays an important role in regards to the discrepancies. Particle image velocimetry (PIV) is a technique very sensitive to experimental conditions. In the case of the MEXICO experiment, the light path close to the hub of the wind turbine can potentially disturb and induce the oscillations in the velocity profile observed in the traverse at R = 1.4 m. The problem with light reflection caused by the blade or the nacelle was also described by Carrión et al. [33], however the numerical discrepancies found in this study could be related to numerical reasons. Wimshurst and Willden [38] mentioned that the upstream axial free-stream velocity is lower in the MEXICO experiment than the computational simulation, arguing that the open tunnel configuration caused expansion of the streamtube between the wind tunnel nozzle and the collector, consequently causing smaller axial induction downstream of the rotor. The computed axial velocity was lower than the experimental axial velocity, which was explained by the greater force applied to the flow by the rotor. Shen et al. [35] observed that the computed axial free-stream velocity upstream of the rotor was 2.5% lower than the experimental (15 m·s^{-1}), and the discrepancies in the near wake were attributed to smaller thrust prediction. A potential contribution to discrepancies is attributed to the type of experiment (PIV measurements), which does instantaneous measurements containing fluctuations. Additionally, the wake fluctuation caused by the tip vortex could not be captured by the computational physical model employed in that study. The type of mesh refinement from Shen's study was claimed to be dependent on the upstream velocity, where a coarse mesh causes excessive dissipation. The sudden drop in velocity for the radial traverse at x = 0.3 m was attributed to the vortex shedding from the transition between the airfoils DU and Riso, and the intensity of the vortex was related to the change of circulation on the blade. Nilsson et al. [37] attributed the slightly overestimated axial velocity to the thrust, which was underestimated for all flow configurations. Furthermore, the light in the tunnel might have reflected on the turbine hub, affecting the experimental PIV measurements at the blade inboard radial position 0.62R (closer to the hub). Garcia et al. [36] found under-prediction of the thrust close to the blade root, attributed to rotational Coriolis effects and centrifugal forces in the boundary layer. Sorensen [41] found that the size of the nacelle influenced the inboard blade flow for yawed cases, so that the nacelle must also be included for accurate CFD modeling at the inboard region.

Our work, unlike previous efforts in literature, simulated the near wake of the MEXICO rotor within an extended downstream region including three diameters, while considering other TSR, free-stream velocity and pitch angle operating conditions. The same trend between axial induction and rotor loading was observed in other studies [27,31], in which the axial induction significantly increased from TSR = 4.2 to TSR = 10. Furthermore, the rotor loading influences the shape of the velocity profile at several downstream positions (Figure 10, Figure 11). While little perturbation to the velocity curves is observed for lower rotor loading, unsteady behavior/oscillation is present for higher rotor loading. The dependence of the velocity deficit on the streamwise distance is clearly more significant for higher TSR. These results agree with other studies in literature [46]. Figure 11 shows the radial traverse in the wake immediately behind the MEXICO rotor at x = 0.3 m, confirming the trend between loading and velocity deficit, even immediately adjacent to the rotor. Moreover, the tip vortices cause the region close to the blade tip to present the highest velocity deficit in comparison to the other blade radial locations; this will determine the wake expansion. Tari et al. [47] also found that the axial induction of horizontal axis wind turbines increases with the TSR, in which a maximum axial velocity deficit occurs between 0.75 < r/R < 0.9.

The TI aerodynamic behavior in the near and far wake was also characterized in previous studies. For instance, Shives & Crawford [48] found that the oscillating/fluctuating behavior is less significant for x/D > 5 in comparison to the near wake, and the curve shape becomes more similar to a Gaussian distribution. This trend was different in comparison to the velocity curve behavior, where the velocity curve starts to define its shape at x/D > 3. Chamorro et al. [49] investigated the effect of the Reynolds number on the wake characteristics, finding that the TI profile in the near wake is dependent on the

Reynolds number, and independent at approximately x/D = 4. It is pointed that the non-uniformity of the boundary layer influences the TI profiles to present relatively asymmetric distribution, which could also explain the asymmetric shape of the velocity profile (Figure 10). Additionally, the effect of the TI could still be observed even up to 12 rotor diameters downstream. Xie and Archer [50] found that the streamwise component of the Turbulence Kinetic Energy (TKE) is dominant for horizontal axis wind turbines. Turbulence Intensity contours showed that the streamwise component of the TI reaches a maximum at 5D, which extends up to approximately 15D, when it starts decaying. A low TI region happens immediately behind the rotor, which contradicts the TI trend behavior found in our study (Figure 9). Zhou et al. [23] investigated the influence of the inflow characteristics on the near wake of the NREL Phase IV, finding that the combination of inflow turbulence and wind shear can also have an impact on the turbulence generation in the near wake.

The pitch angle is proven to have impact on wind farms and wind many researchers. Markou et al. [51] showed that individual-pitch controllers allowed fatigue load reductions for offshore applications, while not significantly influencing the far wake behavior. Tests for a wake compensator resulted in a minimal reduction in average output power of 0.05% for 10D downstream distance. Kanev et al. [52] showed the benefits of using a pitch-based system for wind farms with turbine distances from 6D to 7D, in which 1% to 4% of the wake losses were regained yearly. Additionally, a lifetime extension of 1% was achieved by reducing fatigue loads. In the referred study, the wake loss reduction was found to be insensitive to a particular farm layout. Even higher benefits could be achieved by combining pitch-based and yaw-based wind direction wise systems, in which a pitch-based system would be operated for wind directions well-aligned with the rows of turbines, while yaw system would act as the wind comes at an angle in respect to the rows. Symmetrical layouts combining both systems could achieve almost the sum of the power production benefits of the two separate strategies.

Wake characteristics are closely related to the aerodynamic behavior of the blades. In this research, a numerical CFD model was developed based on MRF approach, which is a CFD technique where a reference frame rotates instead of the body itself. The MRF technique models the blade loading effect by applying an axial induction through the central disc. As a consequence of the axial induction exerted on the central disc, the velocity in the wake has a deficit in comparison with the free-stream velocity. Essentially, this is the same effect that blade loading induces on the wake, producing a velocity-deficit by extracting kinetic energy from the free-stream wind. The order of magnitude of the loading on the blades is a function of the thickness of the moving reference frame (rotational central disc), meaning that a thicker frame will have higher axial induction.

There are many different CFD modelling techniques suitable to mimic experiments. In this work, the main objective is to implement a steady-state CFD model for a quick evaluation of wake effects, aiming to create a computational tool to propose further improvements on the design of wind farm layout. It is important to emphasize that a Fluid-Solid Interaction (FSI) model of the MEXICO rotor is out of the scope of this work. A FSI model would elevate the computational expenses to the point of preventing the applicability of the model to evaluate multiple operational conditions in a reduced time. As we previously stated, the main objective of this research is to develop a computational tool capable of simulating wind turbine operation under variable operating conditions. The MRF technique is capable of representing the axial induction of the wake, allowing the simulation of variable wind operating conditions in a faster way. However, an accurate computation of the torque (consequently mechanical power) would require a FSI model. In these cases when steady state models using MRF approach are implemented, it is recommended to use a hybrid approach: the validated CFD results are utilized to evaluate the wake in terms of velocity deficit and turbulence flow field, and a computational model based on a BEM code could be used to compute the output power.

The modelling technique implemented in this study is the MRF. The model itself is an adaptation of the actuator-surface method: even though the full blade geometry was resolved using the CFD model, the solid blade geometry was suppressed from the rotating disc centrally located at the physical domain. The remaining model has the blade surface geometry represented as solid walls located in the central actuator disc, but the interior part of the blade (solid) was suppressed. A high number of cells (10 million) is necessary to achieve a validation within 5% agreement (see Appendix A) between computational and experimental data for the mesh layout designed for this simulation. Other studies in literature confirmed the necessity for a high number of cells to achieve validation within reasonable agreement [31,33,35,38,41,45]. When solving the full rotor blade geometry, the most efficient way of reducing computational expenses while still keeping a good level of agreement between computational and experimental data is to improve the layout of the mesh. For part I of this study, we focused on developing and validating a wind turbine CFD model, evaluating near wake characteristics under variable operating conditions other than the ones analyzed in the MEXICO experiment. In the second part of this research (Rodrigues and Lengsfeld [53] in review), the model from this study was adapted to analyze an extended wake region while still keeping a similar number of cells. The objective of part II is to develop a CFD model to analyze wind turbines interaction in wind farms. In order to do that, we broke the physical domain into smaller parts to locally design the mesh. This by itself represents an improvement of the mesh layout because there is a reduction in the number of cells relatively to the area analyzed. Further work could achieve an even better mesh layout design. Considering the extension of our analysis to a wind farm, a CFD technique was introduced in the second part of this research ([53], in review) in which a profile is created for the outlet of a first simulation (representing the first row of turbines), and then plugged into a new simulation (as an inlet) to model a hypothetical downstream row of turbines. Even though there is a need for running multiple simulations to analyze interaction effects, we eliminated the need for simulating two turbines rows at once. This could potentially allow users to simulate multiple rows of turbines while still using reduced resources in terms of computational capabilities. Therefore, to study wind farm layout, it would not be required an exorbitant number of cells, reducing the computational cost of these types of simulations.

Even though RANS model using k-ω SST is suitable for complex boundary layer flows under adverse pressure gradient and separation such as turbomachinery and external aerodynamics, separation is typically predicted to be excessive and early. This can reduce the suitability of the model for free shear flows, such as wakes. According to Sanderse et al. [22], RANS is more prevalent to engineering for modeling turbulence in the wake because of the computational expenses, even though eddy viscosity-based models are proved to be diffusive. A Reynolds-stress model based on LES would capture the rotational behavior of the wake, however a LES has to be run for a sufficiently long flow-time to obtain stable statistics of the flow being modeled. LES requires substantially finer meshes than those typically used for RANS models [54]. The computational cost with LES is typically orders of magnitude higher than the costs for steady RANS calculations in terms memory RAM and CPU time. Usually, high-performance computing (for instance, parallel computing) is necessary for LES applications. The main shortcomings of LES is the high resolution requirements for wall boundary layers, where the large eddies become relatively small, limiting LES for wall bounded flows to very ow Reynolds number (Re~10^4–10^5) ([54]).

The CFD technique implemented in this work considers k-ω SST to numerically model wake effects of the MEXICO rotor. The model does not take into consideration transient effects such as the LES model. Another distinctive capability of LES models is the possibility of introducing wind fluctuations for the inlet by using Reynolds Stress components. In despite of that, according to Rodrigo et al. [55] the application of CFD in wind resource assessment is still largely based on RANS models since LES or Detached Eddy Simulation (DES) models are still computationally expensive. According to Rodrigo et al. [55], the compromise between fidelity and cost for wind resource assessment is found with CFD models based on RANS simulations. Moriarty et al. [56] presented a comparison between several LES and RANS models in terms of accuracy on predicting wake deficit, considering the Sixberium and the Horns Rev and Lillgrund wind farms. The comparison showed that there is no apparent winner, as sometimes LES or RANS models have lower normalized average error than lower fidelity models, but often their error is higher. They concluded that more detailed insight into individual models and more experimental observations are required to provide better information about model accuracy. Delineated observations of wind farms under different operating and atmospheric conditions are required for providing better validation data instead of using averaged data over long periods of time. Such information will give the best practices regarding wake modelling and wind farm design, helping to quantify uncertainty bounds for different modelling tools and to determine useful quantities for validation in order to guide future measurement campaigns.

Reducing computational requirements for wake simulations is a tremendously demanding topic for wind energy research applicable for wind farms. The physical modeling implemented in this work essentially represents the forces using a rotating central disc, but careful work has been taken to correctly represent these forces. The model validation process is described throughout Section 3.1, including blade loading and near wake flow field. The full blade surface geometry was modeled, including twist angle, pitch angle and variable chord. This approach is similar to the actuator surface method. According to Sanderse et al. [22], the actuator surface approach is more accurate than the actuator disc and actuator line models. Still according to the same reference, even though the actuator disc method still remains the most widely used method for multiple wake simulation because of its reduced computational requirements, current research approaching the use of actuator surface technique has been evolved because of its relatively higher accuracy on wake modeling.

Extremely valuable work has already been carried out in literature to solve validation of near wake flow field using the MEXICO experimental dataset. In this current research, an extensive numerical effort has been performed to provide new insights related to near wake aerodynamics, which are crucial to understand wake characteristics and consequently to propose improvements to wind farm layout. The influence of some important design parameters on near wake aerodynamics has been determined, providing numerical estimates of wake profile. Such an extensive numerical effort specifically on near wake modeling had not been addressed in literature yet. The simulation of a range of pitch angle values provided a numerical estimate on the influence of this design parameter on the velocity deficit in the near wake. Additionally, a detailed study provided numerical estimates on the impact of TSR on velocity deficit and turbulence intensity/turbulent kinetic energy on the near wake. Furthermore, the near wake analysis in this work considered an extended near wake region: the original MEXICO experiment covered a near wake region up to 1.33 rotor diameters behind the rotor, while in this work the near wake analysis considers a length up to three rotor diameters. A previous work carried out by Bechmann et al. [31] considered an extended near wake region up to three diameters downstream of the MEXICO rotor, simulating the same operating conditions of the original experiment (one TSR value for each of the three velocities tested). Here in this work, the analysis considered different loading and pitch angle conditions, not only analyzing velocity flow field but additionally evaluating turbulent characteristics of the wake. Furthermore, the validation in this work implemented a particular type of lateral boundary condition never before applied for wind turbine CFD analysis: pressure-far-field. The implementation of pressure-far-field boundaries prevent the need for modeling tunnel lateral wall effects, allowing for coarse mesh at lateral boundaries, which are not meaningful for modeling experiments such as the MEXICO rotor (performed in open jet wind tunnel) or even to model turbines in natural field.

In part II of this work [53], an adaptation of CFD model validated in part I was carried out by extending the wake region to numerically model far wake effects. One of the novel aspects of part II is the application of a validated wind turbine CFD model to propose improvements for wind farm layout. The majority of previous works in this topic (wind farm layout optimization) rely on analytical models or non-validated CFD models. As pointed out by Rodrigo et al. [55], "Wind turbine wake aerodynamics is a topic of study that has attracted many researchers, which are divided into the ones studying rotor aerodynamics (near wake) and wind farm array efficiency (far wake). It is common sense that a more realistic description of the wake generation mechanisms in the near wake allows to understand and improve far wake models." In part II, a CFD technique that has never been applied before to solve wind turbine wakes interaction is introduced: we separately implement each wind farm row, creating a profile from each outlet. This profile is implemented as the inlet of a new simulation, allowing to simulate wake interaction effects. The technique allows to simulate multiple wind turbine rows with relatively reduced computational resources in terms of processers, since there is no need to simulate multiple turbines at once. Researchers may take benefit by using this technique to model wind farm rows while still considering wake interaction effects.

5. Conclusions

In this present work, a computational system was designed to analyze and optimize the operational conditions of a wind turbine and the flow field surrounding the rotor wake region. This work is intended to establish a computational framework from which to investigate wind farm layout, and to validate the simulation identifying parameters influencing the wake. The computational results match the selected experimental data for the radial and axial traverse in axial flow conditions. Even though there are minor numerical discrepancies, this CFD model is suitable to determine how design parameters (such as TSR, velocity, and pitch angle) affect the wake aerodynamic behavior. The level of agreement is very similar in comparison to those found in literature. Further improvements in the model could be achieved by refining the near wall resolution y+ for higher velocities, even though there is no variation of y+ in orders of magnitude for the velocity values tested in this work.

An extensive numerical effort has been performed in this research to provide new insights related to near wake aerodynamics, which are crucial to understand wake characteristics and consequently to propose improvements to wind farm layout. Such an extensive numerical effort specifically on near wake modeling had not been addressed in literature before. A detailed study provided numerical estimates on the impact of TSR on velocity deficit and turbulence intensity/turbulent kinetic energy on the near wake. CFD simulation demonstrates that the TSR and the pitch angle greatly influence the near wake behavior, affecting the velocity deficit and the turbulence intensity profile in this region. In the near wake region, the velocity deficit increases as the TSR increases, revealing an increase of 20% in the velocity deficit as the TSR varies from 6 to 10. This corresponds to an increased ratio of approximately 5% m·s^{-1} per dimensionless unit of TSR. The velocity in the wake increases at a rate of approximately 15% of the free-stream velocity per rotor diameter at the wake, regardless of the free-stream velocity applied. The TI peak increases as the free-stream velocity increases. Considering TSR = 6.6, a downstream position at 1D behind the rotor shows an increase of around 85% in the TI peak from U = 10 m·s^{-1} to U = 15 m·s^{-1}, and 40% from U = 15m·s^{-1} to U = 24 m·s^{-1}. This shows that there is a dependence of the TI behavior according to the free-stream velocity. The Pitch Angle can significantly influence the near wake aerodynamic behavior; however, the far wake will not be significantly affected if the pitch angle is close to the designed condition. Wake characteristics such as velocity deficit and TI could also be affected by the pitch angle, the TSR, and at further downstream distances. Our results give support to the notion that the near wake analysis is extremely relevant for the optimal positioning of wind turbines in a wind farm.

Author Contributions: Conceptualization, R.V.R. and C.L.; methodology, R.V.R and C.L.; software, R.V.R.; validation, R.V.R.; formal analysis, R.V.R.; investigation, R.V.R. and C.L.; writing—original draft preparation, R.V.R.; writing—review and editing, R.V.R. and C.L.; funding acquisition, R.V.R.

Funding: This research was funded by Conselho Nacional de Desenvolvimento Científico e Tecnológico (CNPq), grant number 249258/2013-7.

Abbreviations

TSR	Tip Speed Ratio
λ	Tip Speed Ratio
LES	Large Eddy Simulation
MEXICO	Model Experiments in Controlled Conditions
RANS	Reynolds Averaged Navier-Stokes
MRF	Moving Reference Frame
BEM	Blade Element Momentum
TI	Turbulence Intensity
VD	Velocity Deficit
θ	Pitch Angle
ω	Rotational Speed
r	Radius

Appendix A

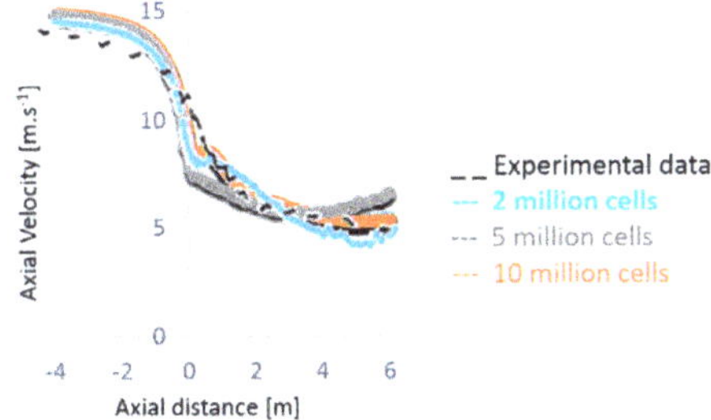

Figure A1. Mesh Sensitivity Study.

References

1. Park, J.; Law, K.H. Layout optimization for maximizing wind farm power production using sequential convex programming. *Appl. Energy* **2015**, *151*, 320–334. [CrossRef]
2. Son, E.; Lee, S.; Hwang, B.; Lee, S. Characteristics of turbine spacing in a wind farm using an optimal design process. *Renew. Energy* **2014**, *65*, 245–249. [CrossRef]
3. Wu, Y.-T.; Porté-Agel, F. Simulation of Turbulent Flow Inside and Above Wind Farms: Model Validation and Layout Effects. *Bound.-Layer Meteorol.* **2012**, *146*, 181–205. [CrossRef]
4. Stevens, R.J.A.M.; Hobbs, B.F.; Ramos, A.; Meneveau, C. Combining economic and fluid dynamic models to determine the optimal spacing in very large wind farms. *Wind Energy* **2017**, *20*, 466–477. [CrossRef]
5. Meyers, J.; Meneveau, C. Optimal turbine spacing in fully developed wind farm boundary layers. *Wind Energy* **2012**, *15*, 305–317. [CrossRef]
6. Park, J.; Law, K.H. A data-driven, cooperative wind farm control to maximize the total production. *Appl. Energy* **2016**, *165*, 151–165. [CrossRef]
7. González, J.S.; Payán, M.B.; Santos, J.R.; Rodríguez, Á.G.G. Maximizing the overall production of wind farms by setting the individual operating point of wind turbines. *Renew. Energy* **2015**, *80*, 219–229. [CrossRef]
8. Lee, J.; Son, E.; Hwang, B.; Lee, S. Blade pitch angle control for aerodynamic performance optimization of a wind farm. *Renew. Energy* **2013**, *54*, 124–130. [CrossRef]
9. Kazda, J.; Zendehbad, M.; Jafari, S.; Chokani, N.; Abhari, R.S. Mitigating adverse wake effects in a wind farm using non-optimum operational conditions. *J. Wind Eng. Ind. Aerodyn.* **2016**, *154*, 76–83. [CrossRef]
10. De-Prada-Gil, M.; Alías, C.G.; Gomis-Bellmunt, O.; Sumper, A. Maximum wind power plant generation by reducing the wake effect. *Energy Convers. Manag.* **2015**, *101*, 73–84. [CrossRef]
11. Chowdhurry, S.; Messac, J.Z.; Castillo, L. Unrestricted wind farm layout optimization (UWFLO): Investigating key factors influencing the maximum power generation. *Renew. Energy* **2012**, *38*, 16–30. [CrossRef]

12. Crespo, A.; Hernández, J.; Frandsen, S. Survey of Modelling Methods for Wind Turbine Wakes and Wind Farms. *Wind Energy* **1999**, *2*, 1–24. [CrossRef]

13. Adamarola, M.S.; Krogstad, P.-A. Experimental investigation of wake effects on wind turbine performance. *Renew. Energy* **2011**, *36*, 2078–2086. [CrossRef]

14. Chamorro, L.P.; Porté-Agel, F. A Wind-Tunnel Investigation of Wind-Turbine Wakes: Boundary-Layer Turbulence Effects. *Bound.-Layer Meteorol.* **2009**, *132*, 129–149. [CrossRef]

15. Mo, J.; Choudhry, A.; Arjomandi, M.; Lee, Y. Large eddy simulation of the wind turbine wake characteristics in the numerical wind tunnel model. *J. Wind Eng. Ind. Aerodyn.* **2013**, *112*, 11–24. [CrossRef]

16. Sturge, D.; Sobotta, D.; Howell, R.; While, A.; Lou, J. A hybrid actuator disc—Full rotor CFD methodology for modelling the effects of wind turbine wake interactions on performance. *Renew. Energy* **2015**, *80*, 525–537. [CrossRef]

17. Sarlak, H.; Meneaveau, C.; Sorensen, J.N. Role of subgrid-scal modeling in large eddy simulation of wind turbine wake interactions. *Renew. Energy* **2015**, *77*, 386–399. [CrossRef]

18. Houssain, M.Z.; Hirahara, H.; Nonomura, Y.; Kawahashi, M. The wake structure in a 2D grid installation of the horizontal axis micro wind turbines. *Renew. Energy* **2007**, *32*, 2247–2267. [CrossRef]

19. Vermeer, L.J.; Sorensen, J.N.; Crespo, A. Wind turbine wake aerodynamics. *Prog. Aerosp. Sci.* **2003**, *39*, 467–510. [CrossRef]

20. Medici, D.; Alfredsson, P.H. Measurements on a Wind Turbine Wake: 3D Effects and Bluff Body Vortex Shedding. *Wind Energy* **2006**, *9*, 219–236. [CrossRef]

21. Krogstadl, P.-Å.; Lund, J.A. An experimental and numerical study of the performance of a model turbine. *Wind Energy* **2011**, *15*, 443–457. [CrossRef]

22. Sanderse, B.; Van der Pijl, S.P.; Koren, B. Review of computational fluid dynamics for wind turbine wake aerodynamics. *Wind Energy* **2011**, *14*, 799–819. [CrossRef]

23. Zhou, N.; Chen, J.; Adams, D.E.; Fleeter, S. Influence of inflow conditions on turbine loading and wake structures predicted by large eddy simulations using exact geometry. *Wind Energy* **2016**, *19*, 803–824. [CrossRef]

24. Hsu, M.C.; Akkerman, I.; Bazilevs, Y. Finite element simulation of wind turbine aerodynamics: Validation study using NREL Phase VI experiment. *Wind Energy* **2014**, *17*, 461–481. [CrossRef]

25. Gundling, C.; Sitaraman, J.; Roget, B.; Masarati, P. Application and validation of incrementally complex models for wind turbine aerodynamics, isolated wind turbine in uniform inflow conditions. *Wind Energy* **2015**, *18*, 1893–1916. [CrossRef]

26. Mo, J.-O.; Choudhry, A.; Arjomandi, M.; Kelso, R.; Lee, Y.-H. Effects of wind speed changes on wake instability of a wind turbine in a virtual wind tunnel using large eddy simulation. *J. Wind Eng. Ind. Aerodyn.* **2013**, *117*, 38–56. [CrossRef]

27. Choudhry, A.; Mo, J.-O.; Arjomandi, M.; Kelso, R. Effects of wake interaction on downstream wind turbines. *Wind Eng.* **2014**, *38*, 535–548. [CrossRef]

28. Schepers, J.G.; Boorsma, K.; Cho, T.; Gomez-Iradi, S.; Schaffarczyk, P.; Jeromin, A.; Shen, W.Z.; Lutz, T.; Meister, K.; Stoevesandt, B.; et al. Final Report of IEA Task 29, Mexnet (Phase 1): Analysis of Mexico Wind Tunnel Measurements. Available online: https://www.um.edu.mt/library/oar//handle/123456789/24132 (accessed on 11 March 2019).

29. Yang, H.; Shen, W.Z.; Sorensen, J.N.; Zhu, W.J. Extraction of airfoil data using PIV and pressure measurements. *Wind Energy* **2011**, *14*, 539–556. [CrossRef]

30. Xudong, W.; Shen, W.Z.; Jin, C. Shape optimization of wind turbine blades. *Wind Energy* **2009**, *12*, 781–803. [CrossRef]

31. Bechmann, A.; Sørensen, N.N.; Zahle, F. CFD simulations of the MEXICO rotor. *Wind Energy* **2011**, *14*, 677–689. [CrossRef]

32. Micallef, D.; Bussel, G.; Ferreira, C.S.; Sant, T. An investigation of radial velocities for a horizontal axis wind turbine in axial and yawed flows. *Wind Energy* **2013**, *16*, 529–544. [CrossRef]

33. Carrión, M.; Steijl, R.; Woodgate, M.; Barakos, G.; Munduate, X.; Gomez-Iradi, S. Computational fluid dynamics analysis of the wake behind the MEXICO rotor in axial flow conditions. *Wind Energy* **2015**, *18*, 1023–1045. [CrossRef]

34. Herraez, I.; Medjroubi, W.; Stoevesandt, B.; Peinke, J. Aerodynamic Simulation of the MEXICO Rotor. *J. Phys. Conf. Ser.* **2014**, *555*, 012051. [CrossRef]

35. Shen, W.Z.; Zhu, W.J.; Sørensen, J.N. Actuator line/Navier–Stokes computations for the MEXICO rotor: Comparison with detailed measurements. *Wind Energy* **2012**, *15*, 811–825. [CrossRef]

36. Ramos-García, N.; Hejlesen, M.M.; Sørensen, J.N.; Walther, J.H. Hybrid vortex simulations of wind turbines using a three-dimensional viscous-inviscid panel method. *Wind Energy* **2017**, *20*, 1871–1889. [CrossRef]

37. Nilsson, K.; Shen, W.Z.; Sørensen, J.N.; Breton, S.-P.; Ivanell, S. Validation of the actuator line method using near wake measurements of the MEXICO rotor. *Wind Energy* **2014**, *18*, 499–514. [CrossRef]

38. Wimshurst, A.; Willden, R.H.J. Extracting lift and drag polars from blade-resolved computational fluid dynamics for use in actuator line modelling of horizontal axis turbines. *Wind Energy* **2017**, *20*, 815–833. [CrossRef]

39. Zhong, H.; Du, P.; Tang, F.; Wang, L. Lagrangian dynamic large-eddy simulation of wind turbine near wakes combined with an actuator line method. *Appl. Energy* **2015**, *144*, 224–233. [CrossRef]

40. Guntur, S.; Sørensen, N.N. A study on rotational augmentation using CFD analysis of flow in the inboard region of the MEXICO rotor blades. *Wind Energy* **2015**, *18*, 745–756. [CrossRef]

41. Sorensen, N.N.; Bechmann, A.; Rethore, P.-E.; Zahle, F. Near wake Reynolds-averaged Navier-Stokes predictions of the wake behind the MEXICO rotor in axial and yawed flow conditions. *Wind Energy* **2014**, *17*, 75–86. [CrossRef]

42. Tsalicoglou, C.; Jafari, S.; Chokani, N.; Abhari, R.S. RANS Computations of MEXICO Rotor in Uniform and Yawed Inflow. *J. Eng. Gas Turbines Power* **2013**, *136*, 01102. [CrossRef]

43. Grasso, F.; Garrel, A. Near Wake Simulation of Mexico rotor in Axial and Yawed Flow Conditions with Lifting Line Free Wake Code. In Proceedings of the Wake Conference, Visby, Sweden, 8–9 June 2011.

44. Shen, W.Z.; Zhu, W.J.; Yang, H. Validation of the Actuator Line Model for Simulating Flows past Yawed Wind Turbine Rotors. *J. Power Energy Eng.* **2015**, *3*, 7–13. [CrossRef]

45. Réthoré, P.-E.M.; Zahle, F.; Sørensen, N.N.; Bechmann, A. CFD Simulations of the Mexico Wind Tunnel and Wind Turbine. In Proceedings of the European Wind Energy Association (EWEA), Brussels, Belgium, 14–17 March 2011; pp. 1–7.

46. Krogstad, P.-Å.; Adaramola, M.S. Performance and near wake measurements of a model horizontal axis wind turbine. *Wind Energy* **2012**, *15*, 743–756. [CrossRef]

47. Tari, P.H.; Siddiqui, K.; Hangan, H. Flow characterization in the near-wake region of a horizontal axis wind turbine. *Wind Energy* **2016**, *19*, 1249–1267. [CrossRef]

48. Shives, M.; Crawford, C. Adaped two-equation turbulence closures for actuator disk RANS simulations of wind & tidal turbine wakes. *Renew. Energy* **2016**, *92*, 273–292.

49. Chamorro, L.P.; Arndt, R.E.A.; Sotiropoulos, F. Reynolds number dependence of turbulence statistics in the wake of wind turbines. *Wind Energy* **2012**, *15*, 733–742. [CrossRef]

50. Xie, S.; Archer, C. Self-similarity and turbulence characteristics of wind turbine wakes via large-eddy simulation. *Wind Energy* **2015**, *18*, 1815–1838. [CrossRef]

51. Markou, H.; Andersen, P.B.; Larsen, G.C. Potential load reductions on megawatt turbines exposed to wakes using individual-pitch wake compensator and trailing-edge flaps. *Wind Energy* **2011**, *14*, 841–857. [CrossRef]

52. Kanev, S.K.; Savenije, F.J.; Engels, W.P. Active wake control: An approach to optimize the lifetime operation of wind farms. *Wind Energy* **2018**, *21*, 488–501. [CrossRef]

53. Rodrigues, R.V.; Lengsfeld, C.S. Development of a computational system to improve wind farm layout, Part II: Wind Turbine Wakes Interaction. *Energies* **2019**. under review.

54. ANSYS Fluent User Guide. Available online: http://support.ansys.com (accessed on 19 January 2019).

55. IEA Wind Task 31 WAKEBENCH 2 (2015–2018)—Wakebench Model Evaluation Protocol for Wind Farm Flow Models. Available online: https://windbench.net/wakebench2 (accessed on 11 February 2019).

56. Moriarty, P.; Rodrigo, J.S.; Gancarski, P.; Chuchfield, M.; Naughton, J.W.; Hansen, K.S.; Machefaux, E.; Maguire, E.; Castellani, F.; Terzi, L.; et al. Iea-task 31 wakebench: Towards a protocol for wind farm flow model evaluation. Part 2: Wind farm wake models. *J. Phys. Conf. Ser.* **2014**, *524*, 012185. [CrossRef]

Article

Wind Energy Prediction in Highly Complex Terrain by Computational Fluid Dynamics

Daniel Tabas [1], Jiannong Fang [2,*] and Fernando Porté-Agel [2]

[1] Department of Mechanical Engineering, The Johns Hopkins University, Baltimore, MD 21218, USA; dtabas1@jhu.edu

[2] Wind Engineering and Renewable Energy Laboratory (WiRE), School of Architecture, Civil and Environmental Engineering (ENAC), École Polytechnique Fédérale de Lausanne (EPFL), 1015 Lausanne, Switzerland; fernando.porte-agel@epfl.ch

* Correspondence: jiannong.fang@epfl.ch; Tel.: +41-21-693-57-01

Received: 18 February 2019; Accepted: 31 March 2019; Published: 5 April 2019

Abstract: With rising levels of wind power penetration in global electricity production, the relevance of wind power prediction is growing. More accurate forecasts reduce the required total amount of energy reserve capacity needed to ensure grid reliability and the risk of penalty for wind farm operators. This study analyzes the Computational Fluid Dynamics (CFD) software WindSim regarding its ability to perform accurate wind power predictions in complex terrain. Simulations of the wind field and wind farm power output in the Swiss Jura Mountains at the location of the Juvent Wind Farm during winter were performed. The study site features the combined presence of three complexities: topography, heterogeneous vegetation including forest, and interactions between wind turbine wakes. Hence, it allows a comprehensive evaluation of the software. Various turbulence models, forest models, and wake models, as well as the effects of domain size and grid resolution were evaluated against wind and power observations from nine Vestas V90's 2.0-MW turbines. The results show that, with a proper combination of modeling options, WindSim is able to predict the performance of the wind farm with sufficient accuracy.

Keywords: wind energy; computational fluid dynamics; complex terrain; model validation

1. Introduction

In recent years, wind flow simulations have gained popularity for wind energy applications, including wind resource assessment, wind power prediction, and wind turbine micro-siting [1]. Compared to field measurements, simulations offer high-resolution three-dimensional wind fields without the need for costly meteorological equipment. Originally, linear models such as the one implemented in the Wind Atlas Analysis and Application Program (WAsP) were used because of their efficiency and their sufficient accuracy over terrain with gentle slopes [2]. However, increased computational capacity combined with a need for more accurate predictions of wind flow over complex terrain have made Computational Fluid Dynamics (CFD) models both practical and necessary. Most simulations solve the steady Reynolds-Averaged Navier–Stokes (RANS) equations, which are time independent and which provide the statistics for wind velocity at each grid point [3]. Other CFD simulation techniques that have higher accuracy, but higher computational cost are also being developed to analyze wind flow patterns and wind farm performance. These time-dependent turbulence-resolving methods include Large-Eddy Simulation (LES) and Direct Numerical Simulation (DNS). LES uses a low-pass spatial filter to average out turbulence at small length scales. In this method, the computationally-expensive calculation of small turbulent structures is replaced by sub-grid-scale modeling. One example of an LES method for wind farm modeling can be found in Porté-Agel et al. [4].

DNS involves solving the full nonlinear Navier–Stokes equations, but is too computationally expensive to be applied in large applications such as wind farms.

WindSim, a CFD package for wind resource assessment and park optimization, has been used and evaluated in both industrial settings and academia. Several groups compared the results of WindSim to WAsP over complex terrain and found better performance in the CFD model WindSim [2,5,6]. Other studies validated the WindSim results against measurements without comparing to linear models [3,7–9]. Castellani et al. [8,10] evaluated turbine wake modeling in wind farms with complex terrain, compared results with on-site measurements, and studied the wake effects together with the terrain effects on the performance of wind farms. Cattin et al. [7] validated the use of WindSim over areas with heterogeneous land cover, but found that implementing a map of roughness lengths did not fully reproduce the effects of forested areas. Dhunny et al. [3] validated the application of WindSim in an island situation using two roughness lengths, one for land and one for sea. Waewsak et al. [9] applied WindSim to a wind resource assessment study in Thailand and found good agreement between simulation results and met mast measurements. Finally, Teneler [11] evaluated the forest model in WindSim and found that modeling the forest as a porous medium improved simulation accuracy in heterogeneous forested regions.

The aim of the present study is to perform a more comprehensive evaluation of the WindSim software taking into account the combination of three complexities: topography, heterogeneous surface cover varying between grassy and forested, and turbine wakes. To accomplish this, we applied WindSim to a case study of a wind farm in the Jura Mountains of Switzerland, for which field data are available. We first performed convergence tests for the simulation domain size and grid resolution. We then investigated WindSim's sensitivity to the forest model, the turbulence model, and the wake model.

2. Methods

2.1. Study Site and Data

The Juvent wind farm in the Jura Mountains of Switzerland contains 16 wind turbines, twelve 2-MW Vestas V90's, and four 3.3-MW Vestas V112's, with a 95-m hub height. The turbines have 90-m and 112-m rotor diameters, respectively. The turbines are situated on two hills, Mont Soleil (alt. 1291 m) and Mont Crosin (alt. 1268 m), where surface cover varies from grassy to forested (Figure 1).

Wind measurements were taken at the nacelle of each turbine from the period 15 January–11 February 2016. Data collected include wind speed, wind direction, power, yaw offset, and temperature. Each measurement was recorded as the mean over a 10-min interval. The standard deviation of wind speed over each interval was also recorded.

For the purposes of this evaluation study, only the predominant wind direction (i.e., 240° as shown in Figure 2) was simulated. Turbines 5–8 were excluded because they were shut down for replacement during the period of analysis. In addition, we focused on the cluster of nine turbines located on the hill of Mont Crosin (Figure 1). Thus, Turbines 9, 15, and 16 were not considered in the simulations because they are far away from the nine turbines and their influences on the flow in the area of interest is negligible. Data of the nine turbines were filtered to an average wind direction of $240 \pm 3°$ and a wind speed range of 8–9 m/s as measured at Turbine 2, the farthest upstream turbine in the cluster. Wind speeds and turbine power outputs were normalized with the measurements at Turbine 2. The normalized results were then averaged over the filtered dataset because, in order to compare simulation results to observed data, we needed a single average measurement for wind speed and power at each turbine. Since Coriolis forces were assumed to be negligible in this study, normalization using linear scaling is valid [2]. Using normalized data from a certain range of wind conditions (hence, a larger dataset) allowed obtaining robust statistical results for a fair comparison.

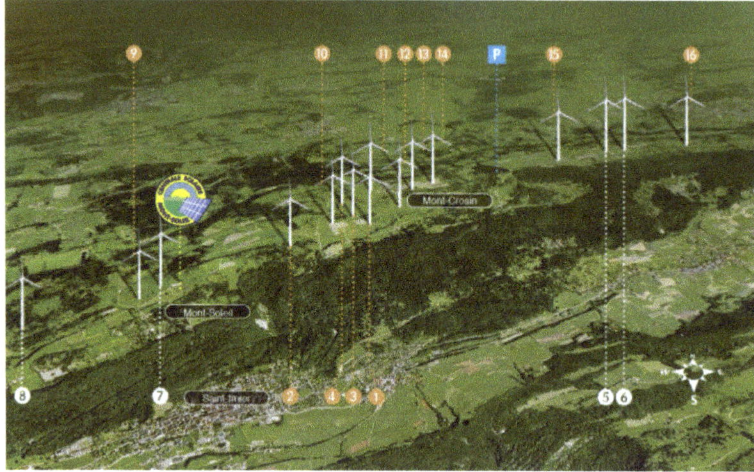

Figure 1. Location of the study site in the Swiss Jura Mountains and numbering of the turbines on the Juvent wind farm (source: www.juvent.ch).

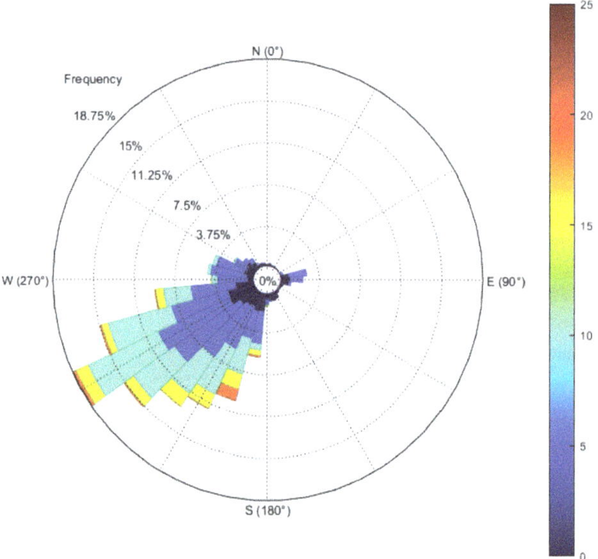

Figure 2. Wind rose at the Juvent wind farm in the Swiss Jura Mountains, 15 January–11 February 2016.

Wind fields in mountainous regions are highly turbulent and are strongly modulated by local, nonlinear interactions with multi-scale surface heterogeneities. The complex land features of interest include both mountainous terrain and heterogeneous vegetation. In this case study, the forest-grassland mosaics of the Jura mountains exhibit land cover whose effects on wind flow are difficult to model accurately. To apply the CFD tools under such complex surface conditions, we needed to feed them with high-resolution data of the relevant surface properties. The high-resolution data of the topography is directly used as input to the CFD tools to determine the surface elevation for the generation of the computational grid. The high-resolution data of the vegetation cover can be used to estimate the surface roughness length in the similarity theory-based wall model and the parameters in the forest modeling. Forest can be modeled either directly through introducing additional forcing terms in the momentum equations or indirectly through the wall model with a high roughness length and displacement height.

In this study, elevation data at 25-m resolution were acquired from the Swiss topographical database through the www.geodata4edu.ch interface developed by the Swiss Federal Institute of Technology in Zurich (ETHZ). Land cover data at 25-m resolution were acquired from the CORINE land cover database, developed by the European Environment Agency, which classifies land cover into 44 different categories and provides the corresponding roughness length for each. Roughness length is a parameter of the vertical log-law profile that models the horizontal mean wind speed near the rough surface. It is equivalent to the height at which the wind speed theoretically becomes zero. As input to the model, we extracted from the elevation and roughness length maps a rectangular domain oriented towards the predominant wind direction (Figure 3). This ensures that the wind profile is allowed to develop over the same distance from every starting point along the inflow boundary. The dimensions of the domain were determined in a convergence test as 19 km × 5 km in the streamwise and spanwise directions, respectively, with 9-km spacing between the upstream border and Turbine 2. The elevation and roughness length presented in Figure 3 show that the Juvent wind farm is located in a highly-complex terrain. Patches with the roughness length value higher than 1 m are identified as forests, which are shown in dark red in the bottom panel of Figure 3.

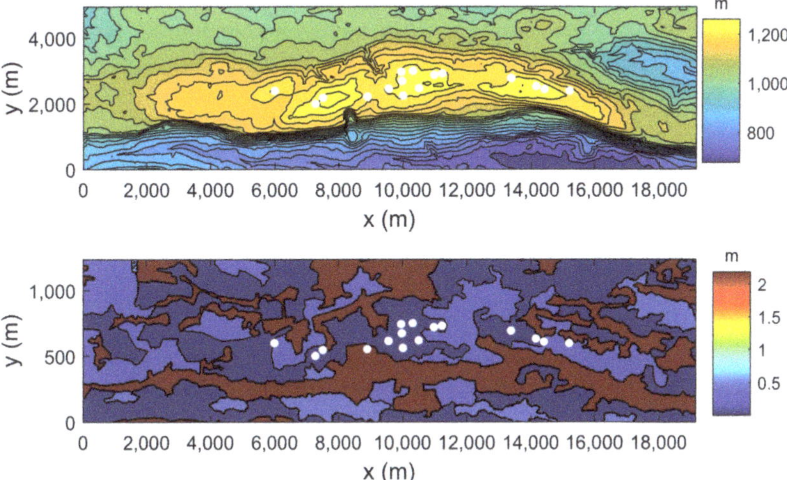

Figure 3. Elevation (**top**) and roughness length (**bottom**) of the area of interest. The turbines are also presented in white circles.

2.2. WindSim

WindSim is a commercial CFD package that simulates flow over wind farms in complex terrain. The program solves the steady Reynolds-Averaged Navier–Stokes (RANS) equations using a two-equation turbulence closure model. Steady RANS simulates time-averaged flow fields assuming a statistically-stationary condition. This study focuses on the evaluation of this approach and its associated models by comparing simulation results with time-averaged field measurements. When a time average is taken, transient phenomena are smoothed out and become invisible. Hence, unsteady effects such as time-varying large-scale atmospheric forcing, topography-induced vortex shedding, and turbine wake meandering cannot be captured by steady RANS. More advanced methods (e.g.,

LES models) capable of capturing the unsteady effects are still too computationally expensive for commercial use in wind energy applications.

The Standard $k - \varepsilon$ model (STD), the modified $k - \varepsilon$ model of the Re-Normalization Group (RNG) version [12], and the $k - \omega$ turbulence model of Wilcox [13] were considered in this study. For flow over complex terrain, some studies showed that the RNG $k - \varepsilon$ model produced promising results, a finding corroborated by Peralta et al. [14], while some other studies showed the superiority of the Wilcox $k - \omega$ model over the STD and RNG models [15]. Capturing the effects of forest is essential for this case study. In WindSim, forest can be modeled by the indirect approach mentioned before or by including porous cells with momentum sinks and turbulence sources [16] in the computational grid for areas that include forest. The latter is called the forest model. Our testing results (not shown) indicated that the forest model yields more realistic results than the indirect approach does. Turbine wake effects can be simulated directly by the use of an Actuator Disc (AD) model [17]. However, in WindSim, the AD model cannot be activated together with the forest model. Hence, in this study, turbine wake effects were modeled through the analytical approach. WindSim has implemented the analytical wake models from Jensen, Larsen, and Ishihara [18]. The accuracy of the three models to predict the observed power production was evaluated.

WindSim can optionally account for atmospheric stability by additionally solving the temperature equation. However, this feature requires several inputs that were not available from the measured data. Instead, we validated the assumption of a neutral boundary layer by examining the measurements. Within the selected range of wind direction and speed, we further filtered by time of day, keeping only wind events from dusk and dawn, when the atmosphere was assumed to be neutral. Comparison with the full dataset showed no significant change in time-averaged wind behavior at any of the turbines. We therefore concluded that the assumption of a neutral boundary layer for the WindSim simulations was valid for the model evaluation study here.

2.3. Boundary Conditions and Numerical Settings

The computational domain, surface elevation data, and turbine locations are shown in Figure 4. For each simulation case, the domain was rotated to make the x-axis along the prevailing wind direction, so there was only one inlet (at $x = 0$) and one outlet (at $x = L_x$). At the inlet, boundary conditions are given as fully-developed flow profiles taking into account the given roughness at the border and the boundary-layer height L_B [19]. For the wind speed, the well-known logarithmic profile is defined from the ground up to L_B, and above this height, the profile is constant. Here, L_B was set to 1000 m above the mean surface elevation, and the constant speed above L_B was set to 15 m/s so that the simulated wind speed at Turbine 2 was around 8.5 m/s, which is the median of the wind speed range applied to filter the data. At the outlet, zero gradient boundary conditions are imposed, meaning that a zero diffusion flux for all flow variables is assumed. On the lateral sides, symmetric conditions are applied. The upper boundary condition is specified as fixed pressure. The bottom boundary condition is no penetration together with the equilibrium log-law wall functions.

WindSim uses a Cartesian grid in the horizontal plane and terrain-following grid points in the vertical direction with tighter spacing closer to the ground level. The number of vertical grid points was set to the maximum (60). Test simulations with four different numerical settings as detailed in Table 1 were performed. Since the purpose of those simulations was to check the convergence of numerical results with regard to grid resolution and domain size, wake effects were not considered, and forest was modeled by the less expansive indirect approach.

For the evaluation simulations, the forest model was used. The height of the forest was set to 20 m, which is the mean height of the trees in the region according to a survey [20]. The number of grid cells in the vertical direction for modeling the forest was set to five, corresponding to $d_z^{min} = 4$ m. According to the table in WindSim, the forest resistive force constant C2 was set to 0.01, twice that of the default value, because the forest at Mont Crosin was sparse, but dominated by *Picea abies* and *Abies alba*, which are evergreen coniferous trees with a higher leaf area index. The influence of C2 on the results is presented in the next section.

Table 1. Information of the numerical settings.

Numerical Setting	Domain Size (L_x, L_y, L_z)	Grid Cells (n_x, n_y, n_z)	Resolution (d_x, d_y, d_z^{min})
S1	19 km, 5 km, 7.5 km	380, 100, 60	50 m, 50 m, 11.4 m
S2	27 km, 7 km, 7.5 km	540, 140, 60	50 m, 50 m, 8.4 m
S3	19 km, 5 km, 7.5 km	760, 200, 60	25 m, 25 m, 6.0 m
S4	19 km, 5 km, 7.5 km	190, 50, 60	100 m, 100 m, 11.4 m

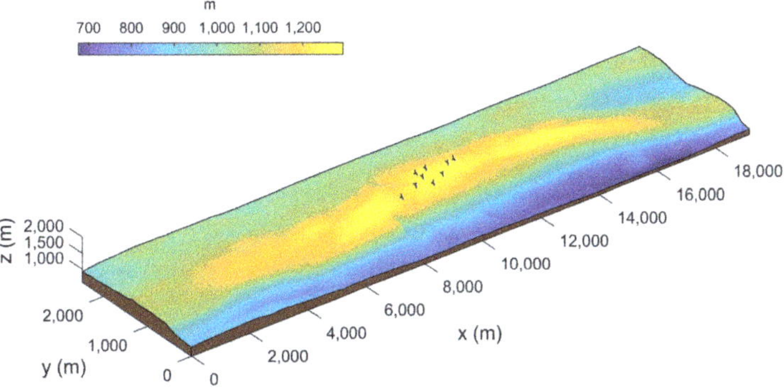

Figure 4. The computational domain for the WindSim simulations. Here, z is the elevation from sea level in m and z_s is the surface elevation.

3. Results and Discussion

Figure 5 shows the normalized wind speeds at the hub height of turbines predicted by WindSim using the standard $k - \varepsilon$ model with the four different numerical settings defined in Table 1. It can be seen that the differences between the results of these numerical tests (except for S4, which had the coarsest grid resolution) were rather small. This indicates that the results presented in the following with the numerical setting S1 did not depend on the domain size and grid resolution. S1 (medium grid resolution) was ultimately chosen because it produced results much faster than using S3 (fine grid resolution). Furthermore, the nesting technique (using the results from a larger outer model with coarser resolution as boundary conditions for flow simulation over a smaller domain with higher resolution) was also tested for S1. The nested simulations did not further change the results (not shown). Therefore, the influence of inaccuracies in the assumed boundary conditions on the results of S1 can be regarded as negligible.

Normalized turbine power outputs predicted by WindSim using three different turbulence models are compared with the wind farm SCADA data in Figure 6. Here, the analytical wake model of Ishihara was used, and the effect of multiple wakes was modeled by the linear superposition of the wake deficits. The predicted power outputs were obtained for three incoming wind directions (237°, 240°, and 243°) and different wind speeds (around 8.5 m/s) at the reference turbine, then averaged to yield the mean values and standard deviations (error bars). It is shown that the results of Wilcox were all within the error bars of the data, while the results of STD and RNG largely under-predicted the power outputs of Turbines 1, 11, and 12. To have a quantitative measure of the model performance, the Root Mean Squared Error (RMSE) and Mean Bias (MB) were calculated as follows:

$$\text{RMSE} = \sqrt{\frac{1}{N} \sum_{n=1}^{N} (P_s^n - P_o^n)^2} \tag{1}$$

$$\text{MB} = \frac{1}{N} \sum_{n=1}^{N} (P_s^n - P_o^n) \tag{2}$$

where P_s^n is the simulated mean power output of turbine n, P_o^n is the observed mean power output of turbine n, and N is the number of turbines used for comparison. RMSE was 0.09 for the Wilcox model, 0.20 for the STD model, and 0.25 for the RNG model. MB was almost zero for the Wilcox model, -0.15 for the STD model, and -0.20 for the RNG model. Overall, it can be concluded that the Wilcox model outperformed the other two models in terms of predicting turbine power outputs in complex terrain.

To have a closer look at the different behaviors of the turbulence models, we plot the fields of predicted wind speed at the height of 95 m for the predominant incoming wind direction in Figure 7. It turns out that at the leeward side of the first hill (marked by the black triangle), where Turbine 2 is located, wind speeds predicted by the Wilcox model were lower than those predicted by the STD model and the RNG model. Since the power of Turbine 2 was used to normalize the results, this explains why the STD and RNG models tended to underestimate the normalized powers at the other turbines. This finding is consistent with other studies showing that the Wilcox model is able to predict mean velocity and turbulent kinetic energy that are closer to the measurements than the other models [15]. The Wilcox model involves the solution of transport equations for the turbulent kinetic energy k and the specific dissipation rate $\omega = \varepsilon / k$ where ε is the dissipation rate of k [13,21]. Compared to the $k - \varepsilon$ models, the $k - \omega$ model has several advantages, namely that: (1) the model is reported to perform better in mildly-separated flows; (2) the model is numerically very stable; (3) the low-Reynolds-number version is more economical and elegant in that it does not require the calculation of wall distances, additional source terms, and/or damping functions based on the friction velocity. It can be inferred from the results that, among those advantages, the first one is mainly responsible for the best performance of the Wilcox model found here.

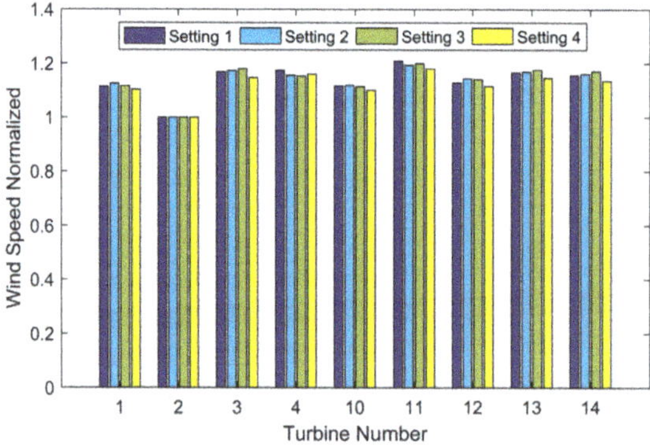

Figure 5. Normalized wind speeds at the hub height of turbines predicted with the four different numerical settings defined in Table 1.

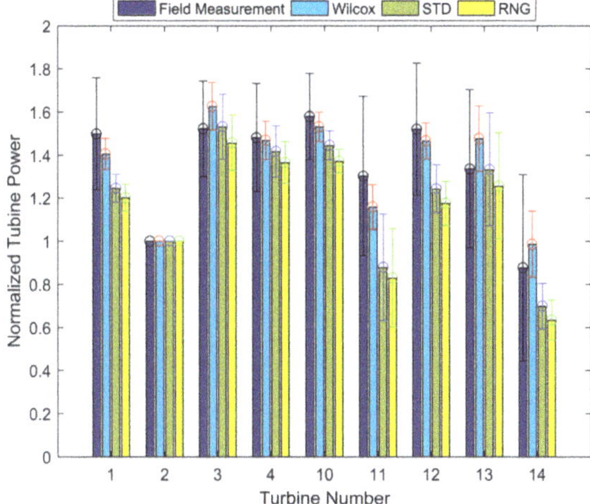

Figure 6. Normalized turbine power outputs observed by field measurement and predicted by WindSim with three different turbulence models: 1. Wilcox; 2. STD; 3. RNG.

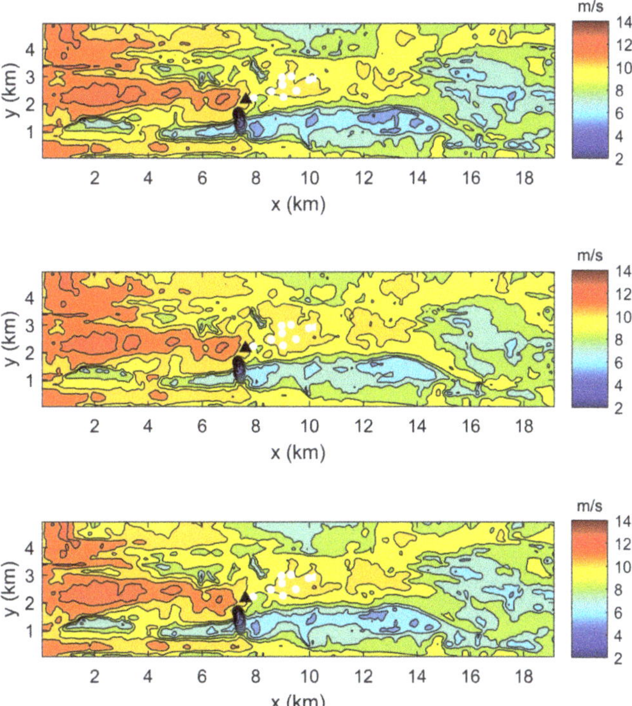

Figure 7. Wind speeds at the height of 95 m predicted by the three turbulence models for the predominant incoming wind direction (from top to bottom: STD, RNG, Wilcox). The first hill is marked by the black triangle.

With the Wilcox turbulence model, the performances of the three analytical wake models implemented in WindSim were evaluated. Again, the linear superposition of velocity deficits was adopted to handle the multiple wakes. As shown in Figure 8, among the three wake models, the Ishihara model yielded the best overall results with an RMSE being 0.09 and an almost zero mean bias, while the Jensen model had an RMSE of 0.15 and an MB of -0.08, and the Larsen model had an RMSE of 0.21 and an MB of 0.10. The better performance of the Ishihara model may be due to the fact that it introduces a turbulence-dependent rate of wake expansion and adopts the Gaussian shape for the velocity deficit. Nevertheless, it is important to note that none of these analytical wake models considers the change of wake growth with topography due to the pressure gradient, which could be significant according to a recent study [22].

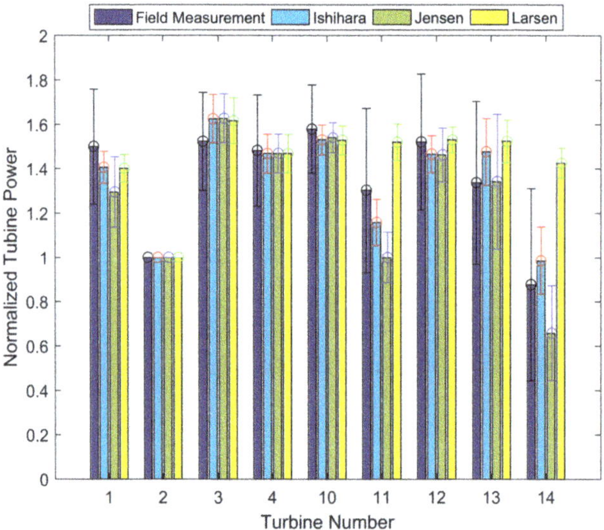

Figure 8. Normalized turbine power outputs observed by field measurement and predicted by WindSim with different analytical wake models: 1. Ishihara; 2. Jensen; 3. Larsen.

Figure 9 compares the results obtained by using two different approaches to calculate the superposition of multiple turbine wakes. It turns out that the linear superposition approach led to stronger multiple wake deficits for the last two downstream turbines (13 and 14) and predicted normalized powers that were in better agreement with the measurements, compared with the other approach that uses the square root of the sum of the squares of the velocity deficits. It is worth mentioning that similar behaviors of the two approaches were found in a study of the Horns Rev offshore wind farm [23].

Table 2 summarizes the prediction errors of the various combinations of modeling options. It is evident that the $k - \omega$ turbulence model of Wilcox together with the analytical wake model of Ishihara and the linear superposition of multiple wake deficits yielded the best performance. Some other combinations of turbulence and wake models were also tested (results not shown), and none of them outperformed the one recommended above. Nevertheless, for this case study, the forest modeling played a key role, and the results were sensitive to the choice of the forest resistive force constant C2, as shown in Figure 10.

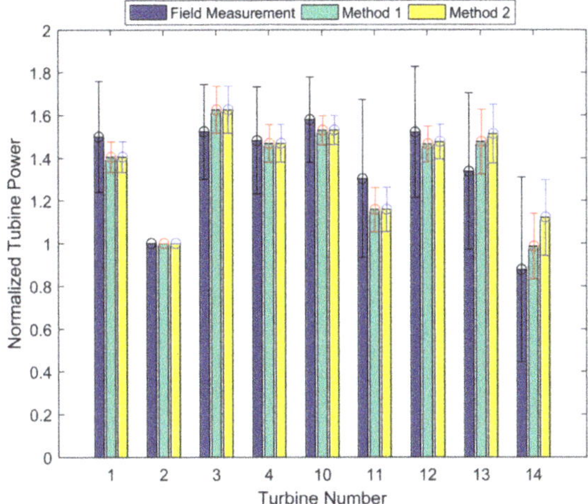

Figure 9. Normalized turbine power outputs observed by field measurement and predicted by WindSim with the multiple wake effects modeled by (1) the linear superposition of velocity deficits and (2) the square root of the sum of the squares of velocity deficits.

Table 2. Data of the RMSE and Mean Bias (MB) of the predicted normalized powers for the model combinations.

Combination	Turbulence Model	Wake Model	Multiple Wakes	RMSE	MB
C1	Wilcox	Ishihara	Method 1	0.09	0.00
C2	STD	Ishihara	Method 1	0.20	−0.15
C3	RNG	Ishihara	Method 1	0.25	−0.20
C4	Wilcox	Jensen	Method 1	0.15	−0.08
C5	Wilcox	Larsen	Method 1	0.21	0.10
C6	Wilcox	Ishihara	Method 2	0.12	0.02

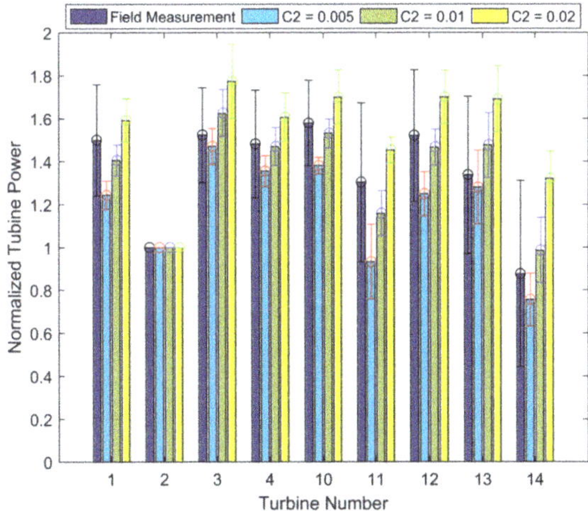

Figure 10. Normalized turbine power outputs observed by field measurement and predicted by WindSim using the forest model with different $C2$ values: 1. 0.005; 2. 0.01; 3. 0.02.

4. Conclusions

The capability of the CFD software WindSim to predict the power outputs of the wind turbines of a wind farm over complex terrain was evaluated in this study. The site of the case study featured the co-presence of three complexities: topography, heterogeneous vegetation with a woodland-grassland mosaic, and interactions between wind turbine wakes. Hence, it allowed an in-depth evaluation of CFD models. The outcome of this study can be concluded as follows:

1. The WindSim modeling setup using the $k - \omega$ turbulence model of Wilcox together with the analytical wake model of Ishihara and the linear superposition of multiple wake deficits was able to simulate turbine power outputs that were in good agreement with the measurements in this case study.
2. Simulation results were sensitive to the choice of modeling schemes and parameters, especially the analytical wake model and the resistive force constant $C2$ in the forest model. Therefore, more validations at different sites of complex terrain are needed before generalizing the optimal modeling setup found in this study.

Comparison with more advanced models such as large-eddy simulation together with actuator disk model would help to verify that the good agreement was not due to the offset of various modeling errors discussed in the paper. Moreover, for forested mountainous regions, high-resolution terrain and vegetation data such as canopy height and density are needed to estimate accurately the relevant parameters for numerical wind energy prediction.

Author Contributions: Conceptualization, J.F. and F.P.-A.; Methodology, J.F. and D.T.; Software, D.T. and J.F.; Validation, D.T. and J.F.; Formal Analysis, D.T. and J.F.; Investigation, D.T. and J.F.; Resources, F.P.-A.; Data Curation, D.T. and J.F.; Writing—Original Draft Preparation, D.T. and J.F.; Writing—Review and Editing, J.F. and F.P.-A.; Visualization, D.T. and J.F.; Supervision, J.F.; Project Administration, J.F. and F.P.-A.; Funding Acquisition, F.P.-A. and J.F.

Funding: This research was funded by the Swiss Innovation Agency (Innosuisse) grant number 1155002544 within the frame of the Swiss Centre for Competence in Energy Research on the Future Swiss Electrical Infrastructure (SCCER-FURIES).

Acknowledgments: The authors thank Johannes Vogel for providing the Juvent wind farm data. The first author wishes to thank Charles Meneveau for providing the scholarship.

Conflicts of Interest: The authors declare no conflict of interest.

References

1. Murthy, K.; Rahi, O. A comprehensive review of wind resource assessment. *Renew. Sustain. Energy Rev.* **2017**, *72*, 1320–1342. [CrossRef]
2. Palma, J.; Castro, F.; Ribeiro, L.; Rodrigues, A.; Pinto, A. Linear and nonlinear models in wind resource assessment and wind turbine micro-siting in complex terrain. *J. Wind Eng. Ind. Aerodyn.* **2008**, *96*, 2308–2326. [CrossRef]
3. Dhunny, A.; Lollchund, M.; Rughooputh, S. Wind energy evaluation for a highly complex terrain using Computational Fluid Dynamics (CFD). *Renew. Energy* **2017**, *101*, 1–9. [CrossRef]
4. Porté-Agel, F.; Wu, Y.T.; Lu, H.; Conzemius, R.J. Large-eddy simulation of atmospheric boundary layer flow through wind turbines and wind farms. *J. Wind Eng. Ind. Aerodyn.* **2011**, *99*, 154–168. [CrossRef]
5. Llombart, A.; Talayero, A.; Mallet, A.; Telmo, E. Performance analysis of wind resource assessment programs in complex terrain. *Renew. Energy Power Qual. J.* **2006**, *1*, 301–306. [CrossRef]
6. Wallbank, T. WindSim Validation Study: CFD Validation in Complex Terrain. Master's Thesis, Victoria University, Melbourne, Australia, 2008.
7. Cattin, R.; Schaffner, B.; Kunz, S. Validation of CFD Wind Resource Modeling in Highly Complex Terrain. In Proceedings of the European Wind Energy Conference 2006, Athens, Greece, 27 February–2 March 2006.
8. Castellani, F.; Astolfi, D.; Piccioni, E.; Terzi, L. Numerical and Experimental Methods for Wake Flow Analysis in Complex Terrain. *J. Phys. Conf. Ser.* **2015**, *625*, 012042. [CrossRef]

9. Waewsak, J.; Kongruang, C.; Gagnon, Y. Assessment of wind power plants with limited wind resources in developing countries: Application to Ko Yai in southern Thailand. *Sustain. Energy Technol. Assess.* **2017**, *19*, 79–93. [CrossRef]

10. Castellani, F.; Astolfi, D.; Mana, M.; Piccioni, E.; Becchetti, M.; Terzi, L. Investigation of terrain and wake effects on the performance of wind farms in complex terrain using numerical and experimental data. *Wind Energy* **2017**, *20*, 1277–1289.

11. Teneler, G. Wind Flow Analysis on a Complex Terrain: A Reliability Study of a CFD Tool on Forested Area including Effects of Forest Module. Master's Thesis, Gotland University, Visby, Sweden, 2011.

12. Yakhot, V.; Orszag, S.; Thangam, S.; Gatski, T.; Speziale, C. Development of turbulence models for shear flows by a double expansion technique. *Phys. Fluids* **1992**, *4*, 1510–1520. [CrossRef]

13. Wilcox, D. Reassessment of the scale-determining equation for advanced turbulence models. *AIAA J.* **1988**, *26*, 1299–1310. [CrossRef]

14. Peralta, C.; Nugusse, H.; Kokilavani, S.; Schmidt, J.; Stoevesandt, B. Validation of the simpleFoam (RANS) solver for the atmospheric boundary layer in complex terrain. *ITM Web Conf.* **2014**, *2*, 01002. [CrossRef]

15. Ramechecandane, S.; Gravdahl, A. Numerical investigations on wind flow over complex terrain. *Wind Eng.* **2012**, *36*, 273–296. [CrossRef]

16. Katul, G.; Mahrt, L.; Poggi, D.; Sanz, C. One- and two-equation models for canopy turbulence. *Bound. Layer Meteorol.* **2004**, *113*, 81–109. [CrossRef]

17. Crasto, G.; Gravdahl, A.; Castellani, F.; Piccioni, E. Wake modeling with the actuator Disc concept. *Energy Procedia* **2012**, *24*, 385–392. [CrossRef]

18. Seim, F.; Gravdahl, A.; Adaramola, M. Validation of kinematic wind turbine wake models in complex terrain using actual windfarm production data. *Energy* **2017**, *123*, 742–753. [CrossRef]

19. Leroy, J. *Wind Field Simulations at Askervein Hill*; Technical Report, VECTOR AS; VECTOR Company: Tønsberg, Norway, 1999.

20. Mathys, L. A Discrete Forest in a Continuous Landscape. Ph.D. Thesis, Faculté de Biologie et de Médecine, Université de Lausanne, Lausanne, Switzerland, 2007.

21. Wilcox, D. Formulation of the $k - \omega$ turbulence model revisted. *AIAA J.* **2008**, *46*, 2823–2838. [CrossRef]

22. Shamsoddin, S.; Porté-Agel, F. Wind turbine wakes over hills. *J. Fluid Mech.* **2018**, *855*, 671–702. [CrossRef]

23. Niayifar, A.; Porté-Agel, F. Analytical Modeling of Wind Farms: A New Approach for Power Prediction. *Energies* **2016**, *9*, 741. [CrossRef]

Article

Development of a Computational System to Improve Wind Farm Layout, Part II: Wind Turbine Wakes Interaction

Rafael V. Rodrigues * and Corinne Lengsfeld

Department of Mechanical and Materials Engineering, University of Denver, Denver, CO 80210, USA;
Corinne.lengsfeld@du.edu
* Correspondence: Rafael.rodrigues@du.edu; Tel.: +1-720-810-8349

Received: 23 January 2019; Accepted: 3 April 2019; Published: 6 April 2019

Abstract: The second part of this work describes a wind turbine Computational Fluid Dynamics (CFD) simulation capable of modeling wake effects. The work is intended to establish a computational framework from which to investigate wind farm layout. Following the first part of this work that described the near wake flow field, the physical domain of the validated model in the near wake was adapted and extended to include the far wake. Additionally, the numerical approach implemented allowed to efficiently model the effects of the wake interaction between rows in a wind farm with reduced computational costs. The influence of some wind farm design parameters on the wake development was assessed: Tip Speed Ratio (TSR), free-stream velocity, and pitch angle. The results showed that the velocity and turbulence intensity profiles in the far wake are dependent on the TSR. The wake profile did not present significant sensitivity to the pitch angle for values kept close to the designed condition. The capability of the proposed CFD model showed to be consistent when compared with field data and kinematical models results, presenting similar ranges of wake deficit. In conclusion, the computational models proposed in this work can be used to improve wind farm layout considering wake effects.

Keywords: wind turbine aerodynamics; wake aerodynamics; computational fluid dynamics; MEXICO experiment; wind farms; wind turbines interaction; wind farm modeling

1. Introduction

The necessity for improving wake models has become more apparent over the last decade with the continuous growth of the wind energy market. Literature shows several analytical wake models: Infinite wind farm boundary layer model, Jensen wake model, Larsen model, dynamic wake meandering model, FUGA (Linearized RANS Model), and EllipSys3D. All these models are excellent tools to estimate wake effects, but there is still room for improvement. Usually, analytical models do not consider wake characteristics according to variable operating conditions. However, Computational Fluid Dynamics (CFD) models have the capability to model wake velocity deficit and Turbulence Intensity (TI) according to variable operating conditions. Although computationally expensive, CFD models are powerful tools that can be applied to solve some of the most complex problems in engineering. This work describes how operational parameters affect the aerodynamic behavior of the near wake of a wind turbine up to 5 diameters downstream of the rotor. Moreover, this study proposes a CFD modeling technique to characterize three-dimensional far wake effects, and numerically quantify the influence of some important wind farm design parameters on the far wake aerodynamic behavior. The literature shows that there is a gap in attempting to solve the Wind Farm Layout Optimization Problem (WFLOP) while still considering a rigorous evaluation of the wake effects. The objective of this work is to develop a CFD model with such capabilities, applicable for future applications related to the WFLOP.

1.1. Review: Wind Farm Aerodynamics

1.1.1. Wake Aerodynamics

Wake models are usually divided in literature [1–4] in two categories: (1) Analytical/empirical/ explicit wake models; and (2) computational/implicit wake models. The analytical models solve a set of equations based on the conservation of mass and empirical relations of wake decay, characterizing the energy content in the flow field, and ignoring the details of the exact nature of the flow field. Kinematic models such as Jensen, Larsen, and Frandsen's model assume self-similar velocity deficit profiles, not solving the turbulence field but only the momentum equation [2]. The velocity deficit is derived from global momentum conversation, using thrust coefficient of the turbine as an input [1]. The computational models solve the fluid flow equations for the wake velocity and turbulence field, whether simplified or not [2].

1.1.2. Wind Energy Computational Fluid Dynamics (CFD) Review

Although there are many CFD studies in the literature approaching wind energy, this is a field of study still in development. CFD modeling techniques applicable for wind turbines significantly vary in literature, showing that there is no well-stablished standard approach. This section presents a comprehensive literature review in CFD models applicable to wind energy, providing an overview on what has been done prior to this work. In regards to CFD techniques for modeling wind turbine flow field, the goal is to investigate what possibilities have not been explored yet, seeking to develop a novel wind turbine CFD model capable of evaluating far wake aerodynamics characteristics. As previously mentioned, a correct evaluation of such characteristics can help to achieve better solutions for the WFLOP.

NREL (National Renewable Energy Laboratory) Phase VI

Several studies utilized the NREL (National Renewable Energy Laboratory)/NASA (The National Aeronautics and Space Administration) Ames Phase VI experimental data campaign to validate their computational models, all of them using pressure coefficient on the blades and aerodynamic torque data for comparison. However, it is difficult to validate wake flow field since no wake measurements were performed in these experiments. Zhou et al. [5] performed Large-Eddy Simulation (LES) of the NREL phase VI, evaluating the effect of different inflow conditions (using user-defined functions) on aerodynamic loading and near wake characteristics. A structured multi-block mesh (with sliding mesh zone) was implemented with refinement on leading and trailing edges. They found that the wind shear and turbulence effects destroyed the uniform and symmetric wake profile in the far wake. Hsu et al. [6] validated a finite-element (Lagrangian–Eulerian) model of the NREL Phase VI using a non-structured rotating mesh. Wake characterization was not the focus of the study, which explains the wake made out of coarse non-structured cells with no refinement. Gundling et al. [7] evaluated low and high fidelity models using the NREL Phase VI for predicting wind turbine performance, aeroelastic behavior, and wakes: (1) The Blade Element Method (BEM) with a free-vortex wake; (2) the Actuator Disc Model (ADM); and (3) the Full Rotor Method (FRM). No specific information or sketch of the wake was provided or described. The FRM showed the largest wind deficits and the slowest dissipation rate for the far wake. Mo et al. [8] developed a study in more depth to understand wake aerodynamics performing a LES of the NREL Phase VI using the dynamic Smagorinsky model. Additionally, verification of the average TI was performed against an analytical model. They found that the downstream distance where instability and vortex breakdowns occur is dependent on wind free-stream inlet conditions: 7 m·s^{-1} happens at four rotor diameters, while 15.1 m·s^{-1} between 11 and 13 diameters. A decrease of the TI happened after instability and vortex breakdowns. The strategy for meshing the physical domain consisted of a virtual wind tunnel with the same dimensions of the NASA Ames with the rotor located at 2 diameters downstream of the inlet with a downstream domain of 20 rotor diameters in length. Choudhry et al. [9] performed a very similar CFD study of the NREL Phase VI using the same computational methods of the study conducted by Mo et al. [8], finding that regions of velocity deficit and high TI are within the high vorticity region. Choudry's study did not specify if the mesh is structured or unstructured.

NREL 5 MW

Many studies have developed CFD models considering the NREL 5 MW wind turbine. Among these studies, Troldborg et al. [10] developed a wake CFD (EllipSys3D) study for the NREL 5MW considering three different models: (1) A fully resolved rotor geometry; (2) the Actuator Line Model (ALM); and (3) the ADM. A comparison for wake properties in uniform and turbulent inflows was performed. All the models correctly predict mean axial velocity within 4 radii downstream of the turbine for laminar inflow. The agreement between ADM and ALM methods is acceptable for the wake deficit. They found that the ADM/ALM model is sufficient to simulate turbines under Atmospheric Boundary Layer (ABL) conditions. Storey et al. [11] implemented a CFD model using a modified actuator technique to develop transient simulations, considering the NREL 5MW turbine. They achieved reduction in the computational time for the simulation while still keeping flow solution fidelity compared to the standard ADM. Seydel et al. [12] performed a Reynolds Averaged Navier-Stokes (RANS) k–ω simulation of the NREL 5 MW to study wake effects between two wind turbines. Réthoré et al. [13] investigated CFD techniques based on permeable body forces including: ADM, ALM, and the Actuator Surface Model (ASM). These approaches can potentially reduce the necessity for mesh refinement next to the rotor. Verification for the ADM in comparison with analytical solution for heavily-loaded turbines demonstrated that the ADM can be a cost-effective way to model wind turbine wake. The verification of the ADM showed that 10 cells per diameter are adequate to describe the near wake flow characteristics, and the cell size becomes less critical in the far wake. The computational domain extends 10 diameters laterally and 25 diameters horizontally, and the wake computational grid is uniformly spaced with cells of the same size. Heinz et al. [14] developed a fluid-structure interaction simulation using EllipSys3D and aero-elastic HAWC2 for the NREL 5 MW considering yaw and standard conditions. Miao et al. [15] developed an unsteady CFD (STAR-CCM+) model for the NREL 5 MW rotor considering yawed flow to investigate wake deviation. The full rotor geometry was modeled considering the NREL 5MW wind turbine, under neutral ABL conditions. Wilson et al. [16] developed a CFD model based on the RANS (OpenFoam and ANSYS Fluent) equations, considering k–ε and k–ω SST (Shear Stress Transport) turbulence model to investigate interactions between wind turbines in neutral ABL conditions. The ADM, the ALM, and the FRM were compared considering the NREL 5 MW. Weipao et al. [17] considered the tilt and cone angle to maximize the power generation of a wind farm for the NREL 5 MW.

Other Topics

CFD modeling techniques have been applied for designing and the analysis of floating offshore wind farms. Wu et al. [18] developed a CFD for an offshore floating wind turbine. The near-wake domain is defined as 3D downstream, whereas a 0.5 D distance upstream of the rotor is maintained with constant size mesh cells. Two different approaches for blade meshing were implemented: unstructured tetrahedral and unstructured hexahedral. Theunissen et al. [19] developed a computational and experimental study to optimize the layout of an offshore wind farm array with 80 turbines. Tran et al. [20] developed an unsteady CFD model for a floating offshore, using the software FAST (Fatigue, Aerodynamics, Structure and Turbulence) and Unsteady BEM equations for the analysis.

RANS techniques have been widely implemented in the literature. Zhale et al. [21] performed an unsteady yaw description for a 500 kW rotor modeling the RANS equations using EllipSys3D. A pressure-based incompressible flow was setup, considering an iterative SIMPLE (Semi-Implicit

Method for Pressure-Linked Equations) and PISO (Pressure-Implicit with Splitting of Operator) second-order accurate scheme, the turbulence k–ω SST model (good performance for wall-bounded adverse pressure gradient flows). The computational mesh was generated using the software Gridgen, with structured elements. Prospathopoulos et al. [22] developed a RANS k-ω model modified for atmospheric flows, finding that CFD models underestimate near wake deficit even for single-wind turbine wake predictions especially under neutral atmospheric conditions. The accuracy was better for the far wake, and this study also considered the multi-wake interaction considering the case of five turbines in a row. AbdelSalam et al. [23] performed experimental procedure and numerical simulation considering a FRM, RANS k–ε modified for atmospheric flows, 2 MW wind turbine SODAR upstream measurements, and wake LIDAR (Light Detection and Ranging) measurements at downstream distances from 2 to 7 diameters. Boudreau et al. [24] studied the axial-flow and cross-flow configurations operating at respective optimal efficiency, with Reynolds' number around 10^7, 3D DES (Detached-Eddy Simulation), and Unsteady RANS. Ammara et al. [25] developed a RANS steady CVFEM (Control Volume Finite-Element Method) model, considering a two-row periodic wind farm in a neutral ABL. Frau et al. [26] developed an unsteady CFD (ANSYS CFX) k–ω SST model to compare downwind and upwind configurations for offshore applications, using 9 million to 25 million cells. They concluded that the downwind turbine configuration is better suited for multimegawatt offshore wind turbines. Lann et al. [27] developed a new k–ε model consistent with Monin–Obukhov similarity theory (MOST), comparing it to other k–ε models. Lann et al. [28] developed a k–ε-fP viscosity model applied to one on-shore and two off-shore wind farms, and the results were compared with power measurements. The k–ε model underpredicts the power deficit of the first downstream wind turbine, while the k–ε-fP eddy viscosity shows good agreement with the measurements. The difference becomes smaller for wind turbines further downstream.

Computational models based on ADM and ALM have also been widely implemented in the literature. More recently, the ADS has been developed for some researchers. Models based on ADM and ALM are relatively less computationally demanding than computational models for the FRM, such as RANS and LES. Sarmast et al. [29] developed an ALM using a new vortex code on the Biot–Savart law, and by considering two different wind turbines: Constant and variable circulation along the blades. They concluded that a simplex vortex code has similar results to the ALM and a lower computational cost. Ivanell et al. [30] developed a CFD (EllipSys3D) ALM using 5 million mesh points to evaluate downstream wake flow field characteristics and the tip vortices positioning. Masson et al. [31] developed a RANS k–ε ADM to assess impacts of the variation of operational parameters influencing the turbulent flow around a wind turbine nacelle. Troldborg et al. [32] developed an unsteady RANS ALM to analyze wake interaction between two wind turbines under different degrees of ambient TI: Laminar, offshore, and onshore conditions. The results show the influence of the upstream turbine wakes on external blade loading of the downstream turbines. Makridis et al. [33] developed a CFD model in ANSYS Fluent solving the RANS equations, assuming ADM (based on BEM) and considering complex terrain and neutral atmospheric wind flow. A validation was performed against wake data over flat terrain. Neutral atmospheric flow conditions over a hill were tested and validated.

LES and DES models have been studied and implemented for wind energy applications over the last years. Although computationally more expensive, these models are capable of modeling the transient behavior of wind turbines. Schulz et al. [34] developed a CFD (FLOWer) study of the yawed flow ($-50°$ to $+50°$) on a generic 2.4 MW using DES. Ivanell et al. [35] studied stability properties of wind turbine wakes using a CFD model based on the LES ALM on the tip vortices of the Tjaereborg wind turbine. Bromm et al. [36] investigated the impact of directionally sheared inflow in the wake development, and analysis of the impact of wakes on energy production and loading on a downstream turbine. A LES was performed using the ALM representation. Storey et al. [37] developed a technique coupling transient wind simulation with an aero-elastic simulation to dynamically model turbine operation and wake structures. A LES with an ADM was performed for that study. Troldborg et al. [38] developed a LES with an ALM technique using 8.4 million grid points to study the near and far wake of a wind turbine at various Tip Speed Ratios (TSR). Lann et al. [39] achieved an improvement for the k–ε model, comparing this model with the original k–ε eddy viscosity model, the LES, and a total of eight field test case measurements. The results showed a better agreement with measurements and LES in comparison to the original k–ε. Transient unsteady models such as LES can account for velocity fluctuations by setting perturbation components using Reynolds stress components. This is important for modeling the fluctuations inherently present at the atmospheric wind. Examples of LES models that simulate wind turbines operating in the ABL can be found in the literature ([40–45]). A full review of LES simulations of wind farm aerodynamics can be found in the literature [46]. According to Rodrigo et al. [47], challenges for ABL modeling include relation between enhanced mixing in operational models, role of land surface heterogeneity, development of LES models with interactive land-surface, and climatology of boundary-layer parameters such as stability.

Moreover, on the topic of wind energy CFD techniques, some researchers have incorporated one-dimensional codes based on BEM to their models, developing a combined hybrid approach CFD-BEM. For instance, Choi et al. [48] developed a CFD model using ANSYS CFX for 2 MW wind turbines, using BEM theory for the blade design. The distance from upstream and downstream wind turbines changed from three to seven times the diameter, and obviously power output was affected. Esfahanian et al. [49] developed a CFD model of the NREL Phase II using ANSYS Fluent and BEM improved methodology. Furthermore, in CFD techniques, Gopalana et al. [50] developed a coupled mesoscale-microscale model (WINDWYO) coupled with WRF (weather research and forecasting) model and CFD codes of different complexity in order to assess the power predictions and wake visualization at the Lillgrund wind farm. Rosenberg et al. [51] extended efforts of the Vortex Lattice Method (VLM) to analyze aerodynamics of dual-rotor wind turbines. Sreenivas et al. [52] studied the interaction between two wind turbines (NREL S826 airfoils) operating in tandem for TSR of 2.5, 4, and 7 in a wind tunnel speed at $10\ \mathrm{m \cdot s^{-1}}$. Larsen et al. [53] reviewed several studies in wake aerodynamics. Mittal et al. [54] developed a CFD model (Tenasi: Finite Volume unstructured flow solver) of a wind turbine at various tip-speed ratios, evaluating the effect of temporal convergence on the predicted thrust and power coefficient. Three turbulence models were evaluated: Spalart–Allmaras, Menter SST two equations, and the DES version of the Menter SST. The results pointed that the DES model is significantly better for predicting velocity components in the wake. AbdelSalam et al. [55] modeled the near and far wake using the RANS rotating reference frame, k–ε turbulence model. A FRM and an ADM were compared, and two additional k–ε previously studied in the literature. Wake results were validated against the 180 kW Danwin (three-bladed), showing good agreement.

1.2. Gaps in the Literature

Basically the gap existent in the literature is related to CFD models capable of simulating a whole wind farm. The vast majority of the methods simulate single turbines, and only a few of them simulate more than one rotor. The computational resources may be a limiting factor for that, however the gap related to lack of CFD models to simulate whole wind farms can be overcome in other ways. Section 2.3 shows a novel approach of this work as an attempt to overcome the main gap identified in the literature. In regards to other aspects, there is no well-established approach to computationally model wind farms. The choice for boundary conditions and turbulence models widely vary in research and any pattern was identified. Moreover, lack of experimental data in controlled environments for the far wake do not allow researchers to validate their data and improve wake aerodynamics knowledge. Consequently, it is not possible to accurately evaluate wake CFD models found in the literature. The majority of the experimental data for far wake characterization comes from field experimental data, which are difficult to replicate in computational models.

2. Methods: Wind Farm CFD Modeling

2.1. Wake Effects

The wake of a wind turbine is characterized by decreased velocity and increased TI. There are many analytical methods to estimate the velocity-deficit in the wake, but models based on CFD are robust and reliable. In this work, a CFD model was developed to determine the wake velocity deficit and consequently its influence on the wind farm output power. The TI profile in the wake is also characterized using a CFD solver. A very important design parameter for wind farms is the TSR, which is defined as the ratio between the blade tip speed velocity and the free-stream velocity (Equation (1)). The TSR and other parameters such as free-stream velocity are critical to determine wake behavior:

$$\lambda = \frac{\omega \cdot R}{U_{freestream}}.$$

(1)

where ω is the rotor rotational speed, R is the blade radius, and U is the free stream velocity.

Another important design parameter is the *TI*. This parameter can be calculated using Equation (2):

$$TI = \frac{\sigma_U}{U_{freestream}}$$

(2)

2.2. CFD Model

The wind turbine modeled in this work was adapted from the previously validated wind turbine CFD model from part I [56], the MEXICO (Model Experiments in Controlled Conditions) rotor (4.5 m diameter) [57] tested in wind tunnel. The wind turbine blade geometry (MEXICO rotor) including twist angle was built using SolidWorks, and then imported to the ANSYS Design Modeler to build the other turbine components (tower, hub) and the physical domain (Figure 1). The geometry of the MEXICO rotor blades is shown in Figure 1c, the three-bladed model has three types of airfoil: DU91-W2-250 (20% to 45%), Riso-A1-21 (54% to 65%), and NACA 64-418 (75% until the blade tip). The blade is also twisted, and a pitch angle of $-2.3°$ was applied for the measurements. Since some of the airfoil data are not publicly available, a reverse engineering process was performed to find the airfoil coordinates. A rectangular physical domain was built, and it was broken into smaller pieces, allowing local wake mesh sizing. The largest rectangle in Figure 1a is an exterior part, and the first rectangle corresponds to the near wake until 2 diameters downstream of the rotor. The wake was simulated with a domain extending 13 diameters downstream of the rotor. The CFD model of this study was adapted from part I of this research [56], which is a validation and near wake analysis of the MEXICO rotor. In part I [56], the wind tunnel inlet is located 7 m upstream of the rotor. In the current study, the same distance was adopted as the length upstream of the wind turbine. The solution of the continuity equation generates fewer amounts of residuals for shorter upstream distances, resulting in a better convergence for the CFD solution.

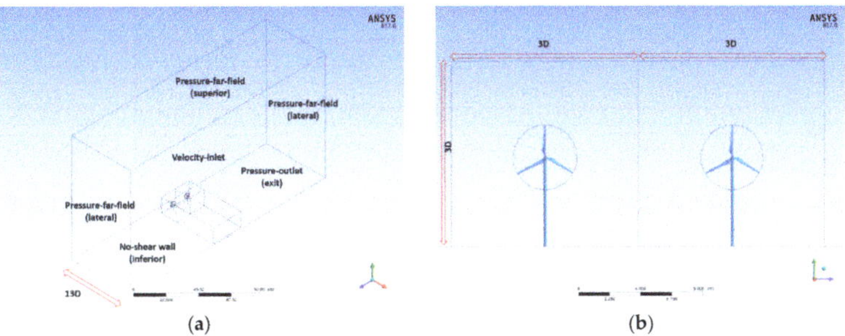

Figure 1. *Cont.*

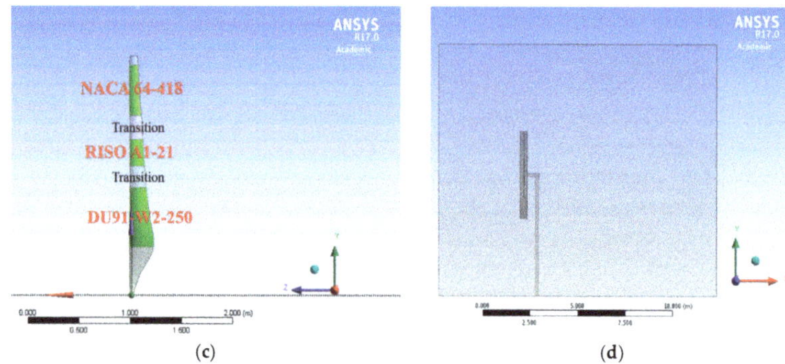

(c)

(d)

Figure 1. (**a**) Physical domain with two rotors and boundary conditions; (**b**) front view of the physical domain, showing the MEXICO (Model Experiments in Controlled Conditions) rotor; (**c**) MEXICO rotor geometry, a three-bladed rotor with a 4.5 m diameter; and (**d**) lateral view of the wind turbine, showing the rotating reference frame.

The strategy for meshing (Figure 2) the physical domain is to build a sphere of influence surrounding each rotor, and break the physical domain into smaller rectangles defining them as the same part in the ANSYS Design Modeler. The sphere of influence option allows for a better convergence of the flow field solution. The smaller rectangles allow the mesh element sizing of the near and far wake to be controlled locally, avoiding gradients in the mesh sizing in the interface of each sub-domain. The flow field solution is determined using the CFD solver ANSYS Fluent 17 (ANSYS, Canonsburg, PA, USA), housed in two computers with 64 GB RAM and 8 processes for each machine. ANSYS Fluent solves the equations of fluid flow and heat transfer by default using a stationary (or inertial) reference frame. However, a Moving (or non-inertial) Reference Frame (MRF) can bring advantages in solving the equations for some problems involving moving parts, such as rotating blades. In those problems, the flow around the moving parts is the variable of interest to be determined. In the case of this work, the region behind the wind turbine corresponding to the wake flow field is the region of interest. The MRF technique models the flow around the moving part as a steady-stead problem with respect to the moving frame, allowing to activate reference frames in selected cell zones. The ANSYS Fluent MRF modeling modify the equations of motion to incorporate additional acceleration terms that occur due to the transformation from the stationary to the moving reference. The main reason for employing a MRF is to solve a problem that is unsteady in the stationary (inertial) frame but steady with respect to the moving frame. In this work, the simulation was performed using a steady state MRF approach, and setting the rotational speed to match experimental conditions. Unlike the ADM, ALM, and ASM approaches, the CFD model of this work is a FRM approach which considers the exact 3D blade geometry, including variable chord length, local twist angle, and blade pitch angle. The boundary layer was solved using 10 inflation layers with a ratio of 1.1 to ensure y+ < 1 next to the blade surface. Even though the full blade geometry was resolved using the CFD model, the solid blade geometry was suppressed from the rotating disc, centrally located at the physical domain. The MRF approach essentially consists in building a central disc (a fluid zone) surrounding the solid three-dimensional blades (solid zone) inside the disc. At this point, there are two physical domains: (1) A solid zone representing the blades; and (2) a fluid zone (central disc) surrounding the blades, which is the central disc. The next step is to subtract the blade domain (solid zone) from the fluid zone corresponding to the central disc. After the subtraction operation, there is no more solid body (blade) inside the central disc, but only the external surfaces (walls) of the full three-dimensional blade geometry, meaning that the interior of the blade is now an empty space. The exterior blade surfaces (three-dimensional blade surface including chord, twist, and pitch) remains in the central disc (fluid zone), behaving exactly in the same way as if the blades had not been suppressed: External walls. This procedure is performed because the remaining central disc is the rotating frame in the MRF approach. The rotating speed is set up to the frame and not the blade itself. The disc evolving the three-blade wind turbine shown in Figure 1b is the reference frame, which is set to rotate at the desired operating condition. The loading on the blades is represented using the rotating central disc from Figure 1b, but careful work has been taken to correctly represent these forces. The model validation process can be found in the first part of this research [56], including blade loading, pressure coefficient on the blades, and near wake velocity flow field. Additionally, more details about the numerical modeling process can be found in part I [56]. The process to adapt the geometry from part I [56] to the extended geometry in this work included the use of Ansys Design Modeler functions. The operations utilized to build the physical domain shown in Figure 1 include extrusion, skin, Boolean, pattern (to duplicate the turbines and wake domain), construction of primitives (cylinder), slicing, rotation, and translation.

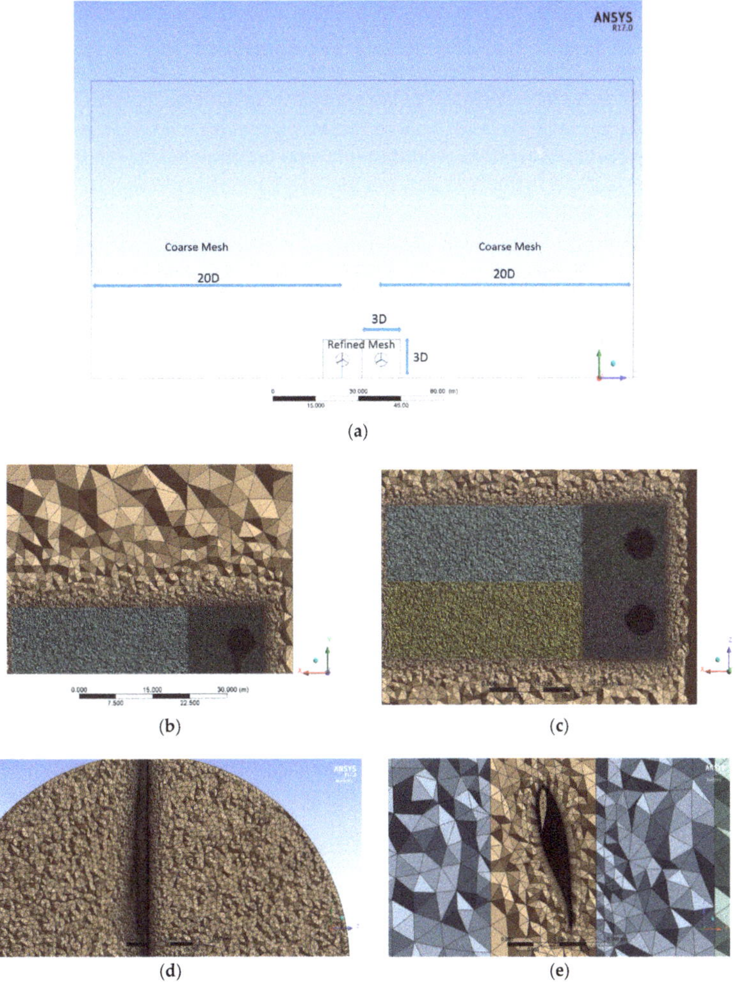

Figure 2. (**a**) Details of the meshing process for the physical domain; (**b**) lateral view of the mesh showing internal details of the sphere of influence; (**c**) top view of the mesh showing internal details of the sphere of influence; (**d**) sectional plane showing details of the mesh on the blade surface; and (**e**) details of the mesh close to the blade surface, showing inflation layers.

Moreover, the turbulence model selected was the k–ω SST, which is suitable for swirl flow and was used in the literature studies as their main turbulence modeling technique. Since there is no public information from the reports of the MEXICO experiment regarding the inlet inflow conditions, default values of 5% and 10% were assumed for the inflow TI and the viscosity ratio (VR) at the inlet, respectively. The Reynolds number based on the average chord length is approximately 1.5×10^5. Pressure-far-field boundaries, which require the larger exterior rectangle to achieve convergence, were applied for the lateral and superior boundaries. The turbines were both rotating in a clockwise direction, which prevented the implementation of symmetry boundary conditions to simulate the two-turbine case. Pressure-far-field boundary conditions are suitable to model the lateral boundaries of the physical domain. If the lateral boundaries are placed far away from the region perturbed by the wind turbine fluid flow, the streamlines have a straight direction. Pressure-far-field boundaries have a good numerical convergence and stability for the case of straight streamlines. We also apply a pressure-outlet for the exit boundary, and a special type of wall with no shear for the inferior boundary. The use of a no shear wall intends to reduce the complexity of the problem by eliminating the need for modeling the surface roughness, which would require a much more refined mesh at the bottom of the physical domain. Essentially, the goal of the study was to develop a preliminary model capable of simulating a wake interaction effect in a wind farm. Originally, a no shear wall condition for the bottom was implemented for the wake validation presented in part I [56] of this research because the validation of the wake velocity field was not affected by the roughness of the bottom. In part II, the validated model from part I was adapted keeping the same type of boundary conditions. The implementation of a shear wall through the definition of ground surface roughness is an intended further improvement of the model. The mesh sensitivity study can be found at Appendix A, showing the need for using 10 million cells. Additionally, the dimensions of the cell elements in each of the fluid cell zones from Figure 2 can be found in the Appendix A.

2.3. Second and Third Rows Simulation

In this work, we developed a new method to evaluate the second and third rows of turbines where the outlet of the first row becomes the inlet of the second row. This results in a significant reduction in the computational expenses, since there is no need to simulate multiple turbines at once. Multiple turbines would require a mesh with a significant higher number of elements. For instance, the three first rows would require three times more elements in comparison with our approach. The goal of this approach was to propose a method to overcome the challenges pointed out in the section: The vast majority of the methods simulate single turbines. This method has never been applied to solve wind farm before in the literature. It is worth mentioning that there is not necessarily an improvement in terms of computational time, since three sequential simulations to simulate three rows take the same amount of time of the conventional simulation with three times more elements. On average, each simulation for case 3 of Table A1 (Appendix A) takes approximately 10 h. The referred reduction in computational expenses comes from the fact that less expensive computational resources are required to perform such simulations. One of the biggest challenges on wind farm computational modeling is the expensive computational resources required to simulate several rows in a wind farm. The use of the technique introduced in this work allows researchers to simulate the wake interaction effect without the need for expensive computational resources. In other words, there is a reduction in the capabilities (processors) required to develop wake interaction simulations.

2.4. Wake Similarity

The wind turbine modeled in this work (the MEXICO rotor) has an extensive wake flow field dataset, which allowed the validation performed in part I [56] of this research. The Reynolds number of a utility-scale turbine is higher than a small wind turbine prototype (such as the MEXICO rotor) mainly because of the differences in the chord length. Matching the Reynolds number of a utility-scale turbine and the MEXICO rotor was not achievable because of the extremely high velocities required to counter balance the difference in chord length. Wake characteristics are highly dependent on the TSR though, as shown in part I [56] of this work. This work relies on the assumption that similarity between large and small-scale turbines in regards to wake characteristics can be ensured by matching TSR operating conditions. Examples of small-scale experiments to study large scale wake aerodynamics can be found in the literature [58–61], supporting the assumptions made in this work.

2.5. Cases of Study and Motivation

In the study case of this work, two turbines were located side by side in the first row (Figure 3). Then, a second row of turbines was placed in a position which was totally aligned with the location of the first row. The downstream distance between the first and second rows was 10 rotor diameters. The operating conditions simulated in this work correspond to wind velocities of 10 m·s^{-1} and 15 m·s^{-1}, pitch angle (θ) within $-1°$ and $-3°$, and a TSR within 4 and 10. The main motivation for positioning the second row aligned with the first row was to study the effect of wake interaction on the velocity and turbulence flow field (Figures 4–6). The motivation for positioning the turbines side by side in the first row was to study the effect of designing staggered rows. For instance, an increase in output power could be achieved in the second row of turbines by staggering the second row rotors out of the region affected by the wake of upstream rows. However, there could be consequences regarding increased turbulence levels at these locations because of wake expansion effects for the turbulence flow field. A further discussion in Figure 6 explains the importance of studying side-by-side distances in upstream rows and its effect on turbulence flow field. Moreover, the motivation for selecting the operating conditions in this work had to do with the MEXICO experiment and typical wind farm conditions: (a) The velocities (10 m·s^{-1} and 15 m·s^{-1}) correspond to values tested in the MEXICO experiment; (b) TSR within 4 and 10 is typically experienced in commercial wind farms; (c) the designed condition for the MEXICO rotor is U = 15 m·s^{-1}, TSR = 6.6, and pitch angle (θ) = $-2.3°$; and (d) part I [56] of this research analyzed a positive value of 2.3°, confirming that pitch angle values much different from the designed condition strongly influence the wake axial induction.

3. Results

3.1. Wind Turbine Wake in the First and Second Rows

3.1.1. Velocity and Turbulence Intensity (TI) Contours

The intensity of the velocity-deficit decayed along the axial distance downstream of the rotor, however the velocity in the wake did not fully recover its free-stream value even after more than 10 diameters downstream of the rotor. Figure 3 shows time-averaged velocity contours for the two-turbine case when considering the designed aerodynamic condition for this specific wind turbine (U = 15 m·s^{-1}, λ = 6.6, ω = 424.5 rpm, θ = $-2.3°$). The region in red (15 m·s^{-1}) represents the area where the velocity was not affected by wake effects. On the other hand, the velocity-deficit in the wake of the wind turbine is represented by green and yellow contours.

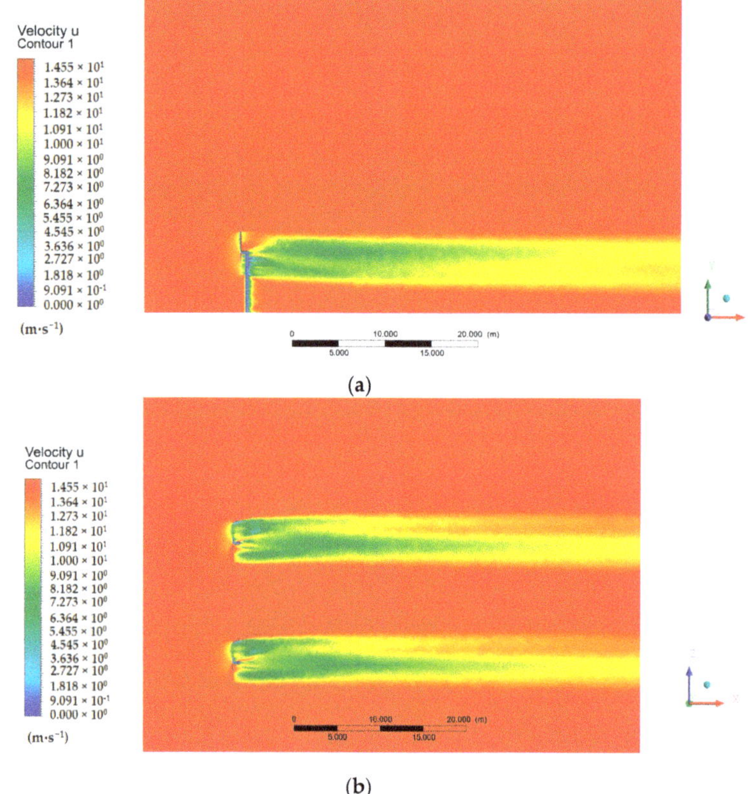

Figure 3. Axial time-averaged velocity contours for the two-turbine case, representing the first row of wind turbines. (**a**) Lateral view of the wake; and (**b**) top view of the wake.

Wind farms experience effects from the interaction between wakes from the different rows, which changes the velocity and turbulence flow field. The region free of wake effects became smaller after each row of turbines. Figure 4 shows the velocity contours for a hypothetical second row of wind turbines, while Figure 5 shows TI contours. These simulations considered the designed operational conditions (U = 15 m·s^{-1}, θ = −2.3°, and TSR = 6.6) for both the first and second row of turbines. Instead of simulating 4 turbines, the methodology applied used data from the previous simulation (Figure 3) for the velocity inlet. Basically, the pressure-outlet of Figure 3 became the velocity-inlet profile for the simulation from Figure 4. This procedure significantly improved the computational efficiency of the simulation with regards to computational time and convergence, since two turbines were simulated instead of four. The second row of wind turbines were not staggered from the first row of turbines, this way occupying a region affected by wake effects from the upstream first row. The wake velocity contours in Figure 4 show a smaller region of unaffected velocity in comparison with Figure 3, meaning that the region free of wake effects becomes smaller after each row of turbines.

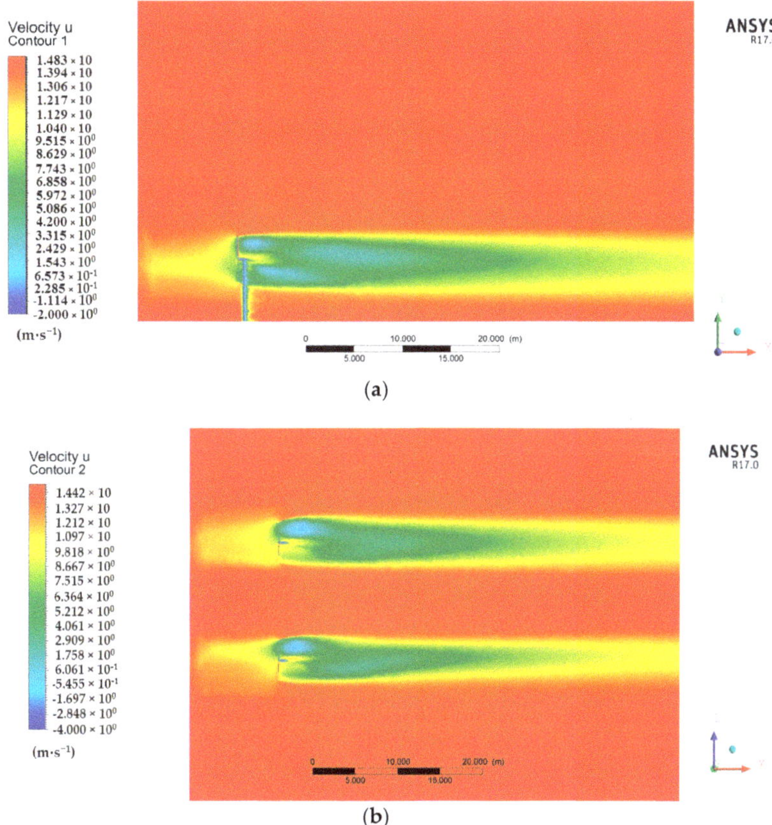

Figure 4. Time-averaged axial velocity contours for the two-turbine case in a hypothetical second row of wind turbines. (**a**) Lateral view of the wake; and (**b**) top view of the wake.

3.1.2. Velocity and TI Plots

Wake data plots for the wake of the first and second turbine rows are shown in Figure 6, comparing the behavior of the velocity deficit (Figure 6a) and the TI (Figure 6b) at a radial traverse at 10 D (diameters) in the wake. The velocity deficit existing in the wake of the second row was slightly higher than the velocity deficit found in the wake of the first row (Figure 6a). The TI of the second row of turbines was considerably higher compared to the same downstream position (10 D) of the first row (Figure 6a). Moreover, interestingly there was an increase of the TI (Figure 6b) in the region between the two turbines (at r = 0), which can be attributed to wake expansion of the turbulence flow field. This can be extremely relevant in the context of wind farm layout optimization, since there is need for improving the turbine packing factor in a wind farm to take the highest benefit/output out of the windiest sites. For instance, the region between the two turbines would not be locally affected with reduced velocities, which could lead to a misleading decision of installing turbines at this position. However, the TI would have increased levels which could have an impact on the components (blades, tower, and turbine) fatigue lifetime. The increase in TI caused by wake interaction effects becomes much more significant as the lateral distance between turbines in the same row decreases. Figure 6b shows that there was a severe increase in TI for the first and second rows when the lateral spacing between turbines was too small (2 D), and Figure 6a shows that even the wake velocity profile was affected. Such effects tend to dissipate for larger lateral distances, as shown by Figure 6e,f which considered 4 D of lateral spacing. Further studies on staggered wind farms should address the influence of the spacing between upstream turbines on turbulence flow field characteristics at the wake, aiming to determine the areas where the level of TI is reduced. If chosen correctly, the side by side distance (from upstream rows) could result in a wake region in which TI levels are reduced, consequently these spots would be more suitable to place downstream turbines.

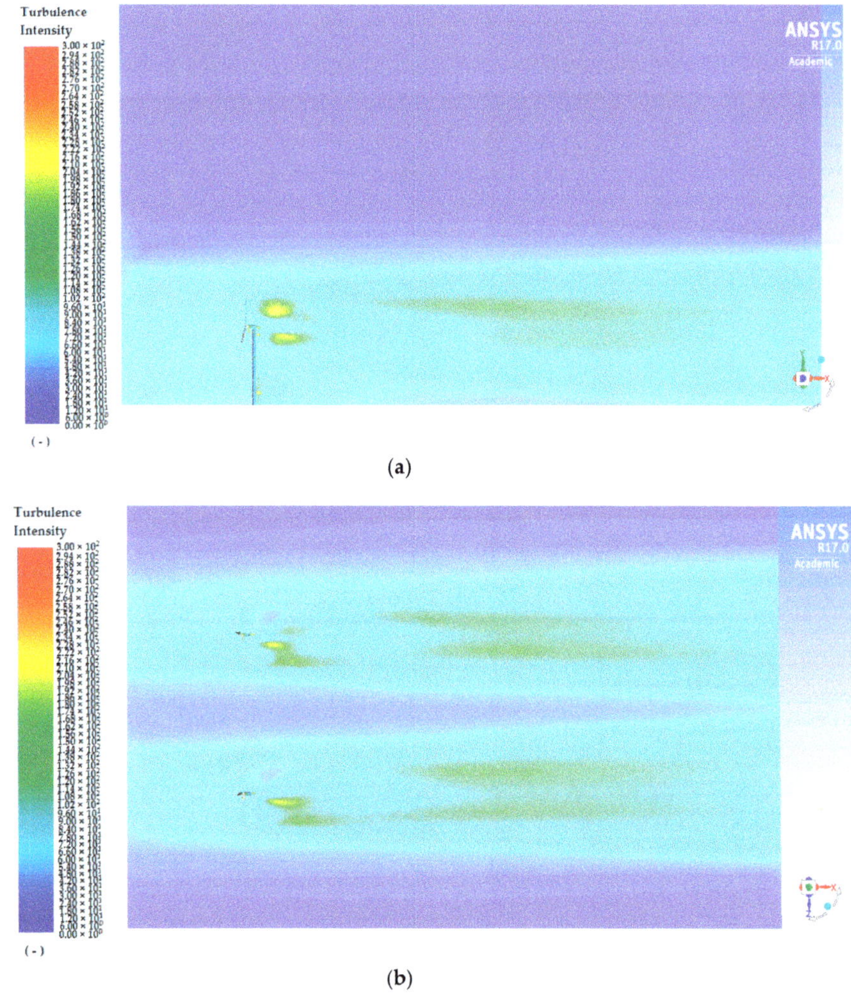

Figure 5. Turbulence Intensity (TI) contours of a second row of turbines, using a profile from a simulation from a first row of turbines. (**a**) Lateral view of the wake; and (**b**) top view of the wake.

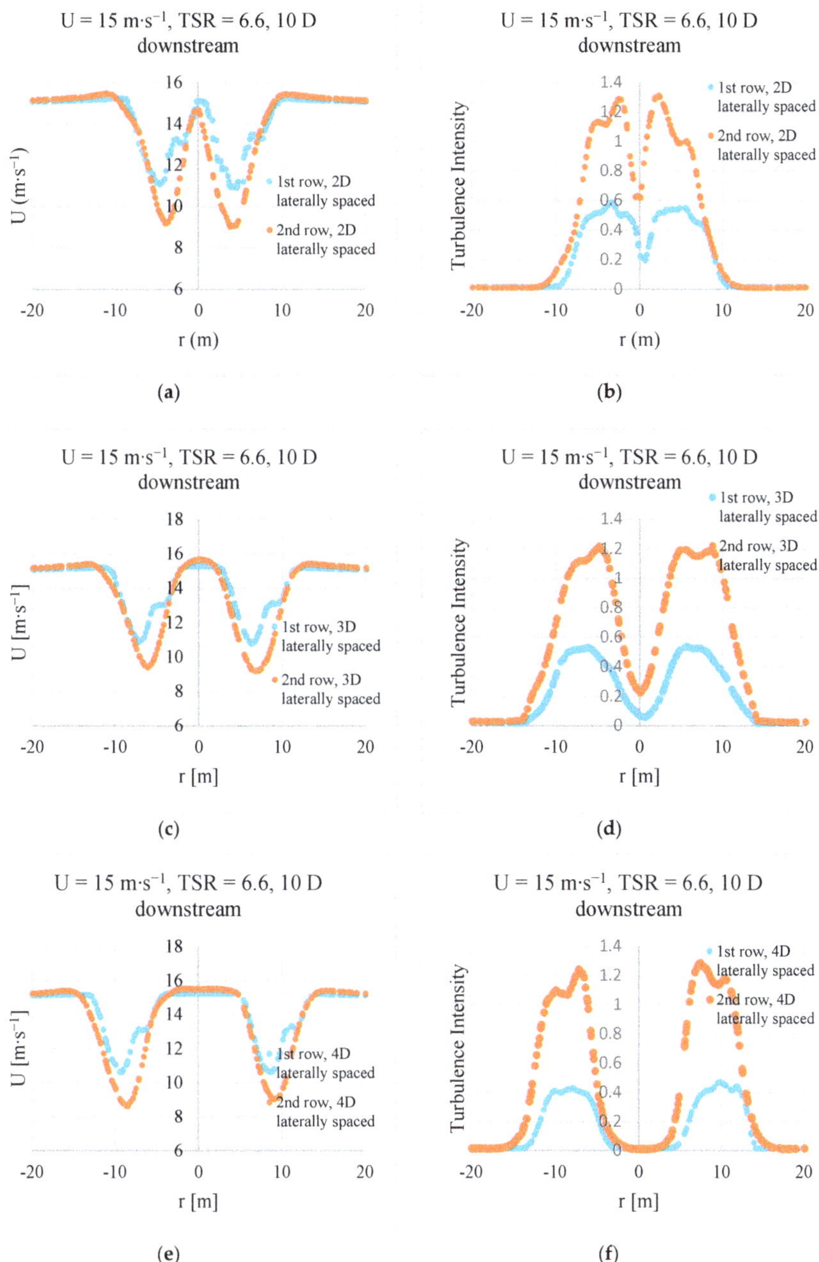

Figure 6. Wake interaction effect, showing wake data plots in the wake of the first and second rows for different lateral spacing in terms of rotor diameter (D): (**a**) Axial velocity for 2 D of lateral spacing; (**b**) TI for 2 D of lateral spacing; (**c**) axial velocity for 3 D of lateral spacing; (**d**) TI for 3 D of lateral spacing; (**e**) axial velocity for 4 D of lateral spacing; (**f**) TI for 4 D of lateral spacing. The spacing distances refer to hub rotor distances.

The evolution of the near wake (up to 6 D) of a single turbine is shown in Figures 7 and 8 for two different free-stream and TSR values (U = 10 m·s^{-1}, U = 15 m·s^{-1}, TSR = 4 and 6.6). The velocity-deficit increased as the TSR increased from 4 to 6.6 for all the positions considered in the wake (Figure 7). In regards to a TSR = 4 and considering U = 10 m·s^{-1}, the wake velocity deficit had a peak of approximately 15% at x/D = 3 in the near wake, and the velocity deficit decreased at x/D = 6 to approximately 11%. The case of TSR = 6.6 and U = 10 m·s^{-1} presents a velocity deficit peak of 25% at x/D = 3 and 17.25% at x/D = 6, which was 9% and 6.25% smaller than the values for U = 10 m·s^{-1} and TSR = 4. The values of velocity deficit for the case of U = 15 m·s^{-1} and TSR = 4 were the same of the case U = 10 m·s-1 and TSR = 4, and so were the other two cases (U = 10 m·s^{-1} TSR = 6.6, and U = 15 m·s^{-1} and TSR = 6.6) as suggests the self-similar theory.

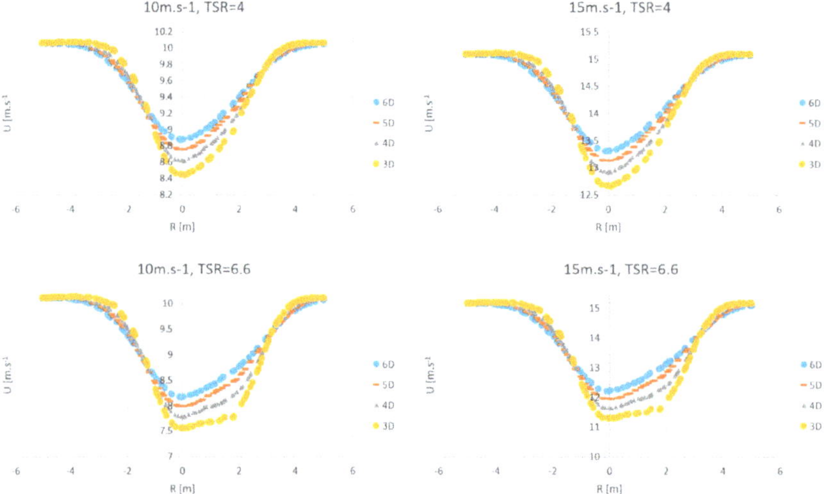

Figure 7. Velocity deficit for two different values of Tip Speed Ratio (TSR) and free-stream velocity.

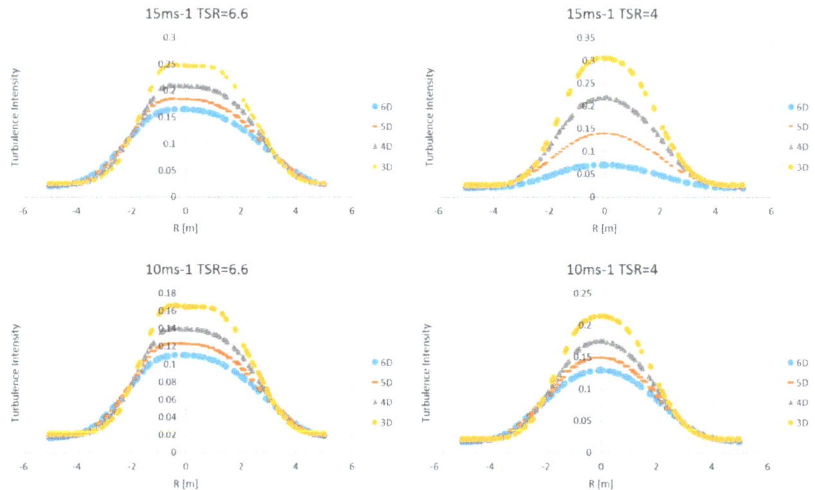

Figure 8. TI for several downstream radial positions and TSR, considering U = 10 m·s^{-1} and U = 15 m·s^{-1}.

3.2. Far Wake Aerodynamics: Influence of Operating Conditions

The problem of optimizing a wind farm layout is very complex, therefore assumptions for the operational conditions are important to allow finding a solution to this type of problem. This explains the importance of this section; it is very important to verify the range of validity of the solution from the optimization routine. In this section, the influence of some important operating design parameters on the velocity deficit and the TI profile in the far-wake development was analyzed including: TSR, pitch angle (θ), and free-stream velocity (U).

3.2.1. Influence of the Tip Speed Ratio (TSR)

The TSR (or λ) critically influenced the far wake behavior. The velocity deficit increased as the TSR increased from 4 to 10, according to the plots from Figure 9 for axial velocity for a radial traverse in the wake at 10 D (diameters) axial location downstream the rotor. Comparing the two values of TSR from Figure 9a, the highest TSR value ($\lambda = 10$) presented the highest velocity-deficit in the far wake behavior for the downstream position considered. Consequently, the TSR was a critical design parameter affecting the three-dimensional extension of the wake. This parameter must be considered to determine the minimal distances between rotors, since a wind farm experiences several different operational conditions with regards to TSR. The TSR (λ) also critically influences the TI in the far wake (Figure 9b), increasing the TSR means that the TI will increase too.

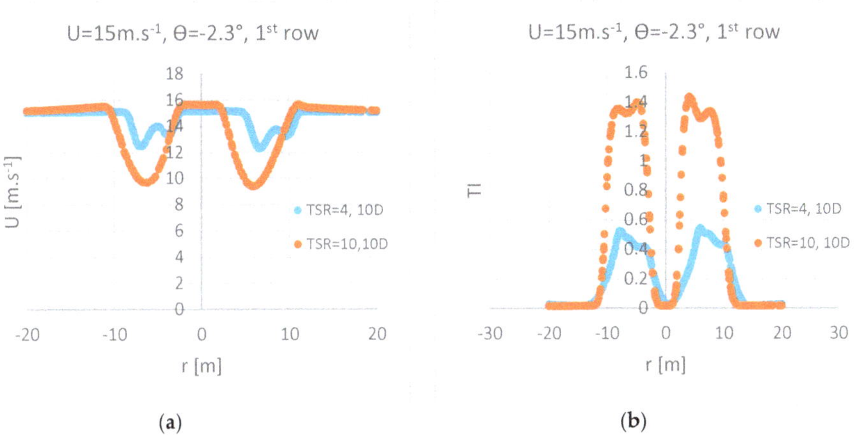

Figure 9. (a) Axial velocity profile at 10D (diameters) downstream the rotor in the wake of the first row; and (b) TI profile at 10D (diameters) downstream the rotor in the wake of the first row. Different line colors represent the TSR (or λ) of 4 (blue) or 10 (orange).

3.2.2. Influence of the Pitch Angle

The pitch angle (θ) had little influence on the velocity and TI profiles in the far wake if the values were kept close to the designed condition. On the other hand, the wake profile was influenced by pitch angle values beyond the designed one. The MEXICO rotor designed condition ($\theta = -2.3°$, $U = 15$ m·s⁻¹, and TSR = 6.6) would result in the best aerodynamic performance when the rotor operates under this specific condition. Part I of this research [56] simulated the MEXICO rotor for pitch angle values ranging from +2.3° to −3°, showing that the velocity deficit in the near wake was significantly higher for pitch angle values close to the designed condition. Particularly, a pitch angle of −3° would result in a velocity deficit three times higher than a pitch angle of +2.3°. These results are expected, since the axial induction of the rotor was higher for the designed condition because more energy was being extracted from the incident wind. For a value of +2.3°, the near wake velocity deficit was lower because of the lower rotor axial induction. In this work, three different values of pitch angle were tested (Figure 10), considering the same free-stream velocity and TSR conditions for all of them. The idea was to check the effect of the variation of the pitch angle on the far wake profile. All three pitch angle values tested were close to the designed condition ($\theta = -2.3°$). The velocity profile (Figure 10a) remained the same at 10 diameters downstream the rotor for all the pitch angles values, whereas there was no significant variation between $\theta = -2.3°$ and $\theta = -3°$ for the TI profile (Figure 10b). Still considering the TI profile (Figure 10b), the case of $\theta = -1°$ showed little deviation from the designed condition $\theta = -2.3°$. This means that the pitch angle may be disregarded for an optimization routine. At least in a preliminary analysis, the pitch angle of individual rotors could be set to the designed condition in order to have the best aerodynamic performance. This could be very important tackling such a complex problem of optimizing wind farm layout, since it is desired to reduce the associated number of variables as much as possible. Figure 10a shows that the velocity wake profile would not be severely affected by doing that, and Figure 10b shows that the effect on the TI profile would be limited to less than a 10% increase. It is important to emphasize again that pitch angle values considerably different than the designed condition would severely affect the wake by altering the velocity and turbulence wake profiles, as shown in part I of this research [56].

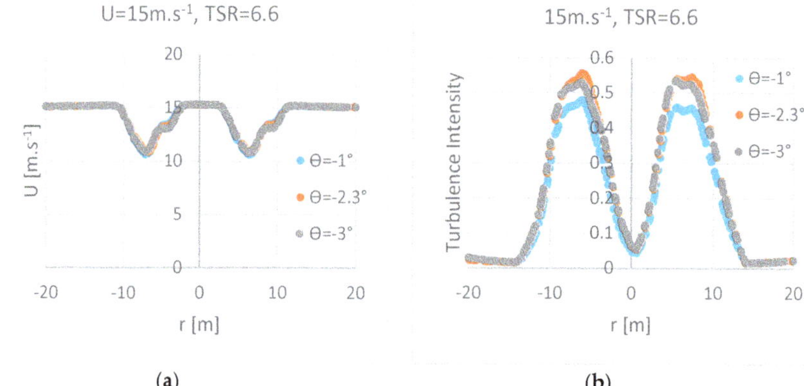

Figure 10. Influence of the pitch angle (θ) on the: (**a**) velocity profile at 10D (diameters) downstream the rotor in the wake; and (**b**) TI profile at 10D (diameters) downstream the rotor in the wake.

3.2.3. Influence of the Free-Stream Velocity

Increasing/decreasing the free-stream velocity value did not affect the magnitude (percentage) of the velocity deficit (Figure 11a). On the other hand, increasing the free-stream velocity value greatly affected the magnitude of the TI (Figure 11b). Consequently, it is important to consider variable free-stream velocity conditions to verify that the optimal wind farm layout solution is not sensitive to the variation of the velocity. Since the turbine components lifetime was closely related to the TI conditions, the variation of the free-stream velocity could be a critical factor to determine the payback of a wind farm.

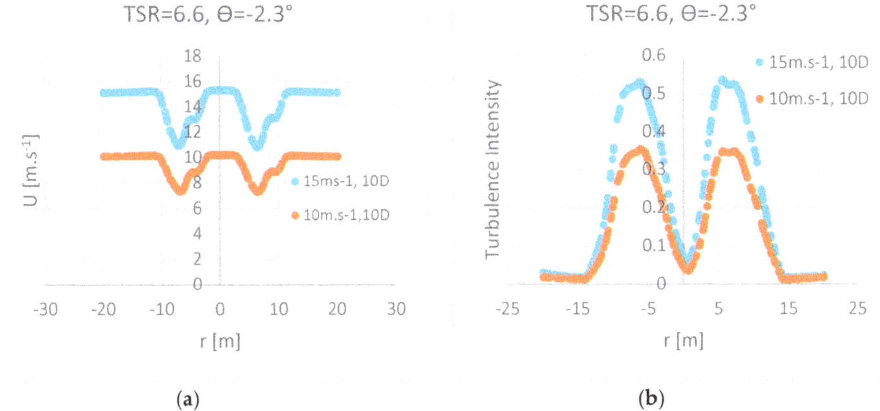

Figure 11. Influence of the free-stream velocity on the: (**a**)velocity-deficit at 10D (diameters) downstream the rotor in the wake; and (**b**) TI at 10D (diameters) downstream the rotor in the wake.

4. Discussion

The MEXICO experiment tested specific operating conditions, providing experimental PIV (Particle Image Velocimetry) measurements for the velocity flow field in the near wake. Far wake measurements in a controlled (wind tunnel) was not viable because of technical constraints, which explains why the MEXICO experiment did not provide such measurements. Although it is impossible to compare the far wake data from the simulations performed in this work against experimental data from the MEXICO rotor, the results of the velocity in the far wake can be verified by comparing other relevant works in the literature. Such comparisons show that the results of the CFD model developed in this work consistently agree with previous literature studies. It is worth mentioning that the following studies cited in this discussion neither computationally implemented the MEXICO rotor nor the exact same operating conditions than the ones implemented in this current work. Despite that, the comparisons presented in this discussion intend to verify an agreement with an acceptable range of consistency.

For instance, the range of velocity deficit found in this study agreed with other works in literature. Some of these studies specified operating conditions in which the results were obtained. Mittal et al. [54] analyzed operating conditions of TSR = 6 at $x/D = 5$, finding 40% of velocity deficit and a not completely symmetrical radial curve profile. Asymmetry was attributed to interaction with the tower. For the off-design condition of TSR = 3, the velocity deficit had a peak of 30% at $x/D = 5$. For the off-design condition of TSR = 10 showed a peak of 80% of the velocity deficit at $x/D = 1$, and 40% at $x/D = 5$. Although the velocity deficit values were slightly greater than the ones found in this work, there is consistency between the results found in this work (Figure 7), in which the estimated velocity deficit showed a value of approximately 25% for a TSR = 6.6 and 15% for a TSR = 4. Storey et al. [37] found that as the free-stream velocity increased and the TSR decreased (rpm maintained constant), the overall velocity deficit decreased. These results confirm the trend found in Figure 7, Figure 8, and Figure 9: A decrease of the TSR, results in a decrease of the velocity deficit. Additionally, Storey et al. [37] found that the shape and magnitude of the velocity deficit vary significantly with the wind speed and TSR. The expansion of the wake varies with wind speed, confirming the trend observed in this work in Figure 11b: The wake presents more pronounced expansion as the velocity varied from 10 m·s⁻¹ to 15 m·s⁻¹.

Moreover, even though some of the works in the literature did not specify operating conditions regarding TSR, their data consistently agree with the results for the velocity deficit found in this work. Prospathopoulos et al. [22] considered a downstream spacing of 5 D between the turbines, finding a velocity deficit in the wake of 40% at 2.5 D and 30% at 3.5 D for the stable stratification case of the Energy Research Centre of the Netherlands (ECN) measurements. Those results are similar to the values found in this present work (Figure 7). Gundling et al. [7] modeled wake wind speed deficits for different wake models and compared them. The UWAKE (Free-vortex Wake with a BEM) model has a maximum velocity deficit of 67% at 4R for 5 m·s^{-1}, and a maximum velocity deficit of 81% at 7R for 10 m·s^{-1}. The FLOWYO (ALM and ADM) LES has a maximum velocity deficit of 70% at 3R for 5 m·s^{-1}, and a maximum velocity deficit of 83% at 6R for 10 m·s^{-1}. The wake deficit is similar for the FLOWYO and UWAKE, but little diffusion of the wake was found when using FLOWYO RANS. The diffusion in the wake is similar using UWAKE and FLOWYO LES, while the FLOWYO RANS had not enough turbulent eddy-viscosity produced by the ADM to result in a similar wake diffusion compared to FLOWYO LES and UWAKE. The HELIOS DES model has a velocity deficit of 58% at 4R for 5 m·s^{-1}, and 65% at 9R for 10 m·s^{-1}. All those values are within an acceptable range when compared to the values found in this work (Figures 7 and 8). Troldborg et al. [10] analyzed different turbulent inflow conditions for a FRM, an ALM, and an ADM approach. The ALM and the ADM showed the same results for velocity deficit in the wake. A maximum peak of approximately 60% was found at 2 R, which remained almost constant in the same value up to the 10 R analyzed. The FRM showed the same 60% velocity deficit at 2 R, but decreasing the peak value to approximately 50% at 10 R. Those results are similar to the ones found in this research for the wake characteristics (Figures 7 and 8).

Furthermore, a similarity in the shape/format of the wake profile was identified. Mo et al. [8] determined velocity profiles at the wake for several downstream positions. A near-symmetrical but not completely at the blades location, the vertical wake velocity profile had a clear W shape at 1 D and 2 D, and overall the velocity deficit decreased as the free-stream velocity increased from 5 m·s^{-1} to 15.1 m·s^{-1}. This was attributed to the state of the completed attached flow in the turbine blade for smaller velocities, and not for more extracted power from the incident wind. Wake shape was not well defined for the further downstream positions. The W shape of the velocity deficit curves is similar to the curve shape found in this present work in Figure 9a for TSR = 4.

The TI behavior showed an increase of its peak value as the TSR increased (Figure 9b). For the cases of wake interaction (Figure 6a,b), there was a considerable increase of TI in the wake of the second row. A verification analysis consistently agrees with the literature results. Troldborg et al. [38] analyzed two turbines case for wake interaction and found that a spacing of 7 D was large enough to allow the wake profile to reach a steady state after the second turbine. AbdelSalam et al. [55] found 65% velocity deficit peak for x/D = 2, 60% velocity deficit peak at x/D = 4, 50% peak velocity deficit at x/D = 6, and 30% velocity deficit peak at x/D = 8. Those results are within an acceptable agreement with the results found in this work (Figures 9–11). Wilson et al. [16] modeled ADM, ALM, and FRM (full rotor). The wake interaction case showed a strong interaction of wakes when spacing 5 D, which is not easy to compare with our study because the downstream distance implemented is not the same. For a single turbine case, the velocity deficit found by Wilson et al. [16] was slightly higher for ADM than ALM, and MR presented the highest velocity deficit in the wake. The TI was significantly higher for ADM and ALM when compared with FRM.

In regards to field experimental data, Barthelmie et al. [62] studied the influence of the downstream spacing between the turbine rows in the normalized power for the case of the Horns Rev offshore wind farm. Considering 8 m·s^{-1} and the 2° sector and a downstream spacing of 7 D, the ratio between the output power of the second and first turbine row was approximately 58%. The output power ratio between the second and third row was 56%. For a downstream spacing of 9.4 D, the ratio between the output power of the second and first turbine row was approximately 70%. The output power ratio between the second and third row was 68%. For a downstream spacing of 10.5 D, the ratio between the output power of the second and first turbine row was approximately 75%. The output power ratio between the second and third row was 70%. Considering 8 m·s^{-1} and the 30-degree sector, the downstream spacing of 7 D had a ratio between the output power of the second and first turbine row of approximately 80%. The output power ratio between the second and third row was 79%. For a downstream spacing of 9.4 D, the ratio between the output power of the second and first turbine row was approximately 85%. The output power ratio between the second and third row was 80%. For a downstream spacing of 10.5 D, the ratio between the output power of the second and first turbine row was approximately 88%. The output power ratio between the second and third row was 83%. As stated at the beginning of the discussion, operating conditions as well as turbines were not the same of the ones analyzed by Barthelmie et al. [62], such that no direct comparison can be made. However, all those results consistently agree within acceptable levels with the results found in this work (Figures 6 and 9–11). In this work, a downstream distance (spacing) between rows was 10 D, and the estimated ratio between output power of the first and second row was within the range of 73% (for a TSR = 4) and 35% (TSR = 10). A TSR = 4 was a more descriptive operating condition for commercial turbines than a TSR = 10, since turbines are more likely to operate in higher TSR conditions (such as TSR = 10) and only in extreme events such as a wind gust. Figure 6 allows a comparison between the second and third row for a TSR = 6.6, showing a power ratio estimate of 56%. The verification against Barthelmie et al. [62] results is especially important since these are wake field data, showing that the CFD model developed in this work is consistent even when compared against atmospheric measurements.

The discussion of Figure 6 gives a good sense on why to simulate two turbines instead of just one. Figure 6 shows the wake interaction effect that the two-turbine side-by-side case can cause on the TI profile in the wake region between these turbines. The simulation of only one turbine would not allow the detection in the increase in TI levels in the referred region. Moreover, a major objective of this research was to create a computational tool to be implemented in future research aiming to improve wind farm land use by investigating the ratio between output power and area for both aligned and staggered configurations. In such a study, TI effects caused by wake interaction need to be pointed out when proposing improvements for land use. This explains the need for simulating at least two turbines.

There are different approaches to model wind turbines, and some of the most common ones are ADM, ALM, and ASM. The ALM represents the blade by a line, and the ASM represents the blade by a planar surface. The ADM does not model the blades surface, and the rotor is modeled by an infinitesimal disc that represents a discontinuity in pressure. All the actuator approaches (ADM, ALM, and ASM) require knowledge of tabulated lift and drag on the blades, and they require corrections for Coriolis, centrifugal, and tip effects when 2D airfoil data are used [3]. According to Vermeer et al. [63], there are many reasons for 2D airfoil data to be corrected to better represent 3D cases. For instance, rotational effects limit the growth of the boundary layer at separation, resulting in increased lift for 3D cases in comparison with 2D cases. Moreover, the drag coefficient can largely differ for 3D cases in comparison with 2D because airfoil characteristics depend on the aspect ratio of the blade. Finally, airfoils under large temporal variations of the angle of attack present hysteresis that changes the static airfoil data. There are some critiques of actuator models. According to Sanderse et al. [3], the ASM requires more accurate airfoil data, as well as knowledge of pressure and skin-friction on the airfoil. According to Churchfield et al. [64], one critique of the ALM is that it does not model the surface of the blade, and in this way this technique is not capable of replicating finer flow features such as boundary layer and separation at high angle of attack. In spite of that, these techniques are still widely used because of the reduced required computational costs.

Previous works in the literature discussed the accuracy of ADM, ALM, and ASM models. Even though the referred models generally represent aerodynamic performance in a satisfactory way, there is a consensus about the need for improvements for properly representing wake flow field under more complex fluid flow configurations. Gundling et al. [7] found that the aerodynamic performance for the NREL Phase VI is well predicted at pre-stalling conditions by low complexity models (BEM and actuator approaches), presenting a similar level of accuracy as higher complexity methods (FRM CFD). The results for high and low fidelity models differ for stalled conditions, but they generally present a good agreement for other less complex situations. The aerodynamic performance at the transition regime was not well predicted by neither low complexity nor high complexity models, but high complexity models accurately predict performance at higher wind speeds when stalling is dominant at the blade surface. The use of a high complexity model (FRM CFD) with adaptative mesh accurately solved the far wake flow field up to a distance of 20 radii downstream of the rotor, but did not show remarkable benefits for performance prediction, in which only the near wake up to 1 radius downstream of the rotor was resolved. Troldborg et al. [10] found that the wake predicted by the ALM and the ADM have very close agreement for uniform inflow conditions, but there is a significant difference when compared with the full resolved rotor method. Additionally, the fully resolved rotor presented higher turbulence levels in the wake. At turbulent inflow, the three methods (ADM, ALM, and FRM) present a close agreement. Troldborg et al. [32] also found good agreement between computational and field data for the turbulence flow field and mean wake deficit. Réthoré et al. [13] compared improved ADM approaches with a CFD FRM, finding significant differences in the turbulence flow field at the wake. The blades and the nacelle from the FRM approach showed a production of turbulence several orders of magnitude higher than the turbulence produced by the ADM. Theunissen et al. [19] accurately performed power calculations using ADM but they found differences in comparison with experimental data for the wake velocity profiles. Storey et al. [37] showed that correct modeling of the ABL and turbulence inflow conditions are important to determine the stability of wind turbine wakes, which may suggest that FRM approach is more suitable for more accuracy on wake prediction. Laan et al. [39] showed that there can be improvements to the actuator approaches by introducing new methodologies to better represent the forces on the rotor.

Even though actuator approaches have still been widely applied in research because of its reduced computational costs, as Sanderse et al. [3] points out, there is need for a detailed study on the influence of exact blade geometry on far wake characteristics. In order to account for the influence of blades geometry on wake characteristics, the FRM CFD approach has been proved as a powerful tool for wind farm design. Unlike the ADM, ALM, or ASM, the CFD model of this work is a direct modeling approach of the rotor which considers the exact 3D blade geometry, including variable chord length, local twist angle, and pitch angle. The boundary layer was solved using 10 inflation layers with a ratio of 1.1 to ensure y+ < 1 next to the blade surface. Therefore, the CFD model of this work does not require the use of tabulated lift and drag data as does the ADM, ALM, and ASM. The capabilities of the model developed in this research allow the evaluation of wind turbine wakes interaction for multiple turbines while still using reduced computational resources. Even though further improvements of the modeling technique implemented in this research might be necessary to account the influence of downstream rows on the wake of upstream rows, the technique is promising towards reducing the computational costs for wind farms simulation. The model is a steady state approach (RANS) aiming to save computational time but can easily be improved to a transient simulation by using a sliding mesh. Moreover, the model of this work has been validated in part I [56] against experimental data for the near wake velocity field. This brings more confidence that the model can properly represent aerodynamics characteristics of the wake flow field. As pointed out by Rodrigo et al. [47], a more realistic description of the wake generation mechanisms in the near wake allows to understand and improve far wake models.

The method implemented in this work applied the outlet from upstream rows as the inlet of downstream rows, with the aim of studying wake interaction effects. This has been performed in this work in a one-way coupling, meaning that only the wake effects from an upstream row will be experienced by downstream rows. A possible effect on upstream rows coming from the interaction with a downstream row (e.g., a second row influencing the far wake of the first row) has not been considered in this work. This might be particularly important for the case of staggered wind farms, which are more densely packed than wind farms with aligned rows. In spite of that, the results presented in this manuscript considered spacing between rows (10 rotor diameters) large enough to dissipate the influence of downstream rows on the wake of an upstream row. A further improvement of the model developed in this research could include the development of a two-way coupling method capable of simulating the effects of downstream rows on the wake of an upstream row, which, as previously mentioned, might matter for the case of staggered farms. Aligned configurations will typically have greater spacing between rows, which potentially dissipates the referred wake interaction effects.

Inflation layers (Figure 2f) were implemented next to the blade surface, keeping y+ < 1 at this location for accurately characterizing boundary layer effects. In ANSYS Fluent, inflation layers can be automatically generated next to selected surfaces, regardless of the surface length (scaled or full-size blades). As specified in the methods section, 10 inflation layers were sufficient to keep y+ < 1 next to the prototype blade surface. In the case of a full-scale wind turbine which operate at higher Reynolds number in comparison with scaled prototypes, a different mesh resolution would be necessary to resolve boundary layer effects. The method implemented in this work would require a larger quantity of inflation layers to achieve mesh resolution small enough to keep y+ < 1 close to the blade surface of a full-scale turbine. The increase in the quantity of inflation layers would require an extra spent, making it more computationally expensive to keep a low y+ value for full-scale rotors. Even though the challenges previously discussed could require more computational costs, they would not prevent the method to be applicable for characterizing full-scale turbines. Moreover, wind turbine wake characteristics have been successfully studied using scaled prototypes such as the one implemented in this work (MEXICO rotor). Whale et al. [65] performed PIV measurements for an untwisted prototype with a flat plate airfoil profile, operating at TSR within 3 and 8 and Reynolds number within 6400 and 16,000. The study showed that wake behavior is insensitive to blade chord Reynolds number, as long as similarity of the TSR is maintained. This is the same hypothesis assumed in this current work (described in Section 2.4), where we assumed that wake similarity can be achieved by matching TSR values of scaled and full-scale turbines. Hossain et al. [61] performed wake measurements of a 500 mm four-bladed turbine using a PIV system, ultrasonic anemometers, and hot-wire anemometers. They performed the same measurements for a 1/10 scaled prototype aiming to study geometry similarity, finding that the wake decays at almost the same levels for the full-scale and the prototype turbines. Ivanell et al. [30] implemented an ADM model for a 0.2 m diameter turbine, showing that the power coefficient is independent of Reynolds number if the Reynolds number is greater than 1000. Additionally, they found that the Reynolds number does not affect the strength of the wake vortex, but only the radial vorticity distribution. Sturge et al. [60] performed an experiment using a wind turbine prototype in which the blade Reynolds number was two order of magnitude lower than the Reynolds number of common full-scale turbines. They presented an interesting discussion about scaled turbines, showing that the Reynolds number became less important for far wake modeling when the ADM was implemented. All these studies mentioned above are examples in the literature that suggest that wake characteristics can be understood by using scaled turbines.

5. Conclusions

In this work, a CFD model based on the MRF approach was developed to assess wind turbine far wake characteristics according to operating conditions typically experienced in commercial wind farms. The influence of the TSR and free-stream wind speed on wake characteristics such as velocity deficit and TI was discussed and compared with the existing literature on this topic.

This paper reviewed most of the wind turbine wakes studies and wind farm CFD techniques from the literature. Overall, we found that the existing literature studies use different turbulence modeling techniques, as well as CFD solvers with different assumptions and boundary conditions. The wake results vary according to the approach adopted in each work. A gap was identified in the literature review of this work, showing that there is a need for more development of CFD models capable of simulating a whole wind farm. The vast majority of the CFD studies simulate single turbines, and only a few of them simulate more than one rotor. The computational resources may be a limiting factor for that, representing one of the biggest challenges on wind farm computational modeling: The expensive computational resources required to simulate several rows in a wind farm. In order to address this need, this work presented a novel methodology to analyze wind turbine wakes interaction with relatively reduced computational resources. The technique had never been applied before in the context of wind farm numerical modeling. Even though multiple simulations are required for studying the interaction effect between upstream and downstream rows in a wind farm, this work successfully achieved a reduction in computational capabilities (processors) required to perform wake interaction simulations. This represents an advance for wind farm modeling, and many researchers could benefit using such techniques to improve wind energy CFD models.

The model presented in this work was previously validated in part I [56] with regards to near wake data. In part II, a verification of the model against other studies in the literature showed consistency in the wake results within acceptable levels. In regards to the velocity deficit and TI assessment, the values found in this work were similar to other CFD wake studies in the literature. This demonstrates the ability of the proposed CFD model in predicting wake characteristics, and this way the model is ready to be applied for determining the optimal spacing between turbines in a wind farm. The capability of the proposed CFD model showed to be consistent when compared with field data, kinematical models, and CFD results from the literature, showing similar ranges of wake deficit.

Further improvement of the model will include a transient approach modeling to determine wake characteristics according to variable rotor operating conditions. This will extend the capabilities of the proposed model by adding a more realistic modeling approach to derive the aerodynamic behavior of the turbine rows. Moreover, a FSI (fluid solid interaction) model would be relevant to determine how the structural behavior of the blades is affected by variable wind conditions. Although the deformation of the blades will have an impact on the blade fatigue lifetime, no study has previously shown that far wake aerodynamics is significantly impacted by the level of blade deformation. Furthermore, there is still room for improvement of the mesh layout in order to reduce even more the computational resources required for simulating wakes interaction effects.

Author Contributions: Conceptualization, R.V.R. and C.L.; methodology, R.V.R. and C.L.; software, R.V.R.; validation, R.V.R.; formal analysis, R.V.R.; investigation, R.V.R. and C.L.; writing—original draft preparation, R.V.R.; writing—review and editing, R.V.R. and C.L.; funding acquisition, R.V.R.

Funding: This research was funded by Conselho Nacional de Desenvolvimento Científico e Tecnológico (CNPq), grant number 249258/2013-7.

Abbreviations

TSR	Tip Speed Ratio
TI	Turbulence Intensity
RANS	Reynolds Averaged Navier-Stokes
BEM	Blade Element Method
CFD	Computational Fluid Dynamics
LES	Large-Eddy Simulation
ABL	Atmospheric Boundary Layer
MRF	Moving Reference Frame
MEXICO	Model Experiments in Controlled Conditions
θ	Pitch Angle
ω	Rotational Speed
R	Blade Radius

Appendix A

Mesh sensitivity study

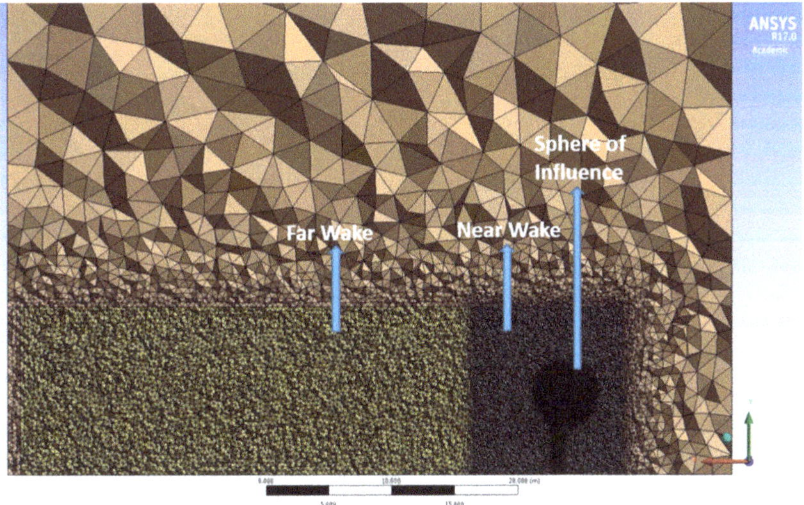

Figure A1. Physical domain showing the refinement of the mesh at different regions in the wake.

Table A1. Four cases of mesh sensitivity study.

	Elements	Nodes	Rotor (m)	Central Disc (m)	Sphere of Influence (m)	Near Wake (m)	Far Wake (m)
Case 1	19,602,483	3,325,671	0.025	0.05	0.075	0.15	0.3
Case 2	10,424,238	1,782,980	0.025	0.05	0.075	0.2	0.4
Case 3	7,055,590	1,215,712	0.025	0.05	0.075	0.25	0.5
Case 4	4,025,982	705,040	0.025	0.05	0.075	0.5	0.75

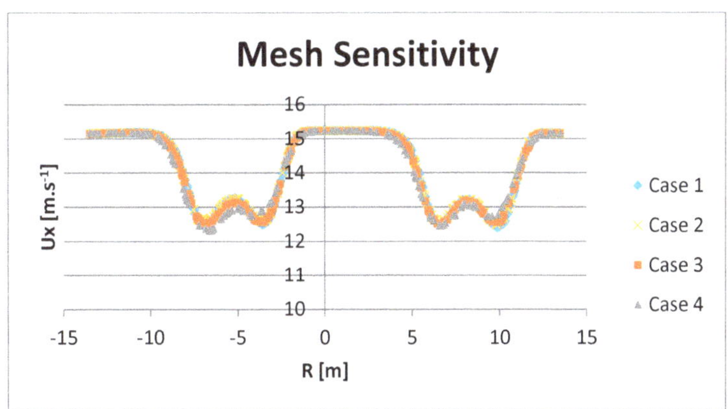

Figure A2. Mesh sensitivity study.

References

1. Crespo, A.; Hernández, J.; Frandsen, S. Survey of Modelling Methods for Wind Turbine Wakes and Wind Farms. *Wind Energy* **1999**, *2*, 1–24. [CrossRef]
2. Shakoor, R.; Hassan, M.Y.; Raheem, A.; Wu, Y.-K. Wake effect modeling: A review of wind farm layout optimization using Jensen's model. *Renew. Sustain. Energ. Rev.* **2016**, *58*, 1048–1059. [CrossRef]
3. Sanderse, B.; van der Pijl, S.P.; Koren, B. Review of computational fluid dynamics for wind turbine wake aerodynamics. *Wind Energy* **2011**, *14*, 799–819. [CrossRef]
4. Ainslie, J.F. Calculating the flow field in the wake of wind turbines. *J. Wind Eng. Ind. Aerod.* **1988**, *27*, 213–224. [CrossRef]
5. Zhou, N.; Chen, J.; Adams, D.E.; Fleeter, S. Influence of inflow conditions on turbine loading and wake structures predicted by large eddy simulations using exact geometry. *Wind Energy* **2016**, 803–824. [CrossRef]
6. Hsu, M.-C.; Akkerman, I.; Bazilevs, Y. Finite element simulation of wind turbine aerodynamics: Validation study using NREL Phase VI experiment. *Wind Energy* **2014**, *17*, 461–481. [CrossRef]
7. Gundling, C.; Sitaraman, J.; Roget, B.; Masarati, P. Application and validation of incrementally complex models for wind turbine aerodynamics, isolated wind turbine in uniform inflow conditions. *Wind Energy* **2015**, *18*, 1893–1916. [CrossRef]
8. Mo, J.-O.; Choudhry, A.; Arjomandi, M.; Kelso, R.; Lee, Y.-H. Effects of wind speed changes on wake instability of a wind turbine in a virtual wind tunnel using large eddy simulation. *J. Wind Eng. Ind. Aerod.* **2013**, *117*, 38–56. [CrossRef]
9. Choudhry, A.; Mo, J.-O.; Arjomandi, M.; Kelso, R. Effects of wake interaction on downstream wind turbines. *Wind Eng.* **2014**, *38*, 535–548. [CrossRef]
10. Troldborg, N.; Zahle, F.; Réthoré, P.-E.; Sorensen, N.N. Comparison of wind turbine wake properties in non-sheared inflow predicted by different computational fluid dynamics rotors models. *Wind Energy* **2015**, *18*, 1239–1250. [CrossRef]
11. Storey, R.C.; Norris, S.E.; Cater, J.E. An actuator sector method for efficient transient wind turbine simulation. *Wind Energy* **2015**, *18*, 699–711. [CrossRef]
12. Seydel, J.; Aliseda, A. Wind turbine performance in shear flow and in the wake of another turbine through high fidelity numerical simulations with moving mesh technique. *Wind Energy* **2013**, *16*, 123–138. [CrossRef]
13. Réthoré, P.; Laan, P.; Troldborg, N.; Zahle, F.; Sørensen, N.N. Verification and validation of an actuator disc model. *Wind Energy* **2014**, *17*, 919–937.
14. Heinz, J.C.; Sørensen, N.N.; Zahle, F. Fluid–structure interaction computations for geometrically resolved rotor simulations using CFD. *Wind Energy* **2016**, *19*, 2205–2221. [CrossRef]
15. Miao, W.; Li, C.; Pavesi, G.; Yang, J.; Xie, X. Investigation of wake characteristics of a yawed HAWT and its impacts on the inline downstream wind turbine using unsteady CFD. *J. Wind Eng. Ind. Aerod.* **2017**, *168*, 60–71. [CrossRef]
16. Wilson, J.M.; Davis, C.J.; Venayagamoorthy, S.K.; Heyliger, P.R. Comparisons of Horizontal-Axis Wind Turbine Wake Interaction Models. *J. Sol. Energy Eng.* **2015**, *137*, 3–8. [CrossRef]
17. Weipao, M.; Chun, L.; Jun, Y.; Yang, Y.; Xiaoyun, X. Numerical Investigation of Wake Control Strategies for Maximizing the Power Generation of Wind Farm. *J. Sol. Energy Eng.* **2016**, *138*, 1–7. [CrossRef]
18. Wu, C.-H.K.; Nguyen, V.-T. Aerodynamic simulations of offshore floating wind turbine in platform-induced pitching motion. *Wind Energy* **2017**, *20*, 835–858. [CrossRef]
19. Theunissen, R.; Housley, P.; Allen, C.B.; Carey, C. Experimental verification of computational predictions in power generation variation with layout of offshore wind farms. *Wind Energy* **2015**, *18*, 1739–1757. [CrossRef]
20. Tran, T.T.; Kim, D.-H.; Nguyen, B.H. Aerodynamic Interference Effect of Huge Wind Turbine Blades with Periodic Surge Motions Using Overset Grid-Based Computational Fluid Dynamics Approach. *J. Sol. Energy Eng.* **2015**, *137*, 1–16. [CrossRef]
21. Zahle, F.; Sorensen, N.N. Charactertization of the unsteady flow in the nacelle region of a modern wind turbine. *Wind Energy* **2011**, *14*, 271–283. [CrossRef]
22. Prospathopoulos, J.M.; Politis, E.S.; Rados, K.G.; Chaviaropoulos, P.K. Evaluation of the effects of turbulence model enhancements on wind turbine wake predictions. *Wind Energy* **2011**, *14*, 285–300. [CrossRef]

23. AbdelSalam, A.M.; Boopathi, K.; Gomathinayagam, S.; Kumar, S.S.H.K.; Ramalingam, V. Experimental and numerical studies on the wake behavior of a horizontal axis wind turbine. *J. Wind Eng. Ind. Aerod.* **2014**, *128*, 54–65. [CrossRef]
24. Boudreau, M.; Dumas, G. Comparison of the wake recovery of the axial-flow and cross-flow turbine concepts. *J. Wind Eng. Ind. Aerod.* **2017**, *165*, 137–152. [CrossRef]
25. Ammara, I.; Leclerc, C.; Masson, C. A viscous three-dimensional differential/Actuator-Disk Method for the Aerodynamic Analysis of Wind Farms. *J. Sol. Energy Eng.* **2002**, *124*, 345–356. [CrossRef]
26. Frau, E.; Kress, C.; Chokani, N.; Abhari, R.S. Comparison of Performance and Unsteady Loads of Multimegawatt Downwind and Upwind Turbines. *J. Sol. Energy Eng.* **2015**, *137*, 1–8. [CrossRef]
27. Van der Laan, M.P.; Kelly, M.C.; Sørensen, N.N. A new k-epsilon model consistent with Monin–Obukhov similarity theory. *Wind Energy* **2017**, *20*, 479–489. [CrossRef]
28. Van der Laan, M.P.; Sørensen, N.N.; Réthoré, P.E.; Mann, J.; Kelly, M.C.; Troldborg, N.; Hansen, K.S.; Murcia, J.P. The k–ε-fP model applied to wind farms. *Wind Energy* **2015**, *18*, 2065–2084. [CrossRef]
29. Sarmast, S.; Segalini, A.; Mikkelsen, R.F.; Ivanell, S. Comparison of the near-wake between actuator-line simulations and a simplified vortex model of a horizontal-axis wind turbine. *Wind Energy* **2016**, *19*, 471–481. [CrossRef]
30. Ivanell, S.; Sørensen, J.N.; Mikkelsen, R.; Henningson, D. Analysis of Numerically Generated Wake Structures. *Wind Energy* **2009**, *12*, 63–80. [CrossRef]
31. Masson, C.; Smaïli, A. Numerical Study of Turbulent Flow around a Wind Turbine Nacelle. *Wind Energy* **2006**, *9*, 281–298. [CrossRef]
32. Troldborg, N.; Larsen, G.C.; Madsen, H.A.; Hansen, K.S.; Sørensen, J.N.; Mikkelsen, R. Numerical simulations of wake interaction between two wind turbines at various inflow conditions. *Wind Energy* **2011**, *14*, 859–876. [CrossRef]
33. Makridis, A.; Chick, J. Validation of a CFD model of wind turbine wakes with terrain effects. *J. Wind Eng. Ind. Aerod.* **2013**, *123*, 12–29. [CrossRef]
34. Schulz, C.; Letzgus, P.; Lutz, T.; Krämer, E. CFD study on the impact of yawed inflow on loads, power and near wake of a generic wind turbine. *Wind Energy* **2017**, *20*, 253–268. [CrossRef]
35. Ivanell, S.; Mikkelsen, R.; Sørensen, J.N.; Henningson, D. Stability analysis of the tip vortices of a wind turbine. *Wind Energy* **2010**, *13*, 705–715. [CrossRef]
36. Bromm, M.; Vollmer, L.; Kühn, M. Numerical investigation of wind turbine wake development in directionally sheared inflow. *Wind Energy* **2017**, *20*, 381–395. [CrossRef]
37. Storey, R.C.; Norris, S.E.; Stol, K.A.; Cater, J.E. Large eddy simulation of dynamically controlled wind turbines in an offshore environment. *Wind Energy* **2013**, *16*, 845–864. [CrossRef]
38. Troldborg, N.; Sorensen, J.N.; Mikkelsen, R. Numerical simulations of wake characteristics of a wind turbine in uniform inflow. *Wind Energy* **2010**, *13*, 86–99. [CrossRef]
39. Van der Laan, M.P.; Sørensen, N.N.; Réthoré, P.E.; Mann, J.; Kelly, M.C.; Troldborg, N. The k–ε-fP model applied to double wind turbine wakes using different actuator disk force methods. *Wind Energy* **2015**, *18*, 2223–2240. [CrossRef]
40. Abkar, M.; Porté-Agel, F. The Effect of Free-Atmosphere Stratification on Boundary-Layer Flow and Power Output from Very Large Wind Farms. *Energies* **2013**, *6*, 2338–2361. [CrossRef]
41. Archer, C.L.; Mirzaeisefat, S.; Lee, S. Quantifying the sensitivity of wind farm performance to array layout options using large-eddy simulation. *Geophys. Res. Lett.* **2013**, *40*, 4963–4970. [CrossRef]
42. Calaf, M.; Meneaveau, C.; Meyers, J. Large eddy simulation study of fully developed wind-turbine array boundary layers. *Phys. Fluids* **2010**, *22*, 015110. [CrossRef]
43. Mirocha, J.D.; Churchfield, M.J.; Muñoz-Esparza, D.; Rai, R.K.; Feng, Y.; Kosović, B.; Haupt, S.E.; Brown, B.; Ennis, B.L.; Draxl, C.; et al. Large-eddy simulation sensitivities to variations of configuration and forcing parameters in canonical boundary-layer flows for wind energy applications. *Wind Energy Sci.* **2018**, *3*, 589–613. [CrossRef]
44. Nilsson, K.; Ivanell, S.; Hansen, K.S.; Mikkelsen, R.; Sørensen, J.N.; Breton, S.; Henningson, D. Large-eddy simulations of the Lillgrund wind farm. *Wind Energy* **2015**, *18*, 449–467. [CrossRef]
45. Wu, Y.; Porté-Agel, F. Large-Eddy Simulation of Wind-Turbine Wakes: Evaluation of Turbine Parametrisations. *Boundary-Layer Meteorol.* **2011**, *138*, 345–366. [CrossRef]

46. Mehta, D.; van Zuijlen, A.H.; Koren, B.; Holierhoek, J.G.; Bijl, H. Large Eddy Simulation of wind farm aerodynamics: A review. *J. Wind Eng. Ind. Aerod.* **2014**, *133*, 1–17. [CrossRef]

47. Rodrigo, J.S.; Arroyo, R.A.C.; Moriarty, P.; Churchfield, M.; Kosovic, B.; Réthoré, P.; Hansen, K.S.; Hahmann, A.; Mirocha, J.D.; Rife, D. Mesoscale to microscale wind farm flow modeling and evaluation. *Energy Environ.* **2017**, *6*, 1–30. [CrossRef]

48. Choi, N.J.; Nam, S.H.; Jeong, J.H.; Kim, K.C. Numerical study on the horizontal axis turbines arrangement in a wind farm: Effect of separation distance on the turbine aerodynamic power output. *J. Wind Eng. Ind. Aerod.* **2013**, *117*, 11–17. [CrossRef]

49. Esfahanian, V.; Pour, A.S.; Harsini, I.; Haghani, A.; Pasandeh, R.; Shahbazi, A.; Ahmadi, G. Numerical analysis of flow field around NREL Phase II wind turbine by a hybrid CFD/BEM method. *J. Wind Eng. Ind. Aerod.* **2013**, *120*, 29–36. [CrossRef]

50. Gopalana, H.; Gundlinga, C.; Brown, K.; Roget, B.; Sitaraman, J.; Mirocha, J.D.; Miller, W.O. A coupled mesoscale–microscale framework for wind resource estimation and farm aerodynamics. *J. Wind Eng. Ind. Aerod.* **2014**, *132*, 13–26. [CrossRef]

51. Rosenberg, A.; Sharma, A. A Prescribed-Wake Vortex Lattice Method for Preliminary Design of Co-Axial, Dual-Rotor Wind Turbines. *J. Sol. Energy Eng.* **2016**, *138*, 1–9. [CrossRef]

52. Sreenivas, K.; Mittal, A.; Hereth, L.; Taylor, L.K.; Hilbert, C.B. Numerical Simulation of the interaction between tandem wind turbines. *J. Wind Eng. Ind. Aerod.* **2016**, *157*, 145–157. [CrossRef]

53. Larsen, G.C.; Crespo, A. Wind turbine wakes for Wind Energy. *Wind Energy* **2011**, *14*, 797–798. [CrossRef]

54. Mittal, A.; Sreenivas, K.; Taylor, L.K.; Hereth, L.; Hilbert, C.B. Blade-resolved simulations of a model wind turbine: Effect of temporal convergence. *Wind Energy* **2016**, *19*, 1761–1783. [CrossRef]

55. AbdelSalam, A.M.; Ramalingamn, V. Wake prediction of horizontal-axis wind turbine using full-rotor modeling. *J. Wind Eng. Ind. Aerod.* **2014**, *124*, 7–19. [CrossRef]

56. Rodrigues, R.V.; Lengsfeld, C.S. Development of a computational system to improve wind farm layout, Part I: Model validation and near wake analysis. *Energies* **2019**, *12*, 940. [CrossRef]

57. Schepers, J.G.; Boorsma, K.; Cho, T.; Gomez-Iradi, S.; Schaffarczyk, P.; Jeromin, A.; Shen, W.Z.; Lutz, T.; Meister, K.; Stoevesandt, B.; et al. *Final Report of IEA Task 29, Mexnet (Phase 1): Analysis of Mexico Wind Tunnel Measurements*; Technical Report; IEA: Petten, The Netherlands, 2012; p. 312.

58. Adamarola, M.S.; Krogstad, P.-A. Experimental investigation of wake effects on wind turbine performance. *Renew. Energy* **2011**, *36*, 2078–2086. [CrossRef]

59. Chamorro, L.P.; Porté-Agel, F. A Wind-Tunnel Investigation of Wind-Turbine Wakes: Boundary-Layer Turbulence Effects. *Boundary-Layer Meteorol.* **2009**, *132*, 129–149. [CrossRef]

60. Sturge, D.; Sobotta, D.; Howell, R.; While, A.; Lou, J. A hybrid actuator disc—Full rotor CFD methodology for modelling the effects of wind turbine wake interactions on performance. *Renew. Energy* **2015**, *80*, 525–537. [CrossRef]

61. Hossain, M.Z.; Hirahara, H.; Nonomura, Y.; Kawahashi, M. The wake structure in a 2D grid installation of the horizontal axis micro wind turbines. *Renew. Energy* **2007**, *32*, 2247–2267. [CrossRef]

62. Barthelmie, R.J.; Hansen, K.; Frandsen, S.T.; Rathmann, O.; Schepers, J.G.; Schlez, W.; Phillips, J.; Rados, K.; Zervos, A.; Politis, E.S.; et al. Modelling and measuring flow and wind turbine wakes in large wind farms offshore. *Wind Energy* **2009**, *12*, 431–444. [CrossRef]

63. Vermeer, L.J.; Sørensen, J.N.; Crespo, A. Wind turbine wake aerodynamics. *Prog. Aerosp. Sci.* **2003**, *39*, 467–510. [CrossRef]

64. Churchfield, M.J.; Schreck, S.J.; Martinez, L.A.; Meneveau, C.; Spalart, P.R. An Advanced Actuator Line Method for Wind Energy Applications and Beyond. In Proceedings of the 35th Wind Energy Symposium, Grapevine, TX, USA, 9–13 January 2017; p. 1998.

65. Whale, J.; Anderson, C.G.; Bareiss, R.; Wagner, S. An experimental and numerical study of the vortex structure in the wake of a wind turbine. *J. Wind Eng. Ind. Aerod.* **2000**, *84*, 1–21. [CrossRef]

Article

Adaptive Nonparametric Kernel Density Estimation Approach for Joint Probability Density Function Modeling of Multiple Wind Farms

Nan Yang [1], Yu Huang [2], Dengxu Hou [1], Songkai Liu [1,*], Di Ye [1], Bangtian Dong [1] and Youping Fan [3]

[1] New Energy Micro-grid Collaborative Innovation Centre of Hubei Province (China Three Gorges University), Yichang 443002, China; ynyyayy@ctgu.edu.cn (N.Y.); hdx270949460@163.com (D.H.); 13657176800@163.com (D.Y.); ljh1017053867@163.com (B.D.)

[2] Yichang Power Supply Company, State Grid Hubei Electric Power Company, Yichang 443002, China; 18971935801@163.com

[3] School of Electrical Engineering, Wuhan University, Wuhan 430000, China; ypfan@whu.edu.cn

* Correspondence: skliu0120@163.com

Received: 26 February 2019; Accepted: 30 March 2019; Published: 9 April 2019

Abstract: The uncertainty of wind power brings many challenges to the operation and control of power systems, especially for the joint operation of multiple wind farms. Therefore, the study of the joint probability density function (JPDF) of multiple wind farms plays a significant role in the operation and control of power systems with multiple wind farms. This research was innovative in two ways. One, an adaptive bandwidth improvement strategy was proposed. It replaced the traditional fixed bandwidth of multivariate nonparametric kernel density estimation (MNKDE) with an adaptive bandwidth. Two, based on the above strategy, an adaptive multi-variable non-parametric kernel density estimation (AMNKDE) approach was proposed and applied to the JPDF modeling for multiple wind farms. The specific steps of AMNKDE were as follows: First, the model of AMNKDE was constructed using the optimal bandwidth. Second, an optimal model of bandwidth based on Euclidean distance and maximum distance was constructed, and the comprehensive minimum of these distances was used as a measure of optimal bandwidth. Finally, the ordinal optimization (OO) algorithm was used to solve this model. The scenario results indicated that the overall fitness error of the AMNKDE method was 8.81% and 11.6% lower than that of the traditional MNKDE method and the Copula-based parameter estimation method, respectively. After replacing the modeling object the overall fitness error of the comprehensive Copula method increased by as much as 1.94 times that of AMNKDE. In summary, the proposed approach not only possesses higher accuracy and better applicability but also solved the local adaptability problem of the traditional MNKDE.

Keywords: kernel density estimation; multiple wind farms; joint probability density; ordinal optimization

---

1. Introduction

In the past decades large-scale wind power integration has become a trend [1]. As a result, a variety of uncertainties have been identified in the power systems [2–7]. The outputs of wind farms are greatly influenced by natural environmental factors such as wind speed, which are random and, therefore, difficult to accurately predict and control [8]. There are many giant wind farms in the northwest of China. When these wind farms are connected to the power grid, a large number of random output generating nodes form in the power system. This brings enormous challenges to the scheduling and planning of the power system because these schemes usually need the accurate prediction data of the generating node outputs. Consequently, it is widely believed that the impacts of wind power uncertainties should be considered in the scheduling and planning of power systems [9–11]. Currently, the best way to describe the uncertainties of wind power is to construct a probability density function (PDF) [12–16].

The outputs of the wind farms maintain a random and probabilistic correlation in the scenario that multiple wind farms are connected to power systems, simultaneously, in the same wind belt [17]. According to the probability theory, when the probability density function is established for multiple subjects with a probability correlation, these subjects cannot be viewed as independent events. Therefore, these wind farms cannot be considered independent [18–21]. That is, the PDF for a single wind farm is not applicable when the uncertainties of these wind farms need to be described. Therefore, it is necessary to construct the joint probability density function (JPDF) for wind farms. Accordingly, it is well-known that the precise construction of the JPDF of wind farms is a foundation for the scheduling and planning of power systems with multiple wind farms. For example, Reference [22] proposes that the probabilistic correlation between multiple wind farms should be considered in the scheduling of power systems. The systematic planning method, considering the probabilistic correlation of multiple wind farms, is studied in Reference [23]. In general, it is of significance to research a method of constructing the JPDFs of multiple wind farms accurately and conveniently.

There exists much literature studying the construction of a JPDF for multiple wind farms. The Copula theory is the most common method used to study this problem, given that it can be used to characterize the probabilistic correlation in multiple wind farms. In Reference [24], the wind farms near the Dutch coastline are equated as two wind farms and a Gaussian-Copula function is introduced to establish their JPDF. In Reference [25], the Copula function is used to build a probabilistic correlation model for the wind speed and the wind power output, and then the model is used to assess the state of the generators. In Reference [26], a variety of two-element Copula functions are utilized to study the dependent structures of wind farms and the goodness of fit of different Copula functions is compared. In Reference [27], a number of basic Copula functions are summed with weights to form a comprehensive Copula function. As a result, compared with the single Copula function model, the JPDF of wind speed can be described more accurately by the comprehensive function [28]. According to previous research, the regular steps are as follows: First, a number of Copula function forms are selected according to the cumulative distribution characteristics of the wind farm outputs, in advance. Second, the unknown parameters are estimated. Finally, the most appropriate Copula function is determined by the optimization method. However, this method, based on Copula function for JPDF modeling of wind farm outputs, is essentially parameter estimation (PE). This kind of method depends on the multiple, prior definition of the JPDF forms. On one hand, once the form selection is wrong, no accurate modeling results can be obtained no matter how accurate the PE process is. Although, for the purpose of improving the accuracy of modeling, some scholars tried to estimate the parameters of all forms of Copula functions, and then selected the most accurate function. However, this kind of thought undoubtedly increases the complexity of the modeling process. On the other hand, a large number of wind farms are scattered throughout China. Consequently, the joint probability characteristics of wind farm clusters on different wind belts may follow different JPDF forms, and it is difficult to ensure the universal applicability of the modeling method based on the Copula function.

Different from the PE method, the probability distributions of objects can be modeled directly, without the prior judgment process of function forms by the nonparametric kernel density estimation (NKDE) method. Accordingly, it has higher accuracy and applicability and has been applied effectively in the field of probabilistic modeling in power systems [2–31]. The main focus of the existing research surrounds PDF modeling of a single random variable [32]. Some literature has begun to study the NKDE method for multidimensional random variables [33–38], but few of them are applied to the field of power systems.

In Reference [39], a JPDF model of grid node loads based on NKDE theory is proposed and the effects of the node load correlation and uncertainty in the aspect of reliability are analyzed. In

Reference [40], a node load conditional probability density modeling method based on NKDE theory is proposed. In these two papers a multivariate nonparametric kernel density estimation (MNKDE) method for the probabilistic correlation modeling of node loads is successfully proposed. However, problems with this method still exist when it is applied to the JPDF modeling for multiple wind farms. The local amplitude of the JPDF for wind farms different from the loads is larger and the bandwidth of the existing MNKDE method is fixed. This fixed bandwidth may be a problem for local applicability because the accuracy of the modeling is high in some intervals but lower in other.

In order to solve this problem, the new idea of modifying the bandwidth, based on the samples themselves and a mathematical model of adaptive univariate NKDE, is proposed in Reference [41]. Based on References [41,42], a new adaptive univariate NKDE model for power system state estimation is proposed. Moreover, a method to determine the bandwidth, discussed in References [43,44], is also proposed. The above references have made significant progress in solving the local applicability problem and provided the idea used in this research. However, the above research was all aimed at the univariate NKDE model. The study of the MNKDE model has not been reported.

In summary, an approach of adaptive multivariate nonparametric kernel density estimation (AMNKDE) is proposed in this paper, and it is utilized to model the JPDF of multiple wind farms. The correctness and effectiveness of the approach is verified by the simulation results, based on the practical operation data of several wind farms in China.

The main contributions of this paper are as follows:

(1) The AMNKDE approach for the JPDF modeling of multiple wind farms is proposed in this paper. Compared with the traditional PE method based on the Copula function, the approach does not require prior judgement of the JPDF forms of multiple wind farms. Consequently, this approach possesses higher modeling accuracy and applicability.

(2) In order to promote the MNKDE in the specific problem of multiple wind farms, an improved adaptive strategy is proposed. Specifically, a model of optimal bandwidth is established and the traditional fixed bandwidth is replaced with the adaptive bandwidth, which can be adjusted automatically according to the samples. The improved strategy in this paper solves the local applicability problem of the existing MNKDE method, and further improves the modeling accuracy.

The rest of the paper is organized as follows: The AMNKDE model for multiple wind farms and bandwidth evaluation indicators are given in Section 2. An optimized method of solving the bandwidth model based on ordinal optimization (OO) is explained in Section 3. The simulation results are compared and analyzed in Section 4. The conclusions are presented in Section 5.

2. Adaptive Multivariate Nonparametric Kernel Density Estimation Model for Multiple Wind Farms

2.1. MNKDE Model for Multiple Wind Farms

Considering m wind farms have n output data samples in each sampling period, the active power vector of the i sampling point is $X_i = [X_{i1}, X_{i1}, \cdots, X_{im}]^T$ $i = 1, 2, \ldots, n$. The random variation of the power output for m wind farms is $x = [x_1, x_2, \cdots, x_m]^T$. The JPDF is $f(x) = f(x_1, x_2, \cdots, x_m)$. The MKDE model of the JPDF is

$$\hat{f}(x) = \frac{1}{n} \sum_{i=1}^{n} \frac{1}{|H|^{1/2}} K\left[H^{-1/2}(x - X_i)\right], \tag{1}$$

where H is the bandwidth matrix, which denotes an $m \times m$ symmetrical positive determined matrix. $K(.)$ is the multivariate kernel function and must satisfy the following conditions:

$$\begin{cases} \int_{R^m} K(x)dx = 1 \\ \int_{R^m} xK(x)dx = 0 \\ \int_{R^m} xx^T K(x)dx = I_m \end{cases}, \tag{2}$$

where R^m is the m-dimensional Euclidean space, I_m denotes the $m \times m$ identity matrix, and x^T is the transpose of x.

According to Reference [45], if the kernel function satisfies Formula (2), its form has little effect on the probability density modeling accuracy. Therefore, the Gauss kernel function was chosen as the kernel function in this paper.

The specific form of the bandwidth matrix H is given in Formula (3),

$$
H = \begin{bmatrix}
h_{11} & h_{21} & \cdots & h_{m1} \\
h_{12} & h_2 & \cdots & h_{m2} \\
\vdots & \vdots & \ddots & \vdots \\
h_{1m} & h_{2m} & \cdots & h_{mm}
\end{bmatrix}.
\tag{3}
$$

In MNKDE modeling, the selection of the bandwidth matrix is the most important factor and can directly affect the modeling accuracy. Generally, the bandwidth matrix is obtained by an optimal model of bandwidth. Due to the large number of bandwidth matrix elements, the computational complexity of the optimal model of bandwidth for MNKDE is much larger than that of univariate NKDE. In order to reduce the computation complexity, the method in Reference [39] was used to simplify the Formula (1) in this paper.

The formula is simplified as follows:

$$
\hat{f}_m(x) = \frac{1}{n} \sum_{i=1}^{n} \frac{1}{h_1 h_2 \cdots h_m} K\left(\frac{x_1 - X_{i1}}{h_1}, \ldots, \frac{x_m - X_{im}}{h_m} \right),
\tag{4}
$$

where $K(x)$ is defined as

$$
K(x_1, x_2, \cdots, x_m) = K(x_1)K(x_2) \cdots K(x_m).
\tag{5}
$$

Here, the Gaussian kernel is used as the kernel function

$$
K(x) = \frac{1}{\sqrt{2\pi}} e^{\left(-\frac{x^2}{2}\right)}.
\tag{6}
$$

According to Formulas (4)–(6), Formula (7) is as follows:

$$
\hat{f}_m(x) = \frac{1}{n} \sum_{i=1}^{n} \frac{1}{h_1 h_2 \cdots h_m} \bullet \frac{e^{-\frac{1}{2}\left(\frac{x_1 - X_{i1}}{h_1}\right)^2}}{\sqrt{2\pi}} \bullet \frac{e^{-\frac{1}{2}\left(\frac{x_2 - X_{i2}}{h_2}\right)^2}}{\sqrt{2\pi}} \cdots \frac{e^{-\frac{1}{2}\left(\frac{x_m - X_{im}}{h_m}\right)^2}}{\sqrt{2\pi}}.
\tag{7}
$$

Further simplification of Formula (7) can be obtained as follows:

$$
\hat{f}_m(x) = \frac{1}{n} \sum_{i=1}^{n} \frac{1}{h_1 h_2 \cdots h_m} \frac{1}{\left(\sqrt{2\pi}\right)^m} e^{-\frac{1}{2} Y(x)},
\tag{8}
$$

where the specific form of $Y(x)$ is shown in Formula (9) as

$$
Y(x) = \left(\left(\frac{x_1 - X_{i1}}{h_1} \right)^2 + \left(\frac{x_2 - X_{i2}}{h_2} \right)^2 + \cdots + \left(\frac{x_m - X_{im}}{h_m} \right)^2 \right).
\tag{9}
$$

2.2. Optimal Model of Bandwidth

In the MNKDE model, H can directly influence the accuracy and smoothness of the model. If the value of H is too large it may lead to high smoothness of the probability density function of $\hat{f}(x)$, which results in a large estimation error. If the value of H is too low the accuracy of estimation can be improved. However, the fluctuation of the probability density function of $\hat{f}(x)$ may be excessively high, especially for the tail of $\hat{f}(x)$.

In conclusion, two kinds of bandwidth evaluation indicators are presented in this paper: the Euclidean distance and the maximum distance. The former is mainly used to evaluate the accuracy of the model and the latter is used to evaluate the smoothness of the model.

Assuming that $f(x)$ is the real JPDF of wind power samples, the Euclidean distance is defined as follows:

$$d_O(H) = \sqrt{\sum_{i=1}^{n} d_{Ji}^2(H)}, \tag{10}$$

where $d_{Ji}(H) = \left| \hat{f}(x_i) - f(x_i) \right|$, which is the geometric distance between the estimation value and the real value for each sample.

The maximum distance is defined as follows:

$$d_M(H) = \max\{d_{Ji}(H)\}. \tag{11}$$

Based on Formulas (10) and (11), an optimal model of bandwidth, considering both accuracy and smoothness of the model, is:

$$\min R(H) = \min[d_O(H) + d_M(H)], \tag{12}$$

where $R(H)$ is the fitness error function of MNKDE.

2.3. Improved Adaptive Strategy Based on the Optimal Bandwidth Adjustment Model

According to Formula (12), a fixed bandwidth H was used in the previous MNKDE theory, which involves obtaining only one H to minimize the fitness error sum of all the samples. However, the fitness error values may be abnormally large for some local sample intervals in that situation. If the adaptive bandwidth matrix, which is adapted to the local sample interval, is solved by modifying H in the sample data and the original fixed bandwidth matrix is replaced with the adaptive bandwidth matrix, the adaptive property of the constructed JPDF in the local sample intervals would be guaranteed. The modeling accuracy would also be further improved. Taking into account the above analysis, based on the MNKDE, the following improved strategies have been used for this paper.

After the bandwidth matrix H is solved by the optimal model of bandwidth (12), we discriminate the fitness of the sample interval. For any local sample intervals, $l \in [l_1, l_2]$ ($l_2 > l_1$ and $l_1, l_2 \in [X_1, X_n]$), we have determined that there exists a local adaptability problem in the local sample interval if the following inequality holds as follows:

$$d_{Jl}(H_{Best}) \geq \lambda \overline{d_J(H_{Best})}, \tag{13}$$

where l denotes any sample intervals, $d_{Jl}(H_{Best})$ is the geometric distance in l, H_{Best} is the result of Formula (12), $\overline{d_J(H_{Best})}$ is the average geometric distance of the entire sample space, and λ is an adjustment factor. If λ is smaller, the screening is more strict and more intervals need to be adjusted. In this scenario, the modeling accuracy is promoted but the complexity of the modeling is higher. In contrast, the complexity of the solution may be reduced but the modeling accuracy will then be declined. The specific value can be determined according to tests.

The $\overline{d_J(H_{Best})}$ is as follows:

$$\overline{d_J(H_{Best})} = \frac{1}{n} \sum_{i=1}^{n} d_{Ji}(H_{Best}), \tag{14}$$

Aiming at the interval with local adaptability problems, a bandwidth adjustment model was built to modify the bandwidth matrix:

$$H_l = \frac{n_l d_{Jl}(H_{Best})_{mid}}{\sqrt{-2 \ln \delta}} H_{Best}, \tag{15}$$

where H_1 is the modified bandwidth in l, n_l is the number of samples in l, $d_{Jl}(H_{Best})_{mid}$ is the median of the geometric distance in l, and δ is the threshold of the kernel function.

Thus, Formula (8) can be modified into Formula (16), which is an AMNKDE model for the JPDF modeling of multiple wind farms:

$$
\begin{aligned}
\hat{f}_m(x) = \quad & \frac{1}{\sum\limits_{i=1}^{l_1} \omega_i} \sum_{i=1}^{l_1} \frac{\omega_i}{\Pi H_{Best}} \frac{1}{(\sqrt{2\pi})^m} e^{-\frac{1}{2}H_{best}(x)} \\
& + \frac{1}{\sum\limits_{i=l_1}^{l_2} \omega_i} \sum_{i=l_1}^{l_2} \frac{\omega_i}{\Pi H_{l_1}} \frac{1}{(\sqrt{2\pi})^m} e^{-\frac{1}{2}H_{l_1}(x)} \\
& + \cdots + \frac{1}{\sum\limits_{i=l_{k-1}}^{l_k} \omega_i} \sum_{i=l_{k-1}}^{l_k} \frac{\omega_i}{\Pi H_{l_{k-1}}} \frac{1}{(\sqrt{2\pi})^m} e^{-\frac{1}{2}H_{l_{k-1}}(x)} \\
& + \frac{1}{\sum\limits_{i=l_k}^{n} \omega_i} \sum_{i=l_k}^{n} \frac{\omega_i}{\Pi H_{l_k}} \frac{1}{(\sqrt{2\pi})^m} e^{-\frac{1}{2}H_{l_k}(x)}
\end{aligned}
\tag{16}
$$

where k is the number of sample intervals that need to be adjusted, H_{l_k} is the modified bandwidth matrix in l_k, and ω_i is the measurement weight. In this paper, the following formula is used for ω_i [42]:

$$
\omega_i = \alpha + \exp\left(-\frac{s_i^2}{\bar{s}^2}\right),
\tag{17}
$$

where α is a small positive number, s_i is the standard deviation of measurement for each sampling interval, and $\bar{s} = \frac{1}{n}\sqrt{\sum\limits_{i=1}^{n} s_i^2}$ is the geometric mean of the standard deviation for all measurements.

3. Solution of the Optimal Model of Bandwidth Based on Ordinal Optimization

For the proposed AMNKDE in this paper, the bandwidth was transformed from a traditional single parameter matrix, which contributed to the increasing difficulty of the solution. In order to solve this problem, a solving approach of the optimal model of bandwidth, based on OO, was proposed.

OO is an effective method for solving complex optimization problems. According to the previous research in Reference [30], this method was successfully applied to solve the optimal model of bandwidth of univariate NKDE and achieved positive results. In this research, the OO was used to solve the bandwidth optimization problem of AMNKDE, the solution is shown in Figure 1 and the detailed steps are as follows:

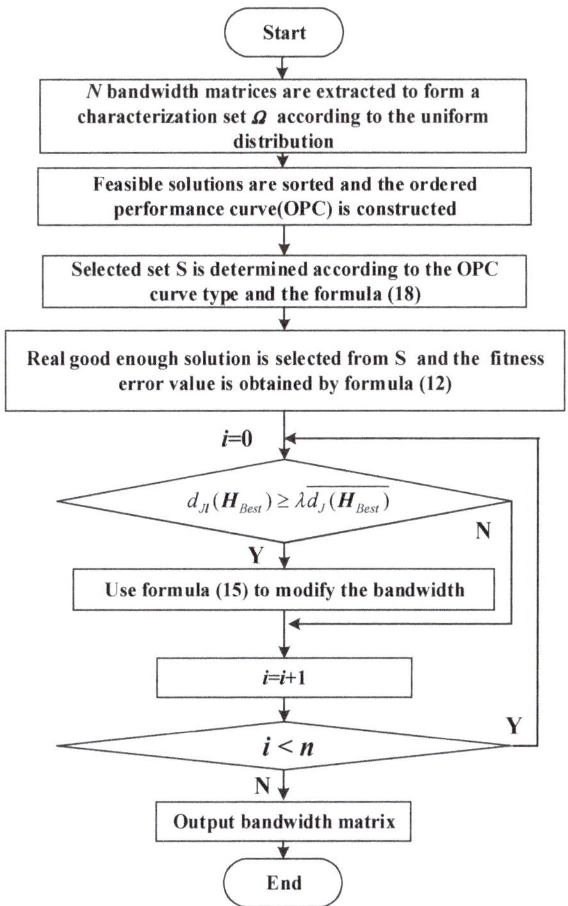

Figure 1. Flow Chat for the Solution.

(1) In the solution space of the bandwidth matrix H, the N bandwidth matrices were extracted to form a characterization set Ω according to the uniform distribution. The N was closely related to the size of the solution space. When the solution space was less than 10^8, $N = 1000$ was recommended by Reference [46].

(2) The N feasible solutions were selected by the rough model of Formula (10). Then, the feasible solutions were sorted according to the assessment results. In addition, the ordered performance curve (OPC) was constructed. The types of OPC are given in Reference [30].

(3) Formula (18) was used to determine the number of solutions in the selected set S,

$$S = e^{\varepsilon} t^{\mu} g^{\omega} + \eta, \tag{18}$$

where S is the number of solutions in the selected set S, t represents that there exist at least t good enough solutions in the selected set S, g represents the size of the good enough solution subset, $\varepsilon, \mu, \omega, \eta$ are the parameters associated with the type of OPC, and the values are 8.1378, 0.8974, 1.2058, 6.00, respectively [29].

(4) Taking the objective function of Formula (12) as the exact model, the order of comparison of solutions in the solution set S is made and the top t solutions will be selected as real, good enough, solutions.

(5) Utilizing Formula (13), the local sample intervals with low accuracy in the model were found. The bandwidths in these intervals were adjusted according to Formula (15).

4. Scenario Study

In this paper, 4773 sampling sequences of wind power outputs from six wind farms in the Hubei province of China were selected as examples. The sampling time interval was 10 min. The sampling period was from 19:40 on 17 March, 2009 to 23:00 on 19 April, 2009. For the frequency histogram of two wind farms, the straight interval chosen for this paper was 30 kW. For the frequency histogram of three wind farms, the straight interval was 100 kW. For the two wind farms, the total probability density of the samples was 1.1×10^{-3}. For the three wind farms, the total probability density of the samples was 1×10^{-6}. When $\lambda = 6$, the comprehensive performance of the proposed model was best. The model could improve the overall modeling accuracy by approximately 10% compared with the traditional MNKDE model and the corresponding calculation time was only 63 s. Accordingly, $\lambda = 6$ was chosen for this research. According to Reference [47], δ was 0.79655.

Program simulation was achieved in the MATLAB platform and related computing was completed on a computer with an Intel Core i5-4460 (3.20-GHz) CPU with 8 G of RAM. The computer time of the OO in this paper was 63.350 s. To verify the validity and applicability of the proposed approach, three-dimensional and four-dimensional JPDFs were obtained from two wind farms and three wind farms for comparison and analysis. The active power output sampling sequence of six wind farms is listed in Figure 2.

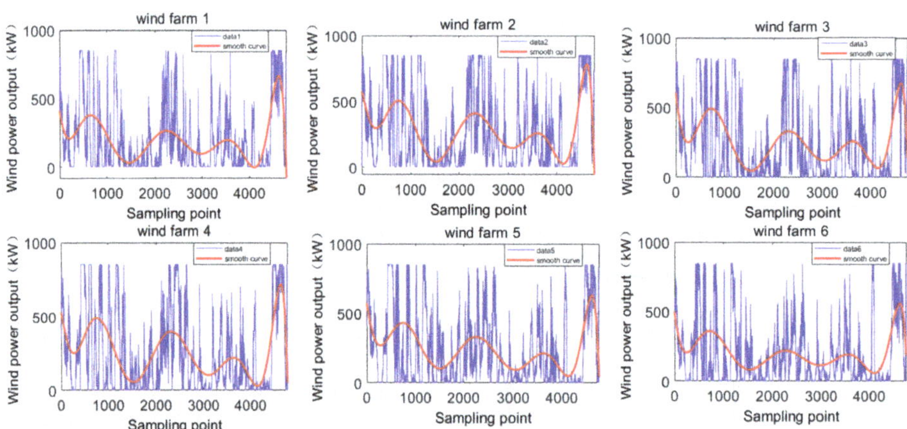

Figure 2. Historical data of six wind farms.

Figure 2 shows the differences between the output trends of the former three wind farms and those of the latter three wind farms. The differences are most obvious for the sampling points between 1300 and 2300. We concluded that the JPDFs of the former three wind farms and the latter three wind farms are different.

4.1. Joint Probability Density Function Modeling of Two Wind Farms

The JPDF of Wind Farms 1 and 2 were obtained via the approach described previously. The frequency histogram is of Wind Farm 1 and Wind Farm 2, based on the sample data. It is shown in Figure 3b. The comparison between them is shown in Figure 3.

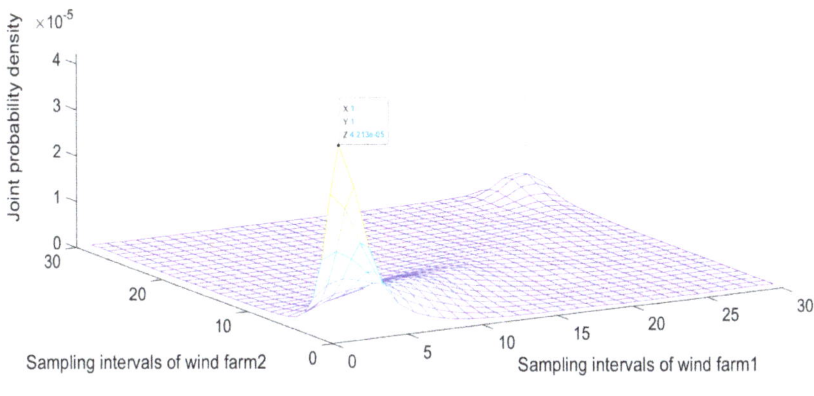

(a) Joint probability density function.

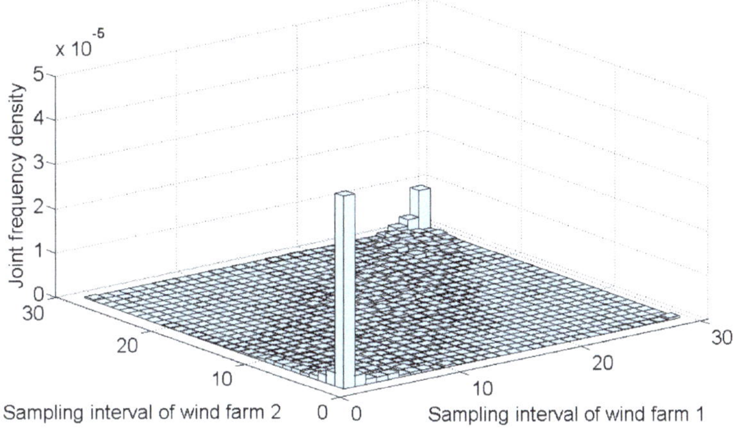

(b) Frequency Histogram.

Figure 3. Joint probability density function and frequency histogram of Wind Farms 1 and 2.

From Figure 3, we found that the outputs of Wind Farms 1 and 2 had a tail correlation. The correlation in the upper tails was stronger, which meant that both wind farms were more likely to produce larger outputs. From the function curve of the model, the modeled JPDF fit well with the real joint distribution of Wind Farms 1 and 2. The detailed calculation results are listed in Table 1.

Table 1. Detailed calculation results of JPDF for Wind Farms 1 and 2 by AMNKDE.

Model	H	Sample Interval		d_O	d_M	$R(H)$
		Wind Farm 1	Wind Farm 2			
AMNKDE	[38.7,38.7] [34.5,34.5] [34.7,34.7]	[30,60] [90,900] [0,30] Other interval	[30,60] [90,900] [0,30]	6.09×10^{-5}	4.1×10^{-6}	6.50×10^{-5}

From the results of Table 1, the modeling error was relatively low and the overall fitness error was only 6.50×10^{-5}.

4.2. Multipart Figures

To guarantee the generalizability of the results, the four-dimensional JPDF of three wind farms is presented. Based on this scenario, a comparative study was carried out. Models 1, 2, and 3 are the traditional MNKDE model, the AMNKDE model and the comprehensive Copula model, respectively.

4.2.1. Validity Analysis of the Improved Adaptive Strategy for MNKDE

To verify the differences between the proposed AMNKDE and the traditional MNKDE, the JPDFs of Wind Farms 1, 2, and 3 were constructed in these methods. The results are given in Table 2.

Table 2. Comparison of AMNKDE and MNKDE.

Model	H	Sample Interval			d_O	d_M	$R(H)$
		Wind Farm 1	Wind Farm 2	Wind Farm 3			
1	[40.7, 40.7, 40.7]	[0,100]	[0,100]	[0,100]	3.60×10^{-8}	3.60×10^{-9}	7.15×10^{-8}
	[40.7, 40.7, 40.7]	[800,1000]	[800,1000]	[0,100]	3.60×10^{-8}		
	[40.7, 40.7, 40.7]		Other interval		2.83×10^{-8}		
2	[37,37,37]	[0,100]	[1,100]	[0,100]	3.00×10^{-8}	1.60×10^{-10}	6.52×10^{-8}
	[38,38,38]	[800,1000]	[800,1000]	[0,100]	1.60×10^{-8}		
	[40.7,40.7,40.7]		Other interval		3.20×10^{-8}		

Compared with Model 1, the Euclidean distance, the maximum distance and the overall fitness error of Model 2 were reduced by 6.76%, 55.5% and 8.81%, respectively, as shown in Table 2. This suggests that the modeling accuracy of MNKDE was effectively improved by the new adaptive strategy. The proposed AMNKDE achieved an adaptive improvement for the bandwidths of the sample interval in [0,100] [0,100] [0,100] and [800,1000] [800,1000] [0,100]. The elements of the bandwidth matrices in the sample intervals [0,100] [0,100] [0,100] and [800,1000] [800,1000] [0,100] were changed from 47 to 32 and 33, respectively.

The rest of the interval elements remained as 40.7 and the above matrix as an adaptive bandwidth matrix. The resulting decline of the Euclidean distance for the corresponding sample intervals was 16.7% and 55.6%, respectively. This improvement resulted in a rise of 13.1% in the Euclidean distance of the other sample intervals, but, for the entire sample interval, the overall Euclidean distance and the maximum distance was evidently reduced and the overall fitness error was cut down by 8.81%. We summarized that the overall modeling accuracy of the MNKDE was effectively facilitated by the adaptive bandwidth improvement strategy of the sample intervals with the local adaptability problem.

4.2.2. Accuracy Comparison between AMNKDE and Copula Parameter Estimation

To verify the accuracy of the proposed AMNKDE approach, the JPDF of Wind Farms 1, 2 and 3 were established using the comprehensive Copula method from Reference [29]. The compared results are shown in Table 3. The optimal Copula function was composed of Gumbel Copula, Clayton Copula and Frank Copula.

Table 3. Accuracy comparison between AMNKDE and Copula parameter estimation.

Model	H	Sample Interval			d_O	d_M	$R(H)$
		Wind Farm 1	Wind Farm 2	Wind Farm 3			
1	[40.7, 40.7, 40.7]	[0,100]	[0,100]	[0,100]	3.60×10^{-8}	3.60×10^{-9}	7.15×10^{-8}
	[40.7, 40.7, 40.7]	[800,1000]	[800,1000]	[0,100]	3.60×10^{-8}		
	[40.7, 40.7, 40.7]		Other interval		2.83×10^{-8}		
2	[37,37,37]	[0,100]	[0,100]	[0,100]	3.00×10^{-8}	1.60×10^{-10}	6.52×10^{-8}
	[38,38,38]	[800,1000]	[800,1000]	[0,100]	1.60×10^{-8}		
	[40.7,40.7,40.7]		Other interval		3.20×10^{-8}		

From Table 3, Model 3, based on the comprehensive Copula method, the Euclidean distance, maximum distance, and overall fitness error of Model 2, based on the proposed AMNKDE approach, were compared and shown to be reduced by 7.8%, 57.9% and 11.6%, respectively. It can be seen that the proposed AMNKDE approach has higher modeling accuracy than the comprehensive Copula method. The reason is that the proposed AMNKDE approach directly models the JPDF based on the sample data. Accordingly, it does not need to choose the specific form of the JPDF in advance and the modeling accuracy is only related to the selection of bandwidth, rather than the prior definition of the JPDF forms.

4.2.3. Comparison of Applicability between AMNKDE and Copula Parameter Estimation

To verify the applicability of the proposed AMNKDE approach, the wind farms were changed for the comparison. The JPDFs of Wind Farms 4, 5, and 6 were established using the AMNKDE and the comprehensive Copula method from Reference [29]. The optimal Copula function still consisted of Gumbel Copula, Clayton Copula and Frank Copula. The detailed results are presented in Table 4.

Table 4. Comparison of Applicability between AMNKDE and Copula Parameter Estimation.

Model	H	Sample Interval			d_O	d_M	$R(H)$
		Wind Farm 4	Wind Farm 5	Wind Farm 6			
2	[38,38,38] [39,39,39] [41.7,41.7,41.7]	[0,100] [800,1000]	[0,100] [800,900] Other interval	[0,100] [0,100]	6.78×10^{-8}	1.1×10^{-10}	6.89×10^{-8}
3	Copula function Gumbel Clayton Frank	λ 0.386 0.403 0.211		θ 5.60 4.940 8.586	7.83×10^{-8}	2.50×10^{-9}	8.08×10^{-8}

From Table 4, the proposed AMNKDE approach still maintained high modeling accuracy for the different wind farms. Compared with Table 3, the overall fitness error of the AMNKDE increased by 5.67%. In contrast, the overall fitness error increase of the comprehensive Copula method was larger, 10.99%, and the increase was 1.94 times that of the AMNKDE. It can be concluded that the proposed AMNKDE approach possesses high applicability compared with the Copula PE method when the modeling object is changed. The reason is that the latter method needs to judge the form of the JPDF, and the joint probability distribution of different wind farms may follow different function forms. Consequently, it may cause a large error if the same function form is used to model different wind farms.

4.2.4. Comparison of Algorithms

To analyze the validity of the OO algorithm in this paper, the calculation efficiency was proposed. The GA, PSO and OO algorithms were used to solve the optimal model of bandwidth in this paper. The optimal bandwidth matrices of these three algorithms, H_{best}, were [48.7,48.7,48.7], [55,55,55], [52,52,52], respectively. The results of the fitness error $R(H)$ and the computation time are shown in Figure 4.

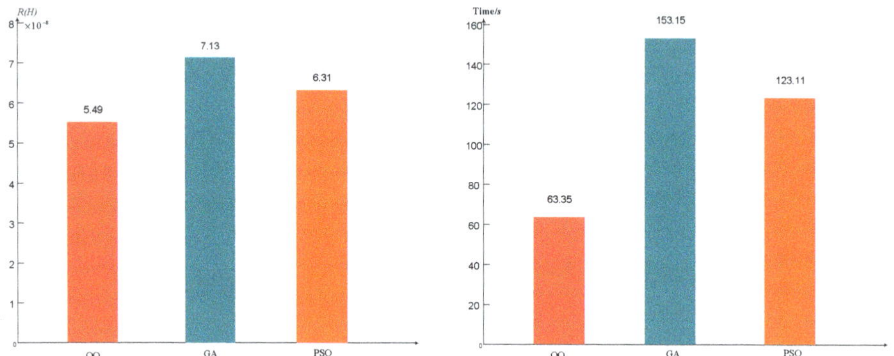

Figure 4. Comparison of Accuracy and Operation Time between Different Algorithms.

Figure 4 compares the traditional genetic algorithm and the PSO algorithm. The proposed OO algorithm was relatively limited in terms of improving the computational accuracy. However, the OO possessed a significant advantage in computational efficiency. It can be concluded that the proposed OO algorithm can effectively guarantee the computational efficiency and accuracy.

5. Conclusions

As the accuracy and applicability of many JPDF modeling methods for multiple wind farms need to be improved, it is of significance to promote the accuracy and applicability of the modeling method of JPDF. This is exactly the purpose of this study.

The specific steps of AMNKDE are as follows. First, the model of AMNKDE was constructed using the optimal bandwidth. Second, an optimal model of bandwidth based on Euclidean distance and maximum distance was constructed and the comprehensive minimum of these distances was used as a measure of optimal bandwidth. Finally, the OO algorithm was used to solve this model.

The specific conclusions of this paper are as follows:

(1) The adaptive bandwidth improvement strategy proposed in this paper replaces the traditional fixed bandwidth of MNKDE with the adaptive bandwidth. It effectively facilitates the overall modeling accuracy of MNKDE by adjusting the bandwidth of local sample interval adaptively. Simulation results in this paper indicate that the overall fitness error of AMNKDE was 8.81% lower than that of traditional MNKDE.

(2) Based on the above strategy, an AMNKDE approach was proposed and utilized to build the JPDF model for multiple wind farms. The simulation results in this paper indicated that the overall fitness error of AMNKDE was 11.6% lower than that of Copula-based PE method. After replacing the modeling object the overall fitness error of the comprehensive Copula method increased by as much as 1.94 times that of AMNKDE. Consequently, the accuracy and applicability of the AMNKDE approach were better than that of the traditional Copula PE method.

In summary, the proposed AMNKDE approach clearly performed better than MNKDE and Copula hybrid models and was suitable for building a multi-wind farm joint probability density model.

This study can be further extended. The proposed approach in this paper can be further applied to many other fields of uncertain modeling, such as the JPDF modeling problems of photovoltaic power systems and wind-solar combined power systems.

Author Contributions: Conceptualization, N.Y. and Y.H.; Data curation, Y.H., D.H., S.L. and D.Y.; Formal analysis, Y.H. and B.D.; Project administration, B.D.; Software, Y.H., D.H. and D.Y.; Supervision, S.L.; Visualization, D.Y.; Writing – original draft, N.Y. and Y.H.; Writing – review & editing, N.Y. and Y.F.

Funding: This research was funded by the National Natural Science Foundation of China No. 51607104.

Conflicts of Interest: The authors declare no conflict of interest.

Nomenclature

$\hat{f}(x)$	probability density function
H	bandwidth matrix
$K(.)$	multivariate kernel function
x^T	transpose of x
d_O	Euclidean distance
d_M	maximum distance
$R(H)$	fitness error function
l	any sample intervals
$d_{Jl}(H_{Best})$	geometric distance in l
$\overline{d_J(H_{Best})}$	average geometric distance of the entire sample space
H_1	modified bandwidth in l
$d_{Jl}(H_{Best})_{mid}$	median of the geometric distance in l
n_l	number of samples in l
δ	threshold of the kernel function
k	number of sample intervals needed to be adjusted
H_{l_k}	modified bandwidth matrix in l_k
ω_i	measurement weight
α	a small positive number
s_i	standard deviation of measurement for each sampling interval
S	number of solutions in the selected set S
t	there exist at least t good enough solutions in the selected set S
g	size of the good enough solution subset

References

1. Wu, J.; Zhang, B.; Wang, K. Optimal economic dispatch model based on risk management for wind-integrated power system. *IET Gener. Transm. Distrib.* **2015**, *9*, 2152–2158. [CrossRef]
2. Hu, D.; Ryan, S.M. Stochastic vs. deterministic scheduling of a combined natural gas and power system with uncertain wind energy. *Int. J. Electr. Power Energy Syst.* **2019**, *108*, 303–313. [CrossRef]
3. Yan, J.; Liu, Y.; Han, S. Reviews on uncertainty analysis of wind power forecasting. *Renew. Sustain. Energy Rev.* **2015**, *52*, 1322–1330. [CrossRef]
4. Zhang, Y.; Jian, L.; Zheng, F.; Zhang, Y.; Liu, K. Two-stage distributionally robust coordinated scheduling for gas-electricity integrated energy system considering wind power uncertainty and reserve capacity configuration. *Renew. Energy* **2019**, *135*, 122–135. [CrossRef]
5. Wu, J.; Zhang, B.; Jiang, Y.; Bie, P.; Li, H. Chance-constrained stochastic congestion management of power systems considering uncertainty of wind power and demand side response. *Int. J. Electr. Power Energy Syst.* **2019**, *107*, 703–714. [CrossRef]
6. Dvorkin, Y.; Lubin, M.; Backhaus, S. Uncertainty Sets for wind power generation. *IEEE Trans. Power Syst.* **2016**, *31*, 3326–3327. [CrossRef]
7. Wang, B.; Fang, B.; Wang, Y.; Liu, H.; Liu, Y. Power System Transient Stability Assessment Based on Big Data and the Core Vector Machine. *IEEE Trans. Smart Grid* **2016**, *7*, 2561–2570. [CrossRef]
8. Liu, R.; Peng, M.; Xiao, X. Ultra-Short-Term Wind Power Prediction Based on Multivariate Phase Space Reconstruction and Multivariate Linear Regression. *Energies* **2018**, *11*, 2763. [CrossRef]
9. Li, J.; Wang, S.; Ye, L.; Fang, J. A coordinated dispatch method with pumped-storage and battery-storage for compensating the variation of wind power. *Prot. Control Mod. Power Syst.* **2018**, *3*, 21–34. [CrossRef]
10. Zheng, D.; Eseye, A.; Zhang, J.; Li, H. Short-term wind power forecasting using a double-stage hierarchical ANFIS approach for energy management in microgrids. *Prot. Control Mod. Power Syst.* **2017**, *2*, 136–145. [CrossRef]
11. Liao, S.; Xu, J.; Sun, Y. Control of Energy-intensive Load for Power Smoothing in Wind Power Plants. *IEEE Trans. Power Syst.* **2018**, *33*, 6142–6154. [CrossRef]
12. Yang, H.; Xie, K.; Tai, H.M. Wind farm layout optimization and its application to power system reliability analysis. *IEEE Trans. Power Syst.* **2016**, *31*, 2135–2143. [CrossRef]

13. Tian, P.C. Estimation of wind energy potential using different probability density functions. *Appl. Energy* **2011**, *88*, 1848–1856.

14. Altunkaynak, A.; Erdik, T.; Dabanlı, İ.; Şen, Z. Theoretical derivation of wind power probability distribution function and applications. *Appl. Energy* **2012**, *92*, 809–814. [CrossRef]

15. Tina, G.; Gagliano, S. Probabilistic analysis of weather data for a hybrid solar/wind energy system. *Int. J. Energy Res.* **2011**, *35*, 221–232. [CrossRef]

16. Wang, C.; Li, X.H.; Tian, T.; Xu, Z.R.; Chen, R. Coordinated control of passive transition from grid-connected to islanded operation for three/single-phase hybrid multimicrogrids considering speed and smoothness. *IEEE Trans. Ind. Electron.* **2019**. [CrossRef]

17. Liu, T.H.; Wei, H.K.; Zhang, K.J. Wind power prediction with missing data using Gaussian process regression and multiple imputation. *Appl. Soft Comput.* **2018**, *71*, 905–916. [CrossRef]

18. Yin, H.; Zivanovic, R. Using probabilistic collocation method for neighbouring wind farms modeling and power flow computation of South Australia grid. *IET Gener. Transm. Distrib.* **2017**, *11*, 3568–3575. [CrossRef]

19. Zhang, L.; Luo, Y. Combined Heat and Power Scheduling: Utilizing Building-level Thermal Inertia for Short-term Thermal Energy Storage in District Heat System. *IEEE Trans. Electr. Electron. Eng.* **2018**, *13*, 804–814. [CrossRef]

20. Olauson, J.; Bergkvist, M. Correlation between wind power generation in the European countries. *Energy* **2016**, *114*, 663–670. [CrossRef]

21. Luo, G.; Chen, J.; Cai, D. Probabilistic assessment of available transfer capability considering spatial correlation in wind power integrated system. *IET Gener. Transm. Distrib.* **2013**, *7*, 1527–1535.

22. Xie, Z.Q.; Ji, T.Y.; Li, M.S. Quasi-Monte carlo based probabilistic optimal power flow considering the correlation of wind speeds using Copula function. *IEEE Trans. Power Syst.* **2018**, *33*, 2239–2247. [CrossRef]

23. Li, C.X.; Dong, Z.Y.; Chen, G. Flexible transmission expansion planning associated with large-scale wind farms integration considering demand response. *IET Gener. Transm. Distrib.* **2015**, *9*, 2276–2283. [CrossRef]

24. Hu, B.Q.; Wu, L.; Marwali, M. On the robust solution to scuc with load and wind uncertainty correlations. *IEEE Trans. Power Syst.* **2014**, *29*, 2952–2964. [CrossRef]

25. Papaefthymiou, G.; Kurowicka, D. Using Copulas for modeling stochastic dependence in power system uncertainty analysis. *IEEE Trans. Power Syst.* **2009**, *24*, 40–49. [CrossRef]

26. Gill, S.; Stephen, B.; Galloway, S. Wind Turbine Condition Assessment Through Power Curve Copula Modeling. *IEEE Trans. Sustain. Energy* **2012**, *3*, 94–101. [CrossRef]

27. Louie, H. Evaluation of bivariate archimedean and elliptical Copulas to model wind power dependency structures. *Wind Energy* **2014**, *17*, 225–240. [CrossRef]

28. Zhang, J.; Chowdhury, S.; Messac, A. A Multivariate and Multimodal Wind Distribution model. *Renew. Energy Int. J.* **2013**, *51*, 436–447. [CrossRef]

29. Wang, J.W.; Zhou, B.X.; Li, H.B. Modeling of wind farm output correlation based on comprehensive Copula function. *Electr. Meas. Instrum.* **2016**, *53*, 100–105.

30. Yang, N.; Cui, J.Z.; Zhou, Z. Research on nonparametric kernel density estimation for modeling of wind power probability characteristics based on fuzzy ordinal optimization. *Power Syst. Technol.* **2014**, *40*, 335–340.

31. Yang, N.; Ye, D.; Zhou, Z.; Cui, J.Z.; Chen, D.J.; Wang, X.M. Research on modelling and solution of stochastic SCUC under AC power flow constraints. *IET Gener. Transm. Distrib.* **2018**, *12*, 3618–3625.

32. Soleimanpour, N.; Mohammadi, M. Probabilistic load flow by using nonparametric density estimators. *IEEE Trans. Power Syst.* **2013**, *28*, 3747–3755. [CrossRef]

33. Ren, Z.Y.; Yan, W.; Zhao, X. Chronological probability model of photovoltaic generation. *IEEE Trans. Power Syst.* **2014**, *29*, 1077–1088. [CrossRef]

34. Carbone, P.; Petri, D.; Barbé, K. Nonparametric probability density estimation via interpolation filtering. *IEEE Trans. Instrum. Meas.* **2017**, *66*, 681–690. [CrossRef]

35. Zhu, B.X.; Ren, L.L.; Hu, X. Kind of high step-up dc/dc converter using a novel voltage multiplier cell. *IET Power Electron.* **2017**, *10*, 129–133. [CrossRef]

36. Kristan, M.; Leonardis, A.; Skoc, D. Multivariate online kernel density estimation with Gaussian kernels. *Pattern Recognit.* **2011**, *44*, 2630–2642. [CrossRef]

37. Li, Z.H.; Tao, Y.; Abu-Siada, A. A New Vibration Testing Platform for Electronic Current Transformers. *IEEE Trans. Instrum. Meas.* **2019**, *68*, 704–712. [CrossRef]

38. Pulkkinen, S.; Mäkelä, M.; Karmitsa, N. A continuation approach to mode-finding of multivariate Gaussian mixtures and kernel density estimates. *J. Glob. Optim.* **2013**, *56*, 459–487. [CrossRef]
39. Chang, M.S.; Wu, X.M. Transformation-based nonparametric estimation of multivariate densities. *J. Multivar. Anal.* **2015**, *135*, 71–88. [CrossRef]
40. Gramacki, A.; Gramacki, J. FFT-based fast computation of multivariate kernel density estimators with unconstrained bandwidth matrices. *J. Comput. Graph. Stat.* **2017**, *26*, 459–462. [CrossRef]
41. Zhao, Y.; Zhang, X.F.; Zhou, J.Q. Load Modeling Utilizing Nonparametric and Multivariate Kernel Density Estimation in Bulk Power System Reliability Evaluation. *Proc. CSEE* **2009**, *29*, 27–33.
42. Zougab, N.; Smail, K.; Célestin, C. Bayesian estimation of adaptive bandwidth matrices in multivariate kernel density estimation. *Comput. Stat. Data Anal.* **2014**, *75*, 28–38. [CrossRef]
43. Sreevani, N.; Murthy, C.A. On bandwidth selection using minimal spanning tree for kernel density estimation. *Comput. Stat. Data Anal.* **2016**, *102*, 67–84. [CrossRef]
44. Zhao, Y.; Yang, J.G.; Li, S.X. Nonparametric disaggregation load model in power system reliability evaluation incorporating the additive correlation. *IEEE Trans. Power Syst.* **2008**, *23*, 6039–6047.
45. Van Kerm, P.; Mohammadi, M. Adaptive kernel density estimation. *Stata J.* **2003**, *3*, 148–156. [CrossRef]
46. Liu, Y.S.; Lin, J.K.; Guo, L.X. A robust state estimation method based on adaptive kernel density estimation Theory. *Proc. CSEE* **2012**, *19*, 4937–4946.
47. Epanecnikov, V.A. Nonparametric estimation of a multidimensional probability density. *Theory Probab. Appl.* **2003**, *14*, 156–161.

Article

Reactive Power Capability Model of Wind Power Plant Using Aggregated Wind Power Collection System

Moumita Sarkar [1],*, Müfit Altin [2], Poul E. Sørensen [1] and Anca D. Hansen [1]

[1] Department of Wind Energy, Technical University of Denmark, 4000 Roskilde, Denmark; posq@dtu.dk (P.E.S.); anca@dtu.dk (A.D.H.)
[2] Energy Systems Engineering Department, Izmir Institute of Technology, Urla, 35430 Izmir, Turkey; mufitaltin@iyte.edu.tr
* Correspondence: mosar@dtu.dk

Received: 20 March 2019; Accepted: 23 April 2019; Published: 27 April 2019

Abstract: This article presents the development of a reactive power capability model for a wind power plant (WPP) based on an aggregated wind power collection system. The voltage and active power dependent reactive power capability are thus calculated by using aggregated WPP collection system parameters and considering losses in the WPP collection system. The strength of this proposed reactive power capability model is that it not only requires less parameters and substantially less computational time compared to typical detailed models of WPPs, but it also provides an accurate estimation of the available reactive power. The proposed model is based on a set of analytical equations which represent converter voltage and current limitations. Aggregated impedance and susceptance of the WPP collection system are also included in the analytical equations, thereby incorporating losses in the collection system in the WPP reactive power capability calculation. The proposed WPP reactive power capability model is compared to available methodologies from literature and for different WPP topologies, namely, Horns Rev 2 WPP and Burbo Bank WPP. Performance of the proposed model is assessed and discussed by means of simulations of various case studies demonstrating that the error between the calculated reactive power using the proposed model and the detailed model is below 4% as compared to an 11% error in the available method from literature. The efficacy of the proposed method is further exemplified through an application of the proposed method in power system integration studies. The article provides new insights and better understanding of the WPPs' limits to deliver reactive power support that can be used for power system stability assessment, particularly long-term voltage stability.

Keywords: reactive power capability; wind power plant; wind power collection system; aggregated, modelling; wind integration studies; long term voltage stability

---

1. Introduction

Growing concerns for climate change, energy security, increasing fuel prices for non-renewable generation sources, price reduction for renewable sources like wind and solar power are driving power systems to have a larger share of renewables all over the world. Due to large onshore and offshore developments, wind power is set to become the leading source of electricity in Europe after 2030 [1]. Around 52.6 GW of wind power capacity was installed globally in 2017, increasing the net installed capacity to 539.6 GW [1]. Increase in the share of renewables is also phasing out the conventional generations like coal based power plants, which brings many new challenges in operation and stability of the power system. Some of these challenges include a decrease in inertia, active and reactive power fluctuations, network congestion, etc. This article deals with reactive power reserve and support from wind power plants (WPP). The reactive power reserves conventionally provided by exciter of synchronous generator reduces when replaced by WPPs. This can cause voltage stability issues. This issue is further pronounced in weak grids where WPPs are connected to the grid through long lines. The need for analysis of reactive power support from WPP is especially essential when the grid is in a stressed condition. However, integration of hundreds of WPPs in large power system analysis is very complex and computationally intensive. Therefore, simplified representations of WPPs accurate enough to reflect capabilities and limitations of the converter based wind turbines (WTs) are required to analyze future power systems.

In power system stability analysis, long-term voltage stability is defined as a slow phenomenon involving slow acting equipment like tap-changing transformers, thermostatically controlled loads, generator limiters etc., such that the network is unable to provide adequate reactive power support (at least at certain nodes or areas in the power system) [2,3]. Traditionally, realistic representations of synchronous generators along with automatic voltage regulators have been used to model the capabilities and limits of reactive power resources in long-term voltage stability studies [2]. A similar reactive power resource model for WPPs needs to be developed for future power systems dominated by converter connected power generations. In this article, WPP reactive power capability is developed for an accurate representation of maximum reactive power generation and absorption capability of IEC 61400-27-1 [4] Type 4 WT (full rated converter based WT) based WPPs.

Several studies have been carried out over the years, where WPP reactive power capability has been used for power system analyses. Reactive power capability has been used to determine the voltage dependent reactive current limitation for modelling of WT by Bech [5] and Sørensen et al. [6]. In power system operation studies, reactive power capabilities of WPPs have been used for load flow studies in [7,8]. Zhang et al. have applied reactive power capability curves of IEC 61400-27-1 Type 3 (also known as doubly fed induction generator (DFIG)) WT for loss minimisation in WPP [9]. Inclusion of Type 3 WTs in optimal power flow for loss minimisation of distribution network have been investigated by Meegahapola et al. [10]. System flexibility studies have been performed by Stankovíc and Söder to determine the reactive power capability of distribution systems with distributed wind generations [11]. Network planning studies including the reactive power capability of WPPs have been done by Ugranli and Karatepe [12]. Reactive power reserve management of WPPs considering the maximum capability of WTs have been proposed by Martínez et al. [13]. Voltage control at the point of common coupling (PCC) considering reactive power capability of WPPs have been investigated by Kim et al. [14] and Karbouj et al. [15]. Reactive power capability of WPPs have also been applied for voltage stability studies. Dynamic voltage stability studies incorporating capability curves have been done by Meegahapola et al. [16]. Londero et al. [17] and Amarasekara et al. [18] have considered WT capability curves to analyze the long-term voltage stability of a power system with wind power generation. Vijayan et al. [19] have developed a voltage stability assessment method depicting that the inclusion of a WT capability curve can result in a larger power transfer margin of the system. Reactive power capability curves have also been applied for studies on ancillary services. For example, Ullah et al. have developed a generalized reactive power cost model for WPPs based on Type 4 WTs [20]. Voltage support as an ancillary service from WPPs using capability curves (for both dynamic and steady-state) have been studied by Karbouj and Rather [15].

Modelling of capability curves can be broadly categorized into: (i) WT capability curves and (ii) WPP capability curves. Lund et al. [21] have derived the steady-state capability of Type 3 WT considering rotor current, rotor voltage and stator current limitation as well as the effect of switching of coupling of DFIG stator on the capability curve. Engelhardt et al. [22] have derived capability of Type 3 WTs considering the generator and converter current limitation, losses in the machine and converter, saturation of flux, converter output voltage limitation, etc. Ullah et al. [20] have derived an analytical expression to compute reactive power capability of Type 4 WTs.

There has been limited work for representing capability curve of a WPP. Generally, there are two methods used in literature for modelling a WPP capability curve:

1. Scaled WT model: The WPP capability curve is derived by scaling up the WT capability curve with number of WTs. Kayikçi and Milanovic have used reactive power capability of WT for reactive power control of WPP, where only a single WT is modelled [23]. Konopinski et al. have modelled the reactive power capability of WPP assuming that the capability of one WT can be scaled to represent the accurate aggregate behaviour of WPP [24]. Ullah et al. have derived the reactive power capability of an aggregated WPP by scaling the output with number of wind turbines in the plant [20]. Meegahapola et al. [16,25] and Londero et al. [17] have used scaled reactive power capability of a WT for WPP representation. Meegahapola et al. [16] and Konopinski et al. [24] have developed capability curve of Type 3 based aggregated WPPs, while Ullah et al. [20] have developed it for Type 4 based aggregated WPPs.
2. WPP detailed model: This involves using a detailed WPP model (including WT transformers and WPP collection system cables' parameters). Kim et al. [26,27] have derived a reactive power capability of Type 3 based WPP based on detailed model of the WPP collection system. Karbouj and Rather [15] have modelled capability curves for Type 4 based WPPs using ABCD parameters of a detailed power collection system.

All these existing methodologies described in literature cannot be applied for simulating large power systems with numerous WPPs because of the following reasons:

- Scaled WT models do not consider WPP collection system parameters; hence, losses in the collection system are neglected. This reduces the accuracy of the reactive power capability estimation.
- For WPPs consisting of large number of WTs, using a detailed model of wind power collection system requires large computational time and resources. This is further worsened when system studies are performed with multiple WPPs in the network. Capability curves need to be computed in real time to utilize the full potential of WPPs in case of stressed system conditions, since reactive power capability of WPPs is dependent on active power production as well as on grid voltage conditions.
- Detailed parameters of WPP collection systems may not be always available to system operators for estimation of reactive power reserve from WPPs.

Therefore, authors have developed a new reactive power capability model in this article which estimates reactive power close to the detailed model while requiring less parameters and computation time. The objective of this article is to develop a reactive power capability model of WPP considering the WPP collection system. The developed model considers active power generation from the WPP as well as voltage dependency at the PCC. Inclusion of the collection system assures that the active and reactive power losses in the collection system are taken into account while computing WPP capability curves. Reduced the number of parameters enables fast real-time calculation of reactive power availability of any WPP. The capability curve of WPPs is dependent on various parameters such as the number of WTs, collection system configuration and length of array cables. Sensitivity studies are performed in order to realize the impact of aforementioned parameters on the WPP capability curve. The accuracy of the proposed model is compared against the WPP detailed model and scaled WT model for different simulated case studies of real WPPs. Furthermore, all these methodologies are applied on a simulated power system model to exemplify the efficacy of the proposed model.

Organisation of the article is as follows: Section 2 describes the methodology for modelling of WPP reactive power capability. In Section 3, case studies are presented and discussed to understand effects of various parameters on WPP reactive power capability. Application of the reactive power capability model for power system studies is also shown in this section. Finally, conclusive remarks are reported in Section 4.

2. Modelling

In this section firstly, the reactive power capability model of Type 4 WT is extended to include both resistance and reactance in the system. Then this model together with the aggregated WPP collection system impedance is further used to calculate WPP reactive power capability.

2.1. Extension of WT Reactive Power Capability Model

The Type 4 WT consists of a generator connected to the grid through full-scale back-to-back converters—machine side converter (MSC) and grid side converter (GSC) [28]. Schematic representation of Type 4 WT with permanent magnet synchronous generator (PMSG) is shown in Figure 1.

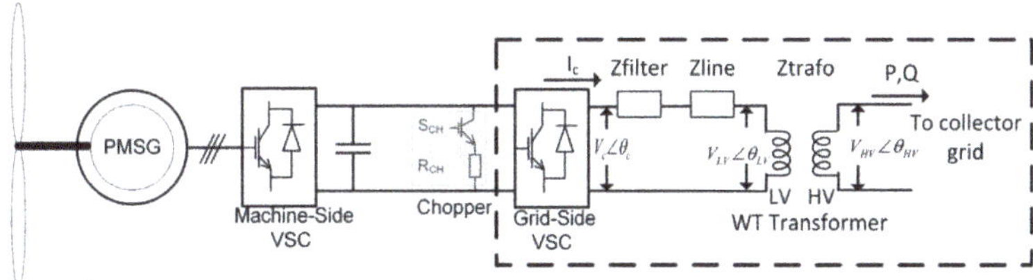

Figure 1. Schematic representation of Type 4 (full rated converter based) wind turbine (WT) connected to wind power plant (WPP) collection system through back-to-back converters and step-up transformer.

Back-to-back converters decouples the WT generator from the grid as well as allowing independent control of active and reactive power. Reactive power is supplied by GSC and is determined by GSC design parameters, namely, current and voltage limitations. The current limitation is due to the maximum current carrying capacity of GSC. Characteristics of power electronics set the maximum and minimum voltage limitation on GSC.

As seen in Figure 1, GSC is connected to the WT transformer through a filter and a short line. Since WT generator generates power at a low voltage level (typically 0.69 kV), the WT transformer is used to step up voltage to medium voltage level (typically 33 kV or 66 kV) to connect to the WPP collection system.

By aggregating the filter and line impedance, Figure 1 can be simplified into GSC with an equivalent impedance in series as shown in Figure 2.

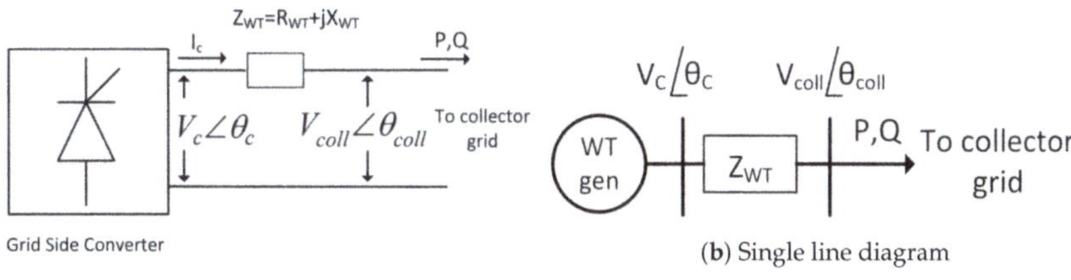

Figure 2. Equivalent representation of grid side converter (GSC) connected to WPP collection system through WT transformer.

In Figure 2, V_C and θ_C represent the converter voltage magnitude and angle, respectively. V_{coll} and θ_{coll} denote the voltage magnitude and angle at the WPP collection system, that is, at the high voltage (HV) side of the WT transformer. The equivalent impedance, Z_{WT} represents impedance from GSC up to the HV side of the WT transformer. Reactive power capability is dependent on two limiting factors—converter voltage limitation and converter current limitation.

2.1.1. Converter Voltage Limitation

Ullah et al. [20] have derived converter voltage limited reactive power capability based on an analogy with the field current limit of a synchronous generator. The relation between P and Q given by Ullah et al. [20] is shown in (1).

$$P^2 + \left(Q + \frac{V_{coll}^2}{X_{WT}}\right)^2 = \left(\frac{V_c V_{coll}}{X_{WT}}\right)^2. \tag{1}$$

Equation (1) from [20] is limited as it consists of only reactance and neglects resistance. However, in this article, WT reactive power capability model is extended for WPP including all resistance and reactance in the circuit.

The relationship between active power, reactive power, voltage and current at the HV side of the WT transformer is given by (2).

$$P + jQ = V_{coll}\angle\theta_{coll}\mathbf{I}^*, \tag{2}$$

where

P	=	active power measured at HV side
Q	=	reactive power measured at HV side
V_{coll}	=	voltage at HV side
$\angle\theta_{coll}$	=	voltage angle at HV side
$\mathbf{I}$	=	complex current flowing into the WPP collection system

Current flowing in the circuit of Figure 2a can be expressed as (3)

$$\mathbf{I} = \frac{V_C\angle\theta_C - V_{coll}\angle\theta_{coll}}{Z_{WT}} \tag{3}$$

where

Z_{WT}	=	equivalent impedance of line, filter and WT transformer
	=	$R_{WT} + jX_{WT}$
V_C	=	converter voltage magnitude
$\angle\theta_C$	=	converter voltage angle

Replacing the current in (2) by (3):

$$P + jQ = V_{coll}\angle\theta_{coll}\left(\frac{V_C\angle\theta_C - V_{coll}\angle\theta_{coll}}{R_{WT} + jX_{WT}}\right)^*. \tag{4}$$

Applying mathematical operation and separating real and imaginary parts, equations for active and reactive power can be written as:

$$P = \frac{1}{R_{WT}^2 + X_{WT}^2}\left[V_{coll}V_C(R_{WT}\cos\theta - X_{WT}\sin\theta) - V_{coll}^2 R_{WT}\right], \tag{5}$$

$$Q = \frac{1}{R_{WT}^2 + X_{WT}^2}\left[V_{coll}V_C(X_{WT}\cos\theta + R_{WT}\sin\theta) - V_{coll}^2 X_{WT}\right], \tag{6}$$

where, $\angle\theta = \angle\theta_{coll} - \angle\theta_C$. Rearranging (5) and (6) and squaring both sides,

$$P^2(R_{WT}^2 + X_{WT}^2)^2 + V_{coll}^4 R_{WT}^2 + 2P(R_{WT}^2 + X_{WT}^2)V_{coll}^2 R_{WT} = V_{coll}^2 V_C^2(R_{WT}\cos\theta - X_{WT}\sin\theta)^2, \tag{7}$$

$$Q^2(R_{WT}^2 + X_{WT}^2)^2 + V_{coll}^4 X_{WT}^2 + 2Q(R_{WT}^2 + X_{WT}^2)V_{coll}^2 X_{WT} = V_{coll}^2 V_C^2(X_{WT}\cos\theta + R_{WT}\sin\theta)^2. \tag{8}$$

By adding (7) and (8), applying mathematical manipulations and simplifying, (9) is obtained.

$$\left(P + \frac{V_{coll}^2 R_{WT}}{R_{WT}^2 + X_{WT}^2}\right)^2 + \left(Q + \frac{V_{coll}^2 X_{WT}}{R_{WT}^2 + X_{WT}^2}\right)^2 = \left(\frac{V_{coll} V_C}{\sqrt{R_{WT}^2 + X_{WT}^2}}\right)^2. \tag{9}$$

Equation (9) can be rearranged and the reactive power limited by converter voltage can be written as a function of active power, converter voltage, voltage at WPP collection system and equivalent impedance as given by (10)

$$Q_V = \sqrt{\left(\frac{V_{coll} V_C}{\sqrt{R_{WT}^2 + X_{WT}^2}}\right)^2 - \left(P + \frac{V_{coll}^2 R_{WT}}{R_{WT}^2 + X_{WT}^2}\right)^2} - \frac{V_{coll}^2 X_{WT}}{R_{WT}^2 + X_{WT}^2}. \tag{10}$$

The maximum injection, $Q_{V,inj}$ and absorption, $Q_{V,abs}$ of reactive power limited by converter voltage can be obtained from (10) by replacing V_C by the maximum and minimum allowable converter voltage, V_{Cmax} and V_{Cmin}, respectively.

2.1.2. Converter Current Limitation

The relation between active and reactive power obtained at the HV side of the WT transformer when limited by the maximum current of GSC can be written as:

$$P^2 + Q^2 = S^2 = (V_{coll} I_{Cmax})^2, \tag{11}$$

where I_{Cmax} is the maximum converter current. Therefore, the reactive power limited by converter current is given by:

$$Q_I = \pm\sqrt{(V_{coll} I_{Cmax})^2 - P^2}. \tag{12}$$

Maximum injection, $Q_{I,inj}$ and absorption, $Q_{I,abs}$ of reactive power limited by converter current can be obtained from positive and negative roots of (12) respectively.

Voltage-limited and current-limited reactive power for different values of active power are plotted in Figure 3. For this illustration, WPP collection system voltage, V_{coll}, is assumed to be 0.95 p.u. Other parameters used are given in Table 1.

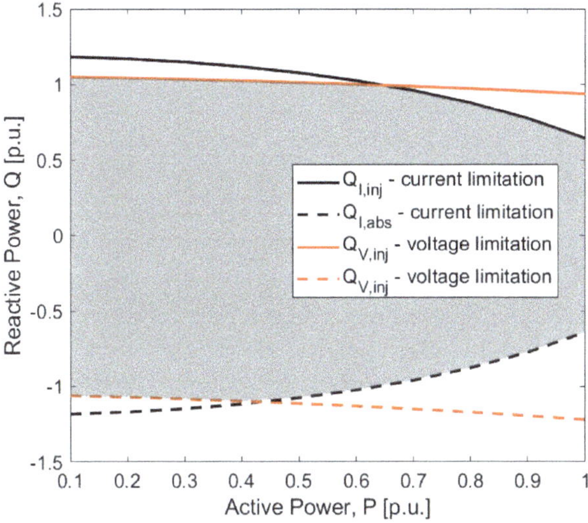

Figure 3. Illustration of voltage-limited and current-limited reactive power capability.

Table 1. Parameters used to plot reactive power capability of WT.

Parameter	Value	Units
I_{Cmax}	1.25	p.u.
V_{Cmax}	1.1	p.u.
V_{Cmin}	0.8	p.u.
R_{WT}	0.0084	p.u.
X_{WT}	0.135	p.u.

Figure 3 shows that WT reactive power capability curves are non-linear. Depending on the active power production and WPP collection system voltage, reactive power capability curves are limited by either voltage or current limitation as represented in the grey shaded region. At a certain operating point, the WT reactive power capability is the minimum of voltage-limited and current-limited reactive power. Maximum reactive power injection, $Q_{inj,max,WT}$, and absorption, $Q_{abs,max,WT}$ capability of WT can be calculated as

$$Q_{inj,max,WT} = min(Q_{V,inj}, Q_{I,inj}),$$
(13)

$$Q_{abs,max,WT} = max(Q_{V,abs}, Q_{I,abs}).$$
(14)

2.1.3. WT Reactive Power Capability Diagrams

The developed model is used to plot the reactive power capability of Type 4 WT at the point of connection (PoC) with the WPP collection system for different values of active power production and different voltages at the WPP collection system terminal. Parameters used are given in Table 1.

It is assumed that the GSC is 25% over dimensioned as compared to the WT generator. Maximum and minimum voltage limitation of GSC, V_{Cmax} and V_{Cmin}, are taken as 1.1 p.u. and 0.8 p.u. respectively. The WT transformer is assumed to have 0.84% resistance and 5% reactance. Reactance of the filter and line taken together is 8.5%, while resistance is taken as zero.

Figure 4 shows the WT reactive power capability at the PoC to WPP collection system, that is, at the HV side of the WT transformer.

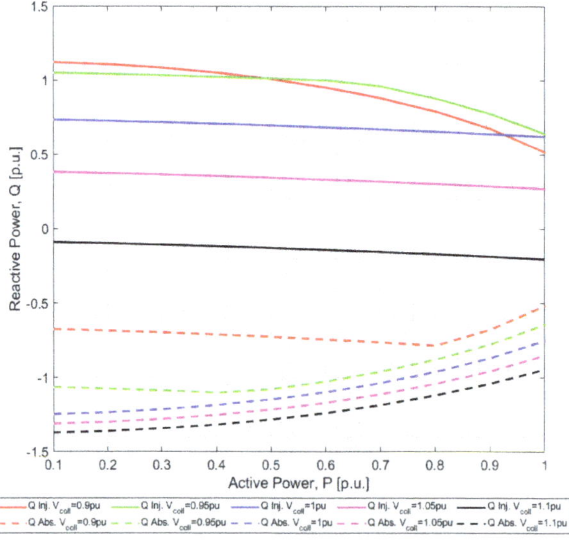

Figure 4. Wind turbine capability curve at the WPP collection system terminal.

According to Danish grid codes [29], WPPs are not required to provide reactive power support when WTs are not producing any active power. Therefore, active power is varied between 0.1 p.u. to 1 p.u. in this study. To illustrate voltage dependency, reactive power capability curves are drawn for different voltage levels of V_{coll} between 0.9 and 1.1 p.u.

In Figure 4, positive reactive power denotes that WT is injecting reactive power into the WPP collection system (denoted by solid lines). Negative reactive power implies that WT is absorbing reactive power (denoted by dotted lines). It can be observed from Figure 4, the WT reactive power injection capability is converter current limited for lower voltages. For higher voltages, voltage limitation of the converter determines the reactive power capability of WT. However, this trend is reversed in the case of reactive power absorption. Reactive power absorption capability is voltage limited for lower voltages and current limited for higher voltages. It can also be observed from Figure 4, for voltage of 1.1 p.u., the upper limitation of reactive power capability becomes negative (denoted by solid black line in Figure 4). This implies that the converter starts to absorb reactive power. This is designed in order to support the system when voltages become too high.

For validation, capability curves obtained using the above mentioned model have been compared with voltage dependent capability diagrams of Type 4 WTs illustrated in [5,30].

2.2. WPP Reactive Power Capability Model

WPPs consist of several WTs on a feeder. One or more of these feeders are then connected to the step-up WPP transformer. Usually a tap-changing transformer maintains the voltage at the WPP end to a constant value under normal operating conditions. In this work, WPP reactive power capability is determined at the low voltage (LV) end of the WPP transformer. The proposed method for modelling WPP reactive power capability considers equivalent impedance of the WPP collection system. To calculate the equivalent impedance of the WPP collection system, the methodology formulated by Muljadi et al. [31] is used. This method of equivalencing the WPP collection system uses circuit analysis to determine equivalent impedance, which is calculated from apparent power loss in the WPP collection system. The equivalent WPP collection system represents both impedance of cables and shunt capacitance of the collection system. Using this equivalent WPP collection system model, reactive power capability of any type of WPP can be obtained because the equivalencing method can be applied to any type of WPP. Since aggregation of the WPP collection system is done assuming that all WTs produce same power, any spatial variation in wind speed (due to variability in wind and wake effect) is neglected in the proposed WPP reactive power capability model.

Figure 5 represents the single line diagram for the proposed WPP reactive power capability model.

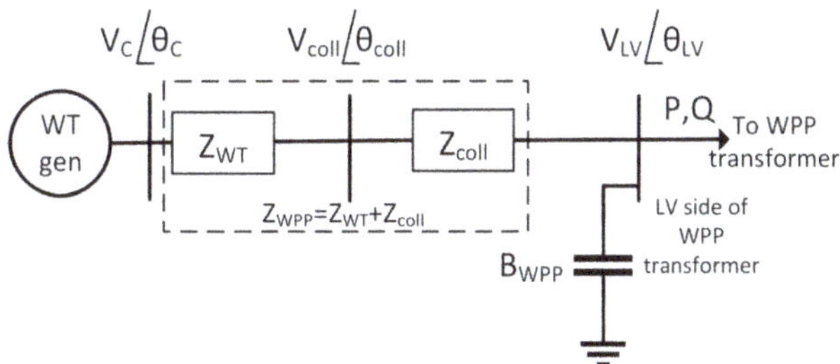

Figure 5. Reactive power capability model of WPP with aggregated WPP collection system.

It is a single WT representation of WPP, while preserving the losses incurred in the WPP collection system as well as incorporating reactive power generated by WPP collection system cables. In Figure 5, Z_{WT} represents the equivalent impedance of filter, line and WT transformer. Z_{coll} represents the equivalent impedance of the WPP collection system. Z_{WPP} represents the combined impedance of WT model and WPP collection system model. B_{WPP} represents the equivalent shunt susceptance of WPP collection system. Using the parameters in Figure 5, Equations (10) and (12) can be modified as

$$Q_{V,WPP} = \sqrt{\left(\frac{V_{LV}V_C}{\sqrt{R_{WPP}^2 + X_{WPP}^2}}\right)^2 - \left(P + \frac{V_{LV}^2 R_{WPP}}{R_{WPP}^2 + X_{WPP}^2}\right)^2 - \frac{V_{LV}^2 X_{WPP}}{R_{WPP}^2 + X_{WPP}^2}}, \tag{15}$$

$$Q_{I,WPP} = \pm\sqrt{(V_{LV}I_{Cmax})^2 - P^2}. \tag{16}$$

Similar to the WT reactive power capability model, the maximum injection ($Q_{V,inj,WPP}$) and absorption ($Q_{V,abs,WPP}$) of WPP reactive power which is limited by converter voltage can be obtained from (15) by replacing V_C by maximum and minimum allowable converter voltage, V_{Cmax} and V_{Cmin}, respectively. For current limitation, the maximum injection ($Q_{I,inj,WPP}$) and absorption ($Q_{I,abs,WPP}$) of reactive power of WPP can be obtained from positive and negative roots of (16) respectively. Reactive power injected by the cables due to the equivalent WPP collection system susceptance, are added to the maximum injection and absorption capability obtained at the LV side of the WPP transformer. Therefore, the maximum reactive power injection, $Q_{inj,max,WPP}$, and absorption, $Q_{abs,max,WPP}$ capability of WPP are calculated as

$$Q_{inj,max,WPP} = min(Q_{V,inj,WPP}, Q_{I,inj,WPP}) + B_{WPP}V_{LV}^2, \tag{17}$$

$$Q_{abs,max,WPP} = max(Q_{V,abs,WPP}, Q_{I,abs,WPP}) + B_{WPP}V_{LV}^2. \tag{18}$$

This WPP reactive power capability model is used to derive capability diagrams of a Type 4 based WPP with the parameters as given in Table 2. The corresponding WPP reactive power capability curves are shown in Figure 6. A summary of parameters required for the proposed model is presented in Appendix A.

Table 2. Parameters used to plot WPP reactive power capability.

Parameter	Value	Units
I_{Cmax}	1.25	p.u.
V_{Cmax}	1.1	p.u.
V_{Cmin}	0.8	p.u.
R_{WT}	0.0084	p.u.
X_{WT}	0.135	p.u.
R_{WPP}	0.0114	p.u.
X_{WPP}	0.0096	p.u.
B_{WPP}	0.0210	p.u.

To illustrate voltage dependency, WPP reactive power capability curves are drawn for different voltage levels of V_{LV} between 0.9 and 1.1 p.u. Comparing Figure 4 and 6, it can be seen that WPP reactive power capability shows similar trends as that of WT reactive power capability. It is to be noted that since WPP is connected to the grid through a tap-changing transformer, the voltage at the LV end of the WPP transformer is maintained at 1 pu. Therefore, as long as the tap-changing WPP transformer is not saturated, WPP reactive power capability is only dependent on active power generated by WTs.

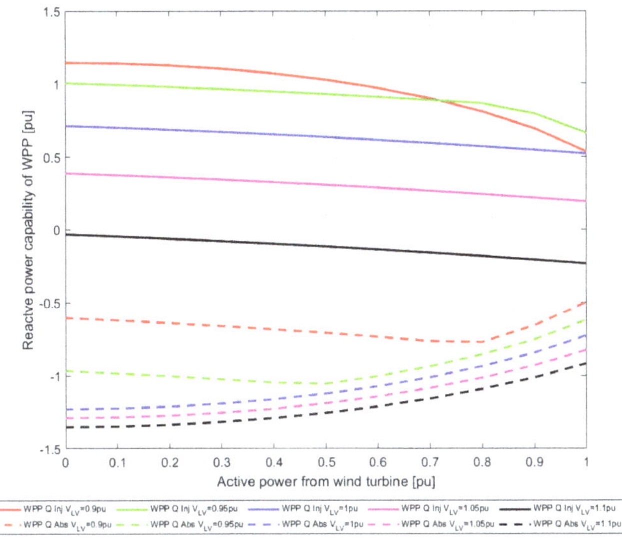

Figure 6. WPP capability curve at the low voltage (LV) end of WPP transformer.

3. Case Studies

In this section, various case studies are performed to determine the behaviour and accuracy of the proposed WPP reactive power capability model as compared to methodologies from literature. In addition, results are compared for different WPP topologies from real WPPs. Finally, an application is shown to demonstrate the difference in performance of bus voltages during increased system stress in a simulated power system, where the WPP reactive power support from the proposed model is compared to that of scaled WT model and detailed WPP model.

3.1. Case Study: Comparison of WPP Reactive Power Capability Curves for Different Models

As mentioned before, there are two existing methodologies to model WPP reactive power capability: Scaling up of the WT reactive power model with the number of WTs in WPP and a detailed WPP model.

3.1.1. Scaled WT Model

In this method, output of a single WT is scaled up with number of WTs in the WPP. WPP collection system is neglected. It is simple and easy to implement, as it requires less parameters. Equations (19) and (20) describe the scaled WT model representation of WPP reactive power capability.

$$Q_{inj,max,WPP} = N * Q_{inj,max,WT},$$ (19)

$$Q_{abs,max,WPP} = N * Q_{abs,max,WT}.$$ (20)

where N = number of WTs in a WPP.

Reactive power capability of the scaled WT model is equal to the reactive power capability of a single WT in per unit, assuming nominal capacity of the WPP is taken as base MVA.

3.1.2. WPP Detailed Model

In this method, the WPP is modelled with WT transformers, WPP collection system impedance and susceptance. Reactive power capability of the WPP can be computed using powerflow studies for different active power and voltage set points. This method is used as the base case for comparison of results in this study. Though this method provides accurate an WPP reactive power capability, the disadvantages of this method are: (i) Many parameters required, (ii) can have high computation time for large WPPs, and (iii) when simulating large power systems with numerous WPP, including detailed model of each WPP may not be efficient.

A method to perform powerflow on the detailed model is described in the flowchart illustrated in Figure 7.

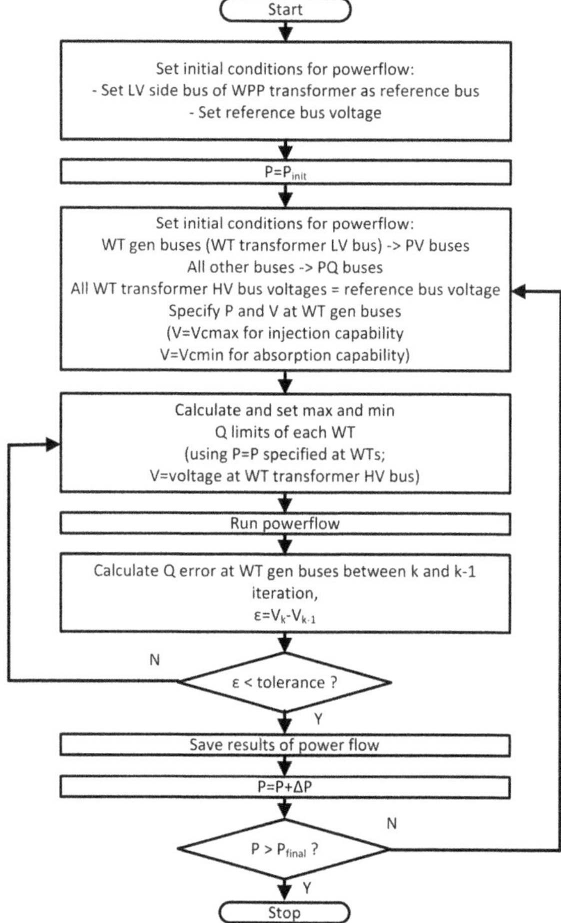

Figure 7. Flowchart to determine WPP reactive power capability using detailed model.

The LV end of the WPP transformer is considered as the reference bus and this reference bus voltage is varied to obtain voltage dependent WPP reactive power capability. Initially, WT generator buses are set as PV buses with voltages equal to V_{Cmax} (to calculate maximum reactive power injection) or V_{Cmin} (to calculate maximum reactive power absorption by WPP). Maximum and minimum reactive power limits of WT generators are calculated using a WT reactive power capability curve model. Powerflow is executed enforcing reactive power limits on the WT generators. WT generator buses are converted to PQ buses when the reactive power limit is reached. Initially, all HV end buses of WT transformer are assumed to have the same voltages as the reference voltage. Since these voltages are close enough but not same in practice, powerflow results are used to update the voltages at the HV ends of the WT transformer. This changes reactive power capability at the LV end of the WT transformer. Therefore, reactive powers calculated at the LV ends of the WT transformers are compared with the results of previous iteration to check if the error is within tolerated levels. This process is repeated for all values of the active power; from the initial to final value.

3.1.3. Scaled WT Model vs. Detailed WPP Model vs. Proposed Model

To compare the result of the proposed model with the existing models, a simple WPP with seven WTs in a string as shown in Figure 8 is studied.

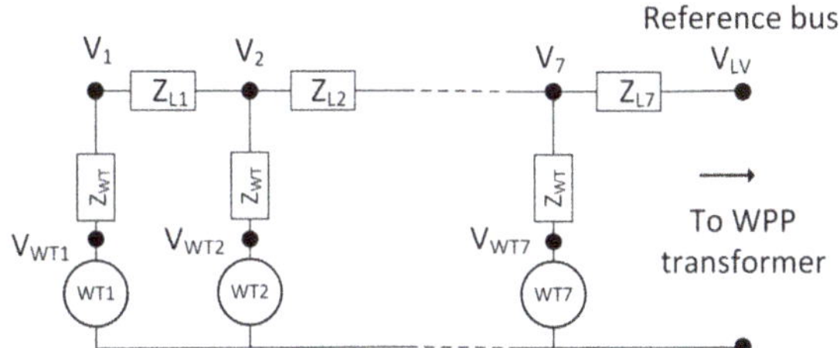

Figure 8. WPP layout with seven WTs on a feeder.

Each WT is assumed to be 2 MW connected to WPP collection system through an impedance $Z_{WT} = 0.0084 + j0.135$ p.u. which includes filter, line and WT transformer. Impedance of collector lines: $Z_{L1} = Z_{L2} = Z_{L3} = Z_{L4} = Z_{L5} = Z_{L6} = 0.0013 + j0.001$ p.u. and $Z_{L7} = 0.0021 + j0.0019$ p.u. Shunt susceptance of collector lines are taken as, $B_1 = B_2 = B_3 = B_4 = B_5 = B_6 = 2.419 \times 10^{-3}$ p.u.; $B_7 = 5.1073 \times 10^{-3}$ p.u.

Figure 9a,b shows reactive power injection and absorption capability of the WPP determined using the three different reactive power capability models. Voltage at the LV end of the WPP transformer is assumed to be 1 p.u. Considering the detailed model as the base case, it can be observed from Figure 9 that the result of the proposed model follows the results from the detailed model.

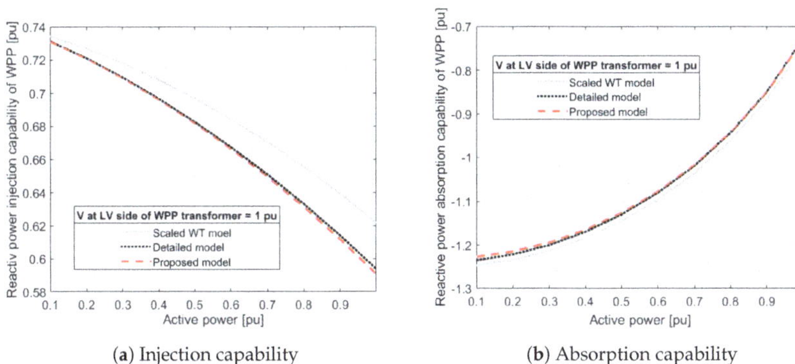

(**a**) Injection capability   (**b**) Absorption capability

Figure 9. Comparison of WPP reactive power capability using scaled WT model, detailed model and proposed model.

On the other hand, the scaled WT model result has considerable difference from the detailed model. It can be therefore concluded that using the proposed model, the WPP reactive power capability is accurately determined using a reduced number of parameters. Thus, a fast and efficient calculation of reactive power availability compared to the reactive power capability of the WPP detailed model can be obtained.

3.2. Case Study: Sensitivity Studies of Different Parameters on the Accuracy of Different Models

In this section, the effect of parameters like the number of WTs in a WPP and the length of WPP collection system cable on the proposed model is investigated. For this purpose, three different WPP topologies are examined: (i) A small WPP represented by seven WTs in a string (as shown in Figure 8), (ii) Horns Rev 2 WPP with 91 WTs which allow us to study the impact of a large number of WTs on the accuracy of the proposed model and (iii) Burbo Bank WPP with 25 WTs but long collection system cables connecting to onshore WPP transformer which allow us to study the impact of length of collection system cables on the accuracy of the proposed model. It is assumed that all WPPs consist of Type 4 WTs.

3.2.1. Horns Rev 2 WPP

Horns Rev 2 WPP is a 210 MW plant with 91 WTs each of 2.3 MW located in the North Sea. Schematic layout of the WPP is shown in Figure 10a. The big bold dots represent the location of WTs, whereas the red square represents the WPP transformer. The dotted line represents the collector cables. The total WPP collection system cable length is 70 km. The nominal voltage of WPP collection system is 33 kV.

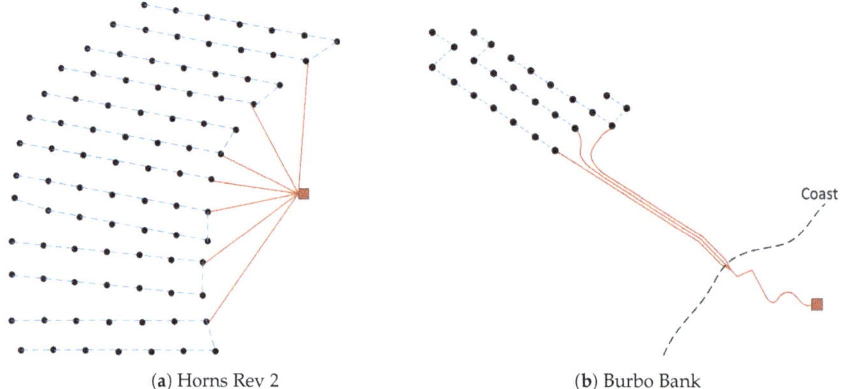

(a) Horns Rev 2   (b) Burbo Bank

Figure 10. Wind power plant layouts.

3.2.2. Burbo Bank WPP

Burbo Bank WPP, situated in the west coast of UK, is a 90 MW wind power plant with 25 WTs, each of 3.6 MW rating. Figure 10b shows the layout of Burbo Bank. The total WPP collection system cable length is 42 km, with three long export cables (total length of approximately 29 km) connecting the WTs with the onshore transformer substation. The nominal voltage of the WPP collection system is 33 kV.

3.2.3. Accuracy of Proposed Reactive Power Capability Model vs. Scaled WT Model

To exemplify the accuracy of the proposed model, the reactive power capability of the above mentioned three WPP topologies are plotted in Figure 11a–c for scaled WT model, detailed WPP model and the proposed model respectively. In this example, voltage at the LV side of the WPP transformer is maintained constant at 1 pu by the tap-changer. Absolute errors between reactive power capability estimation of the scaled WT model and proposed model compared to base case are plotted in Figure 11d–f respectively. It can be observed that for large WPPs, error can be up to 20 Mvar. In this particular example, the scaled WT model is seen to be overestimating the reactive power capability. This can lead to a misinterpretation of the reactive power reserve in the system, which can cause a significant impact on voltage stability analysis, especially when the system is in a stressed condition.

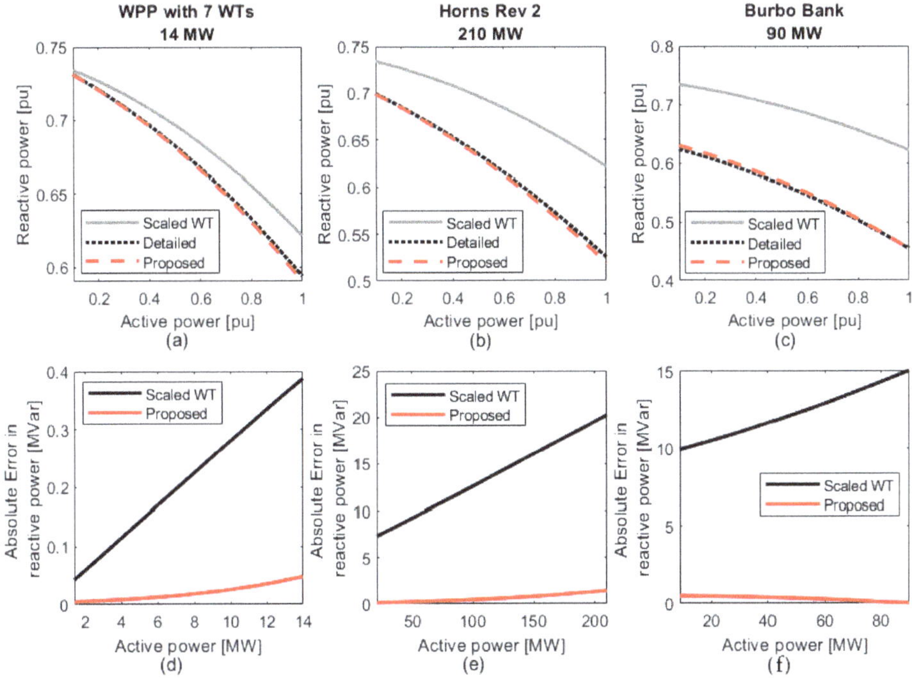

Figure 11. (**a–c**) WPP reactive power capability estimation using different models at 1 p.u. voltage at LV side of WPP transformer. (**d–f**) Error in WPP reactive power capability estimation using scaled WT model and the proposed model compared to detailed WPP model.

Further, the proposed model is used to calculate WPP reactive power capability for each of the WPP topologies at different voltage levels and different active power generation. Results are then compared with the reactive power capability obtained using the detailed WPP model to compute average root mean square error (RMSE) of calculated reactive power as given by

$$\text{Average RMSE} = \frac{\sum_k \sqrt{\frac{\sum_i (Q_{detailed,i} - Q_{calculated,i})^2}{\text{total number of data}}}}{\text{total no. of voltage levels,}} \tag{21}$$

where, $Q_{detailed,i}$ is reactive power calculated using the detailed WPP model for the ith value of active power; $Q_{calculated,i}$ is the reactive power calculated using the proposed reactive power capability model for the ith value of active power. The average RMSE is calculated as the average of RMSE errors across k number of voltage levels simulated. The same process is used to compute the average RMSE of calculated reactive power using a scaled WT model as compared to a detailed model. Table 3 presents the RMSE and average RMSE of calculated reactive power for three WPP topologies for different voltage levels.

A graphical representation of the tabular results are shown in Figure 12. For larger WPPs, losses in WPP collection system can be significant. There can be significant error (5% for Horns Rev 2 WPP and 11% for Burbo Bank WPP) in reactive power capability calculation when using a scaled WT model. However, using the proposed reactive power capability model gives better results (error of 1% for Horns Rev 2 WPP and 4% for Burbo Bank WPP). For smaller WPPs, scaled WT model may be used for simplicity. The error for Burbo Bank is higher than that for Horns Rev 2, though the number of WTs in Horns Rev 2 is higher. Therefore, the error is not directly related to the total number of WTs in a WPP. The length of collector cables causes significant error when the WPP collection system is neglected.

This is evident from the results of Burbo Bank, where using a scaled WT model results in an average RMSE of 0.1 p.u. in the case of absorption and 0.09 p.u. in the case of injection. The error reduces to 0.04 p.u. for absorption and 0.02 p.u. for injection when using the proposed model. From Figure 12b it can be observed that the average error for all three cases reduces to below 4% when using the proposed WPP reactive power capability model as compared to an 11% error for the scaled WT model (Figure 12a).

Table 3. The average root mean square error (RMSE) of calculated reactive power for three different WPPs at different voltage levels when using proposed model and scaled WT model compared to detailed WPP model.

	WPP		V = 0.9	V = 0.95	V = 1	V = 1.05	V = 1.1	MVar	pu
				RMSE (MVar)				**Average RMSE**	
Scaled WT model	WPP with 7 WTs (Cap. = 14 MW)	Inj.	0.20	0.25	0.24	0.14	0.25	0.22	0.015
		Abs.	0.29	0.39	0.22	0.24	0.27	0.28	0.020
	Horns Rev 2 (Cap. = 209.3 MW)	Inj.	6.95	13.21	14.30	9.98	12.03	11.29	0.054
		Abs.	5.76	9.31	3.16	3.64	4.17	5.21	0.025
	Burbo Bank (Cap. = 90 MW)	Inj.	4.92	15.28	12.47	5.52	1.84	8.00	0.089
		Abs.	12.59	16.37	6.03	5.27	5.88	9.23	0.103
Proposed model	WPP with 7 WTs (Cap. = 14 MW)	Inj.	0.03	0.02	0.03	0.02	0.15	0.05	0.003
		Abs.	0.05	0.03	0.07	0.07	0.07	0.06	0.004
	Horns Rev 2 (Cap. = 209.3 MW)	Inj.	3.73	2.28	0.75	0.48	3.87	2.22	0.011
		Abs.	2.41	2.09	1.82	1.65	1.53	1.90	0.009
	Burbo Bank (Cap. = 90 MW)	Inj.	5.17	1.75	0.34	0.50	2.07	1.97	0.022
		Abs.	1.42	2.43	4.04	4.56	4.59	3.41	0.038

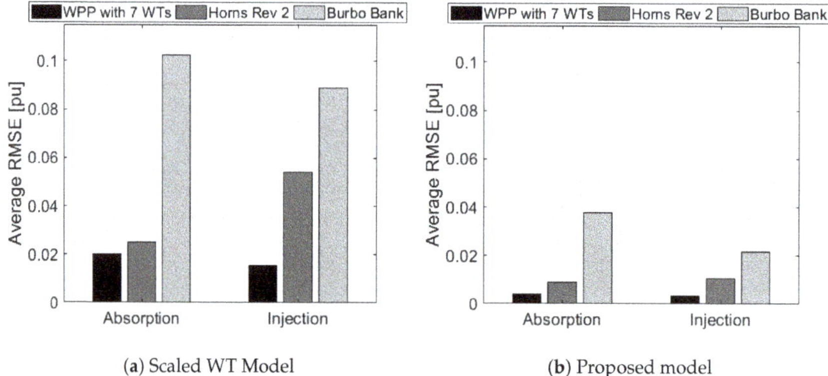

(a) Scaled WT Model

(b) Proposed model

Figure 12. Average root mean square error (in percentage) when using scaled WT model and proposed WPP reactive power capability model as compared to detailed WPP model.

3.2.4. Assessment of Computational Cost

Computational performance of the three different models are assessed in this section. Equations for proposed and scaled WT models are coded in MATLAB. For the WPP detailed model, the algorithm described in Section 3.1.2 is coded in MATLAB and the power flow solution is obtained using MATPOWER. All simulations are performed on a 64-bit Windows OS based computer with 2.6 GHz Intel Core i7-6600U processor. The time required for obtaining maximum and minimum reactive power capability at an operating point (particular active power and voltage) are summarized in Table 4.

Table 4. Time required for three reactive power capability models.

Model	Time (s)		
	WPP with 7 WTs	**Horns Rev 2**	**Burbo Bank**
Scaled WT	0.002	0.002	0.002
Detailed WPP	1.1	3.3	1.5
Proposed	0.002	0.002	0.002

Each value is calculated as the average time required for hundreds of different simulations at a particular operating point. It can be observed that the computational cost of the scaled WT model and the proposed model are the same and 1000 times faster than that of the detailed WPP model. It should be noted that the computation time of the detailed WPP model increases with an increase in the number of WTs in the WPP. Extrapolating these observations, it can be assumed that the computational burden of the power system analysis for a large power systems with numerable WPPs can be significantly improved using the proposed model.

3.3. Case Study: Application of Different Models in Power System Integration

To demonstrate how the proposed reactive power capability curve model can be used for power system studies, a simple power system model as shown in Figure 13 is used.

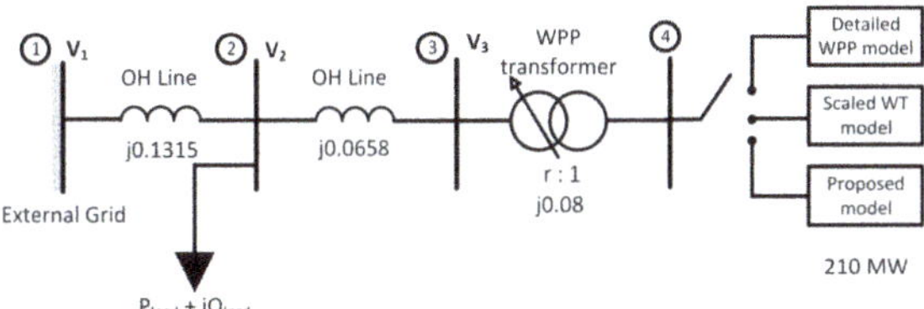

Figure 13. Simple power system model.

The system consists of a 210 MW WPP whose reactive power capability is modelled alternatively as a detailed model, scaled WT model and proposed model. The WPP is connected to the grid through a tap-changing WPP transformer. The transformer maintains the voltage at bus 4 to 1 p.u. The load is modelled as a constant power load, which is increased from 210 MW to 510 MW in order to simulate a voltage stress condition. The load power factor is assumed as 0.9 (lagging). The active power generated from the WPP is kept constant at 1 p.u. (210 MW). For this study, it is assumed that the WPP is controlled such that it provides the maximum available reactive power, that is, the maximum reactive power injection capability at any instant. Figure 14 show profiles of load voltage and voltage at the HV side of the WPP transformer.

Taps of the WPP transformer are not saturated, so voltage at the LV end of the WPP transformer is maintained at 1 p.u. As both active power and voltage remains constant at the LV end of the WPP transformer during the simulation, the WPP reactive power capability remains constant during the simulation. However, the value of the reactive power capability varies according to the capability model used, and the values are given in Table 5.

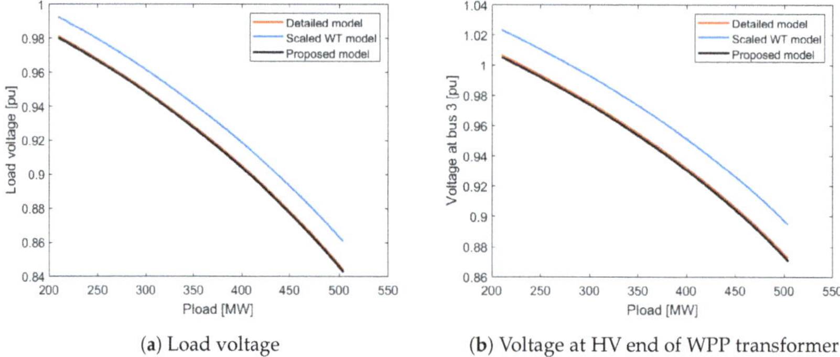

(**a**) Load voltage (**b**) Voltage at HV end of WPP transformer

Figure 14. Simulation results showing how system voltage is affected depending on the type of WPP reactive power capability model used.

Table 5. Values of maximum reactive power injection capability using different models.

Model	Value	Unit
Detailed	110	MVar
Scaled WT	130	MVar
Proposed	109	MVar

The reactive power capability obtained from the detailed WPP model is taken as the base case. From the simulation results shown in Figure 14, it can be observed that the WPP reactive power capability using the proposed model provides a better representation of the reactive power generation capability of WPPs, consequently providing an estimation of system voltages close to the actual values. It should be noticed that using a scaled WT model in the power system studies could lead to a miscalculation of system voltages (overestimation of voltages in this particular example). This study clearly demonstrates that the proposed model should be applied to power system studies.

4. Conclusions

This article proposes and presents a novel approach of modelling WPP reactive power capability using an aggregated WPP collection system parameters for Type 4 based WPPs. The inclusion of a WPP collection system in aggregated form reduces the number of parameters required for simulations, thereby substantially reducing the computational time. Additionally, the accuracy of the proposed model to estimate WPP reactive power capability is much better compared to the scaled WT model predominantly used in literature. WPP reactive power capability depends on the WPP collection system length. For large WPPs with a large collection system, the reactive power capability obtained using the proposed method is close to the actual representation of reactive power generation and absorption limits of WPP. Furthermore, using the reactive power capability of the proposed model in the power system study has shown to be a better estimate of system voltages. Based on the studies and results presented in this article, the proposed model is recommended for power system analysis studies (mainly voltage stability analysis) with large share of converter based generation.

Author Contributions: Conceptualization, M.S. and P.E.S.; Data curation, M.S.; Formal analysis, M.S.; Funding acquisition, P.E.S.; Investigation, M.S.; Methodology, M.S., M.A. and P.E.S.; Project administration, P.E.S.; Resources, M.S. and P.E.S.; Software, M.S.; Supervision, M.A., P.E.S. and A.D.H.; Validation, M.S.; Visualization, M.S.; Writing—original draft, M.S.; Writing—review & editing, M.A., P.E.S. and A.D.H.

Funding: This research was funded by the SARP project, which is funded by Energinet under the Public Service Obligation scheme (Forskel 12427).

Conflicts of Interest: The authors declare no conflict of interest.

Abbreviations

The following abbreviations are used in this manuscript:

WPP	Wind Power Plant
WT	Wind Turbine
DFIG	Doubly Fed Induction Generator
PMSG	Permanent Magnet Synchronous Generator
PCC	Point of Common Coupling
PoC	Point of Connection
MSC	Machine Side Converter
GSC	Grid Side Converter
HV	High Voltage
LV	Low Voltage
RMSE	Root Mean Square Error

Appendix A. Summary of Input Parameters Required for Proposed Wpp Reactive Power Capability Model

Table A1. Description of parameters required for proposed model.

Parameter	Description
V_{Cmax}	Maximum permissible grid side converter voltage
V_{Cmin}	Minimum permissible grid side converter voltage
I_{Cmax}	Maximum permissible current limit of grid side converter
Z_{WT}	Impedance of WT transformer connecting WT to wind power collection system
Z_{coll}	Aggregated equivalent impedance of wind power collection system
B_{WPP}	Aggregated equivalent shunt capacitance of wind power collection system
V_{LV}	Voltage at the low voltage side of WPP transformer (the point at which WPP reactive power capability is calculated)
P	Active power generation from WT

References

1. International Energy Agency. *Global Wind Energy Outlook 2017 Executive Summary*; International Energy Agency: Paris, France, 2017.
2. Lof, P.A.; Hill, D.J.; Arnborg, S.; Andersson, G. On the analysis of long-term voltage stability. *Int. J. Electr. Power Energy Syst.* **1993**, *15*, 229–237. [CrossRef]
3. Kundur, P.; Paserba, J.; Ajjarapu, V.; Andersson, G.; Bose, A.; Canizares, C.; Hatziargyriou, N.; Hill, D.; Stankovic, A.; Taylor, C.; et al. Definition and classification of power system stability IEEE/CIGRE joint task force on stability terms and definitions. *IEEE Trans. Power Syst.* **2004**, *19*, 1387–1401.
4. Das, K.; Hansen, A.D.; Sørensen, P.E. Understanding IEC standard wind turbine models using SimPowerSystems. *Wind Eng.* **2016**, *40*, 212–227. [CrossRef]
5. Bech, J. Siemens experience with validation of different types of wind turbine models. In Proceedings of the IEEE Power and Energy Society General Meeting, Washington, DC, USA, 27–31 July 2014.
6. Sorensen, P.; Fortmann, J.; Buendia, F.J.; Bech, J.; Morales, A.; Ivanov, C. Final draft international standard IEC 61400-27-1. In Proceedings of the 13th Wind Integration Workshop, Berlin, Germany, 11–13 November 2014.
7. Seshadri Sravan Kumar, V.; Thukaram, D. Accurate modeling of doubly fed induction generator based wind farms in load flow analysis. *Electr. Power Syst. Res.* **2018**, *155*, 363–371.
8. Kaempf, E.; Braun, M. Models of reactive power-related wind park losses for application in power system load flow studies. *Wind Energy* **2017**, *20*, 1291–1309. [CrossRef]
9. Zhang, B.; Hou, P.; Hu, W.; Soltani, M.; Chen, C.; Chen, Z. A reactive power dispatch strategy with loss minimization for a DFIG-based wind farm. *IEEE Trans. Sustain. Energy* **2016**, *7*, 914–923. [CrossRef]
10. Meegahapola, L.; Durairaj, S.; Flynn, D.; Fox, B. Coordinated utilisation of wind farm reactive power capability for system loss optimisation. *Eur. Trans. Electr. Power* **2011**, *21*, 40–51. [CrossRef]

11. Stankovic, S.; Soder, L. Analytical Estimation of Reactive Power Capability of a Radial Distribution System. *IEEE Trans. Power Syst.* **2018**, *33*, 6131–6141. [CrossRef]
12. Ugranli, F.; Karatepe, E. Coordinated TCSC allocation and network reinforcements planning with wind power. *IEEE Trans. Sustain. Energy* **2017**, *8*, 1694–1705. [CrossRef]
13. Martínez, J.; Kjær, P.C.; Rodriguez, P.; Teodorescu, R. VAr reserve concept applied to a wind power plant. In Proceedings of the Power Systems Conference and Exposition (PSCE), Phoenix, AZ, USA, 20–23 March 2011; pp. 1–8.
14. Kim, J.; Park, G.; Kang, Y.C.; Lee, B.; Muljadi, E. Voltage control of a wind power plant using the adaptive QV characteristic of DFIGs. In Proceedings of the Power Electronics and Machines for Wind and Water Applications (PEMWA), Milwaukee, WI, USA, 24–26 July 2014; pp. 1–5.
15. Karbouj, H.; Rather, Z.H. Voltage Control Ancillary Service from Wind Power Plant. *IEEE Trans. Sustain. Energy* **2018**, *10*, 759–767. [CrossRef]
16. Meegahapola, L.; Littler, T.; Perera, S. Capability curve based enhanced reactive power control strategy for stability enhancement and network voltage management. *Int. J. Electr. Power Energy Syst.* **2013**, *52*, 96–106. [CrossRef]
17. Londero, R.R.; de Mattos Affonso, C.; Vieira, J.P.A. Long-term voltage stability analysis of variable speed wind generators. *IEEE Trans. Power Syst.* **2015**, *30*, 439–447. [CrossRef]
18. Amarasekara, K.; Meegahapola, L.G.; Agalgaonkar, A.P.; Perera, S. Characterisation of long-term voltage stability with variable-speed wind power generation. *IET Gener. Transm. Distrib.* **2017**, *11*, 1848–1855. [CrossRef]
19. Vijayan, P.; Sarkar, S.; Ajjarapu, V. A novel voltage stability assessment tool to incorporate wind variability. In Proceedings of the Power & Energy Society General Meeting, Calgary, AB, Canada, 26–30 July 2009; pp. 1–8.
20. Ullah, N.R.; Bhattacharya, K.; Thiringer, T. Wind farms as reactive power ancillary service providers—Technical and economic issues. *IEEE Trans. Energy Convers.* **2009**, *24*, 661–672. [CrossRef]
21. Lund, T.; Sørensen, P.; Eek, J. Reactive power capability of a wind turbine with doubly fed induction generator. *Wind Energy* **2007**, *10*, 379–394. [CrossRef]
22. Engelhardt, S.; Erlich, I.; Feltes, C.; Kretschmann, J.; Shewarega, F. Reactive power capability of wind turbines based on doubly fed induction generators. *IEEE Trans. Energy Convers.* **2011**, *26*, 364–372. [CrossRef]
23. Kayikci, M.; Milanovic, J.V. Reactive power control strategies for DFIG-based plants. *IEEE Trans. Energy Convers.* **2007**, *22*, 389–396.
24. Konopinski, R.J.; Vijayan, P.; Ajjarapu, V. Extended reactive capability of DFIG wind parks for enhanced system performance. *IEEE Trans. Power Syst.* **2009**, *24*, 1346–1355. [CrossRef]
25. Meegahapola, L.; Fox, B.; Littler, T.; Flynn, D. Multi-objective reactive power support from wind farms for network performance enhancement. *Int. Trans. Electr. Energy Syst.* **2013**, *23*, 135–150. [CrossRef]
26. Kim, J.; Seok, J.K.; Muljadi, E.; Kang, Y.C. Adaptive Q–V scheme for the voltage control of a DFIG-based wind power plant. *IEEE Trans. Power Electron.* **2016**, *31*, 3586–3599. [CrossRef]
27. Kim, J.; Muljadi, E.; Park, J.W.; Kang, Y.C. Adaptive hierarchical voltage control of a DFIG-based wind power plant for a grid fault. *IEEE Trans. Smart Grid* **2016**, *7*, 2980–2990. [CrossRef]
28. Ackermann, T. *Wind Power in Power Systems*; Wiley Online Library: Hoboken, NJ, USA, 2005.
29. Energinet. *Technical Regulation 3.2.5 for Wind Power Plants with a Power Output Above 11 kW*; 2015. Available online: https://en.energinet.dk/Electricity/Rules-and-Regulations/Regulations-for-grid-connection (accessed on 26 April 2019).
30. North American Electric Reliability Corporation. *Reliability Guideline: Power Plant Model Verification for Inverter-Based Resources*; Technical Report; 2018. Available online: https://www.nerc.com/comm/PC/Documents/4.b_Reliability_Guideline_-_PPMV_for_Inverter-Based_Resources_-_2018-05-17.pdf (accessed on 26 April 2019).
31. Muljadi, E.; Butterfield, C.; Ellis, A.; Mechenbier, J.; Hochheimer, J.; Young, R.; Miller, N.; Delmerico, R.; Zavadil, R.; Smith, J. Equivalencing the collector system of a large wind power plant. In Proceedings of the Power Engineering Society General Meeting, Montreal, QC, Canada, 18–22 June 2006.

Article

Compliance of a Generic Type 3 WT Model with the Spanish Grid Code

Raquel Villena-Ruiz [1], Francisco Jiménez-Buendía [2], Andrés Honrubia-Escribano [1],
Ángel Molina-García [3] and Emilio Gómez-Lázaro [1,*]

[1]   Renewable Energy Research Institute and DIEEAC-ETSII-AB, Universidad de Castilla-La Mancha,
      02071 Albacete, Spain; raquel.villena@uclm.es (R.V.-R.); andres.honrubia@uclm.es (A.H.-E.)
[2]   Siemens Gamesa Renewable Energy, S.A., 31621 Pamplona, Spain; fjimenez@gamesacorp.com
[3]   Department of Electrical Engineering, Universidad Politécnica de Cartagena, 30202 Cartagena, Spain;
      angel.molina@upct.es
[*]   Correspondence: emilio.gomez@uclm.es; Tel.: +34-967-599-200 (ext. 2418)

Received: 20 March 2019; Accepted: 25 April 2019; Published: 29 April 2019

Abstract: The expansion of wind power around the world poses a new challenge that network operators must overcome, namely the integration of this renewable energy source into the grid. Comprehensive analyses involving time-domain simulations must be carried out to plan network operation and ensure power supply. In light of the above, and with the aim of extending the use of the wind turbine models developed by Standard IEC 61400-27-1 and assessing their performance according to national grid code requirements, an IEC Type 3 wind turbine model has been submitted for the first time to Spanish grid code PO 12.3. Indeed, there is a lack of studies submitting generic wind turbine models to national grid code requirements. The model's behavior is compared with field measurements of an actual Gamesa G52 machine and with its detailed simulation model. The outcomes obtained have been comprehensively analyzed and the results of the validation criteria highlight that several modeling modifications, in the cases of non-compliance, must be implemented in the IEC-developed Type 3 model in order to comply with PO 12.3. Nevertheless, the results also show that when the transformer inrush current is not considered, the reactive power response of the generic Type 3 WT model meets the validation criteria, thus complying with Spanish PO 12.3.

Keywords: fault-ride through capability; IEC 61400-27-1; Spanish PO 12.3; Type 3 wind turbine

---

1. Introduction

In 2018, installed wind power capacity in Spain increased by 392 MW to reach a total of 23,484 MW, providing 19% of Spanish electricity consumption, and making it the second-largest wind energy producer in the European Union and fifth in the world [1,2]. This scenario highlights the resurgence of the Spanish wind energy sector and the renewed promotion of its activities, mainly as a result of the three wind energy auctions carried out in 2016 and 2017. Of the total new wind power capacity, 48.5% was installed in the Canary Islands, since the region offers highly suitable wind resources. Thus, Spain currently has around 1123 wind power plants (WPP) and 23,308 wind turbines (WT) installed, spread across 807 municipalities [3].

The contribution of wind power to the Spanish energy demand in 2018 was also reflected in terms of financial savings, benefiting particularly industrial consumers. As an example, for an average industrial annual consumption of 1500 MWh, the total saving amounted to approximately 3500 € [3]. In view of the above achievements, the Spanish Wind Energy Association (AEE) is focused on the development and expansion of the wind energy sector in Spain, and its short-term objective is the commissioning, before March 2020, of the wind power auctioned in 2016 and 2017. This will lead to the growth of both the wind energy market and employment, reduction of greenhouse gases and social and economic development, among other important aspects. Moreover, during 2019, 3000 MW of wind power is expected to be installed.

On the basis of the above information, it is clear that wind energy is a key sector in Spain which will acquire even greater importance in the coming years. However, the unpredictable nature of wind poses a challenge in terms of the integration of the new installed wind power capacity into the grid. Voltage and frequency regulation problems may arise, making planning of network operation activities a pressing need. Thus, knowing the behavior of the grid in advance will allow power system operators to be prepared, overcoming potential power supply problems and forecasting the required power compensations from conventional power plants.

In this regard, dynamic simulation of WT [4] and WPP models representing actual WTs and WPPs connected to the grid is required in order to forecast their active and reactive power responses when subjected to critical situations. Electrical disturbances such as voltage dips are the most important issues, as they cause a voltage reduction of between 10% and 90% and may last up to one minute. In this sense, the Spanish Grid Code developed an Operation Procedure for fault ride-through capability, Operation Procedure 12.3 (PO 12.3), which sets out in detail the response that Spanish WPPs must have under voltage dips. Following a procedure specifically developed for verification, validation and certification (PVVC), the requirements set by PO 12.3 must be complied with by the Spanish WPPs, except for some particular cases. Different adjustments, explained in more detail in Section 2, must then be carried out in the Spanish WPPs to comply with PO 12.3. WTs in operation must therefore follow specific validation criteria, which involves the estimation of validation errors [5].

Furthermore, also driven by the need to provide power system operators with dynamic WT and WPP models to analyze grid integration issues, the International Electrotechnical Commission (IEC) published standard IEC 61400-27-1 in 2015, which defined the so-called generic WT dynamic models [6]. Specifically, four generic WT models are defined, which cover the four main WT typologies currently available in the market. Among them is the generic Type 3 WT model, which represents doubly-fed induction generator (DFIG) WTs. This typology is currently the most widely installed across different countries and, from a technical viewpoint, the most complex one [7]. To validate the generic Type 3 WT model and, in general, all IEC-developed WTs, they must be compared with field measurements, studying their accuracy and testing their performance. To carry out this work, standard IEC 61400-27-1 issued validation guidelines, on which different studies, mentioned later in this document, are based [8,9].

Both the Spanish Grid Code through PO 12.3 and the IEC through standard IEC 61400-27-1 have mapped out the path to be followed in order to regulate the electrical behavior of wind power installations, developing their own validation procedures. However, given that IEC 61400-27-1 is an international standard, and in order to expand the use and scope of application of the originally developed dynamic WT models, this paper submits the generic Type 3 WT, i.e., the DFIG WT, to the requirements of Spanish PO 12.3 on the response of WPP installations in the event of voltage dips, studying its compliance with this grid code. The generic Type 3 WT model is also compared to a detailed model of a DFIG commercial WT, which was previously validated following the PVVC to also comply with the Spanish PO 12.3 [10]. Therefore, this work will allow us to analyze to what extent generic IEC WT models are able to comply with validation criteria established by a national grid code, and to determine their limitations and, in the case of their failure to comply with these criteria, the reasons.

Other key works on this topic, such as [10], focus on the compliance of an actual wind farm composed of Gamesa G52 WTs with Spanish PO 12.3, submitting the entire wind farm to the certification procedure, following the PVVC. In [11], an extension of the work previously performed in [10] is presented. In this case, instead of submitting the complete wind farm to the certification procedure, different voltage dips were applied to a single Gamesa G52 WT, analyzing its compliance with Spanish PO 12.3. Nevertheless, the present contribution goes a step further, studying, for the first time, the compliance of a generic Type 3 WT model developed by standard IEC 61400-27-1 (recently published in 2015) with a national grid code requirement, Spanish PO 12.3. This allows the scope of application of the standard to be extended, facilitating a more widespread use of the IEC-developed generic WT models. Moreover, this work fully implements of the generic model in MATLAB/Simulink with its subsequent dynamic simulation, which provides evidence for the significant differences of the current work with respect to [10,11].

Furthermore, the present work also aims to highlight the strengths and weaknesses of the WT models developed by standard IEC 61400-27-1. On the one hand, for instance, the IEC-developed models are generic enough to represent the wide range of actual WTs developed by different manufacturers. However, on the other hand, given their generic (i.e., simplified) condition, the transient periods of these actual WTs are not accurately represented by the IEC WT models. In this latter case, certain modeling modifications, further detailed in Section 4, must be implemented in the IEC Type 3 WT model to improve its transient behavior. Moreover, the IEC models are clearly specifically designed to represent the different typologies of actual WTs. Hence, the submission of the generic Type 3 WT model to the Spanish grid code is an intermediary step that will allow the wide range of actual WTs in operation to be more rapidly verified, validated and certified according to PO 12.3, without the need for specific detailed WT simulation models.

The paper is structured as follows: Section 2 explains the validation procedure that must be followed to comply with the Spanish grid code. Section 3 presents the WT model studied, which is the Type 3 WT developed by international standard IEC 61400-27-1, while Section 4 shows the results obtained, comparing field measurements and detailed simulation model with responses obtained from the generic IEC model. Finally, Section 5 summarizes the main conclusions of the work.

2. Spanish Grid Code and Procedure for Verification, Validation and Certification

Generic WT models developed by IEC 61400-27-1 have been implemented, simulated and validated using field measurements in several scientific contributions, following the IEC validation procedure [8,9,12,13]. In [8], a Type 3 WT developed by standard IEC 61400-27-1 was validated using the measurements of a real WT following the IEC guidelines. In [9], a Type 4 WT was also validated, although this was done according to both IEC and WECC guidelines (WECC is the Western Electricity Coordinating Council, the other International Organization that has defined generic WT models), while [12] also performed the validation of a generic Type 3 WT according to the IEC Standard. Finally, [13] performs the validation of a Type 1 WT. Moreover, studies such as [11], cited in Section 1, address the validation of a specific-vendor model of DFIG WT following the Spanish grid code, while [14] is based on the improvement of the response of a simplified mechanical model when submitted to fault-ride through capability requirements. However, to the best of the authors' knowledge, there are no studies addressing the simulation process of generic WTs to comply with national grid code requirements. Hence, aiming at a more widespread use of these IEC 61400-27-1 models, the DFIG generic model defined by the IEC is compared to the detailed model of a Gamesa G52 commercial WT, which was, in turn, submitted to the operation procedure for fault ride-through capability within the Spanish Grid Code, PO 12.3 [10]. Furthermore, both the generic and the detailed Type 3 WT models are compared with field measurements, which leads to highly reliable and accurate results. This triple comparison allows the IEC-developed generic WT model to be assessed under different response requirements, thus analyzing its limitations and studying its compliance with the conditions of PO 12.3, based on the PVVC guidelines.

Regarding the development of the PVVC, a specific working group in which WPP owners, WT manufacturers and certification entities and laboratories took part, was created. The corporation that operates the transmission grid in Spain, Red Eléctrica de España, also participated actively in its development. After its completion, this working group was recognized as the technical committee for verification, responsible for monitoring the compliance of WPPs with the Spanish grid code. Hence, based on the active and reactive power responses of WTs, as well as the current reactive ones, which usually define the electrical behavior of the machines during fault and post-fault periods, thus characterizing the fault ride-through capability requirements set by the grid code, the working group developed several editions of the PVVC (the latest released in September 2018). The evolution of the different editions of the PVVC is shown in Table 1 (note that P_n is the nominal power of the WT).

Table 1. Comparison of different editions of the PVVC. $(^{a})\Delta x(\%) = \frac{x_{mea} - x_{sim}}{x_{nom}}$.

PVVC Edition	1, 2	3, 4	5, 6, 7, 8	9, 10, 11
Release date	Jan 2007	Nov 2007, Mar 2008	Jun 2009, Jul 2009, Feb 2010, Sep 2010	May 2011, Jan 2012, Sep 2018
Partial load test, P	10–30% P_n	10–30% P_n	10–30% P_n	10–30% P_n
Full load test, P	$\geq$80% P_n	$\geq$80% P_n	$\geq$80% P_n	$\geq$80% P_n
Power factor, $cos\varphi$	0.95 ind.–0.95 cap.	0.90 ind.–0.95 cap.	0.90 ind.–0.95 cap.	0.90 ind.–0.95 cap.
Reference x_{nom} (^{a})	Measured value	Measured value	Rated value	Rated value
Active power	✓	✓	✓	✓
Reactive power	✓	✓	✓	✓
RMS fundamental phase voltage	✓	✓	-	-
RMS fundamental phase current	✓	✓	-	-
Voltage dip modeling	Fault equipment model	Fault equipment model	Voltage source	Voltage source

It is also worth noting that the Spanish Wind Energy Association (AEE) reported on the problems in adapting the existing WPP installations to the requirements of Spanish national grid code PO 12.3 [15]. Certification forecasts for WPPs were also reflected in the document. The constraints for complying with the PO 12.3 were mainly found in WPPs consisting of WTs equal to or less than 500 kW rated power, since they had insufficient space in the generator to implement the technical solutions required. Moreover, the potential solutions proposed for these types of WPPs to comply with PO 12.3, such as FACTS (flexible alternating current transmission systems), gave rise to administrative and land problems. Old WTs, the manufacturers of which no longer existed, also gave rise to difficulties when looking for specific technical solutions, and were therefore considered beyond the scope of PO 12.3. In this regard, it was also proposed to exclude WPP installations near the end of their expected lifetimes from the certification process, as further financial investment was meaningless. Singular WTs prototypes or machines located in environmentally sensitive areas were also outside the scope of PO 12.3.

2.1. Certification of WTs and WPPs According to the PVVC

Two verifications must be complied with when assessing the response of WTs and WPPs according to Spanish Grid Code PO 12.3: (i) WPPs must remain connected at the point of common coupling (PCC) during voltage dips, which is related to the correct clearance of short-circuits based on the time/voltage curve defined in the grid code, (ii) active and reactive power consumption at the PCC, in case of balanced and unbalanced faults, must be less or equal to the levels specified in the operation procedure.

According to the PVVC, there are two possible ways to certify and verify the response of a WPP installation, described below [5]:

- Particular procedure. This procedure involves compliance of the WTs with PO 12.3, without the need to submit the whole WPP to the certification procedure. Verified WTs following this certification process are known as 'Type Wind Turbines', i.e., WTs which have an accredited testing report. Indeed, a WPP composed of WTs comparable to Type Wind Turbines, i.e., meeting specific conditions, may be directly certified with no extra field tests required, since these WTs have their corresponding verified types of WTs and comply with the requirements set by the PVVC, presenting a certain behavior during voltage dips.
- General procedure. In this case, the simulation of the whole WPP installation is required once the individual WTs and the dynamic compensation systems (FACTS) have been properly tested and their models validated and simulated. For some WT typologies, such as the ones equipped with squirrel cage asynchronous generators, the use of simplified models is allowed, with no additional field tests needed. Moreover, in some cases, the installation of FACTS allows the particular procedure to be conducted. In such situations, the certification entity makes the final decision about the validity of the pertinent accredited report, requiring the general certification procedure to be conducted if necessary.

Drawing on the above, Ref. [15] showed that, in Spain, out of a total of 388 WPPs, 375 were certified following the particular procedure, 9 using FACTS solutions, while the remaining 13 were certified following the general procedure, also using FACTS. Furthermore, approximately 90% of the WTs had to be submitted to specific design modifications to comply with the voltage dip requirements, enabling or disabling protections so they remained connected to the grid during such situations.

2.2. DFIG WT Validation Procedure According to the PVVC

The present work focuses on the achievement of a more widespread use of the IEC-developed generic Type 3 WT model, in addition to the extension of the scope of its applications. This is done by comparing the performance of the generic Type 3 WT with the field tests conducted in a Gamesa G52 commercial WT, as well as with the responses of its detailed simulation model, which was previously verified according to the PVVC to comply with the Spanish Grid Code PO 12.3 [10]. It is, therefore, necessary first to highlight the steps followed to verify, validate and certify the Gamesa G52 WT. Hence, as listed in Section 2.1, two verifications regarding the behavior of WPPs must be complied with according to PO 12.3, and there are two possibilities to certify such compliance with the specified requirements, according to the PVVC. In this particular case, the general verification procedure approach was followed, which consisted of individually validating the WTs and subsequently simulating the WPP by using those validated WT models. As a result, three general steps were followed [10]: (i) wind turbine testing, (ii) wind turbine model validation, (iii) wind farm simulation. Since the first two steps form the basis of this study, the present work focuses on these, paying particular attention to the WT model validation process.

Based on the flowcharts presented in [10,11], once the field tests were conducted following the validity criteria and the equipment specified in the PVVC [5], and the accredited report was received, the model validation process with the field measurements was performed. To carry out this task, the dynamic simulation of the WT model was required. First, based on the data provided by the field tests and the power calculation methodology described in Section 9.2 of the latest edition of the PVVC (Ed. 11), the active and reactive power, as well as the fundamental harmonic of voltage and current Root Mean Square (RMS) values were calculated. Secondly, a voltage source was implemented, along with the detailed WT simulation model, to accurately reproduce the instantaneous voltage measurements corresponding to the field tests, thus obtaining the same instantaneous variables as those recorded during the tests. The time step set during the simulation must be equal to or less than the time interval corresponding to the sample frequency recorded during the field tests [5].

Therefore, both the WT simulation model responses and the WT field measurements can then be compared and analyzed. Based on the PVVC validation criteria, a WT model is considered to be validated when the absolute value of the difference between the field tests' active and reactive power measured values (x_{mea}) and the active and reactive power simulation values (x_{sim}) do not exceed the nominal value (x_{nom}) by 10% in 85% of the data series, (see Equation (1)). Earlier versions of the PVVC (see Table 1) also required the RMS fundamental phase voltage and the RMS fundamental phase current to comply with that criterion.

$$\Delta x(\%) = \left| \frac{x_{med} - x_{sim}}{x_{nom}} \right| \cdot 100 \leq 10\% \tag{1}$$

This validation criterion is applied to the generic IEC-developed Type 3 WT model to study its compliance with Spanish grid code PO 12.3. The RMS values of the measured voltage dip were reproduced at the high voltage side of the transformer -implemented along with the generic IEC WT. The results obtained are analyzed in Section 4.

3. Generic Type 3 Wind Turbine Model Based on Standard IEC 61400-27-1

Based on the current needs of network operators, who must ensure integration of new installed wind power capacity without compromising grid stability, different grid codes have been developed by different countries. Indeed, grid codes were compared and assessed in works such as [16]. However, the European Network of Transmission System Operators for Electricity (ENTSO-E), the aim of which is the regulation of the electricity market in European countries, underlined the need to standardize the technical requirements demanded by the different grid codes. Moreover, some grid codes also included the technical requirements to be complied with by WT and WPP models. In this regard, and with the objective of unifying the technical procedures to assess wind power integration in the grid, the IEC began developing generic, also known as standard, dynamic WT models in October 2009. The tasks conducted resulted in two different parts of standard IEC 61400-27, Part 1 and Part 2, based on the development and validation procedure of the WT and WPP models, respectively. The working group continued the development process with the Final Draft International Standard (FDIS) being issued in 2014, and published one year later, in February 2015. However, the initial structure of the standard was later modified, finally resulting in two different editions: Ed. 1, involving both the WT and the WPP models, and Ed. 2, currently under development, and including their validation procedures.

Regarding WT technologies, Type 3 WT, i.e., the DFIG WT, is the most advanced and currently most widespread model across different countries. It consists of a doubly-fed induction generator with the stator directly connected to the grid and the rotor connected through a back-to-back power converter [17,18]. The IEC-developed Type 3 WT model [6] can be further divided into two sub-models, depending on the generator system implemented: Type 3A and Type 3B. The output signal of both generator systems is a current injected through a current source with parallel impedance, neglecting losses in the generator as the generator air gap power is equal to the power measured at the WT terminals. The main difference lies in the protection system, modeled through a set of dynamic blocks in the case of generic Type 3B WT. Internally, at simulation level, the Type 3B protection system decreases to zero both the active and reactive current signals from the active and reactive control models, respectively, when the voltage differential is above a specific threshold. This whole set of dynamic blocks representing the protection system is therefore not modeled in the case of generic Type 3A WT [6].

In line with the above, since the Gamesa G52 Commercial WT has a break chopper protection, the main function of which is to burn the excess energy to avoid the DC bus voltage increasing outside the set limits, the detailed DFIG WT model simulated using the PSCAD/EMTDC software tool also has a break chopper protection, as it must be capable of representing the fault-ride through capability of the actual WT. Regarding the generic IEC WT, the Type 3A model, which has no specific protection system included, was implemented. This is because, on the one hand, this model is able to control the voltage during the fault and, on the other hand, IEC Type 3B WT is only used to represent actual WTs equipped with active crowbar protection systems [19].

Figure 1 shows a schematic representation of generic Type 3 WT and its control models [20]: *aerodynamic model*, representing the wind turbine rotor (WTR) and providing the value of the wind aerodynamic power; two-mass *mechanical model*, representing both the low and high speed sides of the gear box (GB); *generator system*, which provides the values of the active and reactive current injected into the grid and represents the doubly-fed asynchronous (or induction) generator (DFAG); *pitch control model*, which adjusts the position of the WT blades through calculation of their pitch angle; *active power control model* (P Control Model), the main output of which is the active current command; *reactive power control model* (Q Control Model), which provides, based on the reactive power reference, the reactive current command; *reactive current limitation model* (Q Limitation Model), which calculates the maximum and minimum reactive power allowed; and *current limitation model*, which provides the active and reactive current's limit values. Moreover, the power converter, also shown in Figure 1, consists of the generator side converter (GSC), the direct current link (DCL), the DC capacitor (C), the chopper protection system (CH) and, lastly, the line side converter (LSC). In addition, as mentioned above, some Type 3 WTs include a crowbar protection system (CBR). Finally, the wind turbine terminals (WTT) are connected to the grid through a transformer (TR), and the circuit breaker (CB) may disconnect the WT from the network. The way in which the control models are related to each other may be seen in [6] in more detail.

As will be explained in greater depth in Section 4, the voltage dips were applied to the high voltage side of the WT transformer, so that one of the measurement points to apply the PVVC validation criteria coincides with the testing point. In this way, detailed and generic WTs must also include their transformer models. In the case of generic IEC WT, the transformer model is simulated as an impedance [6,10].

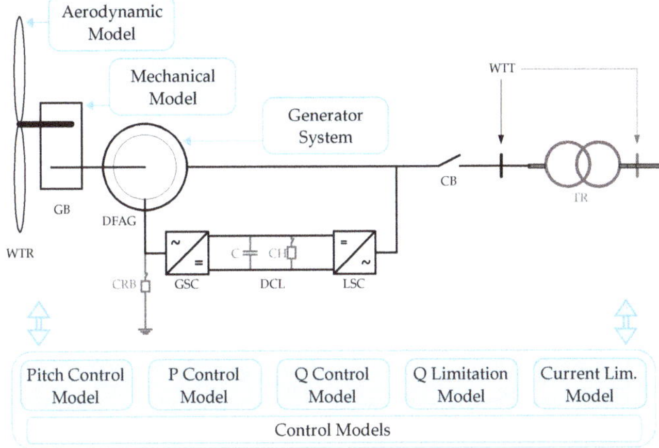

Figure 1. Single-line diagram and control models of generic Type 3 WT based on [6].

4. Results

The voltage dip was first applied on the actual Gamesa G52 WT, the technical specifications of which are shown in Table 2, and measurements were recorded, thus obtaining the positive-sequence values of active and reactive power. The measured voltage dip was then reproduced on both the detailed and the generic WT simulation models. Undoubtedly, the accuracy in the voltage dip's reproduction would affect the accuracy of the models' responses, particularly at the beginning and clearance of the fault. At simulation level, whereas older PVVC editions required the modeling of the physical test bench and the network to reproduce the voltage dip applied, the latest versions establish that a voltage-dependent source is required, setting the time series of data measured as the input signal to the voltage source, thus delivering exactly the same voltage values (see Table 1). This is currently the so-called play-back validation approach [21], which enables an accurate reproduction of the measured voltage dip [22], obtaining highly reliable results.

Table 2. Gamesa G52 WT technical specifications.

Power	
Rated power	850 kW
Cut-in wind speed	4 m/s
Cut-out wind speed	25 m/s
Rated wind speed	15 m/s
Rotor	
Rotor diameter	52 m
Rotor speed	14.6–30.8 rpm
Blade number	3
Blade length	25.3 m
Hub mass with blades	10,000 kg
Blade mass	1900 kg
Top head mass	33,000 kg
Tower	
Tower height	65 m
Tower mass	79,000 kg
Gearbox	
Gearbox stages	3
Gearbox type	1 planetary and 2 helical stages
Gearbox ratio	1:61.74 (50 Hz)
Generator	
Generator type	DFIG
Speed range	900–1900 rpm
Generator voltage	690 V
Control System	
Wind turbine type	Variable speed
Power limitation	Pitch

Both simulation WT models (manufacturer detailed and generic IEC) are therefore submitted to the measured voltage dip, shown in Figure 2, the residual voltage of which is: (i) 19.66% for phase A; (ii) 17.75% for phase B; (iii) 19.91% for phase C. The duration is 0.5705 s. The faults thus fully comply with the characteristics established by the PVVC in the case of three-phase voltage dips, as the residual voltage must be equal to or less than 20% plus the voltage tolerance (+3%), and the dip duration must be higher than, or equal to, 500 ms minus the time tolerance (50 ms) [5]. On the other hand, the time steps used for simulation are 10.024 µs in the case of the detailed model implemented in PSCAD/EMTDC and 1 ms for the generic Type 3 WT modeled in MATLAB/Simulink.

Furthermore, according to Figure 2, which represents the field measurements of the voltage dip and the fault data once applied in both the detailed and generic WT simulation models, it can be stated that the three data series match well and that, therefore, reproduction of the measured voltage in the dynamic simulation models has been performed adequately.

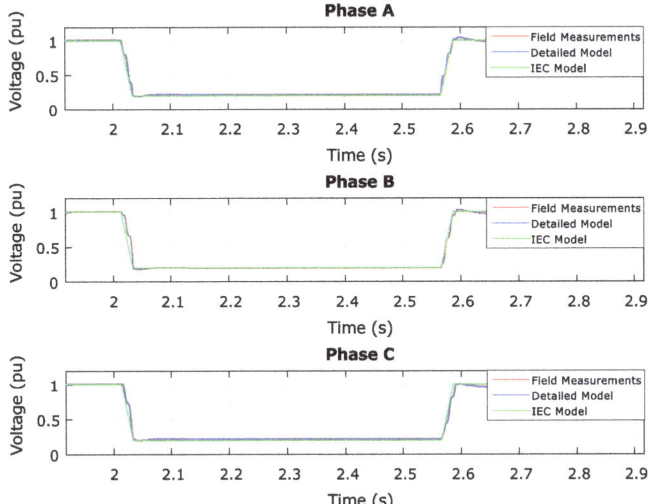

Figure 2. RMS voltage in each phase: measured and simulated.

The duration of the comparison window for the application of the validation criteria is 1000 ms, starting 100 ms before the beginning of the fault. On the basis of the information depicted in Figure 3, this 1000 ms time interval starts at $[T_1 - 100\ \text{ms}]$ and lasts until $[T_1 + 900\ \text{ms}]$. T_1 indicates the time when the fault occurs at one of the phases for the first time, as the voltage drops below the specific threshold (0.85 pu, also shown in Figure 3). T_2 indicates the time when the deepest part of the voltage dip starts (taking into consideration both the reference and residual voltage parameters, U_{ref1} and U_{res1}, see [5]). T_3 indicates the end of the voltage dip's deepest part, and depends on U_{ref2} and U_{res2}. Finally, T_4 is the instant in which the voltage at the three phases is recovered, already above the threshold (0.85 pu).

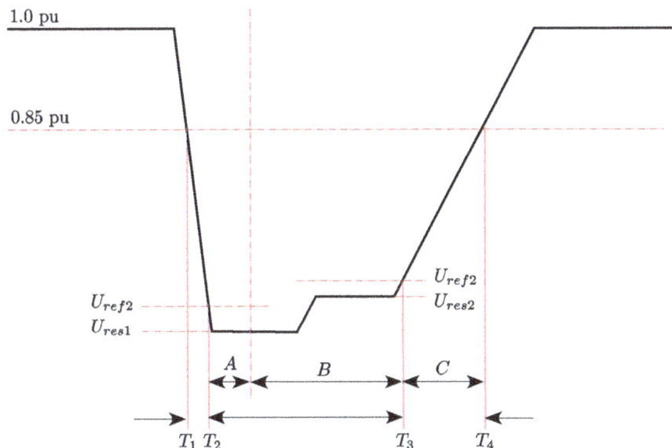

Figure 3. Characterizations of zones during voltage dips according to the PVVC [5].

The validation criteria are therefore applied to the data series obtained over that time interval. As mentioned in Section 2.2, the difference in both three-phase active and reactive powers between measured and simulated data should not exceed 10% for at least 85% of the data series considered.

However, to apply this criterion correctly, it is necessary to know the test conditions. In other words, it is necessary to know which measurement and testing points have been considered.

In the case of WT model validation processes, the measurement point may coincide with the testing point, i.e., with the point at which voltage dips are applied in the actual WT by the voltage dip generator. It is important to note that the dynamic WT model to be validated will comprise all the elements downstream from the measurement point. For instance, in the case that measurement and testing points coincide, both located upstream from the transformer, the transformer and the WT model itself will be considered as validated. If, on the other hand, the testing point is located at the high voltage side of the transformer and the WT model's measurement point is at the low voltage side, only the WT model will be considered as validated. This is explained graphically in Figure 4.

In the present case, the testing point, i.e., the point in which the voltage dip generator is connected, is located at the high voltage side of the transformer, while two different measurement points were considered and therefore two verification procedures were performed, both in the high and low voltage sides. The WT transformer has a transformation ratio of 20 kV/690 V, as indicated in Figure 4.

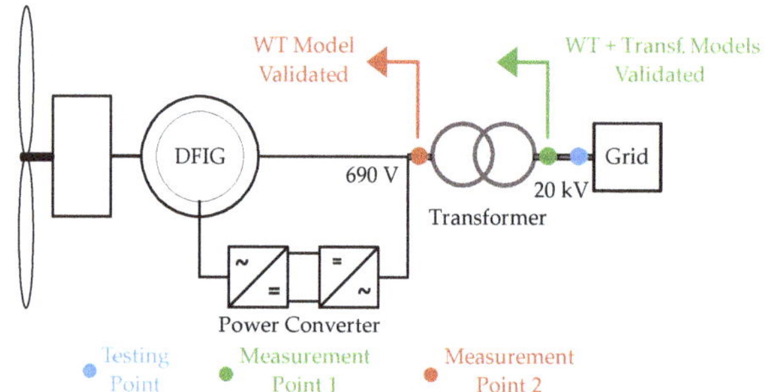

Figure 4. Measurement and testing points for WT validation according to the PVVC.

Based on the information given in Section 2.2, 1 pu is considered as the rated value to apply the PVVC criteria. Hence, 0.1 pu is the maximum deviation allowed for at least 85% of the points within the data series to be compared. Tables 3–5 in Sections 4.1 and 4.2 respectively, estimate these validation parameters, in addition to the percentage of points that comply with this criterion and have a maximum deviation below 0.1 pu.

In this case, the WT is operating under full load conditions of approximately 0.93 pu, also complying with the criteria established by the PVVC (full load test, $\geq$80% P_n, see Table 1).

4.1. PVVC Criteria Applied at the Testing Point: 20 kV, Measurement Point 1

On the one hand, Table 3 presents the IEC model's results for the PVVC criteria applied at the testing point, i.e., at the 20 kV voltage side, which is the so-called *measurement point 1* according to Figure 4. As testing and measurement points coincide in this case, validation results will affect both the transformer and the WT models.

Table 3. Verification of the PVVC validation criteria applied to the IEC generic WT model at the testing point, operating at full load conditions: 20 kV, measurement point 1.

Magnitude	Max. Deviation (pu)	Mandatory Points Below 0.1 pu (%)	Points Below 0.1 pu (%)	Compliance
Active Power, P	0.10	85	80	✗
Reactive Power, Q	0.10	85	56	✗

On the other hand, Figure 5 shows the behavior of the active and reactive power during the voltage dip in the three cases analyzed: field test, detailed WT simulation model and IEC generic WT model.

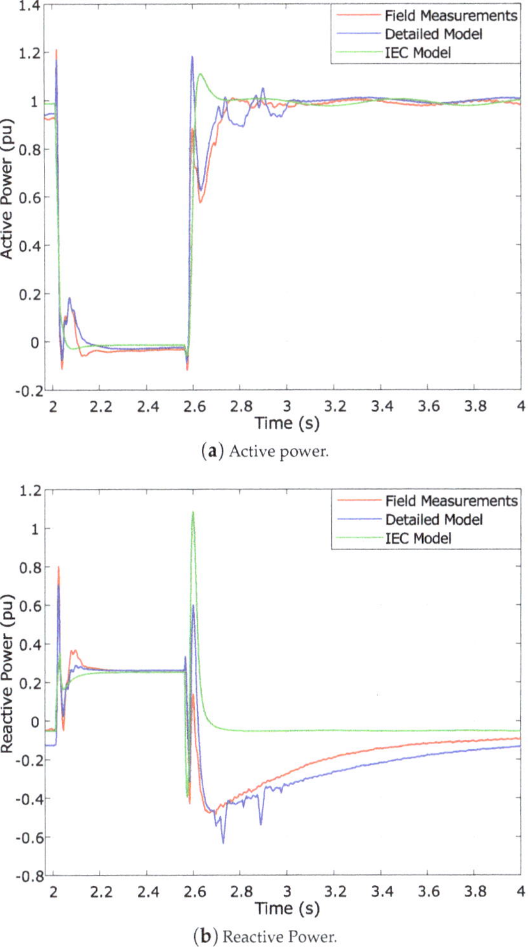

(**a**) Active power.

(**b**) Reactive Power.

Figure 5. Active and reactive power at measurement point 1 (20 kV). WT operating at full load conditions.

Regarding the IEC model results, considering the pertinent time window for evaluation of the data series (see Table 3), only 80% of the points subjected to the analysis comply with the validation criteria for active power. In the case of reactive power, this value is still lower, 56%. Therefore, the validation criteria established by the PVVC are not complied with in either case. However, as will be illustrated in Section 4.2, this non-compliance situation is reversed in one of the cases when the low-voltage measurement point is considered. It can thus be stated that, in this case, neither of the magnitudes subjected to the validation analysis comply with the PVVC criteria when it is applied at the testing point and that, therefore, the transformer model implemented along with the IEC WT model must be considered as non-validated.

Nevertheless, considering the graphic results in Figure 5a, it is clear there exists a reasonably good correlation between the IEC model and the field measurements in the case of active power. The differences are mainly because the generic model is not able to represent the transient periods of the actual WT with great precision. To improve the representation of these transient periods, two modifications should be carried out in the generator system defined by standard IEC 61400-27-1, the modular structure of which is available in [6]:

- The PI control of the generator system should be defined based on the DQ system instead of calculating its real and imaginary components. Moreover, parameters of the active and reactive power PI controllers should have a different value, since the standard considers them equal.

- An electromagnetic transient resistance, Rs, should also be included in the generator system, as in its current form only an electromagnetic transient reactance, named Xs, is considered.

Regarding the reactive power behavior of the IEC WT model (Figure 5b), although there is also non-compliance with the PVVC, transient periods are better represented by the generic WT model in this case. However, the level of accuracy is still low if compared to the transient periods that appear in the detailed model's responses and field measurements. This is also because simplified models such as that developed by the IEC do not represent the fundamental component of the transformer inrush current. This can be understood more easily if we look at the graphic results shown in Section 4.2, specifically at the reactive power graph obtained for the low-voltage side measurement point (Figure 6b), in which reactive power fits much better than in the present case.

The differences between the field measurements and the IEC model, both in the active and the reactive power responses, which are mainly due to the inability of the generic model to accurately represent the transient periods, are described with great precision in [23]. Indeed, the study summarizes the simplifications implemented in the Type 3A WT model, which is the one modeled in the present work, as explained in Section 3.

The validation results for the detailed WT model, also operating at full load conditions, are shown in Table 4. As can be observed, both the active and the reactive power responses fulfill the PVVC criteria when applied at the testing point, since 91% and 90% of the data series analyzed, respectively, are below the maximum deviation allowed. This implies that the detailed transformer and WT models simulated were validated according to the Spanish Grid Code.

Table 4. Verification of the PVVC validation criteria applied to the detailed WT model at the testing point, operating at full load conditions: 20 kV, measurement point 1.

Magnitude	Max. Deviation (pu)	Mandatory Points Below 0.1 pu (%)	Points Below 0.1 pu (%)	Compliance
Active Power, P	0.10	85	91	✓
Reactive Power, Q	0.10	85	90	✓

Compliance of the detailed model with PVVC validation criteria is also supported by Figure 5. On the one hand, Figure 5a shows the detailed model's active power response facing the voltage dip. It can be seen that there is an excellent correlation between this signal and the one provided by the field measurements. The same applies to the reactive power (Figure 5b), the signal of which is also very similar to the field measurements. In this case, transient periods are accurately represented since the detailed WT model considered the actual mechanical, electrical and electronic systems of the Gamesa G52 WT, including their parameters and algorithms.

4.2. PVVC Criteria Applied at the Low Voltage Side: 690 V, Measurement Point 2

The detailed WT model representing the Gamesa G52 WT broadly complies with the validation criteria established by the Spanish Grid Code when analyzed at the testing point, and hence no additional validation analyses are required. However, the generic WT model developed by standard IEC 61400-27-1 does not meet the minimum requirements to be considered a validated WT model according to the same criteria when applied at the testing point. As listed in Section 4.1, there are two main reasons for this situation: the inability of generic WT models to accurately represent, on the one hand, transient responses of actual WTs, and, on the other hand, the fundamental component of transformer inrush current. Indeed, modifications in the structure of the original generator system model developed by the standard have been proposed in order to improve transient responses of the generic Type 3 WT.

Nevertheless, regarding the second reason, the IEC model's reactive power response does comply with the PVVC criteria when the transformer is not taken into account in the validation process, i.e., when the inrush current effects are no longer considered a problem. Therefore, when applying the criteria at the low voltage side, 690 V, *measurement point 2* according to Figure 4, the percentage of data series below the maximum deviation allowed increases to 87%, higher than the 85% set as the target (Table 5). Indeed, Figure 6b shows that the correlation in reactive power between the generic WT model and the field measurements is much better than in the previous case (Figure 5b). Active power, on the other hand, continues to be in non-compliance, with only 80% of data series within the margin established. The graphic results for active power (Figure 6a), are very similar to those obtained at the testing point (Figure 5a, Section 4.1).

It can therefore be stated that the non-compliance situation in the previous case for reactive power is now reversed, since the response of the generic IEC WT model does fulfill the PVCC validation criteria at the low voltage side.

Table 5. Verification of the PVVC validation criteria applied to the IEC generic WT model at the low voltage side, operating at full load conditions: 690 V, measurement point 2.

Magnitude	Max. Deviation (pu)	Mandatory Points Below 0.1 pu (%)	Points Below 0.1 pu (%)	Compliance
Active Power, P	0.10	85	80	✗
Reactive Power, Q	0.10	85	87	✓

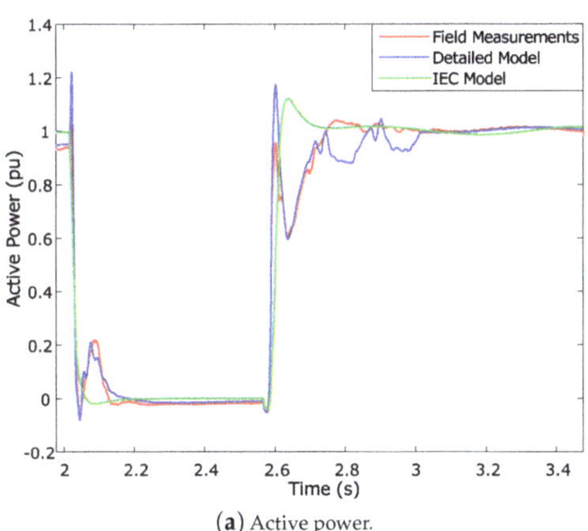

(**a**) Active power.

Figure 6. *Cont.*

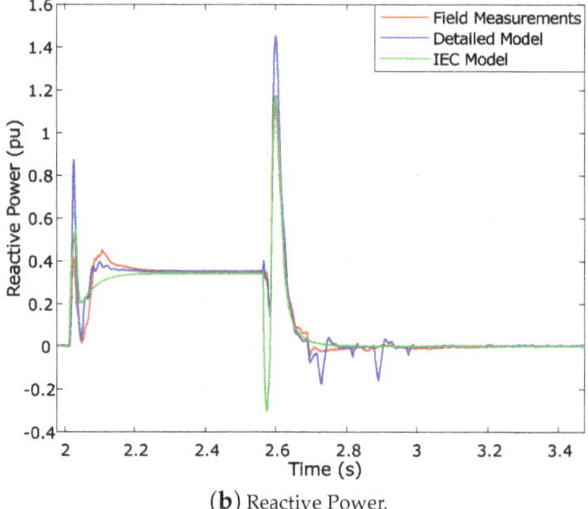

(**b**) Reactive Power.

Figure 6. Active and reactive power at measurement point 2 (690 V). WT operating at full load conditions.

4.3. PVVC Criteria: WT Operating at Partial Load Conditions

The previous sections have presented the analysis of the IEC Type 3 WT model operating under full load conditions. Indeed, such a condition is presented as the worst-case scenario with respect to the compliance of the generic model with Spanish PO 12.3. Thus, the non-compliance cases have been extensively discussed, and the causes analyzed. The modeling modifications that should be implemented within the generic WT model to improve the compliance results have also been presented.

Furthermore, in order to enhance and enrich the current section, a different study case was also analyzed. In this case, the WT is operating at partial load conditions of 0.20 pu, which is in line with the requirements established by the PVVC (partial load test, 10–30% P_n, see Table 1). The voltage dip applied to the IEC WT model working at partial load conditions is the same as that applied to the previous compliance analyses (see Figure 2).

Tables 6 and 7 show the percentage of points that comply with the PVVC validation criteria when applied to the high and low voltage sides of the transformer, respectively. Figures 7 and 8 show the active and reactive power responses in those two measurement points.

A value of 0.1 pu must be the maximum error for at least 85% of points within the data series to be compared. As can be observed in Table 6, the active power response fulfills the PVVC validation criteria at the high voltage side of the power transformer, i.e., at the testing point (or measurement point 1, see Figure 4), since 88% of the data series analyzed are below the maximum deviation allowed. Indeed, if the graphic results in Figure 7a are considered, it can be observed that there is a good correlation between the IEC model response and the field measurements, which justifies the compliance of the active power response at measurement point 1 (see Figure 4) with the PVVC validation criteria. Moreover, the reactive power response fails to comply with PO 12.3, since only 58% of the data series are below 0.1 pu. The differences in this case may be clearly observed in Figure 7b. As in the case of the PVVC criteria applied to the high voltage side with the WT operating at full load conditions (Figure 5b), this non-compliance situation is also due to the inability of the generic WT model to represent the transformer inrush current.

However, in general, the compliance results improved in comparison to the WT operating at full load conditions.

Table 6. Verification of the PVVC validation criteria applied to the IEC generic WT model at the testing point, operating at partial load conditions: 20 kV, measurement point 1.

Magnitude	Max. Deviation (pu)	Mandatory Points Below 0.1 pu (%)	Points Below 0.1 pu (%)	Compliance
Active Power, P	0.10	85	88	✓
Reactive Power, Q	0.10	85	58	✗

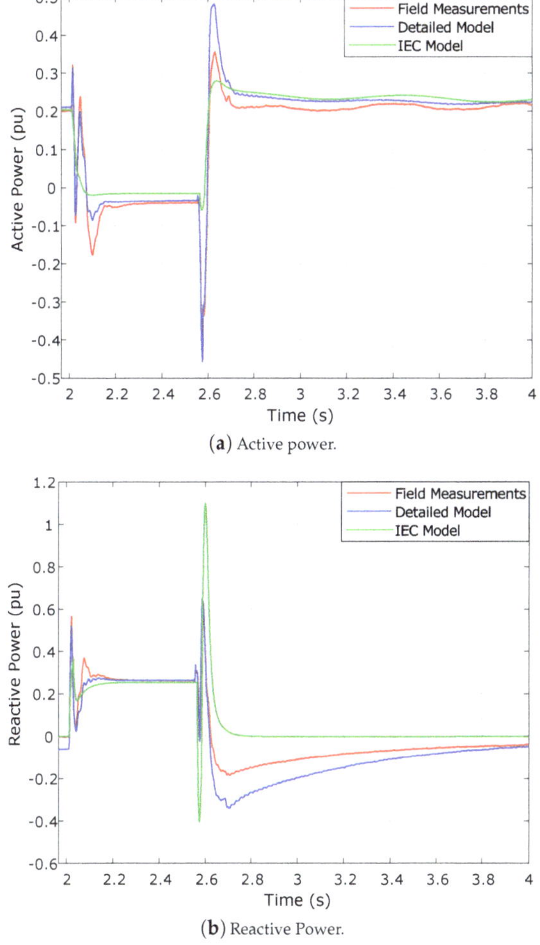

(**a**) Active power.

(**b**) Reactive Power.

Figure 7. Active and reactive power at measurement point 1 (20 kV). WT operating at partial load conditions.

When the validation criteria are applied to the low voltage side (Table 7 and Figure 8), both the active and the reactive power responses of the WT operating at partial load conditions fulfill the PVVC criteria, since 91% and 88% of the data series analyzed, respectively, are below 0.1 pu. The good correlation that exists both in the active and the reactive power between the field measurements and the IEC model is observed in Figure 8a,b, improving as regards the PVVC applied to the low voltage side. The situation for the reactive power has thus reversed, since it complies with the PVVC criteria in this case. This is for the same reasons as those given in Section 4.2: simplified or generic models do not represent the fundamental component of the transformer inrush current.

Table 7. Verification of the PVVC validation criteria applied to the IEC generic WT model at the low voltage side, operating at partial load conditions: 690 V, measurement point 2.

Magnitude	Max. Deviation (pu)	Mandatory Points Below 0.1 pu (%)	Points Below 0.1 pu (%)	Compliance
Active Power, P	0.10	85	91	✓
Reactive Power, Q	0.10	85	88	✓

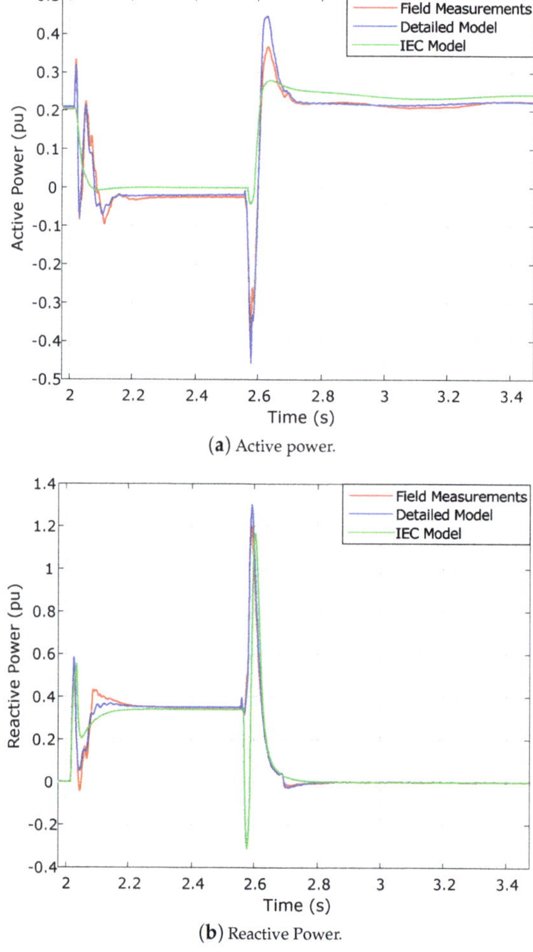

(**a**) Active power.

(**b**) Reactive Power.

Figure 8. Active and reactive power at measurement point 2 (690 V). WT operating at partial load conditions.

5. Conclusions

In Spain, the Spanish Grid Code has developed a specific operation procedure, PO 12.3, which sets out the response that wind power plant installations must have under voltage dips. It also developed the so-called procedure for verification, validation and certification (PVVC), a set of guidelines that establish the steps to follow to comply with PO 12.3. Thus, in order to fulfill the requirements of PO 12.3, voltage dips were applied on actual machines using specialized equipment, and field measurements were compared to the responses of their corresponding detailed simulation models when subjected to those specific voltage dip measurements. Subsequently, errors were calculated.

Based on the concepts mentioned above, since IEC models are precisely intended to represent the wide range of actual WTs available in the market, this research work aimed to extend the functionality of international standard IEC 61400-27-1 by analyzing the compliance of these generic models with national grid code requirements such as the Spanish PO 12.3 for fault ride-through capability. In this way, field measurements from an actual Gamesa G52 WT, in addition to the dynamic responses of its corresponding detailed simulation model, were compared to the responses provided by a generic IEC 61400-27-1 Type 3 WT when they were all subjected to the same voltage dip. The PVVC was applied, and the generic WT's active and reactive power responses were analyzed.

The results obtained are of great interest to network operators as well as other stakeholders concerned with wind power integration. When the WT is operating at full load conditions, if the PVVC is applied at the testing point, the IEC model fails to comply with PO 12.3 in both active and reactive power. This is mainly due to transient periods of the actual WT not being well represented by the generic model as a result of the simplifications introduced in the standard. To improve its electrical behavior, it is proposed that the PI control of the IEC-developed generator system be defined based on the DQ system. It would also be necessary to distinguish between the parameters defining the active and reactive power controllers, since the Standard considers them of equal value. Moreover, regarding specifically the reactive power response, it is also observed that the generic model is unable to represent the fundamental component of the transformer inrush current, unlike in the case of the WT simulation model described. This is better reflected in the reactive power response of the WT at the low voltage side, when the transformer is no longer considered in the validation calculations and hence there is no influence of the inrush current. Indeed, reactive power response does comply with the PVVC in this second case. Concerning the detailed model, its active and reactive power behavior is already validated upstream the transformer, thus broadly complying with PO 12.3. Finally, the IEC WT model operating at partial load conditions only fails to comply with the Spanish grid code in the reactive power response when the validation criteria are applied to the high voltage side. This may be mainly attributed, as in the case of nominal load conditions, to the inability of the WT model to represent the fundamental component of the transformer inrush current.

Wind power is indisputably changing the way electricity networks operate, and this will change even more in the coming years. There is uncertainty about wind lead transmission and distribution system operators thoroughly planning grid activities to ensure power supply. Thus, time-domain analyses such as the ones carried out in the present work can contribute to improving the forecasts required to guarantee proper wind power integration in actual networks. In contrast to other works on this topic, based solely on the study of the certification process of actual WPPs and WTs according to the Spanish grid code, this study analyses the generic Type 3 WT simulation model with the objective of extending the use of Standard IEC 61400-27-1, since the model is subjected for the first time to national grid code validation criteria: Spanish operation procedure PO 12.3. The results show that some modifications should be carried out in the original dynamic sub-models within the generic Type 3 in order to improve its behavior and therefore comply with PO 12.3, thus enhancing the scope of the application of standard IEC 61400-27-1.

Author Contributions: Conceptualization, F.J.-B. and E.G.-L.; Investigation, R.V.-R. and F.J.-B.; Methodology, R.V.-R., F.J.-B. and A.H.-E.; Project administration, E.G.-L.; Supervision, F.J.-B., A.H.-E., Á.M.-G. and E.G.-L.; Writing—original draft, R.V.-R.

Funding: This research was funded by the Spanish Ministry of Economy and Competitiveness and European Union FEDER, grant number ENE2016-78214-C2-1-R, as well as the Agreement signed between the UCLM and the Council of Albacete to foster Research in the Campus of Albacete.

Acknowledgments: The authors would also like to express their gratitude to the wind turbine manufacturer Siemens Gamesa Renewable Energy for the technical support received. Furthermore, the authors would like to thank the TC 88/WG 27 Committee "Wind turbines—Electrical simulation models for wind power generation", to his Convenor, Poul Sørensen, and to the 16 participating countries.

Conflicts of Interest: The authors declare no conflict of interest.

Abbreviations

The following abbreviations are used in this manuscript:

AEE	Spanish wind energy association
DFAG	Doubly-Fed Asynchronous Generator
DFIG	Doubly-Fed Induction Generator
ENTSO-E	European Network of Transmission System Operators for Electricity
FACTS	Flexible Alternating Current Transmission Systems
FDIS	Final Draft International Standard
GB	Gearbox
IEC	International Electrotechnical Commission
PO	Procedure of Operation
PCC	Point of Common Coupling
PVVC	Procedure for Verification, Validation and Certification
RMS	Root Mean Square
WECC	Western Electricity Coordinating Council
WPP	Wind Power Plant
WT	Wind Turbine
WTR	Wind Turbine Rotor
WTT	Wind Turbine Terminals

References

1. GWEC. *Global Wind Statistics 2017*; Technical Report; Global Wind Energy Council: Brussels, Belgium, 2018. Available online: https://gwec.net/wp-content/uploads/vip/GWEC_PRstats2017_EN-003_FINAL.pdf (accessed on 26 April 2019).

2. EWEA. *Wind in Power: 2017 Anual Combined Onshore and Offshore Wind Energy Statistics*; Technical Report; European Wind Energy Association: Brussels, Belgium, 2017. Available online: https://windeurope.org/wp-content/uploads/files/about-wind/statistics/WindEurope-Annual-Statistics-2017.pdf (accessed on 26 April 2019).

3. Asociación Empresarial Eólica (AEE). Datos eólica en España 2019. 2019. Available online: https://www.aeeolica.org/comunicacion/notas-de-prensa (accessed on 17 May 2018).

4. Honrubia-Escribano, A.; Gomez-Lazaro, E.; Fortmann, J.; Sørensen, P.; Martin-Martinez, S. Generic dynamic wind turbine models for power system stability analysis: A comprehensive review. *Renew. Sustain. Energy Rev.* **2017**, *81*, 1939–1952. [CrossRef]

5. AEE. *Procedimientos de Verificación, Validación y Certificación de los Requisitos del PO 12.3 y P.O.12.2. Sobre la Respuesta de las Instalaciones Eólicas y Fotovoltaicas ante Huecos de Tensión*, 11th ed.; Asociación Empresarial Eólica. 2017. Available online: https://www.aeeolica.org/images/Grupos_trabajo/Comite_tecnico_verificacion/PVVC_11.pdf (accessed on 26 April 2019).

6.	IEC 61400-27-1. *Electrical Simulation Models—Wind Turbines*; International Electrotechnical Commission: Geneva, Switzerland, 2015.

7.	Fortmann, J. Modeling of Wind Turbines with Doubly Fed Generator System. Ph.D. Thesis, Department for Electrical Power Systems, University of Duisburg-Essen, Duisburg, Germany, 2014.

8.	Honrubia-Escribano, A.; Jiménez-Buendía, F.; Gómez-Lázaro, E.; Fortmann, J. Field Validation of a Standard Type 3 Wind Turbine Model for Power System Stability, According to the Requirements Imposed by IEC 61400-27-1. *IEEE Trans. Energy Convers.* **2018**, *33*, 137–145. [CrossRef]

9.	Lorenzo-Bonache, A.; Honrubia-Escribano, A.; Jimenez, F.; Gomez-Lazaro, E. Field Validation of Generic Type 4 Wind Turbine Models Based on IEC and WECC Guidelines. *IEEE Trans. Energy Convers.* **2018**. [CrossRef]

10.	Jimenez, F.; Gómez-Lázaro, E.; Fuentes, J.A.; Molina-García, A.; Vigueras-Rodríguez, A. Validation of a double fed induction generator wind turbine model and wind farm verification following the Spanish grid code. *Wind Energy* **2012**, *15*, 645–659. [CrossRef]

11.	Jiménez, F.; Gómez-Lázaro, E.; Fuentes, J.A.; Molina-García, A.; Vigueras-Rodríguez, A. Validation of a DFIG wind turbine model submitted to two-phase voltage dips following the Spanish grid code. *Renew. Energy* **2013**, *57*, 27–34. [CrossRef]

12.	Goksu, O.; Altin, M.; Fortmann, J.; Sørensen, P. Field Validation of IEC 61400-27-1 Wind Generation Type 3 Model with Plant Power Factor Controller. *IEEE Trans. Energy Convers.* **2016**, *31*, 1170–1178. [CrossRef]

13.	Zhao, H.; Wu, Q.; Margaris, I.; Bech, J.; Sørensen, P.E.; Andresen, B. Implementation and validation of IEC generic Type 1A wind turbine generator model. *Int. Trans. Electr. Energy Syst.* **2014**, *25*, 1804–1813. [CrossRef]

14.	Buendia, F.J.; Vigueras-Rodriguez, A.; Gomez-Lazaro, E.; Fuentes, J.A.; Molina-Garcia, A. Validation of a Mechanical Model for Fault Ride-Through: Application to a Gamesa G52 Commercial Wind Turbine. *IEEE Trans. Energy Convers.* **2013**, *28*, 707–715. [CrossRef]

15.	AEE. *Informe Sobre la Situación Actual y la Problemática para la Adecuación de los Parques Existentes a los Requisitos Previstos por el P.O.12.3*; Technical Report, Asociación Empresarial Eólica: Madrid, Spain, 2009.

16.	Jauch, C.; Matevosyan, J.; Ackermann, T.; Bolik, S. International Comparison of Requirements for Connection of Wind Turbines to Power Systems. *Wind Energy* **2005**, *8*, 295–306. [CrossRef]

17.	Lorenzo-Bonache, A.; Villena-Ruiz, R.; Honrubia-Escribano, A.; Gómez-Lázaro, E. Operation of active and reactive control systems of a generic Type 3 WT model. In Proceedings of the 2017 11th IEEE International Conference on Compatibility, Power Electronics and Power Engineering (CPE-POWERENG), Cadiz, Spain, 4–6 April 2017; pp. 606–610.

18.	Lorenzo-Bonache, A.; Honrubia-Escribano, A.; Jiménez-Buendía, F.; Molina-García, Á.; Gómez-Lázaro, E. Generic Type 3 Wind Turbine Model Based on IEC 61400-27-1: Parameter Analysis and Transient Response under Voltage Dips. *Energies* **2017**, *10*, 1441. [CrossRef]

19.	Jiménez Buendía, F.; Barrasa Gordo, B. Generic Simplified Simulation Model for DFIG with Active Crowbar. In Proceedings of the 11th International Workshop on Large-Scale Integration of Wind Power into Power Systems as well as on Transmission Networks for Offshore Wind Power Plants, Lisbon, Portugal, 13–15 November 2012; p. 6.

20.	Lorenzo-Bonache, A.; Honrubia-Escribano, A.; Fortmann, J.; Gómez-Lázaro, E. Generic Type 3 WT models: Comparison between IEC and WECC approaches. *IET Renew. Power Gener.* **2019**. [CrossRef]

21.	Asmine, M.; Brochu, J.; Fortmann, J.; Gagnon, R.; Kazachkov, Y.; Langlois, C.E.; Larose, C.; Muljadi, E.; MacDowell, J.; Pourbeik, P.; et al. Model validation for wind turbine generator models. *IEEE Trans. Power Syst.* **2011**, *26*, 1769–1782. [CrossRef]

22. Honrubia-Escribano, A.; Jiménez-Buendía, F.; Gómez-Lázaro, E.; Fortmann, J. Validation of generic models for variable speed operation wind turbines following the recent guidelines issued by IEC 61400-27. *Energies* **2016**, *9*, 1048. [CrossRef]
23. Fortmann, J.; Engelhardt, S.; Kretschmann, J.; Feltes, C.; Janßen, M.; Neumann, T.; Erlich, I. Generic simulation model for DFIG and full size converter based wind turbines. In Proceedings of the 9th International Workshop on Large-Scale Integration of Wind Power, Quebec, QC, Canada, 18–19 October 2010.

Article

Inertia Dependent Droop Based Frequency Containment Process

Kaushik Das [1,*], Müfit Altin [2], Anca D. Hansen [1] and Poul E. Sørensen [1]

[1] Department of Wind Energy, Technical University of Denmark, 4000 Roskilde, Denmark;
anca@dtu.dk (A.D.H.); posq@dtu.dk (P.E.S.)

[2] Energy Systems Engineering Department, Izmir Institute of Technology, Urla, Izmir 35430, Turkey;
mufitaltin@iyte.edu.tr

* Correspondence: kdas@dtu.dk

Received: 20 March 2019; Accepted: 24 April 2019; Published: 30 April 2019

Abstract: Presently, there is a large need for a better understanding and extensive quantification of grid stability for different grid conditions and controller settings. This article therefore proposes and develops a novel mathematical model to study and perform sensitivity studies for the capabilities of different technologies to provide Frequency Containment Process (FCP) in different grid conditions. A detailed mathematical analytical approach for designing inertia-dependent droop-based FCP is developed and presented in this article. Impacts of different droop settings for generation technologies operating with different inertia of power system can be analyzed through this mathematical approach resulting in proper design of droop settings. In contrast to the simulation-based model, the proposed novel mathematical model allows mathematical quantification of frequency characteristics such as nadir, settling time, ROCOF, time to reach the nadir with respect to controller parameters such as gain, droop, or system parameters such as inertia, volume, of imbalance. Comparative studies between cases of frequency containment reserves (FCR) provision from conventional generators and wind turbines (WTs) are performed. Observations from these simulations are analyzed and explained with the help of an analytical approach which provides the feasible range of droop settings for different values of system inertia. The proposed mathematical approach is validated on simulated Continental Europe (CE) network. The results show that the proposed methodology can be used to design the droop for different technology providing FCP in a power system operating within a certain range of inertia.

Keywords: inertia; wind power; droop; frequency control; primary control; frequency containment process

---

1. Introduction

Non-synchronous generations, such as modern renewable energy sources (RES) like variable speed wind turbine (VSWTs) and solar photovoltaics, are increasingly making larger contributions to electricity generation throughout the world. Unlike conventional synchronous generations, these non-synchronous generations do not inherently contribute to the power system inertia, as they are decoupled from the power system through power electronics. This means that a larger displacement of conventional generations by modern RES without any additional frequency inertia control, the lower the power system inertia and the larger frequency deviations can get during power imbalances following large disturbances such as disconnection of a generator. This aspect might be especially pronounced in island power systems (e.g., Irish power system) or an interconnected power system split into islands following cascading events. For example, the average inertia of Ireland power system in 2020 is prognosed to be reduced by around 25% from the average inertia of 2010 [1]. In [2], it is indicated that the Irish power system inertia can vary from around 14 s to 2 s depending on the degrees of wind power penetration.

Large disturbances in low inertia power systems can cause fast changes in frequency, making the power system vulnerable to short term frequency instability. Frequency reserves are used to prevent frequency emergencies and instability whenever any power imbalance occurs in power system. In the European Network Transmission System Operator (ENTSO-E) Network Code on Load-Frequency Control and Reserves [3], the frequency reserves are classified based on their functionalities as frequency containment reserves (FCR), frequency restoration reserves (FRR) and replacement reserves (RR) [3]. FCR are automatic and expensive reserves activated within seconds following a power imbalance, while FRR and RR are activated within longer time frame from minutes to an hour. The responsibility of FCR is to restrict the sudden rise/decline of frequency, while FRR and RR are activated to release FCR as well as to bring the frequency back to its nominal value [4,5]. A fast release of FCR gives to the power system the ability to handle new possible consecutive power imbalances. Frequency reserves are deployed through different types of frequency control processes, viz. frequency containment process (FCP), frequency restoration process (FRP) and replacement process (RP).

This article only deals with FCP and FCR. FCP is employed to limit the frequency from becoming too high (or low) when exceeding the normal operational limits (50 $\pm$ 0.050 Hz for Continental Europe(CE)) following a large disturbance. FCP is required to be completely deployed within 30 s after a large disturbance and uses FCR from dedicated generators according to ENTSO-E [3]. If FCP is either inadequate or not fast enough to contain the frequency, defense plans such as underfrequency load-shedding (UFLS), or overfrequency generation disconnections are employed to prevent frequency instability [6–8]. However, these defense plans are only considered to be last resorts since they cause economic losses and discomfort to consumers. Therefore, all the available resources in the power system should be employed before activating these defense plans. Generally, in CE network, UFLS starts around 49 Hz, while overfrequency generation disconnection happens at 51.5 Hz [6–8].

It is essential to quantify volume requirements of FCR accurately, more so in continuously changing power system environment. Traditionally volume of FCR is quantified based on n-1 security criterion [9]. A methodology is proposed in [5] to quantify volume of FCR and FRR for future power systems with high penetration of wind power to handle wind power forecast error. Availability of FCR volume is a necessary but not sufficient condition to prevent frequency instability. The technology used to deploy FCR plays a crucial role in FCP especially in power systems with low inertia, as it will be the case of future power systems with large share of renewable generators. FCP also-called primary frequency control in the literature is generally provided by speed-droop governors in conventional generators such as steam, hydro or gas turbine-based generators [10]. These technologies are matured and have been in practice for years; however, the control settings of these technologies such as droop parameters, settling time etc. need to be investigated especially for low inertia power systems, due to fast change in frequency following a disturbance. Recently, several studies have been performed looking at FCP from newer technologies such as demand response, battery storage and wind turbines (WTs). Zhao et al. [11], Molina-Garcia et al. [12] investigates the contribution from demand response to FCP (referred as primary control), while Oudalov et al. [13] and Mercier et al. [14] focus on methods for optimization the battery storage for FCP. FCP support from WTs is studied in Morren et al. [15] and Ullah et al. [16]. A detailed modeling approach for frequency support from WTs is presented in Altin et al. [17], Margaris et al. [18] and Sakamuri et al. [19]. Sun et al. [20] provides a review of WT support for primary control from power systems point of view. Other technologies have also been investigated as viable sources for FCP. For example, Haileselassie et al. [21] and Mu et al. [22] investigates multi-terminal HVDC, electric vehicles for FCP, respectively. All these articles look at the capabilities of these technologies to provide FCR at different grid conditions. Mostly, these articles use simulation-based models for analysis of frequency support. Simulation-based models do not allow for mathematical analytic quantification of frequency response which provides insight about frequency characteristics such as nadir, settling time, ROCOF, time to reach the nadir with respect to controller parameters such as gain, droop, or system parameters such as inertia, volume of imbalance. Catering to the need for a better understanding and extensive quantification of grid stability for different grid conditions and controller settings, this article proposes a novel mathematical methodology to study, compare and perform sensitivity studies for different technologies for different grid conditions; analyzing the stability of the controller.

There have also been papers on the mathematical approach of frequency responses [23,24]. Aik [23] proposed a general-purpose frequency control model considering UFLS. Chavez et al. [25] proposed a simplified model to assess the adequacy of FCR. However, these models do not allow for performing sensitivity studies for different technologies through mathematical analysis. Proposed mathematical model not only allows the choice of proper technology to provide FCR but also allows assessing the relative stability of the controllers for different values of inertia of the system.

Contributions of this article are as follows:

- A novel detailed mathematical analytic approach for FCP is proposed and discussed in this article.
- Analysis of frequency characteristics such as time to reach nadir, attenuation, frequency nadir based on controller parameters and power system characteristics such as inertia and imbalance using the proposed model.
- Capabilities of different technologies of generation sources in providing FCP can be analyzed through this mathematical analytic approach. A set of simulations has been performed in a power system consisting of generic governor and WT models in order to study the impacts of droop settings and system inertia.
- Observations from these simulations are analyzed and explained with the help of analytic approach, which provides a feasible range of droop settings for different values of system inertia.
- The efficacy of the proposed mathematical approach is verified on a case study of simulated realistic CE network.

This article is structured as following. Section 2 derives a mathematical approach for FCP. Several sensitivity studies of different technologies with respect to system inertia and droop settings based on the mathematical model are performed in Section 3. Section 4 applies and validates the mathematical model based on case study on simulated CE network. Finally, conclusive remarks are reported, where the track for future work is also proposed.

2. Mathematical Approach of FCP

Figure 1 depicts block diagram for FCP.

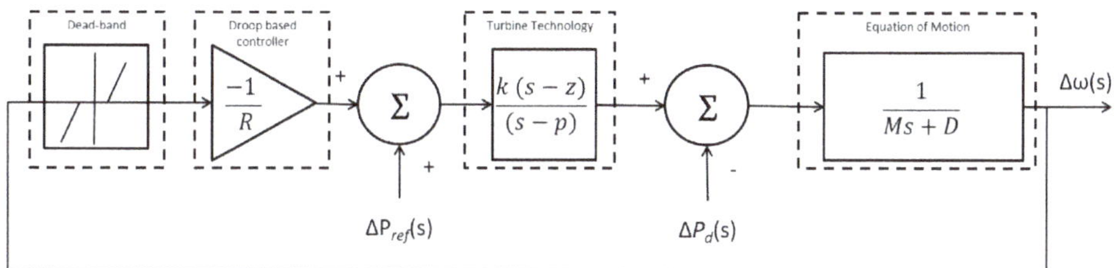

Figure 1. Frequency Containment Process (FCP) model for mathematical analytical approach.

Turbine technology can be either a turbine-governor-controlled power generator or a converter connected WT represented by a zero(z)-pole(p)-gain(k) transfer function model. The FCP controller is further modeled as droop control with slope of $\frac{-1}{R}$. The intention of the studies in this article is to study FCP following a large disturbance. Following a large disturbance, frequency goes much beyond the dead-band, consequently dead-band is neglected for the studies. This makes the system linear. There are two other major non-linearities for these kinds of studies. First kind of non-linearity arises from switching events such as UFLS, overfrequency generation disconnection, pump-storage connection/disconnections etc. All these switching events are part of special protection schemes of defense plans during emergency. The intention of the research presented in this article is to prevent emergency. Therefore, these switching events can be safely ignored. Another non-linearity might arise from the different turbine technologies. For example, hydro generators experience non-linear impacts of backlash and hysteresis in the forward path of the servo systems [26]. Similarly, converter switches in the WTs can also cause non-linearities. However, these non-linearities can also be ignored in lieu of large signal analysis. The advantage of neglecting these non-linearities is that it makes the mathematical model analytically solvable. It should be noted that the methodology is generic for any power system and will give an idea for operators to plan/design/operate their system with low inertia in the event of a large frequency disturbance. Either the non-linearities can be neglected for large signal analysis or parameters of the proposed methodology needs to be tuned for considering the non-linearities. The power system is represented with respect to equation of motion. ΔP_{ref} representing the change in load reference point for the generator set by FRP. In the considered time period of FCP of a few seconds, this value is basically zero. ΔP_d represents the power imbalance seen by the power system due to a disturbance. The disturbance can be disconnection of large generator/load or system separation in the case of large interconnected system. In case of system separation into islanded systems, ΔP_d denotes the loss of import/export to the neighboring islands. Damping of the system can also be neglected since it does not have large and fast impact on FCP, rather has impact on steady-state frequency thereby influencing FRP.

Mathematical formulation for model in Figure 1 is given as

$$\frac{\Delta \omega(s)}{\Delta P_d(s)} = \frac{\frac{-1}{M}s + \frac{p}{M}}{s^2 + \left(\frac{k - pMR}{RM}\right)s + \left(\frac{-kZ}{RM}\right)} \tag{1}$$

It can be observed from (1) that the closed-loop transfer has a zero and 2 poles. The denominator of (1) can be compared with the denominator of the standard second order transfer function given by $s^2 + 2\zeta\Omega_n s + \Omega_n^2$.

Equation (1) can be therefore be rewritten as

$$\frac{\Delta \omega(s)}{\Delta P_d(s)} = \frac{\frac{-1}{M}s + \frac{p}{M}}{s^2 + 2\zeta\Omega_n s + \Omega_n^2} \tag{2}$$

where
natural frequency Ω_n is given by:

$$\Omega_n = \sqrt{\frac{-kz}{RM}} \tag{3}$$

attenuation $\zeta\Omega_n$ is given by:

$$\zeta\Omega_n = \frac{k - pMR}{2RM} \tag{4}$$

damping ratio ζ is given by:

$$\zeta = \frac{k - pMR}{2RM\Omega_n} \tag{5}$$

Generally, a disturbance (such as system separation) occurs instantly, therefore, ΔP_d can be modeled as step response with magnitude A_d. Equation (2) can be thus further written as,

$$\Delta \omega(s) = \frac{-\frac{1}{M}s + \frac{p}{M}}{s^2 + 2\zeta\Omega_n s + \Omega_n^2} \times \frac{A_d}{s} \tag{6}$$

By adding and subtracting $\zeta^2\Omega_n{}^2$ to the denominator and algebraic modifications, the (6) can be written as

$$\Delta\omega(s) = \frac{pA_d}{M\Omega_n{}^2}\left[\frac{1}{s} - \frac{1}{\Omega_d}\frac{(s + \frac{\Omega_n}{p}(\Omega_n + 2p\zeta))\Omega_d}{(s + \zeta\Omega_n)^2 + \Omega_d{}^2}\right] \tag{7}$$

where damped frequency can be written as,

$$\Omega_d = \Omega_n\sqrt{1 - \zeta^2} = \frac{\sqrt{-(k - pMR)^2 - 4kz}}{2RM} \tag{8}$$

Taking inverse Laplace of (7), gives (9)

$$\Delta\omega(t) = \frac{pA_d}{M\Omega_n{}^2}$$
$$\left[1 - e^{-\zeta\Omega_n t}\left(\cos(\Omega_d t) + \frac{\Omega_n{}^2 + p\zeta\Omega_n}{p\Omega_d}\sin(\Omega_d t)\right)\right] \tag{9}$$

Equation (9) shows that $\Delta\omega(t)$ oscillates sinusoidally with an exponential decay, which depends on attenuation $\zeta\Omega_n$.

Main responsibility of the FCP is to contain the frequency peak (or nadir in case of underfrequency) as fast as possible, therefore this peak value $\Delta\omega_{peak}$ and the time to reach this peak value (t_{peak}) are of primal interest.

To find the peak of $\Delta\omega(t)$ given by $\Delta\omega_{peak}$, the derivative of $\Delta\omega(t)$ should be zero, i.e.,

$$\left.\frac{d\Delta\omega(t)}{dt}\right|_{t=t_{peak}} = 0$$

$$t_{peak} = \frac{\tan^{-1}\left(\frac{\Omega_d}{p + \zeta\Omega_n}\right)}{\Omega_d} \tag{10}$$

Notice that time t_{peak} is independent of disturbance ΔP_d and depends on characteristics of technology (k, p, z), droop settings (R) and inertia of the power system (M). This observation is important in designing FCP requirements based on the available energy sources in any specific network. For example, current requirement for full activation time of FCP in CE is 30 s, for Great Britain Network is 10 s, Ireland power system is 15 s and for Northern Europe is 30 s [3].

Substituting the values of t_{peak} from (10) into (9), we get

$$\Delta\omega_{peak} = -\frac{pA_dR}{kz}\left[1 - \frac{e^{-\zeta\Omega_n t_{peak}}}{p}\sqrt{\frac{k(p - z)}{MR}}\right] \tag{11}$$

Please note that in these studies, the peak frequency considered is the first peak value of the frequency. However, there can be many subsequent peaks if the attenuation $\zeta\Omega_n$ is low. If the attenuation is negative, subsequent peaks become even higher than the first peak. Therefore, both first peak and attenuation are taken into consideration in this article.

It can be observed that peak value of $\Delta\omega_{peak}$ is directly proportional to the magnitude of the disturbance, A_d and droop, R. Moreover, it is dependent on generation technology, as $\Delta\omega_{peak}$ is directly proportional to the ratio of pole to gain and zero i.e., $\frac{p}{zk}$. Angular momentum of the system M in p.u. is twice the inertia constant H (M = 2H) and since $\Delta f_p = \Delta\omega_{peak}$, hence the frequency and peak frequency (or nadir) are given by $f = f_{nom}(1 + \Delta\omega)$.

Observations from the mathematical model are as follows:

- Frequency fluctuates sinusoidally with an exponential damping dependent on attenuation $\zeta\Omega_n$

- Peak time t_{peak} is independent of disturbance and dependent on attenuation $\zeta\Omega_n$ and damped frequency Ω_d
- Peak frequency f_{peak} mainly depends on droop, disturbance, generation technology $\frac{pA_dR}{kz}$. When the inertia of the system is low, rate of change of frequency (RoCoF) following a large disturbance can be high. Response of the generation technology should be fast in such system. This can be obtained by reducing the droop (faster droop). Peak frequency is directly proportional to droop. However, capabilities to respond to faster droop depend on type of generation technology. Faster droop can cause reduction in attenuation and damping ratio. This in turn can cause undamped response resulting in oscillatory instability.
- Attenuation $\zeta\Omega_n$ depends on generation technology, droop, system inertia, and independent of disturbance
- Damped frequency Ω_d depends on generation technology, droop, system inertia, and independent of disturbance

The summary of the formulae derived from mathematical model expressed based on inertia constant H are given as (12). Detailed derivation is given in Appendix A.

$$f = f_{nom}\left(1 + \frac{pA_d}{2H\Omega_n{}^2}\right.$$
$$\left.\left[1 - e^{-\zeta\Omega_n t}\left(\cos(\Omega_d t) + \frac{\Omega_n{}^2 + p\zeta\Omega_n}{p\Omega_d}\sin(\Omega_d t)\right)\right]\right)$$
$$f_{peak} = f_{nom}\left(1 - \frac{pA_dR}{kz}\left[1 - \frac{e^{-\zeta\Omega_n t_{peak}}}{p}\sqrt{\frac{k(p-z)}{2HR}}\right]\right)$$
$$t_{peak} = \frac{\tan^{-1}\left(\frac{\Omega_d}{p+\zeta\Omega_n}\right)}{\Omega_d} \tag{12}$$
$$\Omega_n = \sqrt{\frac{-kz}{2RH}}$$
$$\zeta\Omega_n = \frac{k - 2pHR}{4RH}$$
$$\Omega_d = \Omega_n\sqrt{1 - \zeta^2} = \frac{\sqrt{-(k-2pHR)^2 - 4kz}}{4RH}$$

These formulae are used to perform sensitivity studies.

3. Sensitivity Studies

The capabilities of different generation technologies differ by large extent. In this article, 2 main parameters—system inertia and droop are investigated for different generation technologies.

To study different technologies of generation technologies, general-purpose governor block for conventional generators proposed by Anderson and Fouad [27] (Figure D.13. in Appendix D in [27]) is used. This general-purpose governor model basically represents "FCP Controller" of Figure 1. This general-purpose governor has four transfer functions—(i) $\frac{1+T_2s}{1+T_1s}$ representing governor delay (T_1) and pilot valve time (T_2); (ii) $\frac{1}{1+T_3s}$ representing servo or hydro gate time constant; (iii) $\frac{1}{1+T_4s}$ representing steam valve bowl time constant; (iv) $\frac{1+FT_5s}{1+T_5s}$ representing steam reheat time constant, where F is per unit shaft output. The parameters F and $T1 - T5$ vary for different types of generators and affect the output response for change in frequency. In this article, fossil fuel-based steam generator (820 MW) and cross-compound steam generator (436 MW) among conventional generators are consider for studies whose parameters are given in Appendix D in [27]. It should be noted that the methodology is generic for all kinds of generators and accordingly the parameters for specific generators should be used for the studies. Sensitivity studies performed for hydro and nuclear plants can be found in [28].

R is varied from 2% to 6% to study its effects of different R on frequency. $T5$ is the main time constant for the generators. Therefore, T.F.-4 ($\frac{1+FT_5s}{1+T_5s}$) plays the most important role in dictating the output response from these generators. P_{max} is relaxed since volume of FCR is assumed sufficient to handle the disturbance. P_{m0} is assumed constant since it is set by FRP. Considering these assumptions, the generic model is combined with the models presented in Figure 1 to provide the simplified generic delta model for FCP as shown in Figure 2.

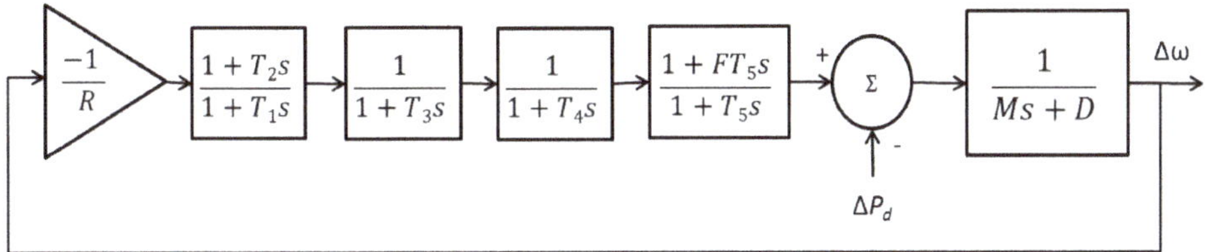

Figure 2. Generic delta model for FCP from conventional Governor-Turbine system.

Generic model of Figure 2 is applicable only for conventional generators and not applicable for WTs. FCP model for WT is shown in Figure 3. Measurement delay is assumed as 100 ms. Measurement delay is assumed comprising of communication delays, delays due to sampling and computation of frequency and measurement delay. Power activation delay is assumed as 50 ms. It should be noted that all these parameters are configurable and varies for different WPPs. 100 ms may be realistic if the WPP is offshore connected through HVDC connection, where communication delay needs to be incorporated. However, if the frequency measurement is obtained locally from PLL, the delay would be much smaller. Maximum and minimum ramp rates are assumed ±0.5 p.u./s. It should be noted that the response capability from WT depend largely on ramp rate and delays of WT control and studied in detail in [28]. These sensitivities are not included in this article because the impact of R & H plays a major role as compared to these other parameters as long as the ramp rates and delay values are within specific limits. However, in future, if technology becomes more flexible, these additional sensitivities must be studied along with R & H.

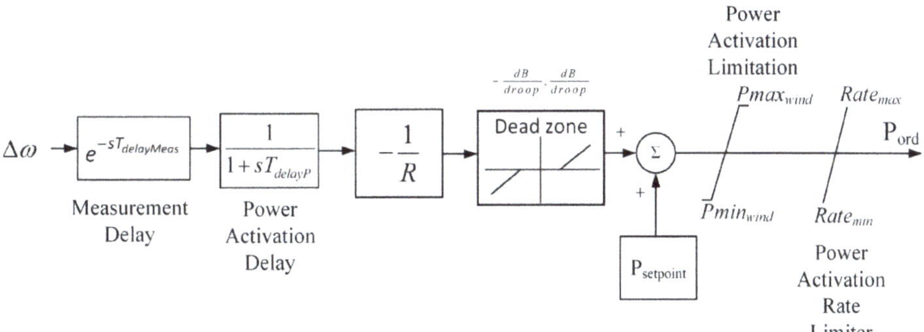

Figure 3. Frequency control model for FCP from WT.

Power imbalance due to fault (A_d) is assumed as large as 0.17 p.u. for the underfrequency studies and −0.17 p.u. for overfrequency studies based on the UCTE 4 November 2006 system separation event when North-Eastern island area had 17% excess generation following the split [29].

To understand the behavior of the generic FCP controller models, equivalent analytical pole-zero-gain model of Figure 1 is identified based on models of Figures 2 and 3 through system identification. This equivalent pole-zero-gain model is used to compute closed-loop poles, attenuation and damping ratio of the frequency response.

To investigate the sensitivities of different technologies of generators following studies are performed:

(A)   Constant droop, constant inertia
(B)   Constant droop, varying inertia
(C)   Varying droop, varying inertia

In all the studies, only one generator technology is considered at a time for FCP model of Figure 1 since the idea is to compare the response of different technologies independently. However, combinations of different generators are studied in the case study involving UCTE disturbance.

3.1. Constant Droop, Constant Inertia

The goal of this study is to understand the individual capabilities of different types of governors and WT to contain frequency and prevent frequency instabilities. In this regard, it is important to find equivalent pole-zero-gain of the analytical model depicted in Figure 1. The parameters of the analytical model for each generator for the response for given $R = 4\%$ and $H = 5$ s is given in Table 1. Fitness of the frequency response for this analytical model with the simulated generic model is also given in Table 1.

Table 1. Analytical model parameters of considered fossil steam generator, cross-compound generator and wind turbine generator.

Parameters	Fossil Steam	Cross-Compound	Wind Turbine
k	5.5349×10^4	-0.1530	-0.4335
p	-1.2553	-2.4096	-7.9687
z	-1.05×10^3	9.8666	18.4136
Fitness [%]	97.6395	99.3584	99.0736
Closed-loop poles	$-0.6283 + 0.9002\mathrm{j}$	$-1.0136 + 0.9567\mathrm{j}$	-6.1596
	$-0.6283 - 0.9002\mathrm{j}$	$-1.0136 - 0.9567\mathrm{j}$	-0.7252
Natural Freq. (Ω_n)	1.2051	1.9425	4.4672
Attenuation $(\zeta\Omega_n)$	0.6283	1.0135	3.4424
Damping Ratio (ζ)	0.5214	0.5218	0.7706

Table 2. Comparison for analytical and generic model.

Generator	Peak Time [s]				Peak Frequency [Hz]			
	Calc.	Obs.	Error		Calc.	Obs.	Error	
			Abs.	Rel. [%]			Abs.	Rel. [%]
Fossil	2.059	2.08	0.020	0.971	50.819	50.821	0.002	0.005
Cross-Compound	1.370	1.377	0.007	0.487	50.586	50.585	0.002	0.003
Wind	0.9062	0.907	8×10^4	0.088	50.3	50.3084	0.0084	0.017

Closed loop poles give information about stability of the power system. It can be observed from Table 1 that the power system is stable for all type of generators since real parts of the poles are negatives. However, their distances from origin (i.e., their absolute values) provide relative stabilities. WT poles have no imaginary component inferring and therefore there is no oscillatory component in the output response from WT. Attenuation $\zeta\Omega_n$ and damping ratio ζ affect the damping of the frequency response. Generally, damping ratio for the controller is chosen between 0.4–0.7 to limit peak overshoot [30]. Remark that all these results are observed based on certain parameters of specific generation technologies and they can vary for different values of parameters. However, the methodology is generic for analyzing different technologies which is purpose of this article.

Table 2 depicts the comparison between proposed mathematical model and generic model. Calculated values are obtained from the mathematical model while observed values are obtained from simulation using the generic models in Figures 2 and 3. It can be noted that error between calculated and observed values is less than 1%. Thus, the confidence on the derived results from the analytical model is quite high.

3.2. Constant Droop, Varying Inertia

With changing penetration of non-synchronous RES replacing conventional generations depending on weather conditions, inertia of the system changes dynamically. Therefore, it is important to choose proper technology for FCP provision in this varying inertia system. In the following studies, it is considered that inertia constant H is decreased from 6 to 1.5 s while R is kept constant at 4%. Closed-loop poles are computed using analytical method as discussed before. Trajectories for closed-loop poles for decreasing inertia constants are shown in Figures 4–6.

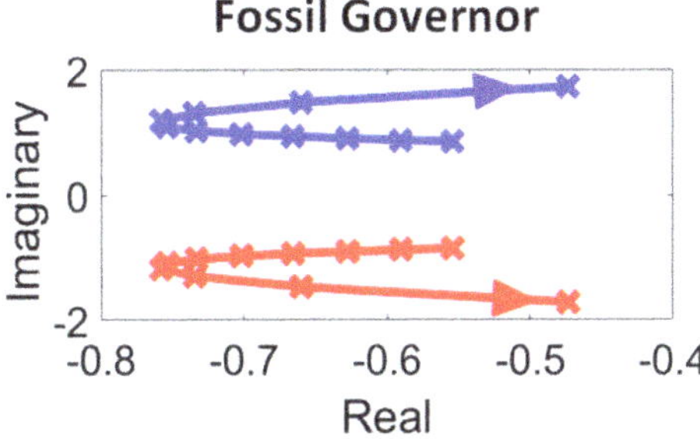

Figure 4. Impact of H on closed loop poles for Fossil Governor.

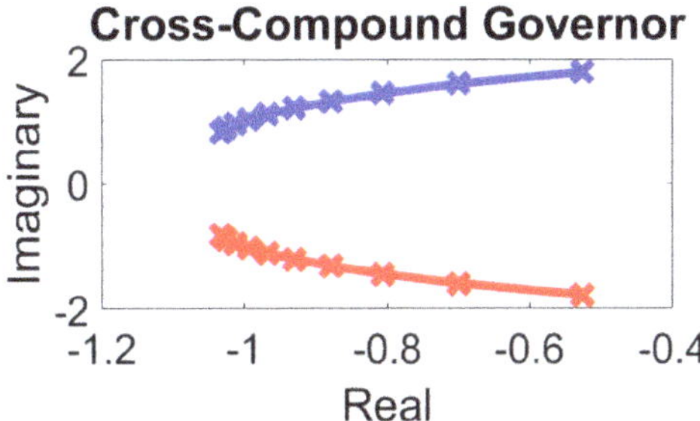

Figure 5. Impact of H on closed-loop poles for Cross-Compound Governor.

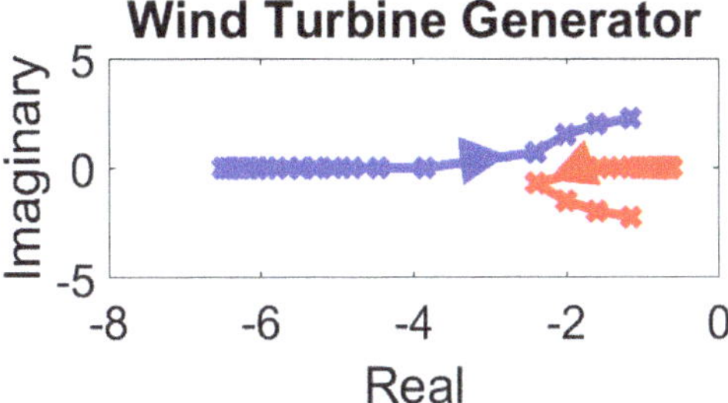

Figure 6. Impact of H on closed-loop poles for Wind Turbine Generator.

Fossil and cross-compound governor systems are quite stable since the real part of the poles is far from origin even with low H i.e., the ending point of the trajectory. However, lowering the inertia reduces relative stability as the poles start moving towards origin. It should also be noted that the imaginary component of the poles starts increasing with reduction in H, indicating increase in frequency oscillations. It should also be noted that these frequency characteristics depend on the controller parameters (as seen from (12)), thereby depending on the size and type of generator. Wind turbine also depicts interesting behavior. For WTs, poles lie on real axis for high inertia. These real poles are quite further from imaginary axis implying fast output response. Since these poles have no imaginary component, there is no oscillation in the output response. However, as inertia is decreased, oscillations begin to appear in the output response as there are imaginary components in the complex poles for lower values of inertia. The values of these complex roots are quite far away from the real axis implying high oscillations in the transient response. This might make WTs incapable of providing FCP for low values of inertia. However, in such situations it might be required to change the droop settings of WTs. This issue is investigated later. Notice that the pole trajectory plots provide information on stability, but not information on peak frequency and attenuation. As peak frequency and attenuation depend both on inertia constant and droop, they should be studied together and not independently. This is especially relevant when system inertia is low and faster responses from generators are required. This can be obtained by reducing the droop. Therefore, impacts on frequency with varying inertia and droop are studied.

3.3. Varying Droop and Varying Inertia

In this study, H is varied from 1.5 s to 6 s while the droop R is varied from 2% to 6%. The success criteria for the FCP is considered to be containment of frequency to less than 51 Hz for overfrequency and greater than 49 Hz for underfrequency events.

Impacts of R and H on f_p (based on first peak) for fossil steam generator is shown in Figure 7. The yellow planes in Figure 7 are the planes of 51 Hz and 49 Hz. Therefore, FCP is deemed successful when the f_p is between these yellow planes. These points are marked with green color while the points outside these planes are marked in red color. If inertia of the system is high (i.e., H = 6 s), droop of around 5% is enough for successful FCP. Decreasing inertia needs to be handled with decreasing droop. However, when the inertia is too low (i.e., $H < 2s$) decreasing droop may not be enough to prevent frequency going outside the range of 49–51 Hz. Similar studies are performed for cross-compound steam generator and WT the result is shown in Figures 8 and 9 respectively. For droop of 4% or lower, frequency always stays within 49–51 Hz for the studied disturbance of ±0.17 p.u. for cross-compound and for any values of R for WT.

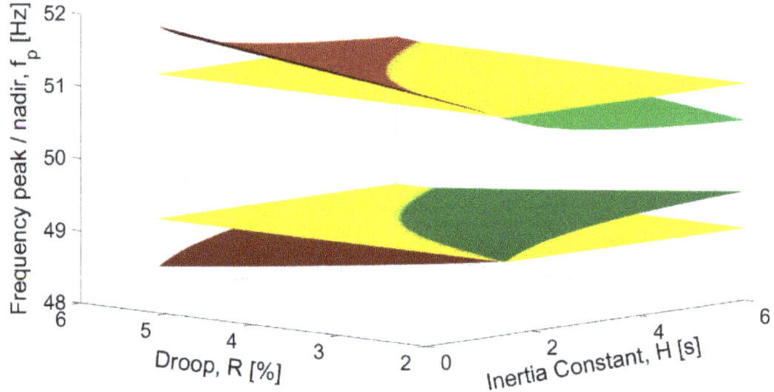

Figure 7. Impacts of R and H on f_p for fossil steam generator.

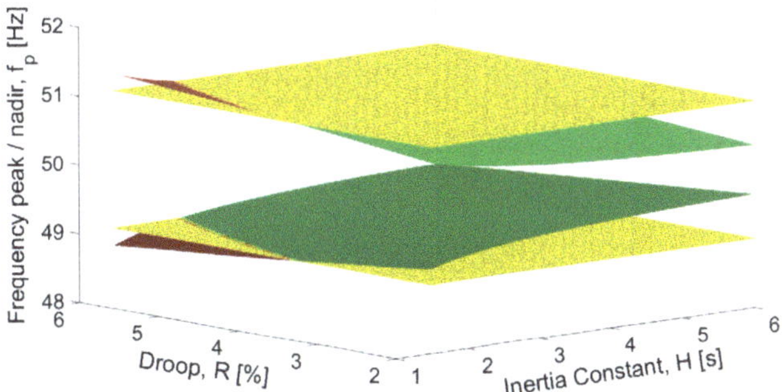

Figure 8. Impacts of R and H on f_p for Cross-compound steam generator.

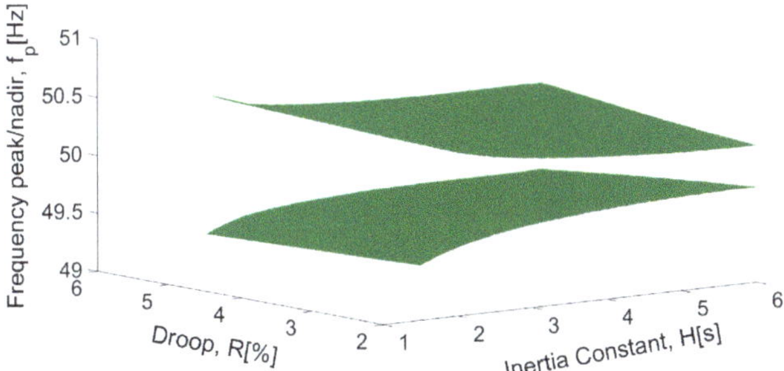

Figure 9. Impacts of R and H on f_p for Wind Turbine.

Impacts of droop and inertia constant on attenuation for cross-compound and fossil steam generators are studied as shown in Figure 10. Attenuation of cross-compound generator are higher than fossil steam generator for higher droop and higher inertia. Meanwhile for lower droop and lower inertia, attenuation of cross-compound generator is lower than fossil steam generator. Remark that attenuation for cross-compound even can be negative when droop and inertia are lower than 3% and 2 s respectively as denoted by gray region in Figure 10.

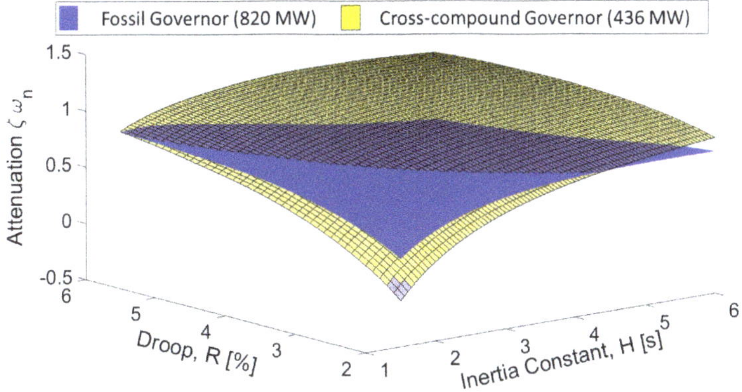

Figure 10. Impacts of R and H on attenuation for Cross-compound and fossil steam.

Figure 11 shows impacts of R and H on attenuation for WT. Notice that lower the droop and lower the inertia, attenuation is lower. Attenuation even becomes negative for very low value of inertia ($H < 3s$) and low droop ($R < 4\%$). These results show that attenuation becomes limiting criteria for deciding droop for different value of inertia for WTs.

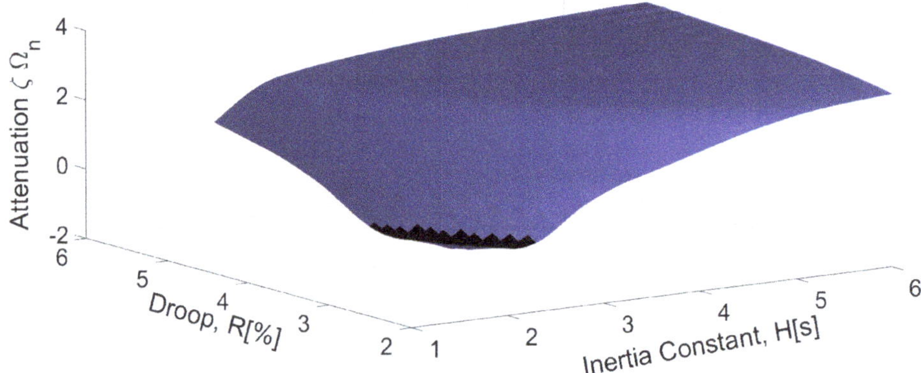

Figure 11. Impacts of R and H on attenuation for Wind Turbine.

From the previous results, it can be understood that peak frequency and attenuation both play crucial role in deciding R for different H. Therefore, feasibility of reducing R with decreasing H is defined as limiting the first peak frequency f_p within 49–51 Hz as well as limiting the damping ratio ζ higher than 0.4. Figures 12–14 show the feasibility of different R for different H for fossil, cross-compound steam generator, and WT. This study is especially important because it gives counter-intuitive result that for lower inertia values the droop should be higher to prevent oscillatory instability. Furthermore, it can be observed that if droop is fixed at 4%, WT can allow for operation with lower H than the other generators. Operating at 4% droop, minimum H possible for fossil steam generator is 2.75 s, while for cross-compound steam generator it is 3.5 s. WT allows operation with H down to 2.35 s for $R = 4\%$. This shows that WT can be attractive choice for providing FCP in future system with low inertia.

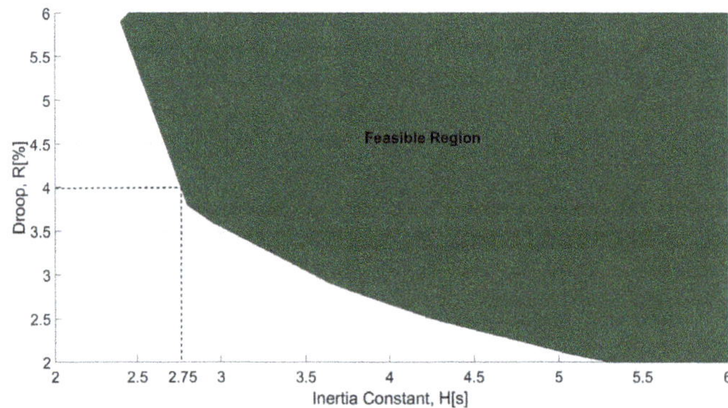

Figure 12. Feasibility of different R for different H for Fossil Steam Generator.

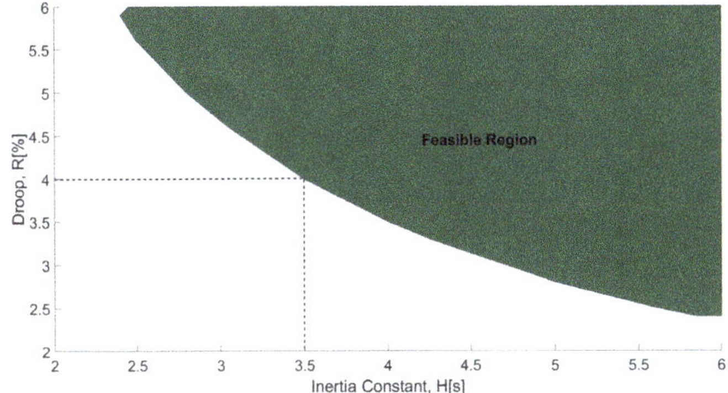

Figure 13. Feasibility of different R for different H for Cross-Compound Fossil Steam.

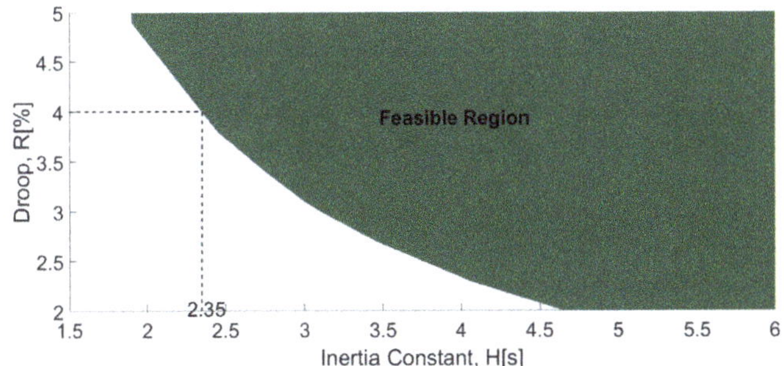

Figure 14. Feasibility of different R for different H for Wind Turbine.

All the above case studies show the importance of mathematical model for designing of FCP and choosing correct type of technology. This mathematical model also helps in specifying the parameters for different generators for provision of FCP.

4. Case Study—UCTE Disturbance on 4 November 2006

The sensitivity studies described above are meant to compare the feasibility of different technologies for providing FCP. However, in a large power system, many different technologies will provide FCP support at the same time. Another simplification used for better understanding was the use of generic model in the sensitivity studies. Therefore, in the considered case study, detailed generator models are used for a large realistic power system to validate the applicability of the proposed methodology.

The disturbance on 4 November 2006 at the "Union for the Co-ordination of Transmission of Electricity" (UCTE) network is one of the most important phenomena seen related to cascading overload phenomena leading to splitting of the network and large frequency deviations. Tripping of a 380 kV line due to overload and other cascading trippings led to the final separation of the entire UCTE network into three islands [29]. The countries in the Western part were in power deficiency situation of about 9 GW. That led to a frequency drop down to about 49 Hz stopped by automatic load-shedding and by tripping of pumping storage units. The countries in the North-Eastern area encountered a surplus of generation. The value of frequency was over 50.5 Hz as shown in Figure 15. This area had around 10% wind power penetration which were being disconnected and reconnected arbitrarily. Conventional generators were mainly responsible for providing frequency support through FCP and emergency control.

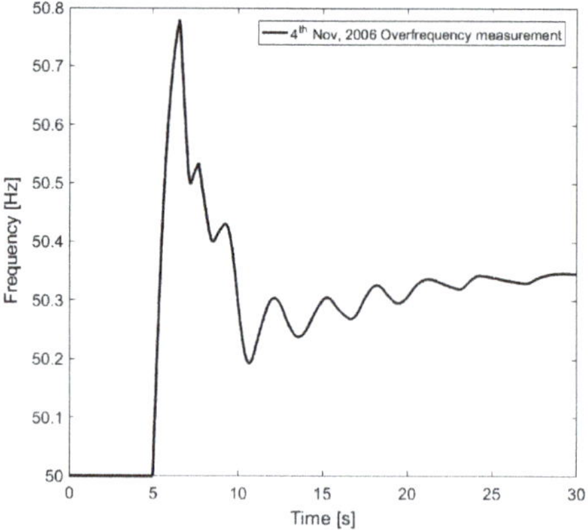

Figure 15. Frequency Response on 4 November Disturbance.

Current grid code requirements require WTs to stay connected up to 51.5 Hz. Therefore, scenario is simulated where WTs are not unintentionally disconnected as shown in Figure 16. Figure 16 shows that response from WTs along with conventional generators can improve the frequency response. Peak frequency reaches 50.8 Hz when FCP support is only provided by conventional generators, while peak frequency reduces 50.68 Hz when FCP support is provided from WT together with conventional generators. The reason for this is that FCP support from WTs is much faster than that of conventional generator. It also depicts that analytical model provides similar results as compared to detailed simulation model (PEGASE model [31]). The difference in peak value is due to the impact of frequency dead-band. To study the impact of frequency support from WTs for future power systems, wind power penetration is increased in these models to 40%. Frequency response in the system with 40% wind penetration is shown in Figure 17. It shows f_p is substantially reduced with additional FCP support from WTs.Peak frequency reaches 51.2 Hz when FCP support is only provided by conventional generators, while peak frequency reduces 50.3 Hz when FCP support is provided from WT together with conventional generators. However, not only f_p is reduced, but also stability margin is improved as evident from the closed-loop poles in Figure 18. Closed-loop poles moves more left in the negative real axis of complex plane thereby improving the relative stability of the system. These results evidently supports and validates the importance of the proposed mathematical analytical model.

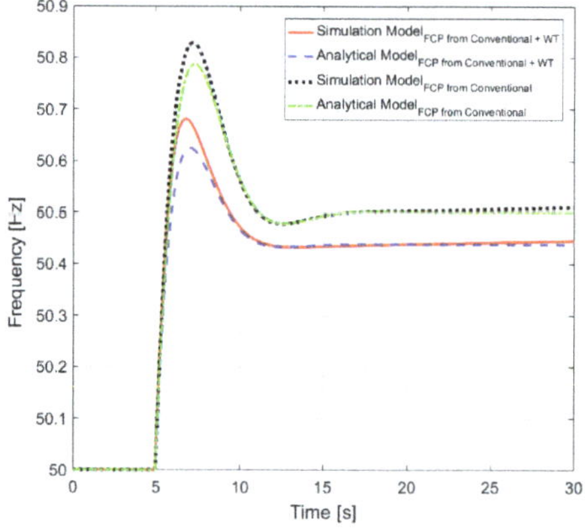

Figure 16. Analytical and Simulation Model for 10% wind penetration.

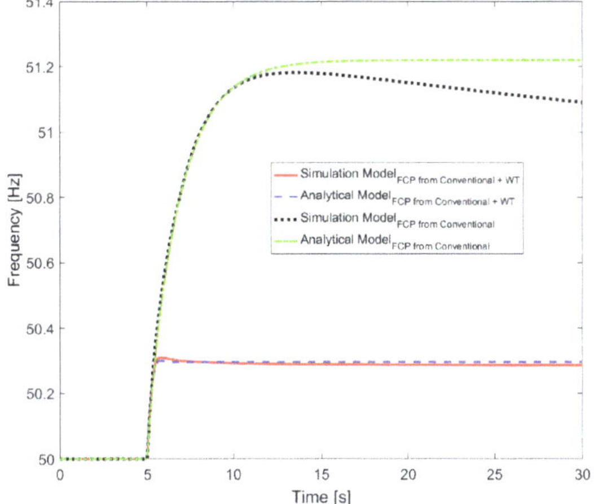

Figure 17. Analytical and Simulation Model for 40% wind penetration.

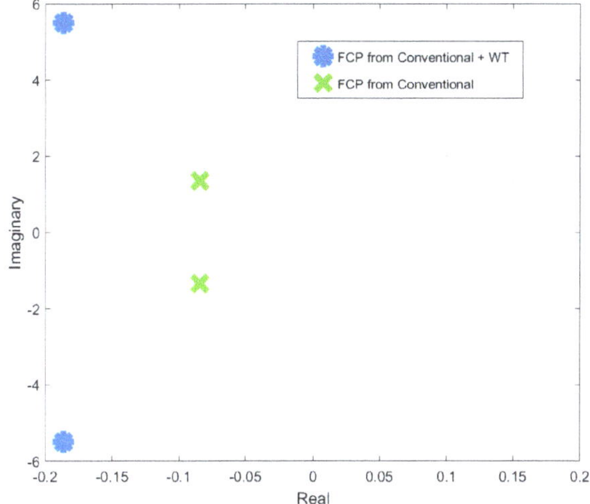

Figure 18. Closed-loop poles for 40% wind penetration.

5. Conclusions

This article presents a developed detailed mathematical approach to study and perform sensitivity studies for the capability of different technologies to provide FCP in different grid conditions. This method allows design of the droop for different technology providing FCP in a power system operating within certain range of inertia. The proposed mathematical approach has been validated on simulated CE network.

Several simulations and comparative studies between FCR provision from conventional generators and WTs have been performed and analyzed in a generic power system with large penetration of WTs in order to study the impacts of droop settings and system inertia. The proposed analytical approach provides a feasible range of droop settings for different values of system inertia.

The results show that providing FCP can become challenging for large disturbances, especially in low inertia systems. The availability of adequate volume of FCR is necessary but not sufficient condition for frequency stability. It has also been observed that time to reach peak response is independent of the size of the disturbance. Furthermore, it has been noticed that the system frequency oscillates sinusoidally with an exponential decay following a large disturbance and that the exponential decay depends on attenuation and damping ratio. Attenuation and damping ratio are independent of size of disturbance and dependent on inertia constant and droop.

The results of this work can be used as a starting base for additional sensitivities studies such as ramp rates and measurement delays and studying future power system scenarios, when the technologies become more flexible.

Author Contributions: Conceptualization, K.D.; Formal analysis, K.D.; Investigation, K.D.; Methodology, K.D.; Resources, K.D. and M.A.; Supervision, M.A., A.D.H. and P.E.S.; Writing—original draft, K.D.; Writing—review & editing, A.D.H. and P.E.S.

Funding: This research received no external funding.

Conflicts of Interest: The authors declare no conflicts of interest.

Abbreviations

The following abbreviations are used in this manuscript:

f	Frequency
s	Laplace Operator
$\Delta\omega$	Change in angular velocity
k	Gain for conventional turbine technology
z	Zero for conventional turbine technology
p	Pole for conventional turbine technology
R	Droop for FCP controller
M	System equivalent angular momentum
D	Equivalent system damping
P_{ref}	Load reference point for the generator
P_d	Power imbalance due to disturbance
Ω_n	Natural frequency
Ω_d	Damped frequency
ζ	Damping coefficient
A_d	Magnitude of the disturbance
ω_{peak}	Peak/nadir value of angular velocity
f_p	Peak/nadir value of frequency
t_{peak}	Time to reach ω_{peak}
H	Inertia constant
f_{nom}	Nominal frequency of the system
$T1\text{–}T5, F$	Parameters of generic governor
P_{m0}	Mechanical power set-point of the generator
P_m	Mechanical power output of the generator
P_e	Electrical power output of the generator
P_a	Accelerating power output of the generator
P_{max}	Maximum power output of generator
$Rate_{max}$	Maximum ramp rate for Wind Turbines(WTs)
$Rate_{min}$	Minimum ramp rate for Wind Turbines(WTs)
$T_{delayMeas}$	Measurement delay for FCP from WTs
T_{delayP}	Power activation delay for FCP from WTs

Appendix A

Detailed Mathematical Formulation

The mathematical formulation for the model in Figure 1

$$\left(-\frac{\Delta\omega(s)}{R} \times \frac{k(s-z)}{(s-p)} - \Delta P_d(s) \right) \frac{1}{Ms+D} = \Delta\omega(s) \tag{A1}$$

$$\frac{\Delta\omega(s)}{\Delta P_d(s)} = \frac{-\frac{1}{M}s + \frac{p}{M}}{s^2 + \left(\frac{k+DR-pMR}{RM}\right)s + \left(\frac{-pDR-kZ}{RM}\right)} \tag{A2}$$

Assuming, D = 0;

$$\frac{\Delta\omega(s)}{\Delta P_d(s)} = \frac{-\frac{1}{M}s + \frac{p}{M}}{s^2 + \left(\frac{k-pMR}{RM}\right)s + \left(\frac{-kZ}{RM}\right)} \tag{A3}$$

Denominator can be compared with $s^2 + 2\zeta\Omega_n s + \Omega_n^2$

$$\frac{\Delta\omega(s)}{\Delta P_d(s)} = \frac{-\frac{1}{M}s + \frac{p}{M}}{s^2 + 2\zeta\Omega_n s + \Omega_n^2} \tag{A4}$$

$$\text{where } \Omega_n = \sqrt{\frac{-kz}{RM}}, \ \zeta = \frac{k-pMR}{2RM\Omega_n} \tag{A5}$$

Modeling disturbance as step response of magnitude A_d

$$\Delta\omega(s) = \frac{-\frac{1}{M}s + \frac{p}{M}}{s^2 + 2\zeta\Omega_n s + \Omega_n^2} \times \frac{A_d}{s} \tag{A6}$$

$$\Delta\omega(s) = \frac{pA_d}{M\Omega_n^2}\left[\frac{1}{s} - \frac{s + \frac{\Omega_n}{p}(\Omega_n + 2p\zeta)}{s^2 + 2\zeta\Omega_n s + \Omega_n^2}\right] \tag{A7}$$

By adding and subtracting $\zeta^2\Omega_n^2$ to the denominator

$$\Delta\omega(s) = \frac{pA_d}{M\Omega_n^2}\left[\frac{1}{s} - \frac{s + \frac{\Omega_n}{p}(\Omega_n + 2p\zeta)}{s^2 + 2\zeta\Omega_n s + \zeta^2\Omega_n^2 + \Omega_n^2 - \zeta^2\Omega_n^2}\right] \tag{A8}$$

$$\Delta\omega(s) = \frac{pA_d}{M\Omega_n^2}\left[\frac{1}{s} - \frac{1}{\Omega_d}\frac{(s + \frac{\Omega_n}{p}(\Omega_n + 2p\zeta))\Omega_d}{(s + \zeta\Omega_n)^2 + \Omega_d^2}\right] \tag{A9}$$

$$\Omega_d = \Omega_n\sqrt{1 - \zeta^2} = \frac{\sqrt{-(k-pMR)^2 - 4kz}}{2RM} \tag{A10}$$

Taking inverse Laplace of (A9)

$$\Delta\omega(t) = \frac{pA_d}{M\Omega_n^2}\left[1 - e^{-\zeta\Omega_n t}\left(\cosh(j\Omega_d t) - \right.\right.$$
$$\left.\left.\frac{j(\frac{\Omega_n^2 + 2p\zeta\Omega_n}{p} - \zeta\Omega_n)\sinh(j\Omega_d t)}{\Omega_d}\right)\right] \tag{A11}$$

$$\cosh(j\Omega_d t) = \frac{e^{j\Omega_d t} + e^{-j\Omega_d t}}{2} = \cos(\Omega_d t), \tag{A12}$$

$$\sinh(j\Omega_d t) = \frac{e^{j\Omega_d t} - e^{j\Omega_d t}}{2} = j\sin(\Omega_d t) \tag{A13}$$

Equation (A11) gets modified to (A14)

$$\Delta\omega(t) = \frac{pA_d}{M\Omega_n^2}\left[1 - e^{-\zeta\Omega_n t}\left(\cos(\Omega_d t) + \frac{\sin(\Omega_d t)}{\Omega_d}\right.\right.$$
$$\left.\left.\left(\frac{\Omega_n^2 + 2p\zeta\Omega_n}{p} - \zeta\Omega_n\right)\right)\right] \tag{A14}$$

$$\Delta\omega(t) = \frac{pA_d}{M\Omega_n{}^2}\left[1 - e^{-\zeta\Omega_n t}\right.$$
$$\left.\left(\cos(\Omega_d t) + \frac{\Omega_n{}^2 + p\zeta\Omega_n}{p\Omega_d}\sin(\Omega_d t)\right)\right] \tag{A15}$$

To calculate the peak of $\Delta\omega(t)$ given by $\Delta\omega_{peak}$,

$$\left.\frac{d\Delta\omega(t)}{dt}\right|_{t=t_{peak}} = 0 \tag{A16}$$

$$\zeta\Omega_n e^{-\zeta\Omega_n t_{peak}}\left[\cos(\Omega_d t_{peak}) + \frac{\Omega_n{}^2 + p\zeta\Omega_n}{p\Omega_d}\sin(\Omega_d t_{peak})\right]$$
$$-e^{-\zeta\Omega_n t_{peak}}\left[-\Omega_d \sin(\Omega_d t_{peak}) + \right.$$
$$\left.\frac{\Omega_n{}^2 + p\zeta\Omega_n}{p}\cos(\Omega_d t_{peak})\right] = 0 \tag{A17}$$

$$\tan(\Omega_d t_{peak}) = \frac{\Omega_d}{p + \zeta\Omega_n} \tag{A18}$$

$$t_{peak} = \frac{\tan^{-1}\left(\frac{\Omega_d}{p+\zeta\Omega_n}\right)}{\Omega_d} \tag{A19}$$

From (A18), $\sin(\Omega_d t_{peak})$ and $\cos(\Omega_d t_{peak})$:

$$\sin(\Omega_d t_{peak}) = \frac{\Omega_d}{\sqrt{p^2 + 2p\zeta\Omega_n + \Omega_n{}^2}} \tag{A20}$$

$$\cos(\Omega_d t_{peak}) = \frac{p + \zeta\Omega_n}{\sqrt{p^2 + 2p\zeta\Omega_n + \Omega_n{}^2}} \tag{A21}$$

$$\text{such that } \sin^2(\Omega_d t_{peak}) + \cos^2(\Omega_d t_{peak}) = 1 \tag{A22}$$

Replacing values of (A20) and (A21) in (A15)

$$\Delta\omega_{peak} = \frac{pA_d}{M\Omega_n{}^2}\left[1 - e^{-\zeta\Omega_n t_{peak}}\right.$$
$$\left.\left(\frac{p + \zeta\Omega_n}{\sqrt{p^2 + 2p\zeta\Omega_n + \Omega_n{}^2}} + \frac{\Omega_n{}^2 + p\zeta\Omega_n}{p\Omega_d}\frac{\Omega_d}{\sqrt{p^2 + 2p\zeta\Omega_n + \Omega_n{}^2}}\right)\right] \tag{A23}$$

$$\Delta\omega_{peak} = \frac{pA_d}{M\Omega_n{}^2}\left[1 - e^{-\zeta\Omega_n t_{peak}}\left(\frac{\sqrt{p^2 + 2p\zeta\Omega_n + \Omega_n{}^2}}{p}\right)\right] \tag{A24}$$

$p^2 + 2p\zeta\Omega_n + \Omega_n{}^2$ calculated from (A5)

$$p^2 + 2p\zeta\Omega_n + \Omega_n{}^2 = \frac{k(p - z)}{MR} \tag{A25}$$

Substituting the values from (A25) into (A24)

$$\Delta\omega_{peak} = -\frac{pA_d R}{kz}\left[1 - \frac{e^{-\zeta\Omega_n t_{peak}}}{p}\sqrt{\frac{k(p - z)}{MR}}\right] \tag{A26}$$

References

1. EirGrid; SONI. Ensuring a Secure, Reliable and Efficient Power System in a Changing Environment. 2011. Available online: http://www.eirgridgroup.com/site-files/library/EirGrid/Ensuring-a-Secure-Reliable-and-Efficient-Power-System-Report.pdf (accessed on 19 March 2019).
2. Bömer, J.; Burges, K.; Nabe, C.; Pöller, M. *All Island TSO Facilitation of Renewables Studies: Final Report for Work Package 3*; EirGrid: Dublin, Ireland, 2010.
3. ENTSO-E. System Operations Code. Available online: https://www.entsoe.eu/network_codes/sys-ops/ (accessed on 19 March 2019).
4. Das, K.; Litong-Palima, M.; Maule, P.; Sørensen, P.E. Adequacy of operating reserves for power systems in future european wind power scenarios. In Proceedings of the Power & Energy Society General Meeting, Denver, CO, USA, 26–30 July 2015; pp. 1–5.
5. Das, K.; Litong-Palima, M.; Maule, P.; Altin, M.; Hansen, A.D.; Sørensen, P.E.; Abildgaard, H. Adequacy of frequency reserves for high wind power generation. *IET Renew. Power Gener.* **2017**, *11*, 1286–1294. [CrossRef]
6. Das, K.; Nitsas, A.; Altin, M.; Hansen, A.D.; Sørensen, P.E. Improved Load-Shedding Scheme Considering Distributed Generation. *IEEE Trans. Power Deliv.* **2017**, *32*, 515–524. [CrossRef]
7. Das, K.; Hansen, A.D.; Sørensen, P.E. Aspects of Relevance of Wind Power in Power System Defense Plans. In Proceedings of the 12th International Workshop on Large-Scale Integration of Wind Power into Power Systems as well as on Transmission Networks for Offshore Wind Power Plants, London, UK, 22–24 October 2013; pp. 416–421.
8. De Boeck, S.; Das, K.; Trovato, V.; Turunen, J.; Halat, M.; Sorensen, P.; Van Hertem, D. Review of defence plans in europe: Current status, strenghts and opportunities. *CIGRE Sci. Eng.* **2016**, *5*, 6–16.
9. Billington, R.; Allan, R.N. *Reliability Evaluation of Power Systems*; Plenum Publishing Corp.: New York, NY, USA, 1984.
10. Wood, A.J.; Wollenberg, B.F. *Power Generation, Operation, and Control*; John Wiley & Sons: Hoboken, NJ, USA, 2012.
11. Zhao, C.; Topcu, U.; Li, N.; Low, S. Design and stability of load-side primary frequency control in power systems. *IEEE Trans. Autom. Control* **2014**, *59*, 1177–1189. [CrossRef]
12. Molina-Garcia, A.; Bouffard, F.; Kirschen, D.S. Decentralized demand-side contribution to primary frequency control. *IEEE Trans. Power Syst.* **2011**, *26*, 411–419. [CrossRef]
13. Oudalov, A.; Chartouni, D.; Ohler, C. Optimizing a battery energy storage system for primary frequency control. *IEEE Trans. Power Syst.* **2007**, *22*, 1259–1266. [CrossRef]
14. Mercier, P.; Cherkaoui, R.; Oudalov, A. Optimizing a battery energy storage system for frequency control application in an isolated power system. *IEEE Trans. Power Syst.* **2009**, *24*, 1469–1477. [CrossRef]
15. Morren, J.; De Haan, S.W.; Kling, W.L.; Ferreira, J. Wind turbines emulating inertia and supporting primary frequency control. *IEEE Trans. Power Syst.* **2006**, *21*, 433–434. [CrossRef]
16. Ullah, N.R.; Thiringer, T.; Karlsson, D. Temporary primary frequency control support by variable speed wind turbines—Potential and applications. *IEEE Trans. Power Syst.* **2008**, *23*, 601–612. [CrossRef]
17. Altin, M.; Kuhlmann, J.; Das, K.; Hansen, A. Optimization of Synthetic Inertial Response from Wind Power Plants. *Energies* **2018**, *11*, 1051. [CrossRef]
18. Margaris, I.D.; Papathanassiou, S.A.; Hatziargyriou, N.D.; Hansen, A.D.; Sorensen, P. Frequency control in autonomous power systems with high wind power penetration. *IEEE Trans. Sustain. Energy* **2012**, *3*, 189–199. [CrossRef]
19. Sakamuri, J.N.; Das, K.; Altin, M.; Cutululis, N.A.; Hansen, A.D.; Tielens, P.; Van Hertem, D. Improved frequency control from wind power plants considering wind speed variation. In Proceedings of the Power Systems Computation Conference (PSCC), Genoa, Italy, 20–24 June 2016; pp. 1–7.
20. Sun, Y.; Zhang, Z.; Li, G.; Lin, J. Review on frequency control of power systems with wind power penetration. In Proceedings of the 2010 International Conference on Power System Technology (POWERCON), Hangzhou, China, 24–28 October 2010; pp. 1–8.
21. Haileselassie, T.M.; Uhlen, K. Primary frequency control of remote grids connected by multi-terminal HVDC. In Proceedings of the Power and Energy Society General Meeting, Providence, RI, USA, 25–29 July 2010; pp. 1–6.

22. Mu, Y.; Wu, J.; Ekanayake, J.; Jenkins, N.; Jia, H. Primary frequency response from electric vehicles in the Great Britain power system. *IEEE Trans. Smart Grid* **2013**, *4*, 1142–1150. [CrossRef]
23. Aik, D.L.H. A general-order system frequency response model incorporating load shedding: Analytic modeling and applications. *IEEE Trans. Power Syst.* **2006**, *21*, 709–717. [CrossRef]
24. Nguyen, N.; Almasabi, S.; Mitra, J. Estimation of penetration limit of variable resources based on frequency deviation. In Proceedings of the North American Power Symposium (NAPS), Charlotte, NC, USA, 4–6 October 2015; pp. 1–6.
25. Chvez, H.; Hezamsadeh, M.R.; Carlsson, F. A simplified model for predicting primary control inadequacy for nonresponsive wind power. *IEEE Trans. Sustain. Energy* **2016**, *7*, 271–278. [CrossRef]
26. Merritt, H.E. *Hydraulic Control Systems*; John Wiley & Sons: Hoboken, NJ, USA, 1967.
27. Anderson, P.M.; Fouad, A.A. *Power System Control and Stability*; John Wiley & Sons: Hoboken, NJ, USA, 2008.
28. Das, K.; Sørensen, P.; Hansen, A.; Abildgaard, H. Integration of Renewable Generation in Power System Defence Plans. Ph.D. Thesis, Technical University of Denmark, Lyngby, Denmark, 2016.
29. Union for the Co-Ordination of Transmission of Electricity (UCTE). Final Report on System Disturbance on 4 November 2006. Available online: https://www.entsoe.eu/fileadmin/user_upload/_library/publications/ce/otherreports/Final-Report-20070130.pdf (accessed on 19 March 2019).
30. Ogata, K. *Modern Control Engineering*; Prentice Hall PTR: Upper Saddle River, NJ, USA, 2009.
31. PEGASE. Available online: http://www.fp7-pegase.com (accessed on 19 March 2019).

Article

Impact of Combined Demand-Response and Wind Power Plant Participation in Frequency Control for Multi-Area Power Systems

Irene Muñoz-Benavente [1], Anca D. Hansen [2], Emilio Gómez-Lázaro [3], Tania García-Sánchez [4], Ana Fernández-Guillamón [1] and Ángel Molina-García [1,*]

[1] Department of Electrical Engineering, Universidad Politécnica de Cartagena, 30202 Cartagena, Spain; irene.munoz.benavente@gmail.com (I.M.-B.); ana.fernandez@upct.es (A.F.-G.)
[2] DTU Wind Energy, Technical University of Denmark, 4000 Roskilde, Denmark; anca@dtu.dk
[3] Renewable Energy Research Institute and DIEEAC-EDII-AB. Universidad de Castilla-La Mancha, 02071 Albacete, Spain; emilio.gomez@uclm.es
[4] Department of Electrical Engineering, Universidad Politécnica de Valencia, 46022 Valencia, Spain; tagarsan@die.upv.es
* Correspondence: angel.molina@upct.es; Tel.: +34-968-32-5462

Received: 15 March 2019; Accepted: 26 April 2019; Published: 4 May 2019

Abstract: An alternative approach for combined frequency control in multi-area power systems with significant wind power plant integration is described and discussed in detail. Demand response is considered as a decentralized and distributed resource by incorporating innovative frequency-sensitive load controllers into certain thermostatically controlled loads. Wind power plants comprising variable speed wind turbines include an auxiliary frequency control loop contributing to increase total system inertia in a combined manner, which further improves the system frequency performance. Results for interconnected power systems show how the proposed control strategy substantially improves frequency stability and decreases peak frequency excursion (nadir) values. The total need for frequency regulation reserves is reduced as well. Moreover, the requirements to exchange power in multi-area scenarios are significantly decreased. Extensive simulations under power imbalance conditions for interconnected power systems are also presented in the paper.

Keywords: wind integration; demand response; frequency control; ancillary services

---

1. Introduction

The integration of renewable energy sources into power system has stressed system operation by causing balancing resources to cycle more frequently, and generating ramps of critical steepness or duration. Flexibility requirements increase strongly in power systems with combined wind and PV (photovoltaics) contribution of more than 30% of total energy and a share of PV in the renewables mix above 20–30% [1]. Nowadays, more than 140 countries currently have renewable energy targets in place. For example, the European Union (EU) has set targets to achieve a 37% renewables share in overall energy use, which could lead to renewable power generation shares in the range of 51–68% [2]. Under this scenario, maintaining a close balance between generated and demanded active power becomes crucial to guarantee power system security and stability, keeping grid frequency within certain intervals—less than $\pm 1\%$ of the nominal value for European power systems [3]. Traditionally, conventional supply-side units are equipped with primary and secondary frequency control systems [4], using the demand-side response to restore the balance only under severe instability conditions [5].

The primary frequency control (PFC) operates locally by means of a governor to modify, around a set-point, the mechanical power input of the supply-side units based on the local frequency deviation [6]. This control system, also known as the droop control, is decentralized with a timescale up to low tens of seconds and an initial rate of change determined by the rotating mass inertia of the power system. A new power balance and frequency grid stabilization is usually achieved, but does not in itself restore the nominal frequency. The main purpose of secondary frequency control (SFC), also called Automatic Generation Control (AGC) from the supply-side, is to balance the total system generation by recovering the global grid frequency and power interchanges among neighboring areas to their set-point values [7,8]. These unintended frequency deviations require reliable and fast-acting controllers to recover the grid frequency. However, many Control Areas (CAs) still adopt a simplified approach in the design of AGC, i.e., the conventional controls—Integral (I), Proportional Integral (PI), and Proportional Integral Derivative (PID). Although these gain controllers are simple to implement, their performance is not always satisfactory, being usually slow and presenting a lack of efficiency in handling system nonlinearities [9,10].

Over the years, the demand-side contribution to the power system started gaining considerable attention as a measure to obtain frequency and voltage regulation. Increased attention has been focused on demand response (DR), strongly motivated by the remarkable penetration of renewables into current power systems, particularly at the distribution level [11]. Loads, such as Thermostatically Controlled Residential Loads (TCRLs), can shift their demand over certain time intervals without compromising their performance and services. In fact, some authors considered that over 40% of residential appliances are compatible with load control strategies [12]. For this reason, TCRLs can be considered as ideal to be used in dynamic DR strategies [13,14]. Moreover, taking into account the large number of consumers and hence small loads connected to the grid, these strategies would improve resource utilization and subsequently would reduce supply-side capacity requirements [15]. Advantages and drawbacks of different load control and dynamic demand can be found in [16].

Most contributions in the last decade have proposed switching-off/on actions applied on TCRLs when frequency variations exceed certain limits [17–24]. However, these works have been mainly focused on PFC. In [17], a simple and optimal control strategy is proposed to modulate the customer load as a linear function of the frequency excursions. Short et al. [18] analyzed how a certain degree of frequency stability could be achieved by integrating dynamic demand controllers into fridges/freezers. These devices monitor the grid frequency and switch-off/on appliances accordingly, while achieving a trade-off between appliance requirements and the grid. Scenarios with high penetration of wind energy are also discussed. Samarakoon et al. [19] described a frequency-based load control scheme for primary frequency response purposes by using smart meters. Loads are grouped according to their relevance for the customer. When the grid frequency falls below the nominal value, each load controller is switched-off for a specific time depending on the frequency excursion. In [20], an experimental platform is proposed by using commercially available smart meters. These appliances are remotely controlled through smart sockets to evaluate the load blocking strategies. In [21], a decentralized approach for using TCRLs is proposed. The authors affirmed that a two-way communication between loads and the control center is not essential when the number of individual loads is considerably large. The value of Dynamic Demand (DD) concept is quantified in [22], enabling domestic refrigeration appliances to contribute to primary frequency regulation through an advanced stochastic control algorithm. In [23], a comprehensive central DR algorithm for primary frequency regulation is described in a smart micro-grid. Contributions for transient studies can be found in [24], where a systematic method to re-balance power and resynchronize bus frequencies after a disturbance with significantly improved transient performance is described. Recently, we discussed DR strategies applied to PFC by including auxiliary frequency control carried out by Wind Farms (WFs) [25]. The work focuses on evaluating the two control actions counteracting frequency deviations as well as their compatibility.

During the last years, the high integration of wind resource into the global energy mix has required an important reformulation of wind power plant services, including their contribution to the frequency control [26,27]. These requirements are regularly updated and often include very rigid criteria, particularly in power systems with a relevant presence of wind power plants, where difficulties in maintaining the grid frequency within an acceptable range emerge as an additional concern under large wind power fluctuations [28,29]. In addition, some authors affirmed that the WT inertia contribution to the total kinetic energy stored in the power systems is considerably less significant than traditional power plants [30,31] and, subsequently, larger frequency deviations will be suffered by the systems after sudden generation or demand variations [32]. Moreover, systems with reduced total inertia experience a sharper immediate frequency drop under imbalance and, thus, are more vulnerable and sensitive to involuntary under-frequency load shedding [33]. Due to this scenario, alternative resources connected to the grid, mainly PV solar installations and wind power plants, are required to provide ancillary services [34]. With this aim, a frequency-dependent control loop is proposed in [35] for Variable Speed Wind Turbines (VSWTs) to improve frequency response and provide an active contribution to the frequency control. This additional controller synthesizes virtual inertia for VSWTs —i.e., kinetic energy stored in their rotating masses—that can be provided at the beginning of a frequency deviation event, diminishing its impact [35]. In this way, Spanish wind power plants are able to participate in ancillary services by a regulation framework issued by the Spanish Secretary of State for Energy [36], which made it legally possible since February 2016.

By considering previous works, this paper analyzes the demand-side contribution to SFC as an additional support to the frequency control strategy proposed in [21,25]. In line with these previous works, a decentralized demand-side solution is proposed to avoid the cost and complexity associated with two-way communications between many loads and the control center. The frequency-responsive load controller is thus extended by considering an additional integral-action function. This function adjusts the thermostat temperature of thermostatically controlled loads based on local frequency estimates. As a result, their power demand profiles are restructured, thus achieving a reduction (or increase) of their energy consumed. In this way, the controller is able not only to modify load's instantaneous power consumption, but also their energy demand during a specific time interval. This innovative load controller operates autonomously and provides a decentralized solution where individual loads are randomly distributed and connected to the grid. A two-area interconnected power system is simulated under severe wind power fluctuations to assess the proposed decentralized solution. An auxiliary frequency controller for VSWTs is also included to combine the contribution of VSWT's inertia to maintain the balance in future power systems with high wind power plant integration.

The rest of the paper is structured as follows: the implemented two-area interconnected power system is described in Section 2, including WF and demand-side modeling. The contribution of demand-side to SFC is discussed in Section 3. Extensive results are provided and widely discussed in Section 4. Finally, the conclusion is given in Section 5.

2. Power System Modeling: Wind Power Plant and Demand-Side Contribution to Frequency Control

2.1. Preliminaries

For studying the dynamic response of power systems dealing with frequency variations, traditional system modeling has been based on the following per unit (pu) expression [5]:

$$\Delta P_G - \Delta P_L = D\Delta f + 2H \frac{d\Delta f}{dt}, \tag{1}$$

where $\Delta P_G - \Delta P_L$ is the power imbalance, Δf is the difference between instantaneous and nominal system frequency and D, referred to as the damping factor, is the load dependence on frequency. Assuming that grid frequency can be considered as constant over large interconnected areas, the generating units can be then combined into an equivalent rotating mass M (being $M = 2H$ and H the inertia constant expressed in seconds). Likewise, loads are grouped and considered as an equivalent load, being D their equivalent damping factor [37].

For multi-area analysis purposes, the previous expression can be extended to different areas. With this aim, Figure 1 shows a general scheme of two-area interconnected power system, including primary and secondary frequency control by conventional generation units. This multi-area general scheme is considered in this paper, where a grid frequency and a system dynamic response characterizes each area. A high-impedance (elastic) transmission line is proposed to connect both areas [5]. Conventional generating units—highlighted with dashed lines—are modeled by considering Generation Rate Constraint (GRC) and speed Governor Dead-Band (GDB). The GRC is modeled by considering an "open loop" method and by adding two limiters bounded by ±0.0017 (pu/s) within the generating units [38]. The GDB is defined as the total magnitude of a sustained speed change within which there is no change in the turbine valve position. The GDB transfer function model can be found in [38], assuming that the GDB of the 0.06% backlash type can be linearized in terms of change and the rate of change in the speed. Under the presence of certain nonlinearities and constraints, such as GRC and GDB, system dynamic responses can present significant overshoots and long settling times for frequency and tie-line power oscillations [39,40].

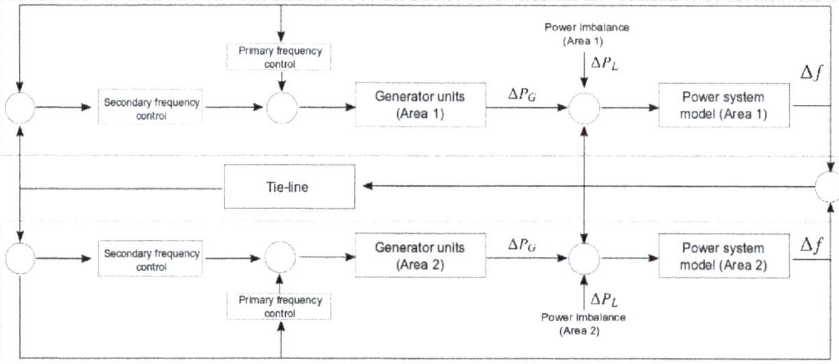

Figure 1. General scheme of two-area interconnected power system.

If one area is under power imbalance, the governor control mechanisms of both areas modify the power system to recover power balance. Grid frequency deviation, Δf, is used as input signal for primary and secondary frequency control. PFC is performed locally at the generator, being the active power increment/decrement proportional to Δf through the speed regulation parameter R, defined as the ratio of ΔP and Δf. Power flow control at the tie-line and damping of tie-line power oscillations are required for effective active power generation and frequency control. In this respect, SFC involves an integral controller modifying the turbine set-point of each area. A linear combination of both tie-line power errors between neighboring areas, $\Delta P_{tie,ij}$, and local frequency variations in each area, Δf_i, is used as the input to the corresponding integral controllers, which is called the Area Control Error (ACE) [41]. ACE in each area is then defined as follows [41]:

$$ACE_i = \sum_{j=1}^{n} \Delta P_{tie,ij} + \beta_i \Delta f_i, \tag{2}$$

where the suffix i refers to the CA and j to the generator number; β_i is the bias coefficient; f_i is the grid frequency of the i-CA; and $\Delta P_{tie,ij}$ is the actual value of the interchange power between i-CA and j-CA. Further information regarding tie-line bias control applicability to load frequency control for multi-area interconnected power systems can be found in [42]. The dynamic performance of the AGC system thus depends on frequency bias factor β_i, in MW/Hz, and the integral controller gain value K_I. Optimal values of K_I and β are estimated by means of an Integral Squared Error (ISE) technique,

$$ISE = \int_0^T \left(\Delta P_{tie}^2 + (\beta_1 \Delta f_1)^2 + (\beta_2 \Delta f_2)^2 \right) dt, \tag{3}$$

where Δf_1 and Δf_2 are frequency deviations in Areas 1 and 2, respectively [43,44]. According to the specific literature, two different power systems were simulated: *(i)* the two-area interconnected power system proposed in [10]; and *(ii)* a two-area wind farm integrated power system based on [43–45]. Figure 2 depicts the initial two-area interconnected power system considered for the simulation by including the corresponding block diagrams.

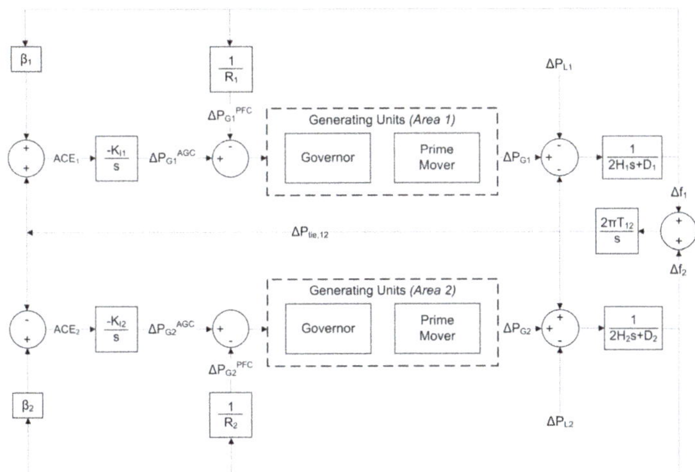

Figure 2. Block diagram of two-area interconnected power system.

2.2. Wind Power Modeling

Accounting for the relevance of wind power integration, two WFs containing VSWTs are explicitly considered in the power system model. Averaged time-varying wind speed profiles are applied to both wind power plants to evaluate wind active power fluctuations on grid frequency stability.

Primary frequency response of WFs is improved by including an additional inertia controller. This controller allows VSWTs to contribute to the system inertia through an active power regulation under frequency deviations [25,35],

$$P_{el}^{correction} = k_{WFs} \cdot \frac{I_{WT}}{2} \frac{df}{dt}, \tag{4}$$

where k_{WFs} is the virtual inertia factor and I_{WT} is the WT inertia (pu). Larger k_{WFs}-factors would improve WT contribution, maintaining WT rotational speed values within acceptable ranges. Figure 3 shows this additional virtual controller depending on the RoCoF (df/dt). An additional control loop that synthesizes virtual inertia is implemented for the VSWTs to contribute to the system's total inertia and thus to the system frequency control. The output power of the WF, $P_{el_{(pu)}}^{ref}$, is modified by an additional power $P_{el_{(pu)}}^{correction}$ depending on the RoCoF, df/dt, [30,46]. The global power of the WFs ($\Delta P_{WFs_{(pu)}}$) depends on the number of WT (N_{WTs}) and the VSWTs power set-point ($P_{0,WT}$) [47,48]. The proposed approach is thus based on a modified inertial control scheme, involving quick response through power-electronic circuits and subsequently providing frequency support from the rotational mass kinetic energy.

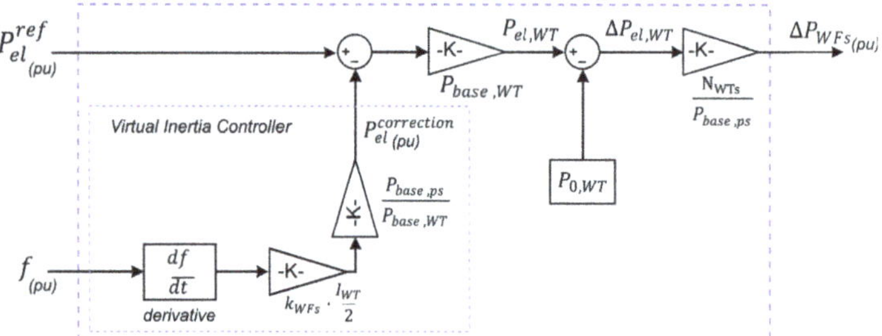

Figure 3. Virtual Inertia Block (VIB) diagram of the virtual frequency controller.

2.3. Demand-Side Modeling

Even thoguh demand-side was initially considered as non-controllable, nowadays, most authors agree with the idea that the devices can be categorized into controllable or uncontrollable [49]. Controllable loads are then defined as the loads suitable for deferral or shifting their power demand to off-peak periods without harming household's convenience, such as heating, ventilation, and air conditioning [50,51]. Because of their thermal properties, these devices can adapt their operation through on/off signals without adversely affecting their temperature requirements. The rest of the loads are considered uncontrollable, i.e., their use is completely determined by the end user. Assuming uncontrollable loads to be modeled by an equivalent load, their global power depends on frequency excursions through parameter D (equivalent damping factor), as discussed in Section 2.1.

In the proposed solution, controllable loads are modeled individually taking into account that their individual power demand can be modified by the following controllers: *(i)* a proportional frequency-dependent controller able to modify the active power demanded by the device and thus emulating PFC from supply-side; and *(ii)* an integral-action controller able to modified the thermostat individual load settings based on the ACE signal. The proportional frequency-dependent controller was previously described by the authors and discussed in [21,25]. This approach modifies the controllable load demand through forced switching-off/on actions. To operate under forced disconnections, frequency excursions exceed a predefined threshold, Δf, for a certain time (see Figure 4). Larger frequency deviations imply faster frequency control responses, thus entering the *Control Region*. Small frequency changes are maintained during longer time intervals, delaying the controller actions. Figure 4 also gives an example of Δf-*time* feature for a given load and its corresponding demand profiles in the event of a linear under-frequency excursion. As can be seen, when Δf-*time* boundaries are exceeded, load is forcibly switched-off for a predefined time period (*OFF-forced*). After a short recovery interval of forced connection (*ON*) to maintain the temperature within an acceptable range, and given that the frequency does not recover its rated value thus continuing within the *Control Region*, the load controller restarts a new cycle of forced disconnection. These *ON-* and *OFF-forced* time periods are designed keeping both appliance requirements in view and considering temperature limits that may be almost imperceptible by the customers. Actually, most controls for conditioning spaces and typical thermostat-controlled devices operate on a time scale of 10–15 min [52], which is in line with the SFC timescales. The proposed integral-action controller is thus based on thermostat temperature changes according to comfort level constraints and the expected demand-side contribution to SFC. Further discussion about thermal inertia, heat–cool flow rates and SFC can be found in Section 3.

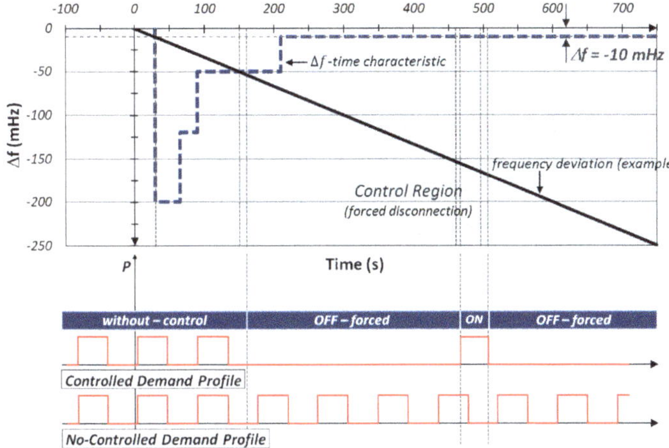

Figure 4. Frequency-responsive load controller: example of demand-side contribution to PFC and Δf-*time* characteristic.

Figure 5 shows an extended block diagram by including wind power generation frequency control (ΔP_{WFs}) and demand-side contribution to PFC and SFC considered in Area 1 (ΔP_{DS}^{SFC}). As can be seen, supplementary power due to wind power generation frequency control, ΔP_{WFs}, is added to the conventional generation response (prime movers), ΔP_{G1}, to provide an additional generation to the supply-side. The demand-side contributions to PFC and SFC are referred to as ΔP_{DS}^{PFC} and ΔP_{DS}^{SFC}, respectively. This multi-area power system model is an extension of the block diagram depicted in Figure 2, by including frequency control to both supply-side—wind power plants—and demand-side.

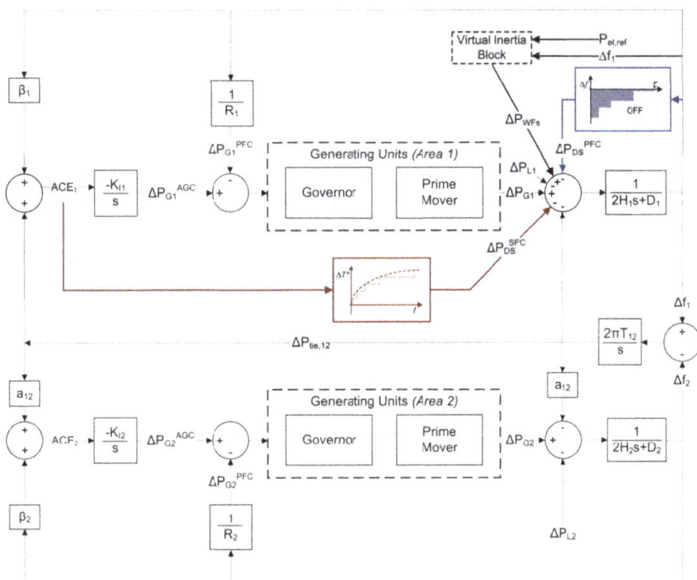

Figure 5. Two-area interconnected power system. WF and demand-side contribution to frequency control.

3. Demand-Side Contribution to SFC: Proposed Solution

3.1. General Description

The implementation of demand-side contribution to SFC is based on adding an integral-action function on the frequency controller proposed by the authors in [25]. By considering two or more areas, this integral-action controller gives a proportional value in line with the accumulated ACE of the corresponding area, i.e., the linear combination of both tie-line power errors (ΔP_{tie}) and local frequency excursions (Δf),

$$\Delta T_{set-point} = k \int (\Delta P_{tie} + \beta \Delta f)dt, \tag{5}$$

where k-parameter (constant of proportionality) is estimated depending on the expected frequency deviations (Δf), the time interval remaining, the frequency excursion and the comfort level ranges allowed by the customers. Figure 6 shows the controller block diagram implemented in *Matlab-Simulink* environment.

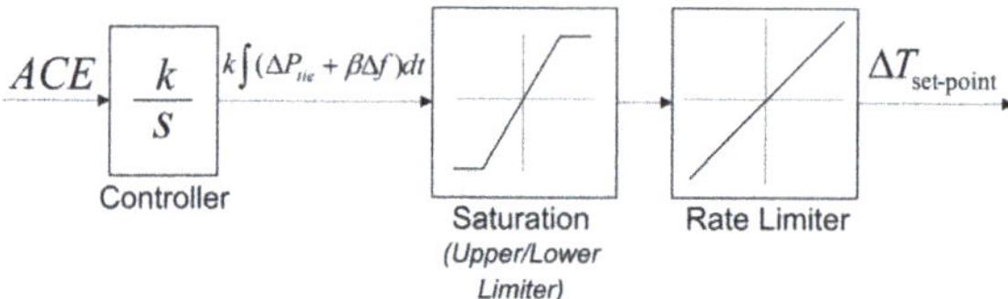

Figure 6. Demand-side contribution to SFC: controller block diagram.

Assuming that controllable loads operate according to their thermostat settings, TCRLs' normal operation patterns are then modified by changing their temperature set-points, $\Delta T_{set-point}$. Recovery to grid frequency deviations can be achieved by removing the residual steady-state error according to both AGC from the supply-side and Equation (5) from the demand-side. Communication is thus needed between the dispatch center and appliances when two or more areas are considered in the power system model. Recent contributions evaluate the impact on power system dynamic of errors in AGC systems [53]. Further discussion can be found in Section 3.2.

3.2. Implementation of Integral-Action Controller

For selecting the optimal value of k-parameter, three variables are considered. (i) *frequency deviation*: According to the European Standard EN–50160 [54], it is stated that the nominal value of grid frequency in most European countries, Asia and Africa is 50 Hz. It also provides the acceptable ranges for frequency variations: for interconnected supply systems under normal operating conditions, the averaged value of grid frequency measured over 10 s must be within a range of ±1% (49.5–50.5 Hz) for 99.5% of a week [55]. The proposed frequency control responds proportionally to the severity of the frequency excursion, and thus a proportional number of loads are called to switch-on/off depending on the frequency excursion to avoid undesirable over-frequency values. Subsequently, the number of loads is proportionally distributed taking into account that 100% of loads are required to support frequency control when the deviation is higher than 500 mHz and then grid frequency is kept within a range of ±1% (49.5–50.5 Hz), as previously commented. (ii) *duration of the frequency deviation*: The frequency deviation is considered to remain for at least 5 s, to be in line with the frequency excursion events emulated in the aforementioned contributions. (iii) *maximum allowable temperature variation for the thermostat set points*: Firstly, controllable loads are categorized into different groups according to their usage patterns and load profiles: *(Load-Group I)* including fridges/freezers; *(Load-Group II)* with air-conditioners/heat pumps; and *(Load-Group III)* corresponding to electric water heaters that have a great potential to store energy in advance [56]. In [57], load profiles of selected major household appliances in the U.S. are discussed in detail. When considering *Load-Group I*, and on the basis of technical specifications from different appliances manufacturers, the maximum allowable temperature variation range for the fridge/freezer thermostat set point is set to +2 °C. For *Load-Group II*, air-conditioners/heat pumps, a variation range of ±3 °C is assumed acceptable for the customers, i.e., +3 °C for cooling mode and −3 °C for heating mode. Examples of cooling devices in demand response can be found in [58]. Finally, −6 °C temperature variation range is considered for *Load-Group III*. Hence, *k*-parameters are determined separately for each individual group of loads as follows:

$$\begin{pmatrix} k_{\text{Load-Group I}} & = -0.8 \,°\text{C}(\text{Hz} \cdot \text{s}), \\ k_{\text{Load-Group II}} & = -1.2 \,°\text{C}(\text{Hz} \cdot \text{s}), \\ k_{\text{Load-Group III}} & = +5.0 \,°\text{C}(\text{Hz} \cdot \text{s}). \end{pmatrix} \tag{6}$$

The value of k is thus estimated taking into account tolerances of the various frequency excursion phenomena that may occur on the mains (almost 95% of cases included). However, given the potential variability of the frequency deviations, in both magnitude and duration, the output of the integral-action controller needs to be limited. For example, in the case of large generation outages, the frequency deviation occurs too rapidly for the frequency containment reserves to be effectively activated and prevent the frequency from reaching values below and above these tolerances. This situation would cause the controllers to order unacceptable changes for temperature set points. For this reason, upper and lower saturation limits and rising and falling slew rates are included in the implemented model. Additionally, to avoid a synchronized and massive response of the loads, and consequently undesired frequency oscillations, a linear density probability function is implemented on each load controller to decide in a distributed manner when an individual load participates (or not) in the demand-side response. Therefore, the percentage of controllable loads varies gradually depending on the severity of the frequency excursion, and thus emulating proportional response of classical turbine-generator governors subjected to under-frequency excursions. The load decision to be involved in frequency control (or not) is then determined individually by each controllable load attending to: (*i*) the frequency excursion along the time; (*ii*) the severity of such frequency deviations (the more severe is the frequency excursion, the higher is the number of controlled loads to participate in the demand response); and (*iii*) the own load thermal characteristics that are in line with the off-time period ranges allowed by the customers. The decentralized solution can thus be used for practical applications, in which it may need a huge number of appliance switches.

Figure 7 shows both thermal and power demand response of controllable loads subjected to an under-frequency event. Their power demand profiles are also included by considering the load controller proposed in Section 3.1. The thermal set-point value is modified according to Equation (5). In this case, we consider $\Delta P_{tie} = 0$, and, thus, $\Delta T_{set-point}$ is estimated proportional to the accumulated deviation between the reference grid frequency and the measured value, according to Equation (5). $\Delta T_{set-point}$ represents the command sent by the individual load controller to the appliance after applying upper/lower and ramp limiters. As can be seen, within the first 25 s after the disturbance, the ramp slope is limited to a maximum rate of 0.08 °C per second. At the end of the simulation, the maximum temperature variation is limited to 2 °C. Moreover, this command has also been discretized in steps of 0.5 °C, i.e., $\Delta T_{set-point}$ *discretized*. The larger is the variation of the temperature set point, the lower is the duty-cycle of the load, thus decreasing power consumption and relieving power system reserves.

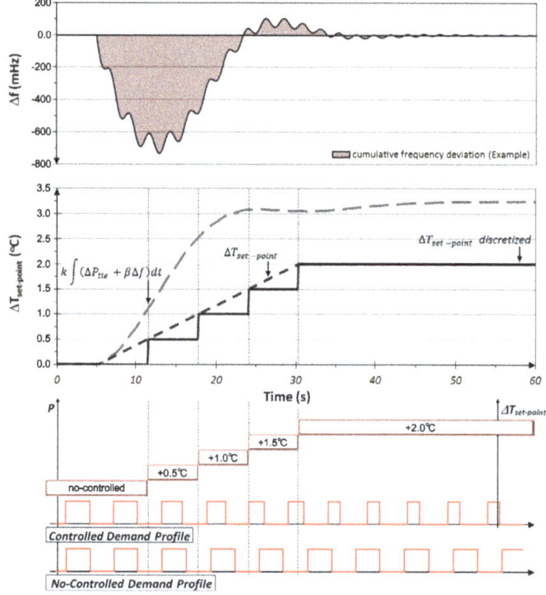

Figure 7. Frequency-response load controller: example of demand-side contribution to SFC.

4. Simulation and Results

4.1. Preliminaries

The most relevant data of the two implemented power systems were referred to an output power of 1200 MW and 1000 MW, with grid frequency nominal values of 50 and 60 Hz, respectively. In reference to wind power generation, both wind power plants had a rated capacity of 100 MW. Further information about the wind turbines can be found in Appendix A (see Table A2). Additional parameters, such as pitch control gains, are also included. Regarding controllable loads, and assuming that the household sector represents about 30% of total final electricity consumption [12,59], the percentage of controllable loads could then be estimated at around 10% of the global power demand. Recent contributions affirm that, averaged across all countries, space cooling accounted for around 14% of peak demand [60].

Controllable loads were divided into different groups attending to their thermal behaviors. Subsequently, different thermal models were used to estimate their frequency control parameters. In this way, fridges/freezers and electric water heaters were modeled according to Shaad et al. [61], providing different approaches for identification of single-zone lumped parameter thermal models. A direct load control algorithm was used to estimate and forecast the temperature and water usage for each individual water heater. Air-conditioners were modeled based on a detailed energy balance proposed by the authors in [62]. The model has been previously assessed for load management applications under different load performances and conditions. By considering these models, as well as k-parameters determined in Equation (6), Table A3 (see Appendix A) summarizes the main configuration parameters according to the different thermal behaviors and customer uses.

4.2. Results

Computer simulations under different operating conditions were carried out using *Matlab-Simulink* environment to evaluate the suitability of both additional control actions counteracting frequency deviations: (i) Two different step load disturbances were considered for Control Area 1

(CA-1), $\Delta P_{L1} = 0.05$ and 0.075 pu; both imbalances remained for 10 min and they were scheduled to disappear after these 10 min of simulation. Control Area 2 remained under balance conditions and non-additional frequency excursions from CA-2 were included in the simulation scenarios. According to the Spanish TSO, different severe situations were collected during the last decade that are in line with the proposed scenarios. Some imbalances were due to wind power curtailments and wind speed oscillations, such as -1.547 MW/min for a 10 min time interval (28 February 2014). Other imbalances were due to special situations, for example a decreasing of 2000 MW for 15 min (3 December 2007) accounting for more than 10% of the power demand. (ii) Different values of the virtual inertia k_{WFs}-factor were also applied: $k_{\mathrm{WFs}} = 0$, 5 and 10. The analysis thus focused on responses under relevant imbalances in one control area, being frequency response under normal operation conditions out of the scope of this paper. A recent discussion including normal operation analysis can be found in [63].

Figure 8 shows the additional wind power contribution to the power system submitted to an under-frequency excursion. The virtual frequency controller made the WTs behave similarly to a conventional synchronous generator during the event, injecting a temporary extra power into the grid. As can be seen, larger k_{WFs}-factor values implied higher amounts of additional power provided by the WTs. Nevertheless, k_{WFs}-factor values had to be within a range of active power temporarily achievable by the WTs according to different constraints. Indeed, excessive k_{WFs}-factor values might lead to additional power values not allowed to be provided by the WTs. Therefore, the k_{WFs}-factor selection process can be considered as a trade-off between an additional inertia provided by the wind resource, i.e., an additional power injected into the power system, and the WT rotational speed deceleration. Contributions focused on low inertia system operation and the relevance of wind turbines to mitigate frequency deviations can be found in [64,65] and in [66] for isolated power systems.

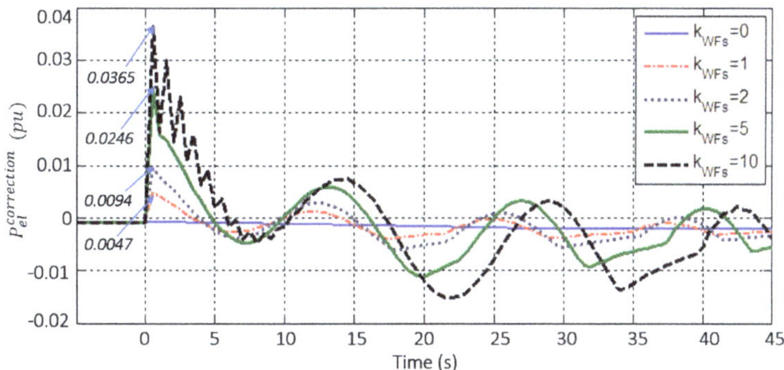

Figure 8. VSWT responses under frequency excursion for different values of the virtual inertia factor (k_{WFs}) and $\Delta P_{L1} = 0.05$.

Figure 9 shows the CA-1 frequency response when demand-side contribution to frequency control was applied (or not), i.e., $\Delta P_{DS} \neq 0$ and $\Delta P_{DS} = 0$, for a series of virtual inertia factor values (k_{WFs}). Results with $\Delta P_{DS} \neq 0$ correspond to simulations where the proposed demand-response frequency control was implemented under different inertia factor values, being $k_{\mathrm{WFs}} = 0$ a scenario where only demand-response was included. According to these results, the maximum value for under-frequency deviation was significantly reduced by 16.6% when only demand-response was considered ($\Delta P_{DS} \neq 0$, $k_{\mathrm{WFs}} = 0$). With regard to virtual inertia factors, larger k_{WFs}-factors implied smaller RoCoF values, as a consequence of the additional inertia provided by the WTs to the power system. In this way, decreases of 22.5% and 23.2% were obtained for $k_{\mathrm{WFs}} = 5$ and 10, respectively, in comparison to other scenarios without demand-side and WT contribution ($\Delta P_{DS} = 0$, $k_{\mathrm{WFs}} = 0$). These aforementioned reduction percentages, i.e., 16.6%, 22.5% and 23.2%, corresponded to a participation share of the controlled load from 40.5% to 44.6%, based on the total available controllable loads. As introduced in

Section 3.2, this percentage of controllable loads varied gradually depending on the severity of the frequency excursion to avoid over-frequency values. In terms of the supply-side frequency response, Figure 10 compares the extra power provided by the CA-1 conventional generating units during the frequency excursion for the different scenarios. This additional active power delivered by conventional generation was significantly reduced when demand and WT frequency response were considered, i.e., maximum peak value was reduced from 0.092 to 0.076 pu. Therefore, this reduction in active power was significant for the presence of demand-side response $\Delta P_{DS} \neq 0$. Additional benefits from virtual inertia factor values ($k_{WFs} \neq 0$) could be achieved by including WT frequency control. This last benefit depended on the wind power capacity considered in the proposed power system. Nevertheless, larger k_{WFs}-factor also reduced the power reserves from the conventional generation units.

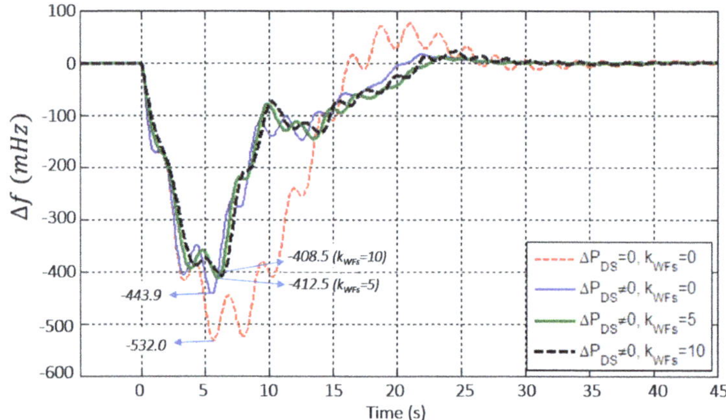

Figure 9. Grid frequency variation for CA-1 (Δf_1) and $\Delta P_{L1} = 0.05$. Comparison of conventional power plant frequency response ($\Delta P_{DS} = 0$, $k_{WFs} = 0$.) vs. demand-side ($\Delta P_{DS} \neq 0$) and VSWT contribution ($k_{WFs} \neq 0$).

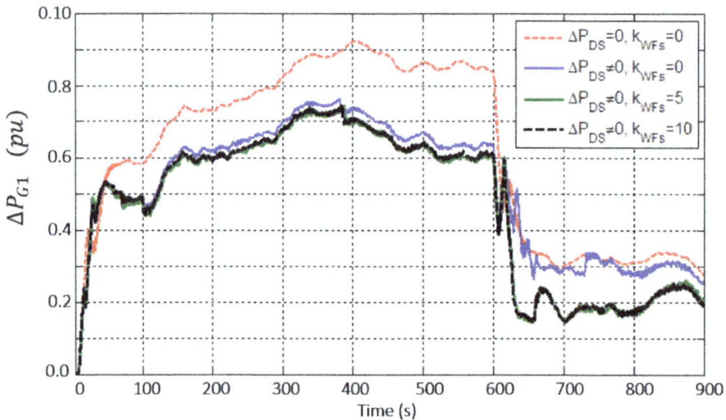

Figure 10. Conventional generation response for CA-1 (ΔP_{G1}) and $\Delta P_{L1} = 0.05$.

With regard to the response of the demand-side frequency control, the participation of the controlled load was linearly distributed from the beginning of the frequency excursion to 10 min. The load controller actions were then spread out along the time of simulation, i.e., 10 min after the disturbance, secondary control response of the controlled load was fully activated. Figure 11 summarizes the contribution to PFC from the controllable loads, ΔP_{DS}^{PFC}, for the different simulated scenarios. In line with the proposed load controller described in Section 2.3 (see Figure 4), the demand response was proportional to the frequency excursion, emulating the natural response of conventional generation units. It is worth pointing out that the demand-side contribution to PFC proportionally disappeared with the frequency deviation, as primary response was progressively replaced by the contribution of demand-side to SFC. Figure 12 shows the controlled load responses for the different frequency excursions as regards demand-side contribution to SFC, ΔP_{DS}^{SFC} according to Section 3. Consequently, SFC of controllable loads was implemented by introducing modifications in their thermostat set-point values through Equation (5). Controlled load thermostats were then readjusted, recovering progressively their initial temperature set point values. Figure 13 shows a box-plot of the temperature refrigerator set points, i.e., *Load Group I*, aiming to demonstrate that the temperature variation remained within acceptable limits during the secondary frequency control load participation.

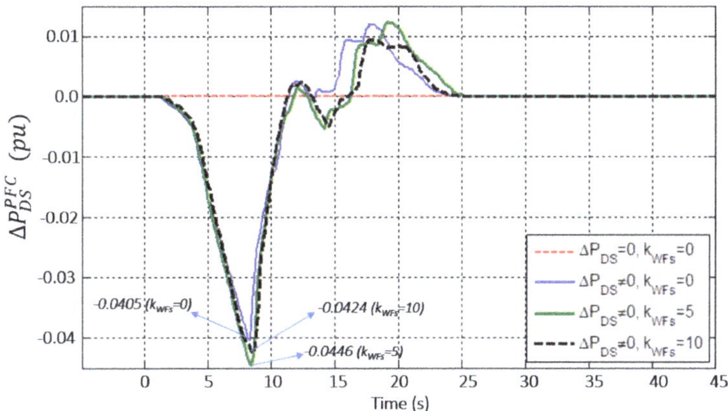

Figure 11. PFC response of controlled loads (ΔP_{DS}^{PFC}) when applied to CA-1 and for $\Delta P_{L1} = 0.05$.

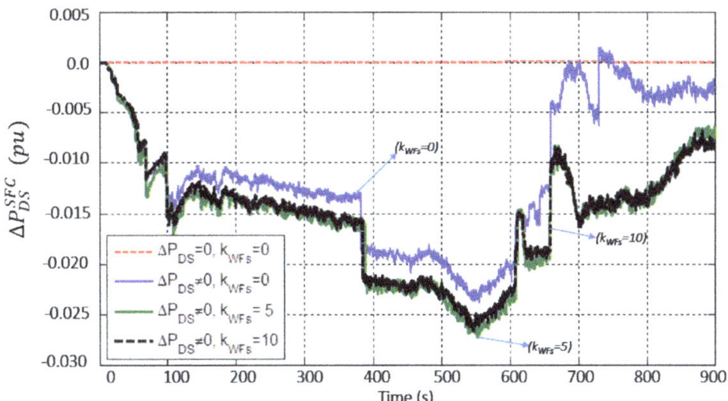

Figure 12. SFC response of controlled loads (ΔP_{DS}^{SFC}) when applied to CA-1 and for $\Delta P_{L1} = 0.05$.

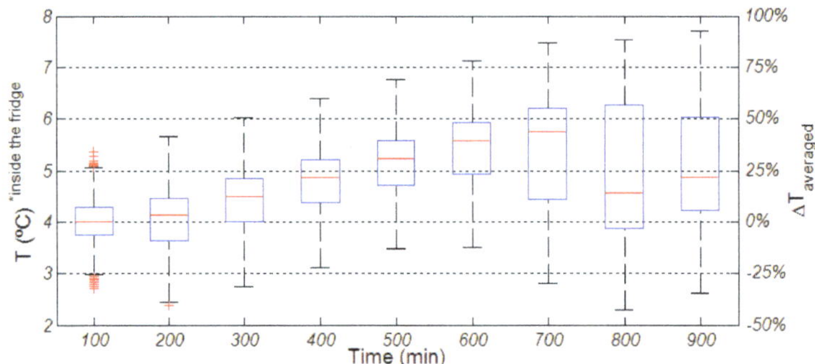

Figure 13. Temperature set point variation for *Load Group I*.

5. Conclusions

An alternative approach to improve dynamic performance of inter-connected power systems with relevant penetration of wind resource is discussed and evaluated. A new frequency control strategy is proposed by introducing demand-side contribution to frequency control, including primary and secondary response under frequency excursions. This demand-side contribution is carried out through the integration of frequency-sensitive load controllers into thermostatically controlled residential loads. Additionally, a supplementary control loop synthesizing virtual inertia for wind power plants is included as well, considering generation rate constraints and a governor dead-band. A combined frequency control strategy is thus proposed and evaluated.

According to the results, frequency deviations were significantly reduced in comparison to classical scenarios by including a combined solution with demand-response and wind power plant participation (peak frequency excursions were reduced by 23%). Similarly, the necessities of both primary and secondary regulation reserves from the supply-side could decrease significantly under power imbalance conditions. From the demand-side, minor effects on the controlled loads were allowed to satisfy minimum comfort levels required by the customers, with set-point temperature variations lower than 2 °C. Consequently, the results show a relevant reduction in supply-side reserve requirements (26.8%) and thus a suitable solution to integrate both controllable loads and wind power plants into the grid frequency stability.

Author Contributions: Data curation, I.M.-B.; Formal analysis, Á.M.-G. and E.G.-L.; Funding acquisition, E.G.-L.; Investigation, A.D.H.; Resources, T.G.-S.; Supervision, Á.M.-G.; Visualization, T.G.-S.; Writing—original draft, A.F.-G.; and Writing—review and editing, Á.M.-G.

Funding: This work was partially supported by by the Regional Seneca Foundation of Spain through the research projects 08747/PI/08 and 19379/PI/14. This work was also supported by project AIM, Ref. TEC2016-76465-C2-1-R (AEI/FEDER, UE).

Acknowledgments: This work was financially supported by the Regional Seneca Foundation of Spain through the research projects 08747/PI/08 and 19379/PI/14.

Conflicts of Interest: The authors declare no conflict of interest.

Abbreviations

The following abbreviations are used in this manuscript:

ACE	Area Control Error
AGC	Automatic Generation Control
CAs	Control Areas
DD	Dynamic demand
DR	Demand response
GDB	Governor Dead-Band
GRC	Generation Rate Constraint
I	Integral
ISE	Integral Squared Error
PFC	Primary frequency control
PI	Proportional Integral
PID	Proportional Integral Derivative
SFC	Secondary frequency control
TCRLs	Thermostatically Controlled Residential Loads
VSWTs	Variable speed wind turbines
WFs	Wind Farms
WT	Wind turbine

Appendix A

Power system and wind turbine parameters for simulations are shown in Tables A1 and A2. Thermal model parameters of controlled residential loads are shown in Table A3.

Table A1. Power system model parameters.

Parameter	Value [PS_1]	Value [PS_2]
Rotational Inertia, H_i [s]	5	7
Damping, D_i [puMW/Hz]	$8.333 \cdot 10^{-3}$	$8.333 \cdot 10^{-3}$
Tie-Line Power Rating, T_{12} [s]	0.0866	0.0866
Speed Regulation, R_i [Hz/puMW]	2.4	9
Integral Controller Gain, K_{Ii} [-]	0.28	0.28
Bias Factor, β_i [puMW/Hz]	0.125	0.150

Table A2. Wind turbine data.

Parameter	Value	Parameter	Value
Nominal power [MW]	2	Rotor inertia [kg $\cdot$ m^2]	$8.6 \cdot 10^6$
Type	VSWT	Generator inertia [kg $\cdot$ m^2]	150
Number of pole pairs	2	Pitch Controller, K_P [-]	2
Gear ratio [-]	96	Pitch Controller, K_I [-]	0.9091
Rotor diameter [m]	80		

Table A3. Thermal model parameters for residential electric loads.

Parameter	Group I	Group II	Group III
T_0, set-point [°C]	2–6	20–26	50–75
T_0, var(0.1) [°C]	4 *appliance*	23.5 *room*	50 *tank*
$\Delta T_{high}/\Delta T_{low}$ [°C]	0.75/-0.75	0.1/-0.1	1.5/-1.5
$t_{on}^{max}/t_{off}^{max}$ [min]	20/20	-/-	28/230

References

1. Huber, M.; Dimkova, D.; Hamacher, T. Integration of wind and solar power in Europe: Assessment of flexibility requirements. *Energy* **2014**, *69*, 236–246. [CrossRef]
2. Simbolotti, G.; Tosato, G.; Kempener, R. *Renewable Energy Integration in Power Grids*; Technical Report; International Renewable Energy Agency (IRENA): Abu Dhabi, UAE, 2015.
3. EURELECTRIC—ENTSO-E Joint Investigation Team. *Deterministic Frequency Deviations—Root Causes and Proposals for Potential Solutions*; EURELECTRIC: Brussels, Belgium, 2011.
4. Wen, T. Load frequency control: Problems and solutions. In Proceedings of the 2011 30th Chinese Control Conference (CCC), Yantai, China, 22–24 July 2011; pp. 6281–6286.
5. Gómez-Expósito, A.; Conejo, A.; Cañizares, C. *Electric Energy Systems: Analisys and Operation*; CRC Press: Boca Raton, FL, USA, 2009.
6. Vijayananda, W.M.T.; Samarakoon, K.; Ekanayake, J. Development of a demonstration rig for providing primary frequency response through smart meters. In Proceedings of the 45th International Universities Power Engineering Conference (UPEC), Cardiff, Wales, UK, 31August–3 September 2010; pp. 1–6.
7. Arghira, N.; Dumitru, I.; Fagarasan, I.; St. Iliescu, S.; Soare, C. Load frequency regulation in power systems. In Proceedings of the IEEE International Conference Automation Quality and Testing Robotics (AQTR), Cluj-Napoca, Romania, 28–30 May 2010; Volume 1, pp. 1–6.
8. European Network of Transmission System Operators for Electricity Union for the Coordination of Transmission of Electricity. *Policy 1—Load-Frequency Control and Performance*; Technical Report; ENTSOE–UCTE: Brussels, Belgium, 2009.
9. Gandhi, N.F.; Mohan, Y.K.; Rao, A.V. Load frequency control of interconnected power system in deregulated environment considering generation rate constraints. In Proceedings of the International Advances in Engineering, Science and Management (ICAESM) Conference, Nagapattinam, Tamil Nadu, India, 30–31 March 2012; pp. 22–25.
10. Chang-Chien, L.R.; Wu, Y.S.; Cheng, J.S. Online estimation of system parameters for artificial intelligence applications to load frequency control. *IET Gener. Transm. Distrib.* **2011**, *5*, 895–902. [CrossRef]
11. Siano, P. Demand response and smart grids—A survey. *Renew. Sustain. Energy Rev.* **2014**, *30*, 461–478. [CrossRef]
12. Cool Appliances. *Policy Strategies for Energy Efficient Homes*; Technical Report; International Energy Agency, IEA: Paris, France, 2003.
13. Ilic, M.D.; Popli, N.; Joo, J.Y.; Hou, Y. A possible engineering and economic framework for implementing demand side participation in frequency regulation at value. In Proceedings of the IEEE Power and Energy Society General Meeting, San Diego, CA, USA, 24–29 July 2011; pp. 1–7.
14. Jay, D.; Swarup, K.S. Frequency restoration using Dynamic Demand Control under Smart Grid Environment. In Proceedings of the IEEE PES Innovative Smart Grid Technologies—India (ISGT India), Kollam, Kerala, India, 1–3 December 2011; pp. 311–315.
15. Douglass, P.J.; Garcia-Valle, R.; Nyeng, P.; Ostergaard, J.; Togeby, M. Demand as frequency controlled reserve: Implementation and practical demonstration. In Proceedings of the 2nd IEEE PES International Innovative Smart Grid Technologies (ISGT Europe) Conference and Exhibition, Manchester, UK, 5–7 December 2011; pp. 1–7.
16. Darby, S. Load management at home: Advantages and drawbacks of some 'active demand side' options. *J. Power Energy* **2012**, *227*, 9–17. [CrossRef]
17. Trudnowski, D.; Donnelly, M.; Lightner, E. Power-System Frequency and Stability Control using Decentralized Intelligent Loads. In Proceedings of the 2006 IEEE PES Transmission and Distribution Conf. and Exhibition, Dallas, TX, USA, 21–24 May 2006; pp. 1453–1459. [CrossRef]
18. Short, J.A.; Infield, D.G.; Freris, L.L. Stabilization of Grid Frequency Through Dynamic Demand Control. *IEEE Trans. Power Syst.* **2007**, *22*, 1284–1293. [CrossRef]
19. Samarakoon, K.; Ekanayake, J. Demand side primary frequency response support through smart meter control. In Proceedings of the 44th International Universities Power Engineering Conference(UPEC), Glasgow, UK, 1–4 September 2009; pp. 1–5.
20. Samarakoon, K.; Ekanayake, J.; Jenkins, N. Investigation of Domestic Load Control to Provide Primary Frequency Response Using Smart Meters. *IEEE Trans. Smart Grid* **2012**, *3*, 282–292. [CrossRef]

21. Molina-García, A.; Bouffard, F.; Kirschen, D.S. Decentralized Demand-Side Contribution to Primary Frequency Control. *IEEE Trans. Power Syst.* **2011**, *26*, 411–419. [CrossRef]
22. Aunedi, M.; Kountouriotis, P.A.; Calderon, J.E.O.; Angeli, D.; Strbac, G. Economic and Environmental Benefits of Dynamic Demand in Providing Frequency Regulation. *IEEE Trans. Smart Grid* **2013**, *4*, 2036–2048. [CrossRef]
23. Pourmousavi, S.A.; Nehrir, M.H. Real-Time Central Demand Response for Primary Frequency Regulation in Microgrids. *IEEE Trans. Smart Grid* **2012**, *3*, 1988–1996. [CrossRef]
24. Zhao, C.; Topcu, U.; Li, N.; Low, S. Design and Stability of Load-Side Primary Frequency Control in Power Systems. *IEEE Trans. Autom. Control.* **2014**, *59*, 1177–1189. [CrossRef]
25. Molina-García, A.; Muñoz Benavente, I.; Hansen, A.; Gomez-Lazaro, E. Demand-Side Contribution to Primary Frequency Control With Wind Farm Auxiliary Control. *IEEE Trans. Power Syst.* **2014**, *29*, 2391–2399. [CrossRef]
26. Tsili, M.; Papathanassiou, S. A review of grid code technical requirements for wind farms. *IET Renew. Power Gener.* **2009**, *3*, 308–332. [CrossRef]
27. Ciupuliga, A.; Gibescu, M.; Fulli, G.; Abbate, A.; Kling, W. Grid Connection of Large Wind Power Plants—A European Overview. In Proceedings of the 8th International Workshop on Large-Scale Integration of Wind Power into Power Systems and Transmission Networks for Offshore Wind Farms, Bremen, Germany, 14–15 October 2009; pp. 349–359.
28. Doherty, R.; Mullane, A.; Nolan, G.; Burke, D.; Bryson, A.; O'Malley, M. An Assessment of the Impact of Wind Generation on System Frequency Control. *IEEE Trans. Power Syst.* **2010**, *25*, 452–460. [CrossRef]
29. Klempke, H.; McCulloch, C.; Wong, A.; Piekutowski, M.; Negnevitsky, M. Impact of high wind generation penetration on frequency control. In Proceedings of the 2010 20th Australasian Universities Power Engineering Conference (AUPEC), Christchurch, New Zealand, 5–8 December 2010; pp. 1–6.
30. Lalor, G.; Mullane, A.; O'Malley, M. Frequency control and wind turbine technologies. *IEEE Trans. Power Syst.* **2005**, *20*, 1905–1913. [CrossRef]
31. Mullane, A.; O'Malley, M. The Inertial Response of Induction-Machine-Based Wind Turbines. *IEEE Trans. Power Syst.* **2005**, *20*, 1496–1503. [CrossRef]
32. Lalor, G.; Ritchie, J.; Rourke, S.; Flynn, D.; O'Malley, M.J. Dynamic frequency control with increasing wind generation. In Proceedings of the IEEE Power Engineering Society General Meeting, Denver, CO, USA, 6–10 June 2004; pp. 1715–1720. [CrossRef]
33. Kou, G.; Till, M.; Bilke, T.; Hadley, S.; Liu, Y.; King, T. Primary Frequency Response Adequacy Study on the U.S. Eastern Interconnection Under High-Wind Penetration Conditions. *IEEE Power Energy Technol. Syst. J.* **2015**, *2*, 125–134. [CrossRef]
34. Johnstone, N.; Hascic, I. Increasing the penetration of intermittent renewable energy: Innovation in energy storage and grid management. In *OECD, Energy and Climate Policy: Bending the Technological Trajectory*; OECD Publishing: Paris, France, 2012; pp. 87–103.
35. Margaris, I.D.; Papathanassiou, S.A.; Hatziargyriou, N.D.; Hansen, A.D.; Sorensen, P. Frequency Control in Autonomous Power Systems With High Wind Power Penetration. *IEEE Trans. Sustain. Energy* **2012**, *3*, 189–199. [CrossRef]
36. *BOE n⁰ 312. Resolution of December 23, 2016, the General Secretariat of Energy, by the Rules for the Electricity Market*; Technical Report; Ministerio de Industria, Turismo y Comercio: Madrid, Spain, 2015. (In Spanish)
37. Wood, A.J.; Wollenberg, B.F. *Power Generation, Operation and Control*; John Wiley & Sons: New York, NY, USA, 1996.
38. Morsali, J.; Zare, K.; Hagh, M. Appropriate generation rate constraint (GRC) modeling method for reheat thermal units to obtain optimal load frequency controller (LFC). In Proceedings of the 5th Conference on Thermal Power Plants (CTPP), Tehran, Iran, 10–11 June 2014; pp. 29–34. [CrossRef]
39. Velusami, S.; Chidambaram, I.A. Decentralized biased dual mode controllers for load frequency control of interconnected power systems considering GDB and GRC non-linearities. *Energy Convers. Manag.* **2007**, *48*, 1691–1702. [CrossRef]
40. Bhatt, P.; Roy, R.; Ghoshal, S.P. GA/particle swarm intelligence based optimization of two specific varieties of controller devices applied to two-area multi-units automatic generation control. *Int. J. Electr. Power Energy Syst.* **2010**, *32*, 299–310. [CrossRef]

41. Rasolomampionona, D.D. A modified power system model for AGC analysis. In Proceedings of the IEEE Bucharest PowerTech, Bucharest, Romania, 28 June–2 July 2009; pp. 1–6.

42. Chen, C.; Zhang, K.; Yuan, K.; Teng, X. Tie-Line Bias Control Applicability to Load Frequency Control for Multi-Area Interconnected Power Systems of Complex Topology. *Energies* **2017**, *10*. [CrossRef]

43. Sheikh, M.R.I.; Muyeen, S.M.; Takahashi, R.; Murata, T.; Tamura, J. Application of self-tuning FPIC to AGC for load frequency control in multi-area power system. In Proceedings of the IEEE Bucharest PowerTech, Bucharest, Romania, 28 June–2 July 2009; pp. 1–7.

44. Sheikh, M.R.I.; Mondol, N. Application of self-tuning FPIC to AGC for Load Frequency Control in wind farm interconnected large power system. In Proceedings of the International Conference on Informatics, Electronics & Vision (ICIEV), Dhaka, Bangladesh, 18–19 May 2012; pp. 812–816.

45. Sheikh, M.R.I.; Takahashi, R.; Tamura, J. Multi-area frequency and tie-line power flow control by coordinated AGC with TCPS. In Proceedings of the International Electrical and Computer Engineering (ICECE) Conference, Dhaka, Bangladesh, 18–20 December 2010; pp. 275–278.

46. Zeni, L.; Rudolph, A.J.; Munster-Swendsen, J.; Margaris, I.; Hansen, A.D.; Sørensen, P. Virtual inertia for variable speed wind turbines. *Wind Energy* **2013**, *16*, 1225–1239. [CrossRef]

47. Morren, J.; Pierik, J.; de Haan, S.W.H. Inertial response of variable speed wind turbines. *Electr. Power Syst. Res.* **2006**, *76*, 980–987. [CrossRef]

48. Ramtharan, G.; Ekanayake, J.B.; Jenkins, N. Frequency support from doubly fed induction generator wind turbines. *IET Renew. Power Gener.* **2007**, *1*, 3–9. [CrossRef]

49. Safdar, M.; Ahmad, M.; Hussain, A.; Lehtonen, M. Optimized residential load scheduling under user defined constraints in a real-time tariff paradigm. In Proceedings of the 17th International Scientific Conference on Electric Power Engineering (EPE), Prague, Czech Republic, 169–18 May 2016; pp. 1–6. [CrossRef]

50. Freris, L.; Infield, D. *Renewable Energy in Power Systems*; John Wiley & Sons, Ltd Registered: Chichester, UK, 2008.

51. Turner, C.; Bining, A.; Gravely, M.; Hope, L.; Oglesby, R.P. *2020 Strategic Analysis of Energy Storage in California*; Technical Report; University of California: Berkeley, CA, USA, 2011.

52. Beil, I.; Hiskens, I.; Backhaus, S. Frequency Regulation From Commercial Building HVAC Demand Response. *Proc. IEEE* **2016**, *104*, 745–757. [CrossRef]

53. Zhang, J.; Domínguez-García, A.D. On the Impact of Measurement Errors on Power System Automatic Generation Control. *IEEE Trans. Smart Grid* **2018**, *9*, 1859–1868. [CrossRef]

54. Masetti, C. Revision of European Standard EN 50160 on power quality: Reasons and solutions. In Proceedings of the 14th International Conference on Harmonics and Quality of Power—ICHQP 2010, Bergamo, Italy, 26–29 September 2010; pp. 1–7. [CrossRef]

55. Tesarova, M.; Nohac, K. Voltage and frequency stability analysis of an island-mode LV distribution network. In Proceedings of the 2014 15th International Scientific Conference on Electric Power Engineering (EPE), Brno, Czech Republic, 12–14 May 2014; pp. 137–142. [CrossRef]

56. Yin, Z.; Che, Y.; Li, D.; Liu, H.; Yu, D. Optimal Scheduling Strategy for Domestic Electric Water Heaters Based on the Temperature State Priority List. *Energies* **2017**, *10*, 1425. [CrossRef]

57. Pipattanasomporn, M.; Kuzlu, M.; Rahman, S.; Teklu, Y. Load Profiles of Selected Major Household Appliances and Their Demand Response Opportunities. *IEEE Trans. Smart Grid* **2014**, *5*, 742–750. [CrossRef]

58. Wai, C.; Beaudin, M.; Zareipour, H.; Schellenberg, A.; Lu, N. Cooling Devices in Demand Response: A Comparison of Control Methods. *IEEE Trans. Smart Grid* **2015**, *6*, 249–260. [CrossRef]

59. Waters, L. *Energy Consumption in the UK*; Technical Report; Dept. for Business, Energy and Industrial Strategy: London, UK, 2017.

60. Motherway, B. *The Future of Cooling Opportunities for Energy-Efficient Air Conditioning*; Technical Report; International Energy Agency: Paris, France, 2018.

61. Shaad, M.; Momeni, A.; Diduch, C.; Kaye, M.; Chang, L. Parameter identification of thermal models for domestic electric water heaters in a direct load control program. In Proceedings of the 25th IEEE Canadian Conf. on Electrical Computer Engineering (CCECE), Montreal, QC, Canada, 29 April–2 May 2012; pp. 1–5. [CrossRef]

62. Molina-Garcia, A.; Gabaldon, A.; Fuentes, J.; Alvarez, C. Implementation and assessment of physically based electrical load models: Application to direct load control residential programmes. *IEE Proc. Gener. Transm. Distrib.* **2003**, *150*, 61–66. [CrossRef]

63. Eriksson, R.; Modig, N.; Elkington, K. Synthetic inertia versus fast frequency response: A definition. *IET Renew. Power Gener.* **2017**, *12*, 514–570. [CrossRef]
64. Gonzalez-Longatt, F. Impact of synthetic inertia from wind power on the protection/control schemes of future power systems: Simulation study. In Proceedings of the 11th IET International Conference on Developments in Power Systems Protection (DPSP 2012), Birmingham, UK, 23–26 April 2012; p. 74.
65. Gonzalez-Longatt, F.; Chikuni, E.; Rashayi, E. Effects of the Synthetic Inertia from wind power on the total system inertia after a frequency disturbance. In Proceedings of the IEEE International Conference on Industrial Technology (ICIT), Cape Town, South Africa, 25–28 February 2013; pp. 826–832. [CrossRef]
66. Delille, G.; Francois, B.; Malarange, G. Dynamic Frequency Control Support by Energy Storage to Reduce the Impact of Wind and Solar Generation on Isolated Power System's Inertia. *IEEE Trans. Sustain. Energy* **2012**, *3*, 931–939. [CrossRef]

Article

Simulation and Protection of Lightning Electromagnetic Pulse in Non-Metallic Nacelle of Wind Turbine

Qibin Zhou [1,*], Yize Shi [2], Xiaoyan Bian [2,*] and Bo Zhou [2]

[1] Shanghai Key Laboratory of Power Station Automation Technology, School of Mechatronics Engineering and Automation, Shanghai University, Shanghai 200444, China

[2] Electric Power College, Shanghai University of Electric Power, Shanghai 200090, China; ezlife@mail.shiep.edu.cn (Y.S.); zhoubo@shiep.edu.cn (B.Z.)

* Correspondence: zhouqibin@shu.edu.cn (Q.Z.); bianxy@shiep.edu.cn (X.B.); Tel.: +86-3530-5938 (X.B.)

Received: 3 April 2019; Accepted: 3 May 2019; Published: 8 May 2019

Abstract: When the nacelle of a wind turbine is struck by lightning, lightning electromagnetic pulse (LEMP) is generated inside the nacelle and consequently impacts inside electronic devices or even seriously destroys them. In order to study the LEMP inside the nacelle, this paper firstly built a full-scale model of a non-metallic nacelle. The lightning electromagnetic environment in the nacelle was simulated and analyzed by the transmission-line matrix method. Then the protective measures of applying metallic shielding mesh on the nacelle were studied, including the mesh size and material of the shielding mesh on the protective effect. The results show that LEMP in the nacelle can be effectively attenuated by metallic shielding meshes. The shielding effect is highly dependent on the conductivity of the shielding mesh material and the mesh size.

Keywords: wind turbine nacelle; lightning electromagnetic pulse (LEMP); magnetic field intensity; shielding mesh

---

1. Introduction

Lightning is a large-scale discharge phenomenon between cloud and cloud or between cloud and ground, as a natural typical electromagnetic hazard source [1,2]. Especially when cloud-to-ground lightning happens, the lightning channel carries hundreds of thousand Ampere current and the rate of rise-time is tens of thousands A/μs. This generates a great thermal effect, dynamic effect and EM effect on sensitive electrical and electronic equipment [3,4].

Lightning strike is an important threat to the safe operation of wind turbine generators [5]. At present, the research on the mechanism of lightning damage to wind turbine mainly focuses on the damaging effect of direct lightning types. Based on the analysis of the development mechanism of "cloud-ground" linear lightning leader, the initial attachment area of the lightning strike was analyzed by the electrostatic field simulation method [6]. The influence of the number and size of blade terminals on the interception effect was analyzed. With the simplified model of the upstream leader development process of blades, a critical length criterion was proposed by using the finite element simulation software [7]. Cooray discussed the capability of some analytical equations, to estimate the lightning channel base current parameters through measured fields and draw a comparison among those analytical equations [8]. Nucci and Rachidi, who measured the lightning induced current of a shielded buried cable and the horizontal magnetic field, discussed the correlation between the horizontal magnetic field and lightning induced current [9]. Through the theory of attraction radius in the space method of lightning induction, the phenomenon of wind turbine shielding was analyzed [10], the maximum shielding failure probability and the probability of shielding failure were calculated, and the effect of adding special down-wire on lightning current discharge was discussed. A miniature model with 1:30 ratio of a typical 2 MW wind turbine generators was used [11], to simulate the tip linear velocity of the actual wind turbine generators, and the effect of blade rotation on clearance breakdown characteristics and lightning initiation ability was studied.

When the large lightning current is attached to the blade or nacelle of a wind turbine, the lightning current generates a strong lightning electromagnetic pulse (LEMP) around the wind turbine. Due to the electromagnetic coupling effect, surges are induced in the electronic equipment and cables in the nacelle, which leads to the malfunction or damage of the nacelle equipment, affecting the safe operation of wind turbines. With the development of "Intelligent Wind Turbine" more and more sensitive electronic devices, which are very vulnerable to LEMP, are used in wind turbine nacelle. In the past, the metallic enclosure of wind turbine nacelle had a good shielding effect on LEMP. However, in order to reduce the weight and cost of nacelles, composite non-metallic materials, such as GFRP (glass fiber reinforced plastics) are widely used as the enclosure of modern wind turbine nacelles. This makes LEMP more dangerous to the sensitive electronic equipment in the nacelle.

The lightning-induced surge in the cable is the main threat to damage electronic equipment. At present, the research on the coupling between the lightning electromagnetic field and cable mainly focuses on the long-distance coupling between overhead transmission line [12,13] and buried cable [14,15]. However, the research on the coupling effect between LEMP and cables in wind turbine nacelle has been little addressed. According to the lightning protection zones (LPZ) of wind turbines classified in IEC 61400-24 [16], the interior area of the wind turbine nacelle is LPZ1, but the electromagnetic field and surge levels in this area are not addressed.

This paper builds a three-dimensional model of the non-metallic wind turbine nacelle and studies the electromagnetic environment when the tail of the metallic nacelle is struck by lightning with the transmission line matrix method. The protective measures of applying metallic shielding mesh on the nacelle were studied, including the influence of the size and material of the shielding mesh on the shielding effect against LEMP.

2. Modeling Method

2.1. Principle of Transmission Line Modeling Method

The transmission-line matrix method (TLM) was first proposed by Johns and Beurle in 1971. Through continuous improvement and development, TLM algorithm has formed a complete time-domain electromagnetic radiation and scattering research method [17–19]. Its core idea is based on the similarity between electromagnetic wave transmission characteristics and voltage and current transmission characteristics in transmission lines. Now it has been extended to three-dimensional space problems.

When the TLM method is used to solve the distribution of electromagnetic field in medium, the medium characteristics are replaced by the TLM matrix, which is composed of intersections (nodes) of transmission lines after spatial discretization. The nodes represent the physical properties of different media, while the transmission lines only assume the distribution and storage of energy. By iterating the computational region from space and time, the distribution of magnetic field in the computational region with time and space can be obtained. The two-dimensional TLM method is composed of parallel transmission lines. The pulse source is incident from four branches to a node, and then scattered to the adjacent node. It can be expressed as:

$$_{k+1}V^r = S_k V^i$$
$$_{k+1}V^i = C_{k+1}V^r$$

(1)

where, S is the impulse scattering matrix of the nodes. C is the connection matrix describing the network topology. The subscripts k and $k + 1$ represent the scattering time interval. The following equation can be obtained as follows:

$$_{k+1}V_n^i = \frac{1}{2}\left[\sum_{m=1}^{4}{}_kV_m^i\right] - {}_kV_n^i \tag{2}$$

In the formula, k means scattering, i means incidence, m and n are port numbers.

In the three-dimensional electromagnetic field simulation, the symmetrical condensed node (SCN) algorithm is used to simulate the electromagnetic propagation in the space element [20]. A three-dimensional SCN node has six branches, each consisting of two transmission lines, which simulate the propagation of electric field and magnetic field, respectively. The cell size is often less than one-tenth of the corresponding wavelength of the highest simulation frequency [21].

In the narrow slot model, the field on the long side of the slot is separated by one-dimensional transmission line and symmetrical condensation node. The calculation formulas of capacitance C_S and inductance L_S per unit length of the slot in the element are as follows:

$$C_s = \frac{2\varepsilon}{\pi}\ln\left(\frac{0.563\Delta z}{w}\right) + \frac{\varepsilon d}{w} \tag{3}$$

$$L_s^{-1} = \left(\frac{\pi\mu}{2\ln\left(\frac{1.591\Delta z}{w}\right)}\right)^{-1} + \left(\frac{\mu w}{d}\right)^{-1} \tag{4}$$

In the formula, ε is the relative dielectric constant in free space, μ is the relative permeability in free space, Δz is the distance of electromagnetic wave propagation in the time step Δt. w is the width of the gap and d is the depth of the gap.

The change of slot transmission line characteristic admittance Y_{LINE} and capacitive stump characteristic admittance Y_{STUB} synchronizes the pulse of slot and stump with the time step of three-dimensional TLM unit, thus establishing the overall scattering model. The calculation formula is as follows:

$$Y_{LINE} = \frac{\Delta t}{L_s\Delta z} \tag{5}$$

$$Y_{STUB} = 2\left(\frac{C_s\Delta z}{\Delta t} - Y_{LINE}\right) \tag{6}$$

$$w \leq 0.3982\Delta z \tag{7}$$

where Y_{LINE} and Y_{STUB} are transmission line characteristic admittance and capacitive stump characteristic admittance, respectively.

According to the above principle, the characteristics of LEMP in the nacelle wind turbine nacelle are studied by using a three-dimensional electromagnetic field simulation program.

2.2. Modeling of the Wind Turbine Nacelle

At present, the shell of most nacelles is made from GFRP (glass fiber reinforced plastics). Copper bars embedded inside the GFRP are used as the down-conductors. When lightning strikes at the tail of the nacelle, the lightning current will flow through the down-conductor copper bar into the MEB (main equipotential bar) located at the bottom of the nacelle and consequently flow into the ground through the tower. When large lightning current flows through the copper bar conductors, intense LEMP is generated inside nacelle, threatening the electric and electronic equipment. This paper built a 3D model of a non-metallic nacelle. In this paper, LEMP inside the nacelle was studied when the lightning current was injected into the lightning rod installed at the tail of the nacelle. The peak value and waveform of the lightning current is chosen as 200 kA and 10/350 µs according to the LPLI (lightning protection level) requirement in IEC 61400-24.

Figure 1 shows the model of real-scale wind turbine nacelle. The shell of the nacelle is made from glass fiber reinforced plastics (GFRP). The center of the nacelle bottom is set as the original point of the Cartesian coordinate system. Seven magnetic probes are placed at seven positions inside the nacelle, as follows, P1 (0, −300, 50), P2 (0, −200, 50), P3 (0, −100, 50), P4 (0, 0, 50), P5 (0, 100, 50), P6 (0, 200, 50), P7 (0, 300, 50).

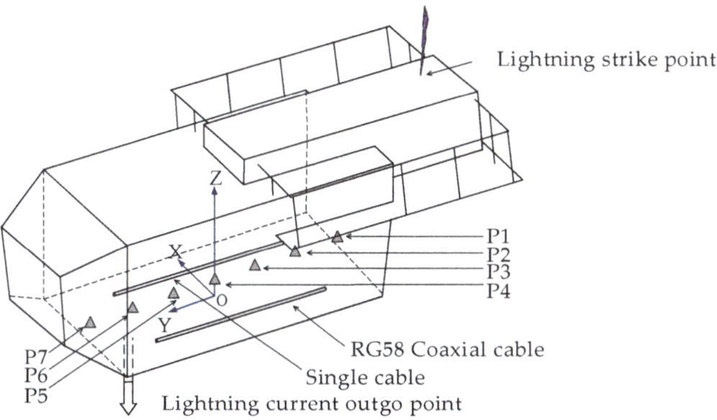

Figure 1. Model of wind turbine nacelle.

Since there are different types of cables in the nacelle. Single-core cables without shielding layer and coaxial cables with shielding layer were selected as typical cables to represent them. One single-core cable and one RG58 coaxial cable of 6 m long were placed at 0.5 m height above the bottom of the nacelle. The induced voltage in the cables was evaluated. Then the inducing current could be easily evaluated along these cables once they were connected to different power sources and loads. In order to study the shielding effect after installing the shielding mesh over the nacelle shell, aluminum meshes with 50 mm, 20 mm and 10 mm size, respectively, were installed individually. In addition, the simulation frequency band was set at 0~30 MHz in the TLM method. The simulation step was 100 µs.

The Heidler model was used to simulate the natural lightning with 10/350 (µs) waveform, as follows [22],

$$I = \frac{I_m}{k} \times \frac{(t/T_1)^{10}}{1 + (t/T_1)^{10}} \times \exp(-t/T_2). \tag{8}$$

where, I_m is the peak value of lightning current (kA), k is the correction factor for the peak current, t is the time (µs), T_1 is the front time constant (µs) and T_2 is the tail time constant (µs). The lightning current is 200 kA, other corresponding parameters can be found in IEC 62305-1, Annex B.

$$k = 0.93, \tag{9}$$

$$T_1 = 19 \ (\text{µs}), \tag{10}$$

$$T_2 = 485 \ (\text{µs}), \tag{11}$$

The lightning current waveform is shown in Figure 2, the current reaches the peak value of 200 kA at 10 µs and decreases to half peak value at 350 µs [23].

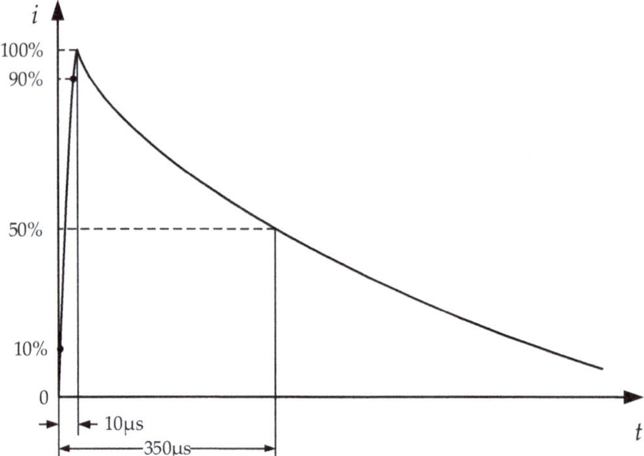

Figure 2. 10/350 lightning current waveform.

3. Simulating Result and Analysis

3.1. Calculated LEMP inside Nacelle

When lightning current flows through the copper bar embedded inside the shell of the nacelle, the induced voltage between the two terminals of the 6 m single-core cable and the coaxial cable is shown in Figure 3.

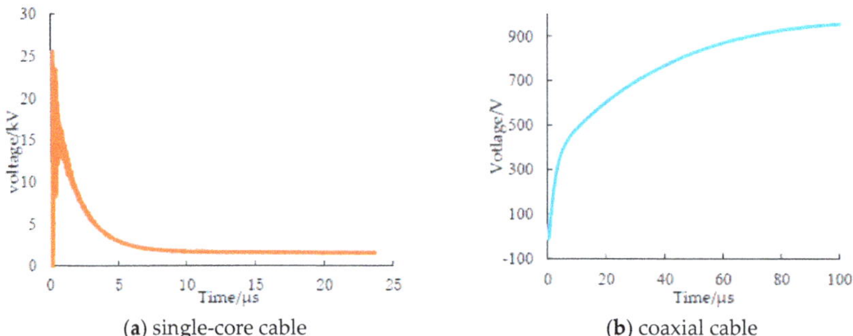

(**a**) single-core cable　　　　　　(**b**) coaxial cable

Figure 3. Induced voltage in the cables in the nacelle.

Figure 3 shows the variation of the induced voltage in different cables. From Figure 3, it is noted that the coupling characteristics of different cable structures have significant differences. Although the coaxial cable has a much better shielding effect than the single-core cable, the induced voltage at the coaxial cable can still reach over a thousand Volts. This is mainly due to the poor shielding ability of GFRP material on LEMP, resulting in the internal magnetic field strength of the nacelle not being well attenuated.

Figure 4 shows the time-domain variation of the magnetic field intensity at seven positions inside the GFRP nacelle. Although there are differences in the magnetic field intensity at different positions in the nacelle, the lightning magnetic field after 20 micros is almost all above 3 kA/m. Relevant studies show that when the magnetic field intensity reaches 190 A/m, some electronic components will be damaged [24,25]. Therefore, when the GFRP nacelle is directly exposed to LEMP, the electronic equipment will be threatened significantly.

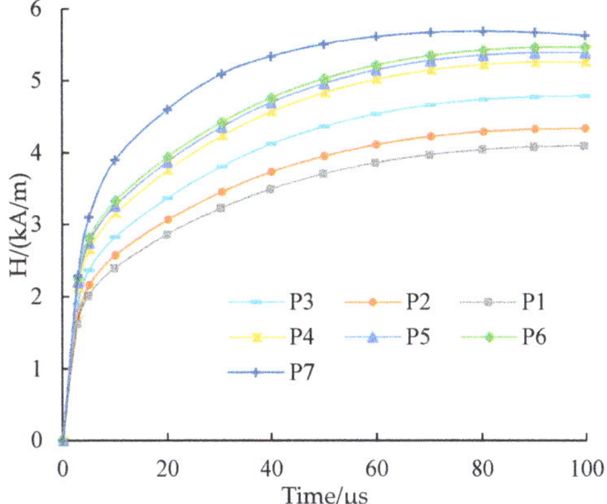

Figure 4. Magnetic field intensity time-domain variation diagrams of probes at different positions in the nacelle.

3.2. Influence of the Mesh Size on the Shielding Effect

IEC61400-24 points out that a metallic mesh can be used in the nacelle with a GFRP cover to shield the external electric field and magnetic field, as well as the magnetic field generated by the current in the metallic mesh. For the nacelle shell made from GFRP, aluminum meshes with small mesh size will protect the nacelle from a direct lightning strike or the leader current. The magnetic field and electric field will be attenuated according to the mesh size and the thickness of the metallic material. In order to analyze the influence of mesh size on the attenuation, diamond aluminum meshes with the side lengths of $l = 2$ cm, 5 cm and 10 cm, respectively, were added to the nacelle shell to calculate the variation of the induced voltage in the cables and magnetic field in the nacelle. Figure 5 shows the time-domain variation of the induced voltage of the coaxial cable for different sizes of meshes. The result shows that adding metallic mesh can greatly reduce the induced voltage in the coaxial cable. After 100 micros, when the aluminum meshes with the side lengths is 2 cm, 5 cm and 10 cm, the induced voltage of coaxial cable is 80 V, 40 V and 16 V. At the same time, the smaller is the mesh size, the more obvious is the attenuation effect on the induced voltage in the cable, the smallest mesh has almost five times the shielding effect of the biggest mesh size. It is proportional to the mesh side length.

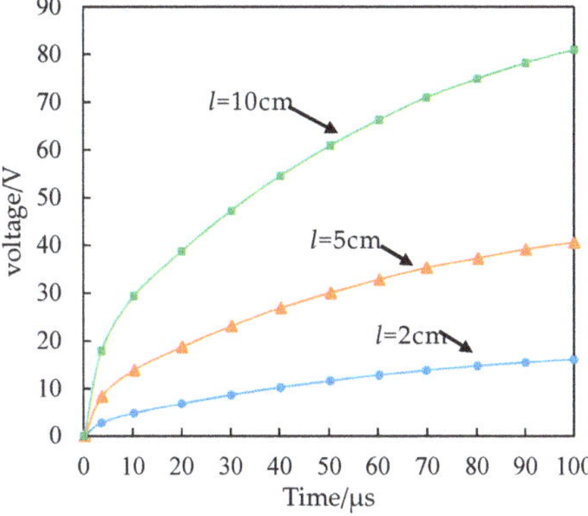

Figure 5. Induction voltage time domain variation diagrams of coaxial cables with different metal meshes.

In order to investigate the distribution of the magnetic field in the nacelle with different mesh sizes, seven probes were located inside the nacelle as sampling points. The distance of each sampling point was 1 m. Figure 6 shows the magnetic field intensity at different points in the nacelle when the lightning current reached its peak value ($t = 10$ µs). It can be seen from Figure 6 that the magnetic field intensity increased gradually when the sampling points approached the lightning current outgo point. Comparing different mesh sizes, it is noted that when the mesh size is reduced from 10 cm to 2 cm, the magnetic field inside the nacelle is reduced by 90%, mostly.

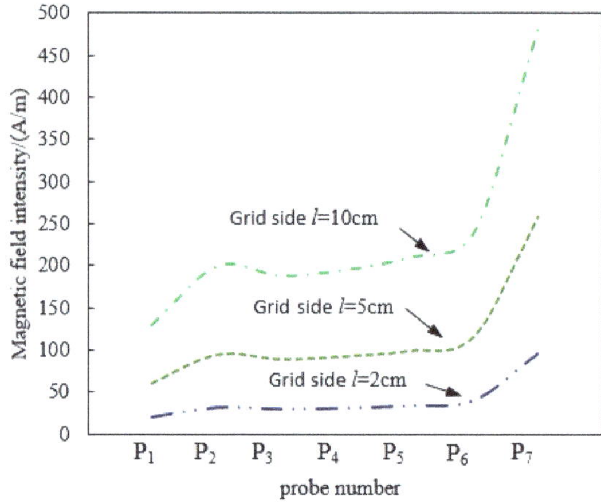

Figure 6. Magnetic field intensity values with different probes inside the nacelle.

3.3. Influence of the Mesh Material on the Shielding Effect

In order to study the influence of mesh material on the electromagnetic effect in the nacelle, the simulation with aluminum and steel was carried out and the simulation results were briefly analyzed. The conductivity of aluminum is 3.56×10^7 S/m. The conductivity of steel is 7.69×10^6 S/m. Figure 7 shows the current distribution on the nacelle surface with two mesh materials when the lightning current reached its peak value ($t = 10$ µs).

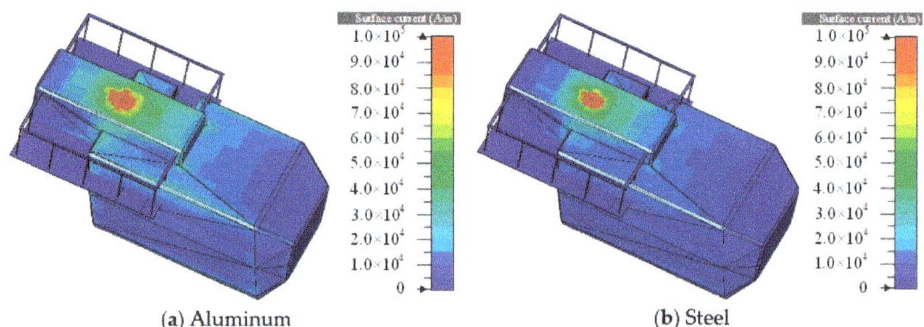

Figure 7. The distribution of surface current on the nacelle with two shielding materials during a lightning stroke.

Figure 7 shows that when the material of shielding mesh is aluminum, the lightning current density can reach 260 kA/m at the lightning strike point and the edge of the nacelles. When the mesh material is steel, the lightning current density in the above area decreases to 225 kA/m. It can be concluded that the lightning current density on the nacelle's surface decreases with the decrease of electrical conductivity, which is mainly due to the increase of Joule heat caused by the decrease of electrical conductivity, leading to a larger current attenuation, which shows that the lightning current intensity on the nacelles surface of the turbine is positively correlated with the electrical conductivity of the shell material.

The attenuation of magnetic field intensity was further studied with two mesh materials. Figure 8 shows the distribution of shield effect against the magnetic field at four sampling points (H1 (0, 0, 50), H2 (0, 0, 150), H3 (0, 0, 250), and H4 (0, 0, 350) with aluminum and steel materials. It is noted that the shielding ability of aluminum is much better than that of steel by nearly 30 dB.

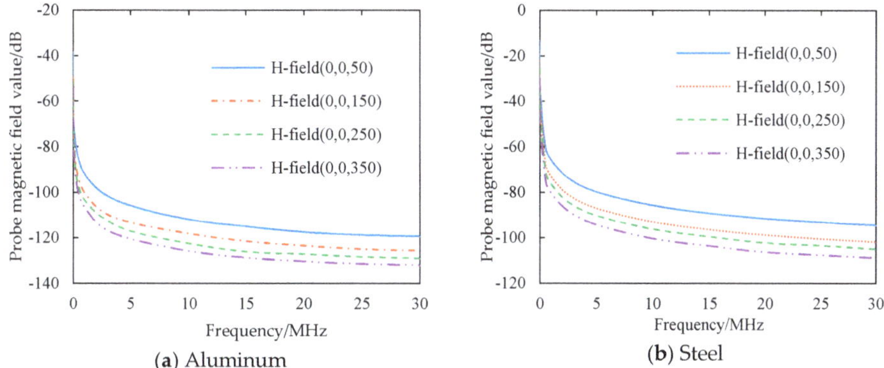

(a) Aluminum　　　　　　　　(b) Steel

Figure 8. The attenuation value of the magnetic field on the probes.

By comparing the results in Figure 8, it shows that the shielding effect of the mesh on a nacelle against LEMP is reduced if the conductivity of the material is reduced. Especially, the wide application of composite materials in the wind turbine nacelle shell at present will inevitably bring serious risks to the internal electronic equipment. Therefore, in order to ensure the reliable and safe operation of the wind turbine, the metallic mesh is highly recommended to be installed over the composite shell to improve the anti-interference ability against LEMP.

4. Conclusions

A three-dimensional model of wind turbine nacelle with internal cables is established in this paper. The induced voltage in cables and the magnetic field distribution in the nacelle are studied by using the transmission line matrix method. The influence of shielding mesh size and material on the attenuation effect again magnetic field in the nacelle is also analyzed. Based on the above analysis, following conclusions are drawn:

(1)	When lightning strikes at the air terminal of the wind turbine nacelle, high transient voltage is induced in the cables inside the nacelle. Applying metallic mesh to the nacelle shell can effectively reduce the magnetic field inside the nacelle as well as the transient voltage in the cable.

(2)	The mesh material will obviously influence the shielding effect. The shielding effect of an aluminum mesh nacelle is nearly 30 dB higher than that of a steel mesh nacelle because of the higher conductivity of the aluminum mesh material.

(3)	The shielding effect is proportional to the mesh side length and the smaller mesh size has better shielding effect against the magnetic field and generates lower transient voltage in the cable. For example, the mesh with the side lengths is 2 cm has a five times shielding effect than the mesh with 5 cm side lengths.

With above study conclusion, it is suggested to apply metallic mesh with a higher conductive metallic material and smaller side length on the nacelle in order to attenuate LEMP inside the nacelle, e.g., aluminum mesh with 2 cm side length is preferred.

Energies **2019**, *12*, 1745

Author Contributions: All the authors contributed equally to this research work.

Funding: This work was financially supported by Shanghai Science and Technology Project under grant (16020501000).

Conflicts of Interest: The authors declare no conflict of interest. The funders had no role in the design of the study; in the collection, analyses, or interpretation of data; in the writing of the manuscript, or in the decision to publish the results.

References

1. Zeng, R.; Zhuang, C.; Zhou, X.; Chen, S.; Wang, Z.; Yu, Z.; Jinliang, H. Survey of recent progress on lightning and lightning protection research. *High Volt.* **2016**, *1*, 2–10. [CrossRef]
2. Wang, H.; Chen, Y.Z.; Wan, H.J.; Wang, X.J. Study on Coupling between Lightning Electromagnetic Field and Twisted Pair. *J. Microw.* **2014**, *40*, 1605–1613.
3. Liu, S.H.; Liu, W.D. Progress of relevant research on electromagnetic compatibility and electromagnetic protection. *High Volt. Eng.* **2014**, *40*, 1605–1613.
4. Zhou, Q.; Cheng, Y.; Bian, X.; Liu, F.; Zhao, Y. Analysis of Restrike Overvoltage of Circuit Breakers in Offshore Wind Farms. *IEEE Trans. Appl. Supercond.* **2016**, *26*, 1–5. [CrossRef]
5. Yamamoto, K.; Noda, T.; Yokoyama, S.; Ametani, A. Experimental and analytical studies of lightning overvoltage in wind turbine generator systems. *Electr. Power Syst. Res.* **2009**, *79*, 436–442. [CrossRef]
6. Zhou, Q.B.; Liu, C.X.; Bian, X.Y.; Lo, K.L.; Li, D.D. Numerical analysis of lightning attachment to wind trubine blade. *Renew. Energy* **2018**, *116*, 584–593. [CrossRef]
7. Ma, Y.F.; Zhang, L.; Yan, J.Y.; Guo, Z.; Li, Q.; Fang, Z.; Siew, W.H. Inception mechanism of lightning upward leader from the wind turbine blade and a proposed critical length criterion. *Proc. CSEE* **2016**, *36*, 5975–5982.
8. Gomes, C.; Cooray, V. On Theoretical Approaches to Estimate the Lightning Return Stroke Current Parameters. In Proceedings of the 11th International Symposium on High-Voltage Engineering (ISH 99), London, UK, 23–27 August 1999; p. v2-401.
9. Petrache, E.; Paolone, M.; Rachidi, F.; Nucci, C.A.; Rakov, V.A.; Uman, M.; Jordan, D.; Rambo, K.; Schoene, J.; Cordier, A.; Verhaege, T. Measurement of Lightning-Induced Currents in an Experimental Coaxial Buried Cable. In Proceedings of the 2003 IEEE Power Engineering Society General Meeting, Toronto, ON, Canada, 13–17 July 2003.
10. Zhou, C.Y. Analysis and protection of lightning shielding failure of wind turbine. *Insul. Surge Arresters* **2017**, *1*, 77–81.
11. Wen, X.S.; Qu, L.; Wang, Y.; Si, T.; Xu, J.; Lan, L. Experimental study of the influence of the blade rotation on triggered lightning ability of wind turbine's blades. *Proc. CSEE* **2017**, *37*, 2151–2158.
12. Chen, W.J.; He, H.X.; He, J.J.; Zhao, X.; Yu, H.; Shi, W.; He, T. On the 3-dimentional leader progression model for the lightning shielding failure performance estimation of overhead transmission lines. *Proc. CSEE* **2014**, *34*, 6601–6612.
13. Liu, G.; Xi, Y.; Tang, J.; Xi, Y.; Li, H.; Deng, L.; Xu, Z. Influence of triggered lightning of high voltage overhead transmission lines on 10 kV overhead distribution lines lightning trip characteristics. *High Volt. Eng.* **2014**, *40*, 690–697.
14. Yu, H.; Dong, W.S.; Chen, S.D.; Wang, J.G.; Zhang, Y.J.; Zhang, Y.; Zhou, M.; Sun, Z.; Deqing, C.M. Observation and analysis of lightning induced overvoltage on buried cables. *High Power Laser Part. Beams* **2010**, *22*, 2373–2377. [CrossRef]
15. Yang, B.; Zhou, B.H.; Chen, B.; Wang, J.B.; Meng, X. Numerical Study of Lightning-Induced Currents on Buried Cables and Shield Wire Protection Method. *IEEE Trans. Electromagn. Compat.* **2012**, *54*, 323–331. [CrossRef]
16. *Wind Turbines, Part 24 Lightning Protection: IEC 61400-24:2010*; International Electrotechnical Commission: Geneva, Switzerland, 2010.
17. Jojns, P.B.; Beurle, R.L. Numerical solution of 2-dimensional scattering problems using a transmission-line matrix. *Proc. Inst. Electr. Eng. IET* **1971**, *118*, 1203–1208.
18. Hoefer, W.J.R. The Transmission-Line Matrix Method—Theory and Applications. *IEEE Trans. Microw. Theory Tech.* **1985**, *33*, 882–893. [CrossRef]

19. Trenkic, V.; Christopoulos, C.; Benson, T.M. New symmetrical super-condensed node for the TLM method. *Electr. Lett.* **1994**, *30*, 329–330. [CrossRef]
20. Johns, P.B. A Symmetrical Condensed Node for the TLM Method. *IEEE Trans. Microw. Theory Tech.* **1987**, *MTT-35*, 370–377. [CrossRef]
21. Johns, D.P.; Mallik, A.; Wlosarcyzk, A.J. TLM enhancements for EMC studies. In Proceedings of the 1992 Regional Symposium on Electromagnetic Compatibility, Tel-Aviv, Israel, 2–5 November 1992.
22. *Protection Against Lightning, Part 1 General Principles: IEC 62305-1:2010*; International Electrotechnical Commission: Geneva, Switzerland, 2010.
23. Zhang, X.Q. *Lightning Protection and Ground Connection of Wind Turbines*; China Electric Power Press: Beijing, China, 2009; pp. 20–22.
24. Gong, X.H.; He, J.L.; Li, Y. Analysis on transient electromagnetic environment of substations. *High Volt. Appar.* **2009**, *45*, 39–43.
25. Wang, X.H. *Investigations on the Electromagnetic Effects in Wind Turbines Struck by Lightning*; Jiao Tong University: Beijing, China, 2010.

Article

Analysis of Wind-Turbine Main Bearing Loads Due to Constant Yaw Misalignments over a 20 Years Timespan

Martin Cardaun *, Björn Roscher, Ralf Schelenz and Georg Jacobs

Center for Wind Power Drives, RWTH Aachen University, 52062 Aachen, Germany;
bjoern.roscher@cwd.rwth-aachen.de (B.R.); Ralf.Schelenz@cwd.rwth-aachen.de (R.S.);
georg.jacobs@imse.rwth-aachen.de (G.J.)
* Correspondence: martin.cardaun@cwd.rwth-aachen.de

Received: 20 March 2019; Accepted: 8 May 2019; Published: 10 May 2019

Abstract: The compact design of modern wind farms means that turbines are located in the wake over a certain amount of time. This leads to reduced power and increased loads on the turbine in the wake. Currently, research has been dedicated to reduce or avoid these effects. One approach is wake-steering, where a yaw misalignment is introduced in the upstream wind turbine. Due to the intentional misalignment of upstream turbines, their wake flow can be forced around the downstream turbines, thus increasing park energy output. Such a control scheme reduces the turbulence seen by the downstream turbine but introduces additional load variation to the turbine that is misaligned. Within the scope of this investigation, a generic multi body simulation model is simulated for various yaw misalignments. The time series of the calculated loads are combined with the wind speed distribution of a reference site over 20 years to investigate the effects of yaw misalignments on the turbines main bearing loads. It is shown that damage equivalent loads increase with yaw misalignment within the range considered. Especially the vertical in-plane force, bending and tilt moment acting on the main bearing are sensitive to yaw misalignments. Furthermore, it is found that the change of load due to yaw misalignments is not symmetrical. The results of this investigation are a primary step and can be further combined with distributions of yaw misalignments for a study regarding specific load distributions and load cycles.

Keywords: wake steering; yaw misalignment; multi body simulation; main bearing loads; rain flow counts

---

1. Introduction

Wind turbines are mainly clustered as wind farms. Due to the limited space, the turbines are densely placed to obtain the full potential of the available space and to avoid unnecessary cabling costs especially at offshore sites. The turbine in the wake flow experiences lower wind speeds and increased turbulence intensity [1]. As a result maximizing the output of the individual turbine does not always lead to a global maximum wind farm output. Furthermore, the increased turbulence intensity leads to greater loads on the drive train and its structural components [1]. This results in accelerated damage accumulation and shortened maintenance intervals. However, there are two approaches to reduce these effects.

During power curtailment the power of an upstream turbine is reduced. This means that the wind speed in the wake is less reduced. Therefore, more power can be extracted by the downstream turbines. However, the turbine in the wake still experiences an increased turbulence intensity. This control scheme was the topic of a significant amount of research, see [2–5].

Wake-steering offers a promising approach. The upstream turbines are misaligned against the inflow direction, which directs the wake flow past the downstream turbine. Simulations showed that the average power capture could be increased by such a control scheme, see [6,7]. In [8] the simulation results were compared with real park measurements and the increase in energy capture matched the predicted values. Further, in [9] it was shown that a wind farm already optimized with respect to wake losses could increase its annual energy supply by up to 3.7% through the use of wake steering. In addition the non-torque loads of the downstream turbine are reduced because of the lower turbulence intensity as shown in [10,11]. In [12,13], the power output and blade root moments of the waked turbine were measured. It was found that the turbine power drops significantly when the yaw misalignment exceeds 10°. Furthermore, a correlation between the yaw angle and flapwise bending moment was made. In [14] aerodynamic loads were compared between measurement and simulation. The results showed that aerodynamic loads could be calculated accurately, even for highly yawed inflows. With respect to these results, a significant influence of the yaw misalignment on bending and tilt moment at the main bearing is expected. Therefore, this study will address the torque and non-torque loads at the main bearing of the turbine in the front that is misaligned.

The main contribution of this work is the structure focussed approach. Many studies regarding wake-steering focus on the flow field and the ones that take the structure into account mostly observe the blade root moments. In this study, the transmission of the aerodynamic loads into the drive train and support structure of the turbine are studied as a function of the yaw misalignment. The work is limited in its consideration of the flow field since the Blade-Element-Momentum Theory is used to calculate the wind loads on the rotor blades instead of a complex CFD simulation which would be necessary to study the flow around the blade profiles. Because of that, the yaw angle is limited to a range of $-10°$ to $+10°$ due to the limitation of the utilised code. As an initial classification of the occurring load changes and the detection of further effects, load calculations are carried out on a generic wind turbine model with flexible structural components. For this purpose, a multi body simulation model of a 3 MW turbine with a rotor diameter of 126 m (C3 × 126 [15]) is simulated at Design Load Case 1.1 according to DIN EN 61400-1 (production operation) [16] with wind class 2B at multiple yaw misalignments. The loads at the main bearing are investigated, since this is where changes in the aerodynamic loads by an inclined flow will mainly be reflected and, for the most part, be introduced into the structure. The resulting understanding can be used both in the design process and in operation. On the one hand, the findings could be taken into account in the design of the bearings, and on the other hand, the park regulation could be adapted to prevent an uneconomical accumulation of damage in favour of energy production.

2. Simulation Model and Setup

Within the scope of this work the resulting turbine loads are determined by a co-simulation between SIMPACK, AERODYN [17] and MATLAB. The multi body simulation software SIMPACK Version 2019 is used to formulate the mechanical structure of the turbine including flexible tower and blades. The AERODYN-code delivers the aerodynamic loads acting in the blades and the controller of the turbine is formulated in MATLAB. For a wind speed below the rated wind speed the turbine is controlled by a generator torque and the above rated wind speed the turbine is pitch-controlled. The drive train of the turbine is modelled as a 2-mass-oscillator with equivalent stiffness and inertia to a gearbox with a ratio of 92.28. Relevant turbine parameters can be found in Table 1. The turbine delivers power in a wind speed range from $v_{hub} = 3$ m/s to $v_{hub} = 25$ m/s (wind speed on hub height in front of the rotor). For this range, three-dimensional turbulent wind fields are generated in steps of 1 m/s with TurbSim [18]. According to the design load case (DLC) 1.1 of the industry standard DIN EN 61400-1 [16], wind fields of wind class 2B are generated with a normal turbulence model ($I_{ref} = 0.14$). The standard deviation of the longitudinal wind speed at the hub height results from Equation (1). The inflow data is arranged in a matrix representing a grid in front of the rotor. For each point on the grid the wind speed data for all three dimensions is stored as a time series. The frequency of the time series is 20 Hz.

$$\sigma_1 = I_{ref}(0.75\, v_{hub} + b); b = 5.6\ \text{m/s} \tag{1}$$

Table 1. C3 × 126 turbine parameters [15].

Variable	Value	Unit
Rated power	3.0	MW
Tower height	112.0	m
Rotor diameter	126.0	m
Hub height	115.0	m
Cut-in wind speed v_{in}	3.0	m/s
Cut-out wind speed v_{out}	25.0	m/s
Rated wind speed v_{rated}	11.0	m/s
Tilt angle	6	°
Gearbox ratio	1:92.28	-
Main suspension	3-Point	-

The loads due to yaw misalignments of $-10°$, $-5°$, $0°$, $+5°$ and $+10°$ are investigated within the scope of this work. The potential weaknesses of the methodology are as follows. The Blade-Element-Momentum Theory doesn't take the interaction between the neighbouring blade elements into account. In addition, it neglects the wake expansion. Since a detailed examination of the flow field around the blades is not intended in this work, these assumptions will have a negligible influence on the results. The used AERODYN-Code utilizes the Blade-Element-Momentum Theory with Glauert Correction for yawed rotors [19], so the simulation is not valid beyond the chosen yaw angle values. However, the overall effects and trends resulting from the misalignment will be detectable within the chosen yaw angle range. In addition, there are the structural assumptions that were made to model the rotor blades which are built as shell elements. The natural frequencies of the flexible blades and tower are considered up to 11 Hz. This ensures that the most important deformations can be mapped but also allows for manageable computation times. For the blades, bending modes up to the first order and torsional modes up to the fourth order are considered. The tower model includes bending modes up to the fifth order and the torsional mode of the first order.

Figure 1 shows the nacelle position relative to the wind direction. The multi body simulation model of the entire turbine is also shown in Figure 1. A simulation of 10 min is performed at each wind speed. The calculated loads are then cumulated over a period of 20 years. The wind speeds are weighted with respect to the probability of their occurrence. The cumulative frequencies of the wind speeds result from the reference location described in the German renewable energies act (EEG 2017) [20]. The height profile of the wind is calculated according to the Hellmann power law with a Hellman exponent of $\alpha = 0.25$ and the reference wind speed and reference hub height (Equation (2); $v_{ref} = 6.45$ m/s, $h_{ref} = 100$ m). This results in the mean wind speed at hub height with $v_{Hub} = v(h = h_{Hub})$. Using the mean wind speed at hub height v_{Hub}, the cumulative frequencies are to be determined by a Rayleigh distribution (Equation (3)). In order to determine the relative frequency of the respective wind speed, the difference of the sum frequencies of v_i and v_{i-1} is calculated (Equation (3)). The distribution of the relative frequencies is shown in Figure 2. Identical wind fields are used for the load calculation of each yaw misalignment.

$$v = v_{ref}\left(\frac{h}{h_{ref}}\right)^{\alpha} \tag{2}$$

$$F(v_i) = 1 - \exp\left[-\frac{\pi}{4}\left(\frac{v_i}{v_{hub}}\right)^2\right] \tag{3}$$

$$H(v_i) = F(v_i) - F(v_{i-1}) \tag{4}$$

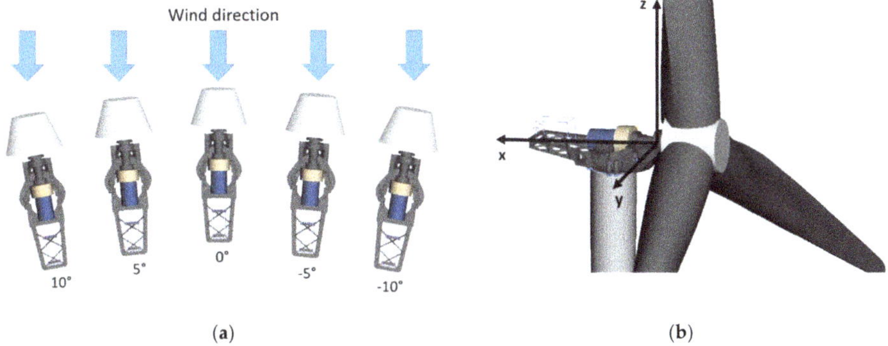

Figure 1. (**a**) Nacelle positions relative to wind direction as seen from above; (**b**) multi body simulation model of the C3 × 126 with coordinate system [15].

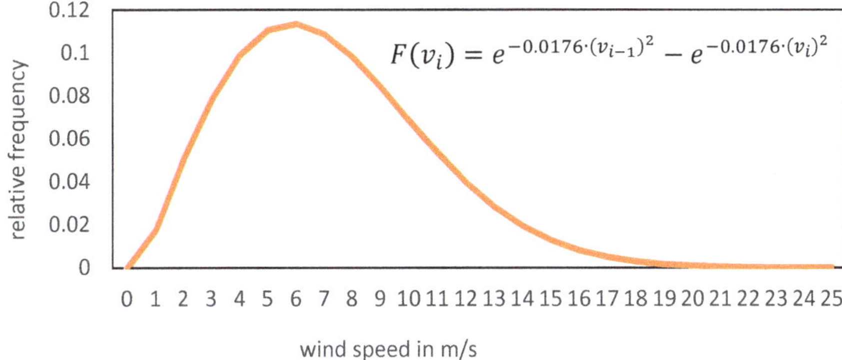

Figure 2. Relative frequencies of occurrence for various wind speeds used for load duration calculation.

3. Results and Discussion

The evaluation of the different inflow angles will be performed for five different yaw angles as mentioned before. Therefore six 10-min simulations with six different random seeds are carried out for each wind speed and each yaw misalignment. This results in a total number of 138 inflow simulations and 690 load simulations. To arrange the results in a compact and observable manner, load duration distributions are calculated in a first step. Building on this, rain flow counts are carried out which will be mainly discussed. To further condense the results, the damage equivalent load (DEL) is calculated with Equation (5) according to the industry standard DIN 50100 [21] for each yaw misalignment.

$$\text{DEL} = \sum_i S_i^k \cdot \left(\frac{N_i}{N_{eq}}\right)^{\frac{1}{k}}, \text{ with } k = 3.3 \tag{5}$$

This parameter is used to retrieve a force or torque for an equivalent load cycle where S_i the load of the corresponding load cycle is N_i. There $N_{eq} = 175,000$ load cycles are used, representative for a 20 year lifetime and an exponent of $k = 3.3$ as it is recommended for the lifetime calculation of bearings with a line contact in [22]. Afterwards, all combinations of load cycles and amplitudes can be summed up to one amplitude with an equivalent amount of load cycles.

In the following part of this work the loads are defined according to the hub coordinate system from Figure 1 (x-axis is coaxial with the shaft axis of rotation). The rain flow count diagrams, shown in Figures 3–5, contain relevant information about the load behaviour of the turbine and will be discussed.

A rain flow count is utilised to determine the number of load cycles from a load time series. The x- and y-axis represent the starting (x-axis) and ending (y-axis) load of the respective load cycle while the logarithmic colour scale gives the total number of load cycles of this type.

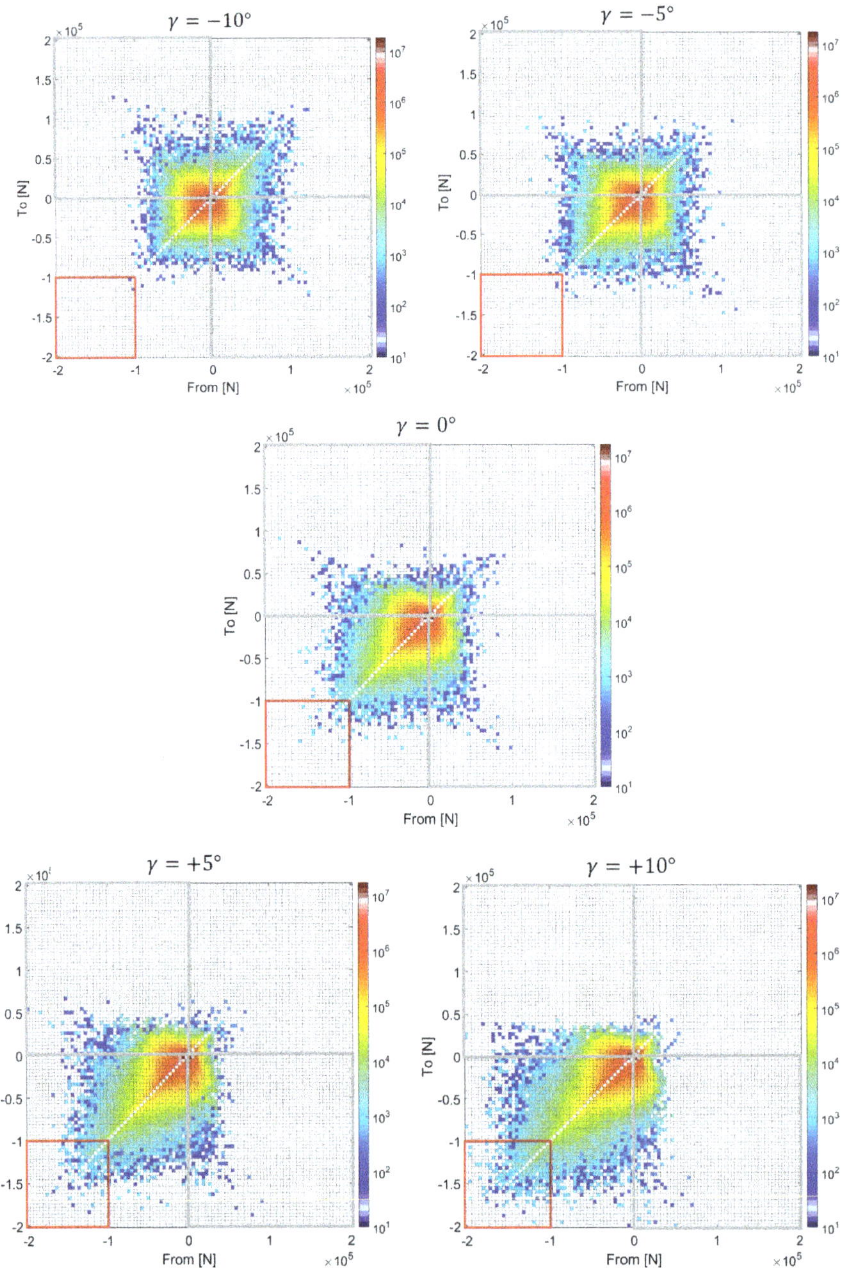

Figure 3. Rain flow count of the in-plane side force (y-axis) for various nacelle positions and a timespan of 20 years (**top left**: $-10°$; **top right**: $-5°$; **middle**: $0°$; **bottom left**: $5°$; **bottom right**: $10°$).

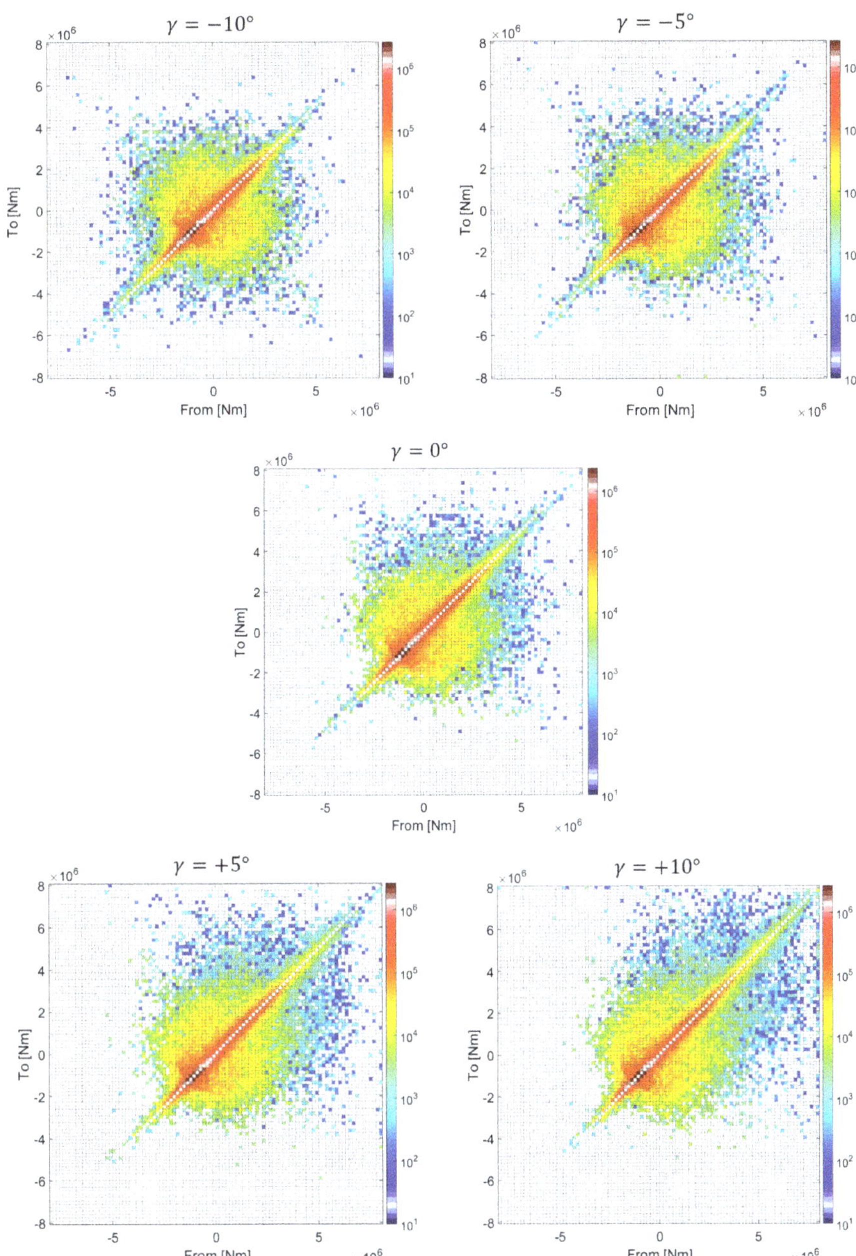

Figure 4. Rain flow count of the tilt moment (around y-axis) for various nacelle positions and a timespan of 20 years (**top left**: $-10°$; **top right**: $-5°$; **middle**: $0°$; **bottom left**: $5°$; **bottom right**: $10°$).

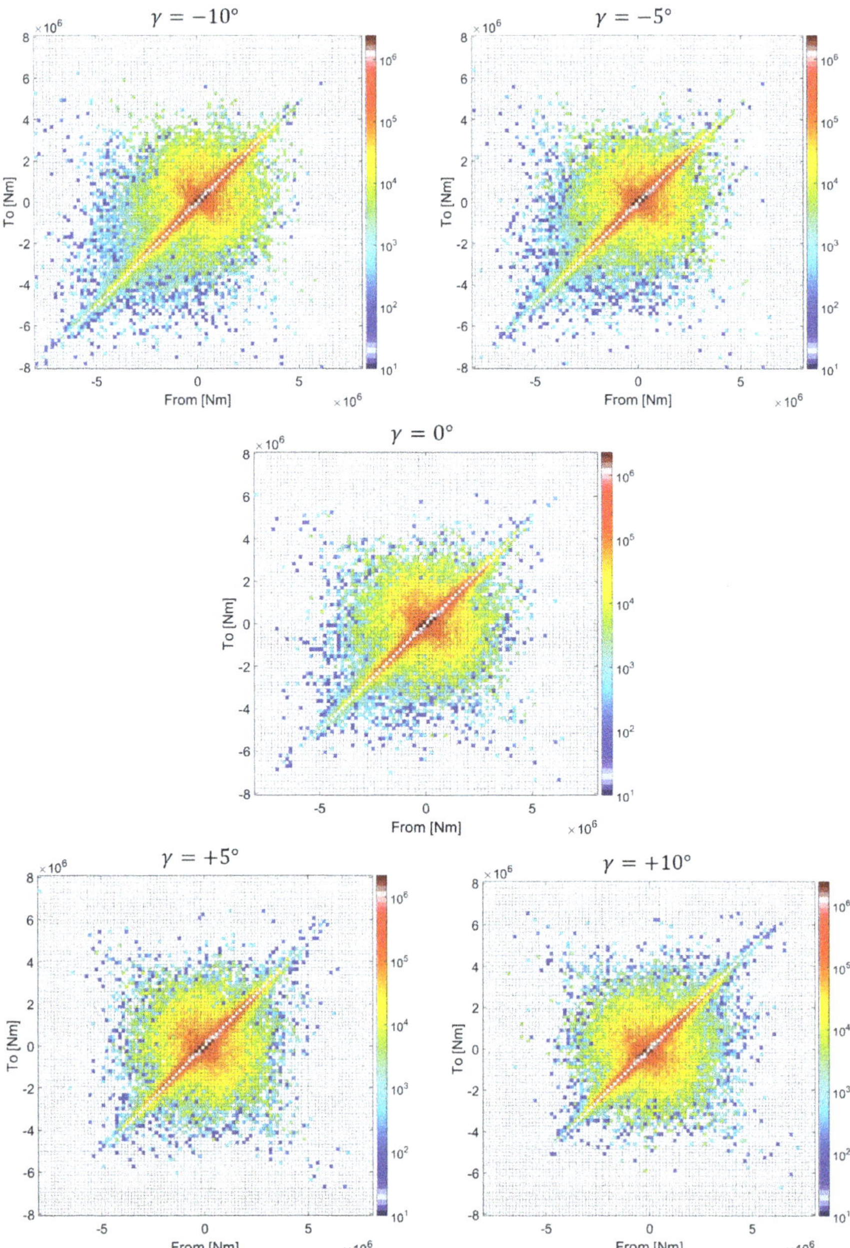

Figure 5. Rain flow count of the bending moment (around z-axis) for various nacelle positions and a timespan of 20 years (**top left**: $-10°$; **top right**: $-5°$; **middle**: $0°$; **bottom left**: $5°$; **bottom right**: $10°$).

Rain flow count diagrams of the torsional torque as well as the axial force (x-axis) and the in-plane vertical force (z-axis) were analysed. The torsional torque and the axial force show no qualitative dependency on the yaw misalignment which matches the results from the literature [13].

The comparable small changes can be attributed to the reduced projected rotor area perpendicular to the inflow. In addition, the vertical in-plane force suffers only marginal changes because it is dominated by the rotor weight which is constant for all simulations.

The in-plane horizontal force is known to have one main direction resulting from the tangential forces acting on the rotor blades. The tangential forces on the blades point towards the rotational direction. However, in the upper half of the rotor they are greater due to the wind shear and the tower shadow leading to a horizontal force on the main bearing pointing to the right when looking downstream. When utilizing the coordinate system shown in Figure 1, the main direction of the side force is the negative y-direction. Figure 3 shows rain flow counts for the in-plane side force. It is observable that the changes are not only dependent on the absolute angle but on the sign as well. Negative angles evoke smaller maximum loads in the main direction. This can be well observed by comparing the load cycle numbers in the interval [−2:−1;−2:−1] (red square). At the same time more load cycles with a changing force direction occur (load cycles in top left or bottom right quadrant, surrounded by grey squares). The tilt moment shows the same qualitative behaviour as the in-plane side force as it can be seen in the top right quadrants of Figure 4. Summarizing, this means that for the side force and bending moment the mean loads become greater with increasing angles but fewer zero-crossings occur. For the bending moment around the z-axis the previously observed effects are inverted (Figure 5). Smaller maximum loads, but more zero-crossings, occur with positive yaw misalignments.

Figure 6 shows the normalized damage equivalent loads for all six degrees of freedom and all yaw misalignments considered in this work. As found before the torsional moment, the axial force and the in-plane vertical force are not affected by yaw misalignments to the same extent as the other loads. As expected from the previous observations, the DEL of the in-plane side force (Fy) increases with the yaw angle. If the value at $0°$ is taken as a reference, the DEL decreases by −15.1% and increases by +18.7% in the investigated range. The same is true for the tilt moment (My) with a relative decrease of −1.2% for a −10° yaw misalignment and an increase of +5.1% for +10° yaw misalignment. Although the in-plane side force experiences the highest relative changes, its absolute values are comparably small to the other loads. The bending moment (Mz) shows a local minimum at 5° yaw angle. The DEL decreases slightly to −0.8%. A broader range of yaw misalignments would be necessary to observe if this is a global minimum. For a negative rotation the DEL is increased by +8.1% for −10° yaw angle.

To better understand these results one can look into the details of the aerodynamic coefficients that are used for calculating the aerodynamic forces. Figure 7 shows the lift and drag coefficients of the NACA64 airfoil taken from [23] which is used in the upper 33% of the blade length. The airfoils used in the rest of the blade differ only insignificantly from the NACA64. The effect of yaw misalignments on the local inflow at the blade elements can be derived from Figure 8. If the nacelle is rotated towards negative angles, the angle of attack decreases for the blade elements in the upper half of the rotor disk and increases for the blades in the lower half. This leads to a change of lift forces which results in a smaller in-plane side force at the main bearing. Conversely, the angle of attack in the upper half increases with a positive rotation of the nacelle, which leads to a greater side force. Similar observations can also be made for the tilt moment. It is mainly affected by the difference between the drag forces acting on the blades in the upper and in the lower half of the rotor disk. Rotating the nacelle in a negative direction will decrease the angle of attack in the upper half and increases it in the lower half. Thereby the difference of the drag forces becomes smaller. This finally leads to a smaller tilt moment at the main bearing. Again similar to the tilt moment, the bending moment is determined by the difference of the drag forces on the left and right rotor disk half. When rotating the nacelle against the inflow the projected wind speed as seen from the blade elements becomes smaller on both sides. This leads to a decrease of angle of attack on the right rotor half (blades that are moving towards the ground) when looking downstream and an increase of angle of attack on the left rotor half (blades that are moving upwards). This effect increases the difference between the drag forces and thus the bending moment and it occurs for both directions of rotation. Therefore, the minimum that was found for the bending moment is very likely to be the global minimum on the basis of these conclusions.

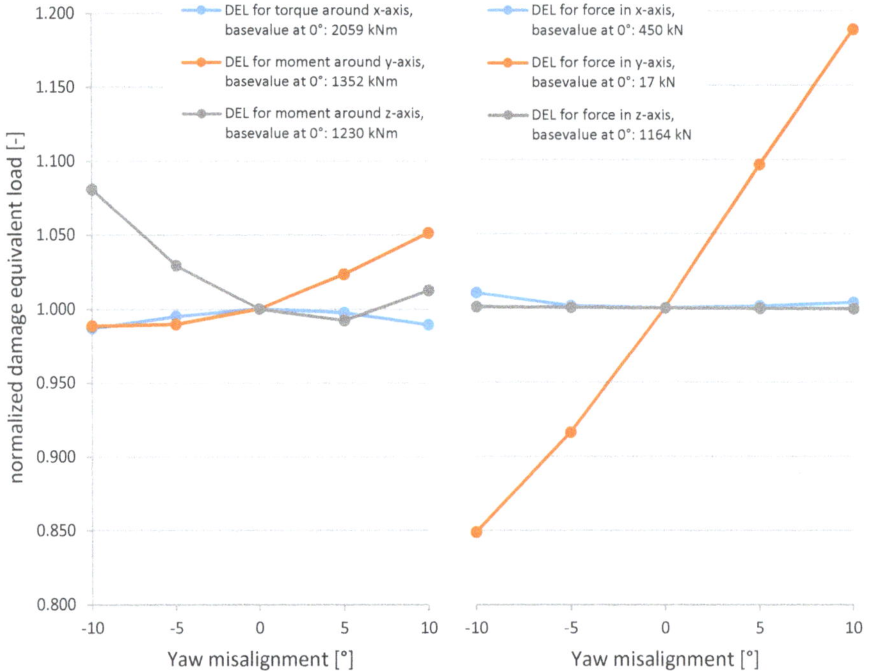

Figure 6. Damage equivalent load for the loads in all six degrees of freedom, various nacelle positions and a timespan of 20 years for an exponent of $k = 3.3$ (**left**: moments; **right**: forces).

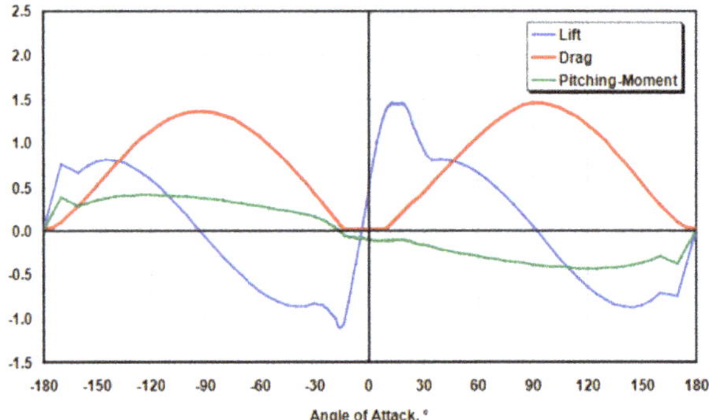

Figure 7. Coefficients of the NACA64 airfoil [23] shown as the respective dimensionless coefficient (y-axis) vs. angle of attack (x-axis).

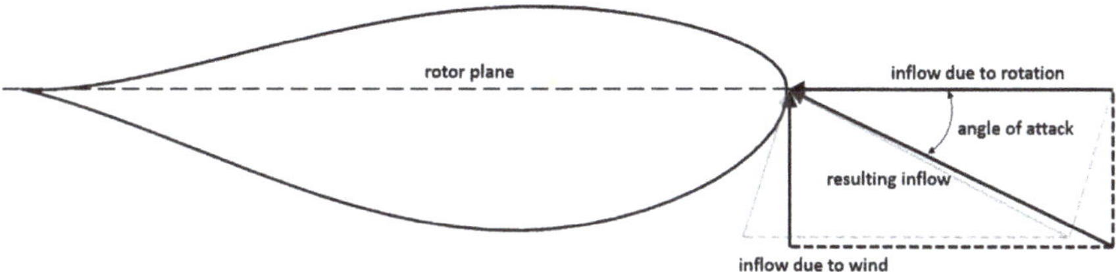

Figure 8. Schematic of the inflow conditions for a yaw misalignment of 0° (black lines) and a positive yaw misalignment (grey lines) for the blade elements of a blade at highest position.

It can be stated that the principle of wake steering offers good potential to increase the general park output as the literature shows. However, it is always to be taken into account that a possibly higher damage accumulation rate is a result. For validation of the real load distribution under wake steering in-field data of a steered turbine e.g., from condition monitoring is necessary.

4. Conclusions

The work carried out in this paper showed the effects of yaw misalignment on loads for a generic turbine. The active misalignment of turbines against the inflow direction is a tool to increase overall power output of the park as well as to decrease overall non-torque loads. However, the turbine that is misaligned itself will not necessarily benefit from this control scheme. A change in non-torque loads is to be expected. The influence of a yaw misalignment over a total timespan of 20 years is examined and analysed in this study. It is found that the differences in loads induced by yaw misalignments aren't symmetrical but dependent on the direction of the inflow. A clockwise rotation of the turbine, as seen from above, leads to smaller maximum loads as well as smaller average load cycle amplitudes for the horizontal in-plane force and the moment around the y-axis. If the turbine is rotated counter clockwise it shows the mentioned above effects for the moment around the z-axis up until 5°. Beyond that the effect seems to be reversed. It can be stated that wake steering does not necessarily have a negative impact on each drive train load, but the influence is to be taken into account. A basic understanding of the load behaviour due to wake steering is achieved. It is absolutely necessary to consider the non-torque-loads when implementing such a control scheme rather than solely focussing on the power output. In further studies, the load distributions calculated in this work can be merged with distributions of yaw misalignments due to wake steering to predict the load spectrum under real conditions. In addition, a cost function for weighting the various loads could be propagated to determine a preferential rotational direction when wake steering is necessary.

Author Contributions: M.C. and B.R. conceived and designed the experiments; M.C. performed the experiments; M.C. and B.R. analyzed the data; R.S. and G.J. contributed materials and analysis tools; M.C. wrote the paper.

Funding: This research was funded by VGB PowerTech.

Conflicts of Interest: The authors declare no conflict of interest.

References

1. Roscher, B.; Werkmeister, A.; Jacobs, G.; Schelenz, R. Modelling of Wind Turbine Loads nearby a Wind Farm. *J. Phys. Conf. Ser.* **2017**, *854*, 012038. [CrossRef]
2. Johnson, K.; Thomas, N. Wind farm control: Addressing the aerodynamic interaction among wind turbines. In Proceedings of the 2009 American Control Conference, St. Louis, MO, USA, 10–12 July 2009.
3. Park, J.; Kwon, S.; Law, K. Wind farm power maximization based on a cooperative static game approach. In Proceedings of the SPIE Active and Passive Smart Structures and Integrated Systems Conference, San Diego, CA, USA, 10 April 2013.

4. Gebraad, P.M.O.; Teeuwisse, F.W.; Van Wingerden, J.W.; Fleming, P.A.; Ruben, S.D.; Marden, J.R.; Pao, L.Y. Wind plant power optimization through yaw control using parametric model for wake effect—A CFD simulation study. *Wind Energy* **2014**, *19*, 95–114. [CrossRef]

5. Marden, J.R.; Ruben, S.D.; Pao, L.Y. A model-free approach to wind farm control using game theoretic methods. *IEEE Trans. Control Syst. Technol.* **2013**, *21*, 1207–1214. [CrossRef]

6. Fleming, P.; Gebraad, P.M.O.; Lee, S.; van Wingerden, J.W.; Johnson, K.; Churchfield, M.; Michalakes, J.; Spalart, P.; Moriarty, P. Simulation comparison of wake mitigation control strategies for a two-turbine case. *Wind Energy* **2015**, *18*, 2135–2143. [CrossRef]

7. Fleming, P.; Gebraad, P.; Lee, S.; van Wingerden, J.; Johnson, K.; Churchfield, M.; Michalakes, J.; Spalart, P.; Moriarty, P. Evaluating techniques for redirecting turbine wakes using SOWFA. *Renew. Energy* **2015**, *70*, 211–218.

8. Fleming, P.; Annoni, J.; Shah, J.J.; Wang, L.; Ananthan, S.; Zhang, Z.; Hutchings, K.; Wang, P.; Chen, W.; Chen, L. Field test of wake steering at an offshore wind farm. *Wind Energ. Sci.* **2017**, *2*, 229–239. [CrossRef]

9. Gebraad, P.; Thomas, J.; Ning, A.; Fleming, P.; Dykes, K. Maximization of the annual energy production of wind power plants by optimization of layout and yaw-based wake control. *Wind Energy* **2017**, *20*, 97–107. [CrossRef]

10. Gebraad, P.M.O. Data-driven wind plant control. Ph.D. Thesis, Delft University of Technology, Delft, The Netherlands, 2014.

11. Santoni, C.; Ciri, U.; Rotea, M.; Leonardi, S. Development of a high fidelity CFD code for wind farm control. In Proceedings of the 2015 American Control Conference (ACC), Chicago, IL, USA, 1–3 July 2015.

12. Ennis, B.L.; White, J.R.; Paquette, J.A. Wind turbine blade load characterization under yaw offset at the SWiFT facility. *J. Phys. Conf. Ser.* **2018**, *1037*, 052001. [CrossRef]

13. White, J.; Ennis, B.E.; Herges, T.G. Estimation of Rotor Loads Due to Wake Steering. In Proceedings of the 2018 Wind Energy Symposium, Kissimmee, FL, USA, 8–12 January 2018.

14. Lee, H.; Lee, D. Wake impact on aerodynamic characteristics of horizontal axis wind turbine under yawed flow conditions. *Renew. Energy* **2019**, *136*, 383–392. [CrossRef]

15. Werkmeister, A.; Schelenz, R.; Jacobs, G. Calculation of the design loads with SIMPACK, In Proceedings of the Simpack User Meeting Conference, Augsburg, Germany, 7 October 2015.

16. International Electrotechnical Commission (IEC). *Wind Turbines—Part 1: Design Requirements*, 3rd ed.; International Electrotechnical Commission (IEC): Geneva, Switzerland, 2014.

17. Zierath, J.; Rachholz, R.; Woernle, C. Field test validation of Flex5, MSC.Adams, alaska/Wind and SIMPACK for load calculations on wind turbines. *Wind Energy* **2016**, *19*, 1201–1222. [CrossRef]

18. Available online: https://nwtc.nrel.gov/TurbSim (accessed on 3 July 2019).

19. Moriarty, P.J.; Hansen, A.C. *AeroDyn Theory Manual*; Technical Report NRELlEL-500-36881; National Renewable Energy Laboratory: Golden, CO, USA, 2005.

20. Internation Electrotechnical Commission standard DIN 50100. *Load Controlled Fatigue Testing—Execution and Evaluation of Cyclic Tests at Constant Load Amplitudes on Metallic Specimens and Components*; Deutsches Institut fur Normung E.V. (DIN): Pforzheim, Germany, 2016.

21. Renewable Energy Act of 21 July 2014 (BGBl. I S. 1066), last amended by Article 1 of the Act of 17 July 2017 (BGBl. I S. 2532). Available online: http://www.gesetze-im-internet.de/eeg_2014/EEG_2017.pdf (accessed on 9 May 2019).

22. Jacobs, G.; Corves, B. *Maschinengestaltung II*; Druck & Verlagshaus Mainz: Aachen, Germany, 2018.

23. Jonkman, J.; Butterfield, S.; Musial, W.; Scott, G. *Definition of a 5-MW Reference Wind Turbine for Offshore System Development*; Technical Report, No. NREL/TP-500-380060; National Renewable Energy Laboratory (NREL): Golden, CO, USA, 2009.

Article

Aeroelastic Analysis of a Coplanar Twin-Rotor Wind Turbine

Amr Ismaiel [1,2,*] and Shigeo Yoshida [3]

[1] Interdisciplinary Graduate School of Engineering Sciences, Kyushu University, 6-1 Kasugakoen, Kasuga, Fukuoka 816-8580, Japan

[2] Faculty of Engineering and Technology, Future University in Egypt (FUE), 5th Settlement, New Cairo 11835, Egypt

[3] Research Institute for Applied Mechanics, Kyushu University, 6-1 Kasugakoen, Kasuga, Fukuoka 816-8580, Japan; yoshidas@riam.kyushu-u.ac.jp

* Correspondence: amr.mohamed@fue.edu.eg or amrmetwally@riam.kyushu-u.ac.jp; Tel.: +81-80-5978-3397

Received: 29 March 2019; Accepted: 14 May 2019; Published: 17 May 2019

Abstract: Multi-rotor system (MRS) wind turbines can be a competitive alternative to large-scale wind turbines. In order to address the structural behavior of the turbine tower, an in-house aeroelastic tool has been developed to study the dynamic responses of a 2xNREL 5MW twin-rotor configuration wind turbine. The developed tool has been verified by comparing the results of a single-rotor configuration to a FAST analysis for the same simulation conditions. Steady flow and turbulent load cases were investigated for the twin-rotor configuration. Results of the simulations have shown that elasticity of the tower should be considered for studying tower dynamic responses. The tower loads, and deformations are not straightforward with the number of rotors added. For an equivalent tower, an additional rotor will increase the tower-top deflection, and the tower-base bending moment both in the fore-aft direction will be more than doubled. The tower torsional stiffness becomes a crucial factor in the case of a twin-rotor tower to avoid a severe torsional deflection. Tower natural frequencies are dominant over the flow conditions in regards to the loads and deflections.

Keywords: aeroelasticity; multi-rotor system; wind turbine

---

1. Introduction

With the world's high demand of energy, and the limitation of the amount of fossil fuels, renewable energies have become a field of interest for many researchers. Wind energy is one of the most growing renewable sources of energy, in terms of usage and research topics. The global cumulative installed wind capacity has increased 2200% from the year 2001 to 2017 [1]. At present, horizontal axis wind turbines remain the dominant wind energy conversion technology. In the past decades, the trend was to increase the diameter of the rotor, since the power produced is proportional to the rotor area. Currently, the world's largest wind turbine has a 12 MW capacity, and a 107-meter blade length [2]. However, with this large size comes big challenges, such as the huge transportation and installation cost of extremely large wind turbines and the severe structure dynamic loads on the blades and the tower, as well as the need to develop each component, including the blade, bearing, generator, gearbox, etc. to be suitable for the large-scale single rotor turbines, which includes risk in cost and quality. Also, it includes a risk if a failure occurs; then the whole wind turbine will shut down and no power will be produced until the failed part is fixed or replaced [3].

The multi-rotor systems (MRS) is a technology with a long history that goes back to 1930, but it has fallen out of consideration for its structure complexity, while large scale single rotor wind turbines have become technically feasible [4]. However, with current advanced materials technology, the materials used to construct the rotors have a higher strength to weight ratio. With those advances in the materials technology, MRS is a promising alternative to large-scale wind turbines. The main advantages of MRS are the standardization of the wind turbine components, ease of transportation (since the rotors are of a small scale compared to large-scaled wind turbines), ease of installation, and the cost and the reliability of wind turbines, since multi rotors ensure that if there is a failure in one rotor, then the other rotors will still produce power. The major challenges in an MRS are the complexity of the supporting structure, the yawing system, and the aerodynamic interaction between rotors placed closely to each other.

As MRS has become an interesting field of research, researchers from many countries have made attempts to issue very interesting research points. Some attempts were interested in studying aerodynamics and the aerodynamic interaction between the rotors of an MRS. Experiments made by Goltenbott et al. [5] have shown that two and three diffuser augmented rotor configurations can increase the power produced per rotor by 5% and 9% respectively, compared to a single rotor. Also, the computational fluid dynamic (CFD) simulations made by Chasapogiannis et al. [6] on a seven rotor system have shown a power increase of 3% per rotor. The coherence effect on the produced power and tower loads on a seven rotor MRS has been studied by Yoshida et al. [7]; Wind models with three different coherences were used in the simulation and showed that larger coherence implies higher power production yet increases the collective loads. MRS was also found to improve the wake recovery; the wakes were found to recover faster for MRS compared to a single-rotor configuration and showed a smaller turbulence intensity in the wake [8].

One of the leading research institutes showing great interest in MRS is the Technical University of Denmark (DTU). The DTU constructed a four rotor wind turbine at the Risø campus. They conducted experiments as well as simulations for the four rotor wind turbine, and both agreed that the interaction between the rotors improved the power performance by 1.8 ± 0.2%, which can increase the annual power production by 1.5 ± 0.2% [9].

Downscaling the design and cost of wind turbine rotors to replace a single large rotor with multiple smaller rotors has been done by Verma et al. [10]. A 5 x 1 MW multi-rotor turbine was compared to the National Renewable Energy Laboratory (NREL) 5MW single rotor turbine. The scaled down multi-rotor configuration has shown a 37% reduction in weight and a 25% reduction in cost compared to a single rotor producing the same amount of power. As an extension of Verma's work, Mate et al. [11] have designed a support structure for the five rotor configuration proposed in their colleague's work, in addition to other configurations that Mate proposed himself. A finite element approach was used for modeling the support structure. However, these simulations did not include a study of the aeroelastic behavior of either the blades or the support structure and the wind conditions.

Aeroelastic analyses for wind turbines are doable for single rotors using either a Computaional Fluid Dynamics (CFD) model like the Fluid Structure Interaction (FSI) made by Bazilevs et al. [12] or Halawa et al. [13] or using deterministic models. NREL's tool, FAST, is one of the most used aeroelastic tools for modeling wind turbines, it is based on models derived from fundamental theory of aerodynamics and structure analysis, which are more time-efficient compared to CFD models [14].

So far there has been no research or open-source tools proposed that introduce aeroelastic analysis for the multi-rotor concept. In this work, the support structure of an MRS is being aero-elastically analyzed, so that structural problems can be addressed in further research studies. It is the first attempt to develop an in-house aeroelastic tool for an MRS support structure. In this work, the present tool is used to model a twin-rotor wind turbine, with two coplanar rotors placed on a T-shaped tower. This tool can be later extended to model support structures for different configurations of MRS.

The theory used in this work is the Blade Element Momentum (BEM) theory, which is used to calculate the aerodynamic loads, and the virtual work method with a modal approach to calculate the structural deformations of blades and tower. Combining the two theories creates an aeroelastic interface between the blades and tower on one hand, and wind on the other.

The results of the present tool are verified by comparing results of a single rotor wind turbine, to NREL's FAST results of the same turbine model. Then, the results for the twin-rotor configuration are introduced.

2. Mathematical Model

2.1. Aerodynamic Model

Unsteady BEM was used in this work to estimate the aerodynamic loads tangential to and normal on each section along the blade. The blade coordinate system was used for the governing equations, as shown in Figure 1.

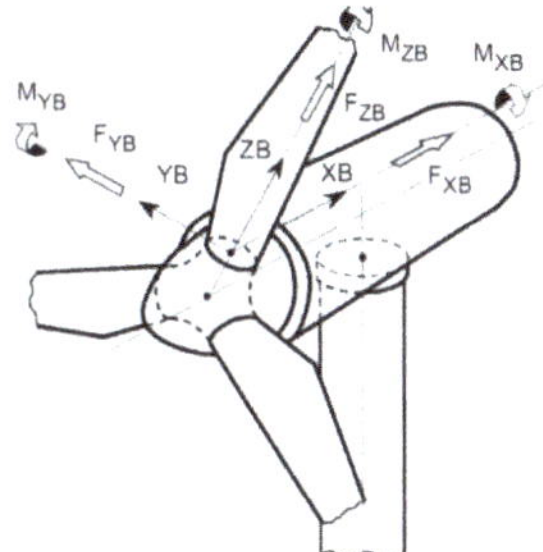

Figure 1. Blade Coordinate System [15].

Initially, the relative wind velocity components on each blade section, in the rotor plane, tangential to the blade width (y_B-axis), and normal to it (x_B-axis) as shown in Figure 2 are as follows:

$$V_{rel,x_B} = V_{0,x_B} + w_{x_B} \tag{1}$$

$$V_{rel,y_B} = V_{0,y_B} - \omega r + w_{y_B} \tag{2}$$

where V_0 is the inflow velocity, ω is the rotational speed of the rotor, r is the blade section position, and w_{x_B}, w_{y_B} are the induced velocities in x_B and y_B directions.

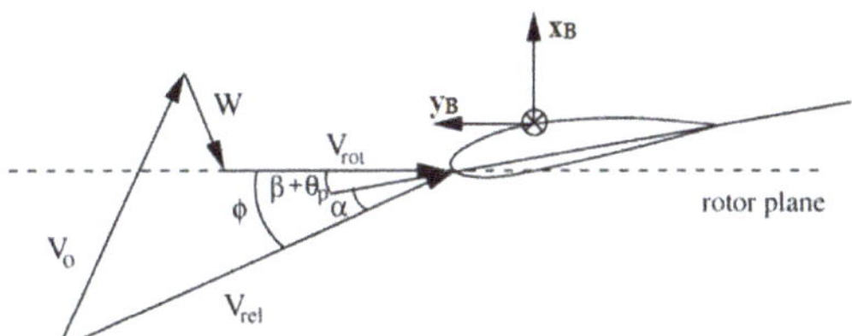

Figure 2. Velocity Triangle on a Blade Section [16].

Prandtl's tip loss factor and Glauert correction are used in the calculation of induced velocities; then, depending on the relative velocity components on each blade section, for every time step, the coefficients of lift and drag forces can be interpolated from the blade airfoils' data tables. Next, lift and drag forces can be calculated for each section, and hence the aerodynamic loads in the plane of rotation ($L_{yB,aero}$) and normal to it ($L_{xB,aero}$) can also be calculated. Dynamic wake, dynamic stall, and yaw misalignment effects were ignored in this study. With the tangential and normal load distributions known, the rotor aerodynamic thrust, torque, and power can be calculated [16].

2.2. Structure Model

The principle of virtual work was used to calculate the structure dynamics parameters. This principle is a method that helps to set up the correct matrices for a discretized mechanical system, such as Newton's second law shown in Equation (3). This method is well suited for a chained multi-body system like wind turbines:

$$M\ddot{x} + C\dot{x} + Kx = F_g \tag{3}$$

where; M is the mass matrix, C is the damping matrix, K is the stiffness matrix, and F_g is the generalized forces array. x, $\dot{x}$, and $\ddot{x}$, are the generalized coordinates for the three modes used in the model, their first, and second time derivatives, respectively.

The degrees of freedom used in this model are the uncoupled modal shapes for both the blades and the tower. For the blades, the first and second flap-wise, and the first edge-wise modes are used, denoted by u^{f1}, u^{f2}, and u^{e1} respectively. For the tower, the first and second fore-aft, and the first side-side modes are used, denoted by u^{fa1}, u^{fa2}, and u^{ss1}. These modes were chosen because the orthogonality constraint between the eigen modes turns the mass matrix into a diagonal matrix, and hence into the damping and stiffness matrices that are dependent on the mass matrix. This results in uncoupled differential equations that can be solved by a fourth order Runge-Kutta numerical technique.

Solving this system of equations results in the values of x, $\dot{x}$, and $\ddot{x}$, on which the deformation of either the blade or the tower is assumed to be dependent, as a linear combination of the three modes. For instance, in the case of the blade, with a stiff tower, the deformation distribution along the blade section position (r), denoted by u_{xB} in the flap-wise direction and u_{yB} in the edge-wise direction, will be as follows:

$$u_{xB}(r) = x_1 u_{xB}^{f1}(r) + x_2 u_{xB}^{e1}(r) + x_3 u_{xB}^{f2}(r) \tag{4}$$

$$u_{yB}(r) = x_1 u_{yB}^{f1}(r) + x_2 u_{yB}^{e1}(r) + x_3 u_{yB}^{f2}(r). \tag{5}$$

The velocity and acceleration distribution along the blade are calculated the same way, except for using $\dot{x}$, and $\ddot{x}$ instead of x [17].

2.3. Aeroelastic Coupling

The deformation of the blade, together with the velocity and acceleration of its vibration, will result in change in the loads and hence structural deformation in the next time-step. The relative wind velocity components on the blade sections can now be updated to include the blade vibrations, with the blade velocity distribution always opposing the wind direction. Equations (1) and (2) can be updated as follows:

$$V_{rel,xB} = V_{0,xB} + w_{xB} - \dot{u}_{xB} \tag{6}$$

$$V_{rel,yB} = V_{0,yB} - \omega r + w_{yB} - \dot{u}_{yB} \tag{7}$$

where $\dot{u}_{xB}$, and $\dot{u}_{yB}$ are the blade sections' vibrational velocities in the flap-wise and edge-wise directions.

The loads on the blade sections are also updated to include gravitational loads and inertia loads due to the blade vibrations. The total load distribution along the blade sections in the normal and tangential directions to the plane of rotation will be as follows:

$$L_{xB}(r) = L_{xB,aero}(r) - m'(r)\ddot{u}_{xB}(r) + m'(r)g\sin(\theta_t + sin\theta_c)\cos\theta_A \tag{8}$$

$$L_{yB}(r) = L_{yB,aero}(r) - m'(r)\ddot{u}_{yB}(r) + m'\sin\theta_A \tag{9}$$

where m' is the mass density distribution along the blade length, $\ddot{u}_{xB}$ and $\ddot{u}_{yB}$ are the blade sections' vibrational accelerations, θ_t is the tilt angle, θ_c is the cone angle, and θ_A is the azimuth angle. The loads in the direction of the blade length are neglected.

The updated relative wind velocity and loads are used for the next time step, guaranteeing that both the aerodynamics and structure models affect each other to create an aeroelastic model.

The tower is modelled independently from the blades, and the rotor loads are transmitted to the tower, including rotor thrust, weight, and torque, taking into consideration the aeroelastic behaviour of those loads. Also, the aerodynamic load on the tower itself is calculated, considering the tower vibrations in the relative wind speed to the tower, and the tower inertial loads.

3. Verification of the Single Rotor Model

The single rotor wind turbine simulated for validation is the NREL 5MW wind turbine. In order to ensure the reliability of the developed aeroelastic tool, the results of aerodynamic analysis, dynamic structural analysis, and aeroelastic analysis are compared to results of the NREL 5MW results in the definition report of the wind turbine [18] and FAST simulation for the aeroelastic case.

3.1. Model Outline

The geometric and material properties of the turbine blades and tower are fully defined in the NREL definition report. General rotor specifications are shown below:

- Rating: 5 MW.
- Rotor orientation, configuration: Upwind, 3 Blades.
- Rotor Diameter: 126 m.
- Hub Height: 90 m.
- Cut-in, Cut-out Wind Speed: 3 m/s, 25 m/s.
- Rated Wind Speed: 11.4 m/s.
- Rotor Mass: 110,000 kg.
- Nacelle Mass: 240,000 kg.
- Tower Mass: 347,460 kg.

Figure 3 shows a CAD model for the turbine blade.

Figure 3. NREL 5MW wind turbine blade [19].

The blade and tower structural specifications are shown in Figure 4.

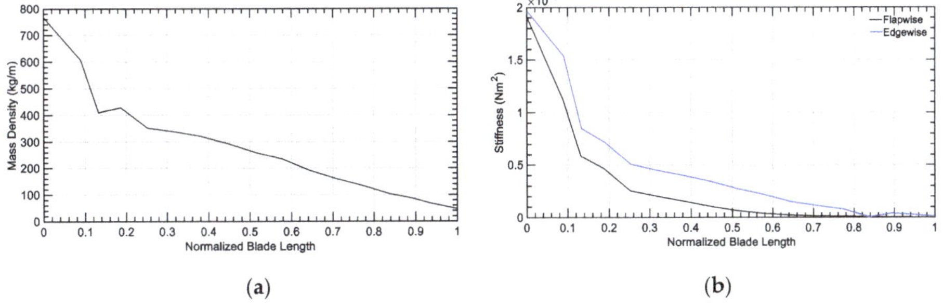

Figure 4. *Cont.*

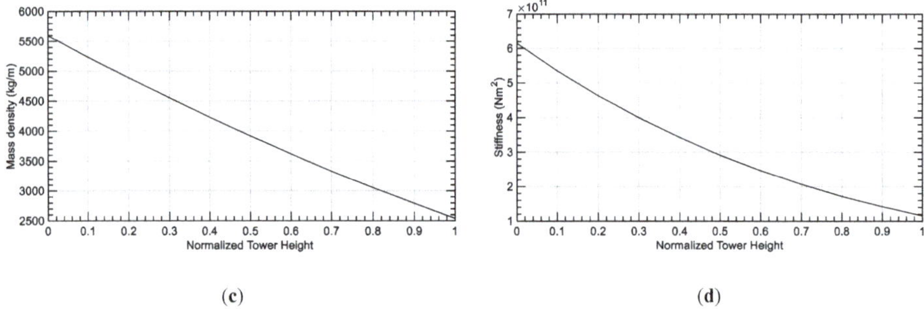

(**c**)   (**d**)

Figure 4. Blade structural properties: (**a**) Blade mass density distribution; (**b**) Blade stiffness distribution; (**c**) Tower mass density distribution; and (**d**) Tower stiffness distribution.

3.2. Rotor Aerodynamics

The NREL 5MW wind turbine is a pitch-controlled turbine. Over the rated wind speed of 11.4 m/s, the pitch control is applied to maintain the rated power output and rotor speed. So far, pitch control has not been applied in the present tool. Accordingly, the full power region cannot be simulated in the current version of the tool, but it will be added later. The BEM in the developed tool is run for wind speeds starting from the cut-in speed of 3 m/s, to the rated speed of 11.4 m/s. The rotor power, thrust, and torque are compared. The results are shown in Figure 5.

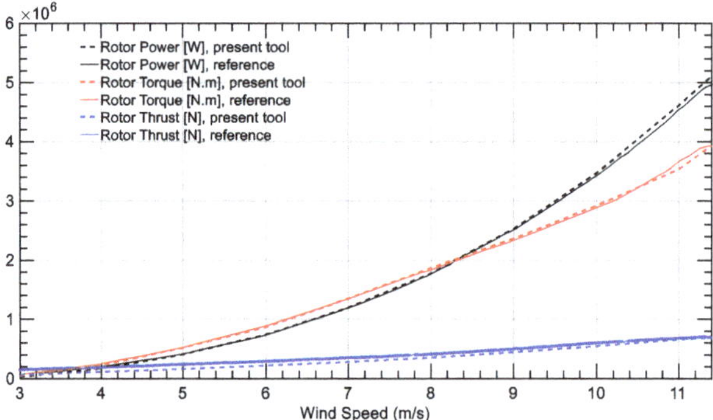

Figure 5. Steady state responses: comparison between results of the developed tool to NREL reference file.

Figure 5 shows good agreement between the calculated steady state values of the power, torque, and thrust of the developed tool, compared to the turbine definition document's results. This agreement proves that the first part of the tool concerned with calculating the aerodynamic responses, without considering the aeroelastic behavior of the blades, is reliable, and the next step would be the dynamic responses of the blades and tower.

3.3. Modal Analysis

In this section, the free vibration modes of the blades and tower are investigated. The structural model is based on the uncoupled modes of vibrations, and it is important that the mode shapes and natural frequencies will be correct, before proceeding to the aeroelastic coupling between the aerodynamics and structural behavior of both the blade and the tower.

The NREL tool Modes [20] was used to calculate the uncoupled mode shapes and natural frequencies of the blades and the tower. Mass and stiffnesses distribution along the blade and tower are available in the NREL 5MW definition document and were used as input for Modes. The mode shapes and natural frequencies are also available in the definition document and are shown for comparison with the results of simulation.

3.3.1. Blades

The mode shapes used in the structural model are the first and second flap-wise modes, and the first edge-wise mode. Natural frequencies and mode shape comparisons are shown in Table 1.

Table 1. Natural frequencies of the blade.

Mode	Calculated (Hz)	Reference (Hz)	Deviation (%)
First flap-wise	0.71	0.70	1.4
Second flap-wise	2.02	2.02	0.0
First edge-wise	1.08	1.08	0.0

Results of the natural frequencies from the Modes simulation agree with the reference values

In Figures 6 and 7, the mode shapes calculated by Modes are compared to the mode shapes in the NREL 5MW definition file. The results are almost identical for both the flap-wise and edge-wise modes.

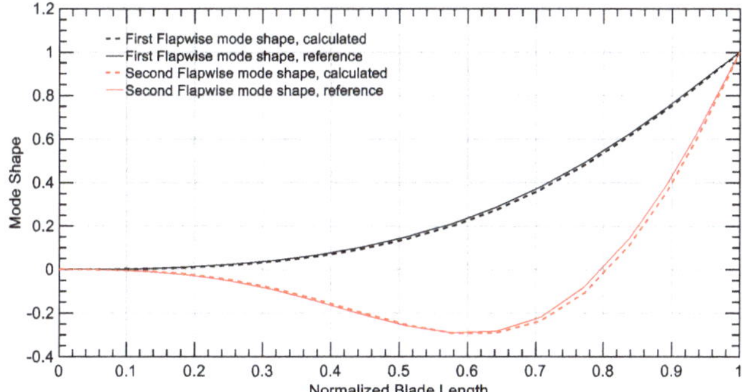

Figure 6. Blade flap-wise mode shapes.

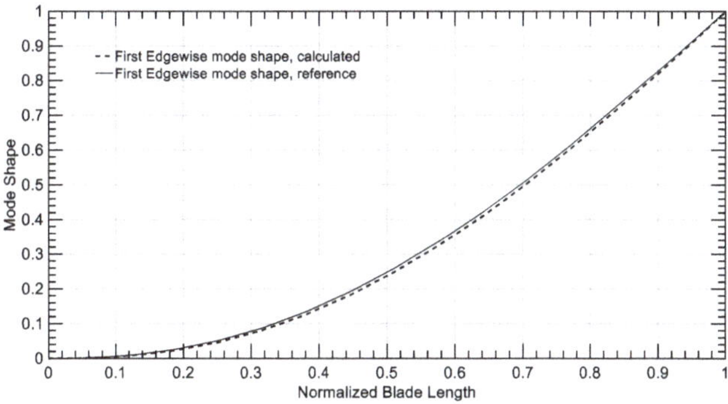

Figure 7. Blade edge-wise mode shape.

3.3.2. Tower

Similarly, the free vibrations of the tower are compared. The modes considered in the aeroelastic tool are the first and second fore-aft modes and the first side-side mode. Natural frequencies are shown in Table 2 and mode shapes are shown in Figures 8 and 9.

Table 2. Natural frequencies for the tower: single rotor configuration.

Mode	Calculated (Hz)	Reference (Hz)	Deviation (%)
First fore-aft	0.32	0.32	0.0
Second fore-aft	3.06	2.90	5.2
First side-side	0.32	0.31	3.1

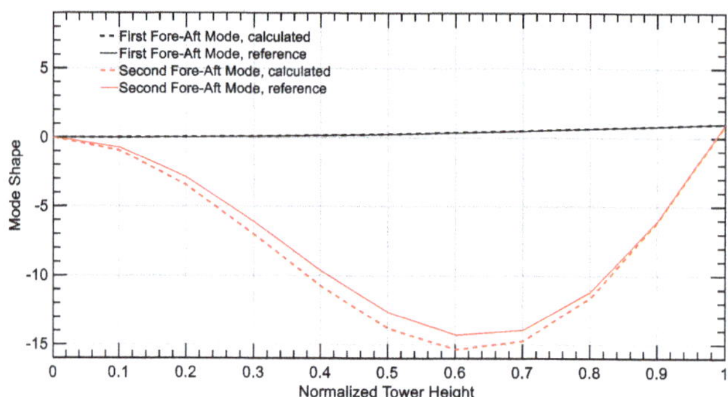

Figure 8. Tower fore-aft mode shapes: single rotor configuration.

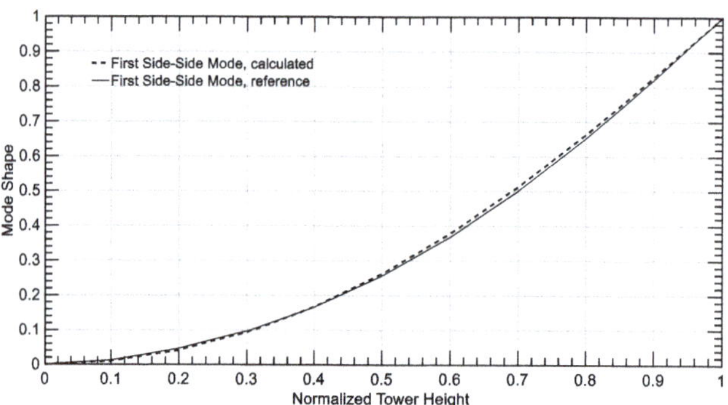

Figure 9. Tower side-side mode shape: single rotor configuration.

Natural frequencies as well as mode shapes of the tower show agreement between the results of Modes compared to the reference.

With the aerodynamic steady state response and the structure dynamic response verified separately, the next step is validation of the aeroelastic tool, where both aerodynamics and the structure of the turbine are coupled.

3.4. Aeroelastic Analysis

In this subsection, the coupling between the aerodynamic loads and the structural behavior is introduced. The aerodynamic loads affecting the blades cause deformation, and hence the relative velocities on the blade sections are changed. Inertial loads generated from the blade vibration also affect the structural behavior. The effects of blade vibrations are considered every time step in the simulation to catch the aeroelastic behavior for both the blades and the tower.

For validation of the results, a FAST simulation is made for the NREL 5MW wind turbine blades and tower, using the same simulation conditions as in the in-house tool. The blades are subject to a constant wind speed and are rotating at a constant angular speed. Simulation parameters are as follows:

- Wind speed: 11.4 m/s (Rated wind speed)
- Rotational velocity: 12.1 rpm
- Pitch angle: 0°
- Time step: 0.01 s
- Simulation time: 50 s

3.4.1. Blades

For the simulation of the blades, the tower is considered as a stiff body. The dynamics are generated from the rotation of the blades in the azimuth direction and the vibrations in the flap-wise and edge-wise directions. Further, the tower is considered to be a stiff body in the FAST simulation to create a similar simulation environment for comparison. In Figures 10 and 11, bending moments at the blade root in the flap-wise and edge-wise directions are shown with the azimuth position of the blade.

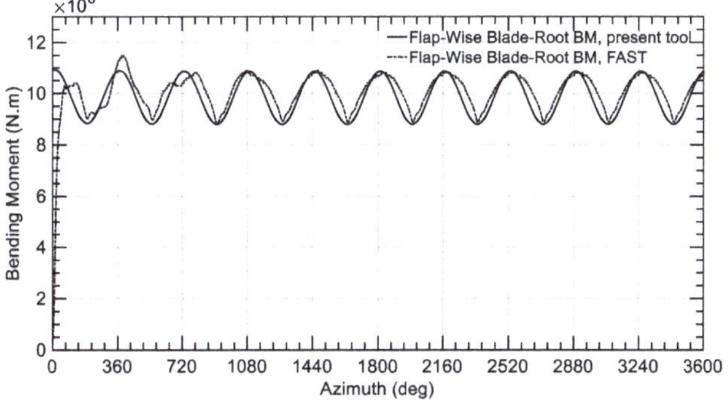

Figure 10. Blade-root, flap-wise bending moment.

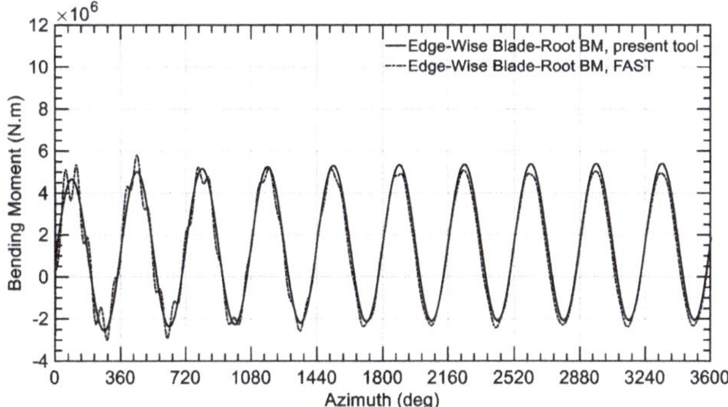

Figure 11. Blade-root, edge-wise bending moment.

The periodic behaviour of the dynamic response of the blade is due to the effect of gravity and inertia on the blades while in different azimuth positions, as well as the shear layer of the flow. As the blade rotates, considering the rotor's tilt angle, the mass centred along the blade length moves such that the blade moves towards or away from the plane of rotation.

Figures 12 and 13 show the blade tip deflections.

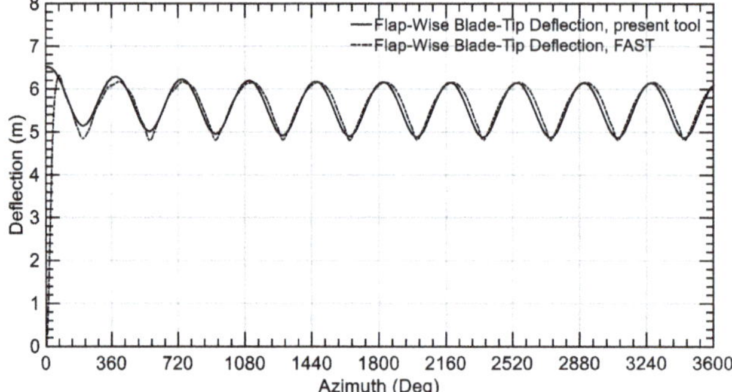

Figure 12. Blade-tip, flap-wise deflection.

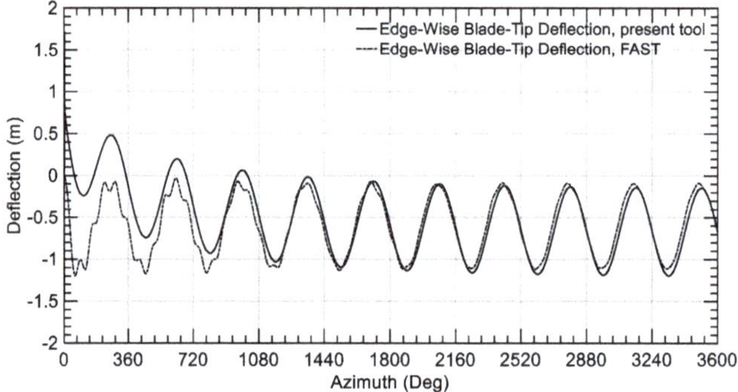

Figure 13. Blade-tip, edge-wise deflection.

There are some discrepancies between the results of the present tool and FAST, especially in the initial runs of the simulation. Those discrepancies are due the difference in the natural frequencies and mode shapes between the present tool and FAST. The aeroelastic model might as well be modeled in a different way in FAST than in the present tool. Other than the initial condition, there is a very good agreement between the results of the present tool and FAST results in terms of the ranges of values, mean values, and frequencies of each time series. Consequently, the proposed tool has proved its ability to describe the blade dynamics and hence its ability to model the tower.

3.4.2. Tower

The tower is now considered for study. The same conditions of simulation were set for both the proposed tool and FAST. The loads were transmitted from the rotor to the tower, considering the dynamic behavior of the rotor loads. The elasticity of the tower together with the aerodynamic and gravitational loads of the tower itself were also considered. As observed from the dynamic responses of the blade, the out-of-plane quantities are more significant than the in-plane ones, and hence, the out-of-plane properties for the tower are shown for comparison.

It is observed in Figures 14 and 15 that there is a very good agreement between the results of the present tool compared to FAST. Discrepancies appear in the tower dynamics in the initial runs as well. This is due to difference in the rotor loads, which appeared in the blade's results, as well as differences in the natural frequencies of the tower itself.

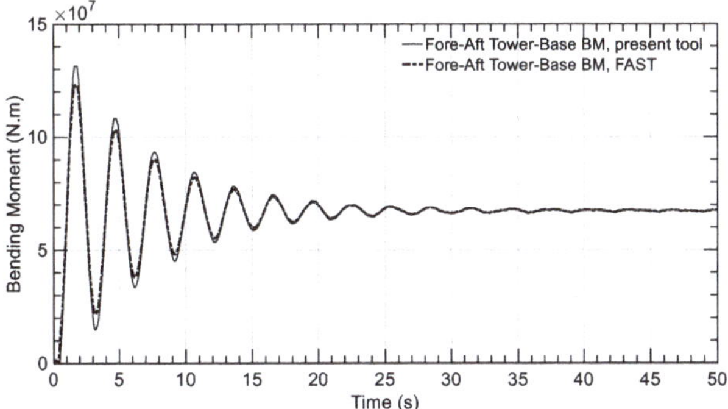

Figure 14. Tower-base, fore-aft bending moment.

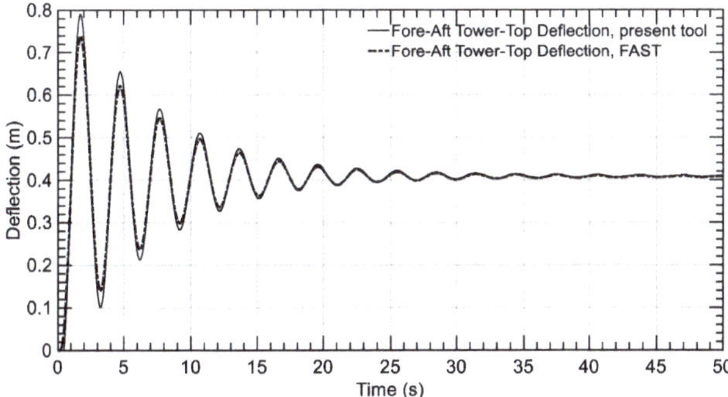

Figure 15. Tower-top, fore-aft deflection.

For both the blade and tower dynamic responses, the results of the proposed tool have shown credibility in modeling the aeroelastic behavior of a wind turbine. So far, the tool is able to generate loads and deflections' time series for the turbine parts in case of a steady wind. In the next section, another rotor is added on the same tower, and the tower dynamics are studied.

4. Simulation of the Twin-Rotor Model

4.1. Twin-Rotor Wind Turbine Model

In this section, the tower of a twin-rotor turbine with two NREL 5MW rotors is modeled. The support structure is assumed to be a T-shaped structure, with the main tower and two side booms connecting the rotors. The side booms are assumed to be a scaled-down structure of the NREL 5MW main tower, each of 63.5 m in length from the main tower center point, such that the two rotors are distanced at 127 meters from hub to hub. These booms' lengths were chosen such that the tips of each of the 126 m diameter rotors are 1 m apart.

It is also assumed that there is no aerodynamic interaction between the two rotors. This assumption can be only accepted as preliminary study, as in reality, rotors affect each other. However, to account for that assumption, a pitch misalignment of 0.2° and −0.2° is added to each rotor's second and third blades, respectively, which is common to generate aerodynamic imbalance. Figure 16 shows a sketch of the proposed twin-rotor configuration.

For comparison of the tower's structural behavior between single and twin rotor configurations, the main tower's geometry and structural properties are changed such that the natural frequency of the first fore-aft mode is the same for both towers. The outer diameter of the tower is changed while the thickness is kept constant. It was found that an outer diameter 1.25 times the diameter of the single-rotor configuration turbine will cause the first fore-aft natural frequency of both configurations' towers to be equal. Table 3 shows the differences in the geometry between both configurations.

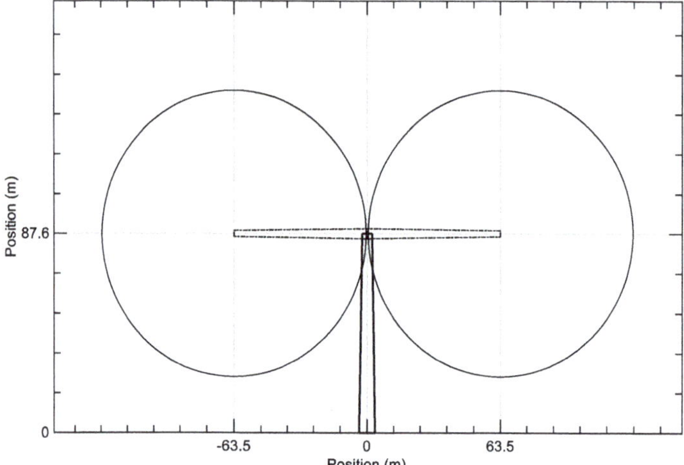

Figure 16. Sketch of the twin-rotor configuration.

Table 3. Tower geometry.

Property (m)	Single rotor	Twin rotor
Tower base diameter	6.000	7.500
Tower base thickness	0.027	0.027
Tower top diameter	3.870	4.840
Tower top thickness	0.019	0.019
Tower height	87.600	87.600

Figure 17 shows the difference in the mass and stiffness distributions between the single rotor and twin rotor configurations' towers.

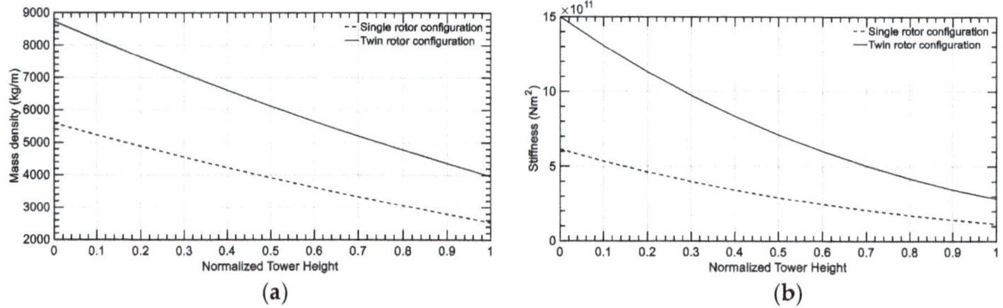

Figure 17. Tower structural properties: single rotor vs. twin rotor: (**a**) tower mass density distribution; (**b**) tower stiffness distribution.

The natural frequencies in case of the twin rotor configuration are shown in Table 4.

The natural frequencies other than the first fore-aft mode are changed compared to the single rotor configuration. This means that the tower stiffness and structural damping matrices in the mathematical model in the case of the twin rotor are different. Moreover, the loads are increased significantly due to the addition of another rotor, and hence the deformations are expected to change in a non-straightforward way.

Table 4. Natural frequencies for the tower – Single Vs. Twin rotor configuration.

Mode	Single rotor (Hz)	Twin rotor (Hz)	Deviation (%)
First fore-aft	0.32	0.32	0.0
Second fore-aft	3.06	3.20	4.4
First side-side	0.32	0.25	21.8

4.2. Simulation Conditions

Two load cases were investigated for the twin rotor configuration; one is in a steady wind condition, and the other is in turbulent wind conditions.

In the steady wind case, the rotors are subject to the same conditions as in the case of single rotor: a steady wind velocity of 11.4 m/s, rotating the rotors at 12.1 rpm. This case studies the aeroelastic properties when the two rotors are rotating simultaneously, such that the rotor loads are superimposed in all the loads' value ranges, and then investigates when the rotors have a 60° phase change in the blades' azimuth positions.

In the turbulent wind condition case, the rotors are subject to a turbulent wind field created by NREL's tool TurbSim [21]. The rotors are subject to different turbulence classes according to IEC 61400-1 standards [22] at an average wind speed at the hub height of 8 m/s and turbulence intensities of class A (high turbulence), B (moderate turbulence), and C (low turbulence). A variable speed control algorithm is used in the turbulent case.

In all cases, the out-of-plane dynamic responses—deflection and bending moment—of the tower are shown.

4.2.1. Case 1: Steady Flow Condition

In this case, the two rotors' loads superimpose in the whole range of values, in the peak values at the beginning of rotation and until it settles for the nominal value of the load, while the blades of both rotors are in the same azimuth position.

First, the tower was modeled as a stiff tower, and the bending moment at the tower base is calculated. Then this model is compared to an elastic tower model to see the differences in results. The results of this simulation are shown in Figure 18.

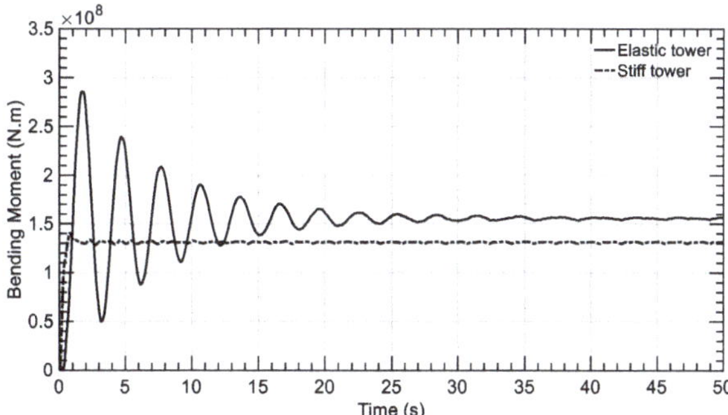

Figure 18. Tower-base fore-aft bending moment: stiff vs. elastic tower.

The stiff tower model doesn't show the dynamic behavior of the load; moreover, the values are less than those in the elastic tower model. The gravity effects due to the vibration of the tower are eliminated in the case of the stiff tower, and hence the loads are far from the real values. This proves that it is not proper to consider the tower to be a stiff for load calculations and it is important to model it as an elastic tower.

Dynamic responses for the elastic tower model were then calculated, Figures 19 and 20 show the tower-base bending moment and tower-top deflection in the fore-aft direction.

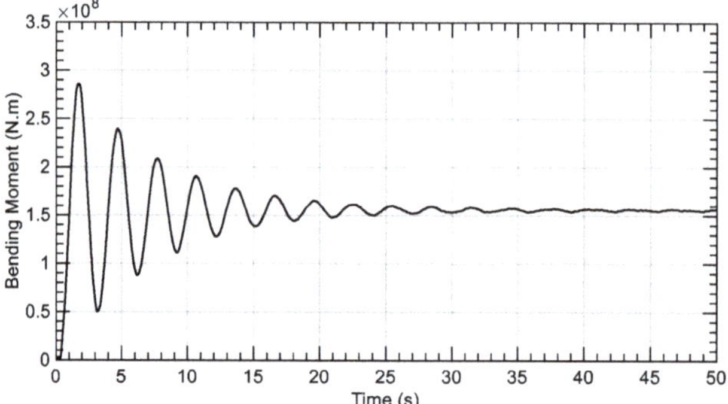

Figure 19. Tower-base, fore-aft bending moment.

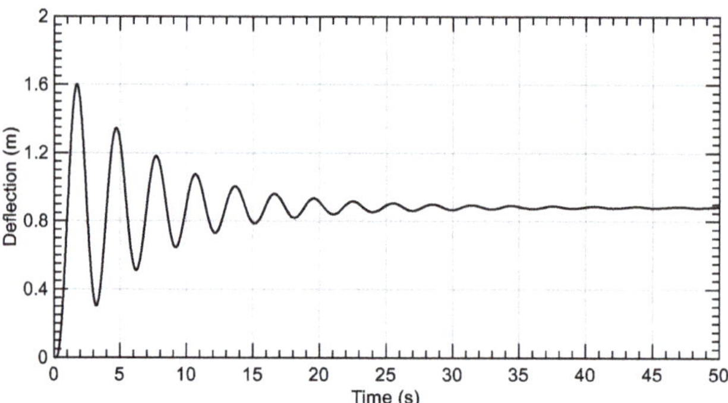

Figure 20. Tower-top, fore-aft deflection.

As was expected, the change in the values of deflection of the tower top and bending moment of the tower base is not linear with the addition of an extra rotor. The change is affected by both the added load and the change in the natural frequencies of the new tower's geometrical properties and hence the stiffness and damping matrices in the mathematical model. The difference is elaborated clearly in Figure 21, where the results of the single-rotor and twin-rotor are shown together.

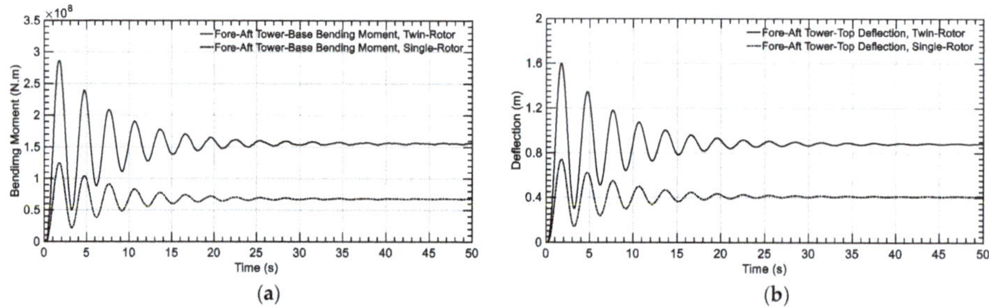

Figure 21. Dynamic responses of the tower: single-rotor vs. twin-rotor load case 1: (**a**) tower-base, fore-aft bending moment, and (**b**) tower-top, fore-aft deflection.

For a tower with the same first fore-aft natural frequency as the single-rotor configuration, the effect of adding one more rotor on the dynamic responses is not straightforward. Two simultaneously rotating rotors on the same tower increase the tower loads and deflections are more than doubled. This is due to the change in the structure of the mathematical model and the added weight and rotor inertias on the top of the tower, which change the natural frequencies in the second fore-aft and first side-side directions.

Then, a phase difference in the initial azimuth position of the first blade of each rotor of 60° is investigated for comparison. The dynamic responses of the tower are shown in Figures 22 and 23.

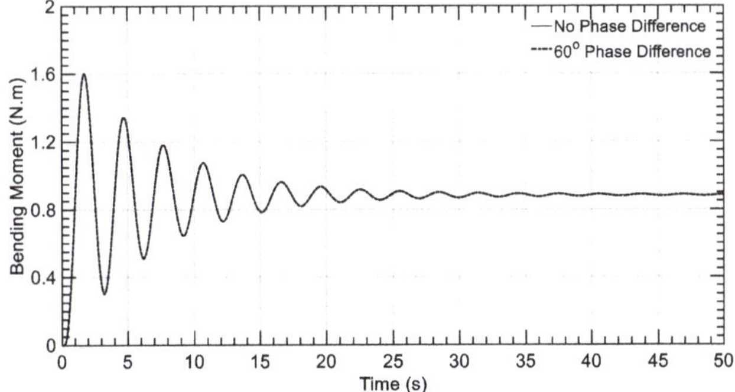

Figure 22. Tower-base, fore-aft bending moment.

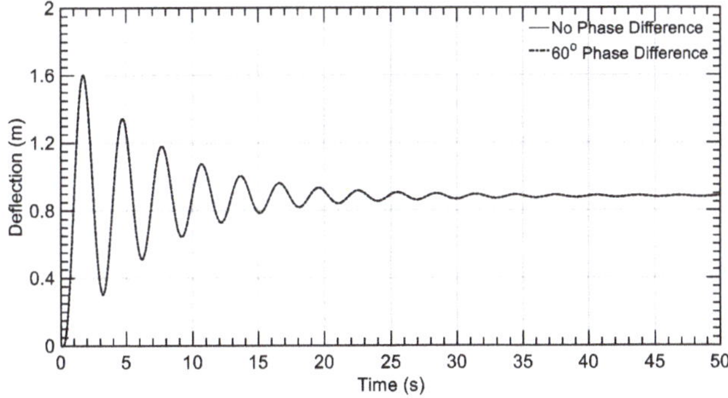

Figure 23. Tower-top, fore-aft deflection.

Here, the dynamic responses of the tower are almost identical with the no phase difference case, but there is a very slight difference which can only be seen only by zooming into the graph. However, with the phase change between the azimuth position of the rotors' blades, there is a slight phase change in the rotor loads, and hence a twisting moment is generated on the tower causing a yawing deflection, which should be studied to anticipate its effect. The yawing deflection of the tower-top is shown in Figure 24.

When the two rotors were rotating simultaneously and under the same aerodynamic conditions, there was no twisting moment for the tower and hence no deflection. When only a phase change between the rotors occurred, a twisting moment was generated causing angular deflection. For different wind conditions the effect of twist can be severe and cause torsional fatigue and hence failure. So, for a twin-rotor configuration, torsional stiffness should be carefully considered in the tower design.

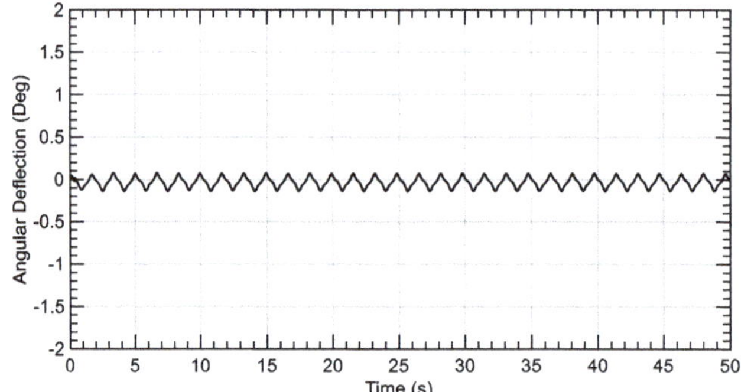

Figure 24. Tower top yawing deflection - Load case 2.

4.2.2. Case 2: Turbulent Flow Condition

In this case, the rotors are subject to a turbulent flow field, using the IEC Kaimal spectral model. Fields having turbulence intensities of IEC 61400 classes A, B, and C, with an average wind speed of 8 m/s, were created with TurbSim. The turbulent grid width was doubled to be able to cover both rotors. Variable speed control was applied to control the rotating speed of the rotors. The generator specifications are available in the NREL 5MW definition report [14]. For a gear ratio of 97:1 and a generator efficiency of 94.4%, the optimal constant of proportionality will be 0.0255764 N.m/rpm^2. The simulation was run for 10 minutes, and the tower dynamic responses were calculated. Figure 25 shows the wind speed at the hub height for all the turbulent cases. Figure 26 shows the tower-base for-aft bending moment and the tower-top fore-aft deflection time series.

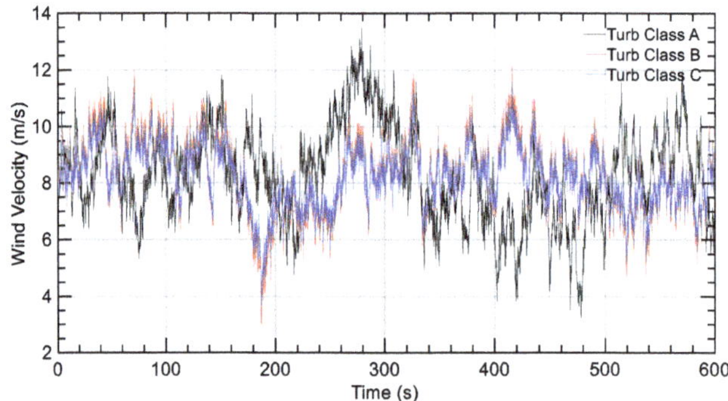

Figure 25. Wind Speed at Hub Height.

Since the turbulent domain covers both rotors, it is expected that each rotor experiences different wind conditions, which indicates the presence of torsional moment over the tower. The yawing deflection over the tower is shown for each turbulence case in Figures 27–29.

It is clear that the turbulent nature of the flow has affected the behavior of the yawing deflection. The deflection is quite random and does not have a general trend, unlike the case of steady wind where the deflection had a periodic nature. This randomness indicates unfavored instability in the dynamics, which affect the lifetime of the turbine, and a thorough fatigue study must be made to avoid sudden failures.

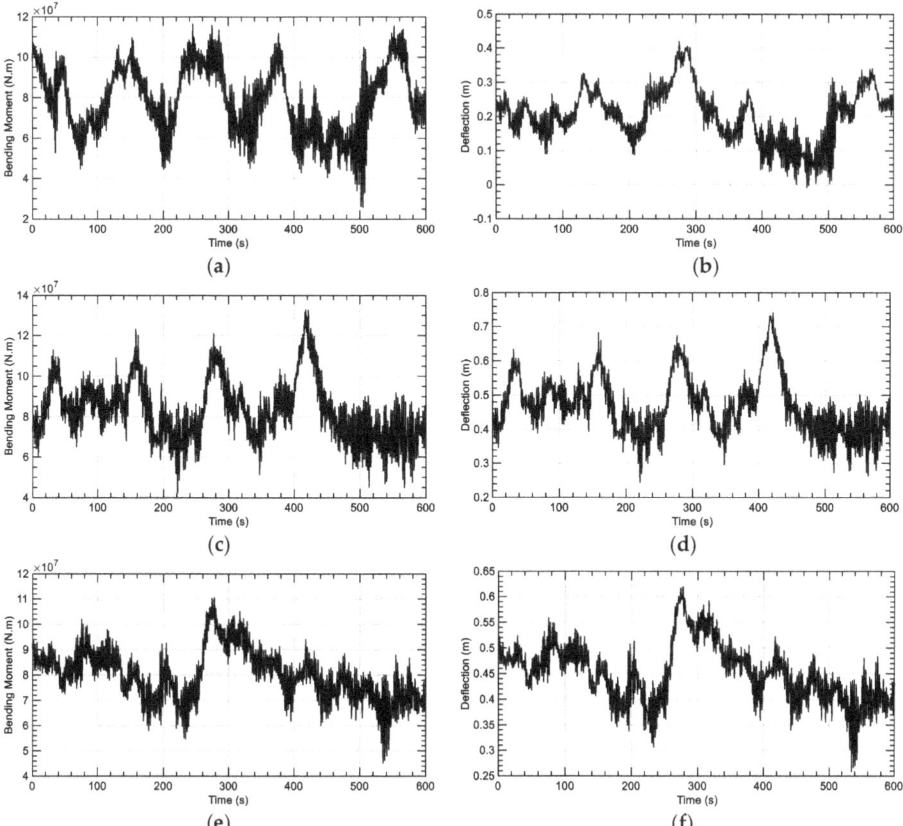

Figure 26. Tower Dynamic Responses; (**a**) Tower-Base Fore-aft Bending Moment, Turb. Class A; (**b**) Tower-Top Fore-aft Deflection, Turb. Class A; (**c**) Tower-Base Fore-aft Bending Moment, Turb. Class B; (**d**) Tower-Top Fore-aft Deflection, Turb. Class B; (**e**) Tower-Base Fore-aft Bending Moment, Turb. Class C; and (**f**) Tower-Top Fore-aft Deflection, Turb. Class C.

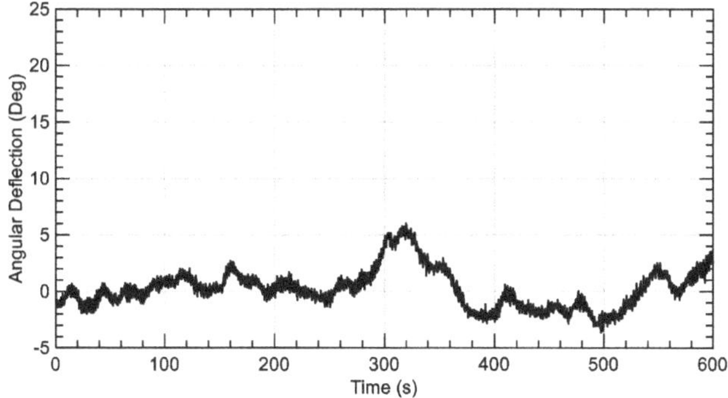

Figure 27. Tower top yawing deflection: turb. class A.

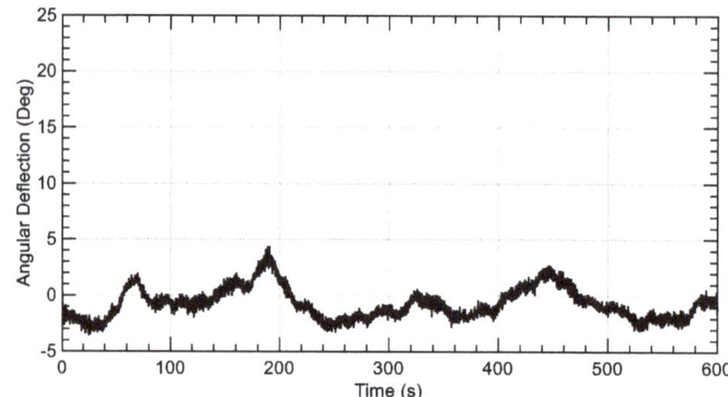

Figure 28. Tower top yawing deflection: turb. class B.

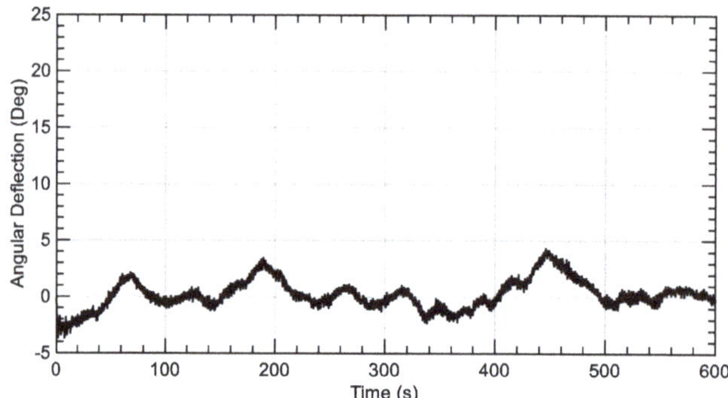

Figure 29. Tower top yawing deflection: turb. class C.

Frequency analysis has been made for the bending moment dynamic response of the tower. Figures 30–32 show the results.

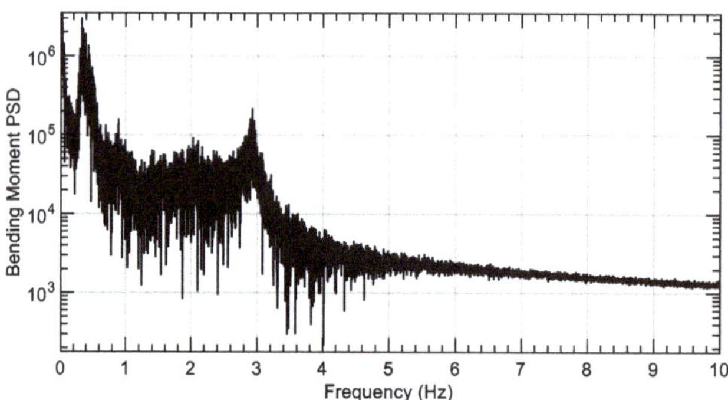

Figure 30. Frequency analysis, tower-base bending moment, turb. class A.

Frequency analyses in all turbulent cases shows that the dominant frequencies are at 0.32 Hz and 3 Hz, which are the same as the free vibration natural frequencies of the tower.

Comparison between dynamic response of the tower-base bending moment for the turbulent flow load cases is shown in Table 5.

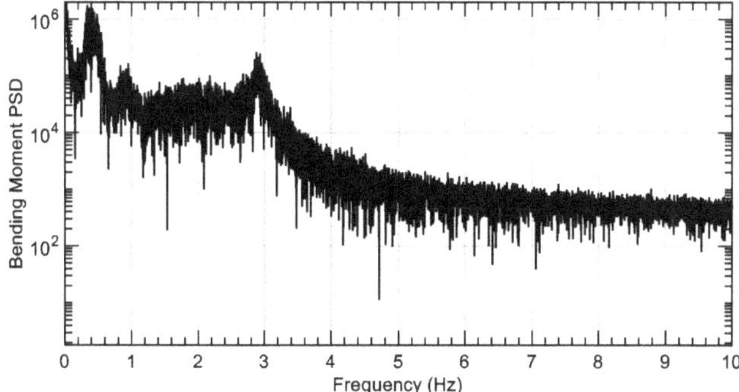

Figure 31. Frequency analysis, tower-base bending moment, turb. class B.

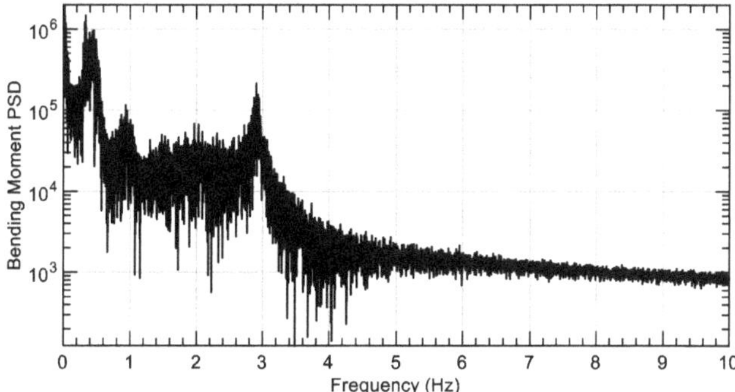

Figure 32. Frequency analysis, tower-base bending moment, turb. class C.

Table 5. Statistical analysis for tower-base bending moment in turbulent cases.

Turbulence Class	Mean Value	Standard Deviation	Dominant Frequencies
A	8.00×10^7	1.79×10^7	0.32 Hz and 3 Hz
B	8.25×10^7	1.53×10^7	0.32 Hz and 3 Hz
C	8.06×10^7	1.14×10^7	0.32 Hz and 3 Hz

Statistical analysis in Table 5 has shown that the tower's natural frequencies are dominant over the flow condition. For the high turbulence intensity, the mean value of the load is less than the other intensities; however a higher standard deviation occurs, which indicates severe oscillation of the loads.

5. Conclusions

In this work, an aeroelastic tool has been developed to provide a two-way FSI model for wind turbine blades and a tower. The tool is validated for a single-rotor configuration and gave very reasonable results. It has been extended to model the aeroelastic behavior of a twin-rotor configuration. Dynamic response of the support tower was investigated for two load cases. The outcomes of this work can be summarized as follows:

- The developed tool has been sufficiently validated for a single-rotor configuration when compared to results of FAST analysis. It can be trusted to model a twin-rotor configuration.
- Tower elasticity should be considered when studying the tower dynamics. The stiff tower model does not count for the vibration and hence the inertial loads of the tower, causing misleading results of the loads in terms of value and behavior.

- For a tower with the same first fore-aft natural frequency, if one more rotor is added; it does not only increase the loads, but also changes the natural frequencies of the rotor and hence the stiffness and structural damping. Accordingly, the change in the tower deflections and loads is not straightforward with the number of rotors.
- Tower torsion is very crucial in case of twin-rotor configuration. The normal case is that the two rotors are not rotating simultaneously, and the results of this work have shown yawing deflections of the tower in case there is a slight phase change in the rotor loads' time series.
- Tower natural frequencies are dominant over the flow conditions for the tower loads and deflections.

The next step to improve the present tool is to implement pitch control, so that high power regions can be modeled. Also, eliminate the assumption of no aerodynamic interaction between the rotors, to study its effect on the dynamics and the optimum distance between the rotors to improve performance.

Author Contributions: A.I. did most of the work for modelling, programming, analyses, discussion, and writing of the manuscript. S.Y. contributed by supervising the work, mentoring, technical advising, and revising the work.

Funding: This work was conducted with the support of Japanese Ministry of Education, Culture, Sports, Science and Technology (MEXT).

Acknowledgments: Special thanks to Professor Martin O. L. Hansen, and Professor Taeseong Kim who helped a lot in understanding the fundamental information for wind turbine aeroelasticity.

Conflicts of Interest: The authors declare no conflict of interest.

Abbreviations

BEM	Blade Element Momentum
CAD	Computer Aided Drawing
CFD	Computational Fluid Dynamics
FAST	Fatigue, Aerodynamics, Structure, Turbulence
IEC	International Electrotechnical Commission
MRS	Multi Rotor System
NREL	National Renewable Energy Laboratory

References

1. Global Wind Energy Council (GWEC). *Global Wind Report*; Global Wind Energy Council: Brussels, Belgium, 2018.
2. GE Renewable Energy. Available online: https://www.ge.com/renewableenergy/wind-energy/turbines/haliade-x-offshore-turbine (accessed on 25 January 2019).
3. Hofmann, M.; Sperstad, I.B. Will 10 MW wind turbines bring down the operation and maintenance cost of offshore wind farms? *Energy Procedia* **2014**, *53*, 231–238. [CrossRef]
4. Jamieson, P. Multi Rotor Systems. In *Innovation in Wind Turbine Design*; Wiley-Blackwell: Chichester, UK, 2011.
5. Goltenbott, U.; Ohya, U.; Yoshida, S.; Jamieson, P. Aerodynamic interaction of diffuser augmented wind turbines in multi-rotor systems. *Renew. Energy* **2017**, *112*, 25–34. [CrossRef]
6. Chasapogiannis, P.; Prospathopoulos, J.M.; Voutsinas, S.G.; Chaviaropoulos, T.K. Analysis of the aerodynamic performance of the multi-rotor concept. *J. Phys. Conf. Ser.* **2014**, *524*, 012084. [CrossRef]
7. Yoshida, S.; Goltenbott, U.; Ohya, Y.; Jamieson, P. Coherence Effects on the Power and Tower Loads of a 7×2 MW Multi-Rotor Wind Turbine System. *Energies* **2016**, *9*, 742. [CrossRef]
8. Ghaisas, N.S.; Ghate, A.S.; Lele, S.K. Large-eddy simulation study of multi-rotor wind turbines. *J. Phys. Conf. Ser.* **2018**, *1037*, 072021. [CrossRef]
9. Van der Laan, M.P.; Andersen, S.J.; García, N.R.; Angelou, N.; Pirrung, G.R.; Ott, S.; Sjöholm, M.; Sørensen, K.H.; Neto, J.X.V.; Kelly, M.; et al. Power curve and wake analyses of the Vestas multi-rotor demonstrator. *Wind Energy Sci.* **2019**, *44*. [CrossRef]

10. Verma, P. Multi Rotor Wind Turbine Design and Cost Scaling. Master's Thesis, University of Massachusetts Amherst, Amherst, MA, USA, September 2013.
11. Mate, G.M. Development of a Support Structure for Multi-Rotor Wind Turbines. Master's Thesis, University of Massachusetts Amherst, Amherst, MA, USA, September 2014.
12. Bazilevs, Y.; Yan, J.; Deng, X.; Korobenko, A. Computer Modeling of Wind Turbines: 2. Free-Surface FSI and Fatigue-Damage. In *Archives of Computational Methods in Engineering*; Springer: Dordrecht, The Netherlands, 2018.
13. Halawa, A.; Sessarego, M.; Shen, W.Z.; Yoshida, S. Numerical Fluid-Structure Interaction Study on the NREL 5MW HAWT. *J. Phys. Conf. Ser.* **2018**, *1037*, 022026. [CrossRef]
14. Jonkman, J. The New Modularization Framework for the FAST Wind Turbine CAE Tool. In Proceedings of the 51st AIAA Aerospace Sciences Meeting, Grapevine, TX, USA, 7–10 January 2013.
15. Vorpahl, F.; Popko, W. *Description of the Load Cases and Output Sensors to be Simulated in the OC4 Project under IEA Wind Annex 30*; Fraunhofer Institute for Wind Energy and Energy System Technology IWES: Bremerhaven, Germany, 2016.
16. Hansen, M.O.L. Unsteady BEM Model. In *Aerodynamics of Wind Turbines*, 2nd ed.; Earthscan: London, UK, 2008; pp. 85–102.
17. Hansen, M.O.L. Dynamic Structural Model of a Wind Turbine. In *Aerodynamics of Wind Turbines*, 2nd ed.; Earthscan: London, UK, 2008; pp. 125–138.
18. Jonkman, J.; Butterfield, S.; Musial, W.; Scott, G. *Definition of a 5-MW Reference Wind Turbine for Offshore System Development*; NREL Technical Report TP-500-38060; National Renewable Energy Lab: Golden, CO, USA, 2009.
19. Bazilevs, Y.; Hsu, M.C.; Akkerman, I.; Wright, S.; Takizawa, K.; Henicke, B.; Spielman, T.; Tezduyar, T.E. 3D Simulation of Wind Turbine Rotors at Full Scale. Part I: Geometry Modeling and Aerodynamics. *Int. J. Numer. Methods Fluids* **2011**, *65*, 207–235. [CrossRef]
20. NWTC Information Portal (Modes). Available online: https://nwtc.nrel.gov/Modes (accessed on 15 May 2019).
21. Jonkman, B.J.; Kilcher, L. *TurbSim's User Guide: Version 1.06.00*; Technical Report; National Renewable Energy Lab (NREL): Golden, CO, USA, 2012.
22. IEC. *Wind Turbines—Part 1: Design RequirementsIEC-61400-1*, 3rd ed.; International Electrotechnical Commission: Geneva, Switzerland, 2005.

Article

A Large-Eddy Simulation-Based Assessment of the Risk of Wind Turbine Failures Due to Terrain-Induced Turbulence over a Wind Farm in Complex Terrain

Takanori Uchida [1,*] and Susumu Takakuwa [2]

[1]   Research Institute for Applied Mechanics (RIAM), Kyushu University, 6-1 Kasuga-kouen, Kasuga, Fukuoka 816-8580, Japan
[2]   Eurus Energy Holdings Corporation, 3-13, Toranomon 4-Chome, Minato-ku, Tokyo 105-0001, Japan; susumu.takakuwa@eurus-energy.com
*   Correspondence: takanori@riam.kyushu-u.ac.jp; Tel.: +81-92-583-7776; Fax: +81-92-583-7779

Received: 3 April 2019; Accepted: 17 May 2019; Published: 20 May 2019

Abstract: The first part of the present study investigated the relationship among the number of yaw gear and motor failures and turbulence intensity (TI) at all the wind turbines under investigation with the use of in situ data. The investigation revealed that wind turbine #7 (T7), which experienced a large number of failures, was affected by terrain-induced turbulence with TI that exceeded the TI presumed for the wind turbine design class to which T7 belongs. Subsequently, a computational fluid dynamics (CFD) simulation was performed to examine if the abovementioned observed wind flow characteristics could be successfully simulated. The CFD software package that was used in the present study was RIAM-COMPACT, which was developed by the first author of the present paper. RIAM-COMPACT is a nonlinear, unsteady wind prediction model that uses large-eddy simulation (LES) for the turbulence model. RIAM-COMPACT is capable of simulating flow collision, separation, and reattachment and also various unsteady turbulence–eddy phenomena that are caused by flow collision, separation, and reattachment. A close examination of computer animations of the streamwise (x) wind velocity revealed the following findings: As we predicted, wind flow that was separated from a micro-topographical feature (micro-scale terrain undulations) upstream of T7 generated large vortices. These vortices were shed downstream in a nearly periodic manner, which in turn generated terrain-induced turbulence, affecting T7 directly. Finally, the temporal change of the streamwise (x) wind velocity (a non-dimensional quantity) at the hub-height of T7 in the period from 600 to 800 in non-dimensional time was re-scaled in such a way that the average value of the streamwise (x) wind velocity for this period was 8.0 m/s, and the results of the analysis of the re-scaled data were discussed. With the re-scaled full-scale streamwise wind velocity (m/s) data (total number of data points: approximately 50,000; time interval: 0.3 s), the time-averaged streamwise (x) wind velocity and TI were evaluated using a common statistical processing procedure adopted for in situ data. Specifically, 10-min moving averaging (number of sample data points: 1932) was performed on the re-scaled data. Comparisons of the evaluated TI values to the TI values from the normal turbulence model in IEC61400-1 Ed.3 (2005) revealed the following: Although the evaluated TI values were not as large as those observed in situ, some of the evaluated TI values exceeded the values for turbulence class A, suggesting that the influence of terrain-induced turbulence on the wind turbine was well simulated.

Keywords: terrain-induced turbulence; complex terrain; computational fluid dynamics (CFD); LES

---

1. Introduction

In recent years, wind power has started to be implemented across the world. In the midst of this movement, preparations for further dissemination of wind power are being advanced in Japan with the passage of the Act on Special Measures Concerning Procurement of Renewable Electric Energy by Operators of Electric Utilities. However, it is still true that there remain a large number of issues to be resolved for further dissemination of wind power. Some of these issues are technical, and they concern noise, lightning, and turbulence (terrain-induced turbulence). Terrain-induced turbulence is the main cause of the reduction of availability factors and wind turbine failures that increase repair costs, significantly affecting the wind power generation industry as a whole. Terrain-induced turbulence originates from the complexity of terrain. In Japan, where a larger proportion of the land is covered by mountains than in many countries abroad, areas for wind farm development are often characterized by complex terrain. Accordingly, the risk of wind turbine failure caused by terrain-induced turbulence is high in Japan. It should be noted that not all the wind turbines deployed on complex terrain break down because of terrain-induced turbulence. In addition, terrain undulations in complex terrain cause local increases of wind speed, which is considered to be an advantage. With the continuingly decreasing availability of flat areas for potential wind farm development, dissemination of wind power in Japan requires detailed business potential assessments, which involve highly accurate assessments of wind over complex terrain on local scales beforehand.

Deployment of a wind vane and an anemometer ensures the most reliable qualitative and quantitative confirmation of the presence of terrain-induced turbulence. However, over complex terrain, wind conditions can be different between two locations that are separated by several hundred meters; thus, the properties of terrain-induced turbulence assessed by a wind vane and an anemometer at a wind turbine site are not necessarily the same as those at another wind turbine site. From the perspective of cost, it is not feasible to deploy wind vanes and anemometers to all proposed wind turbine sites. In order to avoid unanticipated failures of wind turbines after wind farm construction, it is desirable to assess the risk of wind turbine failures, optimize the layout of the wind turbines, and select the most suitable wind turbine models prior to the construction of the wind farm.

Given the above background, the present study investigated the validity of a computational fluid dynamics (CFD)-based method for assessing the risk of wind turbine failures caused by terrain-induced turbulence at the planning stage of wind farm construction. CFD is a numerical simulation technique (software), which can simulate three-dimensional wind flow using computers [1–12]. The use of CFD software allows desktop assessment of wind flow for a potential wind farm prior to its actual construction. The Wind Atlas Analysis and Application Program (WAsP), developed by the Technical University of Denmark (DTU), is one such wind analysis software and has been commonly used in the wind power industry for some time [11,12]. However, because WAsP was developed for linear, steady flow analyses, its applicability for analyses of wind flow over complex terrain is highly limited. In recent years, nonlinear, unsteady flow analyses have become possible as a result of the rapid improvement of computers. A representative approach for such wind flow analyses uses a numerical turbulence model called large-eddy simulation (LES). In the present study, wind over an existing wind farm on complex terrain was analyzed with RIAM-COMPACT software, which was developed based on LES by the first author of the present paper [13–21]. At the wind farm investigated in the present study, failures of the yaw gears and motors that were likely caused by terrain-induced turbulence occurred frequently on only one particular wind turbine. Thus, the failure risk of the yaw gears and motors of this wind turbine was assessed by analyzing numerical simulation data.

2. An Overview of the Investigated Wind Farm, In Situ Data Analysis Results, and Discussions

Figure 1 shows a general view of the wind farm investigated in the present study. The wind farm and its surrounding area are characterized by highly complex terrain, and the wind turbines are deployed along a mountain ridge. At this wind farm, a total of 16 Siemens 1.3 MW, International Electrotechnical Commission (IEC) class IA wind turbines with a rotor diameter of 62 m and a hub height of 60 m have been deployed. The wind farm began operation in February 2004.

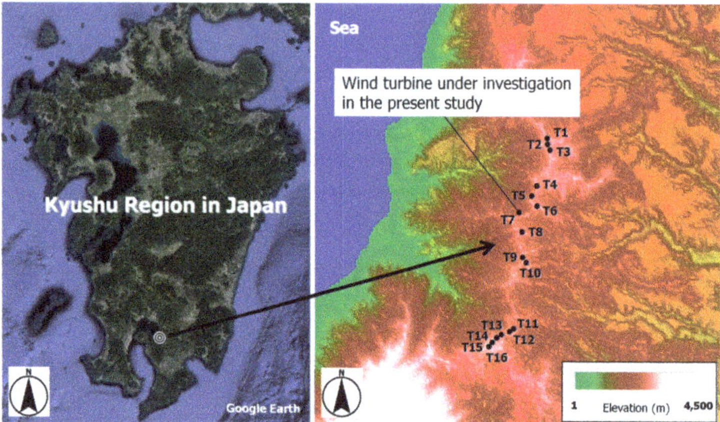

Figure 1. General view of the wind warm under investigation. (T1–T16 indicate the locations of wind turbines #1–#16).

Figure 2 shows the total number of failures of the yaw gears and motors of each of the deployed turbines in the first seven years of the wind turbine operation. Figure 2 reveals that the number of failures at wind turbine #7 (T7) was larger by far than those at the other wind turbines.

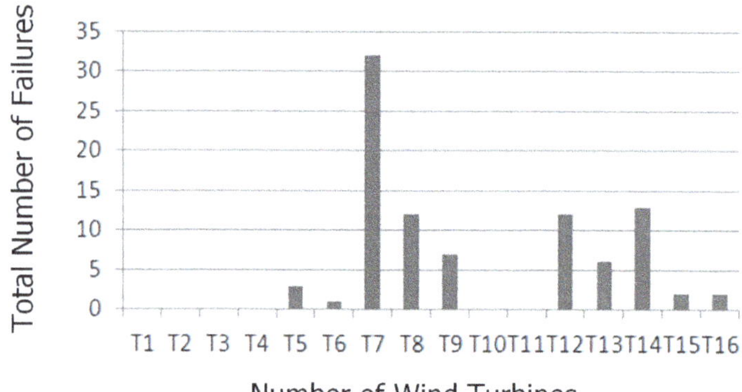

Figure 2. Comparison of the number of yaw gear and motor failures at all the wind turbines.

Figure 3 illustrates actual damage to a yaw gear. Cracks and shaft breakage can be identified, and it can be speculated that the damage occurred as a result of excessive force exerted on the yaw gear. Because the occurrence of cracks and shaft breakages was concentrated at T7, it was hypothesized that the cracks and shaft breakages at this turbine were attributable to a cause that was unique to this turbine. We presumed that a large number of the failures of T7 were caused by terrain-induced turbulence that originated from the terrain features in the area surrounding the wind turbine.

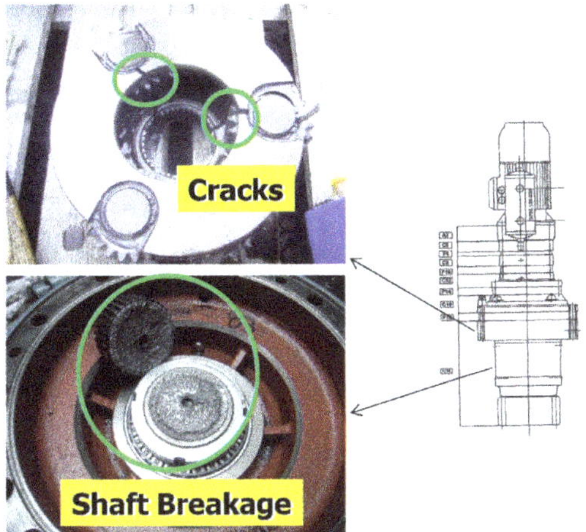

Figure 3. Yaw gear damage at wind turbine #7 (T7).

In order to examine the effect of terrain-induced turbulence on T7 in detail, the turbulence intensity (TI), evaluated from the Supervisory Control and Data Acquisition (SCADA) dataset for this wind turbine, was examined (Figure 4; see Equation (1) for the mathematical definition of TI). For comparison, Figure 4 also shows the TI evaluated from the SCADA dataset for wind turbine #5 (T5), which failed less often than T7. Each data point indicates a 10-min TI value. A close examination of Figure 4 reveals that the TI at T7 was slightly higher than that at T5. The red lines in Figure 4 show the relationship between TI and wind speed for turbulence class A (see Section 3.2 for details) from the Normal Turbulence Model (NTM) in 61400-1 Ed.3, IEC (2005). The white lines in Figure 4, which show the sum of the bin average and one bin standard deviation (σ) of the plotted TI values, indicate that the sum of these two values falls almost on the NTM line in the range of wind speeds greater than or equal to 7 m/s and less than 12 m/s for T7. For wind speeds of 12 m/s or higher, the sum of the two abovementioned values significantly exceeds the value of the TI for IEC turbulence class A from the NTM.

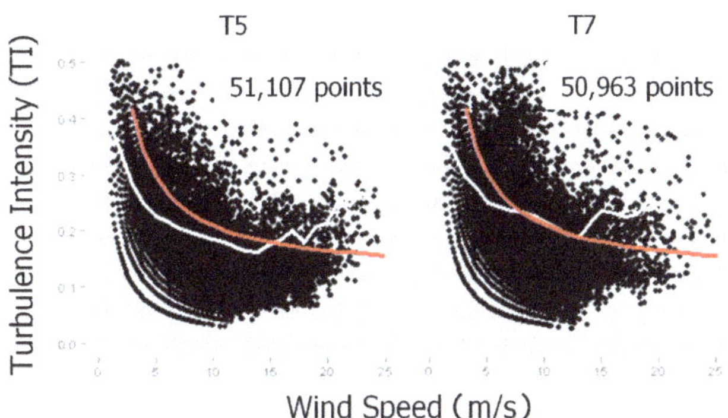

Figure 4. Comparison between the turbulence intensity (TI) at wind turbine #5 (T5) and that at T7. The data plotted here are those that were collected during the first year of the wind turbine operation. The numbers in the upper right indicate the number of plotted data values. The red lines indicate the relationship between TI and wind speed for International Electrotechnical Commission (IEC) turbulence class A. The white lines indicate the sum of the bin average and one bin standard deviation (σ) of the plotted TI values.

Figure 5 shows the number of 10-min periods in which the TI values evaluated from the in situ data exceeded the TI values for IEC turbulence class A at T5 and T7 in the first seven years of wind turbine operation. Figure 5 reveals that such 10-min periods occurred much more frequently at T7 than at T5. The colors used in Figure 5 indicate the amount by which the observed TI exceeded the values for IEC turbulence class A at the two wind turbine sites. The results in Figure 5 indicate that T7 was frequently affected by strong terrain-induced turbulence. Of all the wind turbines, T7 was the most frequently affected by terrain-induced turbulence that had TI values that exceeded the TI value for IEC turbulence class A (not shown due to limited space). From these findings, it can be speculated that terrain-induced turbulence was quite likely the main cause of the failures of T7.

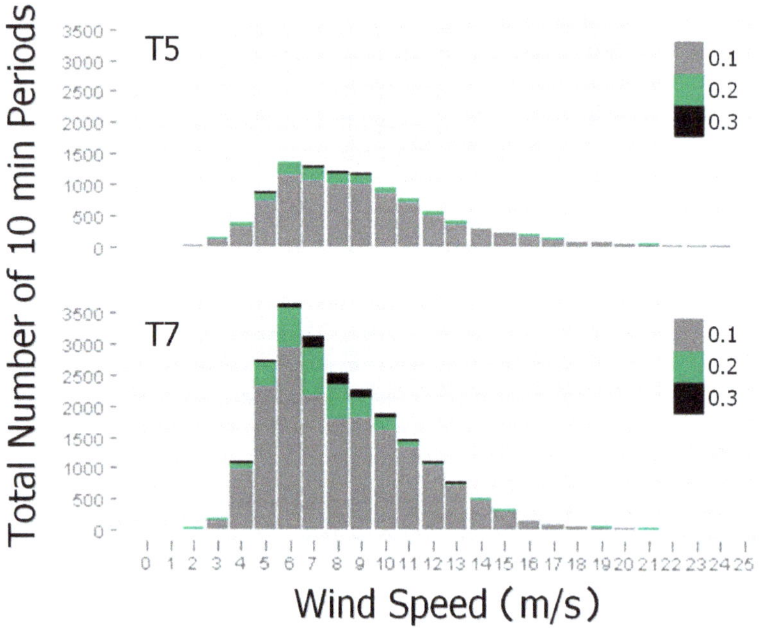

Figure 5. Comparison of the number of 10-min periods in which the TI of terrain-induced turbulence exceeded that for IEC turbulence class A at T5 and T7.

When TI values evaluated for a wind turbine are high only in some of the examined periods, it is likely that upstream terrain mainly accounts for the high values of TI in those periods and that the TI value becomes high in specific wind directions. Therefore, the values of TI for T7 were analyzed according to the wind direction. Figure 6 shows the relationship between the values of TI and wind speed at T7 for six wind directions during the first seven years of wind turbine operation. Figure 6 shows that the values of TI were high in westerly to north-westerly wind. Westerly to north-westerly wind occurred during one-third of the entire period, and such frequent occurrence of north-westerly wind is likely quite significant for T7 in terms of the increased risk of failures. (The prevailing wind direction of the area under investigation is north-westerly.) For comparison, the relationship between the values of TI and wind speed at T7 for east-south-easterly to south-south-easterly wind is shown in Figure 7. (The second most common wind direction in the area under investigation is south-easterly.) The comparisons of the TI values for different wind directions clearly revealed that terrain-induced turbulence occurred quite frequently in westerly to north-westerly wind.

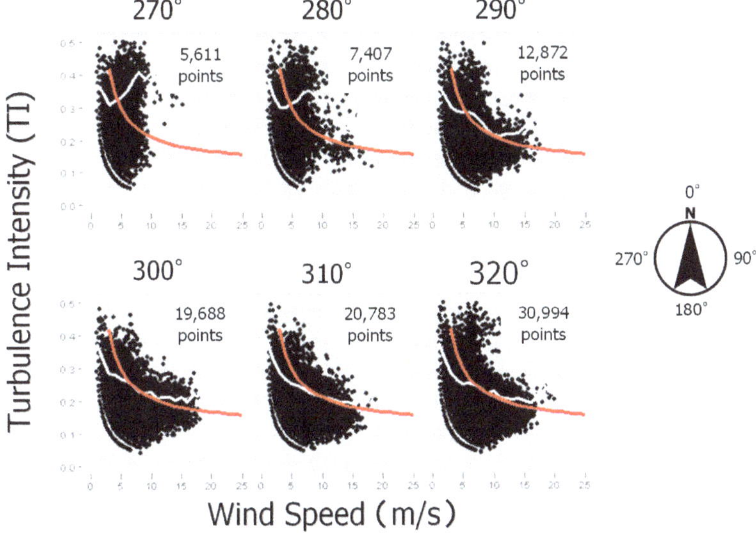

Figure 6. TI at T7 in westerly to north-westerly wind. The plot in the upper left panel is for a wind direction of 270°, and the plot in the lower right panel is for a wind direction of 320°.

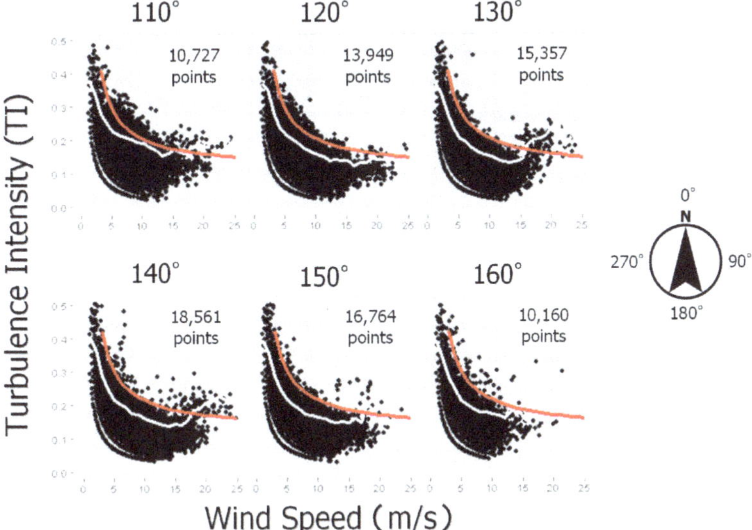

Figure 7. TI at T7 in east-south-easterly to south-south-easterly wind. The plot in the upper left panel is for a wind direction of 110°, and the plot in the lower right panel is for a wind direction of 160°.

Figure 8 shows the topography in the vicinity of T7. Several ridges exist to the west and northwest of this turbine site. Figure 9 shows the ridges in the area where T7 is located. The terrain in this area is complex with large undulations. Thus, it was speculated that wind flows separated due to the ridges, and terrain-induced turbulence was generated as a result. In Section 3, the attempt to simulate this flow separation and turbulence with CFD is discussed.

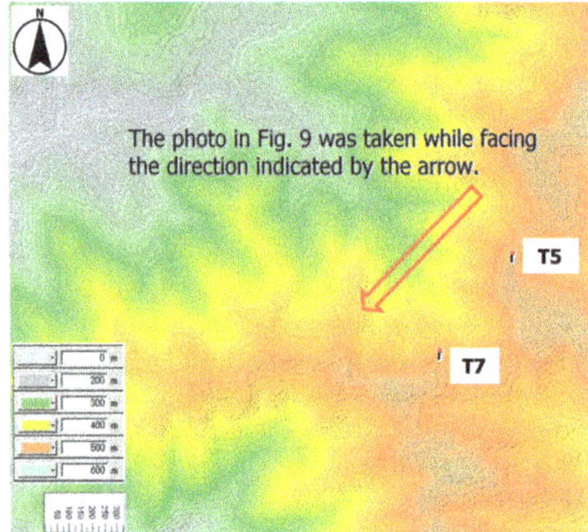

Figure 8. Topography in the vicinity of T7.

Figure 9. Photo of the area where T7 is located.

3. CFD Simulation Overview

In order to closely examine the wind flow characteristics that have caused failures of the yaw gears and motors on T7, a CFD simulation was conducted. The CFD software used in the present study is RIAM-COMPACT, which was developed by the first author of the present paper. Because the details of the numerical simulation methods used in RIAM-COMPACT have been discussed in previous papers [13–21], they will be omitted here. In the present study, the LES is assumed to reproduce the wind tunnel testing. Therefore, the effects of atmospheric stability associated with vertical thermal stratification of the atmosphere and inflow turbulence were neglected. Thus, TI calculated from numerical results is smaller than measured data. The standard Smagorinsky model was used for the subgrid-scale model (SGS) model [22]. The model coefficient was assumed to be 0.1 by using a wall-damping function. In addition, as in [13,16–18], the effects of the surface roughness were taken into consideration by reconstructing surface irregularities in high resolution. A comparison between the Reynolds-averaged modeling (RANS) results and the present LES results is summarized in a recent article [14], and the prediction accuracy of the present LES approach by comparison with wind tunnel experiments is discussed in [21].

Figure 10 illustrates the computational domain and grid used for the present study. For the study, the complex terrain of the wind farm and its surroundings were numerically constructed using the 10-m resolution land surface digital elevation model (DEM) from the Geospatial Information Authority of Japan (GSI). The dimensions of the computational domain were 10.0 km × 4.0 km × 4.0 km in full scale in the streamwise (x), spanwise (y), and vertical (z) directions, respectively. The computational domain was set in such a way that T7 was located in the center of the x-y plane of the computational domain. The grid spacing for all directions varied so that grid density was high in the vicinity of T7. The minimum grid spacing for the x- and y-directions was set to approximately 8 m, and the minimum grid spacing for the z-direction was set to approximately 1.7 m. The number of grid points was $501 \times 201 \times 101$ in the x-, y-, and z-directions, respectively, which resulted in a total number of approximately 10 million grid points. The wind direction considered for the simulation was west-north-westerly. (West-north-westerly wind is wind that flows from an angle of 292.5° clockwise from north, which was set to 0° in the wind direction coordinate system adopted in the present study.) Because inflow boundary was located over the ocean (Figure 10), the vertical wind profile of the inflow streamwise wind velocity was set according to a power law ($\alpha = 0.1$, where α is the power law exponent, i.e., N = 10, where N is the inverse of α). Other boundary conditions are detailed in [13–21].

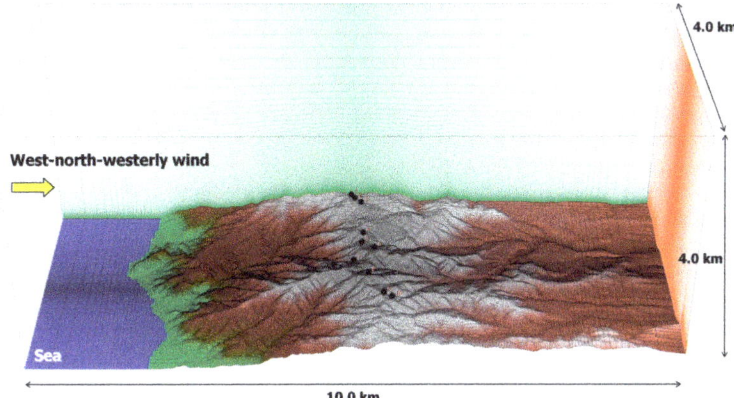

Figure 10. Computational domain and grid used for the present study.

3.1. Results of Non-Dimensional Simulation and Discussions

The governing equations of the flow adopted in the present study (i.e., a filtered continuity equation for incompressible viscous flows and filtered Navier–Stokes equations for incompressible viscous flows) were non-dimensionalized with the use of 1) the difference between the minimum and maximum terrain elevations within the computational domain and 2) the inflow streamwise wind velocity at the height of the maximum surface elevation in the computational domain. Therefore, all the physical variables that are output by the model are non-dimensional quantities. In the current sub-section, the non-dimensional data of the three wind velocity components that were obtained from the simulation will be analyzed, and the results will be shown and discussed.

Figure 11 shows the temporal change of the fluctuating parts of the three wind velocity components at the hub height of T7 (60 m above the ground surface) in the non-dimensional time period from 600 to 800. Specifically, the data values shown for each wind velocity component are those that were obtained by subtracting the period-averaged wind velocity component from the original time series of the wind velocity component. Figure 11 shows that the values of the streamwise (x), spanwise (y), and vertical (z) wind velocity components all fluctuated significantly in time.

Figure 12 shows vertical profiles of statistical quantities of the turbulent flow at the site of T7 from the non-dimensional time period from 600 to 800. Figure 12a shows the vertical profiles of the streamwise wind velocity. Specifically, the red line in Figure 12 indicates the vertical profile of the streamwise inflow velocity, and the blue line indicates the vertical profile of the mean streamwise wind velocity at the site of T7 from the time period under investigation. Figure 12a also includes the values of the speed-up ratio at the bottom of the swept area (29 m above the ground surface), at the hub center

(60 m above the ground surface), and at the top of the swept area (91 m above the ground surface), where the speed-up ratio is defined as the ratio of the streamwise wind velocity at a height of interest above the ground surface at the site of T7 to the inflow streamwise wind velocity at the height of interest. These results show that, due to terrain effects, the streamwise wind velocity increased locally at the wind turbine site, and additionally, there was no significant streamwise wind velocity deficit at the site. As can be presumed from Figures 10 and 13, the locally increased streamwise wind velocity likely occurred as the wind flowed uphill along the terrain and into the wind turbine. Figure 12b shows the vertical profiles of the standard deviations of the streamwise (x), spanwise (y), and vertical (z) wind velocity components. The values of the standard deviations for all three components are relatively large, reflecting the temporal change of the fluctuating parts of the wind velocity components in Figure 11. Examinations of the values of the standard deviations at the hub center (60 m above the ground surface) in Figure 12b reveal that the value of the standard deviation of the vertical (z) wind velocity component is large and that the ratio of the values of the standard deviations of the three wind velocity components at the hub center was $\sigma_1:\sigma_2:\sigma_3 = 1.0:0.7:0.65$, which clearly indicates that there was an influence of terrain-induced turbulence at this site. (This finding will be discussed again in Section 3.2).

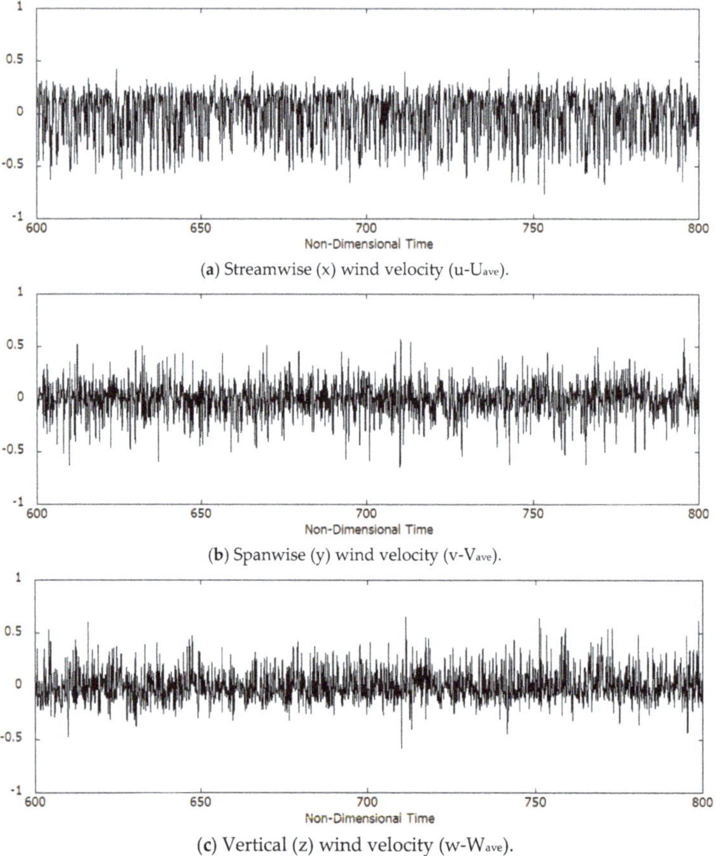

Figure 11. Temporal change of the fluctuating parts of the wind velocity components at the hub height (60 m above the ground surface) of T7.

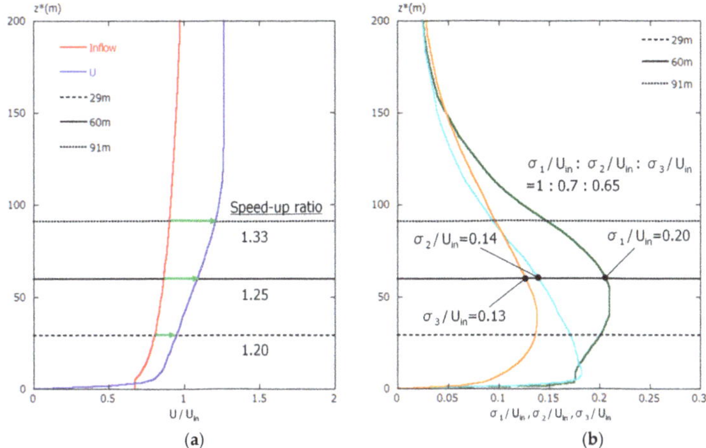

Figure 12. Vertical profiles of statistical quantities of the turbulent flow at the site of T7. (**a**) Normalized mean streamwise wind velocity; (**b**) Normalized standard deviation of the three wind velocity components. (U_{in}: the inflow streamwise wind velocity at the height of the maximum surface elevation in the computational domain. $z^*(m)$: the height above the ground.)

To investigate details of the flow field in the vicinity of T7, Figure 13 illustrates the temporal change of the streamwise (x) wind velocity as contour plots. The visualized streamwise wind velocity field in Figure 13 shows that, as a separation vortex (indicated by the arrows in Figure 13) that was shed upstream of the wind turbine passed through the wind turbine, the wind velocity field surrounding the wind turbine changed significantly.

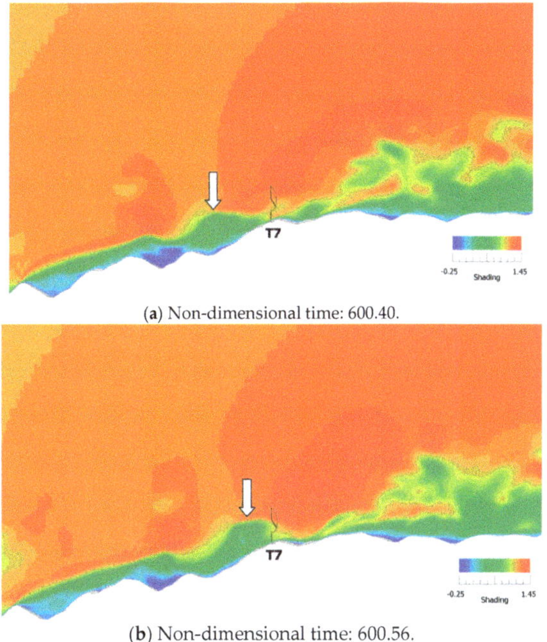

(**a**) Non-dimensional time: 600.40.

(**b**) Non-dimensional time: 600.56.

Figure 13. *Cont.*

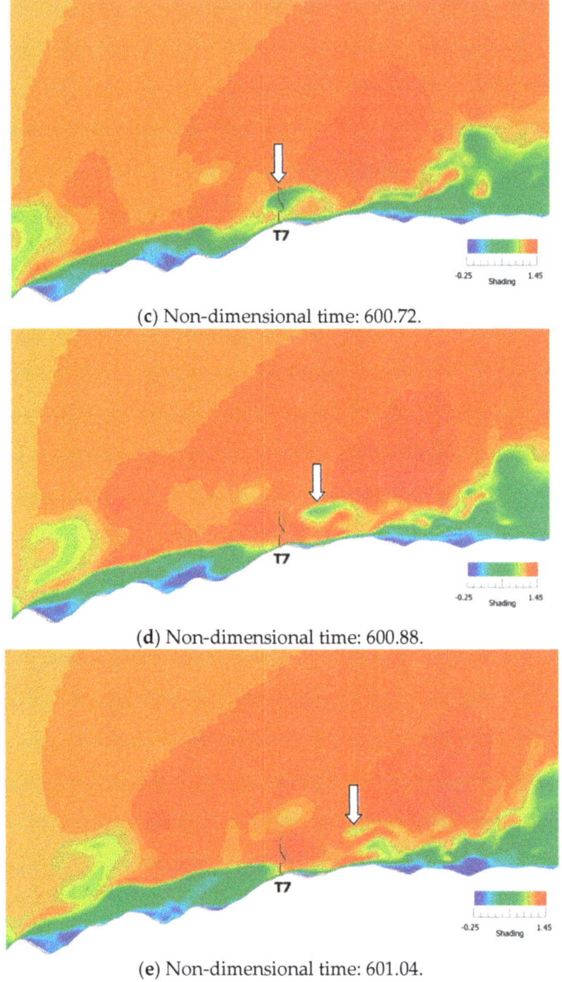

(**c**) Non-dimensional time: 600.72.

(**d**) Non-dimensional time: 600.88.

(**e**) Non-dimensional time: 601.04.

Figure 13. Temporal change of the streamwise (x) wind velocity field in the vertical cross-section that includes the site of T7.

Figure 14 shows the vertical profiles of the streamwise (x) wind velocity at the site of T7 from the same times for which the cross-sectional views of the streamwise (x) wind velocity in Figure 13 were created. Immediately before the separation vortex passed through the wind turbine (Figure 13 (a)), the vertical profile of the streamwise (x) wind velocity showed a local increase of the velocity due to terrain effects and thus showed no significant wind velocity deficit with respect to the power law profile of the streamwise (x) wind velocity (Figure 14a). As the separation vortex that had been located upwind of the turbine approached the turbine, a velocity deficit occurred in the layer between the hub center (60 m above the ground surface) and the bottom of the rotor (Figure 14b). At the time at which the separation vortex arrived at the wind turbine (Figure 14c), negative wind shear was evident between the hub center height (60 m above the ground surface) and heights that were slightly higher than the hub center height. As illustrated in Figure 14d,e, after the passage of the separation vortex, the wind velocity recovered to values predicted by the power law.

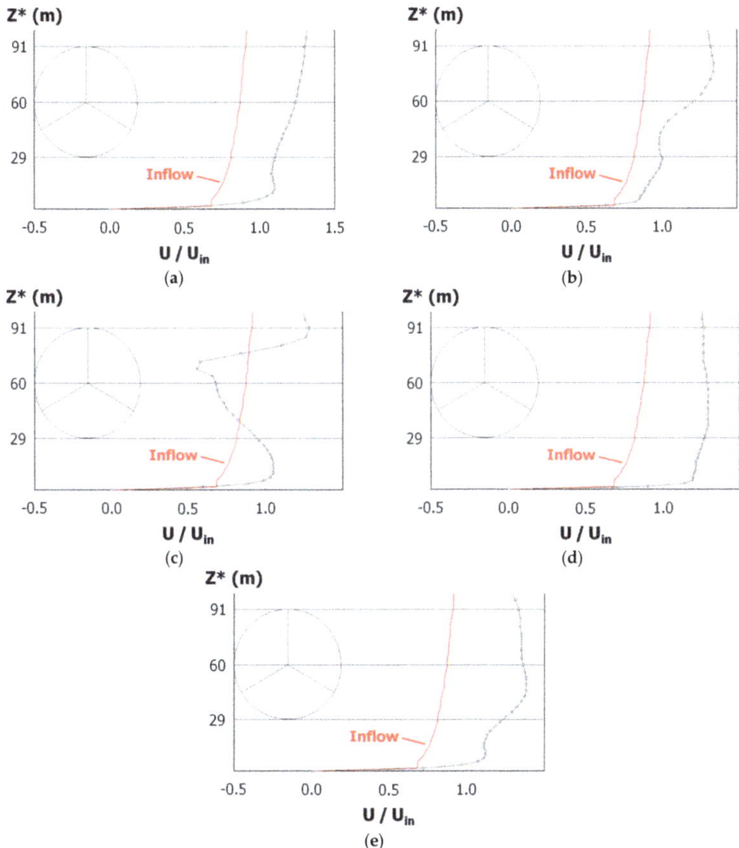

Figure 14. Temporal change in the vertical profile of the streamwise (x) wind velocity at the site of T7. The profiles in (**a–e**) correspond to the vertical cross-sectional views of the streamwise (x) wind velocity field shown in Figure 13a–e. (**a**) Non-dimensional time: 600.40, (**b**) Non-dimensional time: 600.56, (**c**) Non-dimensional time: 600.72; (**d**) Non-dimensional time: 600.88, (**e**) Non-dimensional time: 601.04. (z*(m): the height above the ground.)

A close examination of computer animations of the simulations results in Figures 13 and 14 led to the following finding: The sequence of wind flow patterns described above, that is, large vortex shedding that originated from the micro-topographical features upstream of the wind turbine and its accompanying generation and disappearance of a separation vortex, occurred in a nearly periodic manner. A complex wind flow field with a vertical profile of the streamwise wind velocity such as the one in Figure 14c, which is generally rare, periodically formed in the vicinity of the wind turbine. More specifically, this vertical profile of streamwise wind velocity deviated significantly from the power law and also had negative wind shear. It can be surmised that when complex wind flow with such a profile passes through a wind turbine, it causes a large wind load on the turbine. In addition, because the structure of the abovementioned complex wind flow is three-dimensional, wind loads on the left and right side of the swept area of a wind turbine in such a wind flow are expected to differ. Thus, it can be speculated that such wind loads would exert force on the wind turbine in such a way that they would forcibly rotate the nacelle of the wind turbine, which in turn would cause impact loads on both the yaw gears and motors and result in the failures of the yaw gears and motors in the end. Accordingly, it may be possible to make prior assessments of wind turbine failure risks due to terrain-induced turbulence by studying, with the use of CFD, wind velocity fluctuations in the vicinity of a wind turbine and the vertical profiles of statistical quantities of the three velocity components (i.e., the three-dimensional flow structure) within the swept area.

3.2. Re-Scaled Dimensional Simulation Results and Discussions

The temporal change of the streamwise (x) wind velocity (a non-dimensional quantity) in the period from 600 to 800 in non-dimensional time at the hub height of T7 was re-scaled such that the average value of the streamwise (x) wind velocity at the hub height of this turbine for this period became 8.0 m/s. The abovementioned re-scaling procedure can be summarized as follows:

(1) The average value of the streamwise (x) wind velocity (a non-dimensional quantity) at the hub height of T7 was calculated for the period from 600 to 800 in non-dimensional time. The calculated average value was 1.087 in the present study.

(2) A correction coefficient was calculated so that the average value of the streamwise (x) wind velocity at the hub height of T7 in the period under investigation was 8.0 m/s in full scale. Then, the non-dimensional wind velocity data from the entire simulation time period were multiplied by the calculated correction coefficient. The calculated correction coefficient was 7.36 (= 8.0/1.087) in the present study. With this procedure, the streamwise (x) non-dimensional wind velocity was converted to full-scale wind velocity (m/s).

(3) The time in the period from 600 to 800 in non-dimensional time was converted to full scale using the equation T = t (h/U_{in}), where T is full-scale time (s), t is non-dimensional time, h is the difference between the minimum and maximum terrain elevations within the computational domain (m), and U_{in} is the streamwise wind velocity (m/s) at the height of the maximum surface elevation in the computational domain at the inflow boundary. In the present study, the 200 non-dimensional time period, i.e., the period from 600 to 800 in non-dimensional time, was converted to approximately 15,500 s (approximately 4 h) in full scale. The time step was 0.3 s in full scale.

Figure 15 shows the temporal change of the full-scale streamwise (x) wind velocity (m/s) that was obtained from the rescaling procedure with the method described above (total data points: 50,000; time interval: 0.3 s). The green line in the figure indicates 8.0 m/s, which is the average streamwise (x) wind velocity that the re-scaling procedure was designed to attain for the hub height of T7 in the time period under investigation.

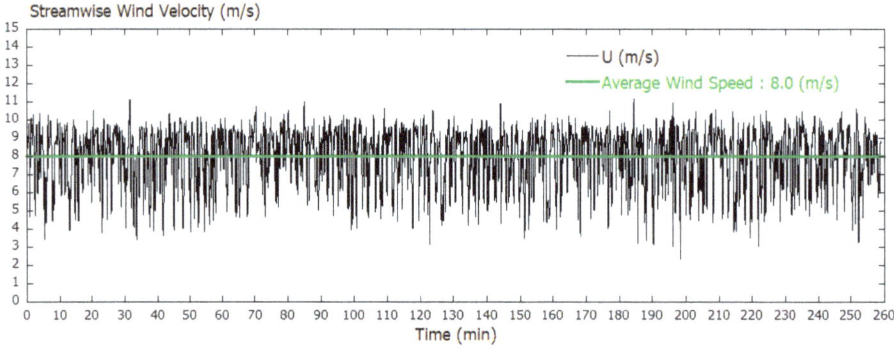

Figure 15. Temporal change of the streamwise (x) wind velocity at the hub height of T7 (60 m above the ground surface) that was rescaled to full scale (m/s).

Figure 16 shows a histogram of the streamwise (x) wind velocity data from Figure 15 with bin widths of 1 m/s. The average streamwise (x) wind velocity that the re-scaling procedure was designed to attain for the time period investigated was 8.0 m/s in the present study. As a result, the re-scaled streamwise (x) wind velocity ranged between approximately 4.0 and 11.0 m/s, and the occurrence frequency of the wind velocity class of 9.0 to 10.0 m/s in particular was large.

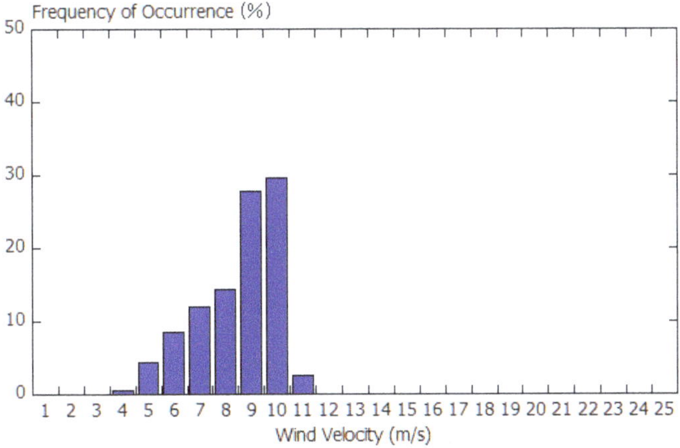

Figure 16. Frequency distribution of the streamwise (x) wind velocity that has been rescaled to full scale (m/s) for the time period shown in Figure 15.

In the present study, following a common statistical processing procedure adopted for in situ data, a 10-min moving average filter (1932-point averaging filter) was applied to the time series of the re-scaled streamwise (x) wind velocity (m/s) in Figure 15 (total data points: 50,000; time interval: 0.3 s) to evaluate the values of the moving-averaged wind velocity and the corresponding TI (Figure 17) (48,068 data points). In Figure 17a, the green line indicates 8.0 m/s, which is the average streamwise wind velocity (x) that the re-scaling procedure in this study was designed to attain for the time period under investigation. Figure 17b shows the evaluated TI values. These values were obtained with Equation (1) below using a moving-averaged filter with a window length of 10-min and the wind velocity data within the window (number of sample data points: 1932).

$$TI = \frac{\sigma_u}{\overline{u}} = \frac{\sqrt{\frac{1}{N}\sum_{i=1}^{N}\left(u_i'\right)^2}}{\overline{u}} \tag{1}$$

where

$$u' = u(t) - \overline{u} \tag{2}$$

where u(t) is the instantaneous streamwise wind velocity, $\overline{u}$ is the average of the instantaneous streamwise wind velocity within the 10-min moving-averaged window, and u' is the fluctuating component of the streamwise wind velocity due to turbulence. Figure 17a,b shows that the values of the average streamwise velocity and TI fluctuate in a correlated manner.

That is, as the average wind velocity in Figure 17a increases, the TI in Figure 17b decreases. Conversely, as the average wind velocity in Figure 17a decreases, the TI in Figure 17b increases. A further examination of the temporal change of the TI in Figure 17b reveals that the TI changes in large amplitude with the increasing and decreasing average wind velocity. The average value of TI was 0.19 (the green line in Figure 17b), which is relatively large. This result also indicates that T7 was strongly affected by terrain-induced turbulence.

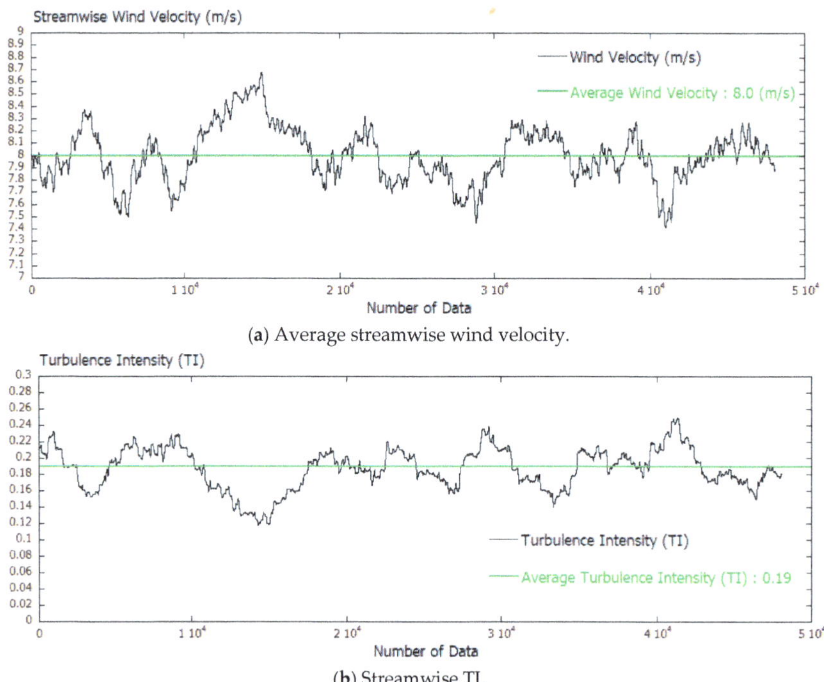

(**a**) Average streamwise wind velocity.

(**b**) Streamwise TI.

Figure 17. Temporal change of the full-scale streamwise (x) wind velocity (m/s) and TI. Both are calculated by applying a moving-averaged window to the time series data in Figure 15. The data shown are for the hub height (60 m above the ground surface) at T7.

Finally, the TI values from Figure 17 were examined by comparing them with those from the NTM in IEC 61400-1 Ed.3 (2005) (Figure 18). The NTM defines wind turbine classes as in Table 1. V_{ref} in Table 1 represents the 50-year return period values of 10-min average wind speed. I_{ref} is the expected value of TI for a wind speed of 15 m/s.

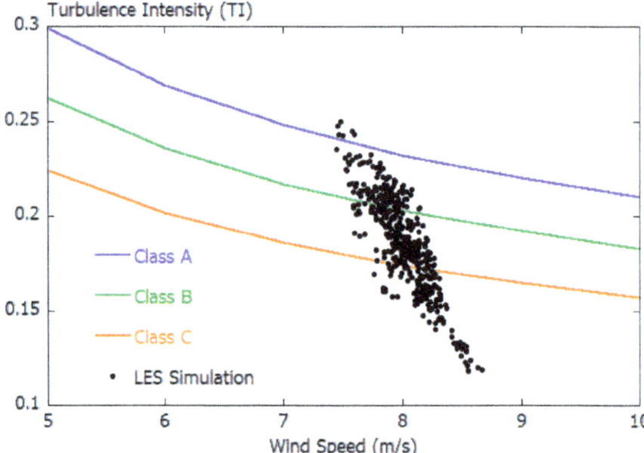

Figure 18. The relationship between TI and wind speed in the location of T7 at hub height (60 m above the ground surface). Dots: Simulation results from the present study. The plotted data were extracted every 100 data points from the time series of TI calculated from the simulation data. Lines: IEC turbulence categories A, B, and C from the NTM, defined in IEC 61400-1 Ed.3 (2005).

Table 1. Wind turbine class according to IEC 61400-1 Ed.3 (2005).

Class		I	II	III	S
V_{ref}		50.0	42.5	37.5	
	A		0.16		Values specified
I_{ref}	B		0.14		by the designer
	C		0.12		

For NTM, the values of and σ_{90q} and TI_{90q} are calculated using the streamwise (x) wind velocity as

$$\sigma_{90q} = I_{ref}(0.75U + 5.6) \tag{3}$$

$$TI_{90q} = \frac{\sigma_{90q}}{U} = \frac{I_{ref}(0.75U + 5.6)}{U} \tag{4}$$

where I_{ref} is the expected value for the turbulence intensity for V = 15 m/s; TI is the turbulence intensity; U is the 10-min average streamwise wind velocity (m/s); σ is wind velocity standard deviation (m/s); and subscript 90q is the 0.9 quantile value.

Wind turbine designers design wind turbines in such a way that they meet both the wind turbine class and turbulence class requirements. Wind power providers are able to reduce their business risks by confirming that the values of TI at their wind turbine sites lie under the curve defined by Equation (4). Figure 18 shows that the values of the streamwise (x) TI simulated with the abovementioned method were not as large as those observed in situ (Figure 6). However, some of the simulated values exceeded the TI values for turbulence class A, suggesting that the influence of terrain-induced turbulence on the wind turbine was well simulated.

Based on turbulence spectral relationships, the spanwise (y) wind-velocity standard deviation, σ_2, and the vertical (z) wind-velocity standard deviation, σ_3, are given with respect to the streamwise (x) wind-velocity standard deviation, σ_1, as in Equations (5) and (6). Both of these equations were derived from turbulence spectra from wind flow over flat terrain.

$$\sigma_2 \geq 0.7\sigma_1 \tag{5}$$

$$\sigma_3 \geq 0.5\sigma_1 \tag{6}$$

In the present study, the value of the standard deviation of the vertical (z) wind velocity, σ_3, at the hub height of T7 in Figure 12b was fairly large as discussed earlier and shown below:

$$\sigma_2 \approx 0.7\sigma_1 \tag{7}$$

$$\sigma_3 \approx 0.65\sigma_1. \tag{8}$$

4. Conclusions

The first part of the present study investigated the relationship among the number of yaw gear and motor failures and TI at all the wind turbines under investigation with the use of in situ data. The investigation revealed that wind turbine #7 (T7), which experienced a large number of failures, was affected by terrain-induced turbulence with TI that exceeded those presumed for the wind turbine design class for which T7 was designed. The frequency of occurrence of such terrain-induced turbulence at this wind turbine was also significantly higher than that at the other wind turbines. When the TI values evaluated for a wind turbine are high only in some of the examined periods, it is likely that upstream terrain mainly accounts for the high values in those periods and that the TI values become high in specific wind directions. Accordingly, the TI values at T7 were examined with respect various wind directions. The examination revealed that the TI values were high in westerly to northwesterly winds. Because complex terrain with ridges and surface undulations of various scales exists in the vicinity of T7, it was speculated that wind flows separated due to these ridges, resulting in terrain-induced turbulence.

Subsequently, a CFD simulation was performed to examine if the abovementioned observed wind flow characteristics could be successfully simulated. The CFD software package that was used in the present study was RIAM-COMPACT, which was developed by the first author of the present paper. RIAM-COMPACT is a nonlinear, unsteady wind prediction model that uses LES for the turbulence model. RIAM-COMPACT is capable of simulating flow collision, separation, and reattachment and also various unsteady turbulence–eddy phenomena that are caused by flow collision, separation, and reattachment. A close examination of computer animations of the streamwise (x) wind velocity revealed the following findings: As we predicted, wind flow that was separated from a micro-topographical feature (micro-scale terrain undulations) upstream of T7 generated large vortices. These vortices were shed downstream in a nearly periodic manner, which in turn generated terrain-induced turbulence, affecting T7 directly. In addition, vertical wind profiles of the (instantaneous) streamwise wind velocity at the wind turbine site were studied from the time at which the abovementioned unsteady turbulence–eddy phenomena occurred. As a result, it was found that a complex wind flow field with unusual vertical profiles of streamwise wind velocity formed at the wind turbine. More specifically, these vertical profiles of streamwise wind velocity deviated significantly from the power law profile and also had negative wind shear. On the other hand, the time-averaged wind flow data showed that the average streamwise wind velocity increased locally at the wind turbine site due to terrain effects and that no significant wind velocity deficit occurred at the site. Thus, even when no significant wind velocity deficit existed in the vertical profile of the time-averaged streamwise wind velocity, significant velocity deficits were present in vertical profiles of the instantaneous wind velocity, and such instantaneous wind velocity deficits should not be overlooked. An examination of the values of the standard deviations of the three wind velocity components at the wind turbine hub center (60 m above the ground surface) revealed that the value for the vertical (z) direction was large. The ratio of the standard deviations of the three wind velocity components at the wind turbine hub height was $\sigma_1:\sigma_2:\sigma_3 = 1.0:0.7:0.65$, which clearly showed the effect of terrain-induced turbulence. When wind flow with such properties passes through a wind turbine, it causes a large wind load on the wind turbine. Furthermore, because the structure of this complex wind flow is three-dimensional, wind loads imposed on the left and right side of the swept area are expected to differ. It was speculated that such wind loads would exert force on the wind turbine in such a way that they would forcibly rotate the nacelle of the wind turbine, which in turn would cause impact loads on both the yaw gears and motors and ultimately result in the failures of the yaw gears and motors.

Finally, the temporal change of the streamwise (x) wind velocity (a non-dimensional quantity) at the hub height of T7 in the period from 600 to 800 in non-dimensional time was re-scaled in such a way that the average value of the streamwise (x) wind velocity for this period was 8.0 m/s, and the results of the analysis of the re-scaled data were discussed. With the re-scaled full-scale streamwise wind velocity (m/s) data (total number of data points: approximately 50,000; time interval: 0.3 s), the time-averaged streamwise (x) wind velocity and turbulence intensity (TI) were evaluated using a common statistical processing procedure adopted for in situ data. Specifically, 10-min moving averaging (number of sample data points: 1932) was performed on the re-scaled data. Comparisons of the evaluated TI values to the TI values from the Normal Turbulence Model in IEC61400-1 Ed.3 (2005) revealed the following: Although the evaluated TI values were not as large as those observed in situ, some of the evaluated TI values exceeded the values for turbulence class A, suggesting that the influence of terrain-induced turbulence on the wind turbine was well simulated.

Author Contributions: The study idea, plan and design were conceived by T.U.; And, all authors prepared the manuscript.

Funding: This work was supported by JSPS KAKENHI Grant Number 17H02053.

Acknowledgments: The present study was supported by collaborative research with Eurus Energy Holdings Corporation: (1) Collaborative research and development on a method for assessing the wind speed ratio for wind turbine sites on complex terrain, July 2013-March 2014, December 2015-March 2018; (2) Collaborative research on land use modeling and its effect on simulated local-scale wind fields, December 2015-June 2016, principle investigator: Takanori Uchida. The authors wish to express their gratitude to all the individuals involved.

Conflicts of Interest: All authors declare no conflict of interest.

References

1. Rodrigues, R.V.; Lengsfeld, C. Development of a Computational System to Improve Wind Farm Layout, Part I: Model Validation and Near Wake Analysis. *Energies* **2019**, *12*, 940. [CrossRef]
2. Gargallo-Peiró, A.; Avila, M.; Owen, H.; Prieto-Godino, L.; Folch, A. Mesh generation, sizing and convergence for onshore and offshore wind farm Atmospheric Boundary Layer flow simulation with actuator discs. *J. Comput. Phys.* **2018**, *375*, 209–227. [CrossRef]
3. Sessarego, M.; Shen, W.Z.; Van der Laan, M.P.; Hansen, K.S.; Zhu, W.J. CFD Simulations of Flows in a Wind Farm in Complex Terrain and Comparisons to Measurements. *Appl. Sci.* **2018**, *8*, 788. [CrossRef]
4. Temel, O.; Bricteux, L.; van Beeck, J. Coupled WRF-OpenFOAM study of wind flow over complex terrain. *J. Wind Eng. Ind. Aerodyn.* **2018**, *174*, 152–169. [CrossRef]
5. Castellani, F.; Buzzoni, M.; Astolfi, D.; D'Elia, G.; Dalpiaz, G.; Terzi, L. Wind Turbine Loads Induced by Terrain and Wakes: An Experimental Study through Vibration Analysis and Computational Fluid Dynamics. *Energies* **2017**, *10*, 1839. [CrossRef]
6. Murthy, K.S.R.; Rahi, O.P. A comprehensive review of wind resource assessment. *Renew. Sustain. Energy Rev.* **2017**, *72*, 1320–1342. [CrossRef]
7. Simões, T.; Estanqueiro, A. A new methodology for urban wind resource assessment. *Renew. Energy* **2016**, *89*, 598–605. [CrossRef]
8. Gopalan, H.; Gundling, C.; Brown, K.; Roget, B.; Sitaraman, J.; Mirocha, J.D.; Miller, W.O. A coupled mesoscale-microscale framework for wind resource estimation and farm aerodynamics. *J. Wind Eng. Ind. Aerodyn.* **2014**, *132*, 13–26. [CrossRef]
9. Ishugah, T.F.; Li, Y.; Wang, R.Z.; Kiplagat, J.K. Advances in wind energy resource exploitation in urban environment: A review. *Renew. Sustain. Energy Rev.* **2014**, *37*, 613–626. [CrossRef]
10. Porté-Agel, F.; Wu, Y.-T.; Chen, C.-H. A Numerical Study of the Effects of Wind Direction on Turbine Wakes and Power Losses in a Large Wind Farm. *Energies* **2013**, *6*, 5297–5313. [CrossRef]
11. Gasset, N.; Landry, M.; Gagnon, Y. A Comparison of Wind Flow Models for Wind Resource Assessment in Wind Energy Applications. *Energies* **2012**, *5*, 4288–4322. [CrossRef]
12. Palma, J.M.L.M.; Castro, F.A.; Ribeiro, L.F.; Rodrigues, A.H.; Pinto, A.P. Linear and nonlinear models in wind resource assessment and wind turbine micro-siting in complex terrain. *J. Wind Eng. Ind. Aerodyn.* **2008**, *96*, 2308–2326. [CrossRef]
13. Uchida, T. Numerical Investigation of Terrain-induced Turbulence in Complex Terrain by Large-eddy Simulation (LES) Technique. *Energies* **2018**, *11*, 2638. [CrossRef]
14. Uchida, T.; Li, G. Comparison of RANS and LES in the Prediction of Airflow Field over Steep Complex Terrain. *Open J. Fluid Dyn.* **2018**, *8*, 286–307. [CrossRef]
15. Uchida, T. Computational Investigation of the Causes of Wind Turbine Blade Damage at Japan's Wind Farm in Complex Terrain. *J. Flow Control Meas. Vis.* **2018**, *6*, 152–167. [CrossRef]
16. Uchida, T. Computational Fluid Dynamics Approach to Predict the Actual Wind Speed over Complex Terrain. *Energies* **2018**, *11*, 1694. [CrossRef]
17. Uchida, T. LES Investigation of Terrain-Induced Turbulence in Complex Terrain and Economic Effects of Wind Turbine Control. *Energies* **2018**, *11*, 1530. [CrossRef]
18. Uchida, T. Computational Fluid Dynamics (CFD) Investigation of Wind Turbine Nacelle Separation Accident over Complex Terrain in Japan. *Energies* **2018**, *11*, 1485. [CrossRef]
19. Uchida, T. Large-Eddy Simulation and Wind Tunnel Experiment of Airflow over Bolund Hill. *Open J. Fluid Dyn.* **2018**, *8*, 30–43. [CrossRef]
20. Uchida, T.; Ohya, Y. Latest Developments in Numerical Wind Synopsis Prediction Using the RIAM-COMPACT®CFD Model. *Energies* **2011**, *4*, 458–474. [CrossRef]

21. Uchida, T.; Ohya, Y. Micro-siting technique for wind turbine generators by using large-eddy simulation. *J. Wind Eng. Ind. Aerodyn.* **2008**, *96*, 2121–2138. [CrossRef]
22. Smagorinsky, J. General circulation experiments with the primitive equations. Part 1, Basic experiments. *Mon. Weather Rev.* **1963**, *91*, 99–164. [CrossRef]

Article

Design and Simulation of an LQR-PI Control Algorithm for Medium Wind Turbine

Kwansu Kim [1], Hyun-Gyu Kim [1], Yuan Song [1] and Insu Paek [2,*]

[1] Department of Advanced Mechanical Engineering, Kangwon National University, Chuncheon-si 24341, Korea; kwansoo@kangwon.ac.kr (K.K.); khg0104@kangwon.ac.kr (H.-G.K.); songwon@kangwon.ac.kr (Y.S.)
[2] Division of Mechanical and Biomedical, Mechatronics and Materials Science and Engineering, Kangwon National University, Chuncheon-si 24341, Korea
* Correspondence: paek@kangwon.ac.kr; Tel.: +82-033-250-6379

Received: 31 March 2019; Accepted: 11 June 2019; Published: 12 June 2019

Abstract: In this paper, a new linear quadratic regulator (LQR) and proportional integral (PI) hybrid control algorithm for a permanent-magnet synchronous-generator (PMSG) horizontal-axis wind turbine was developed and simulated. The new algorithm incorporates LQR control into existing PI control structures as a feed-forward term to improve the performance of a conventional PI control. A numerical model based on MATLAB/Simulink and a commercial aero-elastic code were constructed for the target wind turbine, and the new control technique was applied to the numerical model to verify the effect through simulation. For the simulation, the performance data were compared after applying the PI, LQR, and LQR-PI control algorithms to the same wind speed conditions with and without noise in the generator speed. Also, the simulations were performed in both the transition region and the rated power region. The LQR-PI algorithm was found to reduce the standard deviation of the generator speed by more than 20% in all cases regardless of the noise compared with the PI algorithm. As a result, the proposed LQR-PI control increased the stability of the wind turbine in comparison with the conventional PI control.

Keywords: horizontal-axis wind turbine (HAWT); permanent-magnet synchronous-generator (PMSG); linear quadratic regulator (LQR); PI control algorithm; LQR-PI control

---

1. Introduction

Control algorithms for a wind turbine are generally designed to control both power and load [1]. The power control includes the maximum power region at wind speed lower than the rated wind speed, the rated power region at wind speed higher than the rated wind speed, and the transition region between the two mentioned power control regions. These control regions are named region 2, 3, and 2.5, respectively [2]. The load control is targeted to reduce loads that the wind turbine experiences and is distinguished on the basis of the load that is mostly reduced [3,4]. The tower damper is known to reduce the tower load and uses the acceleration signal of the nacelle to calculate the command to the pitch actuator to reduce loads [4–9]. This is used in region 3. The peak shaving is known to reduce the tower and blade loads at region 2.5 by slightly adjusting the pitch angle of the blade by a pre-designed pitch schedule [10]. The individual pitch control is used in region 3 to reduce the blade load due to imbalance loads caused by wind shear, tower shadow, etc. It uses the signals from strain sensors mounted on the blade roots to calculate the command to the pitch actuator [4,11]. The drivetrain damper is used in region 2 to reduce the low-speed shaft torque due to torsional modes from the drive train [12]. It uses the generator speed signal to calculate the torque command to the generator to cancel out the drivetrain mode in the torque command.

Power control for modern wind turbines is achieved by a combination of open-loop and closed-loop control. In region 2, the control strategy is to maximize the wind turbine power, and this is achieved by an open-loop torque control with a fixed pitch angle (known as fine pitch) to maximize the power coefficient which is the aerodynamic conversion efficiency of the rotor. Either a generator torque–speed lookup table or an optimal mode gain (optimal relationship between generator speed and torque) is used for this [13,14]. The power control in region 2.5 is an extension of the power control in region 2, and the control strategy just performs a smooth transition from region 2 to region 3. The control strategy in region 3 consists in regulating the power so that it does not exceed the rated power of the wind turbine. The generator speed is controlled by either a PI (proportional-integral) or a PID (proportional-integral-differential) control to achieve the rated wind speed. The generator torque is controlled by open-loop control and PI control [4,15].

Although many modern control algorithms, including the linear quadratic regulator (LQR), fuzzy control, and model predictive control (MPC) algorithms, have been proposed by researchers as control algorithms for wind turbines to improve their performance [16–20], no algorithm has been chosen as an alternative to the conventional PI or PID power control by wind turbine manufacturers or companies to provide wind turbine control solutions. This is because the conventional PI or PID control algorithms for wind turbines have been used for a long time as power control algorithms and found to be robust and effective. This practice is not likely to change fast, as manufacturers often adopt a conservative approach towards innovation in control system design.

Efforts have been made to improve the performance of the conventional PI or PID control algorithms by adding extra commands to the calculated pitch command [21,22]. These methods measure the wind speed by Light Detection and Ranging (LIDAR) or by other techniques; they commonly use the partial derivative of the pitch angle with respect to the generator speed to calculate the required pitch angle variation based on the current wind speed variation and add this, multiplied by a suitable proportional gain, to the pitch command from the conventional PI or PID control algorithm. Although these feed-forward controls could not be validated, they are considered to be applicable to the actual wind turbines because they use the conventional PI or PID control algorithms as a basis and integrate the feed-forward control in region 3.

This study was performed to develop a new power control algorithm to be applied to a 100 kW medium-capacity wind turbine to improve its performance using a similar approach to the previous feed-forward control. The target wind turbine is not a multi-megawatt wind turbine and cannot afford a LIDAR to measure the wind speed, therefore a wind speed estimator was chosen for this study. Also, to calculate the feed-forward pitch command signal, contributions from other measured parameters as well as the estimated wind speed were considered for fine adjustment of the pitch angle command that was added to the command from the conventional PI or PID control algorithm. Therefore, an LQR controller was finally selected for this purpose. The LQR control uses wind speed estimators to estimate the representative wind speed experienced by that wind turbines and determine the magnitude of the control command [21,23–26]. Reference [24] constructed a tower and blade state estimator using accelerometers and strain gauges arranged along structural members and used it to estimate the wind state. The demonstration was conducted through an aerosol-servo-elastic simulator, which suggested that the individual blade fatigue and load could be reduced. Reference [25] demonstrated power curve tracking through a model-based control using a wind schedule for 3 MW wind turbines with blade tip speed constraints in simulated environments. In Reference [26], a wind observer was tested using field test data collected from NREL CART3 wind turbines. The results showed that the rotor equivalent wind speed estimated by the proposed observer correlated with the meteorological data and was much more accurate than the speed measured by an onboard wind vane. The wind speed estimator used in this study used a three-dimensional (3D) lookup table based on the two-mass drivetrain model with measured generator speed, torque, and pitch angle [4,21]. In [16], an LQR controller was designed for a megawatt (MW)-class wind turbine, and simulations were performed to test its performance. The simulation results showed that the performance of the wind turbine was improved by the proposed

LQR controller compared to that obtained with the conventional PI control, and the blade and tower loads were also reduced. Reports in the literature show that LQR controllers are effective as wind turbine controllers [16], but their performance relies on the accuracy of the wind speed estimators, so they are vulnerable to the noise or unexpected events influencing the measurement signal that is used for wind speed estimation. The reason is that the sensitivity varies with the wind speed. This issue has not been studied.

The purpose of this paper was to improve the performance of a PI control algorithm by virtue of an LQR controller which has a good control performance but is vulnerable to uncertainties in wind speed estimation. Therefore, a hybrid controller is newly proposed in this study. A PI control was used as the conventional power control, and an LQR control was used as a feed-forward controller to improve the control performance. This new control algorithm minimizes changes in the conventional PI control algorithm so that it could be relatively easily adopted by wind turbine manufacturers as a new control algorithm for modern wind turbines. Also, the proposed algorithm was expected to limit the contribution from the LQR controller which was significantly affected by wind speed estimation errors because the LQR controller was used as a feed-forward controller. For this, a new hybrid controller, which is a combination of the conventional PI and LQR controllers, was designed for a 100 kW wind turbine. It is difficult to validate wind turbine control algorithms in a field test with multi-MW-class wind turbines. Therefore, numerical modeling is generally used to validate the performance of a single wind turbine or in wind farms [27–32]. The target wind turbine had a permanent-magnet synchronous generator (PMSG) without a gearbox and blades with a substantially smaller rotor moment of inertia and faster rotational speed compared to those of MW-class wind turbines. The proposed LQR-PI controller was tested with dynamic simulations, and the performances were compared with those of a PI and an LQR control algorithms with and without noise in the measured generator speed signal.

2. Target Wind Turbine

The target wind turbine used in this study is a PMSG horizontal-axis wind turbine. An overview of the specifications and an image of the target wind turbine are presented in Table 1 and Figure 1, respectively. The wind turbine is installed on an onshore test bed located in Gimnyeong-ri, Jeju-do, South Korea.

Table 1. Specifications of the target wind turbine.

Specifications	Units	Values
Rotor diameter	m	24.25
Hub height	m	30
Rated generator speed	rpm	50
Rated electrical power	kW	100
Cut-in, rated, cut-out wind speed	m/s	4, 10.5, 20

Figure 1. Image of the target wind turbine.

3. Numerical Modeling

The commercial code DNVGL-Bladed (4.6, DNV·GL, Oslo, Norway) was used for numerical modeling. DNVGL-Bladed was used to extract linear models, blade power coefficients, and thrust coefficients for the wind turbine. The in-house code includes control algorithms, wind speed estimators, and wind turbine numerical models. This section describes the wind speed estimator and wind turbine numerical models, and the next section introduces the control algorithm. From a control system perspective, a wind turbine numerical model includes aerodynamics, drive trains, generators, towers, and pitch actuators.

A block diagram of the overall functional scheme of a wind turbine is shown in Figure 2. The blue box indicates the control algorithm, the green box indicates the wind speed estimator, and the yellow box presents the wind turbine numerical model.

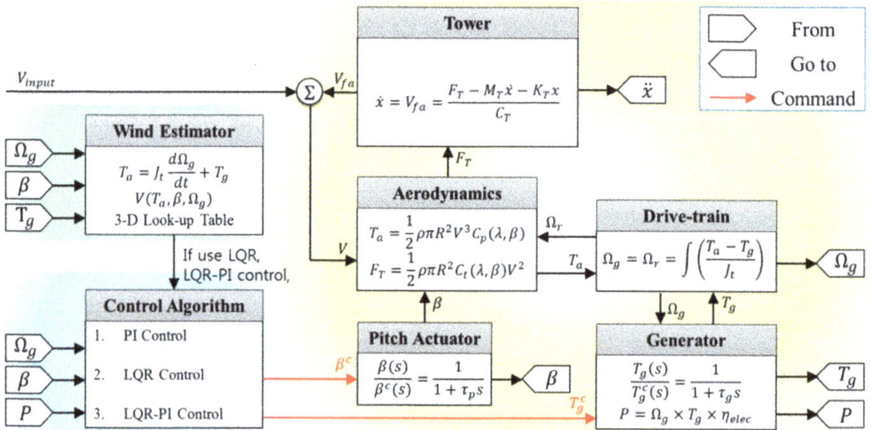

Figure 2. Block diagram of the in-house code.

3.1. Aerodynamics

The aerodynamics model makes use of power coefficient and thrust coefficient lookup tables extracted as a function of the pitch angle and tip speed ratio (TSR) through the aerodynamic analysis of DNVGL-Bladed. In this component, the wind speed, generator speed, and pitch angle are the input. The aerodynamic torque and thrust force are calculated through Equations (1) and (2), respectively, and applied to the drivetrain and tower model, respectively.

$$T_a = \frac{1}{2}\rho\pi R^3\left(\frac{C_p(\lambda, \beta)}{\lambda}\right)V^2 = \frac{1}{2}\rho\pi R^3 C_q(\lambda, \beta)V^2 \tag{1}$$

$$F_T = \frac{1}{2}\rho\pi R^2 V^2 C_t(\lambda, \beta) \tag{2}$$

3.2. Drivetrain

The target wind turbine is a PMSG type without a gearbox, so Equation (3) can be derived from Figure 3. This component receives the generator torque and aerodynamic torque from a generator and aerodynamics model, calculates the generator speed, and delivers it back to the aerodynamics and generator model.

$$\left(J_r + J_g\right)\frac{d\Omega_r}{dt} = J_t\frac{d\Omega_r}{dt} = J_t\frac{d\Omega_g}{dt} = T_a - T_g \tag{3}$$

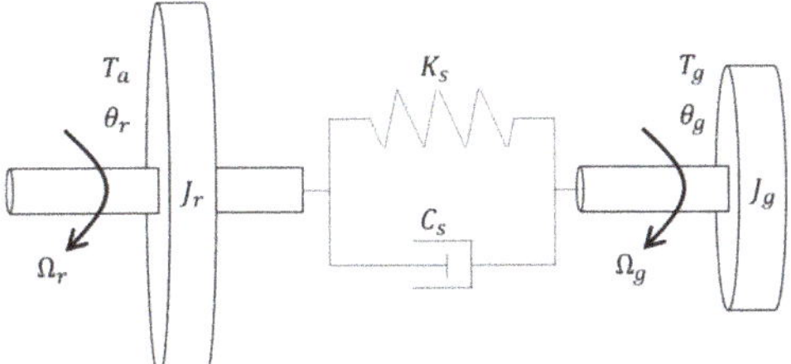

Figure 3. Drivetrain model.

3.3. Generator

The generator can be simplified to Equation (4) from the side of the control system. The torque command and generator speed are input from the control algorithm and the drivetrain model, respectively, to calculate the generator torque and electrical power. The electrical power is calculated using Equation (5).

$$\frac{T_g(s)}{T_g^c(s)} = \frac{1}{1 + \tau_g s} \tag{4}$$

$$P = T_g \Omega_g \eta_g \tag{5}$$

3.4. Tower

The tower model is expressed as the equation of motion given in Equation (6). In this model, the velocity at which the nacelle sways fore and aft because of the wind speed is added to the input wind speed and entered into the aerodynamics model.

$$m_T \ddot{x}_{faft} + c_T \dot{x}_{faft} + k_T x_{faft} = F_T \tag{6}$$

3.5. Pitch Actuator

The dynamic characteristics of the pitch actuator model are expressed by Equation (7). The pitch actuator operates within the limits of Equations (8) and (9) according to the design specification.

$$\frac{\beta(s)}{\beta^c(s)} = \frac{1}{1 + \tau_p s} \tag{7}$$

$$-5° \leq \beta \leq 90° \tag{8}$$

$$-10 \; (°/s) \leq \dot{\beta} \leq 10 \; (°/s) \tag{9}$$

4. Control Algorithms

This section introduces PI control, LQR control, and LQR-PI hybrid control algorithms. The PI control algorithm is a control technique applied to the target wind turbines and in this study, it is presented as a reference control algorithm to compare the performance of LQR control and LQR-PI control.

4.1. PI Control Algorithm

The PI control algorithm adopted and used by the target wind turbine receives feedback on the measured pitch angle, electrical power, and generator speed, and sends pitch angle and torque commands to the pitch actuator and generator [7,10,15]. In practice, mechanical load-reduction control techniques such as tower dampers and peak shaving are usually applied, but in this study, only power control was considered.

Figure 4 shows a block diagram of the pitch PI control algorithm containing the gain schedule. The configuration consists of a pitch PI control, torque schedule, and mode switch. The torque schedule is a lookup table with an input of generator speed and an output of generator torque. It was constructed to perform an open-loop maximum power point tracking (MPPT) control to achieve the optimal tip speed ratio with the measured generator speed in region 2. The optimal values of the generator torque with respect to the input generator speeds were calculated on the basis of the aerodynamic analysis of the rotor using DNVGL-Bladed. The gain selection of pitch PI control and operation of the mode switch are explained in detail below.

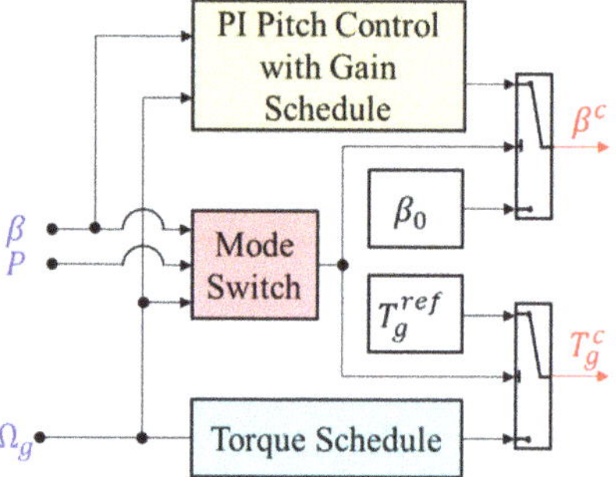

Figure 4. Block diagram of the proportional integral (PI) control algorithm.

Figure 5 shows the block diagram of the mode switch. The mode switch determines the control mode using an internal logic with the measurement values of generator speed, electrical power, and pitch angle. A set-reset (SR) flip-flop is a logic that remembers one bit and remains in the current state until a change in the state signal (clock) is generated. If the measured generator speed or power exceeds the rated values, the mode switch outputs a signal of 1 (switched on). Also, the mode switch outputs a signal of 1 (switched on) if the measured pitch angle is greater than the fine pitch angle.

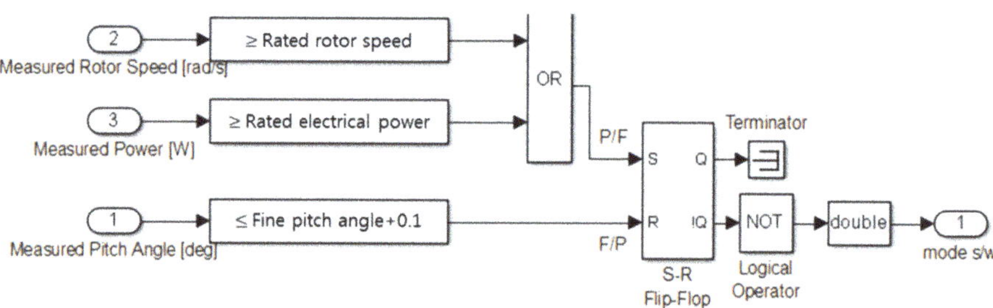

Figure 5. Block diagram of mode switch.

When the mode switch is on, pitch PI control is performed, and the generator torque command is fixed to be the rated value. When the mode switch is off, to perform open-loop MPPT control, the pitch angle command is fixed to be the fine pitch angle, and the torque control is performed through the torque schedule.

Figure 6 shows the frequency response of the pitch control loop gain. Figure 6a shows the frequency response of the open-loop transfer function given in Equation (10). The frequency response was drawn over all the wind speeds from the rated wind speed to the cutout wind speed by 0.5 m/s intervals. As shown in the figure, the frequency response varies depending on wind speed. Therefore, to have uniform pitch sensitivity, gain scheduling should be applied to maintain a constant value of the cross frequency of the pitch control loop. In this study, the cross frequency was set to 1 rad/s, taking into account the fact that most energy components in wind speed exist at frequencies lower than 1 rad/s based on the power spectrum of wind speed [4].

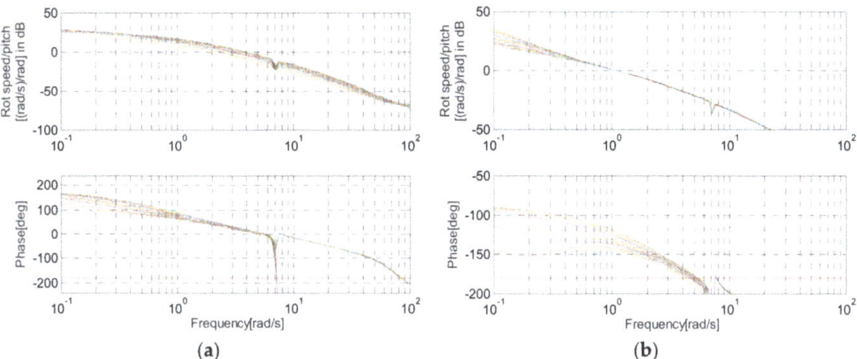

(a)   (b)

Figure 6. Frequency response of the pitch loop. (**a**) Open-loop transfer function for the pitch input; (**b**) pitch control loop gain transfer function with gain scheduling.

Figure 6b shows the frequency response of the pitch control loop gain transfer function of Equation (11). The pitch control loop consists of gain scheduling, PI control, pitch actuator dynamics (Equation (7)) and open-loop transfer function (Equation (10)). As shown in the figure, all the frequency responses (magnitude plot) at different wind speeds had a cross frequency of 1 rad/s, and the phase margin (phase plot) of at least 30 degrees was achieved for system stability.

$$G(s) = \frac{\delta\Omega_g(s)}{\delta\beta(s)} \tag{10}$$

$$L(s) = k_G(\beta)\left(k_p + \frac{k_i}{s}\right)\left(\frac{1}{\tau_p s + 1}\right)\left(\frac{\delta\Omega_g(s)}{\delta\beta(s)}\right) \tag{11}$$

4.2. LQR Control Algorithm

Figure 7 shows the control structure of the LQR control algorithm. The pitch control was replaced by LQR control. The LQR control received the generator speed, torque, and pitch angle as inputs. In addition, the wind speed estimator used the current generator speed, torque, and pitch angle to deliver the estimated wind speed to the designed LQR control. The design of the wind speed estimator is presented in detail in Section 5.

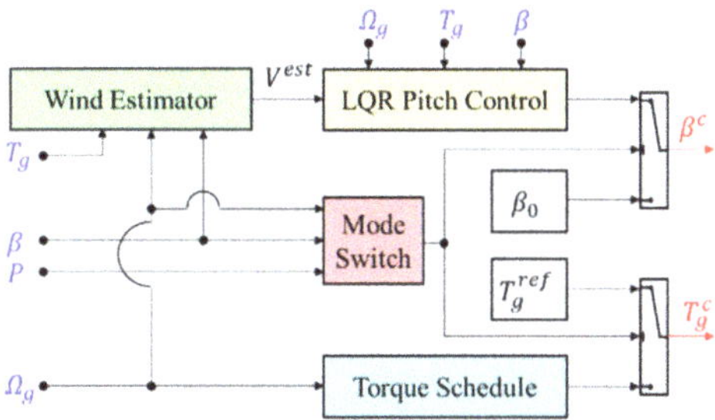

Figure 7. Block diagram of the linear quadratic regulator (LQR) control algorithm.

Linearization models are required to select the optimal gain for LQR control. In this study, a linearization model was acquired through DNVGL-Bladed. In Equation (12), the state matrix A and input matrix B are stabilizable. The state vectors and control vectors are presented in Equation (13). In state vector x, the pitch angle actually reflects only one result because the target wind turbine performs collective pitch control (CPC). In order to stabilize the system Equation (12), the optimum gain K in Equation (17) that minimizes the quadratic cost function (Equation (15)) through the state feedback method (Equation (14)) must be selected. In Equation (17), S is the symmetric positive semidefinite solution of the Riccati Equation (16). Since the size of the matrix was large, K was obtained using the LQR function of MATLAB (R2014a, The MathWorks, Inc, Natick, MA, USA).

Also, the weight matrices Q and R that met the conditions in Equation (18) were chosen randomly and simulated by solving the Riccati Equation. Then Q and/or R were re-selected if transient response specifications and/or size constraints were not met. This means that the weight matrix was selected as a tuning process until a satisfactory performance was achieved. The Q matrix was weighted more heavily for the state variables, so that the objective was achieved in a short time, and the R matrix was chosen through the simulation response.

$$\dot{x} = Ax + Bu \tag{12}$$

$$x = \left[x_{faft}\ \dot{x}_{faft}\ x_{side}\ \dot{x}_{side}\ \theta_g\ \Omega_g\ T_g\ \beta_1\ \dot{\beta}_1\ \beta_2\ \dot{\beta}_2\ \beta_3\ \dot{\beta}_3 \right]^T, u = \left[\beta^{cmd} \right] \tag{13}$$

$$u = -Kx \tag{14}$$

$$J = \int_0^\infty \left(x^T Q x + u^T R u \right) dt \tag{15}$$

$$A^T S + SA - SBR^{-1}B^T S + Q = 0 \tag{16}$$

$$K = R^{-1}B^T S \tag{17}$$

$$Q = Q^T \geq 0, R = R^T > 0 \tag{18}$$

4.3. LQR-PI Control Algorithm

Figure 8 shows a block diagram of the LQR-PI control algorithm proposed in this paper. The LQR-PI control combines the control output of the PI pitch control and that of the LQR control to transmit the combined pitch command to the pitch actuator. As shown in the figure, the LQR control was applied in this structure as a feedforward term for pitch control.

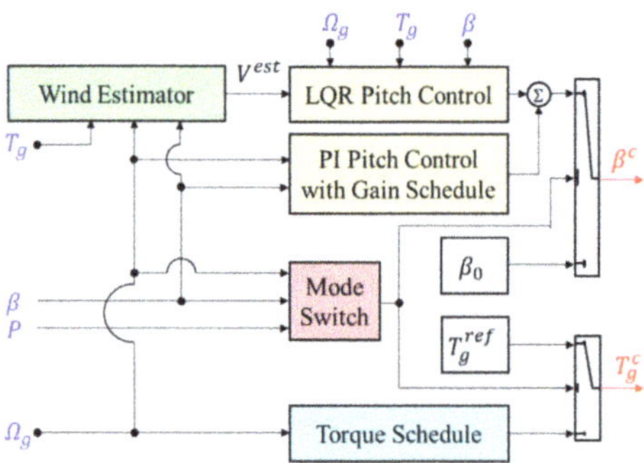

Figure 8. Block diagram of the LQR-PI hybrid control algorithm.

Although it is simply a combination of LQR control and PI control, both controls can complement each other in the pitch control domain to improve the operating stability of the wind turbine. If the LQR control delivers the optimum pitch angle command to the pitch actuator and performs a sufficiently stable control, the contribution of the PI control will be small, because the pitch PI control will intervene in control when the measured generator speed exceeds the rated value. However, because of noise or unexpected circumstances, the LQR control may send incorrect pitch commands to the pitch actuator, thus not achieving its original purpose of maintaining the rated generator speed. In this case, PI control takes the lead in pitch control.

LQR-PI control was configured to perform a PI control also when LQR control was removed (or disconnected by a switch) from the control structure. The advantage of this feature is that when the LQR-PI control is applied to actual wind turbines, only a feed-forward loop of LQR control can be added to the pitch loop of the existing PI controller without modifying existing control algorithms.

5. Wind Speed Estimator

The nacelle wind speed is not suitable for feeding a control-scheduling logic because it is disturbed by the rotation of the rotor, which introduces a periodic decrease with multiples of the rotor frequency, as well as higher frequency disturbances due to wake turbulence [33]. Therefore, a wind speed estimator was designed for LQR control and used to calculate the rotor average wind speed.

Figure 9 shows a block diagram of the wind speed estimator. It consists of an aerodynamic torque estimator, a 3D look-up table for wind speed, and a low-pass filter. The aerodynamic torque estimator is just a Simulink representation of Equation (3), which is the two-mass drivetrain model. The aerodynamic torque was firstly estimated from the aerodynamic torque estimator using the rotor speed and the generator torque and then was supplied to the 3D lookup table as an input. Two more inputs of pitch angle and rotor speed were provided to the 3D lookup table to get the wind speed as an output. The wind speed from the 3D lookup table was finally low-pass filtered to remove high-frequency components [21].

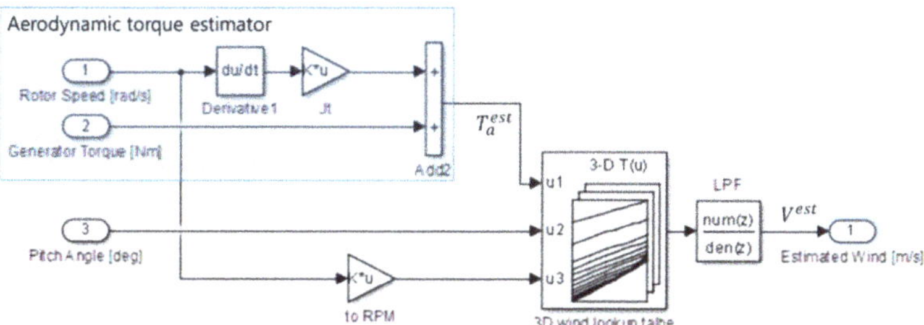

Figure 9. Block diagram of the wind speed estimator. Low pass filter (LPF); Revolutions per minute (RPM).

The 3D lookup table in Figure 9 was created using the fminsearch function of MATLAB. The fminsearch is a function minimization algorithm based on the Nelder-Mead simplex method [34–36]. It can be applied to nonlinear functions whose derivatives are not known and is one of the most widely used function minimization algorithms for a direct search method. This method uses a simple value known as a polytope with $n + 1$ vertices (or $n + 1$ test point) in the n variable of the objective function. To find a value that can minimize objective function values, compare the function values at the $n + 1$ test point, avoid the test points that provide the worst function values, and repeat the reflection, contraction, and extension of the variables [36].

That is, the fminsearch finds the wind speed to minimize the error between the power calculated and the rated power, as shown in Figure 10. To calculate the electrical power, the aerodynamic torque was firstly calculated using Equation (1) with inputs of wind speed, rotor speed, and pitch angle. The wind speed, in this case, was a trial value from the fminsearch algorithm, and the others were the given inputs. The aerodynamic torque was finally multiplied by the rotor spee, and the generator loss and converted into electrical power. The error between the calculated power and the rated power was used as a cost function in the fminsearch algorithm to be minimized. At the function minimum, the wind speed could be obtained. The inputs were varied to construct a 3-D lookup table whose output was the wind speed. The inputs in the lookup table were rotor speed, pitch angle, and aerodynamic torque.

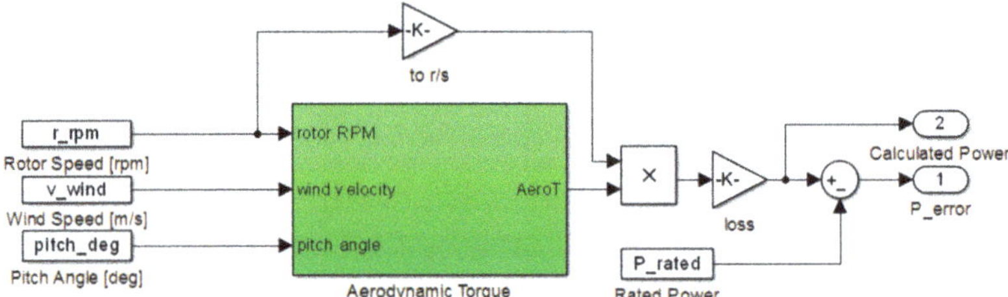

Figure 10. Block diagram of a function to produce a 3D lookup table.

Figure 11 shows the results of the 3D lookup table constructed through this process. The 3D lookup table was constructed for the operation range as a function of rotor speed, pitch angle, and estimated aerodynamic torque. For example, if the estimated aerodynamic torque for a given rotor speed of 40 RPM, as in Figure 11 was 25 kNm and the pitch angle was 14°, the wind speed would be 15 m/s.

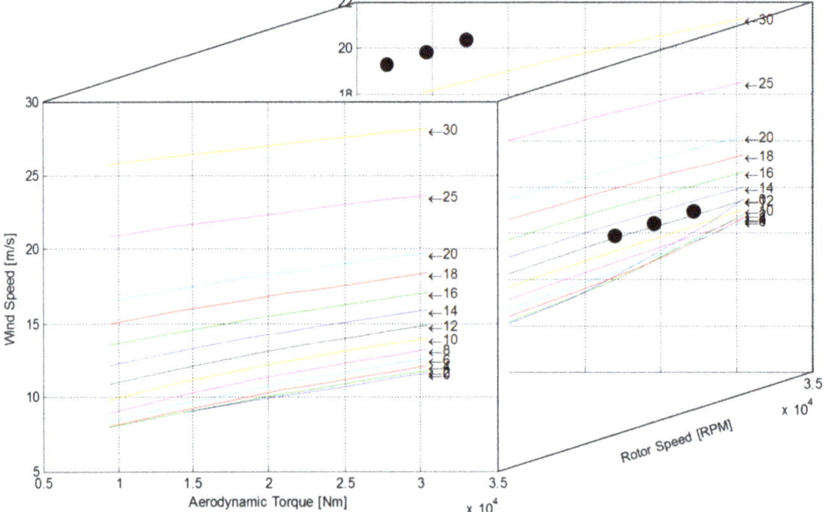

Figure 11. 3D lookup table for the estimated rotor averaged wind speed.

To validate the wind speed estimator using simulation, the wind speed estimated from the wind speed estimator was compared with the rotor averaged wind speed from DNVGL-Bladed (Figure 12). A dynamic simulation at a mean wind speed of 14 m/s was performed with the target wind turbine, and the rotor averaged wind speed was obtained from DNVGL-Bladed. The generator torque, rotor speed, pitch angle from DNVGL-Bladed at a time interval of 10 ms were used as inputs to the wind speed estimator, and the turbulent wind speeds were obtained as outputs. Although a delay of less than 1 second was found, the wind speed estimator considered appeared valid because the mean and the standard deviation of the two wind speed data were almost identical.

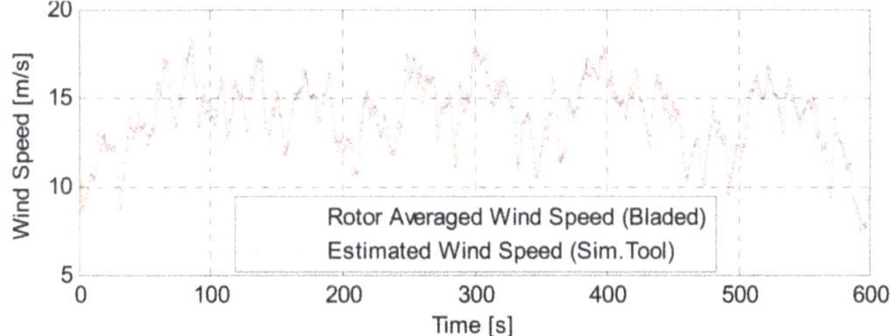

Figure 12. Comparison of the wind speed estimator results with those obtained with DNVGL-Bladed.

6. Simulation

6.1. Method

The simulation was performed for the transition region and the rated power region where pitch control was used. The wind speed calculated through the wind speed estimator was used as the input wind speed for the simulation. For this, the measured generator speed, generator torque, and pitch angle from the target 100 kW wind turbine were used.

Figure 13 shows the wind speed estimated by the measured data, and the nacelle wind speed measured by an anemometer on top of the nacelle. Figure 13a shows the wind speed of the transition region, and Figure 13b shows the wind speed of the rated power region. In the case of the nacelle wind speeds, although the speeds actually had a higher frequency, they appeared to be similar to the estimated wind speed because the data measuring device collected data with a sample rate of 1 Hz. Although the nacelle wind speed was affected by the rotor rotation, the estimated wind speed was found to be similar to the nacelle wind speed, and as expected, it was found to be slightly higher than the nacelle wind speed. The mean value and the standard deviation of the input wind speed were 10.58 m/s and 0.98 m/s, respectively, for the transition region, and they were 15.21 m/s and 1.73 m/s, respectively, for the rated power region.

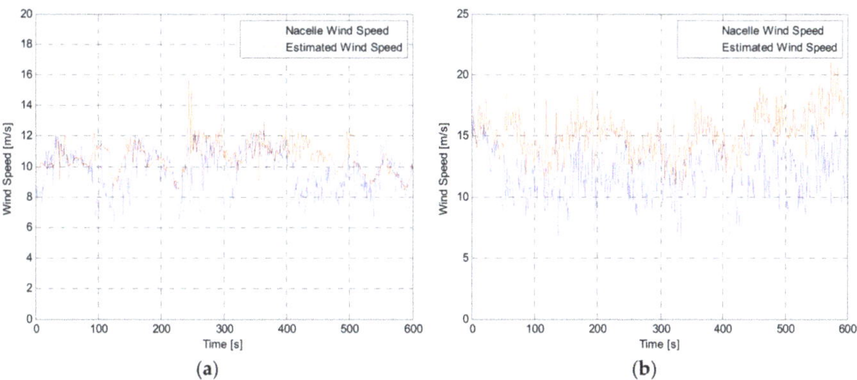

Figure 13. Comparison between actual nacelle wind speed and estimated wind speed: (**a**) transition region; (**b**) rated power region.

The simulation was performed with three different control algorithms including the conventional PI, the LQR, and the proposed LQR-PI. The simulation was also performed with and without noise to evaluate the controller performance in the presence of noise in the measured signal. In the simulation with noise, randomly mixing Gaussian noise was added to the generator speed.

6.2. Results without Noise

Figure 14a,b show the simulation results without considering noise in two different wind speed regions, i.e., the transition region and the rated power region. They compare the results with the conventional PI, the LQR, and the proposed LQR-PI controller. The simulation was performed for 600 s, but for visibility purposes, only the results from 0 to 100 s are presented. In Figure 14, the black line for wind speed represents the input wind speed obtained from the previous section. The subplot of wind speed also includes the wind speeds obtained from the wind speed estimators in the simulations with three different controllers. The estimated wind speed obtained by three different controllers showed a difference of less than 1% in mean wind speed compared with input wind speed, but the standard deviations were 4.08% and 3.47% higher for transition and rated power regions, respectively.

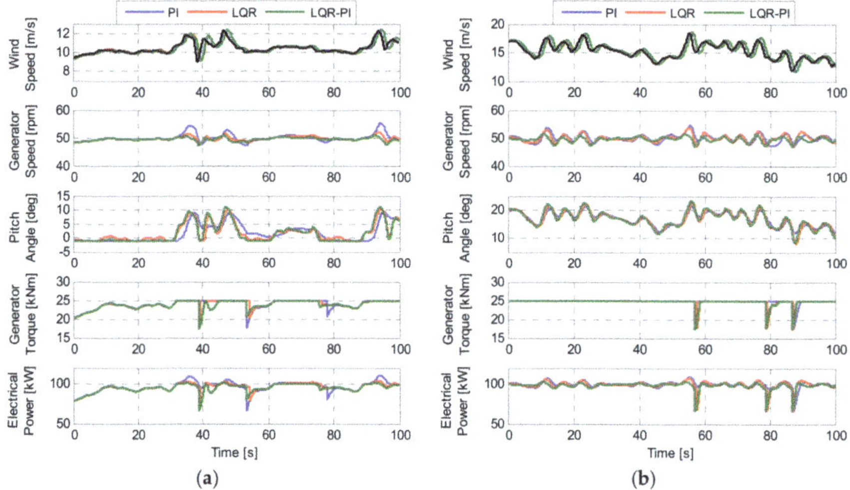

Figure 14. Simulation results according to the control method applied. (**a**) Transition region; (**b**) rated power region.

In the transition region, the PI control showed the largest overshoot of generator speed. This can be explained in relation to the pitch angle. The PI control regulates the pitch angle from the moment it exceeds the rated generator speed, but the LQR and the LQR-PI controls started the pitch angle in advance to attenuate the generator speed increase. Although the LQR and the LQR-PI controls used the estimated wind speed delayed by about 1 second for their control command calculation, the standard deviation of the generator speed was reduced compared with that obtained with the PI control. Sudden dips in the generator torque and power were observed in the simulation with all three controllers, but the greatest one was obtained with the PI controller. The results with the LQR controller were the best, and that those the LQR-PI were intermediate. On the basis of these results, it was concluded that the overshoot of the power was mostly due to the generator speed, and the dip was mostly due to the generator torque.

For the rated power region, the results with three different controllers were similar, but the lowest standard deviation of the generator speed was achieved when the LQR-PI control was used. Quantitative comparisons of the simulation results are shown in Tables 2 and 3.

Table 2. Quantitative comparison of performance data in the transition region without noise.

Mean	Performance Data			Difference (%)		
	PI (A)	LQR (B)	LQR-PI (C)	(B−A)/A	(C−A)/A	(C−B)/B
V^{est} (m/s)	10.61	10.61	10.60	0.00	0.00	−0.09
Ω_g (rpm)	49.53	49.43	49.26	−0.20	−0.55	−0.35
β (°)	3.02	3.05	2.88	0.80	−4.91	−5.66
T_g (kNm)	23.43	23.48	23.43	0.21	0.00	−0.21
P (kW)	97.41	97.40	96.85	−0.01	−0.58	−0.57
Std. Dev.	**PI (A)**	**LQR (B)**	**LQR-PI (C)**	**(B−A)/A**	**(C−A)/A**	**(C−B)/B**
V^{est} (m/s)	1.03	1.02	1.03	−0.97	0.00	0.98
Ω_g (rpm)	1.57	1.15	0.96	−26.58	−38.85	−16.70
β (°)	4.00	3.93	4.23	−1.55	5.76	7.42
T_g (kNm)	2.21	2.20	2.24	−0.53	1.19	1.73
P (kW)	11.11	10.71	10.62	−3.65	−4.41	−0.78

Table 3. Quantitative comparison of performance data in the rated power region without noise.

Mean	Performance Data			Difference (%)		
	PI (A)	LQR (B)	LQR-PI (C)	(B−A)/A	(C−A)/A	(C−B)/B
V^{est} (m/s)	15.27	15.27	15.27	0.00	0.00	0.00
Ω_g (rpm)	50.00	49.99	49.71	0.00	−0.58	−0.57
β (°)	16.61	16.52	16.56	−0.57	−0.29	0.28
T_g (kNm)	24.50	24.51	24.51	0.03	0.03	0.00
P (kW)	98.82	98.85	98.29	0.03	−0.54	−0.57
Std. Dev.	**PI (A)**	**LQR (B)**	**LQR-PI (C)**	**(B−A)/A**	**(C−A)/A**	**(C−B)/B**
V^{est} (m/s)	1.79	1.79	1.79	0.00	0.00	0.00
Ω_g (rpm)	1.52	1.51	1.12	−1.15	−26.86	−26.01
β (°)	3.14	3.69	3.87	17.63	23.13	4.68
T_g (kNm)	1.24	1.33	1.38	7.41	11.63	3.94
P (kW)	6.48	6.80	6.58	4.90	1.53	−3.21

Tables 2 and 3 show the simulation results for 600 seconds. The most notable performance indicators in the results presented are the standard deviations of the generator speed, which can represent the operating stability of the wind turbine. The estimated wind speed in Tables 2 and 3 represents the estimated wind speed from the wind speed estimator. These were about the same with three different controllers, although the operating points were slightly different.

The results given in Table 2 indicate that the LQR control reduced the standard deviation of the generator speed by 26.58% compared to the PI control. In the case of the LQR-PI control, the standard deviation of the generator speed was even 16.7% lower than that for the LQR control, and 38.85% less than that for the PI control. However, as a side effect, the mean power with the LQR-PI control was reduced by 0.57% with respect to that measured with the LQR control.

As can be seen in Table 3, the LQR control and LQR-PI control had less than a 1% difference in all average performance indices compared with the PI control. For the standard deviation, the LQR control had a lower generator speed of 1.15% and a higher pitch angle of 17.63% compared to the PI control. A higher standard deviation of the pitch angle means that the pitch control was busier. The generator torque and power generation increased by 7.41% and 4.9%, respectively. The LQR-PI control reduced the standard deviation of the generator speed by 26.86% compared with the PI control, and the standard deviations of the pitch angle, generator torque, and power increased by 23.13%, 11.63%, and 1.53%, respectively.

As a result, the LQR control was able to increase the stability of wind turbines by reducing the standard deviation of the generator speed. However, in regions where the pitch control was continually used, the effect was reduced. On the other hand, the LQR-PI control was able to reduce the standard deviation of the generator speed in the two wind speed regions compared with the PI control, and its effect was the greatest in the rated control region, where the pitch control was used continually.

6.3. Results with Noise

The LQR control can improve the stability of the generator speed, but a practical problem is that it relies on the accuracy of the wind speed estimators. Noise in the feedback signal causes the wind speed estimator to become inaccurate, which causes the controller to send abnormal commands to the actuator. Figure 15a,b show the simulation results in the transition and power controlled regions, respectively, when noise was taken into consideration. To simulate the noise, white noise was introduced into the generator speed. Compared with Figure 14a,b, the dip in the generator torque by mode switch occurred more frequently, and the pitch angle movement was more active.

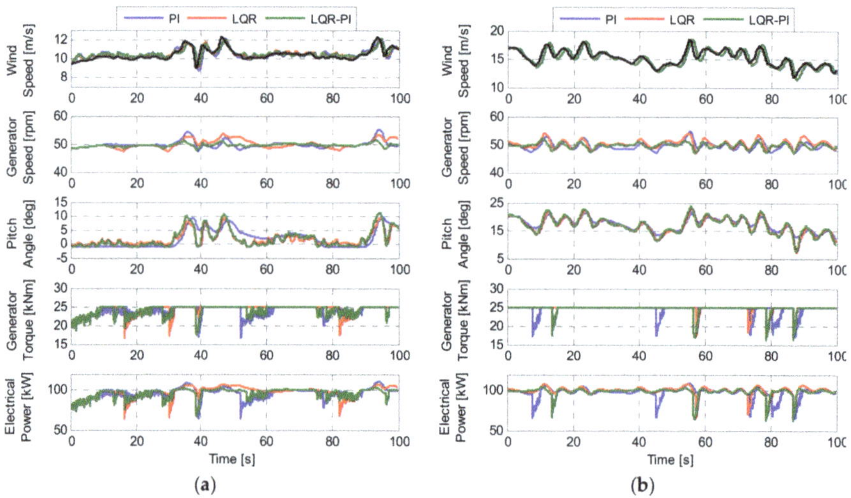

Figure 15. Simulation results according to the control method applied in the presence of noise. (**a**) Transition region; (**b**) rated power region.

Figure 15a shows the input wind speed (black line) as well as the wind speeds estimated in the simulations with three different controllers. Unlike the results without noise, the estimated wind speeds were now oscillatory with high-frequency components. However, this oscillation in the estimated wind speed is not visible in Figure 15b.

In the transition region, the PI control had the largest overshoot in the generator speed, similar to the results obtained without noise. However, the LQR control showed unstable behavior, much differently from the results obtained without noise. This is because the input wind speed to the LQR control which was obtained from the wind speed estimator was distorted by the noise of the generator speed. In addition, the oscillations of the LQR and LQR-PI controls were reflected in the behavior of pitch angle and were more clearly detected than when using the PI control. The generator torque command was determined using the generator speed, so the noise component was still present, and showed unstable behavior, which also affected the electrical power.

In the rated power region, the LQR-PI control yielded the lowest standard deviation in the generator speed, similar to the results without noise. The difference in the pitch angles with the three different control techniques was not significant.

The dip in the generator torque affected the overall electrical power. The LQR control reduced the frequency of the dip in the generator torque and resulted in an increase of the electrical power.

Tables 4 and 5 show a quantitative comparison of the simulation results in the presence of noise. The wind speed estimated for the three different controllers was found to differ more compared to the estimated wind speed without noise as a consequence of the noise added to generator speed. Similar to the condition without noise, the mean wind speed did not show a significant difference, but the standard deviation decreased or increased by 23.70% and 2.98% for the transition and rated power regions, respectively.

Table 4. Quantitative comparison of the performance data in the transition region in the presence of noise.

Mean	Performance data			Difference (%)		
	PI (A)	**LQR (B)**	**LQR-PI (C)**	**(B−A)/A**	**(C−A)/A**	**(C−B)/B**
V^{est} (m/s)	10.59	10.60	10.62	0.09	0.28	0.19
Ω_g (rpm)	49.40	49.68	49.38	0.57	−0.04	−0.60
β (°)	3.29	3.14	3.02	−4.56	−8.21	−3.82
T_g (kNm)	23.27	23.53	23.43	1.12	0.69	−0.42
P (kW)	92.87	94.44	93.43	1.69	0.60	−1.07
Std. Dev.	**PI (A)**	**LQR (B)**	**LQR-PI (C)**	**(B−A)/A**	**(C−A)/A**	**(C−B)/B**
V^{est} (m/s)	0.80	0.78	0.77	−2.50	−3.75	−1.28
Ω_g (rpm)	1.53	1.57	0.99	2.61	−35.29	−36.94
β (°)	4.01	3.70	4.10	−7.73	2.24	10.81
T_g (kNm)	2.31	2.24	2.26	−3.03	−2.16	0.89
P (kW)	10.97	10.87	10.39	−0.91	−5.29	−4.42

Table 5. Quantitative comparison of the performance data in the rated power region in the presence of noise.

Mean	Performance Data			Difference (%)		
	PI (A)	**LQR (B)**	**LQR-PI (C)**	**(B−A)/A**	**(C−A)/A**	**(C−B)/B**
V^{est} (m/s)	15.02	15.03	15.03	0.07	0.07	0.00
Ω_g (rpm)	49.72	50.04	49.71	0.64	−0.02	−0.66
β (°)	16.83	16.55	16.59	−1.66	−1.43	0.24
T_g (kNm)	24.16	24.34	24.41	0.75	1.03	0.29
P (kW)	96.95	98.28	97.93	1.37	1.01	−0.36
Std. Dev.	**PI (A)**	**LQR (B)**	**LQR-PI (C)**	**(B−A)/A**	**(C−A)/A**	**(C−B)/B**
V^{est} (m/s)	1.84	1.85	1.84	0.54	0.00	−0.54
Ω_g (rpm)	1.49	1.58	1.17	6.04	−21.48	−25.95
β (°)	3.02	3.81	3.93	26.16	30.13	3.15
T_g (kNm)	1.72	1.64	1.61	−4.65	−6.40	−1.83
P (kW)	8.36	8.16	7.56	−2.39	−9.57	−7.35

Based on Table 4, the LQR and LQR-PI controls used average pitch angles smaller than those of the PI control by 4.56% and 8.21%, respectively, and achieved power increases of 1.69% and 0.60%, respectively. For the standard deviation in the generator speed, it increased by 2.61% with the LQR control, while it decreased by 35.29% with the LQR-PI control.

Table 5 lists the simulation results in the rated power region. The average values show differences within 2%. However, the standard deviation of the generator speed increased by 6.04% with the LQR compared with the PI and decreased by 21.48% with the LQR-PI. When the noise was taken into consideration, the standard deviation in the generator speed increased with the LQR control compared with the PI control for both transition and rated power regions. In the case of the LQR-PI control, on the other hand, the standard deviation of the generator speed was reduced compared with that of the PI control, even though noise was introduced.

Overall, the LQR control was better in performance compared with other controllers without any noise; however, when noise was considered, the LQR-PI was the best. Also, the LQR-PI controller showed better performances than the PI controller in both situations, with and without noise. Especially, the target 100 kW wind turbine in this study has a much lower rotor inertia compared with MW wind turbines, and power shutdowns are often encountered because of the generator overspeeding. The proposed LQR-PI controller reduced the standard deviation of the generator speed substantially and is expected to reduce the occurrence of shutdowns in the target wind turbine.

7. Conclusions

In this study, a new LQR-PI control algorithm was designed and proposed to improve the performance of conventional PI control. For this, numerical modeling of a target 100 kW horizontal-axis PMSG-type wind turbine was performed, and an LQR-PI control algorithm using an LQR controller as a feedforward controller to the conventional PI control was introduced. To verify the proposed control algorithm by simulation, a conventional PI and an LQR controller were also designed for the target wind turbine, and comparisons of the simulation results for the three different controllers were carried out. The simulations were performed with and without noise.

The results showed that the LQR control improved the performance only in the rated power region where the noise was not considered, but the proposed LQR-PI control was able to maintain the stability by reducing the standard deviation of the generator speed in all cases, with and without considering noise in the generator speed signal. With the proposed LQR-PI, the standard deviation of the generator speed was reduced by 38.85% in the transition region and by 26.86% in the rated power region when the noise was not considered. Also, it was reduced by 35.29% in the transition region and by 21.48% in the rated power region when the noise was considered. Therefore, it can be concluded that the LQR-PI control was effective in improving the stability of the wind turbine with a minimal change to the existing PI control. In particular, the proposed LQR-PI control is expected to improve the annual energy production of the target 100 kW wind turbine because it can significantly reduce the standard deviation of the generator speed and, finally, the frequency of shutdowns due to overspeed in the generator.

Author Contributions: All authors equally contributed to the data analysis and the simulation, the results analysis, the writing, and review.

Funding: This work was partly supported by a Korea Institute of Energy Technology Evaluation and Planning (KETEP) grant funded by the Korean government (MOTIE) (20173010025010, Advancing of micrositing technology for wind farm development, and 20184030201940, Graduate Track for Core Technologies of Wind Power System Engineering).

Conflicts of Interest: The authors declare no conflict of interest.

Nomenclature

Acronyms

CPC	Collective Pitch Control
DFIG	Doubly Fed Induction Generator
LQR	Linear Quadratic Regulator
MIMO	Multi-input Multi-output
MPPT	Maximum Power Point Tracking
MW	Multi-megawatt
PMSG	Permanent Magnet Synchronous Generator
TSR	Tip Speed Ratio

Symbols

T_a	Aerodynamic torque	θ_g	Drivetrain axis torsional angle
Ω_g	Generator speed	Ω_r	Rotor speed
T_g	Generator torque	T_g^{ref}	Reference generator torque
T_g^c	Torque command	T_a^{est}	Estimated Aerodynamic torque
V	Wind speed	V^{est}	Estimated wind speed
V_{fa}	Nacelle fore-aft velocity	V_{input}	Input wind speed
P	Electrical power	β^c	Pitch command
β	Pitch angle	β_0	Fine pitch angle
$\ddot{x}_{faft}$	Nacelle fore-aft acceleration	$\dot{x}_{faft}$	Nacelle fore-aft velocity
x_{faft}	Nacelle fore-aft displacement	$\dot{x}_{side}$	Nacelle side-side velocity
x_{side}	Nacelle side-side displacement	F_T	Thrust force
M_T	Nacelle mass	K_T	Tower stiffness
H	Hub height	ρ	Air density
R	Rotor radius	λ	Tip speed ratio
C_p	Power coefficient	C_t	Thrust coefficient
C_q	Torque coefficient	J_r	Rotor moment of inertia
J_t	Total moment of inertia	J_g	Generator moment of inertia
τ_g	Generator time constant	η_{elec}	Electrical efficiency
τ_p	Pitch actuator time constant	C_T	Tower damping coefficient
K_s	Drivetrain axis torsional modulus	C_s	Drivetrain axis torsional damping

References

1. Bossanyi, E.A. The design of closed loop controllers for wind turbines. *Wind Energy* **2000**, *3*, 149–163. [CrossRef]
2. Fingersh, L.; Johnson, L. Baseline results and future plans for the NREL controls advanced research turbine. In Proceedings of the 42nd AI/AA Aerospace Sciences Meeting and Exhibit, Reno, NV, USA, 5–8 January 2004. [CrossRef]
3. Bossanyi, E.A. Wind turbine control for load reduction. *Wind Energy Int. J. Prog. Appl. Wind Power Convers. Technol.* **2003**, *6*, 229–244. [CrossRef]
4. Nam, Y. *Wind Turbine System Control*, 1st ed.; GS Intervision: Seoul, Korea, 2013.
5. Lio, W.H.A. *Blade-Pitch Control for Wind Turbine Load Reductions*; Springer: New York, NY, USA, 2018.
6. Oh, Y.; Kim, K.; Kim, H.; Paek, I. Control algorithm of a floating wind turbine for reduction of tower loads and power fluctuation. *Int. J. Precis. Eng. Manuf.* **2015**, *16*, 2041–2048. [CrossRef]
7. Kim, K.; Paek, I.; Kim, C.; Kim, H.; Kim, H. Design of power and load reduction controller for a medium-capacity wind turbine. *J. Korean Sol. Energy Soc.* **2016**, *36*, 1–12. [CrossRef]
8. Kim, C.; Kim, K.; Paek, I. Design of tower damper gain scheduling algorithm for wind turbine tower load reduction. *J. Korean Sol. Energy Soc.* **2018**, *38*, 1–13.
9. Kim, C.; Kim, K.; Song, Y.; Paek, I. Tower load reduction control by pitch loop individual gain scheduling. *J. Wind Energy* **2018**, *9*, 25–32.
10. Kim, K.; Kim, H.; Paek, I. Application and validation of peak shaving to improve performance of a 100 kW wind turbine. *Int. J. Precis. Eng. Manuf. Green Technol.*. under review.
11. Bossanyi, E.A. Individual blade pitch control for load reduction. *Wind Energy Int. J. Prog. Appl. Wind Power Convers. Technol.* **2003**, *6*, 119–128. [CrossRef]
12. Dixit, A.; Suryanarayanan, S. Towards pitch-scheduled drive train damping in variable-speed, horizontal-axis large wind turbines. In Proceedings of the 44th IEEE Conference on Decision and Control, Seville, Spain, 12–15 December 2005; pp. 1295–1300. [CrossRef]
13. Nam, Y.; La, Y.; Son, J.; Oh, Y.; Cho, J. The effect of torque scheduling on the performance and mechanical loads of a wind turbine. *J. Mech. Sci. Technol.* **2014**, *28*, 1599–1608. [CrossRef]
14. Lim, C. Design and manufacture of small-scale wind turbine simulator to emulate torque response of MW wind turbine. *Int. J. Precis. Eng. Manuf. Green Technol.* **2017**, *4*, 409–418. [CrossRef]
15. Kim, K.; Kim, H.; Paek, I.; Kim, H.; Son, J. Field validation of demanded power point tracking control algorithm for medium-capacity wind turbine. *Int. J. Precis. Eng. Manuf. Green Technol.*. accepted. [CrossRef]

16. Pham, T.; Nam, Y.; Kim, H.; Son, J. LQR control for a multi-MW wind turbine. *World Acad. Sci. Eng. Technol.* **2012**, *62*, 670–675.

17. Park, S.; Nam, Y. Two LQRI based blade pitch controls for wind turbines. *Energies* **2012**, *5*, 2028–2046. [CrossRef]

18. Das, S.; Pan, I.; Halder, K.; Das, S.; Gupta, A. LQR based improved discrete PID controller design via optimum selection of weighting matrices using fractional order integral performance index. *Appl. Math. Model.* **2013**, *37*, 4253–4268. [CrossRef]

19. Gupta, A.; Chauhan, Y.K.; Singh Pal, N. Constant torque control schemes for PMSG based wind energy conversion system. In Proceedings of the 2018 IEEE International Conference on Power Energy, Environment and Intelligent Control (PEEIC), Greater Noida, India, 13–14 April 2018. [CrossRef]

20. Jeon, G.; No, T. A Design of wind turbine control system using nonlinear model predictive control. *J. Wind Energy* **2016**, *7*, 14–21. [CrossRef]

21. Nam, Y.; Kim, J.; Paek, I.; Mun, Y.; Kim, S.; Kim, D. Feedforward pitch control using wind speed estimation. *J. Power Electron.* **2011**, *11*, 211–217. [CrossRef]

22. Wang, N.; Johnson, K.E.; Wright, A.D.; Carcangiu, C.E. LIDAR-assisted preview controllers design for a MW-scale commercial wind turbine model. In Proceedings of the 52nd IEEE Conference on Decision and Control, Florence, Italy, 10–13 December 2013; pp. 1678–1683. [CrossRef]

23. Østergaard, K.Z.; Brath, P.; Stoustrup, J. Estimation of effective wind speed. *J. Phys. Conf. Ser.* **2007**, *75*, 012082. [CrossRef]

24. Bottasso, C.L.; Croce, A.; Riboldi, C. Computing spatial estimates of the over-the-rotor wind distribution for advanced wind turbine active control. In Proceedings of the 5th European & African Conference on Wind Engineering, Florence, Italy, 19–23 July 2009; pp. 1000–1004.

25. Bottasso, C.L.; Croce, A.; Nam, Y.; Riboldi, C. Power curve tracking in the presence of a tip speed constraint. *Renew. Energy* **2012**, *40*, 1–12. [CrossRef]

26. Bottasso, C.L.; Riboldi, C. Validation of a wind misalignment observer using field test data. *Renew. Energy* **2015**, *74*, 298–306. [CrossRef]

27. Kim, K.; Lim, C.; Oh, Y.; Kwon, I.; Yoo, N.; Paek, I. Time-domain dynamic simulation of a wind turbine including yaw motion for power prediction. *Int. J. Precis. Eng. Manuf.* **2014**, *15*, 2199–2203. [CrossRef]

28. Kim, H.; Kim, S.; Ko, H. Modeling and control of PMSG-based variable-speed wind turbine. *Electr. Power Syst. Res.* **2010**, *80*, 46–52. [CrossRef]

29. Kim, H.; Kim, K.; Paek, I. Power regulation of upstream wind turbines for power increase in a wind farm. *Int. J. Precis. Eng. Manuf.* **2016**, *17*, 665–670. [CrossRef]

30. Kim, H.; Kim, K.; Paek, I.; Bottasso, C.L.; Campagnolo, F. A study on the active induction control of upstream wind turbines for total power increases. *J. Phys. Conf. Ser.* **2016**, *753*, 32014. [CrossRef]

31. Kim, H.; Kim, K.; Paek, I. Model Based Open-loop wind farm control using active power for power increase and load reduction. *Appl. Sci.* **2017**, *7*, 1068. [CrossRef]

32. Kim, K.; Kim, H.; Kim, C.; Paek, I.; Bottasso, C.L.; Campagnolo, F. Design and validation of demanded power point tracking control algorithm of wind turbine. *Int. J. Precis. Eng. Manuf. Green Technol.* **2018**, *5*, 387–400. [CrossRef]

33. Allik, A.; Uiga, J.; Annuk, A. Deviations between wind speed data measured with nacelle-mounted anemometers on small wind turbines and anemometers mounted on measuring masts. *Agron. Res.* **2014**, *12*, 433–444.

34. Olsson, D.M.; Nelson, L.S. The Nelder-Mead simplex procedure for function minimization. *Technometrics* **1975**, *17*, 45–51. [CrossRef]

35. Lagarias, J.C.; Reeds, J.A.; Wright, M.H.; Wright, P.E. Convergence properties of the Nelder-Mead simplex method in low dimensions. *SIAM J. Optim.* **1998**, *9*, 112–147. [CrossRef]

36. Nelder, J.A.; Mead, R. A Simplex Method for Function Minimization. *Comput. J.* **1965**, *7*, 308–313. [CrossRef]

Article

New Assessment Scales for Evaluating the Degree of Risk of Wind Turbine Blade Damage Caused by Terrain-Induced Turbulence

Takanori Uchida [1,*] and Yasushi Kawashima [2]

[1] Research Institute for Applied Mechanics (RIAM), Kyushu University, 6-1 Kasuga-kouen, Kasuga, Fukuoka 816-8580, Japan

[2] West Japan Engineering Consultants, Inc., Denki Building Kyosokan 7F, 2-1-82 Watanabe-dori, Chuo-ku, Fukuoka 810-0004, Japan

* Correspondence: takanori@riam.kyushu-u.ac.jp; Tel.: +81-92-583-7776; Fax: +81-92-583-7779

Received: 4 June 2019; Accepted: 2 July 2019; Published: 8 July 2019

Abstract: The present study scrutinized the impacts of terrain-induced turbulence on wind turbine blades, examining measurement data regarding wind conditions and the strains of wind turbine blades. Furthermore, we performed a high-resolution large-eddy simulation (LES) and identified the three-dimensional airflow structures of terrain-induced turbulence. Based on the LES results, we defined the Uchida-Kawashima Scale_1 (the U-K scale_1), which is a turbulence evaluation index, and clarified the existence of the terrain-induced turbulence quantitatively. The threshold value of the U-K scale_1 was determined as 0.2, and this index was confirmed to not be dependent on the inflow profile, the influence of the horizontal grid resolution, and the influence of the computed azimuth. In addition, we defined the Uchida-Kawashima Scale_2 (the U-K scale_2), which is a fatigue damage evaluation index based on the measurement data and the design value obtained by DNV GL's Bladed. DNV GL (Det Norske Veritas Germanischer Lloyed) is a third party certification body in Norway, and Bladed has been the industry standard aero-elastic wind turbine modeling software. Using the U-K scale_2, the following results were revealed: the U-K scale_2 was 0.86 < 1.0 (within the designed value) in the case of northerly wind, and the U-K scale_2 was 1.60 > 1.0 (exceeding the designed value) in the case of easterly wind. As a result, it was revealed that the blades of the target wind turbine were directly and strongly affected by terrain-induced turbulence when easterly winds occurred.

Keywords: wind turbine blade; complex terrain; terrain-induced turbulence; large-eddy simulation; turbulence evaluation index; fatigue damage evaluation index

---

1. Introduction

The adoption and promotion of renewable energy has gained widespread interest worldwide, including in Japan. For this reason, the introduction of wind farms with wind turbines has accelerated. However, there have also been great concerns. Unfortunately, an increasing trend in the number of serious accidents has been reported, such as the accidental fall of a wind turbine nacelle, particularly in wind power stations built on complex terrain in mountainous areas. Utmost caution is required, especially for wind power stations with complicated topography, in planning an optimum arrangement of wind turbines and controlling both maintenance and management. Diversified tasks are increasingly important to avoid accumulated fatigue of wind load in wind turbines due to terrain-induced turbulence, to reduce malfunctions and accidents inside and outside wind turbines, and to improve the availability of wind turbines [1–11]. Considering social and engineering requests at present, the crucial purpose of the present study is to establish a system with a numerical diagnostic technique for wind status, which contributes to the proper operation of wind farms, an adequate understanding of the indigenous wind environments of each site, including terrain-induced turbulence, and a reduction of malfunctions and accidents associated with wind turbines [12–21].

We conducted research with a demonstration not only to examine the impacts of terrain-induced turbulence on flapwise fatigue damage of wind turbine blades, but also to clarify the mechanism of terrain-induced turbulence generation. Specifically, we targeted a large-scale wind turbine built on complex mountainous terrain. Electric strain gauges were installed at the base of three blades of the wind turbine. We developed a measurement system to automatically obtain time-series data of strain fluctuation via the gauges. Subsequently, the damage equivalent load (DEL) [22,23] of the flapwise bending and vibration of wind turbine blades was calculated on the basis of the collected data. Furthermore, we examined the relationship with the output results of wind direction and speed sensors, which were installed on the wind turbine nacelle for the purpose of wind turbine control.

Firstly, the time period of the maximum DEL on the blades during the measurement period in this study was identified, as well as the time period of the maximum load regarding fatigue damage of the blades, and the situation was quantitatively examined. Simultaneously, a high-resolution simulation of numerical wind conditions was performed on the basis of large-eddy simulation (LES) to closely examine three-dimensional airflow structures when the wind turbine blades experienced terrain-induced turbulence. Secondly, the present study derived a formula, a relational expression indicating a correlation between the DEL calculated on the basis of actual measured data and the values of nacelle wind speed, which were obtained by the nacelle anemometer, that is, the fatigue damage of the blades. Using this formula and measurement data of wind conditions over a one-year period, the influence of terrain-induced turbulence on the age-related degradation of wind turbine blades was quantitatively assessed.

Lastly, through the analysis of the series of measurement data and numerical wind condition simulations, we proposed two types of new assessment scales to evaluate the impacts of terrain-induced turbulence on wind turbine blades and demonstrated the operation methods of these generalization indexes.

2. Overview of the Kushikino Reimei Wind Farm

The present study was conducted, with cooperation of Kyudenko New Energy Co., Ltd.(Fukuoka, Japan), focusing on the Kushikino Reimei Wind Farm (established in November 2012) in Hashima, Ichikikushikino City, Kagoshima Prefecture, Japan (Figure 1). This wind farm was installed with ten 2 MW (megawatt) downwind wind turbines (Hitachi, Ltd. (Tokyo, Japan)). The target wind turbine was wind turbine #10. The present study focused on the impacts of terrain-induced turbulence on the blades of wind turbine #10. Turbulence was generated when easterly winds passed over Mt. Benzaiten (elevation 519 m) (see Figures 2 and 3, and Table 1). Figure 4 provides an outline of the wind turbine. Figure 5 illustrates the power curve of the wind turbine. Figure 6 shows a vane anemometer which was installed on the wind turbine nacelle. The pitch and yaw control of the wind turbine was performed based on the sensor information, as shown in Figure 5. The present study analyzed the airflow field generated around wind turbine #10, utilizing output results of wind direction and speed sensors (research data results are described later).

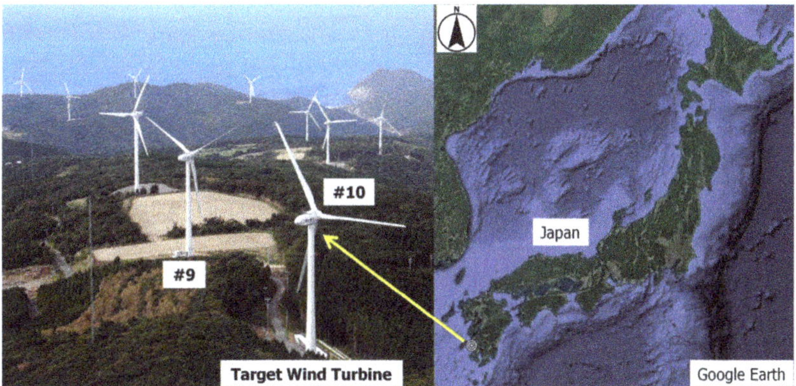

Figure 1. Map of the Kushikino Reimei Wind Farm and the surrounding area.

Figure 2. Photo of wind turbine #10.

Figure 3. Relative locations of Mt. Benzaiten (elevation 519 m) and wind turbine #10.

Table 1. Elevation information for wind turbine #10 and distance between Mt. Benzaiten (elevation 519 m) and wind turbine #10.

Elevation at Base of Wind Turbine #10	Maximum Blade Tip Elevation (Above Sea Level)	Distance Between Mt. Benzaiten and Wind Turbine #10
418 m	518 m	Approx. 300 m

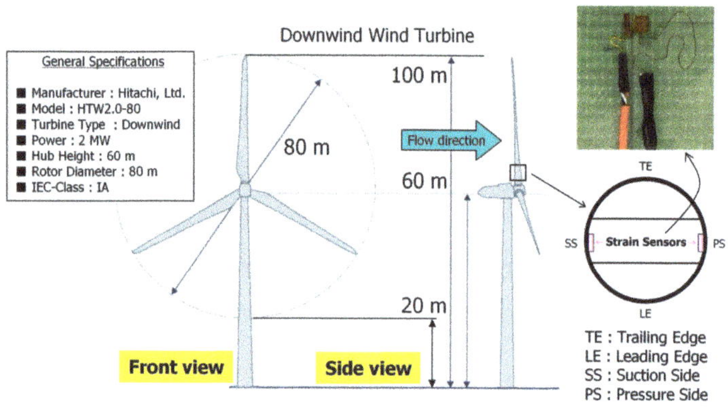

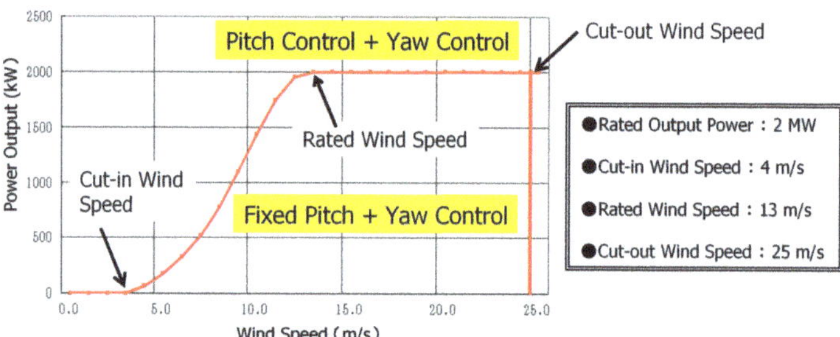

Figure 4. Outline of a wind turbine.

Figure 5. Power curve of a wind turbine.

Figure 6. Nacelle propeller-vane anemometer.

3. In-Situ Data Analysis

3.1. Analysis of Wind Turbine Power Output Data

The theoretical power output was compared with the measured data for wind turbine #10. The results are shown in Figure 7. Figure 7a,b show measured values of northerly and easterly winds, respectively; the number of measured data values was 1578 for the northerly wind, obtained in 10-min periods in January of 2013, and similarly, 601 data values were obtained for the easterly wind in June of 2013. The easterly wind pattern shows a high dispersion compared with that of the northerly wind.

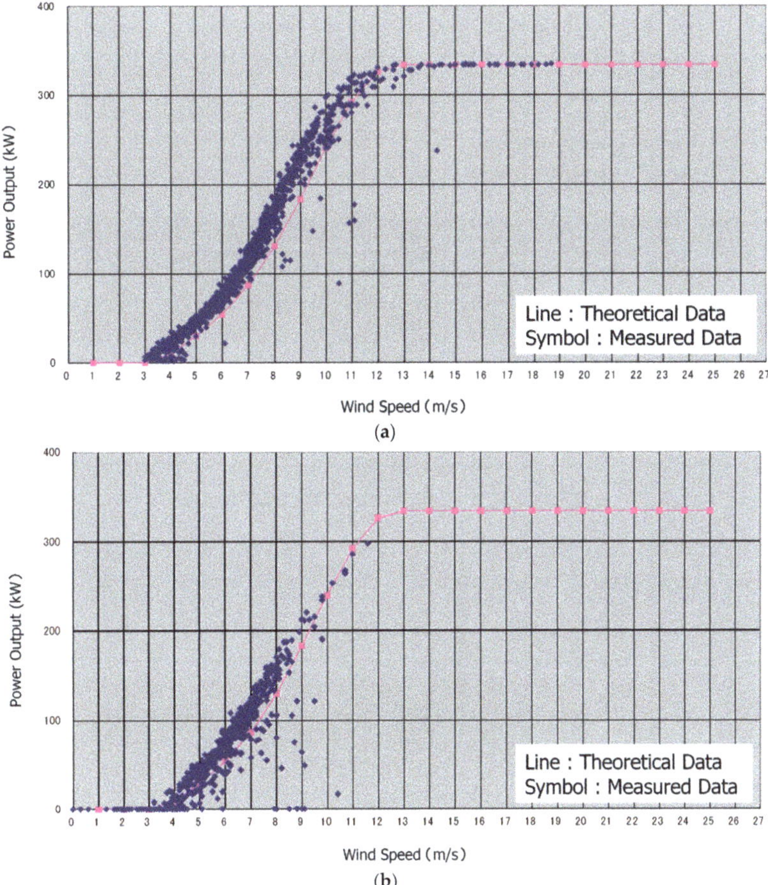

Figure 7. Comparison between theoretical data and measured data for wind turbine #10. (**a**) Northerly wind. (**b**) Easterly wind.

3.2. Analysis of Wind Turbine Alarm Data

A variety of alarm conditions exist for a wind turbine. For this research, the following two items regarding wind conditions were our focus, and the operation status of wind turbine #10 was inspected. The target period of the data analysis was 14 months (November 2012–January 2014).

(1)   Shutdown due to excessive yaw error (definition: malfunctions occur due to deviation of the nacelle direction and the anemoscope direction).

(2) Discordance in wind directions of sensors for wind direction and speed (definition: malfunctions occur due to disagreement in the values of the two sensors).

Table 2 provides the number of alarm occurrences due to the above alarm items for all wind directions. The mean values of other wind turbines are shown for a comparative examination. This table indicates that the alarm of wind turbine #10 occurred frequently due to the two items mentioned above. In Table 3, alarm occurrences of Table 2 are shown depending on the wind directions. Figure 8 shows a graph generated from Table 3. Table 3 and Figure 8 clearly indicate that wind turbine #10 had an extremely high number of alarm occurrences when wind blew from the east compared with other wind directions. This phenomenon suggested that wind turbine #10 was affected over time by unsteady wind direction fluctuations when easterly winds occurred.

Table 2. Number of alarm occurrences for wind conditions under all wind directions.

Alarm Item	Wind Turbine #10	Other Wind Turbines (Average)
Shutdown due to excessive yaw error	1448	530
Discordance in wind directions of sensors	308	80

Table 3. Number of alarm occurrences due to wind conditions for each wind direction.

Alarm Item	N	NNE	NE	ENE	E	ESE	SE	SSE
Shutdown due to excessive yaw error	39	12	130	150	560	176	58	18
Discordance in wind directions of sensors	5	2	33	35	146	45	16	10

Alarm Item	S	SSW	SW	WSW	W	WNW	NW	NNW	Total
Shutdown due to excessive yaw error	11	7	2	2	8	2	158	115	1448
Discordance in wind directions of sensors	6	0	0	1	1	2	3	3	308

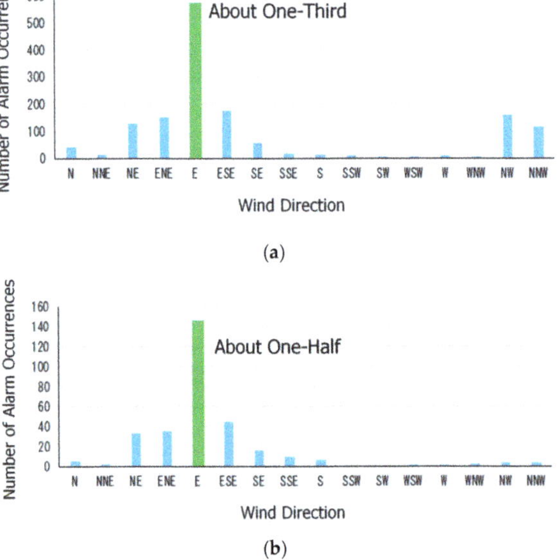

Figure 8. Result of graphing the values in Table 3. (**a**) Number of alarm occurrences for yaw misalignment. (**b**) Number of alarm occurrences for wind direction mismatch of the wind vane. Note: Wind turbine systems resort to shutting down the wind turbine if the yaw misalignment exceeds a threshold.

3.3. Analysis of Nacelle Propeller-Vane Anemometer Data

Wind energy is input via wind turbine blades. Thus, it is absolutely essential to monitor the behavior of the flapwise bending and vibration at the base of the blades for the assessment of the durability of the entire wind turbine. In the present study, two types of electric strain gauges were installed at the base of three blades on wind turbine #10 (the base: the position at a distance of approximately 1.3 m from the hub-connected point, referred to in Figure 4). A measurement synchronization system was developed to measure both of these values and basic information on the wind turbine operation (eight items: wind direction, wind speed and azimuth of the nacelle, pitch angle, rotational speed of the generator, active power of the power conditioning system (PCS) system, azimuth angle, and longitudinal acceleration of the nacelle). Actual measurement data were collected at 50 Hz (0.02 interval, 50 cycles per second) via this measurement synchronization system. The average values of the two sets of wind direction and speed sensors for wind turbine control, which were installed on the wind turbine nacelle shown in Figure 6, were used as measurement data regarding the wind direction and speed of the nacelle. The measurement period was from 3 November 2015, 0:00 a.m. JST to 17 March 2016, 7:00 a.m. JST.

Wind roses for all of the wind velocity classes are shown in Figure 9, where the above-measured data of wind direction and wind speed of the nacelle were classified into 16 wind directions. Further, the numerical data are shown in Table 4. Regarding frequency distribution, the highest frequency, 22.5%, was for the northerly wind (Total: 19237, N: 4331) and the frequency of the easterly wind was 4.4% (Total: 19237, E: 856). The mean speed of the easterly wind was lower than that of the northerly wind; the northerly wind mean speed was 6.1 m/s, while the easterly mean speed was 4.5 m/s.

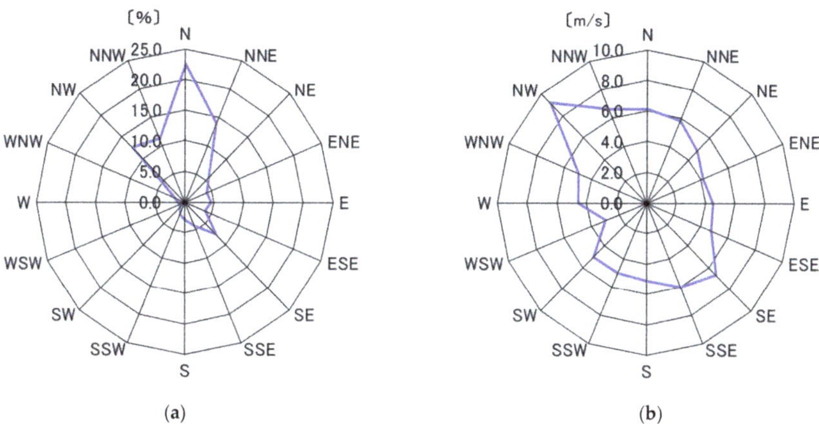

(a) (b)

Figure 9. Frequency distribution of the direction of the 10-min average wind (%): (**a**) wind rose (frequency distribution) and the average of the 10-min average wind speed observed for 16 wind directions (m/s): (**b**) wind speed by direction (wind measurement height: hub height (60 m), analysis period: 3 November 2015, 0:00 a.m. JST–17 March 2016, 7:00 a.m. JST).

Table 4. Frequency distribution of the direction of the 10-min average wind (%) and the average of the 10-min average wind speed observed for 16 directions (wind measurement height: hub height (60 m), analysis period: 3 November 2015, 0:00 a.m. JST–17 March 2016, 7:00 a.m. JST).

Height	Item	N	NNE	NE	ENE	E	ESE	SE	SSE	S	SSW	SW	WSW	W	WNW	NW	NNW	Total
60 m	Frequency Distribution (%)	22.5	13.8	5.6	4.0	4.4	3.6	7.5	4.3	3.0	2.2	1.2	0.9	1.3	1.8	12.6	11.2	100.0
	Average Wind Speed (m/s)	6.1	5.8	4.8	4.1	4.5	4.7	6.7	6.0	5.1	5.0	5.0	3.0	4.6	5.0	9.2	6.6	6.1

Data groups were analyzed for nacelle wind direction and nacelle wind speed, and the standard deviation was calculated for nacelle wind speed values on the basis of 10-min intervals for the data measurement period of wind turbine #10 from 3 November 2015, 0:00 a.m. JST to 17 March 2016, 7:00 a.m. JST. Furthermore, measurement data were classified into 12 wind directions and were arranged with analysis results of the damage equivalent load (DEL), which is described later (see Table 5). Figures 10 and 11 illustrate the analysis results of the wind velocity standard deviation and turbulence intensity for both northerly and easterly winds. To examine the status of wind turbine generation, analysis data targets corresponded to winds of 4 m/s or higher.

Table 5. Wind direction range and total number of data values.

	Wind Direction Range	**Total Number of 10-min Periods for Which Wind Statistics are Calculated**
Northerly Wind	$0° \pm 15°$	4036 (Total: 12,567; 32.1%)
Easterly Wind	$90° \pm 15°$	496 (Total: 12,567; 4.0%)

Note: includes only data from 10-min periods with an average wind speed of 4 m/s (cut-in wind speed) or higher.

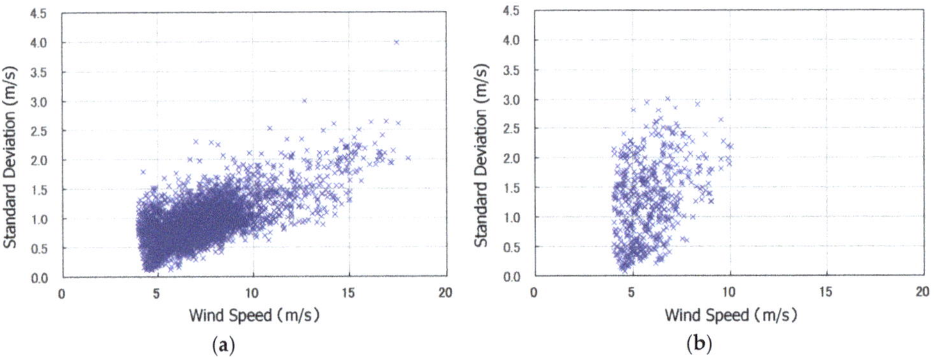

Figure 10. Relationship between the standard deviation and the average of the wind speed in 10-min periods for two wind directions (wind measurement height: hub-height (60 m), analysis period: 3 November 2015, 0:00 a.m. JST–17 March 2016, 7:00 a.m. JST). (**a**) Northerly wind. (**b**) Easterly wind.

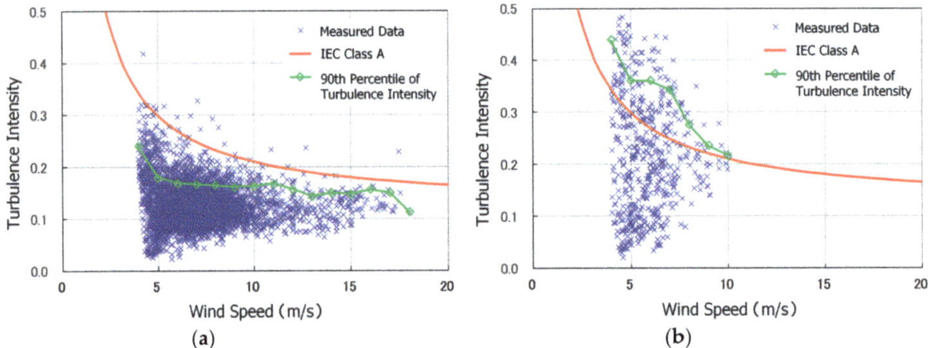

Figure 11. Relationship between turbulence intensity and the average of the wind speed in 10-min periods for two wind directions (wind measurement height: hub-height (60 m), analysis period: 3 November 2015, 0:00 a.m. JST–17 March 2016, 7:00 a.m. JST). (**a**) Northerly wind. (**b**) Easterly wind.

Wind velocity standard deviations and turbulence intensity were calculated by formula (1):

$$TurbulenceIntensity(TI) = \frac{\sigma_u}{\overline{u}} = \frac{\sqrt{\frac{1}{N}\sum_{i=1}^{N}\left(u_i'\right)^2}}{\overline{u}} \tag{1}$$

where:

$$u' = u(t) - \overline{u} \tag{2}$$

In Figures 10 and 11, values of standard deviations of easterly wind velocity and values of their corresponding turbulence intensity were notably larger in comparison to those of the northerly wind. In addition, it was confirmed that the turbulence intensity of the easterly wind, in a wind velocity class 10 m/s and lower, frequently exceeded the turbulence intensity category of the International Electrotechnical Commission (IEC). It was inferred that the influence of Mt. Benzaiten (elevation 519 m) at a distance of 300 m from wind turbine #10 was a major cause of the increased turbulence intensity when easterly winds occurred. This is examined later, employing numerical wind condition simulations.

3.4. Analysis of Wind Turbine Blade Strain Data

As mentioned above, wind energy is input via wind turbine blades. Thus, it is absolutely essential to monitor the behaviors of flapwise bending and vibration at the base of blades for the assessment of the durability of the entire wind turbine. In the present study, electric strain gauges were installed at the base of three blades on wind turbine #10, as shown in Figure 4. Figure 12 presents the blade strain data of two types of winds in the measurement period. They were compared between two time-history data waveforms when the mean values of the strain gauge of nacelle wind speeds were approximately 9 m/s. The northerly wind showed the highest occurrence frequency, and the easterly wind showed notably large values in turbulence intensity, as was described previously. Data that are framed by a continuous line represent the time zones of two wind patterns, which showed approximately 9 m/s mean values of the nacelle wind velocity (Figure 12). A comparison of the results of measurement data between the easterly wind in Figure 12b and the northerly wind in Figure 12a reveals that variable amplitudes of strain gauges were notably large for the easterly wind, and thus, the blades of the target wind turbine #10 vibrated due to the large flapwise wind loads.

This study conducted a further quantitative examination of the strain data of wind turbine blades. From strain data, time history data of flapwise bending moments were extracted with the cooperation of a wind turbine manufacturer. Upon applying the rainflow counting algorithm [22,23], the damage equivalent load (DEL) was calculated (see Formula 3). DEL is the most commonly applied index in the wind power industry in arguments regarding the fatigue damage of wind turbines. In the present research, the obtained values of DEL were normalized by a design value for a wind velocity of 12 m/s, employing the aero-elastic analysis software Bladed.

$$DEL = \left\{ \frac{\sum_{i=1}^{n}\left(F_i^m \cdot n_i\right)}{N} \right\}^{\frac{1}{m}} \tag{3}$$

where

F_i is the load of the *i*-th class of the fatigue load spectrum;

n_i is the number of cycles in the *i*-th class of the fatigue load spectrum;

N is the equivalent of cycles;

m is the S-N (stress-number of cycles to failure) curve slope for relevant material.

$N = 600$ and $m = 10$ with fiber-reinforced plastic (FRP) blades in this study.

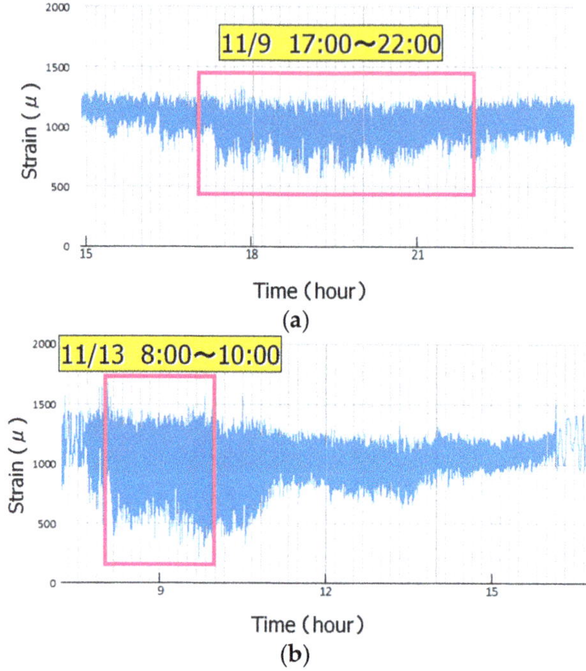

Figure 12. Blade strain data (blade flapwise bending raw data). (**a**) Northerly wind. (**b**) Easterly wind. Note: interval: 0.02 seconds, average wind speed: approx. 9 m/s.

Figure 13 shows time variations at 10-min intervals for nacelle wind velocity, its standard deviation and DEL. Similar to Figure 12, time zones where the 10-min mean value of nacelle wind velocity was approximately 9 m/s are the focus of this discussion. In the time period of the northerly wind (November 9, 2015, 19:30–19:40 p.m. JST), the wind velocity value was 9.4 m/s. Corresponding to this time period, the standard deviation was 1.3 m/s and DEL was 0.99. Contrarily, the time period for which 9 m/s was intended to be a mean speed was from November 13, 2015, 9:40 to 9:50 a.m. JST for the easterly wind, showing approximately 9.1 m/s as the 10-min mean values of the nacelle wind velocity. The value of standard deviation for the easterly wind is worth mentioning in particular. The mean value of the standard deviation of the nacelle wind velocity (November 13, 2015, 9:40–9:50 a.m. JST) was 2.3 m/s, which was approximately 1.8 times higher than that of the northerly wind. Accordingly, the DEL was also notable at 2.03, which was approximately twice as large as that of the northerly wind. Conclusively, the easterly wind status, which had the issue of terrain-induced turbulence, was proven to have distinctive differences in comparison with the northerly wind status, which was scarcely affected by terrain-induced turbulence, while it had the highest frequency of occurrence. Based on these findings, it was revealed that for easterly wind patterns, the standard deviation was large. Accordingly, the DEL was large, and thus terrain-induced turbulence directly influenced the turbine blades. The DEL value of 2.03 for the easterly wind (November 13, 2015, 9:40–9:50 a.m. JST) was the highest in the entire present study; that is, this period was the time period where the largest load was generated regarding the fatigue damage of the blades of the wind turbine. This scale of 2.03 in DEL means that if airflow with a property of DEL = 2.03 continued for 5.88 years, the total load on the wind turbine blades would reach the design load for the designed service life.

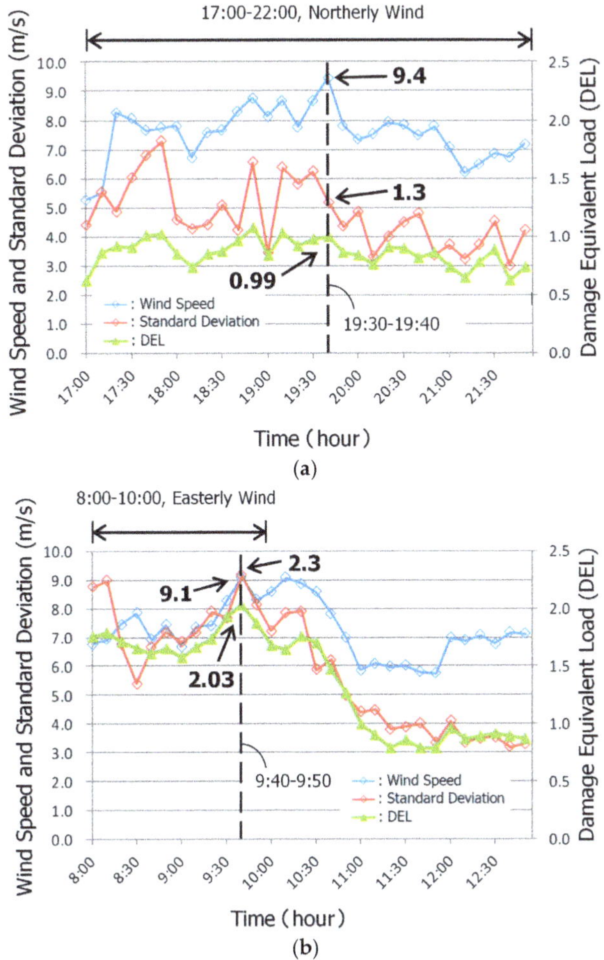

Figure 13. Time series of wind speed, standard deviation and normalized damage equivalent load (DEL) for flapwise blade bending. Plotted values were evaluated from instantaneous data values within 10-min periods. (**a**) Northerly wind. (**b**) Easterly wind.

4. Numerical Simulation of Airflow with the WRF Mesoscale Model

As described in Section 3.4, the wind vanes on the nacelle of WT#10 indicated the presence of an easterly wind from 9:40 to 9:50 a.m. JST, November 13, 2015. To investigate the wind flow from this time period closely, a numerical simulation was performed using the Weather Research and Forecasting (WRF) mesoscale model [24] in the present study. This mesoscale model was developed cooperatively by a number of institutions, including NCAR (National Center for Atmospheric Research), the University of Oklahoma, NCEP (US National Center for Environment Prediction), NOAA (Forecast System Laboratory of the National Oceanic and Atmospheric Administration), and AFWA (Air Force Weather Agency). WRF is a three-dimensional, fully compressible, non-hydrostatic model that has been used both operationally and for research worldwide. WRF is considered a successor to the non-hydrostatic MM5 model, which was developed principally by NCAR. A number of physics models are included in WRF, such as radiation models to calculate solar and atmospheric radiation, a turbulence model that simulates turbulence in the mixed-layer, a cloud physics model that takes into account water vapor, cloud water, rainwater, snow, and hail, and a land surface model that calculates the surface temperature, soil temperature, soil water content, snowfall, and surface flux. Furthermore, WRF allows the use of the latest physics models and a data assimilation system; thus, WRF is suitable for predicting and simulating localized heavy rainfall, gusts, and other meteorological phenomena.

Figure 14 illustrates the computational domain used for the present simulation. In the present study, four-layer nesting was adopted. The spatial resolution of the smallest domain (d04), in which the wind farm is located, was 333.33 m. The height of the smallest grid cell was approximately 8 m. For the terrain data set, the model used the global digital elevation model GTOPO30 with spatial resolutions of 30 seconds in latitude and longitude provided by the United States Geological Survey (USGS) and higher-resolution data; specifically, the 50-m digital elevation model (GSI50, spatial resolution: 1.5 seconds in the latitude direction and 2.25 seconds in the longitude direction) provided by the Geospatial Information Authority of Japan (GSI). For land use and vegetation data, the USGS data that are pre-installed in WRF (spatial resolution: 1°) were used. The meteorological GPV (grid point value) data that were used for the boundary conditions in the present study were the NCEP Final Analysis (NCEP-FNL) data (spatial resolution: 0.5°; temporal resolution: six hours), which are global analysis data. These data were used every six hours (with no nudging). For sea surface temperature (SST) data, the skin temperature from the NCEP–FNL data was used. The airflow simulation in the present study takes cloud physics and precipitation processes into account.

Figure 15 shows the wind velocity vectors at the wind turbine height (60 m above the ground surface) in the smallest domain, d04 (Figure 10), at 9:40 a.m. JST on November 13, 2015. This figure confirms that the wind was easterly in the entire computational domain.

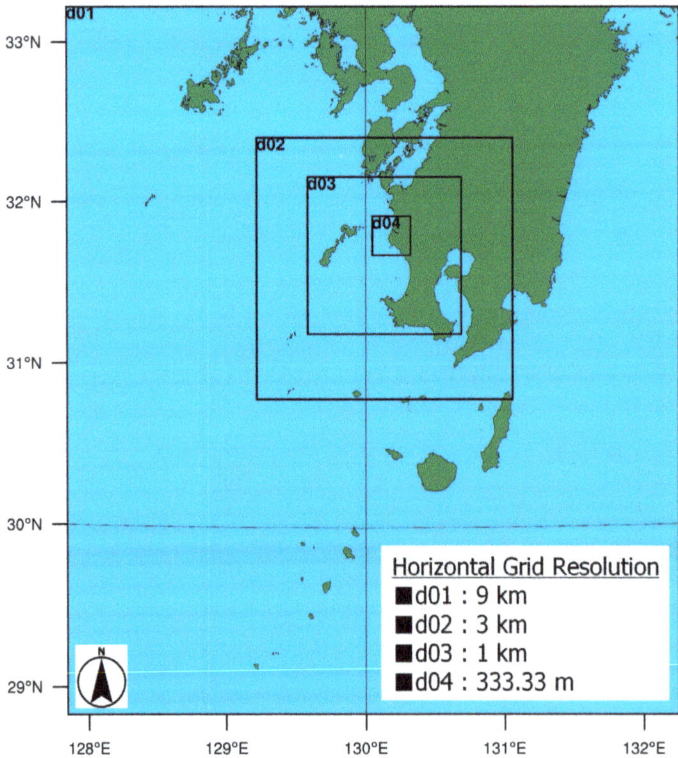

Figure 14. Computational domain used in the Weather Research and Forecasting (WRF) mesoscale model.

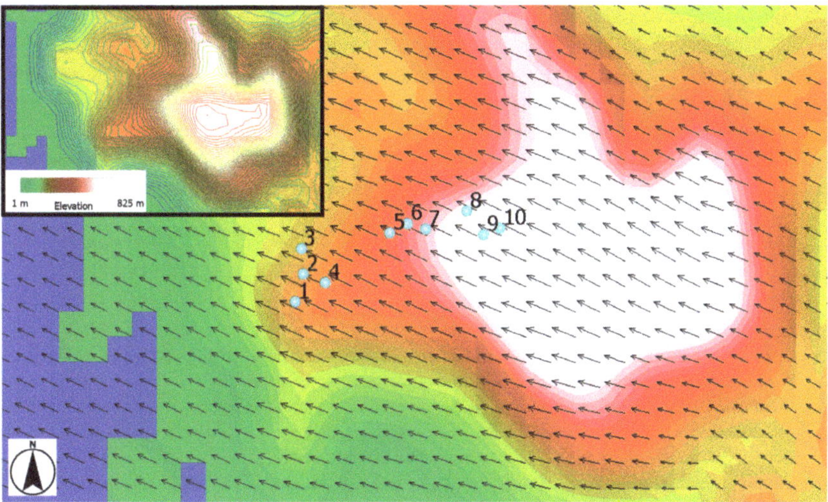

Figure 15. Distribution of the horizontal wind vectors in Domain 4, approximately 60 m above the ground surface. 9:40 a.m. JST, Nov. 13, 2015.

5. Overview of the Numerical Simulation Method Based on Large-Eddy Simulation (RIAM-COMPACT)

5.1. Setting of Numerical Parameters

As mentioned above, when easterly wind occurred, turbulence intensity was high around wind turbine #10, and thus the DEL of the wind blades was high from the viewpoint of actual measurement data analysis. Consequently, we inferred that the major cause was the influence of Mt. Benzaiten (elevation 519 m), located on the eastern side at a distance of approximately 300 m from wind turbine #10 (see Figure 16a). Thus, high-resolution numerical simulations for wind conditions were performed based on large-eddy simulation (LES) to scrutinize three-dimensional airflow structures, which directly affected the wind turbine blades due to terrain-induced turbulence. To compare flow distributions, the northerly wind was also examined. The northerly wind had the highest frequency of occurrence and was scarcely affected by terrain-induced turbulence (see Figure 16b).

For the numerical simulations, the RIAM-COMPACT (Research Institute for Applied Mechanics, Kyushu University, Computational Prediction of Airflow over Complex Terrain) natural terrain version was used, for which a collocated grid in a general curvilinear coordinate system was adopted. The collocated grid system is characterized by functions to define physical velocity components and pressure at the cell center of computational grids and to define variables at the cell faces, which are obtained with a contravariant velocity component multiplied by the determinant of the Jacobian. The numerical calculation method was based on the finite-difference method (FDM), and large-eddy simulation (LES) was employed to examine the turbulence models. The computational algorithm confirmed the fractional step method (F-S method) [25], and the time marching method was based on an explicit scheme of the Euler method. Poisson's equations regarding pressure were solved using successive over-relaxation (SOR). For the discretization of all the spatial terms, with the exception of the convective term in the equation, a second-order accurate central-difference scheme was applied. For the convective term, a third-order upwind difference scheme was applied. The interpolation technique was used for a fourth-order finite-difference that comprised the convective terms [26]. For the weighting of the numerical diffusion term in the convective term discretized by third-order upwind differencing, $\alpha = 0.5$ is used as opposed to $\alpha = 3.0$ [27] from the Kawamura-Kuwahara scheme in order to minimize the influence of numerical diffusion. The LES sub-grid-scale model employed the standard Smagorinsky model [28] combining a wall-damping function, setting the model coefficient as 0.1. In the present study, the LES is assumed to reproduce the wind tunnel testing. Therefore, the effects of atmospheric stability associated with vertical thermal stratification of the atmosphere and inflow turbulence were neglected. In addition, as in [12,16,17], the effects of surface roughness were taken into consideration by reconstructing surface irregularities in high resolution. A comparison between the Reynolds-averaged modeling (RANS) results and the present LES results is summarized in a recent article [13], and the prediction accuracy of the present LES approach by comparison with wind tunnel experiments is discussed in [19].

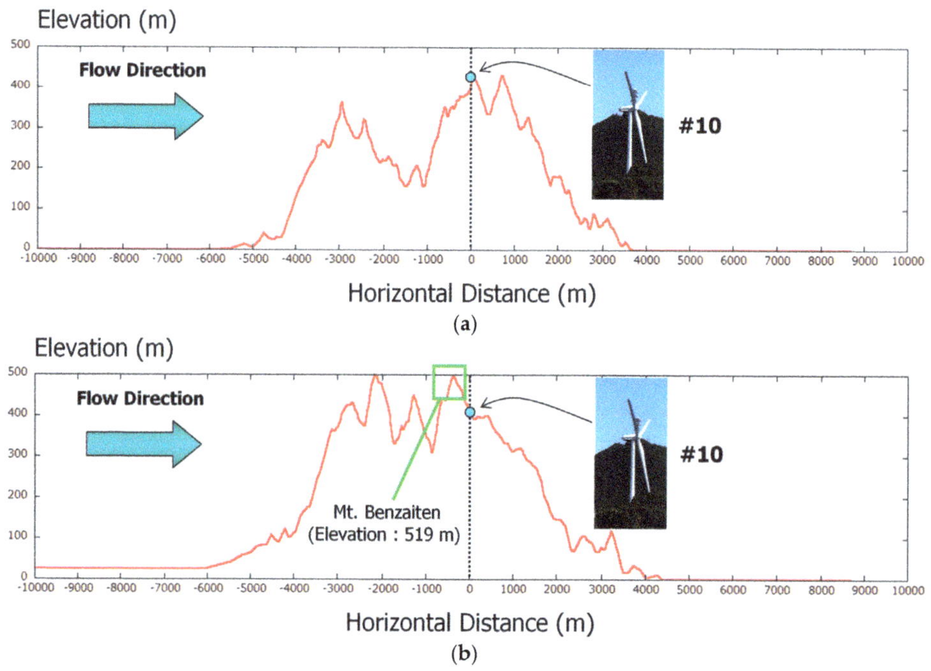

Figure 16. Comparison of the topographic section. (**a**) Northerly wind. (**b**) Easterly wind.

Conditions for numerical wind simulations were as follows: the computational domain held a space of 12.0 (x), 5.0 (y) and 2.5 (z) km for the streamwise direction, spanwise direction, and vertical direction, respectively (see Figure 17). A buffer zone was established in the upstream end of the computational domain, in which the terrain irregularities were reduced by 95% to form flat terrain. Similarly, a buffer zone was added to the downstream end of the domain. In the computational region, the maximum and minimum altitudes were 523.5 m and 0 m, respectively. This simulation utilized terrain elevation data with 10 m surface imagery provided by the Geospatial Information Authority of Japan (GSI). Generated grids, including added grids of the marginal area upstream and downstream of the computational region, were approximately eight million points in number, 496 (x) × 201 (y) × 81 (z) in three-dimensional coordinates. Horizontal grid resolutions in the vicinity of the wind turbine were 10 m in the x- and y-directions. The minimum resolution of the vertical grid was approximately 1.5 m above ground level to enable smooth drawing.

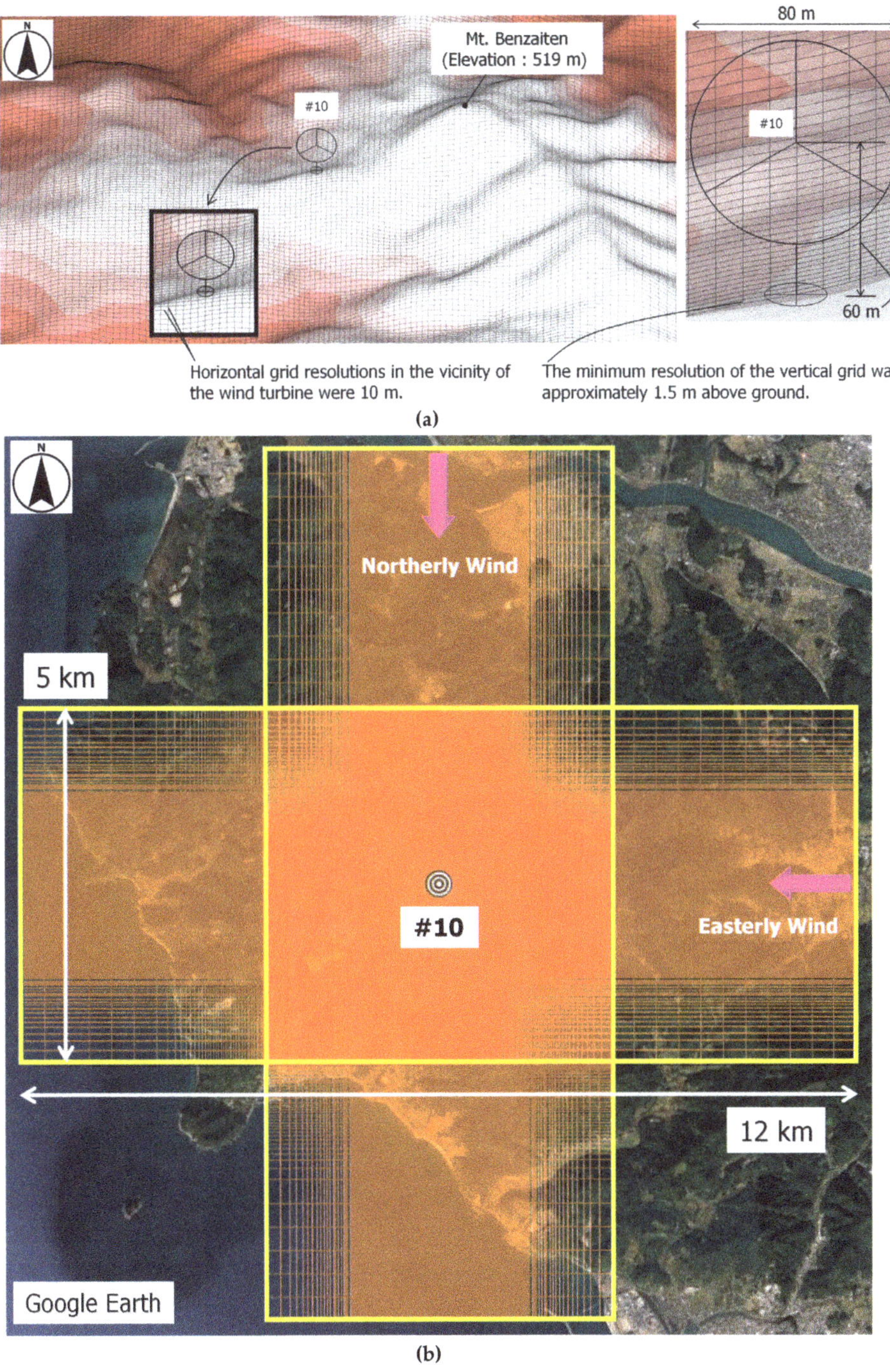

Figure 17. Computational grids and domain. (**a**) Enlarged view. (**b**) Overall view.

The power law distribution was provided with N = 7.0, which is shown in Figure 18, regarding the inlet boundary conditions of numerical wind simulations. The present study focused on the impacts of terrain-induced turbulence for discussion, and thus did not examine fluctuations in the inlet airflow. A free-slip boundary condition was applied to the side and upper walls. A convective outflow condition was applied to the outlet boundary. A no-slip condition was imposed on the ground surface. Characteristics of reference scales in the present study are shown in Figure 19. In the figure, h represents differences in altitude in the computational region, U_{in} represents streamwise wind velocity of the inlet boundary at the maximum altitude point, and v represents the coefficient of kinematic viscosity. On the basis of the two types of reference scales, the dimensionless parameter, Re, is the Reynolds number (= U_{in} h/v). For this simulation, Re = 10^4. The time step was specified as $\Delta t = 2 \times 10^{-3}$ h/U_{in}. Furthermore, identical simulation conditions were applied to both northerly and easterly winds. In the present study, we used a vector-parallel supercomputer system named NEC SX-ACE. The NEC SX-ACE provides four cores per node. When using one node of this system, about 8 million grid points of LES simulation took several hours.

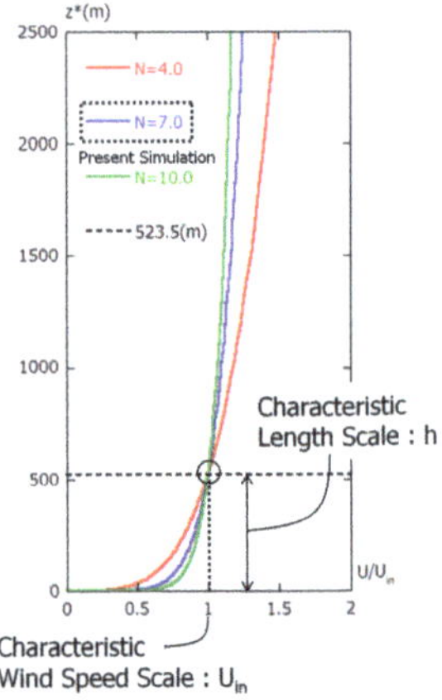

Figure 18. Inflow condition.

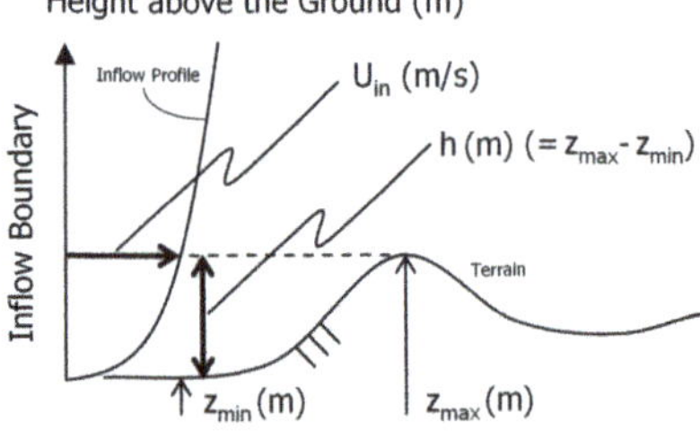

Figure 19. Two characteristic scales (U_{in} and h).

5.2. Flow Visualization of Simulation Results

Figure 20 illustrates the distribution of the streamwise (x) wind velocity component (u) in an instantaneous flow field. With the attentive observation of the two shadings in the figures, it is evident that the airflow pattern that was generated around the wind turbine in Figure 20b was significantly distinct in easterly winds compared to that of the northerly winds in Figure 20a Conclusively, with the visual effects of these illustrations, it was evident that when the wind blew from the east, wind turbine #10 was directly affected by the separated flow from the east, which was the terrain-induced turbulence formed due to Mt. Benzaiten (elevation 519 m), which is located upstream of the wind turbine.

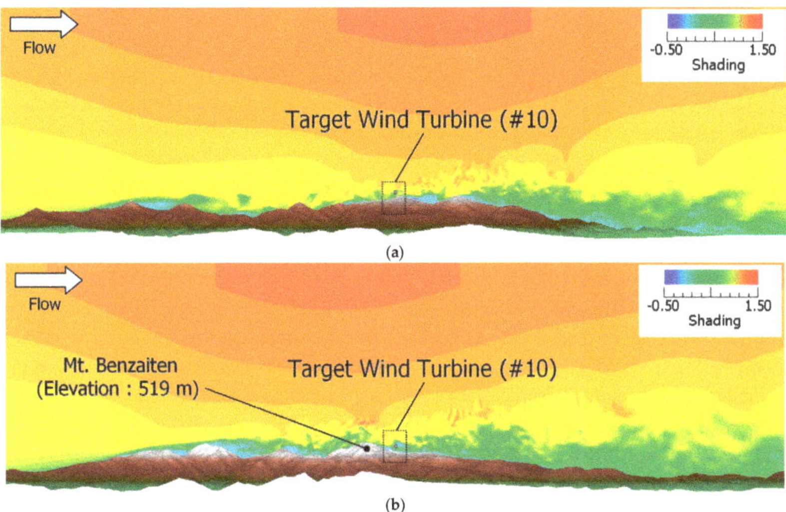

Figure 20. Distribution of the streamwise wind velocity component on a vertical cross-section that includes wind turbine #10 and the instantaneous flow field. (**a**) Northerly wind. (**b**) Easterly wind.

5.3. Proposal of Turbulence Evaluation Index (Uchida-Kawashima Scale_1)

The present study proposes two types of new indexes regarding wind conditions and loads. The former, the Uchida-Kawashima Scale_1 (the U-K scale_1), which is the turbulence evaluation index regarding wind conditions, is defined as formula (4) below; the latter, the Uchida-Kawashima Scale_2 (the U-K scale_2), which evaluates the fatigue damage, is defined as the fatigue damage evaluation index. Its definition is described later. The U-K scale_1, which is the turbulence evaluation index, is obtained as follows: standard deviation is calculated using the streamwise (x) wind velocity component (u) at the hub height, and this is normalized by U_{in}, wind velocity at the maximum height above the ground of the inflow boundary, as shown in Figure 15. Instead of the average wind velocity at the hub height, the U_{in} inflow velocity at the inflow boundary is used for normalization; the formula is generalized and does not depend on wind directions or terrain undulations.

$$\text{U-K Scale\_1} = \frac{\sigma_u}{U_{in}} = \frac{\sqrt{\frac{1}{n}\sum_{i=1}^{n}\left(u_i - \overline{u}\right)^2}}{U_{in}} \tag{4}$$

5.4. Analysis of Turbulence Statistics and Uchida-Kawashima Scale_1 Verification Part 1

Figure 21 shows time-series data (dimensionless time, 100) of the streamwise wind velocity component at the hub height (60 m above the ground) of wind turbine #10. The research data results of wind velocity revealed that the mean speed was lower, and the fluctuation was greater for easterly winds (shown in blue) than for northerly winds (shown in red). The threshold of the U-K scale_1 was determined as follows: standard deviation, which was calculated with the streamwise (x) wind velocity component (u) assessed at the hub height, was normalized by inflow velocity at the maximum height above the ground of inflow boundary; the U-K scale_1 is the turbulence evaluation index defined in formula (4). The value obtained for the northerly wind in this manner was 0.17, and similarly, that for the easterly wind was 0.25. Based on these data results, the present study defines the threshold of the U-K scale_1 as 0.2. The validity of the value 0.2 of the threshold is discussed later.

Vertical profiles of turbulence statistics at the site of wind turbine #10 are shown regarding both northerly and easterly winds in Figures 22 and 23. Firstly, the mean values of the streamwise wind velocity components in Figure 22 were our focus. The profile of inflow wind velocity is also illustrated in green in Figure 22. For the northerly wind (in red), a speed-up effect of about 1.3 times (= 0.97/0.74) was obtained at the hub height (approximately 60 m) due to the topography effect on the wind. On the other hand, in the case of the easterly wind shown in blue, it was decelerated by about 0.4 times (= 0.29/0.74) under the influence of Mt. Benzaiten (elevation 519 m). Furthermore, it was confirmed in the figure that extremely large-scale velocity shears existed in the range of the swept area of the wind turbine ($z^* = 20$–100 m)(The variable z^* is the height above the ground). This large-scale velocity shear induces malfunctions of major parts that make up the wind turbine, such as the main shaft and gearbox. Thus, utmost caution must be exercised.

Secondly, among the three-dimensional components of standard deviations, the vertical profile is observed, as shown in Figure 23. The threshold value of 0.2 in the U-K scale_1, which was previously described, is drawn in Figure 23. Results of the easterly wind, shown in blue, were greater than those of the northerly wind, shown in red, in all three components. Focusing on the wind turbine hub height (60 m above the ground), it is worth mentioning in particular that both northerly and easterly winds had almost the same level of values for the three components.

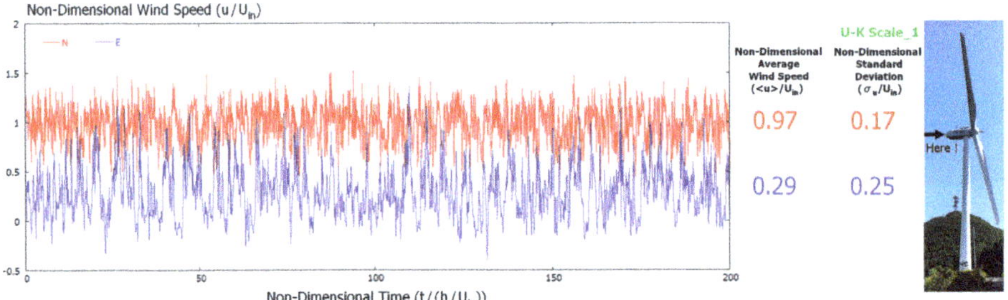

Figure 21. Time-series data of streamwise wind velocity from the numerical simulations. Red: northerly wind, Blue: easterly wind.

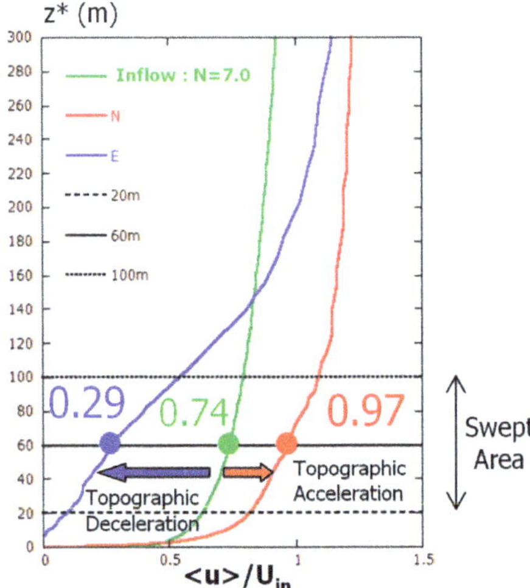

Figure 22. Vertical profiles of the streamwise wind velocity at wind turbine #10, with a time-averaged flow field. Red: northerly wind, Blue: easterly wind.

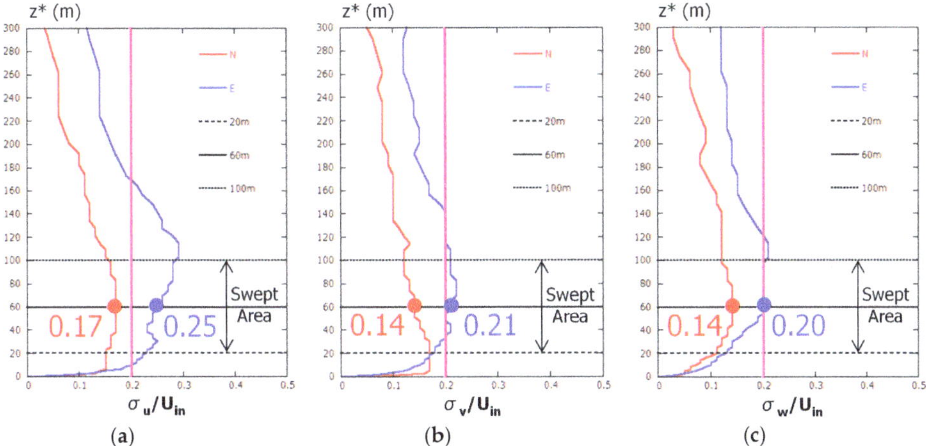

Figure 23. Vertical profiles of the non-dimensional standard deviations at wind turbine #10, with a time-averaged flow field. Red: northerly wind, Blue: easterly wind. (**a**) Streamwise direction. (**b**) Spanwise direction. (**c**) Vertical direction.

Compared with the threshold value of 0.2 in the U-K scale_1, northerly patterns were lower than 0.2 in all three components; however, easterly patterns exceeded 0.2 in all three components. Based on these findings, it was quantitatively shown that the terrain-induced turbulence of the easterly wind had a three-dimensional structure, and turbulence was generated due to the presence of Mt. Benzaiten (elevation 519 m).

To verify the validity of the threshold value of 0.2 in the U-K scale_1, configurations of inflow wind velocity profiles were altered using calculations of both northerly and easterly winds. Specifically, calculations using N = 4.0 and N = 10.0 were performed, and the obtained values were compared with the values of N = 7.0. These results are shown in Table 6. This table also shows the case of the threshold value of 0.2. The following results were obtained: in the case of the easterly wind, all of the calculated values were higher than 0.2 of the threshold values, and in all cases of the northerly wind, the calculated results were lower than 0.2 of the threshold value. It was suggested that the U-K scale_1 did not depend on inflow wind velocity profiles, and thus 0.2 of the threshold level was a pertinent judgment criterion on the whole.

Table 6. Comparison of the values of the U-K Scale_1 at wind turbine hub height (z* = 60 m) under different N values.

	N = 4.0	N = 7.0	N = 10.0	Criteria of the U-K Scale_1
Northerly Wind	0.16	0.17	0.17	0.20
Easterly Wind	0.24	0.25	0.24	

5.5. Mt. Benzaiten Impact Assessment and Uchida-Kawashima Scale_1 Verification Part 2

The present study performed a computational simulation where the effects of Mt. Benzaiten (elevation 519 m) were eliminated from the examined impacts. Specifically, all the terrains higher than the elevation of wind turbine #10 were eliminated from the simulation. Furthermore, the validity of the threshold level in the U-K scale_1 was considered. The grid resolution was changed from 10 m, which was used in previous numerical wind condition simulations in the horizontal direction, to five meters, which was half of the previous resolution. We investigated how this alternation in grid resolution affected the calculation results; that is, how they affected the values of the U-K scale_1.

Figure 24 shows two types of situations for the easterly wind: simulation results of the current situation with Mt. Benzaiten (Figure 24a) and simulation results with Mt. Benzaiten eliminated (Figure 24b). In both simulations, the figures show the streamwise (x) wind velocity component (u) profiles and wind velocity vector in the vertical direction at the wind turbine #10 site for inlet wind to wind turbine #10 in the instantaneous flow field. From Figure 24a, for the current situation results, the separated flow of terrain-induced turbulence, which was generated from Mt. Benzaiten located in the upper stream of wind turbine #10, was precisely observed. Conclusively, wind velocity vectors at the wind turbine #10 site presented intricate distributions in height. Contrarily, the simulation results with Mt. Benzaiten removed (Figure 24b), as expected, showed that terrain-induced turbulence was not generated, and wind velocity vectors at the wind turbine site had an ideal distribution with a gradual increase in wind velocity by height.

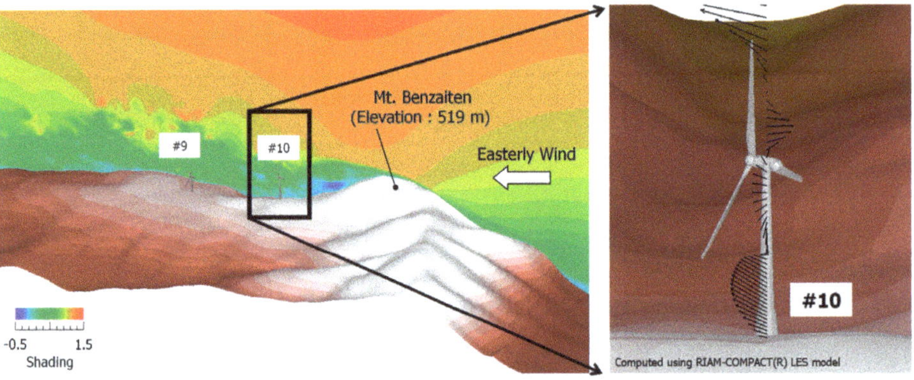

(**a**) Simulation result of the current situation.

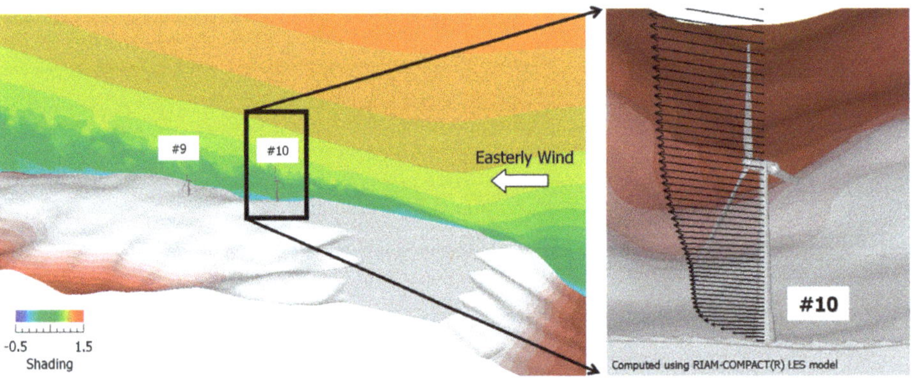

(**b**) Simulation result of removing Mt. Benzaiten (elevation: 519 m).

Figure 24. Distribution of the streamwise wind velocity component on a vertical cross-section, which includes wind turbine #10 and wind velocity vectors at wind turbine #10, easterly wind, and instantaneous flow field.

Table 7 provides simulation results of the U-K scale_1, where the horizontal grid resolution was changed to 5 m in high-resolution simulations to examine numerical wind conditions. The simulation with Mt. Benzaiten eliminated resulted in the U-K scale_1 being 0.01, which was significantly lower than the threshold value of 0.2. In contrast, the simulation with the current situation, where terrain-induced turbulence exists, resulted in the U-K scale_1 being 0.28, which was higher than 0.2 of the threshold level, similar to the cases in Table 6. This meant that the effectiveness of the U-K scale_1 and the validity of 0.2 of the threshold level were verified.

Table 7. Values of the U-K Scale_1 with horizontal grid resolution set to 5 m.

	Easterly Wind, N = 7.0	Criteria of the U-K Scale_1
Case of Removing Mt. Benzaiten (elevation 519 m)	0.01	0.20
Current Situation	0.28	

6. In-situ Data Analysis of Impacts of Terrain-induced Turbulence on Fatigue Damage in Wind Turbine Blades

6.1. Relationship between Wind Speed and its Standard Deviation and Damage Equivalent Load (DEL)

Figure 25 shows the results of the plotted nacelle wind speed and DEL obtained at 4 m/s or higher in wind turbine operation during the data measurement period of wind turbine #10 (3 November 2015, 0:00 a.m. JST–17 March 2016, 7:00 a.m. JST). In this figure, the values of the nacelle wind speed and DEL were in intervals of 10-min. The red symbol represents northerly wind, the blue symbol the easterly wind, and the black symbol represents design values calculated by the aero-elastic wind turbine simulation software Bladed. From these results, the following was clarified: firstly, when the speed of the easterly wind was between 6–10 m/s, the measured value exceeded the design value. This means that the blades of wind turbine #10 experienced wind loads that exceeded the design value due to terrain-induced turbulence, as a result of the proximity of Mt. Benzaiten (elevation 519 m), under 6–10 m/s easterly wind. Secondly, regarding the northerly wind, which had the highest frequency of occurrence during the above data measurement period, the damage was below the level of the design value in any wind speed class. Generally, it is known that wind speed and its standard deviation have a linear correlation. In Figure 26, the horizontal axis represents the standard deviation, which was converted by calculation from the wind speed axis of Figure 25. A strong correlation was revealed between DEL and the standard deviation values, as expected.

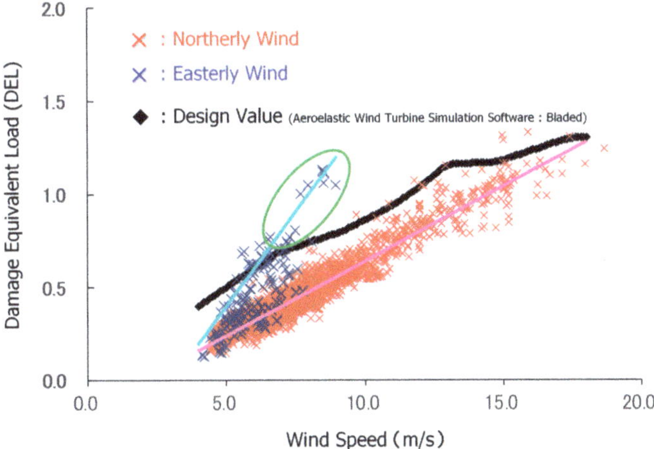

Figure 25. Relationship between wind speed (m/s) and damage equivalent load (DEL).

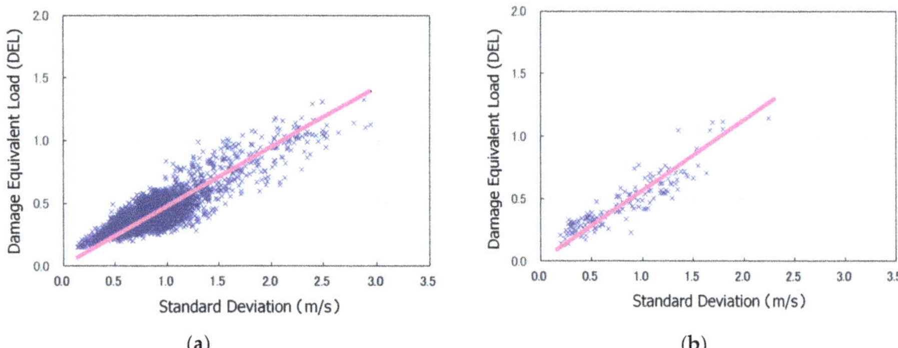

Figure 26. Relationship between standard deviation (m/s) and damage equivalent load (DEL). (**a**) Northerly wind. (**b**) Easterly wind.

6.2. Investigation Regarding Accumulated Fatigue Damage of Wind Turbine Blades

In the previous section, it was revealed that a linear trend was recognized between the nacelle wind speed (its standard deviation) and DEL in wind turbine operations at 4 m/s or higher, and it was possible to approximate to the regression line (see Figures 25 and 26). In other words, by assigning the DEL as a dependent variable (an objective variable) and nacelle wind speed as an independent variable (an explanatory variable), it is possible to apply a linear regression model between them. Figure 27 also shows a regression line of the nacelle wind speed and the DEL. The result of the northerly wind, which was represented by a red symbol, was defined as a low-turbulence flow case, and the results of the easterly wind, which was represented by a blue symbol, was defined as a high-turbulence flow case. At this point, the regression lines were modified at wind speeds of 10 m/s or higher. Similar to Figure 22, a black symbol was drawn as a design value calculated by the aero-elastic analysis software Bladed.

In the present study, as an index regarding the load, the Uchida-Kawashima Scale_2 (the U-K scale_2) was defined as the fatigue damage evaluation index, which was obtained by using two types of regression lines calculated based on the following actual measurement values and the design value on the basis of Bladed: one regression line was the northerly wind result as the low-turbulence flow case, and the other was the easterly wind result as the high-turbulence flow case. The definition of the U-K scale_2 was as follows:

$$\text{U-K Scale\_2} = \frac{\sum_{i=1}^{n} DEL_{Proposal}}{\sum_{i=1}^{n} DEL_{Design}} \tag{5}$$

At this point, we used m (S-N curve slope), which was used to calculate the DEL. The values raised to the *m*-th power were totaled, and the obtained value was raised to the (*1/m*)-th power for the integration of DEL, which is a scalar quantity. For example, the integration of two DELs is obtained with formula (6):

$$DEL_{total} = \left(DEL1^{m} + DEL2^{m}\right)^{\frac{1}{m}} \tag{6}$$

Formula (5) refers to the ratio of the integrated value in the measured DEL to the integrated value in the designed DEL (Bladed). Therefore, the accumulated fatigue damage of wind turbine blades is as follows:

U-K scale_2 > 1.0: more than the design value, impact of the terrain-induced turbulence: large.
U-K scale_2 $\leqq$ 1.0: design value and less, impact of terrain-induced turbulence: small.

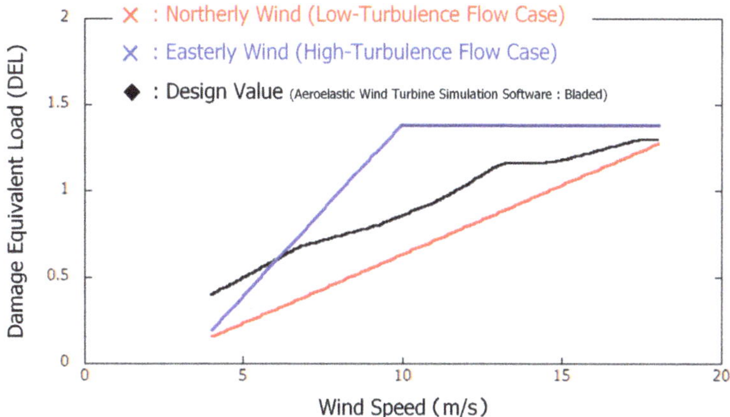

Figure 27. Regression line between wind speed (m/s) and damage equivalent load (DEL).

In the present study, the data of both northerly and easterly winds were extracted from measured data for one year, April 2015–March 2016, corresponding to a 4 m/s or higher speed in the wind turbine's operation. Sequentially, using these extracted data and the U-K scale_2, which was defined in formula (5), how terrain-induced turbulence influenced age-related degradation in wind turbine blades was assessed quantitatively. A total of 7485 data points (14.3%) were obtained for northerly wind and 2342 (4.5%) for easterly wind. In the case of northerly wind,

U-K scale_2 = 0.86 < 1.0: within the design value.

Contrarily, in the case of the easterly wind,

U-K scale_2 = 1.60 > 1.0: over the design value.

The U-K scale_2 exceeded the design value, as was shown. Furthermore, the integrated value of fatigue damage was approximately 1.9 times greater for the easterly wind compared with the northerly wind. Based on this result, it was notably indicated that the blades of the wind turbine #10, which was the target turbine, were directly and intensely affected by the terrain-induced turbulence when easterly winds occurred.

7. A Proposal for the Use of the U-K Scales and Future Research

The present study showed that the effect of terrain-induced turbulence on wind turbine blades remains low as long as the value of the wind condition index, the U-K scale_1, is equal to or smaller than the threshold value, 0.2. Furthermore, it was suggested that by combining this index and the threshold value of 1.0 or smaller for the U-K scale_2, which is an index for wind loads on wind turbine blades, an optimal wind turbine siting can be planned with higher accuracy than with the conventional approach. Figure 28 shows an example flowchart for planning wind turbine siting with the use of the two generalized parameters proposed in the present study (the U-K scale_1 and the U-K scale_2). When planning the optimal wind turbine siting, it is desirable to maximize the output in prevailing winds, while simultaneously minimizing wind turbine failures. It is very likely that the use of the flowchart in Figure 28, even for a few prevailing wind directions, at a site under consideration will be effective. It is also possible to apply the flowchart for the placement of wind observation poles and for the so-called repowering of wind turbines at existing wind farms and individual wind turbine sites. Repowering refers to rebuilding old wind turbines in order to increase the wind power generation capacity and improve the efficiency of wind power generation.

The present study showed regression lines for the relationship between the DEL and wind speed at the nacelle height only for northerly winds as a low-turbulence flow case and easterly winds as a high-turbulence flow case. However, similar regression lines have also been evaluated for other wind directions. Thus, with the use of these regression lines, we are planning to quantitatively evaluate the effect of long-term DEL accumulation, for all wind directions, on the blades of wind turbine #10, which was the turbine studied in the present study, in the future. In addition, since WT #10 is affected by the wakes of other wind turbines in some wind directions, we are also planning to conduct analyses on the effect of wind turbine wakes on the blades of wind turbine #10.

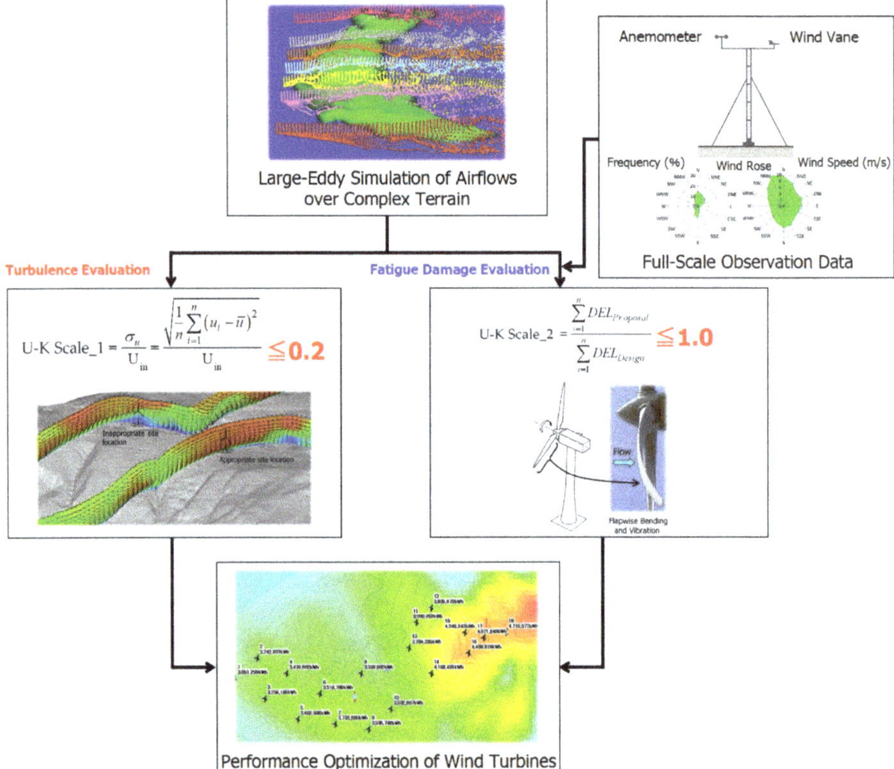

Figure 28. An example of wind energy resource assessment based on the two reference scales (the U-K scales).

Our future studies will expand the present empirical research and aim to develop an advanced micro-siting technique. Specifically, this micro-siting technique will be able to 1) maximize the output from a group of wind turbines both onshore and offshore; and also 2) enable accurate assessments and predictions of fatigue damage and fatigue life of the main shaft, gearbox, and other main components of a wind turbine that result from terrain-induced turbulence and wind turbine wakes.

8. Conclusions

The present study scrutinized the impacts of terrain-induced turbulence on wind turbine blades of wind turbine #10 in the Kushikino Reimei Wind Farm (established in November of 2012) in Hashima Ichikikushikino City, Kagoshima Prefecture, Japan, measuring actual data of wind conditions and strains of wind turbine blades. High-resolution numerical simulations were performed on the basis of large-eddy simulation (LES). The research findings were as follows:

On the basis of measurement data obtained by the above auto-measuring system from 3 November 2015, 0:00 a.m. JST to 17 March 2016, 7:00 a.m. JST, the output results of sensors were extracted using wind speed and direction sensors for the control of the wind turbine, which were installed on the wind turbine nacelle. These data results were analyzed to examine the behavior of airflows that were generated surrounding wind turbine #10. Consequently, it was confirmed that the turbulence intensity of the easterly wind with a speed of 10 m/s or lower frequently exceeded the turbulence intensity (Class A) that was defined in the International Electrotechnical Commission (IEC). Furthermore, it was inferred that the influences of Mt. Benzaiten (elevation 519 m) were a cause of the increased turbulence intensity when easterly wind occurred.

During the measurement period of this investigation (3 November 2015, 0:00 a.m. JST–17 March 2016, 7:00 a.m. JST), the maximum value of 0.23 for DEL on the wind turbine blades under easterly wind was identified for the time period 9:40–9:50 a.m. JST of November 13, 2015. It was revealed that if airflow with a property of DEL 2.03 continued for 5.88 years, the total load on wind turbine blades would reach the design load for the designed service life.

The three-dimensional airflow structures, which were considered to be terrain-induced turbulence formed under the easterly wind from Mt. Benzaiten (elevation 519 m), were closely examined using high-resolution numerical simulations on the basis of large-eddy simulation (LES). The present study defined the U-K scale_1 as the value obtained by calculation where the standard deviation assessed at the wind turbine hub height was normalized by the wind speed obtained at the maximum height above the ground point of the flow boundary layer. Therefore, the existence of the terrain-induced turbulence was quantitatively identified. The threshold value of the U-K scale_1, the wind condition index, was determined as 0.2, and this index was confirmed to not be dependent on the height direction distributions of inflow wind speed, the influences of horizontal grid resolution, and the influences of the computed azimuth. Under these circumstances, the U-K scale_1 is a unique indicator that distinguishes it from the IEC turbulence category. Moreover, this scale is based on the premise of comparison with actual measurement data and was found as a research result of about 15 years, including this research result. The U-K scale_1 is extremely effective as a guideline for solving the problem of generation loss and failure caused by topographical turbulence.

A linear trend was recognized in wind turbine operations at 4 m/s or higher between the nacelle wind speed (and its standard deviation) and damage equivalent load (DEL), and it was possible to approximate to the regression line. For an index regarding load, the U-K scale_2 was defined as the fatigue damage evaluation index, which was obtained by the calculation using two types of regression lines calculated based on actual measurement values and the design value on the basis of Bladed: one regression line was the northerly wind as the low-turbulence flow case, and the other was that of the easterly wind as the high-turbulence flow case. The U-K scale_2 was equal to the ratio of the integrated value for the measured DEL to the integrated value for the design DEL (Bladed).

Data on both northerly and easterly winds were extracted from measurement data for the year April 2015–March 2016, corresponding to wind turbine operations with a speed of 4 m/s or higher. Using these extracted data and the U-K scale_2, the influence of terrain-induced turbulence on the age-related degradation of wind turbine blades was evaluated quantitatively. Consequently, it was revealed that the results of northerly wind were within the design value, with the U-K scale_2 being 0.86 < 1.0. In contrast, for easterly wind, the U-K scale_2 was 1.60 > 1.0 and exceeded the design value. Furthermore, the integrated value of the fatigue damage was approximately 1.9 times greater in the case of easterly wind compared to northerly wind. Based on this result, it was notably indicated that the blades of wind turbine #10 were directly and strongly affected by terrain-induced turbulence when easterly winds occurred.

Author Contributions: The study idea, plan and design were conceived by T.U. And, all authors prepared the manuscript.

Funding: This work was supported by JSPS KAKENHI Grant Number 17H02053.

Acknowledgments: We owe our deepest gratitude to Kyudenko New Energy Co., Ltd., whose encouragement and support in providing the in situ data enabled us to develop our research, and to E-Wind Co., Ltd., whose commitment to data analysis supported us in completing our investigation. This work was supported by both the collaborative research of Kyushu University and West Japan Engineering Consultants, Inc. and the collaborative research of Kyushu University and Hitachi, Ltd. We pay our heartfelt respect and gratitude to all the organizations and individuals for their particular technique and wisdom.

Conflicts of Interest: All authors declare no conflict of interest.

References

1. Tang, X.-Y.; Zhao, S.; Fan, B.; Peinke, J.; Stoevesandt, B. Micro-scale wind resource assessment in complex terrain based on CFD coupled measurement from multiple masts. *Appl. Energy* **2019**, *238*, 806–815. [CrossRef]
2. Gargallo-Peiró, A.; Avila, M.; Owen, H.; Prieto-Godino, L.; Folch, A. Mesh generation, sizing and convergence for onshore and offshore wind farm Atmospheric Boundary Layer flow simulation with actuator discs. *J. Comput. Phys.* **2018**, *375*, 209–227. [CrossRef]
3. Sessarego, M.; Shen, W.Z.; Van der Laan, M.; Hansen, K.; Zhu, W.J. CFD Simulations of Flows in a Wind Farm in Complex Terrain and Comparisons to Measurements. *Appl. Sci.* **2018**, *8*, 788. [CrossRef]
4. Temel, O.; Bricteux, L.; Beeck, J.V. Coupled WRF-OpenFOAM study of wind flow over complex terrain. *J. Wind Eng. Ind. Aerodyn.* **2018**, *174*, 152–169. [CrossRef]
5. Dhunny, A.Z.; Lollchund, M.R.; Rughooputh, S.D.D.V. Wind energy evaluation for a highly complex terrain using Computational Fluid Dynamics (CFD). *Renew. Energy* **2017**, *101*, 1–9. [CrossRef]
6. Murthy, K.S.R.; Rahi, O.P. A comprehensive review of wind resource assessment. *Renew. Sustain. Energy Rev.* **2017**, *72*, 1320–1342. [CrossRef]
7. Simões, T.; Estanqueiro, A. A new methodology for urban wind resource assessment. *Renew. Energy* **2016**, *89*, 598–605. [CrossRef]
8. Blocken, B.; Hout, A.V.D.; Dekker, J.; Weiler, O. CFD simulation of wind flow over natural complex terrain: Case study with validation by field measurements for Ria de Ferrol, Galicia, Spain. *J. Wind Eng. Ind. Aerodyn.* **2015**, *147*, 43–57. [CrossRef]
9. Gopalan, H.; Gundling, C.; Brown, K.; Roget, B.; Sitaraman, J.; Mirocha, J.D.; Miller, W.O. A coupled mesoscale-microscale framework for wind resource estimation and farm aerodynamics. *J. Wind Eng. Ind. Aerodyn.* **2014**, *132*, 13–26. [CrossRef]
10. Ishugah, T.F.; Li, Y.; Wang, R.Z.; Kiplagat, J.K. Advances in wind energy resource exploitation in urban environment: A review. *Renew. Sustain. Energy Rev.* **2014**, *37*, 613–626. [CrossRef]
11. Palma, J.M.L.M.; Castro, F.A.; Ribeiro, L.F.; Rodrigues, A.H.; Pinto, A.P. Linear and nonlinear models in wind resource assessment and wind turbine micro-siting in complex terrain. *J. Wind Eng. Ind. Aerodyn.* **2008**, *96*, 2308–2326. [CrossRef]
12. Uchida, T.; Takakuwa, S. A Large-Eddy Simulation-Based Assessment of the Risk of Wind Turbine Failures Due to Terrain-Induced Turbulence over a Wind Farm in Complex Terrain. *Energies* **2019**, *12*(1925), 1485. [CrossRef]
13. Uchida, T.; Li, G. Comparison of RANS and LES in the Prediction of Airflow Field over Steep Complex Terrain. *Open J. Fluid Dyn.* **2018**, *8*, 286–307. [CrossRef]
14. Uchida, T. Computational Investigation of the Causes of Wind Turbine Blade Damage at Japan's Wind Farm in Complex Terrain, Journal of Flow Control. *Meas. Vis.* **2018**, *6*, 152–167.
15. Uchida, T. Designed Wind Speed Evaluation Technique in Wind Turbine Installation Point by Using the Meteorological Model and CFD Model. *J. Flow Control Meas. Vis.* **2018**, *6*, 168–184. [CrossRef]
16. Uchida, T. LES Investigation of Terrain-Induced Turbulence in Complex Terrain and Economic Effects of Wind Turbine Control. *Energies* **2018**, *11*(6), 1530. [CrossRef]
17. Uchida, T. Computational Fluid Dynamics (CFD) Investigation of Wind Turbine Nacelle Separation Accident over Complex Terrain in Japan. *Energies* **2018**, *11*(6), 1485. [CrossRef]
18. Uchida, T.; Maruyama, T.; Ohya, Y. New Evaluation Technique for WTG Design Wind Speed using a CFD-model-based Unsteady Flow Simulation with Wind Direction Changes. *Model. Simul. Eng.* **2011**. [CrossRef]
19. Uchida, T.; Ohya, Y. Micro-siting technique for wind turbine generators by using large-eddy simulation. *J. Wind Eng. Ind. Aerodyn.* **2008**, *96*, 2121–2138. [CrossRef]

20. Uchida, T.; Ohya, Y. Large-eddy simulation of turbulent airflow over complex terrain. *J. Wind Eng. Ind. Aerodyn.* **2003**, *91*, 219–229. [CrossRef]
21. Uchida, T.; Ohya, Y. Numerical simulation of atmospheric flow over complex terrain. *J. Wind Eng. Ind. Aerodyn.* **1999**, *81*, 283–293. [CrossRef]
22. Kiyoki, S.; Uchida, T.; Kawashima, Y.; Kondo, K. Impact Assessment of Terrain Turbulece to Wind Turbine Fatigue. Presented at the 15th World Wind Energy Conference (WWEC 2016), Tokyo, Japan, 31 October 31–3 November 2016.
23. Freebury, G.; Musial, W. *Determining Equivalent Damage Loading for Full-Scale Wind Turbine Blade Fatigue Tests*; Technical Report for National Renewable Energy Laboratory: Golden, CO, USA, February 2000.
24. Takemi, T.; Kusunoki, K.; Araki, K.; Imai, T.; Bessho, K.; Hoshino, S.; Hayashi, S. Representation and Localization of Gusty Winds Induced by Misocyclones with a High-Resolution Meteorological Modeling. *Theor. Appl. Mech. Jpn.* **2010**, *58*, 121–130.
25. Kim, J.; Moin, P. Application of a fractional-step method to incompressible Navier-Stokes equations. *J. Comput. Phys.* **1985**, *59*, 308–323. [CrossRef]
26. Kajishima, T. Upstream-shifted interpolation method for numerical simulation of incompressible flows. *Bull. Jpn. Soc. Mech. Eng. B* **1994**, *60*, 3319–3326. [CrossRef]
27. Kawamura, T.; Takami, H.; Kuwahara, K. Computation of high Reynolds number flow around a circular cylinder with surface roughness. *Fluid Dyn. Res.* **1986**, *1*, 145–162. [CrossRef]
28. Smagorinsky, J. General circulation experiments with the primitive equations: I. Basic experiments. *Mon. Weather Rev.* **1963**, *91*, 99–164. [CrossRef]

Article

Implementation of IEC 61400-27-1 Type 3 Model: Performance Analysis under Different Modeling Approaches

Raquel Villena-Ruiz [1], Alberto Lorenzo-Bonache [1], Andrés Honrubia-Escribano [1], Francisco Jiménez-Buendía [2] and Emilio Gómez-Lázaro [1,*]

[1] Renewable Energy Research Institute and DIEEAC-ETSII-AB, Universidad de Castilla-La Mancha, 02071 Albacete, Spain
[2] Siemens Gamesa Renewable Energy, S.A., 31621 Pamplona, Spain
* Correspondence: emilio.gomez@uclm.es; Tel.: +34-967-599-200 (ext. 2418)

Received: 16 May 2019; Accepted: 9 July 2019; Published: 12 July 2019

Abstract: Forecasts for 2023 position wind energy as the third-largest renewable energy source in the world. This rapid growth brings with it the need to conduct transient stability studies to plan network operation activities and analyze the integration of wind power into the grid, where generic wind turbine models have emerged as the optimal solution. In this study, the generic Type 3 wind turbine model developed by Standard IEC 61400-27-1 was submitted to two voltage dips and implemented in two simulation tools: MATLAB/Simulink and DIgSILENT-PowerFactory. Since the Standard states that the responses of the models are independent of the software used, the active and reactive power results of both responses were compared following the IEC validation guidelines, finding, nevertheless, slight differences dependent on the specific features of each simulation software. The behavior of the generic models was assessed, and their responses were also compared with field measurements of an actual wind turbine in operation. Validation errors calculated were comprehensively analyzed, and the differences in the implementation processes of both software tools are highlighted. The outcomes obtained help to further establish the limitations of the generic wind turbine models, thus achieving a more widespread use of Standard IEC 61400-27-1.

Keywords: DIgSILENT-PowerFactory; IEC 61400-27-1; MATLAB; model validation; transient stability; type 3 wind turbine

---

1. Introduction

Renewable energy power plants are growing at a spectacular rate all over the world. The International Energy Agency (IEA) states that, specifically in the electricity sector, renewable energies will undergo the fastest growth, providing approximately 30% of the total power demand in 2023 [1]. It also underlines that, although bioenergy will still be the largest source of renewable energy in the years to come, especially due to its consumption in heat and transport, its share will decline as a result of the expansion of both wind power and solar PV. Thus, forecasts for 2023 position wind energy as the third-largest renewable energy source in the world, only surpassed by bioenergy and hydropower [1]. Other entities such as the Global Wind Energy Council (GWEC) [2] or WindEurope annually publish statistics on new onshore and offshore wind power capacity installed across the countries. WindEurope has already uploaded the 2018 report [3], which informs that Europe installed 11.7 GW of new wind power capacity (up to a total of 189 GW), more than any other power generation source, covering 14% of the European Union's electricity demand.

In light of the above, it is highly important for countries to continue developing a clear strategy that allows the objectives set by the European Union with reference to the renewable energy production in 2030 [4] to be achieved. Among other tasks, countries need to foster economic stability in order to attract investment. Furthermore, there is a need to promote high participation of renewable energy in electricity mixes by seeking a broad consensus among the stakeholders in renewable energy production.

Focusing on wind power, it indisputably plays an increasingly important role in current power systems. However, despite all the advantages mentioned above, grid integration of the installed wind power capacity is regarded as a challenge mainly due to the unpredictable nature of wind. Moreover, fluctuations in wind power generation will lead conventional power plants to compensate for these variations, thus forcing them to operate under conditions for which they are not planned. Therefore, the integration of such a large number of wind turbines (WT) into power systems may increase voltage and frequency regulation problems, necessitating forward planning of network operation activities. Thus, optimizing the utilization of wind energy and securing the continuity of the electricity supply is a key issue. Transmission and Distribution System Operators (TSOs and DSOs, respectively) are the entities authorized to manage and maintain power systems and thus they require detailed operation plans to have advance knowledge of the behavior of the power systems. To carry out this work, transient stability analyses of WTs [5] and wind power plants (WPP) [6] dynamic models are required. These types of analyses will allow the electrical responses of the models, once connected to the grid, to be forecasted [7].

Under this scenario, in February 2015, the International Electrotechnical Commission (IEC) published Standard IEC 61400-27-1 [8], the first edition of which includes the description of both generic WT and WPP models. These models are referred to as 'generic' or 'standard' because they are intended to represent any commercial WT model, regardless of the vendor. They consist of a small number of blocks and parameters and may be implemented in any simulation software tool. Moreover, as detailed WT simulation models are technically complex and usually belong to private companies, in addition to being subject to confidential agreements, publicly available generic WT models developed by standard IEC 61400-27-1 [9] are intended to faithfully replicate the behavior of actual WTs when they are connected to the grid and submitted to electrical disturbances. Generic WT models are classified to represent the four main typologies of actual WTs available in the market, including their principal technical features: Type 1, which uses an asynchronous generator directly connected to the grid, Type 2, consisting of an asynchronous generator equipped with a variable rotor resistance; Type 3, which uses a doubly-fed induction generator (DFIG), where the stator is directly connected to the grid and the rotor is connected through a back-to-back power converter [10]; and Type 4, connected to the grid through a full-scale power converter. Therefore, the implementation and dynamic simulation of these WT models [11] will allow TSOs and DSOs to properly plan network operation and secure electricity supply. To fully achieve this objective and guarantee their effective operation, generic WT simulation models must be validated against field measurements of actual WTs [12]. Thus, it is necessary to compare and analyze their responses under the most critical conditions, i.e., under voltage dips [13]. Indeed, in order to conduct a quantitative comparison, specific validation guidelines have been specifically developed by the IEC [14].

IEC 61400-27-1 also states that the responses obtained from the dynamic simulation analyses of the models must be independent of the simulation software used. This study aims to show that, nevertheless, there exist slight deviations between simulation results depending on the implementation software. This is mainly due to differences in the integration algorithms or in the implementation processes of the dynamic sub-systems. The current work presents these differences and addresses the general implementation processes of the generic Type 3 WT model in two different software tools: MATLAB/Simulink (2018, MathWorks, Natick, MA, US) and DIgSILENT-PowerFactory [15]. Indeed, there is a lack of research works related to the study of the DFIG WT developed by Standard IEC 61400-27-1.

In particular, to the best of the authors' knowledge, there are few studies in the scientific literature regarding the implementation and simulation of the generic Type 3 WT model in specialized electrical engineering software tools such as DIgSILENT-PowerFactory. For instance, works such as [16] or [17] analyze the behavior of the active and reactive power control systems of the standard Type 3 WT and its transient response when subjected to voltage dips, respectively. In both cases, MATLAB is the only simulation tool used. Moreover, studies related to the validation of WTs are based on private, specific and detailed WT simulation models developed by specific vendors, such as the one carried out in [18]. Thus, also related to this topic are found works such as [11,13,19]. In [13], two voltage dips are conducted in a 2 MW DFIG WT, and its responses are compared to the simulation responses of a detailed WT simulation model (i.e., not RMS model). The authors in [11] analyzes the WT model that represents the technology used by a specific vendor, while the authors in [19] validate a 3 MW DFIG WT model by comparing its responses with the responses of a detailed DFIG vendor model. Finally, Reference [20] also performs validation tasks during faults, although they compare a WT model with an analytical method. Therefore, in view of the above, Reference [21] is the only work addressing the performance of generic Type 3 WT models using DIgSILENT-PowerFactory. However, although PF is the tool used, the responses at plant level during changes on the reference points are analyzed, so that no voltage dip tests at a WT model level are conducted.

The benefits of using specialized software tools are numerous, since they allow actual, larger and more complex power systems to be simulated, in addition to being tools with which TSOs and DSOs are used to working. On the other hand, simulation tools such as MATLAB/Simulink are highly attractive due to their versatility and ease of use. This work showcases the use of both types of software tools, presenting their advantages and particularities. In addition, it aims to define the modeling processes that must be followed, as well as demonstrating the differences when simulating the same generic WT dynamic model, despite claims for their non-existence.

Therefore, the present work addresses the implementation and validation process of the IEC 61400-27-1 generic Type 3 WT model when it is subjected to different voltage dips. The generic WT model is modeled and simulated in both MATLAB/Simulink and DIgSILENT-PowerFactory. The parametrization and validation are conducted using field data from the WT manufacturer Siemens-Gamesa Renewable Energy (Zamudio, Spain). The simulated active and reactive power responses are compared to the ones measured from the actual WT in operation. The validation error magnitudes—following IEC 61400-27-1 guidelines—are calculated in three different comparisons: (i) MATLAB/Simulink WT model vs. field measurements, (ii) DIgSILENT-PowerFactory WT model vs. field measurements, and (iii) MATLAB/Simulink WT model vs. DIgSILENT-PowerFactory WT model.

Summarizing, the main contributions of the present study are focused on: (i) providing feedback to Standard IEC 61400-27-1 regarding the assumption that the performance of the electrical simulation models are independent of any software simulation tool; (ii) conducting an in-depth analysis of the root causes that lead to the differences in the simulated responses; (iii) expanding the scope of application of the Standard; and (iv) providing the opportunity of conducting analyses involving large power systems that include generic WT models in specialized electrical engineering software tools such as DIgSILENT-PowerFactory.

The paper is structured as follows: Section 2 discusses the limitations of the generic WT models and the validation guidelines in Standard IEC 61400-27-1. Section 3 describes the modeling process of the generic Type 3 WT when implemented in both software tools, while Section 4 analyzes the research results. Finally, Section 5 summarizes the main conclusions obtained.

2. IEC 61400-27-1 Generic Type 3 WT Model

The generic Type 3 WT model can be divided into two sub-models: Type 3A and Type 3B. The main difference lies in the generator system, which includes a protection system in the case of the Type 3B WT. This protection system is modeled through a set of dynamic blocks that decreases the current signals to zero when a voltage dip occurs. In the present work, as the field measurements were recorded on an actual DFIG WT in operation, the generic Type 3B is the model studied.

The generic Type 3B WT model developed by Standard IEC 61400-27-1 (hereinafter referred to as Type 3) consists of several sub-models, as shown in Figure 1: aerodynamic control model [22], which provides the two-mass mechanical model with the aerodynamic power (p_{aero}) coming from wind; pitch control model [23], which calculates the value of the pitch angle (θ)-position angle of the WT blades—required to follow the rotor speed and power generation setpoint (p_{WTref}); two-mass mechanical model [24], which models the actual gearbox representing both the low-speed and the high-speed sides and provides the wind turbine rotor and the generator rotational speed (w_{WTR} and w_{gen}, respectively); active power control model [16], which provides the generator system with the active current command (i_{pcmd}) and also calculates both the reference rotational speed and the active power order; reactive power control model, which controls the reactive power injection through the calculation of the reactive current command (i_{qcmd}) based on the user-defined reactive power reference (x_{WTref}); reactive power limitation model, which provides the reactive power control model with the reactive power injection's maximum and minimum dynamic values allowed at the wind turbine terminals (WTT); current limitation model, which calculates the limit values of both the active and reactive currents (i_{pmax}, i_{qmax} and i_{qmin}); and generator system, equipped with a crowbar model [25–27], which has as output signals the active and reactive currents injected into the grid through a current source (i_{gen}, and the generator air gap power (p_{ag})). Measured values of voltage (u_{gen} and u_{WT}), as well as active and reactive power (p_{WT} and q_{WT}) at the test network are also required as input signals in some control models. The single-line diagram of the generic Type 3 WT model is available in [8].

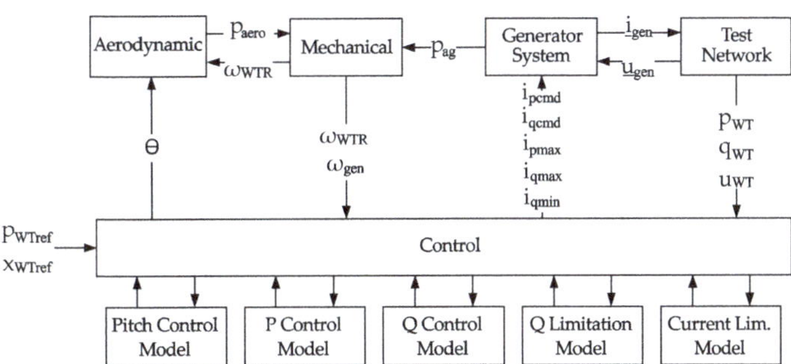

Figure 1. Modular structure of the generic Type 3 WT adapted from [8].

As discussed in Section 1, International Standard IEC 61400-27-1 states that there should be no differences when studying the responses of generic WT models using different software tools. However, there may exist differences in the implementation and initialization processes, as well as in the algorithms used by the simulation tools. Moreover, although the time step used during the simulations may be set as fixed when conducting the comparison between models implemented in different software tools, some of them use variable-step solvers, adjusting the time step depending on the variability of the signals.

IEC 61400-27-1 also describes the limits of the accuracy of the generic WT models [8]. Since the model's accuracy is analyzed comparing simulation responses with measured data, errors in both cases may appear. On the one hand, regarding simulation responses, errors occur, as during the development of the standard, some simplifications were implemented in order to obtain technically simplified dynamic WT models, which are, nevertheless, accurate enough to represent the existing wide range of actual WTs. On the other hand, limits to possible model accuracy may also arise from the measurements. When performing transient stability analyses, electromagnetic phenomena are not of interest, and are thus not emulated and represented by the generic WT models. For instance, if the high voltage side of the transformer is chosen to perform the measurements, electromagnetic phenomena caused by the behavior of the transformer may be reflected in the measured data. The accuracy of the model will thus be lower. Furthermore, some errors and limitations are also generated when performing the measurements at actual WPPs or test benches. Deviations resulting from tolerances or bandwidth limitations of the measurement equipment, as well as errors derived from the measurements' conversion, lead to cumulative errors in some measured signals. When converting from instantaneous values to root mean square (RMS) values to perform the comparisons, some errors also occur.

In view of the above, it is clear that there will exist errors between measurements and simulated data. For this reason, they must be correctly assessed and calculated according to IEC 61400-27-1 guidelines. Deviations between the responses of the two simulated models (PowerFactory and MATLAB) will also be assessed during this work, with the aim of identifying the main factors that cause them. The following subsections describe, on the one hand, the calculation process of the validation errors established by IEC 61400-27-1, and, on the other, the general implementation process followed by both simulation software tools used, also showing some of the dynamic control models that are part of the generic Type 3 WT, highlighting the particularities and adaptations required in each case.

2.1. Validation of Generic Models Based on IEC 61400-27-1 Guidelines

The validation process of the generic models involves several steps. First, the definition of the dynamic simulation model must be performed. Second, field measurements are needed to measure the responses of actual WTs. Thus, in order to perform the field tests, a fault-ride through (FRT) mobile test unit was installed in a Spanish wind farm during a measurement campaign led by the manufacturer Siemens-Gamesa. The connection of the test unit was made between the switchgear and the high voltage side of the power transformer of the WT. Further information regarding the FRT mobile test unit can be found in [25,28], since similar units have been used to analyze the FRT capability of WTs. Once the voltage dips were applied to the WT, the rate of the measurement sample was set to 10 kHz, and the measurements composing each data set began 10 s before the voltage dip was applied and ended 15 s after this point in time. This period is enough to carry out the measurements, since the transient stability responses of the simulation are reflected during that period and can then be properly assessed. After that, measurements of the events conducted are replicated in the WT model, so that the simulation data and the field measurements can be compared. Finally, the validation errors are estimated. In this regard, the most significant disturbances in power systems that lead to integration issues are voltage dips [29]. In terms of RMS values, voltage dips involve severe voltage reductions ranging between 10% and 90% and a duration of up to one minute [19,30]. Residual voltage (minimum value of voltage) and dip duration are the two parameters that characterize these types of disturbances in the grid. However, the complexity of WTs means that not all kinds of voltage dips are suitable to be conducted on actual WTs, and, hence, IEC 61400-21 established a representative set of disturbances in order to validate a WT model [17].

As explained in Section 1, only a few scientific studies have performed validation tasks using field measurements, and even fewer when it comes to the implementation, simulation and validation of the most widely used type of WT, Type 3 (i.e., DFIG), in specialized software tools. Moreover, this study presents a triple comparison, which involves field measurements and simulation results from the same generic

WT model implemented in two different tools. It also provides information, for the first time, about the slight deviations that may exist between those simulation results, despite IEC 61400-27-1 indicating they should not exist. This comparison thus helps to further determine the limitations of the IEC-developed WTs, in addition to achieving a more widespread use of these models.

In order to be able to obtain a response and a well-founded and justified conclusion, and based on the objectives previously mentioned, the criteria used to validate the generic WT models with field measurements, defined by IEC 61400-27-1, were also applied when comparing the simulated models with each other. Three time windows are defined during the simulation: (i) pre-fault window, which starts 1000 ms before the fault initiation, (ii) fault window, which lasts from the onset to the clearance of the fault, (iii) post-fault window, which lasts 5000 ms after the voltage dip clearance. Furthermore, in order not to consider transient periods in the assessment of the accuracy of the models, several time sub-windows are also defined, formerly known as 'quasi-steady state' sub-windows. Hence, the transient responses appearing at the start of both fault and post-fault windows are discarded in the calculation of the maximum errors, as will be seen later in this document. On the one hand, a transient period of 140 ms is not considered in the calculations of the maximum validation errors at the start of the fault window due to the limitations of the generic model in reproducing the DC-component of the generator flux [12]. On the other hand, since the model is unable to faithfully represent the transformer inrush current, thus affecting the active and reactive power responses, a period of 500 ms is not considered at the start of the post-fault window. In addition, it is worth considering that, because of the inability to model the aerodynamic oscillations or nonlinear aerodynamic effects by generic WT models, larger errors may occur during the entire post-fault window. Figure 2 shows both the time windows and the quasi-steady state sub-windows defined during the voltage dip according to [8].

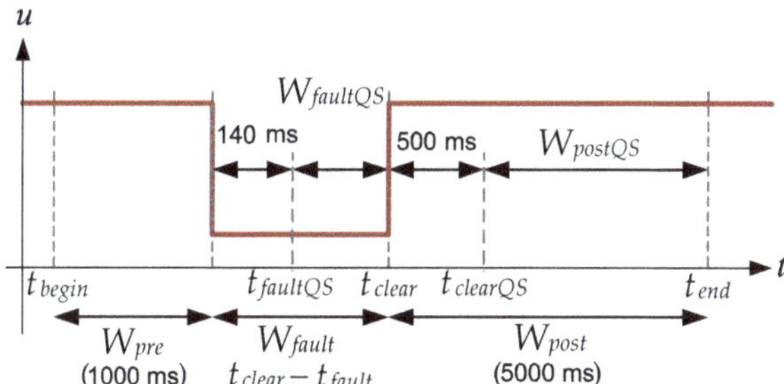

Figure 2. Voltage dip windows adapted from [8].

The calculation of validation errors during the pre-fault window is not critical for the evaluation of the performance of the model. Therefore, error time series are only calculated during fault and post-fault windows for each of the comparisons (x_{error} in Equation (1)), i.e., between the data series of the following cases: (i) MATLAB Model–Field Data, (ii) PowerFactory Model–Field Data, (iii) MATLAB Model–PowerFactory Model. On the basis of this information, three validation errors (also referred to as validation performance indicators) are calculated in both fault and post-fault windows, the calculations of which are presented in Equations (2)–(4): Mean Error (*ME*), Mean Absolute Error (*MAE*) and Maximum Absolute Error (*MXE*) [12].

In Equations (1)–(4), numbers 1 and 2 indicate the data series to be compared in each case, while n indicates the indices of the vectors and N the total number of samples used:

$$x_{error}(n) = x_1(n) - x_2(n), \tag{1}$$

$$x_{ME} = \frac{\sum_{n=1}^{N} x_{error}(n)}{N}, \tag{2}$$

$$x_{MAE} = \frac{\sum_{n=1}^{N} |x_{error}(n)|}{N}, \tag{3}$$

$$x_{MXE} = max(|x_{error}(n)|). \tag{4}$$

ME is defined as the mean value of the error over the corresponding time window, and is related to the steady-state performance of the dynamic model; *MAE* is the mean value of the absolute error, calculated over the entire time window in the case of the post-fault period, and over the corresponding quasi-steady state sub-window in the case of the fault period; and, finally, *MXE* is the maximum value of the absolute error, mainly related to the transient performance of the simulated model. In this last case, the quasi-steady state sub-windows are considered in the calculation. The electromagnetic transient periods measured are thus not considered in the evaluation of the accuracy of the model, since fundamental-frequency generic models are not designed to represent them. As a summary, Table 1 shows the time periods considered for the calculation of each type of error in both the fault and post-fault periods, according to Figure 2.

Table 1. Time periods considered for the calculation of the validation errors based on Figure 2.

Validation Error	Fault Window	Post-Fault Window
Mean error, *ME*	$[t_{fault}\ t_{clear}]$	$[t_{clear}\ t_{end}]$
Mean absolute error, *MAE*	$[t_{faultQS}\ t_{clear}]$	$[t_{clear}\ t_{end}]$
Maximum absolute error, *MXE*	$[t_{faultQS}\ t_{clear}]$	$[t_{clearQS}\ t_{end}]$

Having explained the validation criteria to be applied, the following sections are devoted to describing, in general terms, the generic Type 3 WT modeling process in the two software tools used.

3. Modeling Process of Generic Type 3 WT

The structure of the generic Type 3 WT and the validation process that must be followed to test its electrical performance have been extensively described in previous sections. Moreover, limitations to the model's accuracy included in IEC 61400-27-1 have also been listed. However, with the objective of achieving well-reasoned explanations of the causes of the differences in the model's behavior when it is simulated in the two software tools, the ways in which the dynamic WT model is implemented are described in detail below.

3.1. Implementation in MATLAB/Simulink

MATLAB/Simulink is currently one of the most widely used software tools in engineering. Its flexibility, supported by its large community, allows a wide range of studies to be conducted. Furthermore, with the addition of Simulink, its add-on, which works with block language, the possibilities are unlimited. For this work, the generic Type 3 WT model was first modeled and parametrized in Simulink, as the facilities included permit faster development. Furthermore, the validation work was conducted in MATLAB, the programming language of which permits easy implementation of the IEC 61400-27-1 validation methodology described in Section 2.1.

Regarding the IEC 61400-27-1 generic Type 3 modeling, its general structure is shown in Figure 3. Additionally, the systems included within the 'Control model' are shown in Figure 4. The control of the generic Type 3 model has previously been explained in the literature. In [17], the behavior and parametrization of the systems related to the active power response is explained. Furthermore, in [31], the behavior of the reactive power control, the current limitation system and the generator system are depicted and compared with a simpler model [32].

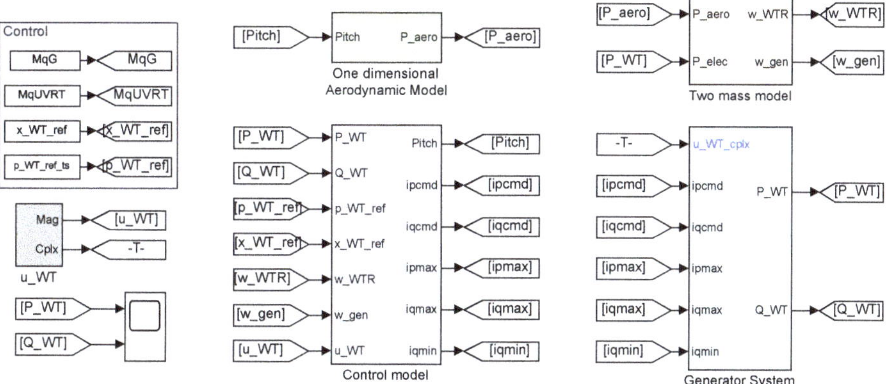

Figure 3. General structure of the IEC 61400-27-1 generic Type 3 model implemented in Simulink.

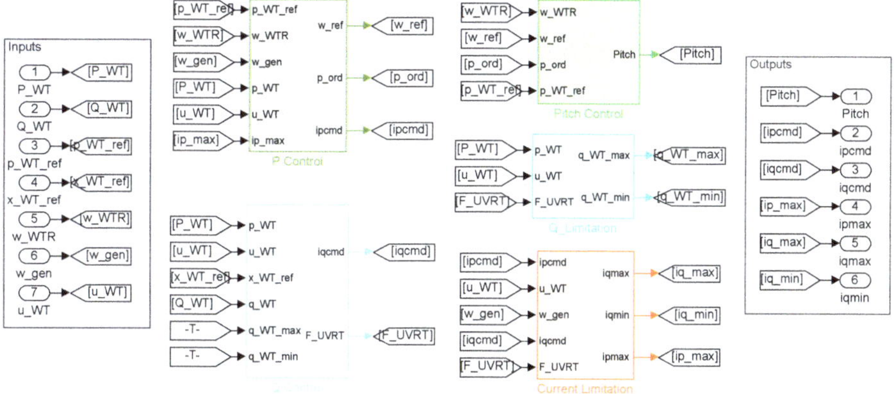

Figure 4. Control systems of the IEC 61400-27-1 generic Type 3 model implemented in Simulink.

The electrical generator system is one of the most complex in this Type 3 WT model. It is based on the physical dynamics of a DFIG, and the theory behind its development is described in [10,15]. For the present work, as explained in Section 2, the authors modeled a generic Type 3B electrical generator system, which includes the crowbar protection system. Its modular structure is shown in Figure 5. The 'reference frame rotational model' coordinates the active and reactive command currents with the grid reference frame. Furthermore, this system includes the dynamics of the generator sub-transient reactance (x_s). Additionally, the crowbar system is modeled as a system that multiplies by 0 the command currents over a short time when the fault occurs and ends, depending on the derivative of voltage. Finally, as shown in Figure 3, the Simulink model is not connected to a grid. The active and the reactive power are calculated based on Equation (5):

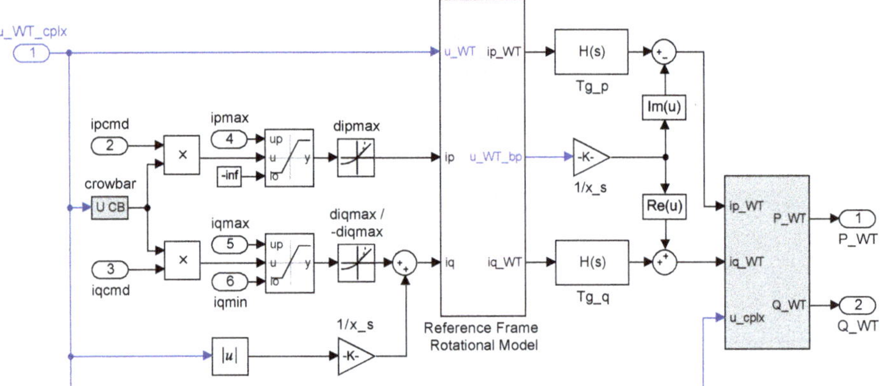

Figure 5. Electrical generator system of the IEC 61400-27-1 generic Type 3B model implemented in Simulink.

$$\bar{S}_{\mathrm{WT}} = P\_WT + jQ\_WT = \bar{U}_{\mathrm{WT}}\bar{I}^*_{\mathrm{WT}} = u_{\mathrm{WT}}\angle\varphi_{u\_\mathrm{WT}} \cdot (ip\_WT + j \cdot iq\_WT) \tag{5}$$

Additionally, it should be mentioned that the fine adjustment of the model parameters was conducted using the Simulink Design Optimization tool. An example illustrating the use of this tool is shown in Figure 6. For this work, the reference signal was either the active or the reactive power response. Then, the parameters that we wish to adjust are selected, as well as their ranges and scale. Finally, the program iterates until reaching the objective selected. The iteration method and the error followed (absolute or relative) can be selected. Finally, the software provides the finely adjusted parameters. As a side note, although this process might seem automatic, the approximately 100 parameters which define the generic Type 3 model means the user can have a deep understanding of the behavior of the model, as well as the parameters which should be adjusted and their logical ranges in which to conduct the tuning.

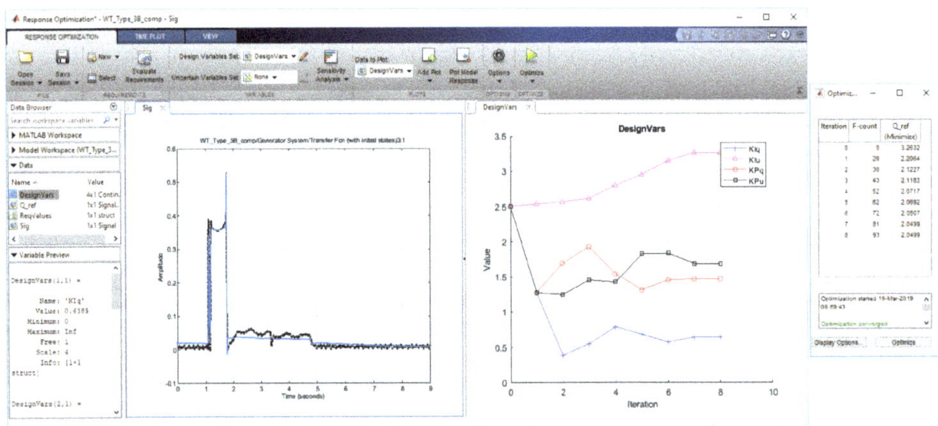

Figure 6. Simulink Design Optimization tool example.

3.2. Implementation in DIgSILENT-PowerFactory

The previous sections highlighted the importance of using specialized electrical engineering software tools, as they permit a wide range of studies to be carried out. In addition, these types of tools allow highly complex studies to be performed, enabling also the consideration of the vast amount of information and the large number of parameters which characterize actual power systems [33]. It is therefore of prime importance to have in-depth knowledge of the dynamic model's implementation process, and an overview of particularities of the software and also of the calculation method or the network representation, as these will help to further understand the causes of the deviations between simulation results provided by the two software tools. The way in which voltage dips are conducted at the WT model implemented in DIgSILENT-PowerFactory (PF) is also an important element.

3.2.1. DIgSILENT Simulation Language

DIgSILENT Simulation Language (DSL) is the PF function used to model dynamic systems [34]. When working with DSL, predefined electrical devices available in the software and user-defined dynamic blocks may be incorporated into the power system, i.e., complete dynamic models are implemented in PF by relating a certain electrical component with its corresponding control models, influencing its electrical behavior. In fact, the WT is modelled as an AC current source that injects active and reactive current into the grid. Active and reactive power are thus measured through a predefined power measurement device connected at the WTTs, while a conversion system has been specifically designed to adapt the output current signals from the generator system model to the input signals' format required by the AC current source.

Based on [8] and the DSL working structure [35], Figure 7 shows the complete power system implemented in PF, which relates all the generic Type 3 WT dynamic models described at the start of Section 2, including the AC current source representing the WT and measurement and auxiliary devices.

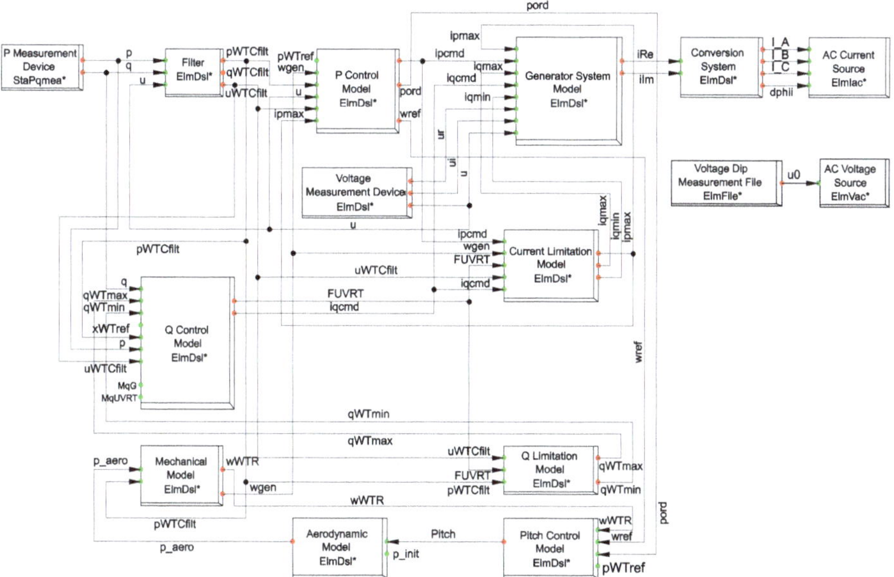

Figure 7. General structure of the IEC 61400-27-1 generic Type 3 model implemented in DIgSILENT-PowerFactory.

3.2.2. Time Domain Simulations in DIgSILENT-PowerFactory

The initialization of the power system is of great importance, since correctly executing the initialization process avoids fictitious electrical transients and allows the system to rapidly reach the steady-state condition [36]. Otherwise, numerical instability may appear [33]. First, a load flow calculation to initialize the predefined electrical components must be executed. Second, the equations of the initial conditions must be set manually in every user-defined DSL dynamic model (based on the information provided by the load flow calculation) [37]. Finally, the 'calculation of initial conditions' command must also be executed.

'AC Load Flow, balanced, positive sequence' is the calculation method followed. It performs the calculation for an equivalent single-phase, i.e., for a network representation of positive sequence. This method is valid in the case of balanced symmetrical networks, which justifies the subsequent choice of the simulation method. Therefore, for the calculation of initial conditions, the balanced RMS simulation method, which considers a steady-state, symmetrical representation of the electrical grid, is selected. Dynamics in electromechanical and control devices are thus taken into account, and only fundamental components of voltages and currents are considered [34]. Under these conditions, only symmetrical faults are allowed. Indeed, generic WT models defined by IEC 61400-27-1 are designed to conduct studies of three-phase symmetrical faults.

3.2.3. Conduction of Voltage Dips at the WTTs

Once the complete power system is initialized, it may generally be studied under two different operating conditions: normal and fault operating conditions. However, as stated in Section 1, electrical disturbances such as voltage dips are the most critical situations to be assessed in order to allow network operators to properly plan system operation. Hence, since this study aims to calculate validation errors between field data and the results of generic Type 3 WT model simulation by reproducing two voltage dips measured on an actual WT in operation, the play-back validation approach was used [38]. Instead of conducting a voltage dip by defining a short-circuit at one of the WTTs [35], an external AC voltage source was connected to the WT model. This voltage source is controllable through a voltage dip measurement file and positive sequence voltage values are used as the input signal. In this way, the power system is forced to behave under the desired conditions. The play-back method thus constitutes the most suitable approach in such cases, as it enables the accurate reproduction of voltage dips that correspond to field measurements (see dynamic sub-models 'Voltage Dip Measurement File' and 'AC Voltage Source' in Figure 7).

3.2.4. Test Network

The AC current source to which the control models are connected and the controllable AC voltage source reading the voltage dip measurement file are part of the same test network. Considering that this work is focused on comparison studies and calculation of validation errors, as well as on the analyses of the Type 3 WT's electrical responses in isolation, strictly equal voltage profiles must be considered in the cases analyzed (in MATLAB and PowerFactory). Therefore, it is not important to go into depth in the test network as it does not represent an actual power system, nor does it affect the performance of the generic WT model, since it has only been defined to be able to reproduce the voltage dip measured at the actual WT and assess, in such a way, the accuracy of the simulation responses.

However, both the current and voltage sources need to be physically connected because they are predefined electrical devices in PowerFactory, with a test network such as the one presented in Figure 8 being the modelled auxiliary power system. It consists only of the sources of the AC current and voltage and a terminal interconnecting them. Figure 8 is also a schematic illustration of how dynamic control models are related to the AC current source through the DSL.

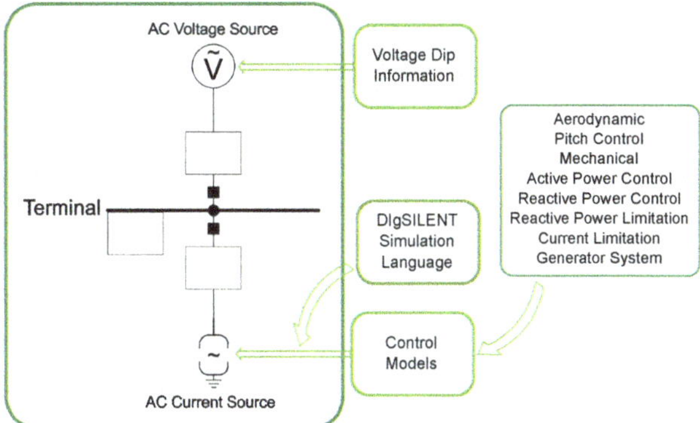

Figure 8. Test network modeled in DIgSILENT-PowerFactory: interconnection of the dynamic control models using DSL.

4. Results

This section presents a comparison of the responses of the generic Type 3 WT model implemented in both software tools during two different voltage dips, and also the calculation of the validation errors according to Section 2.1. Moreover, it performs a comprehensive analysis of the reasons for such errors. The values of the main parameters that are part of the generic Type 3 WT model are shown in Table 2.

Table 2. Main parameters of the generic Type 3 WT model.

Parameter	Description	Model	Value
T_{cw}	Crowbar duration vs. voltage variation		0.05
x_s	Electromagnetic transient reactance	Generator system	0.4
T_{wo}	Time constant for crowbar washout filter		0.5
di_{pmax}	Maximum active current ramp rate		3
M_{qG}	Reactive power control mode		1
M_{qUVRT}	Under Voltage Ride Through (UVRT) reactive power control mode		2
K_{qv}	Voltage scaling factor for UVRT current		5.55
i_{qpost}	Post-fault reactive current injection	Reactive power	0.05
r_{drop}	Resistive component of voltage drop impedance	control	0.01
x_{drop}	Inductive component of voltage drop impedance		0.1
T_{qord}	Time constant in reactive power order lag		0.001
u_{max}	Maximum voltage in voltage PI controller integral term		2
u_{min}	Minimum voltage in voltage PI controller integral term		0
u_{qdip}	Voltage threshold for UVRT detection		0.9
M_{DFSLim}	Limitation of Type 3 stator current	Current limitation	0
M_{qpri}	Prioritisation of reactive power control during UVRT		1

Table 2. *Cont.*

Parameter	Description	Model	Value
$T_{wfiltp3}$	Filter time constant for generator speed measurement		500
w_{offset}	Offset to reference value that limits controller action		0.02
K_{DTD}	Gain for active DTD		0.8834
T_{pfilt}	Time constant in power measurement filter	Active power	0.001
T_{pord}	Time constant in power order lag	control	0.0005
T_{ufilt}	Time constant in voltage measurement filter		0.001
dp_{max}	Maximum WT power ramp rate		2.75
T_{wref}	Time constant in speed reference filter		2
$d_{\tau,UVRT}$	Limitation of torque rise rate during UVRT		0
τ_{uscale}	Voltage scaling factor of reset torque	Torque PI control	0.45
KPp	Proportional constant of torque PI controller		10,000
KIp	Integral constant of torque PI controller		0.3722
Kpx	Pitch cross-coupling gain		0
θ_{max}	Maximum pitch angle		35
θ_{min}	Minimum pitch angle	Pitch control	0
$d\theta_{max}$	Maximum pitch angle rate		10
$d\theta_{min}$	Minimum pitch angle rate		−4

4.1. Validation of the Generic Type 3 WT Model

The WT dynamic model (implemented in MATLAB/Simulink and DIgSILENT-PowerFactory) was submitted to two different voltage dips, the residual voltage (u), and dip duration (t), of which are $u = 0.50$ pu and $t = 920$ ms for the Test Case 1, and $u = 0.25$ pu and $t = 625$ ms for the Test Case 2. In order to do so, the field data series were converted to positive sequence values, since the generic WT models must be studied for fundamental frequency-positive sequence response analyses. Thus, the positive sequence values of the voltage dips were reproduced in the generic WT models.

Figures 9 and 10 show the active and reactive power responses and their validation errors in Test Case 1, while Figures 11 and 12 show those corresponding to Test Case 2. The value of the more than a hundred parameters which are part of the generic Type 3 WT model do not vary from one software to the other. This explains the excellent correlation between the two simulation data series (in blue and black), in both the active power (Figures 9a and 11a) and the reactive power (Figures 10a and 12a), for both Test Cases. In these figures, a red line represents the measured data from the field tests. As might be expected, the differences between measured and simulation data are higher than the differences between both simulation responses, despite the generic Type 3 WT model having been adjusted to the field data series to the maximum extent.

In the graphics of the error data series, Figures 9b–12b, vertical lines and red bars at the top indicate the transient periods of 140 ms and 500 ms that must not be considered in the calculations of errors at the start of fault and post-fault periods, respectively, as explained in Section 2.1. As can be observed in these figures, error lines between simulation models and measured data for all cases (in red and blue) show sharp peaks. In addition, Figure 9 shows how an aerodynamic imbalance may affect the error between the field data and the simulation responses. The real response shows an undershoot during the post-fault period due to a wind speed fluctuation, which can be emulated by neither of the generic WT models. Nevertheless, as will later be explained drawing on Tables 3 and 4, the maximum errors obtained do not exceed 13% in either case (the *MXE* values are marked in the figures as small circles). Finally, the black lines represent the error obtained between both simulation models, which is at around 0% in all cases.

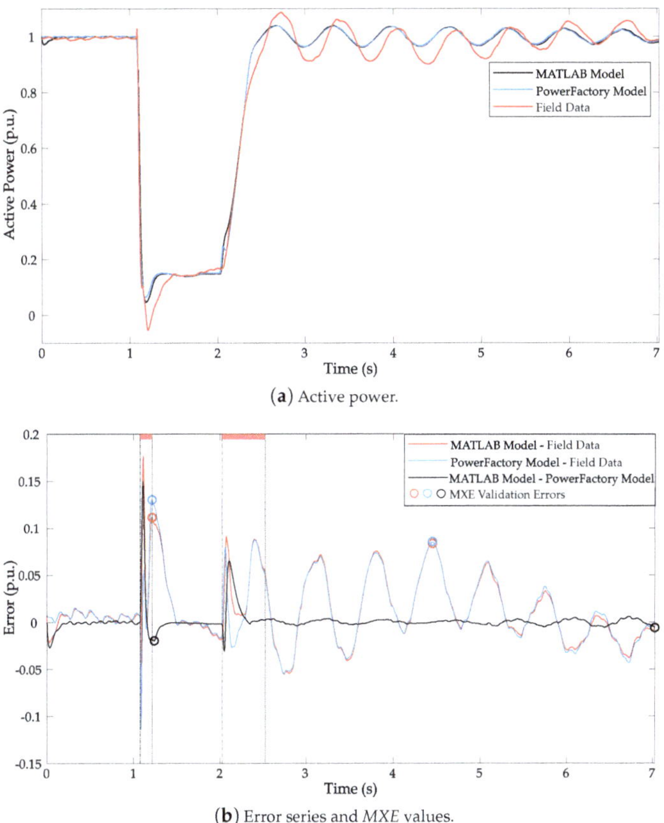

(**a**) Active power.

(**b**) Error series and *MXE* values.

Figure 9. Active power in Test Case 1: $u = 0.50$ pu, $t = 920$ ms.

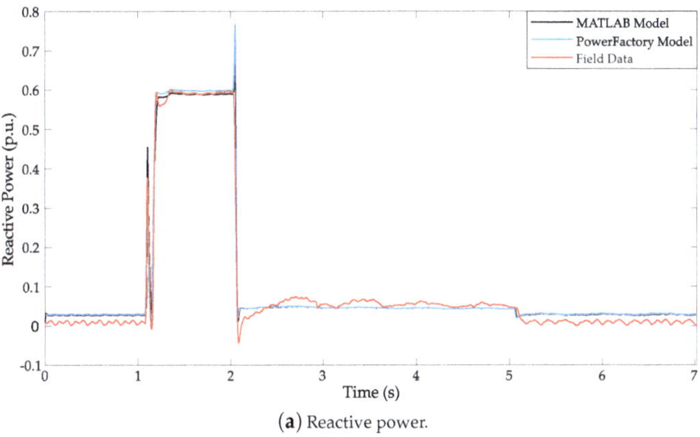

(**a**) Reactive power.

Figure 10. *Cont.*

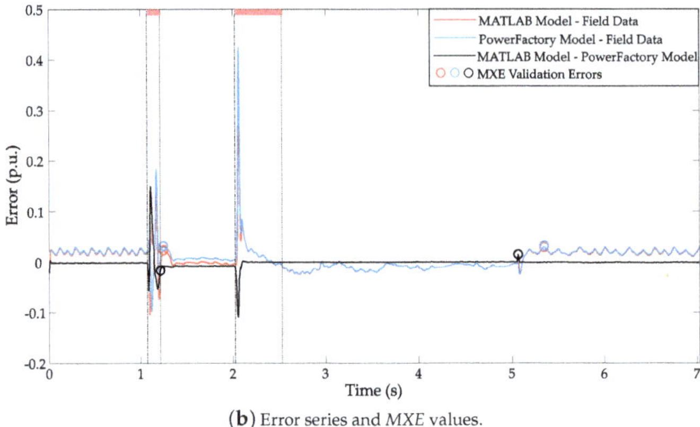

(**b**) Error series and *MXE* values.

Figure 10. Reactive power in Test Case 1: $u = 0.50$ pu, $t = 920$ ms.

In general, the graphic representation of the errors obtained is a good way to provide a general but comprehensive overview of the results. Thus, a quick glance reveals the performance of the simulation models, since the higher the deviation from 0%, the lower is the accuracy of the simulation models compared with each other and with the field measurements.

Furthermore, the numerical results of the validation errors are presented in Table 3 for Test Case 1 and Table 4 for Test Case 2. Different trends may be observed. First, as expected, lower errors are obtained for the 'MATLAB-PowerFactory' comparison (third column of Tables 3 and 4), since the behavior of both simulation models is very similar. In these cases, the *ME* and the *MAE* do not exceed 1% either in the active power or in the reactive power of the two test cases analyzed, in both fault and post-fault periods. Regarding the *MXE*, it does not exceed 2% in either case for the 'MATLAB-PowerFactory' comparison. It can therefore be stated that, despite the slight deviations which will be analyzed in Section 4.2, there is very little difference between the MATLAB and the PowerFactory simulation responses.

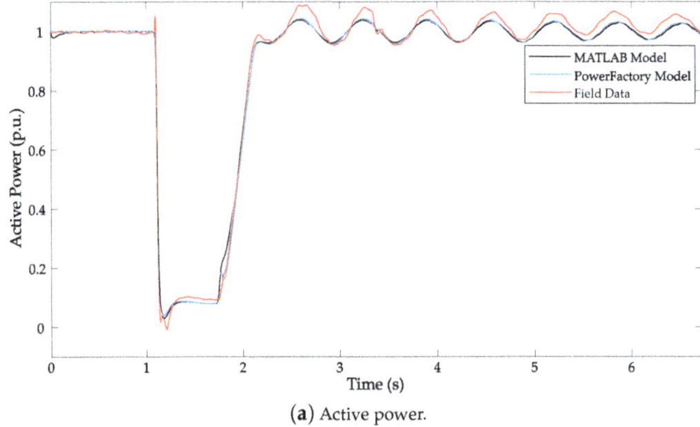

(**a**) Active power.

Figure 11. *Cont.*

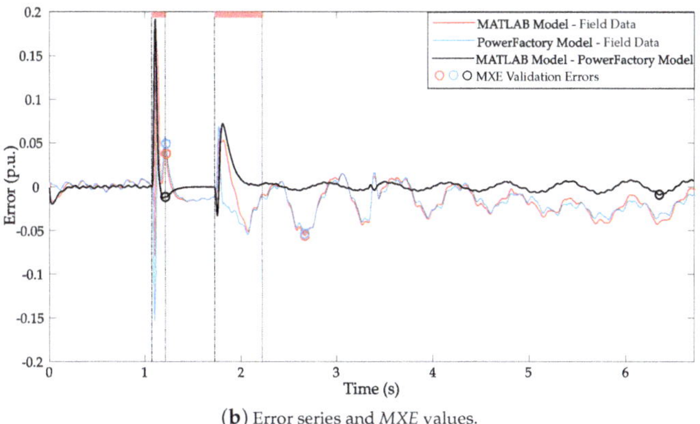

(**b**) Error series and *MXE* values.

Figure 11. Active power in Test Case 2: $u = 0.25$ pu, $t = 625$ ms.

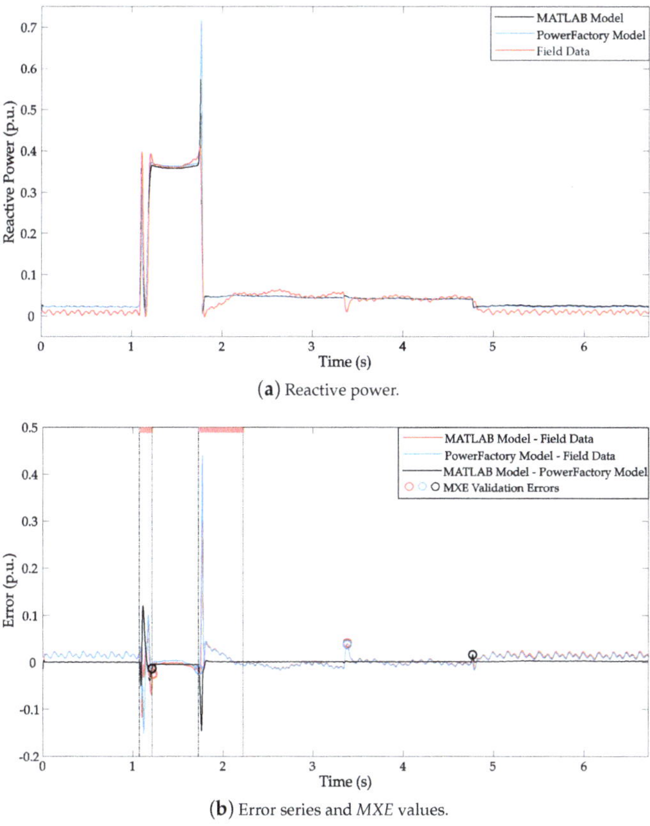

(**a**) Reactive power.

(**b**) Error series and *MXE* values.

Figure 12. Reactive power in Test Case 2: $u = 0.25$ pu, $t = 625$ ms.

Table 3. Validation errors of Test Case 1: u = 0.50 pu, t = 920 ms.

Active Power	MATLAB—Field	PowerFactory—Field	MATLAB—PowerFactory
$ME_{fault}(\%)$	2.34	1.99	0.25
$MAE_{fault}(\%)$	2.48	2.61	0.27
$MXE_{fault}(\%)$	11.12	12.98	1.97
$ME_{post}(\%)$	1.33	1.21	0.13
$MAE_{post}(\%)$	3.04	3.04	0.35
$MXE_{post}(\%)$	8.44	8.43	0.61
Reactive Power	MATLAB—Field	PowerFactory—Field	MATLAB—PowerFactory
$ME_{fault}(\%)$	0.31	0.87	−0.60
$MAE_{fault}(\%)$	0.46	0.82	0.84
$MXE_{fault}(\%)$	2.35	3.11	1.69
$ME_{post}(\%)$	0.41	0.54	−0.10
$MAE_{post}(\%)$	1.60	1.72	0.11
$MXE_{post}(\%)$	2.95	3.02	1.47

Table 4. Validation errors of Test Case 2: u = 0.25 pu, t = 625 ms.

Active Power	MATLAB—Field	PowerFactory—Field	MATLAB—PowerFactory
$ME_{fault}(\%)$	0.08	−0.87	0.77
$MAE_{fault}(\%)$	1.39	1.43	0.21
$MXE_{fault}(\%)$	3.73	4.87	1.18
$ME_{post}(\%)$	−1.65	−1.82	0.18
$MAE_{post}(\%)$	2.07	2.05	0.52
$MXE_{post}(\%)$	5.61	5.39	0.90
Reactive Power	MATLAB—Field	PowerFactory—Field	MATLAB—PowerFactory
$ME_{fault}(\%)$	−0.56	−0.50	−0.14
$MAE_{fault}(\%)$	0.60	0.37	0.49
$MXE_{fault}(\%)$	2.53	1.65	1.34
$ME_{post}(\%)$	0.62	0.57	0.08
$MAE_{post}(\%)$	1.11	1.14	0.23
$MXE_{post}(\%)$	3.90	3.71	1.59

In light of the above, the values of the errors between the 'MATLAB-Field' and 'PowerFactory-Field' data series are very similar, as can be observed in Tables 3 and 4. Moreover, in general, the *ME*, the *MAE* and the *MXE* are lower in the reactive power responses than in the active power responses for both test cases. This means that the reactive power simulation response of the generic Type 3 WT model is better adjusted to the field measurements. Despite this, there also exists a good correlation between the simulation and the field measurements in active power for both test cases.

Hence, as the generic WT models developed by IEC 61400-27-1 are not intended to be studied during the transient periods appearing at the start and the clearance of the faults, higher error values may be obtained. Indeed, most of the *MXE* errors obtained are near the transient periods of the responses (see the small circles in Figures 9b–12b and the red bars at the top, respectively), where the accuracy of the model with regard to the field measurements is usually lower. This is because transient periods of the actual WTs are not adequately represented by the generic models.

The efficient performance of the simulation models is also supported by the good correlation in the amplitude and the phase shift of the active power responses after the voltage dip clearance (Figures 9a and 11a). This is mainly due to the good fit to the parameters of the two-mass mechanical model. Moreover, the reactive power responses provided by the generic WT models are highly accurate (Figures 10a and 12a), since they present a very similar behavior to that of the field measurements (including the reactive power injection period during the voltage dips to stabilize the voltage).

Therefore, in view of the explanations above and the low values of the validation errors obtained, it can be concluded that the simulation responses of the generic Type 3 WT model yield satisfactory results, since, on the one hand, both emulated responses are very similar (with errors around 0%) and, on the other hand, both models are adjusted adequately to the field measurements.

4.2. Analysis of the Limitations of the Software Tools: Causes of the Differences in the Simulated Response

Section 4.1 analyzed the values of the validation errors calculated according to the validation guidelines issued by IEC 61400-27-1 in the three cases considered: (i) MATLAB Model—Field Data, (ii) PowerFactory Model-Field Data, (iii) MATLAB Model—PowerFactory Model. This section aims to explain the main differences in the software tools that may cause the errors found between both simulation responses of the generic WT model. However, it is worth noting that the validation errors are smaller than 2% for all cases. The explanations included in this Section intend to clarify the small differences between both simulation responses (from the two software tools used) at the time of conducting dynamic simulations, but always considering that the results are equally valid for both of them.

Figure 13 shows one of the constraints when comparing both simulation tools. Specifically, the response of a signal passing through a built-in rate limiter in PowerFactory and Simulink is analyzed. The rising rate was set to 5 and the falling rate was set to –999. It can be observed that the PowerFactory response is not the expected behavior, while the Simulink response is correct. Basically, PowerFactory applies a first-order filter to the input signal of the block, the time constant of which can also be adjusted. Thus, setting a lower time constant should dampen the effect of this filter. Nevertheless, this filter has an effect on the signal and, hence, the responses from PowerFactory and Simulink when using these limiters (which are several in IEC 61400-27-1 models) are not the same. More precisely, the generic Type 3 WT model includes a total of five rate limiters, two in the generator system model and three in the active power control model, so that the combined effect of these blocks on the output signals also explains the differences between the responses of both software tools. These filters are not implemented in Simulink since IEC 61400-27-1 does not include them in the models.

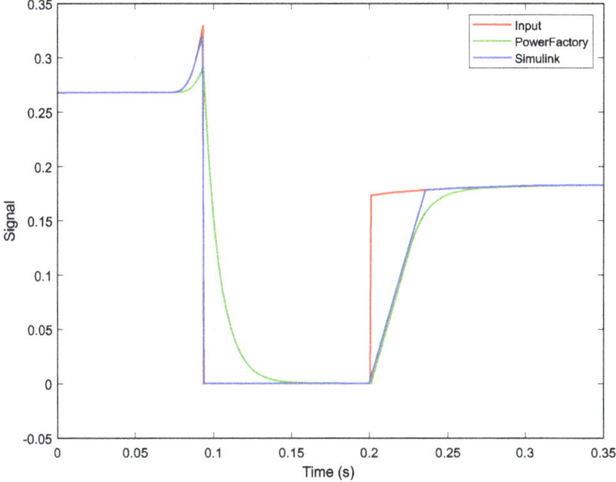

Figure 13. Rate limiters responses for the same input signal.

Additionally, despite the PowerFactory solver being set to 'fixed step' with a time step of 1 ms, the time step varies during the simulations. Regarding Simulink, the solver was set to ode4 (Runge–Kutta) with a fixed time step of 1 ms. This type of solver is widely used due to its balance between accuracy and simulation time. This is shown in Figure 14, which represents the time step at each sample, as well as the active power response for that simulation. When the active power response varies greatly, the time step decreases to improve the accuracy. In fact, this is the appropriate behavior for a variable step solver. However, for comparison purposes, a real fixed-step simulation time would be desirable. Therefore, this variable time step used by DIgSILENT-PowerFactory for the simulation of the WT model was identified as another one of the main causes of the slight differences between the simulation responses of the two software tools.

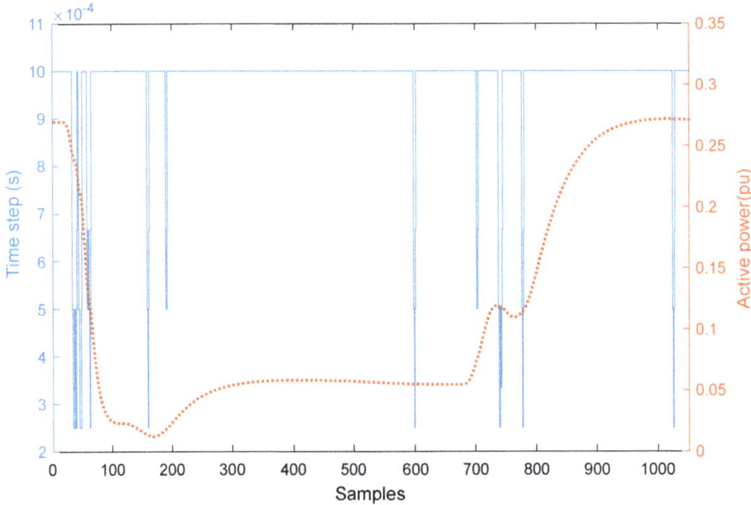

Figure 14. Simulation time steps during a fault in PowerFactory.

Therefore, differences in the rate limiter responses in both software tools, and the variable time step used by DIgSILENT-PowerFactory during the simulation, are identified as the key factors explaining the errors between the responses of the generic Type 3 WT dynamic model.

5. Conclusions

Wind power will be the third largest renewable energy source around the world in 2023, and will provide approximately 30% of the total power demand during that year. However, this scenario gives rise to the problem of integrating the installed wind power capacity into the grid. Since the nature of wind is unpredictable, WPP installations are regarded as non-dispatchable, and hence TSOs and DSOs in different countries must carry out proper planning of network operation, and also forecast the participation of conventional power plants to compensate fluctuations in wind power generation. Moreover, voltage and frequency regulation problems may emerge in these situations.

In light of the above, transient stability analyses of WT and WPP dynamic models are required to forecast the electrical behavior of actual WTs before being connected to the grid. For this purpose, the International Electrotechnical Commission, through IEC 61400-27-1, defined four WT simulation models intended to represent actual WTs of different manufacturers. These generic WTs consist of a small number of dynamic sub-models and parameters, and may be implemented and simulated in any software tool. Moreover, the Standard states that the responses obtained must be the same and independent of the software used. The present work demonstrates that, nevertheless, there may exist deviations between the two sets of results.

In order to analyze these differences, the generic Type 3 WT model, i.e., the DFIG WT, was implemented and simulated in MATLAB/Simulink and DIgSILENT-PowerFactory. Although MATLAB is widely used in the scientific literature, more specialized software tools, such as PowerFactory, are often used by TSOs and DSOs in different countries. Thus, the present work contributes to achieving a more widespread use of the recently developed generic WT models, also benefiting all stakeholders in the wind power industry.

The models were subjected to two different voltage dips, and their active and reactive power responses were compared with field measurements from an actual WT in operation. Validation errors following IEC 61400-27-1 guidelines were also estimated. In this way, it has been demonstrated that both simulation software tools are equally valid for the development of generic WT models. Beyond the specific applications of each tool, numerical differences between them are smaller than 2% for all the validation errors. Thus, it can be concluded that generic WT models can be implemented in the simulation software tool that fits better to the needs of the user.

With regard to the validation study, in general, the validation errors obtained between the field data series and the active and reactive power simulation responses of the models implemented in both software tools are low. The slight differences are due, firstly, to the limits of the model's accuracy, since, during the development of the Standard, some simplifications were introduced to design technically easy WT models (for instance, transient periods are not accurately represented by the generic WT models). Secondly, the errors may also be due to the limitations when the measurements were performed, such as the ones introduced by the tolerances of the measurement equipment, and also due to the conversion from instantaneous values to RMS values.

In addition, the errors between the two sets of simulated responses are near zero in the two Test Cases considered. Nevertheless, after carefully analyzing the differences in the implementation processes of both software tools and the behavior of each dynamic block, the errors between the active and reactive power responses may be attributed to two causes. The first cause could be the differences in the rate limiters used by the simulation tools. In these cases, PowerFactory applies a first-order filter to the input signal, the time constant of which affects its behavior and may dampen the effect of the filter. This affects the overall behavior of the WT, since IEC models use several of these rate limiters. The second cause concerns the time steps used. PowerFactory uses a variable step solver despite the time step being set as fixed. This may distort the dynamic responses of the WT model over time, since a fixed step solver is desirable for comparison purposes.

In summary, the outcomes of this work include the assessment of the advantages and disadvantages of the two power system simulation tools through the analysis of their implementation processes. Furthermore, the dynamic simulation of the generic Type 3 WT model is intended to provide a better understanding of IEC 61400-27-1, assessing its limitations and the accuracy of its originally-developed dynamic WT models, thus extending their field of application.

Author Contributions: Conceptualization, R.V.-R., A.L.-B. and E.G.-L.; Investigation, R.V.-R. and A.L.-B.; Methodology, R.V.-R., A.L.-B. and A.H.-E.; Project administration, E.G.-L.; Resources, F.J.-B.; Supervision, A.H.-E., F.J.-B. and E.G.-L.; Writing—Original draft, R.V.-R.; Writing—Review and editing, A.H.-E.

Funding: This research was funded by the Spanish Ministry of Economy and Competitiveness and European Union FEDER, grant number ENE2016-78214-C2-1-R, as well as the Agreement signed between the UCLM and the Council of Albacete to foster Research in the Campus of Albacete.

Acknowledgments: The authors would also like to express their gratitude to the wind turbine manufacturer Siemens Gamesa Renewable Energy for the technical support received. Furthermore, the authors would like to thank the TC 88/WG 27 Committee "Wind turbines—Electrical simulation models for wind power generation", to his Convenor, Poul Sørensen, and to the 16 participating countries.

Conflicts of Interest: The authors declare no conflict of interest.

Abbreviations

The following abbreviations are used in this manuscript:

AC	Alternating Current
DC	Direct Current
DFIG	Doubly-Fed Induction Generator
DSL	DIgSILENT Simulation Language
DSO	Distribution System Operator
GWEC	Global Wind Energy Council
IEA	International Energy Agency
IEC	International Electrotecnical Commission
MAE	Mean Absolute Error
ME	Mean Error
MXE	Maximum Absolute Error
PF	DIgSILENT-PowerFactory
PV	Photovoltaics
RMS	Root Mean Square
TSO	Transmission System Operator
WPP	Wind Power Plant
WT	Wind Turbine
WTT	Wind Turbine Terminals

References

1. IEA. *Renewables 2018. Analysis and Forecasts for 2023*; Technical Report; International Energy Agency: Paris, France, 2018.
2. GWEC. *Global Wind Statistics 2017*; Technical Report; Global Wind Energy Council: Brussels, Belgium, 2018.
3. EWEA. *Wind Energy in Europe in 2018. Trends and Statistics*; Technical Report; European Wind Energy Association: Brussels, Belgium, 2018.
4. EWEA. *Wind Energy in Europe: Scenarios for 2030*; Technical Report; European Wind Energy Association: Brussels, Belgium, 2017.
5. Honrubia-Escribano, A.; Gomez-Lazaro, E.; Fortmann, J.; Sørensen, P.; Martin-Martinez, S. Generic dynamic wind turbine models for power system stability analysis: A comprehensive review. *Renew. Sustain. Energy Rev.* **2018**, *81*, 1939–1952. [CrossRef]
6. Fortmann, J.; Miller, N.; Kazachkov, Y.; Bech, J.; Andresen, B.; Pourbeik, P.; Sørensen, P. Wind Plant Models in IEC 61400-27-2 and WECC-latest developments in international standards on wind turbine and wind plant modeling. In Proceedings of the 14th International Workshop on Large-Scale Integration of Wind Power into Power Systems as well as on Transmission Networks for Offshore Wind Power Plants, Brussels, Belgium, 20–22 October 2015; p. 5.
7. Villena-Ruiz, R.; Jiménez-Buendía, F.; Honrubia-Escribano, A.; Molina-García, Á.; Gómez-Lázaro, E. Compliance of a Generic Type 3 WT Model with the Spanish Grid Code. *Energies* **2019**, *12*, 1631. [CrossRef]
8. IEC 61400-27-1. *Electrical Simulation Models-Wind Turbines*; IEC: Geneva, Switzerland, 2015.

9. Sørensen, P.; Andersen, B.; Fortmann, J.; Johansen, K.; Pourbeik, P. Overview, Status and Outline of the New IEC 61400-27—Electrical Simulation Models for Wind Power Generation. In Proceedings of the 10th International Workshop on Large-Scale Integration of Wind Power into Power Systems as well as on Transmission Networks for Offshore Wind Power Farms, Roskilde, Denmark, 25–26 October 2011; p. 6.

10. Fortmann, J. Modeling of Wind Turbines with Doubly Fed Generator System. Ph.D. Thesis, Department for Electrical Power Systems, University of Duisburg-Essen, Duisburg, Germany, 2014.

11. Timbus, A.; Korba, P.; Vilhunen, A.; Pepe, G.; Seman, S.; Niiranen, J. Simplified Model of Wind Turbines with Doubly-Fed Induction Generator. In Proceedings of the 10th International Workshop on Large-Scale Integration of Wind Power into Power Systems as well as on Transmission Networks for Offshore Wind Power Farms, Aarhus, Denmark, 18 October 2011; p. 6.

12. Honrubia-Escribano, A.; Jiménez-Buendía, F.; Gómez-Lázaro, E.; Fortmann, J. Field Validation of a Standard Type 3 Wind Turbine Model for Power System Stability, According to the Requirements Imposed by IEC 61400-27-1. *IEEE Trans. Energy Convers.* **2018**, *33*, 137–145. [CrossRef]

13. Seman, S.; Niiranen, J.; Virtanen, R.; Matsinen, J.P. Low voltage ride-through analysis of 2 MW DFIG wind turbine—Grid code compliance validations. In Proceedings of the 2008 IEEE Power and Energy Society General Meeting-Conversion and Delivery of Electrical Energy in the 21st Century, Pittsburgh, PA, USA, 20–24 July 2008; pp. 1–6. [CrossRef]

14. Lorenzo-Bonache, A.; Honrubia-Escribano, A.; Jimenez, F.; Gomez-Lazaro, E. Field Validation of Generic Type 4 Wind Turbine Models Based on IEC and WECC Guidelines. *IEEE Trans. Energy Convers.* **2018**, *34*, 933–941. [CrossRef]

15. Lorenzo-Bonache, A.; Villena-Ruiz, R.; Honrubia-Escribano, A.; Molina-García, A.; Gómez-Lázaro, E. Comparison of a standard type 3B WT model with a commercial build-in model. In Proceedings of the 2017 IEEE International Electric Machines and Drives Conference (IEMDC), Miami, FL, USA, 21–24 May 2017; pp. 1–6.

16. Lorenzo-Bonache, A.; Villena-Ruiz, R.; Honrubia-Escribano, A.; Gómez-Lázaro, E. Operation of active and reactive control systems of a generic Type 3 WT model. In Proceedings of the 11th IEEE International Conference on Compatibility, Power Electronics and Power Engineering (CPE-POWERENG), Cadiz, Spain, 4–6 April 2017; pp. 606–610.

17. Lorenzo-Bonache, A.; Honrubia-Escribano, A.; Jiménez-Buendía, F.; Molina-García, Á.; Gómez-Lázaro, E. Generic Type 3 Wind Turbine Model Based on IEC 61400-27-1: Parameter Analysis and Transient Response under Voltage Dips. *Energies* **2017**, *10*, 1441. [CrossRef]

18. Akhmatov, V.; Andresen, B.; Nielsen, J.N.; Jensen, K.H.; Goldenbaum, N.M.; Thisted, J.; Frydensbjerg, M. Unbalanced Short-Circuit Faults: Siemens Wind Power Full Scale Converter Interfaced Wind Turbine Model and Certified Fault-Ride-Through Validation. In Proceedings of the 2010 European Wind Energy Conference and Exhibition, Warsaw, Poland, 20–23 April 2010; p. 9.

19. Trilla, L.; Gomis-Bellmunt, O.; Junyent-Ferre, A.; Mata, M.; Sánchez Navarro, J.; Sudria-Andreu, A. Modeling and Validation of DFIG 3-MW Wind Turbine Using Field Test Data of Balanced and Unbalanced Voltage Sags. *IEEE Trans. Sustain. Energy* **2011**, *2*, 509–519. [CrossRef]

20. Chang, Y.; Hu, J.; Tang, W.; Song, G. Fault Current Analysis of Type-3 WTs Considering Sequential Switching of Internal Control and Protection Circuits in Multi Time Scales during LVRT. *IEEE Trans. Power Syst.* **2018**, *33*, 6894–6903. [CrossRef]

21. Goksu, O.; Altin, M.; Fortmann, J.; Sørensen, P. Field Validation of IEC 61400-27-1 Wind Generation Type 3 Model with Plant Power Factor Controller. *IEEE Trans. Energy Convers.* **2016**, *31*, 1170–1178. [CrossRef]

22. Fortmann, J. Generic aerodynamic model for simulation of variable speed wind turbines. In Proceedings of the 9th International Workshop on Large-Scale Integration of Wind Power into Power Systems as well as on Transmission Networks for Offshore Wind Power Plants, Quebec City, QC, Canada, 18–19 October 2010; p. 7.

23. Zhang, J.; Cheng, M.; Chen, Z.; Fu, X. Pitch angle control for variable speed wind turbines. In Proceedings of the Third International Conference on Electric Utility Deregulation and Restructuring and Power Technologies, Nanjing, China, 6–9 April 2008; pp. 2691–2696.

24. Buendia, F.J.; Vigueras-Rodriguez, A.; Gomez-Lazaro, E.; Fuentes, J.A.; Molina-Garcia, A. Validation of a Mechanical Model for Fault Ride-Through: Application to a Gamesa G52 Commercial Wind Turbine. *IEEE Trans. Energy Convers.* **2013**, *28*, 707–715. [CrossRef]

25. Buendía, F.J.; Gordo, B.B. Generic Simplified Simulation Model for DFIG with Active Crowbar. In Proceedings of the 11th International Workshop on Large-Scale Integration of Wind Power into Power Systems, Lisbon, Protugal, 13–15 November 2012.

26. Salles, M.B.C.; Hameyer, K.; Cardoso, J.R.; Grilo, A.P.; Rahmann, C. Crowbar System in Doubly Fed Induction Wind Generators. *Energies* **2010**, *3*, 738–753. [CrossRef]

27. Subramanian, C.; Casadei, D.; Tani, A.; Sørensen, P.; Blaabjerg, F.; McKeever, P. Implementation of electrical simulation model for IEC standard Type-3A generator. In Proceedings of the 2013 European Modelling Symposium, Manchester, UK, 20–22 November 2013; pp. 426–431.

28. Jimenez, F.; Gómez-Lázaro, E.; Fuentes, J.A.; Molina-García, A.; Vigueras-Rodríguez, A. Validation of a double fed induction generator wind turbine model and wind farm verification following the Spanish grid code. *Wind Energy* **2012**, *15*, 645–659. [CrossRef]

29. Honrubia-Escribano, A.; Gómez-Lázaro, E.; Molina-Garcia, A.; Fuentes, J. Influence of voltage dips on industrial equipment: analysis and assessment. *Int. J. Electr. Power Energy Syst.* **2012**, *41*, 87–95. [CrossRef]

30. Jiménez, F.; Gómez-Lázaro, E.; Fuentes, J.A.; Molina-García, A.; Vigueras-Rodríguez, A. Validation of a DFIG wind turbine model submitted to two-phase voltage dips following the Spanish grid code. *Renew. Energy* **2013**, *57*, 27–34. [CrossRef]

31. Lorenzo-Bonache, A.; Honrubia-Escribano, A.; Fortmann, J.; Gómez-Lázaro, E. Generic Type 3 WT models: comparison between IEC and WECC approaches. *IET Renew. Power Gener.* **2019**, *3*, 1168–1178 [CrossRef]

32. WECC REMTF. *WECC Wind Power Plant Dynamic Modeling Guidelines*; Technical Report; WECC: Salt Lake City, UT, USA, 2014.

33. Gonzalez-Longatt, F.M.; Rueda, J.L. *PowerFactory Applications for Power System Analysis*; Springer: Cham, Switzerland, 2014.

34. GmbH, D. *DIgSILENT PowerFactory V18–User Manual*; DIgSILENT GmbH: Gomaringen, Germany, 2018.

35. Villena-Ruiz, R.; Lorenzo-Bonache, A.; Honrubia-Escribano, A.; Gómez-Lázaro, E. Implementation of a generic Type 1 wind turbine generator for power system stability studies. In Proceedings of the International Conference on Renewable Energies and Power Quality, Malaga, Spain, 4–6 April 2017; pp. 4–6.

36. Hansen, A.; Sørensen, P.; Iov, F.; Blaabjerg, F. Initialisation of Grid-Connected Wind Turbine Models in Power-System Simulations. *Wind Eng.* **2003**, *27*, 21–38. [CrossRef]

37. Hansen, A.D.; Jauch, C.; Sørensen, P.; Iov, F.; Blaabjerg, F. *Dynamic Wind Turbine Models in Power System Simulation Tool DIgSILENT*; Technical Report; Riso-DTU National Laboratory: Roskilde, Denmark, 2007.

38. Asmine, M.; Brochu, J.; Fortmann, J.; Gagnon, R.; Kazachkov, Y.; Langlois, C.E.; Larose, C.; Muljadi, E.; MacDowell, J.; Pourbeik, P.; et al. Model validation for wind turbine generator models. *IEEE Trans. Power Syst.* **2011**, *26*, 1769–1782. [CrossRef]

Article

Fault-Ride Trough Validation of IEC 61400-27-1 Type 3 and Type 4 Models of Different Wind Turbine Manufacturers

Andrés Honrubia-Escribano [1,*], Francisco Jiménez-Buendía [2], Jorge Luis Sosa-Avendaño [2], Pascal Gartmann [3], Sebastian Frahm [4], Jens Fortmann [5], Poul Ejnar Sørensen [6] and Emilio Gómez-Lázaro [1]

[1] Renewable Energy Research Institute and DIEEAC-ETSII-AB, Universidad de Castilla-La Mancha, 02071 Albacete, Spain
[2] Siemens Gamesa Renewable Energy, S.A., 31621 Pamplona, Spain
[3] WRD Wobben Research and Development GmbH, D-26607 Aurich, Germany
[4] Senvion GmbH, Überseering 10, 22297 Hamburg, Germany
[5] HTW Berlin-University of Applied Sciences, 12459 Berlin, Germany
[6] Wind Energy Systems, Department of Wind Energy, Technical University of Denmark, 4000 Roskilde, Denmark
* Correspondence: andres.honrubia@uclm.es; Tel.: +34-967-599-200 (ext. 8216)

Received: 19 June 2019; Accepted: 2 August 2019; Published: 7 August 2019

Abstract: The participation of wind power in the energy mix of current power systems is progressively increasing, with variable-speed wind turbines being the leading technology in recent years. In this line, dynamic models of wind turbines able to emulate their response against grid disturbances, such as voltage dips, are required. To address this issue, the International Electronic Commission (IEC) 61400-27-1, published in 2015, defined four generic models of wind turbines for transient stability analysis. To achieve a widespread use of these generic wind turbine models, validations with field data are required. This paper performs the validation of three generic IEC 61400-27-1 variable-speed wind turbine model topologies (type 3A, type 3B and type 4A). The validation is implemented by comparing simulation results with voltage dip measurements performed on six different commercial wind turbines based on field campaigns conducted by three wind turbine manufacturers. Both IEC validation approaches, the play-back and the full system simulation, were implemented. The results show that the generic full-scale converter topology is accurately adjusted to the different real wind turbines and, hence, manufacturers are encouraged to the develop generic IEC models.

Keywords: DFIG; field testing; full-scale converter; generic model; IEC 61400-27-1; validation

---

1. Introduction

Wind energy emerged as the most promising renewable energy source (RES) in the world over the past few years. Since 2014, annual wind power installations have surpassed 50 GW each year on a global scale, bringing the total cumulative capacity up to 591 GW at the end of 2018 [1]. China is leading the global market with 206 GW of installed capacity, followed by the US (127 GW) and several EU countries. With a total installed capacity of 179 GW in the EU at the end of 2018, wind power had installed more capacity than any other type of electricity generation in the EU in that year [2], positioning itself as the second largest type of power generation capacity in the region.

In addition to the installed capacity, wind power plays a key role in electricity demand coverage. In the EU, wind power met 14% of the electricity demand in 2018 [2], which is 2% higher than in 2017. Denmark presents the highest share of wind energy in its electricity demand (41%) in the EU. Ireland,

Portugal, Germany and Spain also exhibited a considerable contribution of wind power to demand coverage in 2018: 28%, 24%, 21% and 19%, respectively.

Network operators, either transmission system operators (TSOs) or distribution system operators (DSOs), perform transient stability analysis to correctly integrate the increasing penetration of wind power into the energy mix of current power systems. Dynamic wind turbine (WT) simulation models are required for this purpose [3]. However, in contrast to traditional synchronous generators, most WT models are not standardized or validated [4]. In this sense, the models developed by WT manufacturers are able to reproduce the behavior of their WTs with the greatest accuracy [5]. Nevertheless, the use of WT vendor models for transient stability analysis presents the following challenges: (i) they require specific simulation software [6], (ii) each vendor model is commonly subject to a non-disclosure agreement [7], (iii) each WT has specific controls depending on the manufacturer [8], (iv) increased accuracy is provided at the expense of increased complexity and number of parameters and, as a consequence, high computation time [9].

In light of the above considerations, the International Electrotechnical Commission published the Standard International Electronic Commission (IEC) 61400-27-1 in February 2015 [10]. IEC 61400-27-1 defined four generic WT models to conduct dynamic simulations of power system disturbances such as short-circuits. These generic models, also known as standard or simplified models, involve several assumptions and have several key properties, as follows:

- They are public [11].
- They are independent of the software simulation tool used [12].
- They should be easily parameterized to emulate particular responses from any WT vendor available in the market.
- They are intended for fundamental frequency positive sequence response [13]. Hence, they can be used for balanced short-circuits, i.e., three-phase symmetrical faults.
- Wind speed is assumed to be constant over the simulation. This assumption is acceptable because generic WT models use simulation time steps in the range of 1 ms and 10 ms and the total simulation time is between 10 s and 30 s [14], with both of these conditions being common features for transient stability analysis [15].

Under this framework, the present paper performs the validation of six generic WT models based on the guidelines imposed by IEC 61400-27-1. For the first time in the literature, field campaigns conducted by three WT manufacturers, Siemens–Gamesa, Senvion and ENERCON, are used for the validation of three different WT technologies. Specifically, the variable-speed WT topologies, i.e., the doubly-fed induction generator (DFIG) and the full-scale converter, which represent the largest market share in current power systems, were submitted to voltage dips of different magnitude and duration. The validation methodology defined by the IEC 61400-27-1 was implemented to evaluate the accuracy of the generic WT models.

Following this introduction, the rest of the paper is structured as follows: Section 2 provides an overview of the current state of the art regarding variable-speed WTs, where the lack of field validation works is highlighted. Section 3 describes the methodology and testing procedure implemented in the present work, the results of which are provided in Section 4. Finally, Section 5 summarizes the main conclusions of the paper.

2. Overview of Generic Variable-Speed WTs and Previous Field Validation Works

MW-range WTs may be operated in two different ways: either fixed-speed or variable-speed operation. Fixed rotor speed is the oldest WT technology [15], while variable-speed is the most advanced technology and hence the current choice for every WT manufacturer [16]. Two different WT topologies are identified as variable-speed operation, Figure 1: the DFIG, also known as type 3 (Figure 1a), and the full-scale converter, also known as type 4 (Figure 1b).

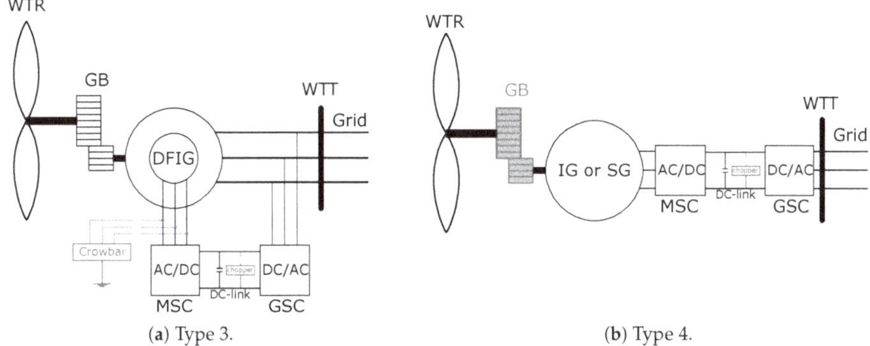

(**a**) Type 3. (**b**) Type 4.

Figure 1. Diagrams of the variable-speed wind turbines (WTs).

As shown in Figure 1, both of these variable-speed WT technologies include a bi-directional AC-DC/DC-AC converter. The main difference between them is the converter rated power: the converter is rated to 25–30% of the WT rated power for type 3 [17]; while the converter evacuates all the energy produced by the generator (either an induction generator, IG, or a synchronous generator, SG) in type 4. Hence, the type 4 generator is completely decoupled from the grid by the converter [18]. The power converter is composed of a machine (or rotor) side converter (MSC), a grid side converter (GSC) and the dc-link. Depending on the fault-ride trough (FRT) capability of the WT, the generic type 3 WT model is divided into two subtypes [10]: type 3A for WTs where the MSC and the chopper are sufficiently dimensioned for FRT without disconnecting the converter; and the type 3B, which is equipped with a crowbar device connected to the MSC in order to short-circuit the rotor when over-currents and over-voltages voltages are detected [19]. In fact, the type 3B WT is transformed into an induction generator with a rotor-connected resistance during crowbar activation [20]. In a similar way, two subtypes are also defined for the generic type 4 WT model: type 4A, which omits the aerodynamic and mechanical components due to the addition of a chopper in the dc-link; and type 4B, where choppers are not included and hence post-fault power oscillations are present.

Due to the complex behavior of variable-speed WTs, and taking into account the particular features of the generic IEC WT models listed in Section 1, there is little previous literature on the validation of these models. Two of the first contributions are found in [20,21], where both generic type 3 models, type 3B and type 3A, respectively, were validated with a 2 MW based WT operating at full-load conditions against one voltage dip test case. A generic type 4B model was validated against one voltage dip in [22], where the post-fault power oscillations were clearly observed. Generic type 3B and type 4A models, both from the same vendor, were validated in [5,8,12] based on the field results obtained from several test cases. It should be noted that the authors of the present work collaborated in most of the previously cited contributions, as well as being members of Working Group 27 of the IEC Technical Committee 88 in charge of the development of IEC 61400-27.

Under this framework, it can be clearly observed that the field validation of generic WT models is a current topic of interest in the wind power industry. Nevertheless, the number of contributions found in the literature is limited. Furthermore, there is a lack of contributions with the involvement of several WT manufacturers and this is the gap the present paper aims to fill. Since each WT vendor has specific controls, the FRT response of each actual WT is different. Hence, the validation of several WT topologies provided by different manufacturers is the key contribution of the present paper.

3. Description of the Validation Methodology and Testing Procedure

Validating a model consists of comparing the emulated response with the measured data from field tests, both referring to the same wind turbine terminals (WTT). According to IEC 61400-27-1 [10], the measured and simulated data should be represented in per unit (pu) values based on the nominal active power and the nominal voltage at the WTT. The results of the validation procedure will include the following parameters:

- Time series of the measured and simulated fundamental frequency positive sequence parameters, such as voltage (u), active power (p) and reactive power (q).
- Error time series for the previous parameters, x_E where x represents the specific parameter to be validated (u, p or q), which are defined by the difference between measured field data (x_{field}) and simulated data (x_{sim}), Equation (1).

$$x_E(n) = x_{field}(n) - x_{sim}(n) \tag{1}$$

- Three key validation errors are estimated for the previous error time series: mean error (x_{ME}), Equation (2), mean absolute error (x_{MAE}), Equation (3) and maximum absolute error (x_{MXE}), Equation (4).

$$x_{ME} = \frac{\sum_{n=1}^{N} x_E(n)}{N} \tag{2}$$

$$x_{MAE} = \frac{\sum_{n=1}^{N} |x_E(n)|}{N} \tag{3}$$

$$x_{MXE} = max\big(|x_E(1)|, |x_E(2)|, ..., |x_E(N)|\big) \tag{4}$$

Three different fault windows (W) are considered for the estimation of each key validation error, as represented with different colors in Figure 2: (i) a pre-fault window lasting 1000 ms before the fault occurs at t_{fault} (this is the first time the voltage dip occurs in one of the phases); (ii) a fault-window that covers a time period from t_{fault} to the fault clearance, t_{clear}; (iii) a post-fault window lasting 5000 ms after t_{clear}. As observed in Figure 2, two quasi-steady state (QS) sub-windows were defined during both fault and post-fault periods. These QS sub-windows are used to avoid a misunderstanding of the validation errors due to electromagnetic transients that could appear in the field but are outside the scope of root mean square (RMS) simulations. The calculation of the final validation errors at each window is summarized in Table 1.

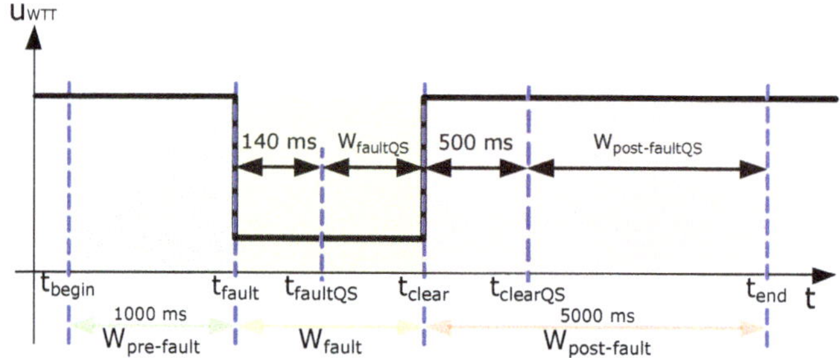

Figure 2. Voltage dip validation windows.

Table 1. Windows used for the estimation of the validation errors.

Error	Pre-Fault	Fault	Post-Fault
x_{ME}	$W_{pre\text{-}fault}$	W_{fault}	$W_{post\text{-}fault}$
x_{MAE}	$W_{pre\text{-}fault}$	$W_{faultQS}$	$W_{post\text{-}fault}$
x_{MXE}	$W_{pre\text{-}fault}$	$W_{faultQS}$	$W_{post\text{-}faultQS}$

Furthermore, the validation methodology defined by IEC 61400-27-1 includes two different approaches to represent the grid model. On the one hand, the full system simulation approach considers the modeling of the whole system composed of the equivalent grid, the interface between the grid and the WT and the generic WT model [23]. On the other hand, the play-back approach involves only the WT being modeled and the measured voltage being directly played-back into the generic WT model. Therefore, the play-back validation methodology is recommended for assessing the accuracy of the generic WT model as the uncertainties related to grid and test equipment models are reduced.

FRT mobile test units were used to perform the field tests and measurements of the actual WTs. Figure 3 shows several photos of the different field campaigns carried out by the manufacturers involved in the present work: Siemens–Gamesa (Figure 3a), Senvion (Figure 3b) and ENERCON (Figure 3c), to perform the field tests used for the validation of the generic IEC WT models.

(**a**) Siemens–Gamesa. (**b**) Senvion. (**c**) ENERCON©.

Figure 3. Photos of the field campaigns carried out by the WT manufacturers.

The validation methodology previously described, as well as the FRT mobile units, were used to perform six different field tests for the validation of the generic WT models, as shown in Table 2. Three different WT topologies (Type 3A, Type 3B and Type 4A) from three WT manufacturers were considered. Siemens–Gamesa implemented the play-back validation methodology, while Senvion and ENERCON deployed the full system simulation approach. A wide range of voltage dip characteristics (residual voltage and dip duration) were also considered. It should be noted that the residual voltage shown in Table 2 is based on the measurement guidelines defined by IEC 61400-21-1 [24]. This means that the field test is defined by a voltage dip without a WT and, subsequently, when the WT is connected and the test is performed, the final residual voltage may increase due to the actual reactive current injection.

Table 2. Description of the validation tests performed.

Test ID	WT Type	WT Capacity	WT Manufacturer	Validation Methodology	WT Load, p (in pu)	Residual Voltage, u (in pu)	Dip Duration, t (in ms)
1	3A	3.46 MW	*Siemens-Gamesa*	play-back	1.00	0.20	550
2	3B	2 MW			0.45	0.35	723
3	3A	2 MW	*Senvion*	full system	0.97	0.23	980
4	4A	3.4 MW			1.02	0.50	500
5	4A	2 MW	*ENERCON*	full system	0.98	0.25	1520
6	4A	6 MW			0.21	0.75	3000

4. Results

This section, which presents the main results obtained in this work, is divided into two subsections in order to differentiate between type 3 and type 4 validation test cases. For each field test shown in Table 2, results will be addressed in two different ways: (i) three figures with the time series of the three measured (in black) and simulated (in blue) key parameters (u, p and q), as well as the error time series (in red); (ii) one table summarizing the three validation errors (x_{ME}, x_{MAE} and x_{MXE}) at each validation window.

4.1. Type 3 WT Validation Test Cases

This subsection discusses the validation results for the DFIG field tests: test ID 1, test ID 2 and test ID 3.

4.1.1. Test ID 1

Figure 4 shows the results of test ID 1, which was performed on a DFIG WT with a dc-link chopper as active protection device, i.e., a type 3A WT. The measured voltage profile shown in Figure 4a was obtained through the connection of a series impedance at the FRT mobile test unit before the measurement starts, which is disconnected at $t = 4.12$ s.

Regarding the active power response, Figure 4b, a considerable constant deviation is observed between field and simulation when the fault was cleared ($t_{clear} = 2.05$ s) and the active power recovery ramp has finished. This deviation is due to the far greater complexity found in the pitch model and torque controller in the actual WT compared to the simplified generic IEC WT model. Therefore, a significant validation error was found for the average value during the post-fault period, as observed in Table 3, $p_{ME} = p_{MAE} = 0.09$ pu. This active power oscillation also occurs because the drive-train model of the real WT is more complex than the two-mass model considered for the generic WT model. Nevertheless, it can be observed that the oscillation frequency fits properly.

Regarding the reactive power response, Figure 4c, the IEC generic WT model generally emulates the behavior of the actual WT with great accuracy. However, a negative reactive power peak appears in the field at the fault clearance due to the transformer inrush current, which is a non-linear effect that cannot be properly represented by transformer RMS models. Therefore, as observed in Table 3, mean reactive power errors are considerably low (≤ 0.01 pu), while the maximum error is large (0.15 pu) due to the disconnection of the series impedance of the FRT test unit.

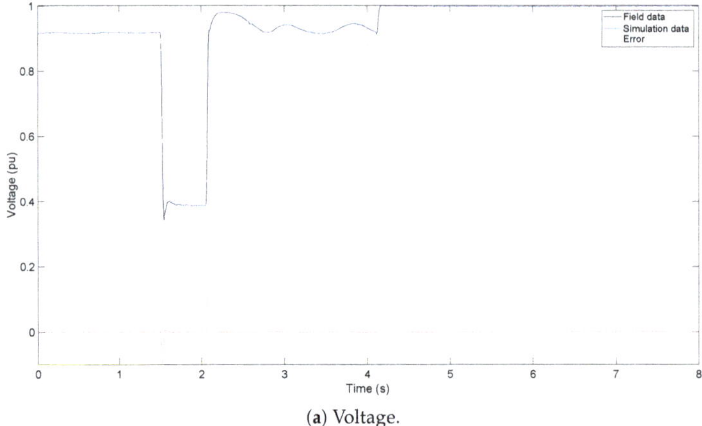

(a) Voltage.

Figure 4. *Cont.*

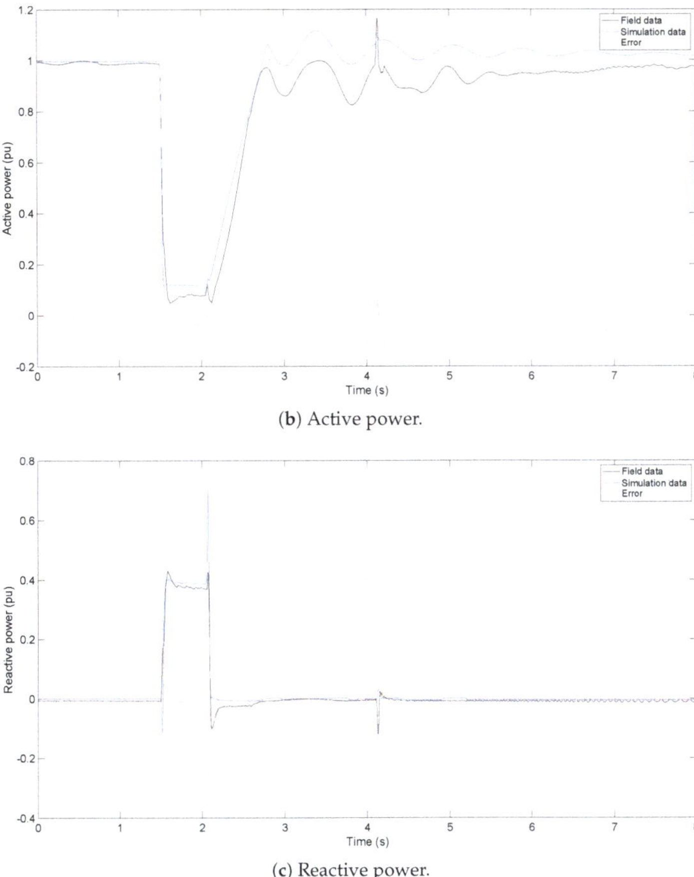

(**b**) Active power.

(**c**) Reactive power.

Figure 4. Test ID 1 results.

Table 3. Validation errors for test ID 1, in pu.

Error	Pre-Fault		Fault		Post-Fault	
	p	q	p	q	p	q
ME	0.01	0.01	0.02	0.00	0.09	0.01
MAE	0.01	0.01	0.04	0.01	0.09	0.01
MXE	0.01	0.01	0.07	0.03	0.16	0.15

4.1.2. Test ID 2

Test ID 2 presents the second field case performed by the vendor Siemens–Gamesa. Figure 5 shows the results of this test, while Table 4 provides the calculation of the validation errors. It is worth noting that the voltage dip characteristics of test ID 2 were quite similar to those of test ID 1, with the main difference being the loading condition of the WT and the WT topology.

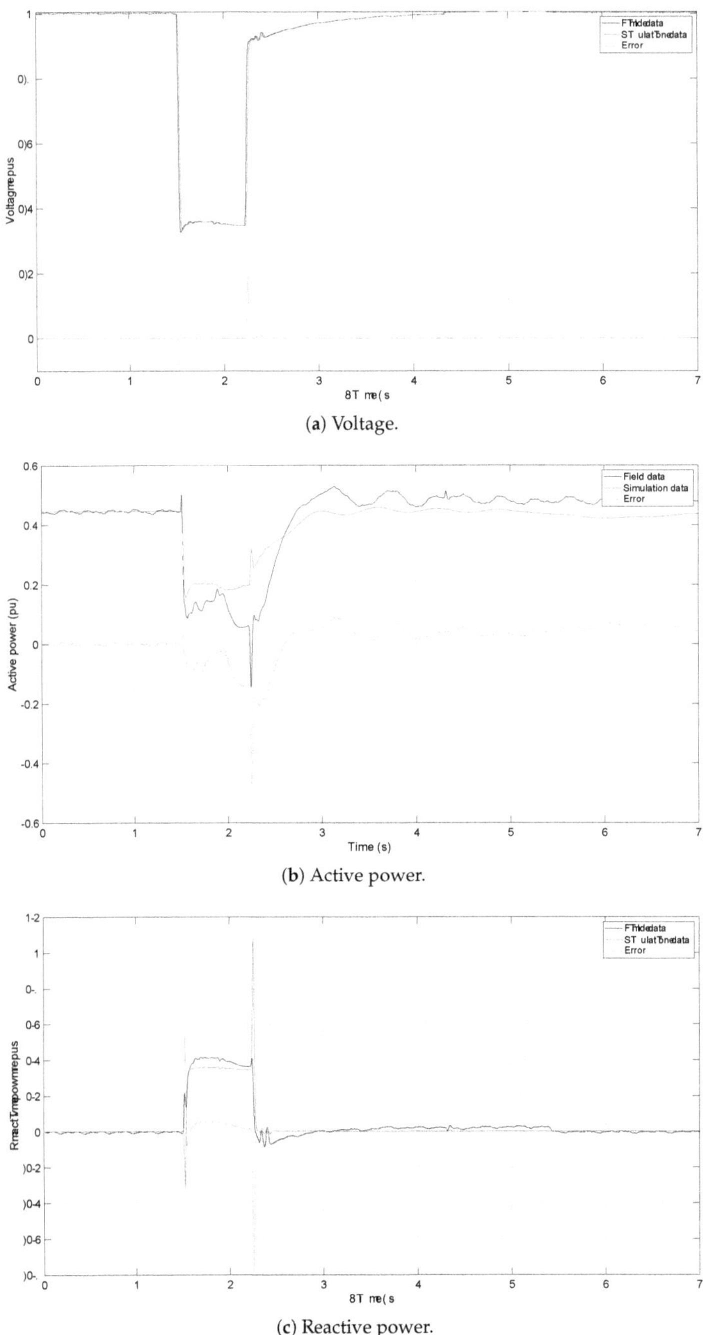

(**a**) Voltage.

(**b**) Active power.

(**c**) Reactive power.

Figure 5. Test ID 2 results.

Table 4. Validation errors for test ID 2, in pu.

Error	Pre-Fault		Fault		Post-Fault	
	p	q	p	q	p	q
ME	0.00	0.00	0.07	0.04	0.03	0.00
MAE	0.00	0.00	0.08	0.04	0.05	0.02
MXE	0.01	0.01	0.14	0.06	0.09	0.03

Firstly, regarding the voltage profile, Figure 5a, the identical response between simulation and field, as also found in Figure 4a, should be highlighted. This is due to the implementation of the play-back validation approach.

The active power response, Figure 5b, shows a slight deviation between field and simulation due to the difficulties of representing exactly the same active power delivery during the voltage dip. This is because the actual WT integrates a particular active power limitation algorithm that cannot be represented with the generic WT model. In addition, the power dynamic in the actual WT during the voltage dip is slower than in continuous operation and this different dynamic behavior cannot be represented by the IEC model. This also has an impact on the ramp-up once the fault is cleared. Furthermore, the real WT absorbs active power at the fault clearance ($t_{clear} = 2.22$ s), which cannot be properly emulated by the generic WT model due to the simplification in terms of transients. Therefore, the active power validation errors have a significant value during both fault and post-fault periods, as shown in Table 4.

Furthermore, reactive power validation errors also present a larger value in comparison to test ID 1, which is directly related to the crowbar dynamics, as observed in Figure 5c at both fault inception and fault clearance.

4.1.3. Test ID 3

Figure 6 shows the results of the last field test performed on a DFIG WT. Specifically, this WT is a Senvion MM series WT that implements the same IEC model type as that used for test ID 1, i.e., Type 3A. In addition, both the WT loading condition and the residual voltage were almost identical. Table 5 summarizes the validation of test ID 3.

In this field test, the grid was modeled by the full system simulation approach of IEC 61400-27-1, as commented in Section 3. Figure 6a compares both the measured and the simulated voltage at the wind turbine terminals, where the three-phase voltage dip occurs at $t = 1.05$ s. Due to the reactive current infeed of the WT during the dip, the voltage level rises. In contrast to tests ID 1 and ID 2, a small hysteresis was observed in the measured voltage at voltage dip clearance ($t_{clear} = 2.02$ s), which cannot be represented by the simulated voltage due to the lack of hysteresis in the transformer model.

The eigenfrequency between active power measurement and simulation shown in Figure 6b is quite similar during both the fault and post-fault period. Specifically, the simulated response is almost identical to the measured one during the fault period, which causes a notably reduced validation error: $p_{ME} = p_{MAE} = 0.01$ pu. However, a delay was identified in the measurement at fault clearance, which is caused by the power converter operation (further details are provided in [16]). This power converter effect was not considered in the generic IEC WT model.

Table 5. Validation errors for test ID 3, in pu.

Error	Pre-Fault		Fault		Post-Fault	
	p	q	p	q	p	q
ME	0.00	0.00	0.01	0.01	0.02	0.04
MAE	0.01	0.00	0.01	0.01	0.05	0.04
MXE	0.03	0.01	0.03	0.02	0.10	0.10

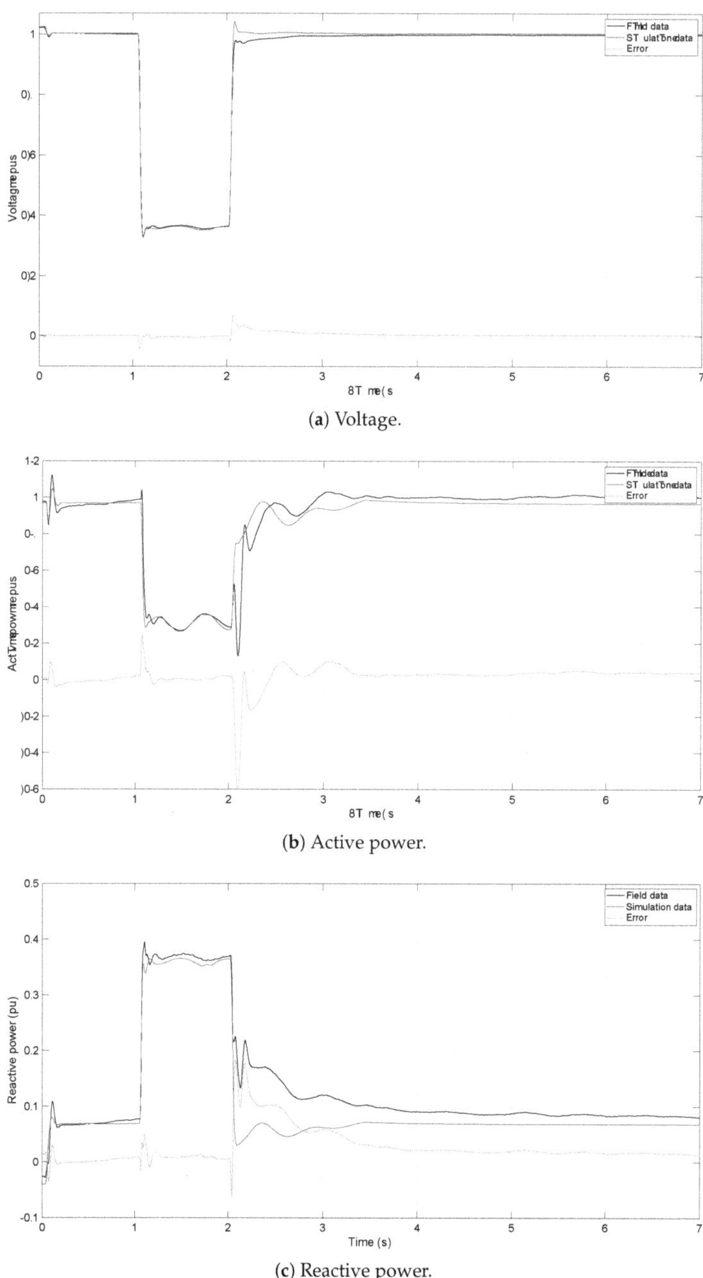

(**a**) Voltage.

(**b**) Active power.

(**c**) Reactive power.

Figure 6. Test ID 3 results.

4.2. Type 4 WT Validation Test Cases

This subsection discusses the validation results for the full-scale converter WT topology: tests ID 4, ID 5 and ID 6.

4.2.1. Test ID 4

The WT used for the test ID 4 was model *3.4M* ($P_n = 3.4$ MW), belonging to the manufacturer *Senvion*. It is represented by IEC 61400-27-1 as a Type 4A WT model. The grid was the same as that used for test ID 3, which was modeled by the full system simulation approach.

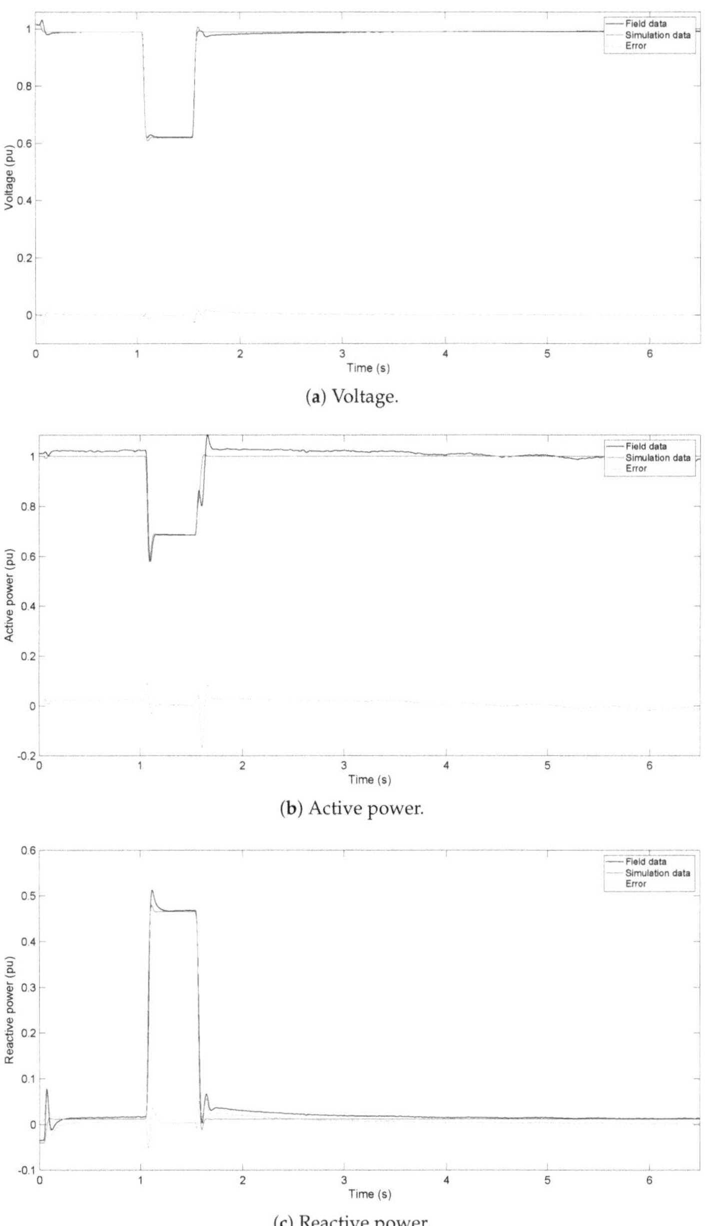

(**a**) Voltage.

(**b**) Active power.

(**c**) Reactive power.

Figure 7. Test ID 4 results.

Figure 7 shows the results of validation test ID 4. A three-phase voltage dip with a residual voltage of 0.5 pu occurred at $t_{fault} = 1.05$ s and ends at $t_{clear} = 1.54$ s, Figure 7a. As previously mentioned, this residual voltage was increased by the reactive current infeed during the voltage dip. In addition, a small hysteresis effect of the transformer is observed on the measured voltage (in black) at fault clearance. However, the hysteresis was not implemented in the transformer model and, hence, this effect cannot be reproduced by the simulation (in blue).

The measured active power, Figure 7b, fluctuates according to the wind speed variations. However, the simulated active power is set to rated power because the wind speed was assumed to be constant, according to IEC 61400-27-1, as stated in Section 1. The fit of the active power during the fault period was reasonably accurate: $p_{ME} = p_{MXE} = 0.01$ pu. However, a small deviation in the active power ramp is observed once the fault is cleared. This was caused by the simplifications included in the generic WT model, which should be neglected for power system stability studies.

Furthermore, the reactive power, Figure 7c, which was controlled according to the voltage level, also presents quite an accurate adjustment between field and simulation. There was a small deviation in the reactive power at fault clearance due to the already mentioned hysteresis effect in the measurement, which was not represented in the generic WT model. This effect was not considered in the IEC validation methodology because it was a question of transients, which were outside the scope of system stability studies, as commented in Section 2.

Finally, Table 6 summarizes the validation errors estimated for test ID 4, where it can be observed that the representation of the generic IEC type 4A model is reasonably accurate before, during and after the voltage dip.

Table 6. Validation errors for test ID 4, in pu.

Error	Pre-Fault		Fault		Post-Fault	
	p	q	p	q	p	q
ME	0.02	0.00	0.01	0.00	0.01	0.01
MAE	0.02	0.00	0.00	0.00	0.02	0.01
MXE	0.03	0.01	0.01	0.01	0.03	0.02

4.2.2. Test ID 5

Figure 8 shows the validation results for test ID 5, which are provided by an ENERCON E-82 WT model with 2 MW rated power. For the simulation case, the generic IEC type 4A model was used, implementing a full system validation approach.

Figure 8A shows a three-phase voltage dip down to 25% of the nominal voltage. The WT was operating at rated active power and zero reactive power. As can be observed, the series impedance of the FRT container was switched on at $t = 1$ s. The short-circuit impedance was switched on at $t_{fault} = 3$ s and the fault duration was 1.5 s ($t_{clear} = 4.5$ s).

Once the FRT series impedance is connected, a small deviation is found in the reactive power response, Figure 8c. This is caused by an additional voltage regulation of the actual WT, which is not represented by IEC 61400-27-1 type 4A model. When the fault occurs, the WT starts injecting reactive power according to the adjusted factor $K = 2$, as detailed in Equation (5), where I_q represents the reactive current, I_n the nominal current, U_+ the positive sequence voltage, U_n the nominal voltage and U_0 the reference voltage.

$$\Delta I_q = K \cdot I_n \cdot \frac{-\Delta U_+}{U_n} \; ; \text{with,} \; \Delta U_+ = U_+ - U_0 \qquad (5)$$

As observed in Figures 8b,c, quite an accurate fit between the simulation and the measurements was observed during the fault period. This implies that the current limitation model was well represented by the generic type 4A model. However, a transient transformer effect is shown when the fault is cleared, which cannot be accurately emulated. In summary, considerably accurate validation results were obtained for both active and reactive power, as summarized in Table 7.

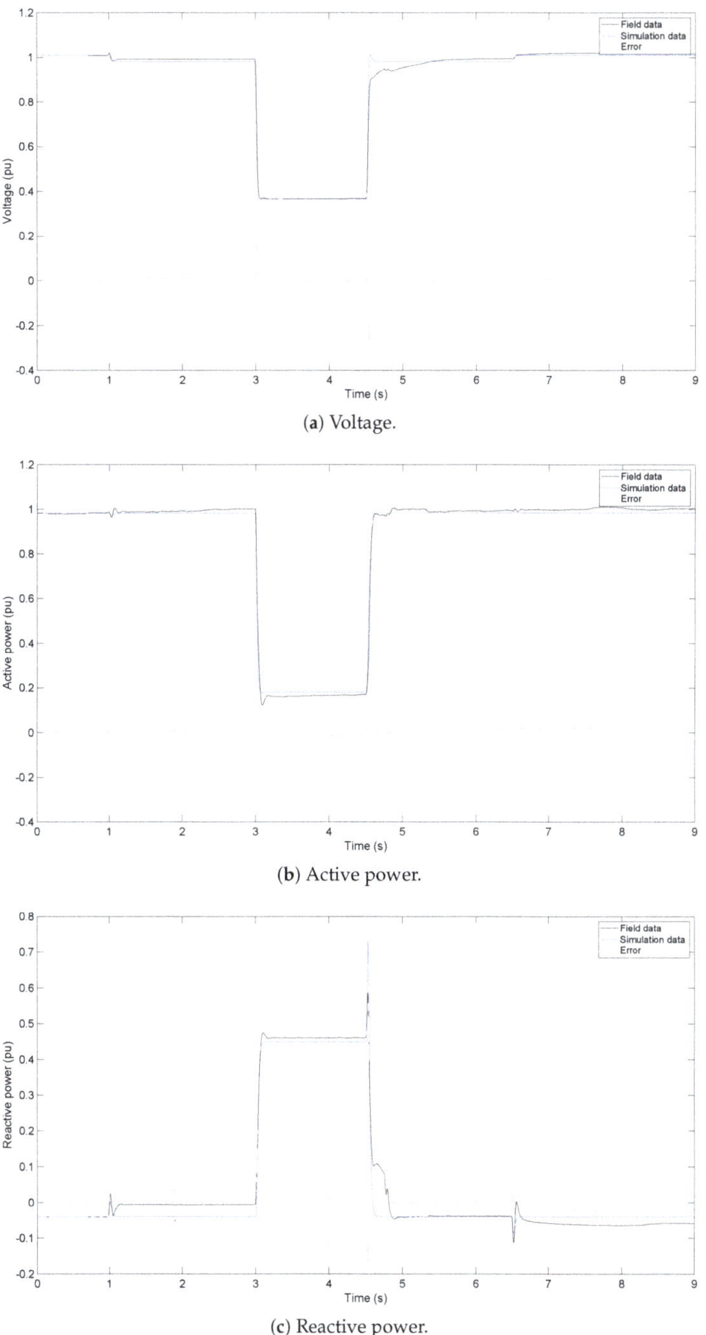

(**a**) Voltage.

(**b**) Active power.

(**c**) Reactive power.

Figure 8. Test ID 5 results.

Table 7. Validation errors for test ID 5, in pu.

Error	Pre-Fault		Fault		Post-Fault	
	p	q	p	q	p	q
ME	−0.02	−0.04	0.01	−0.01	−0.01	0.01
MAE	0.02	0.04	0.02	0.01	0.02	0.02
MXE	0.02	0.04	0.03	0.02	0.03	0.05

4.2.3. Test ID 6

Test ID 6 provides the validation results of an ENERCON E-126 WT with 6 MW rated power, as shown in Figure 9. As for test ID 5, the generic IEC type 4A model was used in a full system validation approach.

Figure 9a shows a three-phase voltage dip down to 75% of the rated voltage. The WT was operating at partial active power and zero reactive power. As observed in Figure 9a, the series impedance of the FRT container was switched on at $t = 1$ s, the short-circuit impedance is switched on at $t_{fault} = 2$ s and the fault duration is 3 s. In fact, ID 6 was the test with the longest dip duration.

Once the short-circuit occurred, quite an accurate response of both active and reactive power was observed, Figure 9b,c, respectively. As in test ID 5, the WT turbine starts injecting reactive power when the fault occurs according to the adjusted factor $K = 2$, as detailed in Equation (6), where $U_{UVRT\ or\ OVRT}$ defines an additional dead band for the reactive current calculation and the other parameters are the same as those defined for Equation (5).

$$\Delta I_q = K \cdot I_n \cdot \frac{-\Delta U_r}{U_n} \; ; \text{with,} \; \Delta U_r = \Delta U \pm (U_n - U_{UVRT\ or\ OVRT}). \tag{6}$$

A small constant deviation between the simulation and the reactive power measurements is found during the fault period, Figure 9c. In contrast, higher deviations are observed for active power, Figure 9b. These deviations were due to the current injection method of the actual WT, which was based on a regulation algorithm that includes the dc-link voltage. In fact, the dc-link part and the regulation algorithm were not represented by the generic IEC type 4A model. Once the fault is cleared, some saturation effects are observed in the measurements, which were not represented by the generic transformer model.

Finally, Table 8 summarizes the key validation errors for test ID 6. As observed, very low validation errors were found for active and reactive power during both fault and post-fault periods.

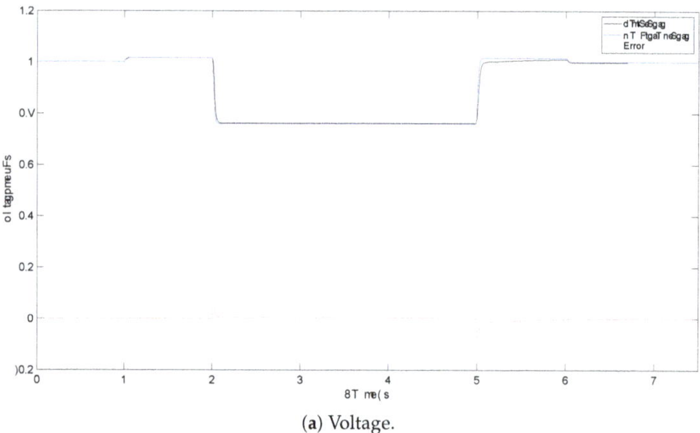

(a) Voltage.

Figure 9. *Cont.*

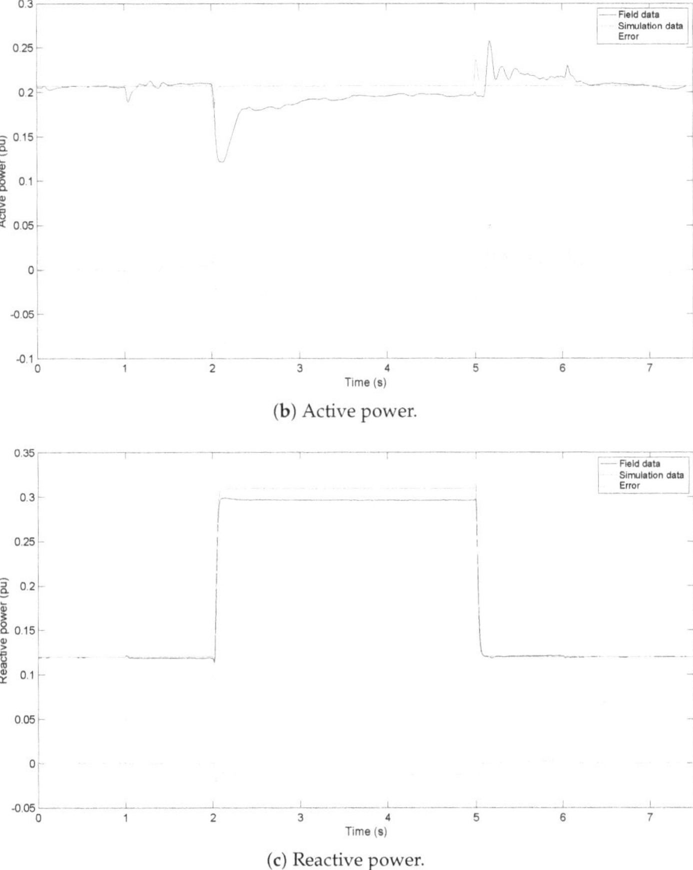

(**b**) Active power.

(c) Reactive power.

Figure 9. Test ID 6 results.

Table 8. Validation errors for test ID 6, in pu.

Error	Pre-Fault		Fault		Post-Fault	
	p	q	p	q	p	q
ME	0.00	0.00	0.02	0.01	−0.01	0.00
MAE	0.00	0.00	0.02	0.01	0.01	0.00
MXE	0.02	0.00	0.08	0.01	0.02	0.00

5. Conclusions

Given the increased penetration of wind power in the energy mix of current power systems, the need for public standard (i.e., generic) WT models to perform transient stability analysis is growing. IEC 61400-27-1, published in February 2015, defined four generic WT dynamic models able to be adapted to any particular WT vendor's commercial model. As it was published relatively recently, very few validation works with field data have been performed. Hence, TSO, DSOs, WT manufacturers and other stakeholders do not currently have evidence of the generic WTs' accurate response. The present work has validated three different generic WT topologies (type 3A, type 3B and type 4A) with six different actual variable-speed WTs from three manufacturers (Siemens–Gamesa, Senvion and ENERCON).

First, both IEC validation approaches, the play-back and the full system, were implemented, finding that when the play-back approach is used, the simulated voltage is identical to the measured voltage. In fact, when using the full system approach, the measured voltage experiences a slight hysteresis at voltage dip clearance, which cannot be represented by the simulations due to the lack of hysteresis in the transformer model.

Regarding the DFIG WT model validations, type 3B presented larger validation errors than type 3A, which is due to the crowbar protection system. In fact, the generic crowbar model implemented in type 3B is a simplification of a quite complex model.

Furthermore, the type 4 WT models provided a highly accurate response, for both active and reactive power, with respect to the three different type 4A WTs considered. In the case of the *ENERCON E-126* WT, a larger deviation between field and simulation was found, which was based on the particular representation of the current injection for this WT. In this sense, if a deeper voltage dip occurs, the active power results will be affected by the current limitation, which would yield a more accurate validation result.

Author Contributions: Conceptualization, A.H.-E., P.E.S. and E.G.-L.; methodology, A.H.-E., F.J.-B., P.G., S.F. and J.F.; software, J.L.S.-A., P.G. and S.F.; validation, A.H.-E., F.J.-B., P.G., S.F. and E.G.-L.; writing—original draft preparation, A.H.-E.; supervision, J.F., P.E.S. and E.G.-L.

Funding: This research was funded by the Spanish Ministry of Economy and Competitiveness and European Union FEDER, which supported this work under Project ENE2016-78214-C2-1-R, as well as the Agreement signed between the UCLM and the Council of Albacete to foster Research in the Campus of Albacete.

Acknowledgments: The authors would like to express their appreciation to the wind turbine manufacturers Enercon, Senvion and Siemens–Gamesa for their technical support, as well as to the rest of the members of Working Group 27 of IEC Technical Committee 88 in charge of the development of IEC 61400-27.

Conflicts of Interest: The authors declare no conflict of interest. The funders had no role in the design of the study; in the collection, analyses, or interpretation of data; in the writing of the manuscript, or in the decision to publish the results.

Abbreviations

The following abbreviations are used in this manuscript:

AC	Alternating current
DC	Direct current
DFIG	Doubly-fed induction generator
DSO	Distribution system operator
EU	European Union
FRT	Fault-ride through
GB	Gearbox
GSC	Grid side converter
IEC	International Electrotechnical Commission
IG	Induction generator (also known as asynchronous generator)
MSC	Machine (or rotor) side converter
pu	per unit
QS	quasi-steady state
RES	Renewable energy source
RMS	Root mean square
SG	Synchronous generator
TSO	Transmission system operator
US/USA	United States of America
W	Window
WECC	Western Electricity Coordinating Council
WT	Wind turbine
WTR	Wind turbine rotor
WTT	Wind turbine terminal

References

1. GWEC. *Global Wind Report. Annual Market Update 2015*; Technical Report; Global Wind Energy Council: Brussels, Belgium, 2016.
2. WindEurope. *Wind Energy in Europe in 2018*; Technical Report; WindEurope: Brussels, Belgium, 2019.
3. Trilla, L.; Gomis-Bellmunt, O.; Junyent-Ferre, A.; Mata, M.; Sánchez Navarro, J.; Sudria-Andreu, A. Modeling and Validation of DFIG 3-MW Wind Turbine Using Field Test Data of Balanced and Unbalanced Voltage Sags. *IEEE Trans. Sustain. Energy* **2011**, *2*, 509–519. doi:10.1109/TSTE.2011.2155685. [CrossRef]
4. Zhao, S.; Nair, N.K. Assessment of wind farm models from a transmission system operator perspective using field measurements. *IET Renew. Power Gener.* **2011**, *5*, 455–464. [CrossRef]
5. Honrubia-Escribano, A.; Jiménez-Buendía, F.; Gómez-Lázaro, E.; Fortmann, J. Validation of Generic Models for Variable Speed Operation Wind Turbines Following the Recent Guidelines Issued by IEC 61400-27. *Energies* **2016**, *9*, 1048. [CrossRef]
6. Lorenzo-Bonache, A.; Honrubia-Escribano, A.; Fortmann, J.; Gómez-Lázaro, E. Generic Type 3 WT models: Comparison between IEC and WECC approaches. *IET Renew. Power Gener.* **2019**, *13*, 1168–1178, doi:10.1049/iet-rpg.2018.6098. [CrossRef]
7. Zhao, H.; Wu, Q.; Margaris, I.; Bech, J.; Sørensen, P.E.; Andresen, B. Implementation and validation of IEC generic Type 1A wind turbine generator model. *Int. Trans. Electr. Energy Syst.* **2014**, *25*, 1804–1813. [CrossRef]
8. Lorenzo-Bonache, A.; Honrubia-Escribano, A.; Jiménez-Buendía, F.; Gómez-Lázaro, E. Field Validation of Generic Type 4 Wind Turbine Models Based on IEC and WECC Guidelines. *IEEE Trans. Energy Convers.* **2019**, *34*, 933–941, doi:10.1109/TEC.2018.2875167. [CrossRef]
9. Lorenzo-Bonache, A.; Honrubia-Escribano, A.; Jiménez-Buendía, F.; Molina-García, A.; Gómez-Lázaro, E. Generic Type 3 Wind Turbine Model Based on IEC 61400-27-1: Parameter Analysis and Transient Response under Voltage Dips. *Energies* **2017**, *10*, 1441. [CrossRef]
10. International Electrotechnical Commission. *IEC 61400-27-1: Electrical Simulation Models-Wind Turbines*; IEC: Geneva, Switzerland, 2015.
11. Goksu, O.; Sørensen, P.; Morales, A.; Weigel, S.; Fortmann, J.; Pourbeik, P. Compatibility of IEC 61400-27-1 Ed 1 and WECC 2nd Generation Wind Turbine Models. In Proceedings of the 15th International Workshop on Large-Scale Integration of Wind Power into Power Systems as well as on Transmission Networks for Offshore Wind Power Plants, Vienna, Austria, 15–17 November 2016.
12. Honrubia-Escribano, A.; Jiménez-Buendía, F.; Gómez-Lázaro, E.; Fortmann, J. Field Validation of a Standard Type 3 Wind Turbine Model for Power System Stability, According to the Requirements Imposed by IEC 61400-27-1. *IEEE Trans. Energy Convers.* **2018**, *33*, 137–145, doi:10.1109/TEC.2017.2737703. [CrossRef]
13. Das, K.; Hansen, A.D.; Sørensen, P.E. Understanding IEC standard wind turbine models using SimPowerSystems. *Wind Eng.* **2016**, *40*, 212–227. [CrossRef]
14. Flynn, D.; Rather, Z.; Ardal, A.; D'Arco, S.; Hansen, A.; Cutululis, N.; Sorensen, P.; Estanquiero, A.; Gómez, E.; Menemenlis, N.; et al. Technical impacts of high penetration levels of wind power on power system stability. *Wiley Interdiscip. Rev. Energy Environ.* **2017**, *6*, e216. [CrossRef]
15. Honrubia Escribano, A.; Gomez-Lazaro, E.; Fortmann, J.; Sørensen, P.; Martín-Martínez, S. Generic dynamic wind turbine models for power system stability analysis: A comprehensive review. *Renew. Sustain. Energy Rev.* **2018**, *81*, 1939–1952. [CrossRef]
16. Fortmann, J. Modeling of Wind Turbines with Doubly Fed Generator System. Ph.D. Thesis, Department for Electrical Power Systems, University of Duisburg-Essen, Duisburg, Germany, 2014.
17. Justo, J.J.; Mwasilu, F.; Jung, J.W. Doubly-fed induction generator based wind turbines: A comprehensive review of fault ride-through strategies. *Renew. Sustain. Energy Rev.* **2015**, *45*, 447–467. [CrossRef]
18. Hu, Y.L.; Wu, Y.K.; Chen, C.K.; Wang, C.H.; Chen, W.T.; Cho, L.I. A Review of the Low-Voltage Ride-Through Capability of Wind Power Generators. *Energy Procedia* **2017**, *141*, 378–382, doi:10.1016/j.egypro.2017.11.046. [CrossRef]
19. Luna, A.; Lima, F.; Santos, D.; Rodriguez, P.; Watanabe, E.; Arnaltes, S. Simplified Modeling of a DFIG for Transient Studies in Wind Power Applications. *IEEE Trans. Ind. Electron.* **2011**, *58*, 9–20. [CrossRef]

20. Jiménez Buendía, F.; Barrasa Gordo, B. Generic Simplified Simulation Model for DFIG with Active Crowbar. In Proceedings of the 11th International Workshop on Large-Scale Integration of Wind Power into Power Systems as well as on Transmission Networks for Offshore Wind Power Plants, Lisbon, Portugal, 13–15 November 2012; p. 6.
21. Fortmann, J.; Engelhardt, S.; Kretschmann, J.; Feltes, C.; Erlich, I. New Generic Model of DFG-Based Wind Turbines for RMS-Type Simulation. *IEEE Trans. Energy Convers.* **2014**, *29*, 110–118. [CrossRef]
22. Bech, J. Siemens Experience with Validation of Different Types of Wind Turbine Models. In Proceedings of the IEEE Power and Energy Society General Meeting, Washington, DC, USA, 27–31 July 2014.
23. Goksu, O.; Altin, M.; Fortmann, J.; Sorensen, P. Field Validation of IEC 61400-27-1 Wind Generation Type 3 Model with Plant Power Factor Controller. *IEEE Trans. Energy Convers.* **2016**, *31*, 1170–1178. [CrossRef]
24. International Electrotechnical Commission. *IEC 61400-21-1: Wind Energy Generation Systems. Part 21-1: Measurement and Assessment of Electrical Characteristics-Wind Turbines*; IEC: Geneva, Switzerland, 2019.

Article

Investigations on the Unsteady Aerodynamic Characteristics of a Horizontal-Axis Wind Turbine during Dynamic Yaw Processes

Xiaodong Wang [1,*], Zhaoliang Ye [1,2], Shun Kang [1] and Hui Hu [3]

[1] Key Laboratory of Condition Monitoring and Control for Power Plant Equipment, Ministry of Education, North China Electric Power University, Beijing 102206, China

[2] China Huaneng Clean Energy Research Institute, Beijing 102209, China

[3] Department of Aerospace Engineering, Iowa State University, Ames, IA 50011, USA

* Correspondence: wangxd@ncepu.edu.cn; Tel.: +86-155-0116-6882

Received: 11 June 2019; Accepted: 7 August 2019; Published: 14 August 2019

Abstract: Wind turbines inevitably experience yawed flows, resulting in fluctuations of the angle of attack (AOA) of airfoils, which can considerably impact the aerodynamic characteristics of the turbine blades. In this paper, a horizontal-axis wind turbine (HAWT) was modeled using a structured grid with multiple blocks. Then, the aerodynamic characteristics of the wind turbine were investigated under static and dynamic yawed conditions using the Unsteady Reynolds Averaged Navier-Stokes (URANS) method. In addition, start-stop yawing rotations at two different velocities were studied. The results suggest that AOA fluctuation under yawing conditions is caused by two separate effects: blade advancing & retreating and upwind & downwind yawing. At a positive yaw angle, the blade advancing & retreating effect causes a maximum AOA at an azimuth angle of 0°. Moreover, the effect is more dominant in inboard airfoils compared to outboard airfoils. The upwind & downwind yawing effect occurs when the wind turbine experiences dynamic yawing motion. The effect increases the AOA when the blade is yawing upwind and vice versa. The phenomena become more dominant with the increase of yawing rate. The torque of the blade in the forward yawing condition is much higher than in backward yawing, owing to the reversal of the yaw velocity.

Keywords: HAWT; aerodynamic characteristics; dynamic yawing process; near wake; start-stop yaw velocity

---

1. Introduction

The use of wind power has grown rapidly over the past few decades. The aerodynamics of the wind turbine are a core subject of wind power generation. Owing to the intermittency of wind, the orientation of a wind turbine must be frequently adjusted, typically using the yaw mechanism, to ensure the rotor disk is perpendicular to the direction of the wind. Yaw control is one of the most important strategies for improving the efficiency of wind energy conversion. Under the yawed condition, velocity component of the upstream wind is parallel to the plane of rotation. Thus, the angle of attack of airfoil along blade spanwise exhibits periodic variation in one revolution [1]. The aerodynamic performance of a wind turbine under yaw, as well as the surrounding flow, show highly unsteady characteristics, resulting in high fatigue loads that can dramatically reduce the operational life of a wind turbine. Therefore, understanding the unsteady aerodynamic characteristics of wind turbines under yaw is important for improving their design [2]. In particular, accurate predictions of unsteady aerodynamic loads under yawed conditions are important for blade design and operational control.

Knowledge of complex three-dimensional (3D) flow phenomena over wind turbine blades and near wake under the yawing condition are still lacking. Thus, improved numerical models that consider the combined effects of stall delay, unsteady flow, and other significant effects such as the retreating & advancing effect and upwind & downwind effect are needed. The aim of present paper is to further elucidate the aerodynamic and 3D flow characteristics of yawed and yawing turbines.

The aerodynamics of wind turbines under yaw have been extensively investigated using both experiments and numerical simulations. Detailed aerodynamic performance measurements along the blade in combination with 3D flow field measurements have helped further our understanding of the flow mechanism. Wind tunnel experiments performed by Hand et al. [3] on a two-blade wind turbine provide a benchmark for numerical simulations. In their study, Hand and colleagues obtained the pressure and aerodynamic force coefficients for an azimuthal distribution at different radial locations under five yawed cases: 0°, 10°, 30°, 45°, and 45°. The angle of attack was found to vary approximately with the blade azimuth angle according to a sinusoidal function. Further to this, Schepers et al. [4] performed wind tunnel experiments on a three-bladed wind turbine, the MEXICO rotor, under three different yaw angles: 15°, 30°, and 45°. They obtained sectional aerodynamic forces and torques for different azimuth angles; however, the sinusoidal variation rule did not correlate well along the inner span of the blade owing to the velocities induced by root vortices. Sant et al. [5] investigated the TUDelft reference rotor in an open jet wind tunnel under both axial and yawed conditions and assessed the blade element moment (BEM) method, which was found to be limited to the skewed wake under yawed conditions.

Based on the work of Schepers et al. [6], Micallef et al., performed stereo particle image velocimetry measurements to study the blade aerodynamic performance and near wake development. Under yaw, flow in the windward region exhibited inboard motion due to rapid motion of the tip vortices away from the blade. In field tests, Ven et al. [7] applied in-field measurements of a turbine to investigate the three dimensional stall phenomena under yawed inflow conditions. Imbalances in the crossflow fraction along the radial direction of the blade were observed between azimuth angles of 90° and 270°. The phenomenon is caused by yawed inflow leading to in-wash and outwash on the blade surface. Furthermore, Dai et al. [8] provided a detailed investigation of the yaw effect based on Supervisory Control and Data Acquisition (SCADA), and presented characteristics of the power coefficient and rotor torque under yaw. Interestingly, many yaw error control solutions based on the smart prediction of the wind direction had been used to maximum power extraction of wind power [9,10]. Further to this, Saenz-Aguirre et al. [11] claim that the drawbacks of the yaw control are the very large time constant and the strict yaw angle rate owing to the high mechanical loads, resulting that yaw control is less studied in the literature than that pitch and speed control. Furthermore, Munters et al. [12] performed wind-farm control research with a new dynamic yaw control based on the large-eddy simulation, and found that with the consideration of the unsteady-control interacting with flow dynamics, dynamic yaw control for the given wind farm setup is better than induction control in the process of power extraction. In fact, the yawing operation of wind turbine is very common, and the aerodynamic characteristics of wind turbine under yaw had been extensively studied in the literature with numerical simulation.

Numerical methods to evaluate the aerodynamic performance of wind turbines mainly include BEM, vortex theory, and computational fluid dynamics (CFD). The most commonly used method is BEM, which is fast with low computational costs. Ryu et al. [13] developed a successive under-relaxation technique based on the classical BEM to compute specific radial locations of the blade and found that yawed inflow leads to periodic changes in the angle of attack on different spanwise sections. Jimenez et al. [14] firstly established an analytical wake model with a simple formula to predicit the wake skew effect based on the momentum conservation and top-hat model offered by Jensen [15] for the velocity deficit effect. However, the experimental validation was not sufficient as mentioned by the author and the model overestimate the wake deflection of wind turbine, owing to the inaccurate top-hat model for velocity deficit. With the goal of optimize the yaw angles of the wind turbines, Gebraad et al. [16] presented a new flow redirection and induction in steady-state model based on internal parametric model for wake effects, which demonstrate the optimization control for increasing the energy production of the wind plant. Recently, Bastankhah and Porte-Agel [17] developed a new analytical model on the assumption of the Gaussian distribution to predict the wake deflection and far wake velocity deficit, which showed a good agreement with the experimental data. However, some parameters of that model have not specified which is needed to be researched. Based on the works of them, Qian Guo-Wei et al. [18] presented a new analytical wake model with Gaussian distribution function of velocity deficit and momentum conservation in the lateral direction, considering the ambient turbulence intensity, thrust coefficient and yawed effect, which enables the model to have good application under various conditions. Yaw not only influences the wake flow behind the wind turbine, but also flows on the surface of the blade and 3D aerodynamic characteristics of the rotor [19]. Castellani [20] analyzed a small three-blade wind turbine and investigated the effects of large yaw angles, using numerical and experimental methods, and found that the 3P blade frequency is clearly visible. In addition, non-uniform flow around the blades was modeled using a simplified wake model modified by Prandtl's tip loss factor.

The time constant used in dynamic inflow models results in an unrealistic drop at the tip region, and therefore, cannot accurately model for the dynamic conditions of wind turbines. For instance, both the tip loss model and dynamic inflow model have several shortcomings when used for non-design load cases but can possibly be improved by free vortex methods (FVM), which combine the blade aerodynamics with wake flow computational models. Qiu et al. [21] investigated the dynamic variation of loads on a wind turbine blade during the yaw process using an improved lift line method and proposed a wake model comprised of the vortex sheet model and tip vortex model. Further, the yaw rate and dynamic wake were shown to significantly influence the shaft thrust and torque of the wind turbine. Micallef et al. [22] performed both experimental and numerical investigations based on an FVM to investigate the tip vortex generation of a horizontal-axis wind turbine (HAWT) under yaw. Vorticity on the suction side under non-yawed conditions was shown to be more concentrated than for the yawed condition, in which vorticity spreads over a small region at the tip of the turbine blade.

Several aerodynamic correction models have been proposed and widely used, including Pitt and Peter's yaw correction [23,24], Suzuki's dynamic inflow model [25], Prandtl's tip loss function [26], Du and Selig's 3D stall delayed model [27], and Buhl's wake correction [28]. Nonetheless, classical BEM and FWM are still not sufficiently reliable for predicting the aerodynamic load distributions on wind turbine blades, especially for the yawed and stalled rotor conditions. Investigations were performed on the complicated yaw aerodynamic problem of wind turbines using more accurate CFD simulations using detailed visualization of the 3D flow [29]. Tongchitpakdee et al. [30] investigated the aerodynamic characteristics of the National Renewable Energy Laboratory (NREL) Phase VI wind turbine modeled by an unstructured grid under different wind speeds and yaw angles using the BL model. At low wind speeds (<7 m/s) and under low yaw angles, the flow remains attached over most regions of the rotor. Further to this, Yu et al. [31] performed time-accurate aerodynamic simulations of the NREL Phase VI wind turbine under yawed flow conditions, which was also modeled with an unstructured mesh, along with the k-ω shear stress transport (SST) turbulent model. They found that the retreating side shown aerodynamic loads (C_n, C_t) higher magnitudes of loads than on the advancing side. However, the three-dimensional unsteady stall effect at different yaw angles was not studied. Schulz et al. [32] modeled and analyzed a generic 2.4-MW wind turbine with over-set structured grids and yaw angles ranging from $-50°$ to $50°$ using the FLOWer CFD solver. The simulation was performed using a detached eddy simulation method and revealed azimuthal non-uniformity of the load variation along spanwise sections of the blade due to the retreating & advancing effects of the blade. Furthermore, the deformation characteristics of the blade were considered in the yawed case of the wind turbine simulation. Jeong et al. [33] investigated the effects of yaw errors on the aerodynamic and aeroelastic behaviors of NREL 5-MW HAWT blades, and showed how yaw misalignment adversely affects the dynamic aeroelastic stability of the blade. Dai et al. [34] performed an aeroelastic analysis on the Tjæreborg wind turbine under yaw, and found that fluid-solid coupling results in higher averaged power and thrust, as well as violent oscillation amplitudes.

Yaw is a continuous rotation process with a constant or variable velocity of rotation. Qiu et al. [21] investigated the shaft torque of the blade and instantaneous wake flow under the dynamic yawing process with a yaw rate between 5 and 20°/s for the NREL Phase VI wind turbine. For larger wind turbines, such as the NREL 5-MW et al., these yaw rate seem to be impossible owing to the larger moment of inertia of the rotor-nacelle subsystem due to enlargement of the blade and tower length. Leble et al. [35] carried out a series of CFD simulations to investigate the performance of the DTU 10-MW wind turbine and modeled the turbine as a structured grid with moving boundaries. Simulations were performed with a yaw rate of 0.3°/s and 3° yaw angle and the results showed that overall power under the dynamic yawing condition was much larger than for the yawed case. Then, Wen et al. [36] investigated the NREL 5-MW yawing dynamic sinuous motion with an averaged yaw rate of 1.2°/s and 2.4°/s for the case of $f = 0.1$ Hz and $f = 0.2$ Hz, respectively. The dynamic yaw motion induced the upwind & downwind yawing effect, which considerably influenced the AOA of blade sections. As the blade is yawing upwind, the AOA increases and vice versa. In summary, the unsteady aerodynamic characteristics of wind turbine blades under yaw considering full 3D rotational effects are still unclear. In the present paper, unsteady dynamic CFD simulations were performed to investigate the aerodynamic characteristics of a wind turbine blade during the rotational revolutions using a full 3D wind turbine model. With the assumption of the rigid body, the main contribution of this paper is the proposal of a new grid methodology for analyzing the overall performance using URANS simulation, aerodynamic loads, and flow field of wind turbines in the yawed and yawing processes. The method can be used to analyze interactions between transient aerodynamic phenomena associated with the wind turbine control system. The other main objective of this study was to investigate the effects of yaw on the dynamic output power, rotor thrust, and the blade sectional aerodynamic characteristics caused by continuous changes in the yaw angle. The main novelty of the current work is the inclusion of the dynamic yawing simulation of wind turbine with the multiple structured domain sliding mesh. The rest of this paper is organized as follows: Section 2 describes the geometry model and computational method. In Section 3, simulation results under several yaw angles are compared to experimental data to validate the model. The discussion mainly focuses on the sectional azimuthal variations of aerodynamic force coefficients, as well as the pressure distribution on the blade. Finally, the conclusions are summarized in Section 4.

2. Numerical Model

The 5-MW reference wind turbine (RWT) designed by the NREL was used in this study [37]. The blade uses the Delft University (DU) airfoil family with a relative thickness ranging from 21 to 40%. The blade chord, relative thickness, and twist angle all have non-linear distributions. The rotor diameter is 126 m and the wind turbine operates at a wind speed of 11.4 m/s with a rotational velocity of 12.1 rpm, resulting in the tip speed ratio of $\lambda = 7$. The definitions of yaw angle and azimuth angle are shown in Figure 1.

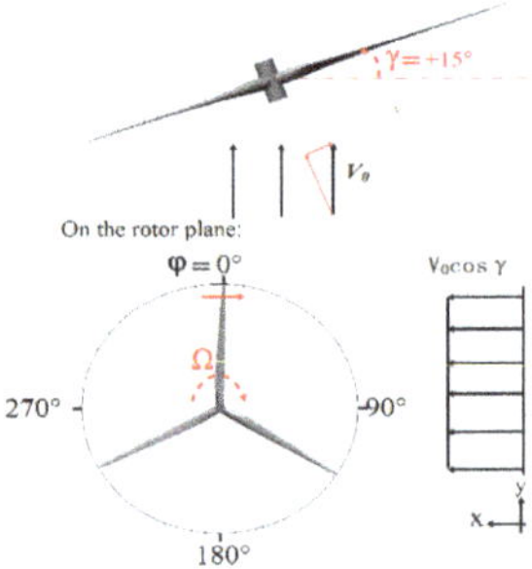

Figure 1. Sketch of the asymmetric relative velocity under yaw.

2.1. Computational Domain and Boundary Conditions

The computational domain, as shown in Figure 2a, is a rectangular zone that is 330*D* in length and 240*D* in width, where *D* is the rotor diameter. The computational domain is made up of two main zones, the far-field zone and internal zone, as depicted in Figure 2b. The far-field is used to set the inflow conditions. The internal zone can be further subdivided into two parts: (1) cylinder yaw zone with a radius of 6*D* and a height of 7.5*D*; (2) rotor roll zone with 1.5*R* radius and 1.6-m height. The cylinder zone was used to configure the dynamic yaw motion. The distance between the inlet and rotational cylinder was 120*D*. The rotor is located at the center of the rotor roll zone.

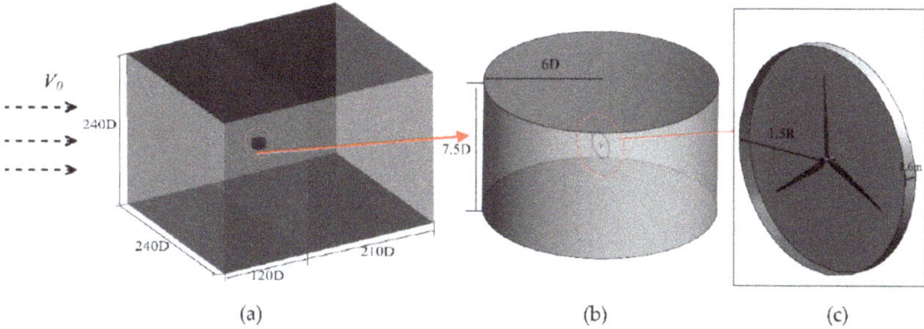

Figure 2. Computational Zones: (**a**) far-field (static); (**b**) yaw zone; (**c**) rotational zone.

Figure 3 shows the boundary conditions used in the simulations. A uniform wind speed of 11.4 m/s was set at the inlet of the domain with a turbulence intensity of 5%. Boundary conditions for the up and bottom planes in the domain are free-slip and no-slip walls. The blade surface was set as a no-slip wall. The pressure outlet condition was assigned to the outlet of the domain. To simplify the static yaw simulation, the rotational axis was fixed, while the inlet velocity direction was varied to generate different angles mimicking different yaw condition. In Fluent, it is possible to select a mesh that can process non-conformal interfaces, namely boundaries interfaces between cell zones, for which the mesh node locations are not identical. In the current simulation case, the matching mesh interface option was applied to three interfaces, for example, the interface between the far-field zone and yawing zone, which is more accurate than the mapped interface option.

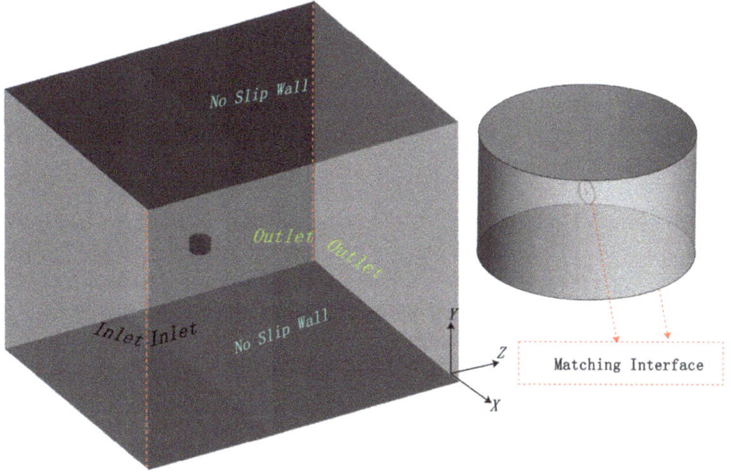

Figure 3. Boundary conditions.

2.2. Grid Setup

Building multiple blocks is recommended for studying the yawing dynamics of coupled effects within two axes (yawing axis and rotational axis) since the relative rotation between zones can be investigated. Figure 4a,b show grids in the three different zones of the turbine (far-field, yaw cylinder, and rotation) generated using the ICEM software. The NREL 5-MW rotor was modeled including the hub (without the tower) [31]. The grid number for the far-field was 3.35 million, which is enough to transition from the inflow condition to the internal zone. Dynamic yaw motion was set, along with the rotational velocity in the cylinder yaw zone, with 3.25 million grids. Figure 4a illustrates the grids of the rotor zone, generated using AutoGrid tool in the NUMECA software. The grid number for this zone was 3.76 million, the first layer wall normal distance is about 10^{-4} m. The total grid number was about 10.27 million. The work of Tran and Kim [38] employed a 6 million-cell grid for a 5-MW wind turbine. The blade surface was resolved with 93 cells along the span. In order to resolve the boundary layer, the grids around the blades were refined to keep $y+$ less than 5 on the blade surface, which satisfies the needs of the T-SST turbulent model.

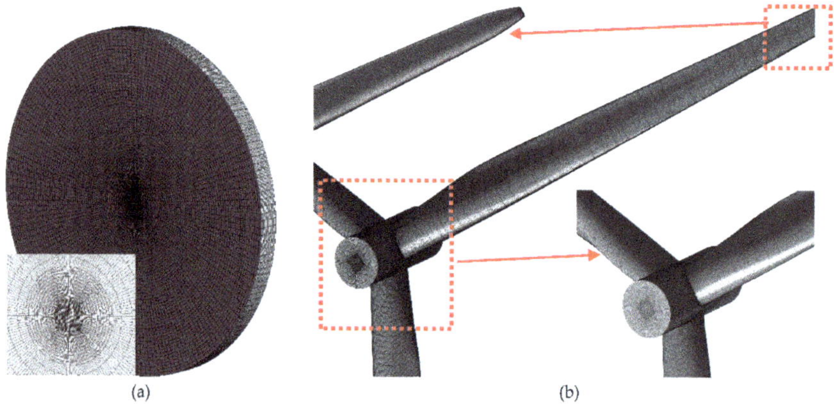

(a)   (b)

Figure 4. Computational mesh for NREL 5-MW wind turbine; (**a**) rotor cylinder; (**b**) blade surface mesh.

2.3. Numerical Methods

The unsteady Reynolds-averaged Navier-Stokes equations (URANS) were solved in ANSYS Fluent. The k-ω transitional SST turbulent model was used for turbulence modeling considering both laminar and transitional effects at the blade surface. The sliding mesh technique was used for data exchange between the rotational zone and the stationary zone. To ensure time-accurate calculations, a dual-time implicit time integration algorithm based on linearized second-order Euler backward differencing was used. The time-step size was equivalent to a rotor azimuthal increment of 5° coupled with 20 pseudo-time sub-iterations. The residual of the continuity equation was reduced by at least three orders throughout the calculations. For comparison, BEM computations using the Fast software were also performed. Work of Tran and Kim [38] employed 2° increments of the azimuth angle. The yawed case was achieved by changing the inlet velocity component to implement yaw inflow.

The yaw velocity can be constant, shown as the black dashed line in Figure 5. In fact, yawing is a process that changes gradually, similar to the definition of 2-s (or 4-s) start or stop duration. The selected variation of angular velocity for studying the influence of start-stop duration on the dynamic aerodynamic characteristics under yawing is shown a solid black line in Figure 6, which can be separated into three part: 1. Start duration with sinusoidal variation law of yaw velocity, 2. the stage with constant yaw velocity of 0.3°/s, 3. Stop duration with sinusoidal law of yaw velocity.

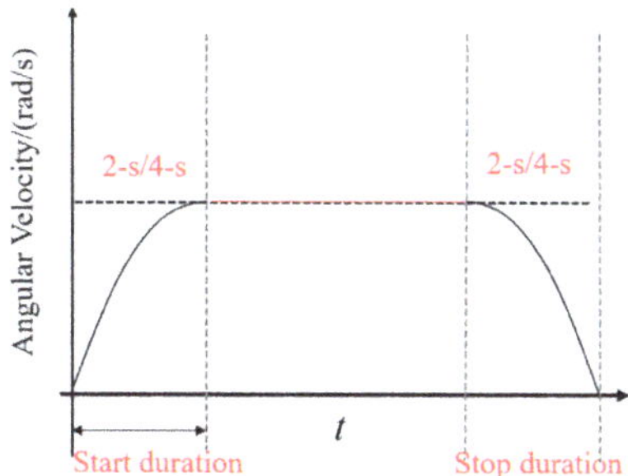

Figure 5. The sketch for start-stop yawing process.

In the present investigation, the yawing process of the NREL 5-MW turbine was calculated under two different start-stop durations, 2-s and 4-s. Variation of the yaw velocity is shown in Figure 6, and changes more gradually with the 4-s duration of the start-stop process compared to the 2-s duration. During the start-stop yawing process, the sinusoidal variation of yaw velocity was applied to achieve transient yawing of the wind turbine, as follows:

$$yawVel(t) = rpm \times \sin(2\pi(1/(duration \times 4) \times t) \tag{1}$$

where rpm = 0.3°/s, according to the description of the NREL5 WM design manual [37], and duration is either 2-s or 4-s. Up to 0.3°/s, the wind turbine changes from transient to a yaw angle of approximately 20° with the maximum rpm.

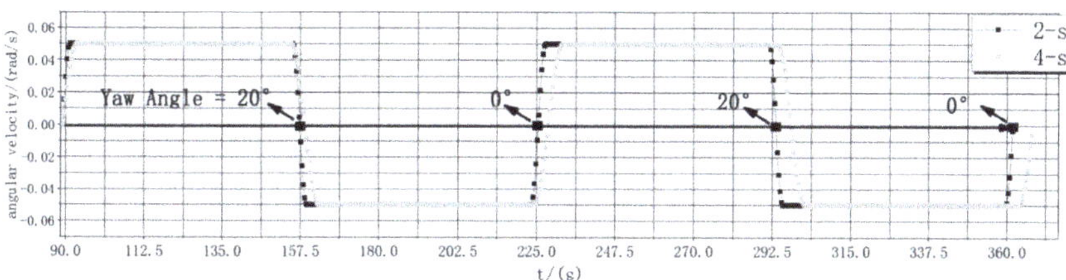

Figure 6. Variation of yaw velocity under different transient start-stop processes.

3. Analysis of Results

The NREL 5-MW reference wind turbine (RWT) was studied [28]. The yawed and yawing wind turbines were considered. The parameters used in the simulation are listed in Table 1 for each case. Two of the cases assumed uniform inflow and the blades were assumed to be rigid in the model.

Table 1. Simulation cases.

Configuration	Angle/Amplitude (°)	Unsteady Computation Time (s)
Aligned	0	89.26
Fixed yaw	5, 10, 15, 20, 25, 30	89.25
Dynamic yawing	0–20–0	126

Physically, the flow-field around a rotating HAWT is significantly influenced by the existence of wind shear, turbulence, gust, and yaw motion of the nacelle. For a yawing wind turbine, flow characteristics are more complex than those of a static yawed wind turbine, and the additional wind contribution effects transmitted to the rotor due to the nacelle motion must be considered.

Therefore, accurate prediction of the unsteady aerodynamics calculated by many conventional numerical approaches is still questionable for a yawing wind turbine. In this study, unsteady CFD simulations based on the sliding structure mesh technique were performed to analyze the yawing motion of the wind turbine due to the nacelle motion. Thus, to investigate the effects of vortex-wake-blade interactions on the aerodynamic performance of the wind turbine, the yawing motion of the rotating turbine blades due to the nacelle motion was considered.

The CFD method was first verification with grid independence, turbulent model studies and time step studies. Then the computational results are validated using both fixed yaw and yawing of the NREL 5-MW wind turbine. The unsteady aerodynamic loads of the yawing wind turbine were more sensitive to change as the yaw angle was varied. Nine simulation conditions were performed based on a fixed yaw angle of $0°$, $5°$, $10°$, $15°$, $20°$, $25°$, and $30°$.

Next, the total aerodynamic performance, which was used to validate the results of the simulation, will be discussed. Thereafter, aerodynamic characteristics and three-dimensional stall characteristics at different span sections will be investigated. Finally, the results of two separate simulations of the yawing start-stop stage will be used to examine the mechanism of the yawing effect.

3.1. Verification of the Computational Results

3.1.1. Grid Independence

Table 2 shows the different torques calculated for five different numbers of cells. Every mesh was refined along the airfoil circumferential direction. The grid numbers were computed in the T-SST turbulent mode and simulations were performed with the following grid numbers: 1.6 million, 3.09 million, 3.67 million, 5.31 million, and 6.05 million. In addition, two grid numbers (1.6 million and 3.67 million) were investigated under different yaw angles ($10°$ and $20°$), as shown in Table 2. For non-yaw condition, the design torque at 11.4 m/s is about 4.08×10^6 N·m, which is taken as a reference value. The relative errors for the results using other four mesh are 1.9~0.49%. The relative error for mesh of 3.67 million cells is smaller than 1% and the computational cost using this mesh is moderate. Thus this mesh was selected for further investigation.

Table 2. Computed torque of different grid size averaged in one revolution.

Torque (N·m)	Number of Cells (Million)					
Yaw Angle (°)	1.6	3.09	3.67	5.31	6.05	Designed Value
$0°$	4.0×10^6	4.05×10^6	4.06×10^6	4.11×10^6	4.10×10^6	4.08×10^6
relative error	1.9%	0.73%	0.49%	0.73%	0.49%	

3.1.2. Turbulent Model Studies

In this section, we present results of the unsteady independence study based on four turbulent models: SST, k-kl-ω, Reynold stress, and T-SST. The rotor torque under axial flow obtained for each turbulent model is presented in Table 3. The computed torque was similar for each of the four turbulent models. The simulation cases using the SST model and Reynold stress model were used to interpret the fully developed flow mechanism without transient processes, resulting less torque compared to using Transient SST turbulent model. When the wind turbine operates under low wind speeds, transient phenomena occur on the suction side of the blade; therefore, the T-SST turbulent model was selected for static and yawing process studies.

Table 3. Time averaged rotor torque calculated using different unsteady turbulence models.

Turbulent Models	k-ω SST	k-kl-ω	Reynold Stress	T-SST	Designed Value
Torque/(N·M)	3.89×10^6	4.17×10^6	3.76×10^6	4.06×10^6	4.08×10^6
relative error	4.65%	2.2%	7.8%	0.49%	

3.1.3. Time Step Studies

In the unsteady numerical simulation, the time step could influence the overall performance of the model, therefore, it is necessary to investigate time-step independence. Table 4 lists the averaged rotor torques during one revolution for non-yawed case obtained using the T-SST turbulent model with three different time steps (72, 144, and 360). The comparisons show slight differences among the results (less than 3%). The result using 72 time steps in one revolution gave a better agreement comparing to the designed value. Considering the computational cost, a time step of 72 during one revolution was selected for subsequent simulations.

Table 4. Rotor torque calculated using different time step sizes averaged over one revolution.

Time Step	72	144	360	Designed Value
Torque/(N·M)	4.06×10^6	4.16×10^6	4.19×10^6	4.08×10^6
relative error	0.49%	1.9%	2.69%	

3.2. The Validation of Numerical Simulation Results for Yawed Wind Turbine

After the verification of the simulation, the subsections below will show some aerodynamic analysis of wind turbine under yawed and yawing simulation.

3.2.1. Overall Performance Analysis for Different Yaw Angles

Figure 7 shows the overall performance of the rotor under yaw. As shown in Figure 7a, only a small amount of variation in rotor power can be observed under small yaw angles ($\leq 5°$). When the yaw angle exceeds $5°$, the rotor torque significantly decreases as the yaw angle increases. The torque value computed using the BEM method, with or without the Beddoes stall model, is higher than the value computed by CFD since the BEM computation does not consider flow separation and flow transient phenomena. Zhu [39] extensively investigated the combined effects of rotational augmentation and dynamic stall, and found that the hysteresis loop of aerodynamic load is much larger compared to 2D simulations. In fact, for the NREL 5-MW wind turbine and a wind speed of 11.4 m/s, flow separation and 3D radial flow mainly occur on the inner board region, resulting in a lower torque and lower thrust. The deviation in the power and thrust between CFD and BEM was 4.1% and 5.91%, respectively. Compared with the axial free inflow, the three functions on $\cos(\gamma)$, $\cos^2(\gamma)$, $\cos^3(\gamma)$ of the power and thrust are shown with dashed, dotted and dash-dotted line, respectively. The variation of averaged power in yaw conditions will decrease by $\cos^2(\gamma)$; the averaged thrust agrees well with $\cos(\gamma)$.

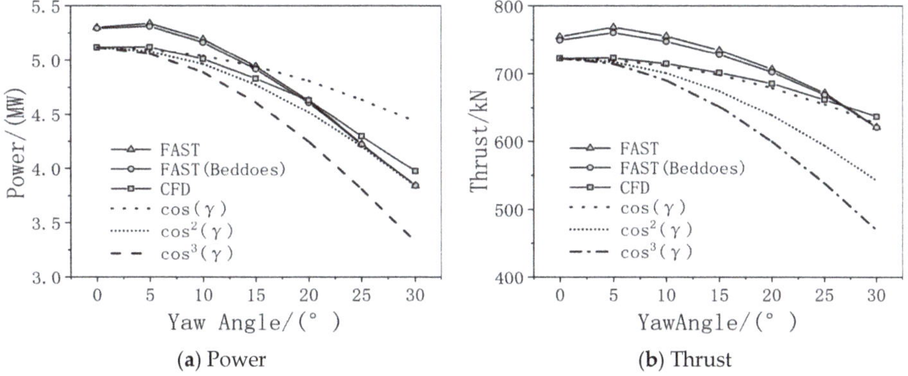

Figure 7. Variation of the power and thrust of the wind turbine under yaw.

3.2.2. Aerodynamic Load Analysis along the Span of the Blade

Theoretical analysis is necessary to determine the mechanism behind the yawed effect. Figure 8 shows the velocity diagram under the yawed condition. A velocity component exists in the blade revolution plane and can be projected into radial and chordwise components denoted V_r and V_c, as shown in Figure 8a. The retreating and advancing effects occur because of V_c, which causes the sectional load and AOA to fluctuate periodically. The maximum AOA and maximum load occur when the blade is in the 12 o'clock direction in the current simulation setup. Additionally, loads on the retreating side are much larger than on the advancing side. Figure 9b,c quantify the influence of V_c, which is calculated by:

$$V_{rel} = \sqrt{\left(V_0(\cos\gamma - a) + \eta V_{dyn}\right)^2 + \left(\omega r(1 + b) - \beta V_0 \sin\gamma \cos\varphi\right)^2} \tag{2}$$

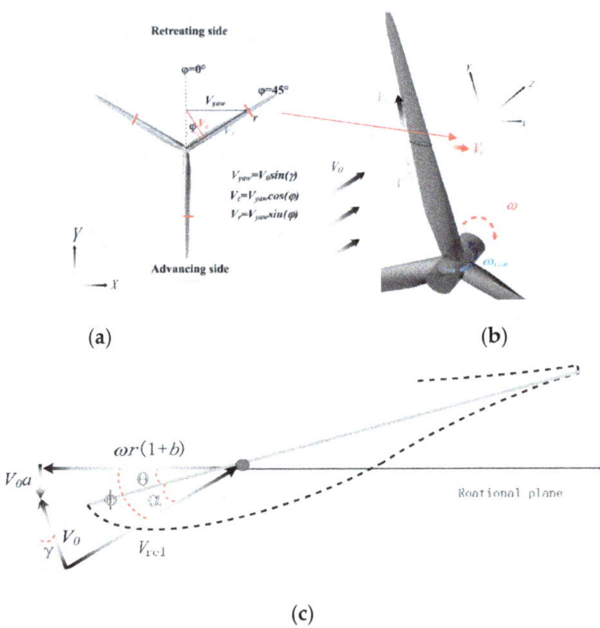

Figure 8. Section airfoil induced velocity V_c under stable yaw: (**a**) magnitude of radial and chordwise velocity; (**b**) direction of radial and chordwise velocity; (**c**) velocity diagram of blade element.

(1) Time-averaged load analysis on spanwise section under yawed condition

Figures 9–11 show the analysis of the variation of averaged AOA, C_n, and C_t under one rotor revolution. The value of C_n and C_t is calculated by:

$$C_n = \int_{\frac{x}{c}=0}^{\frac{x}{c}=1} C_{p-up} d(\frac{x}{c}) - \int_{\frac{x}{c}=0}^{\frac{x}{c}=1} C_{p-down} d(\frac{x}{c}) \tag{3}$$

$$C_t = \int_{y0}^{y1} C_{p-up} d(\frac{y}{c}) - \int_{y0}^{y1} C_{p-down} d(\frac{y}{c}) \tag{4}$$

where C_{p-up} and C_{p-down} denote the pressure coefficients in the upper and down side of the airfoil, respectively; $y0$ and $y1$ mean the leading edge position and trailing edge position of the airfoil.

The AOA distribution presented in Figure 9 takes into account the advancing and retreating effects, as previously shown by Castellani [20] in the yawed simulation of a HAWT using the BEM model. Comparing the simulation results for CFD and FAST, the same trends can be observed along the blade span, whereas the AOA value is much larger than the FAST result. Figures 10 and 11 show the variation of aerodynamic load spanwise along the blade. For the axial flow, some large abnormal fluctuations in the AOA and aerodynamic loads can be observe d. Differences in the results of the CFD and FAST methods mainly occur along the inner board, owing to flow separation, which lead to smaller aerodynamic loads compared to those computed by the BEM model that does not consider flow separation.

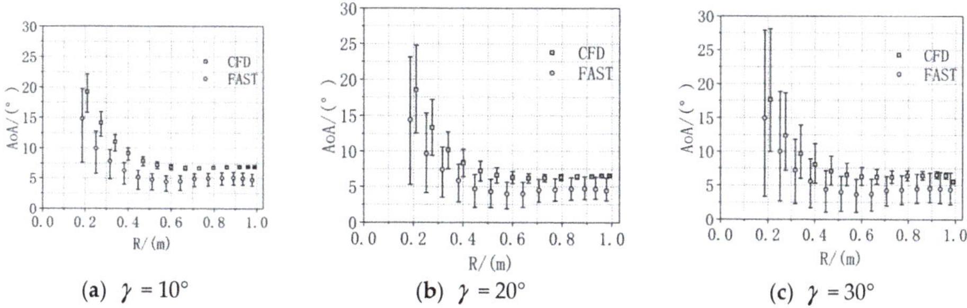

(a) $\gamma = 10°$ (b) $\gamma = 20°$ (c) $\gamma = 30°$

Figure 9. The profile of AOA along the blade span under yawed cases.

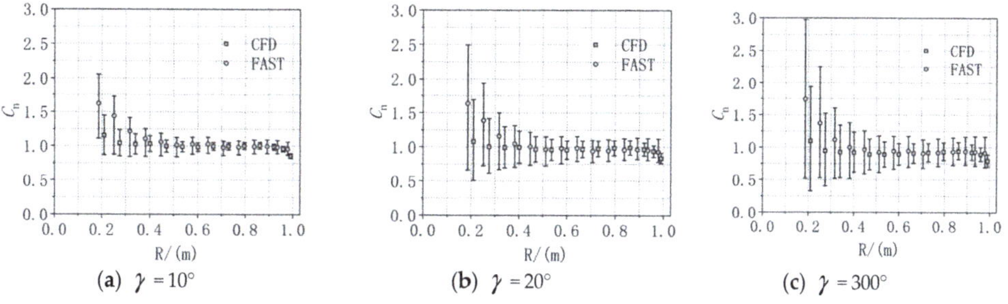

(a) $\gamma = 10°$ (b) $\gamma = 20°$ (c) $\gamma = 300°$

Figure 10. The profile of C_n along the blade span under yawed cases.

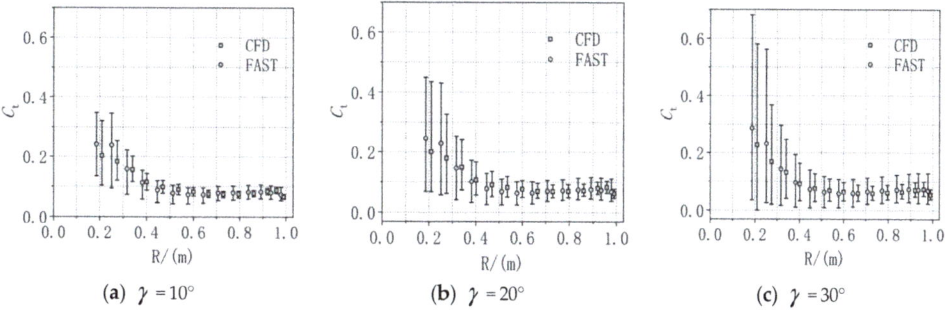

(a) $\gamma = 10°$ (b) $\gamma = 20°$ (c) $\gamma = 30°$

Figure 11. The profile of C_t along the blade span under yawed cases.

(2) The spanwise section aerodynamic load analysis under yawed conditions

Figures 12 and 13 show the distribution of the AOA with respect to the azimuth angle under a yaw angle of 15° and 30°. Fluctuations become larger as the yaw angle increases due to the advancing and retreating effect, which is partly caused by the inflow velocity component under yaw. After post-processing, the AOA using the combination of the CFD sectional airfoil aerodynamic loads and the BEM method, the maximum and minimum AOA during one revolution occur at an azimuth angle of 0° and 180°, respectively. Wen [36] found that under the impact of non-uniform effects (due to variations of the induced factor caused by the radial position and azimuth angle in the rotational plane), the maximum AOA tends to occur at 90° for inboard airfoils. The maximum and minimum aerodynamic loads occur at an azimuth angle of 90° and 270°, respectively. Thus, future computations of the AOA should consider non-uniform effects.

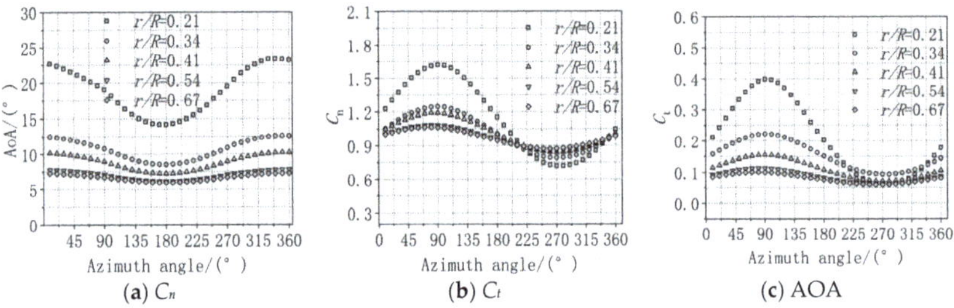

Figure 12. Azimuthal aerodynamic loads at different spanwise sections during one revolution under a yaw angle of 15°.

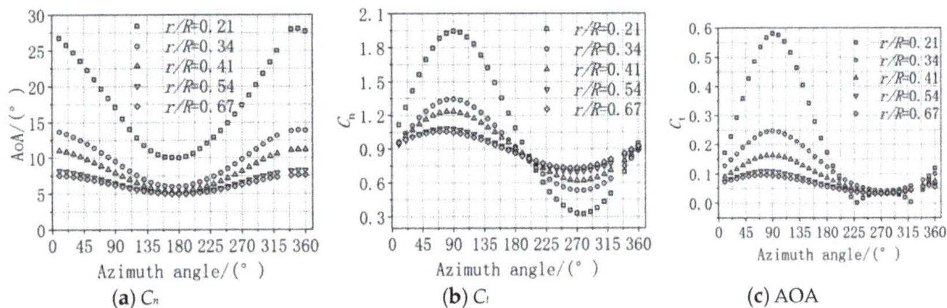

Figure 13. Azimuthal aerodynamic loads at different spanwise sections during one revolution under a yaw angle of 30°.

Figure 14 shows the variation of the averaged AOA with aerodynamic load at five typical spanwise sections with respect to yaw angle. Fluctuations of the aerodynamic performance are clearly observed and become larger as the yaw angle increases.

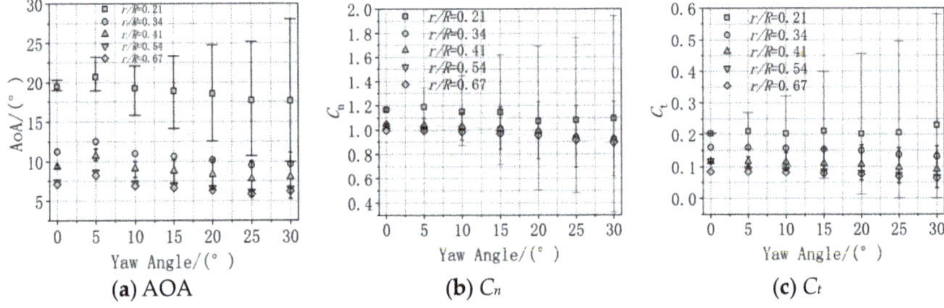

Figure 14. Variation of time-averaged AOA and aerodynamic load under yaw during one revolution.

The mean aerodynamic performance remains relatively constant with less decrease due to the yaw effect. Variation of the AOA and aerodynamic loads exhibit much higher fluctuations in the inner board compared to the middle and outer board. With the increasing of the yaw angle, the circumferential loads and AOA also cause higher maximum loads.

3.3. Numerical Simulation of Dynamic Yawing Wind Turbine

This section presents results of the numerical analysis of the NREL 5-MW RWT under the dynamic yawing process.

3.3.1. Torque Characteristics of Wind Rotor

Figure 15 shows the distribution of the wind rotor torque under different start-stop yaw velocities. During the yawing process, fluctuation of the wind rotor torque is small. At the beginning of dynamic yaw, larger torques occur due to changes in the rotor position. As the yaw angle increases, the torque gradually decreases and is 5×10^5 N·m larger in the case of the 2-s duration compared to the 4-s duration. The 2-s duration may induce a larger rotor torque since the yaw angle changes more quickly than with 4-s duration. The reasons are given below. At the start and stop stage, the yaw velocity is changed with the sinusoidal variation law related to the frequency, the higher frequency of yaw velocity caused much larger of the power and thrust, which is similar to the variation of power under platform pitching [38,40]. When the wind turbine is yawing with the constant yaw velocity, the additional velocity inducing by yawing is the same, and the power is only decreased with the square cosine of yaw angle. When wind turbine yawed to the stage of stop, the power under the 2-s case decrease much faster than under 4-s cases, which is similar to the yaw start period.

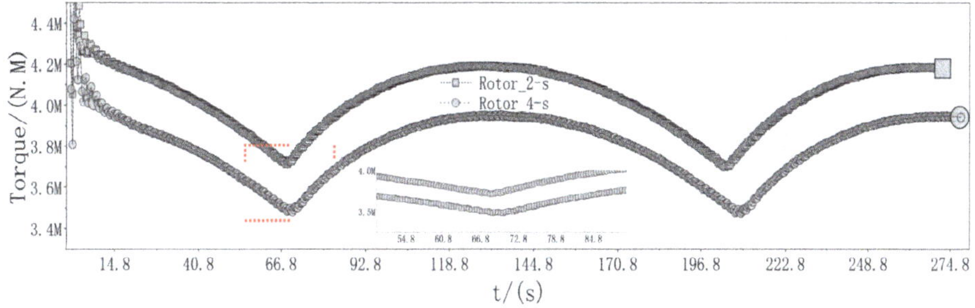

Figure 15. Torque characteristics of rotor under two different yawing rotational velocities.

The results of 58 rotor revolutions were analyzed. During each revolution, 12 torque measurements were collected. Then, the fast Fourier transform was used to obtain the rotor aerodynamic frequency. Figure 16 shows the torque power spectra of the rotor under 2-s and 4-s duration yawing process. The blade passing frequency of the NREL 5-MW turbine is about 0.2017 Hz (1P fluctuation), and is clearly the main frequency of the two yawing processes. Both are 0.6 Hz (which is approximately a 3P rotor fluctuation), which is similar to results presented by Castellani [20]. Figure 16a shows a secondary frequency of 0.2 Hz (1P fluctuation). Due to the less sampling data in the sinusoidal stage, the yawing start-stop frequency (2-s and 4-s duration with a corresponding main frequency of 1/8 Hz and 1/16 Hz, respectively) cannot be captured in the computation of the torque power structure.

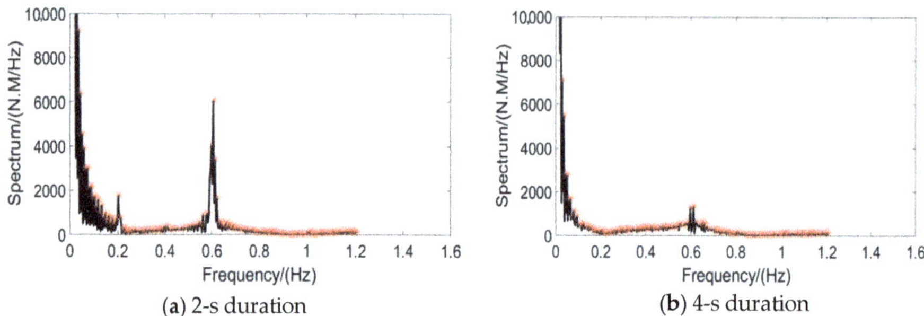

Figure 16. Frequency of rotor torque for a yawing start-stop process with a duration of 2-s and 4-s.

3.3.2. Torque Characteristics of Blade

Figure 17 shows the variation of torque under different yaw rates in two periodic processes of start-stop yawing. Similarly, the torque of the blade under a 2-s duration is larger than the torque of the 4-s duration. Interestingly, the torque of the blade in the forward yaw stage (shown in Figure 18 within 0~66.8 min, yaw angle from 0° to 20°) is larger than that of the backward yaw stage (shown in Figure 18 within 66.8~133.6 min, yaw angle from 20° back to 0°) since the dynamic yawing effect, which generates the dynamic velocity, reduces the relative velocity.

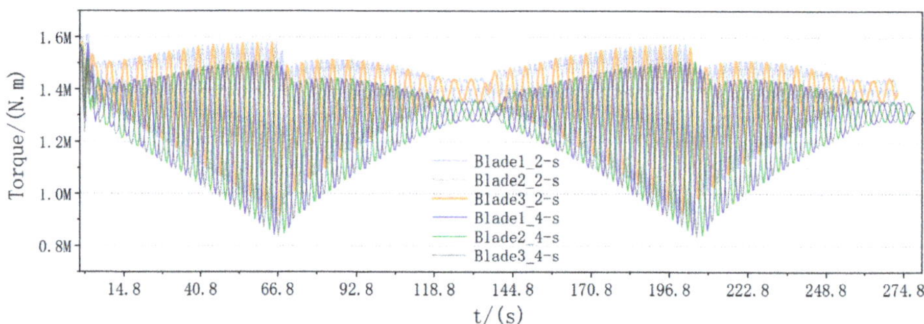

Figure 17. Torque characteristics of blades under two different yaw velocities.

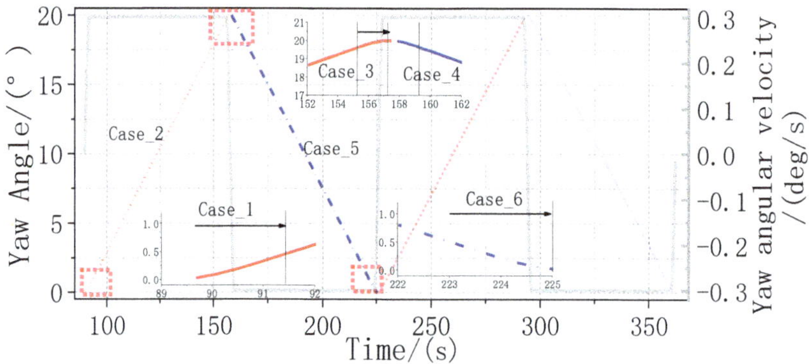

Figure 18. Six cases during the start-stop stage under dynamic yaw.

3.3.3. Wake Flow Characteristics

Wake effects are important in analyzing wind turbine aerodynamics. For convenience, the dynamic yaw was classified into six cases, as illustrated in Figure 18. Cases 1,3,4,6 include the forward-start yaw, forward-stop yaw, backward-start yaw, and backward-stop yaw, respectively. For Cases 2 and 5, the yaw angle is 10° and they represent forward yaw and backward yaw, respectively. The grey line in the figure illustration the variation of yaw angular velocity during the simulation, and shows that except the Cases 2 and 5, other cases are all in the simulation about the variation of yaw angular velocity.

(1) Forward yaw-start stage (Case 1). The velocity contours at t = 1/8T (T = 8 s or 16 s) at the beginning of dynamic yaw are shown in Figure 19. The near wake velocity flow structure shows the same status at the beginning of the yawing process for both the 2-s and 4-s start-stop duration. The different yaw velocities have very little influence on the velocity distribution in the 2D range downstream.

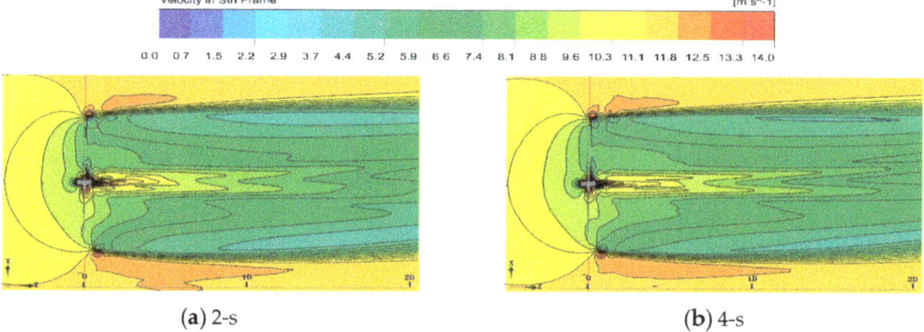

(a) 2-s   (b) 4-s

Figure 19. Velocity contours for Case 1.

(2) Yaw angle of 10° under forward yawing stage (Case 2). Sketches of the velocity streamlines of the two dynamic yawing processes (yaw angle of 10°) are shown in Figure 20a,b are similar to those obtained for the yawed case, as shown in Figure 20c. The dynamic yaw rotates with a fixed yaw velocity of 0.3°/s, and similar results are observed for the dynamic process. More energy intermediate effects can be observed between the wake zone and the main flow zone than in the yawed case. This may be due to the effects of dynamic stall.

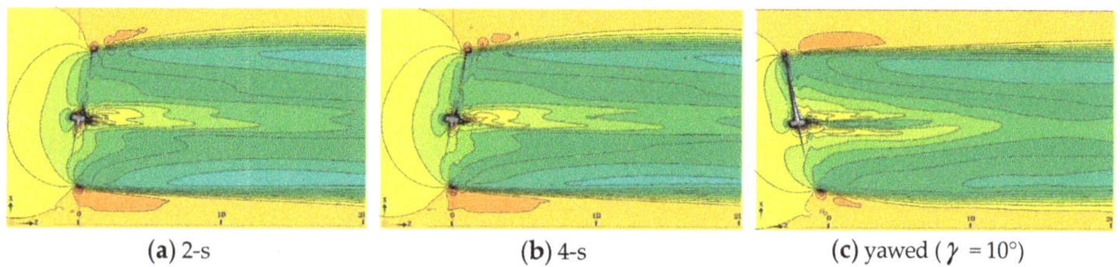

(a) 2-s   (b) 4-s   (c) yawed (γ = 10°)

Figure 20. Velocity contours for Case 2.

(3) Forward-yaw-stop stage (Case 3). Figure 21 shows the instantaneous velocity contours for the dynamic yawing and yawed cases. Both dynamic yawing processes result in a much larger wake zones than under the yawed case. Meanwhile, some deflection occurs in the velocity wake, which is similar to the wake deflection effect reported in the work of Qian [18], which took into account the velocity deficit and turbulent intensity using a Gaussian-based wake model.

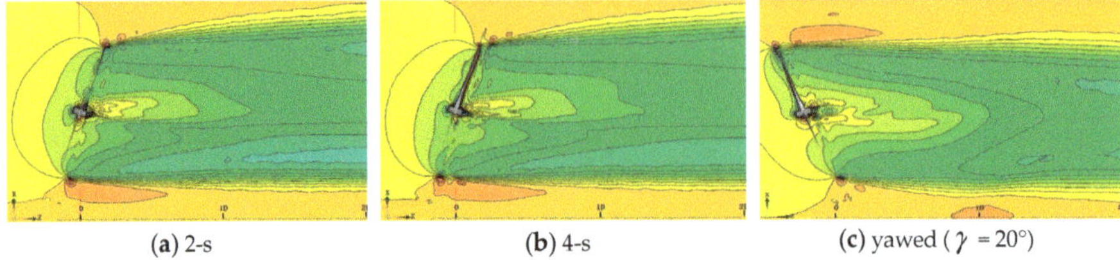

(**a**) 2-s (**b**) 4-s (**c**) yawed ($\gamma = 20°$)

Figure 21. Velocity contours for Case 3.

(4) Backward-yaw start stage (Case 4). Figure 22 illustrates the velocity wake in the backward-yaw start stage. The wake zone velocity field retains almost the same flow structure. To refine the simulation results, large eddy simulations can be used to improve the interpretation of the flow process.

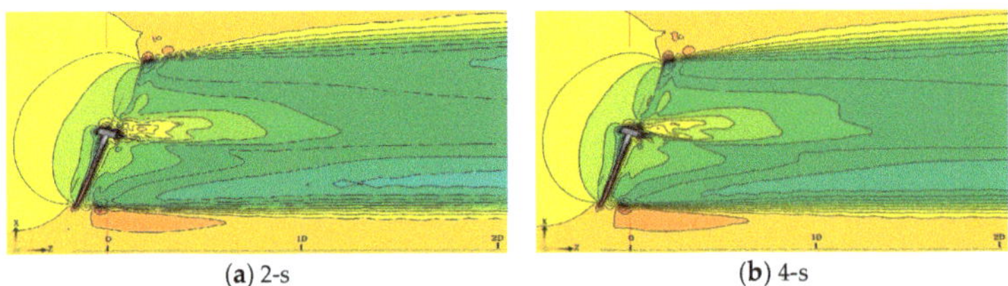

(**a**) 2-s (**b**) 4-s

Figure 22. Velocity contours for Case 4.

(5) Yaw angle of 10° under backward yawing stage (Case 5). When the wind turbine rotates about the Y-axis with a yaw angle of 10°, the velocity wake gradually become symmetrical and the 4-s duration simulation recovers to the symmetry state faster than 2-s duration simulation, as shown in Figure 23.

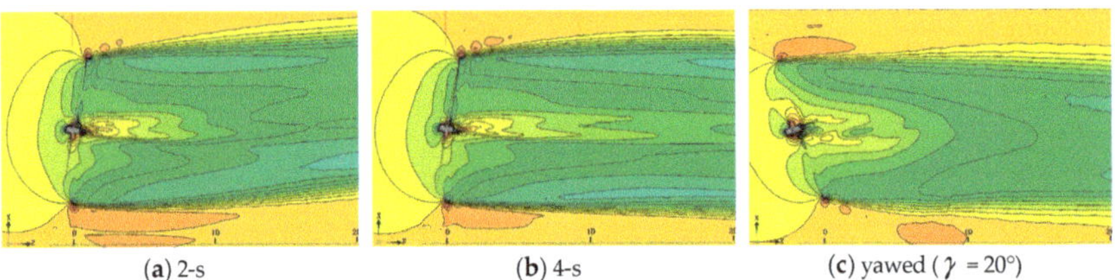

(**a**) 2-s (**b**) 4-s (**c**) yawed ($\gamma = 20°$)

Figure 23. Velocity contours for Case 5.

(6) Backward-yaw stop stage (Case 6). As shown in Figure 24, the wind turbine returns to its initial state, the wind direction is normal to the rotor rotational plane, and the three-dimensional flow structure is symmetrical in streamwise.

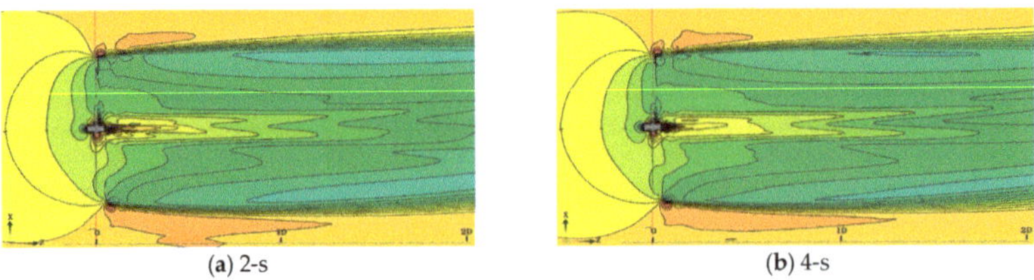

(**a**) 2-s (**b**) 4-s

Figure 24. Velocity contours for Case 6.

In summary, dynamic yawing based on two start-stop durations shows the approximate flow structure, and the upwind and downwind effect induced by yawing process expands the wake zone earlier and much larger than the yawed condition.

3.3.4. Aerodynamic Characteristics along Blade Spanwise Section

To investigate the yawing effect, additional velocity induced by the yawing wind rotor should also be examined. Figure 25 illustrates the velocity triangle along the span of the blade under dynamic yawing. The dashed line of the yawing zone indicates positive yaw (negative yaw was not investigated in the present simulation). Four process variations can be extracted according to the dynamic yawing stage. Note that $R = 0$ means that initial position of the rotational axis of the wind turbine is the rotor hub position, which is different from the platform yawing process used for the offshore wind turbines. The direction of V_{dyn} is different on both sides of the yaw axis. In the process of yawing counterwise, V_{dyn} and the inflow wind speed create an acute angle in the right side of yaw axis, while V_{dyn} and the inflow wind speed create an obtuse angle in the left one of yaw axis, as shown in Figure 26a. The process of yawing clockwise is just the contrary to the case of yawing counterwise. The absolute formulation of V_{dyn} can be written as:

$$V_{dyn} = \left| \omega_{yaw} r \sin(\varphi) \right| \tag{5}$$

where, V_{dyn} become zero at the azimuth angle of $0°$ and $180°$ in the current setup of wind turbine.

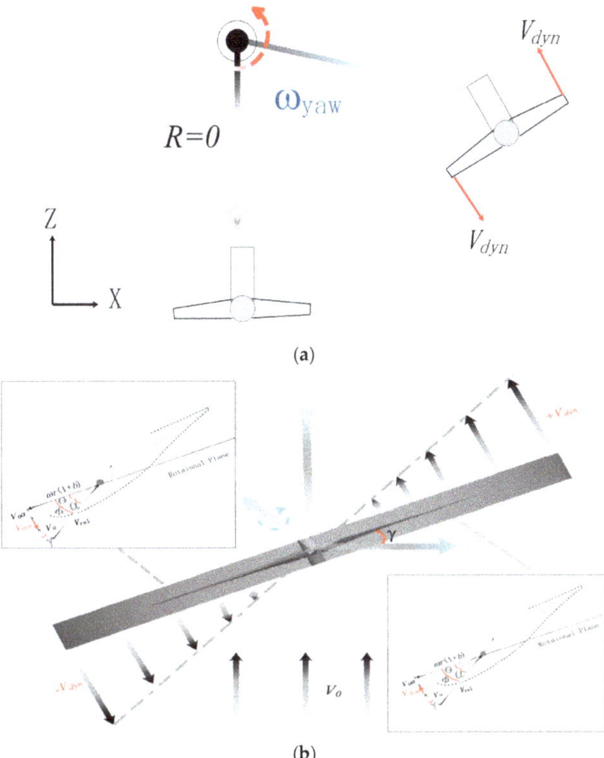

Figure 25. Velocity diagram for a blade section under dynamic yaw at radius r: (**a**) sketch of the yawing process for the velocity dynamic analysis; (**b**) velocity diagram along the span of the blade.

The relative velocity of the section under dynamic yawing can be defined as:

$$V_{rel} = \sqrt{\left(V_o(\cos\gamma - a) + \eta V_{dyn}\right)^2 + \left(\omega r(1 + b) - \beta V_0 \sin\gamma \cos\varphi\right)^2} \tag{6}$$

where $\eta = 1$ if the direction of V_{dyn} and inflow wind velocity creates an acute angle and $\eta = -1$ when the direction of V_{dyn} and inflow wind velocity creates an blunt angle. In the present yawing simulation, if the wind turbine is rotating in the positive yaw direction, $\beta = 1$.

Figure 26 shows the variation of the AOA in three typical section (r/R = 0.21, 0.4, or 0.67), with respect to yaw angles 0–20–0–20–0°. The upper abscissa is the wind rotor yaw angle, while the bottom abscissa is the wind turbine rotational cycle. The vertical coordinate indicates the AOA distribution along the radial direction of the blade. The three sections are indicated by solid lines of different thickness of airfoil. Inner board airfoils are the thickest and outer board one are thinnest.

The AOA gradually decreases along the span of the blade. Due to the effect of dynamic yawing, the AOA oscillates during the rotational period of the rotor, with the combined effects of both retreating & advancing and upwind & downwind. Figure 26a shows the calculated results of 2-s start-stop yaw rate. In this case, the start-stop velocity is faster, and the overall fluctuation are larger than those in the 4-s scenario. The start-stop effect mainly affects the AOA near the outer board but has less influence on other areas. In addition, the start-stop process leads to changes in the AOA.

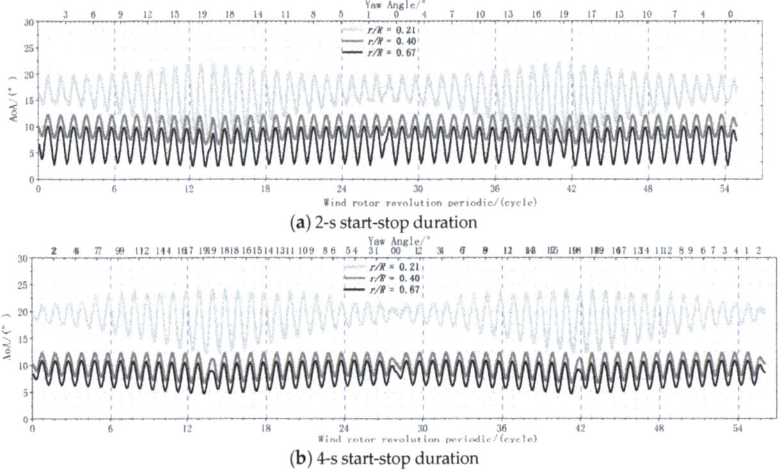

(**a**) 2-s start-stop duration

(**b**) 4-s start-stop duration

Figure 26. Comparison of AOA under different start-stop durations for three different spanwise sections under dynamic yawing.

Figure 27 show the distribution of the normal force coefficient at three typical sections (r/R = 0.21, 0.4, 0.67) within the dynamic process, including two forward yawing and backward yawing states with a yaw angle of 20°. Variation of C_n under dynamic yawing with a 2-s duration causes larger differences at the radial position (r/R = 0.4) than the 4-s duration. Under the yawed case, the average normal force coefficient of the r/R = 0.21 section with a yaw angle of 20° is 1.1, and the maximum and minimum load coefficients are 1.7 and 0.5, respectively (see Figure 14). In the current yawing case, the maximum and minimum load factors are 1.9 and 1.0, which are due to the downwind and upwind effects of the yaw dynamics, similar to the horizontal wind shear effect, but more pronounced than typical horizontal wind shear effects. In the start and stop duration of yawing, the yawing velocity of 2-s case has much higher frequency than 4-s case, resulting to much higher additional velocity and aerodynamic loads. Thus the overall performance of 2-s yawing presents much larger fluctuation in the process of yawing.

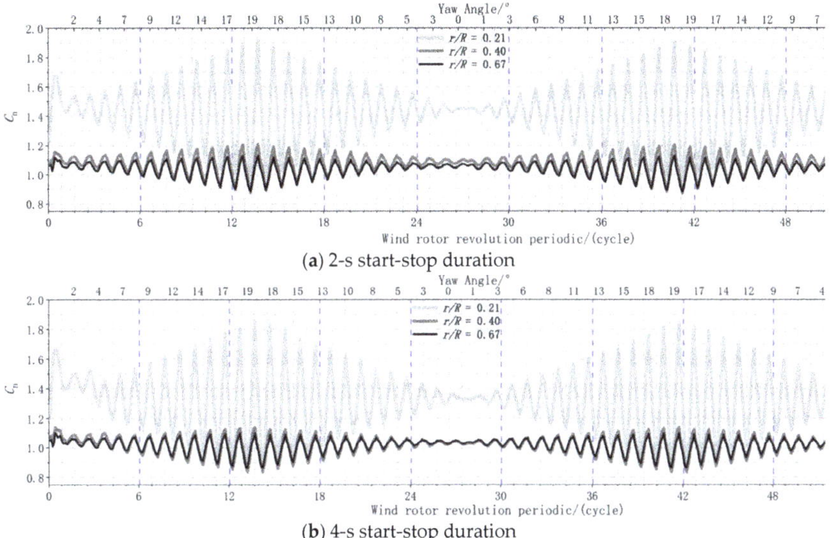

(**a**) 2-s start-stop duration

(**b**) 4-s start-stop duration

Figure 27. Comparison of C_n under different start-stop durations for three different spanwise sections under dynamic yawing.

Figure 28 shows how the tangential force coefficient varies with simulation time at three typical sections ($r/R = 0.21, 0.4,$ and 0.67). The results suggest the aerodynamic load inside the inner board is more influenced under dynamic yawing with the 2-s during than the 4-s duration. Fast shifts of the yaw angle under the yawing start-stop stage influence the aerodynamic load along the inner and middle blade span.

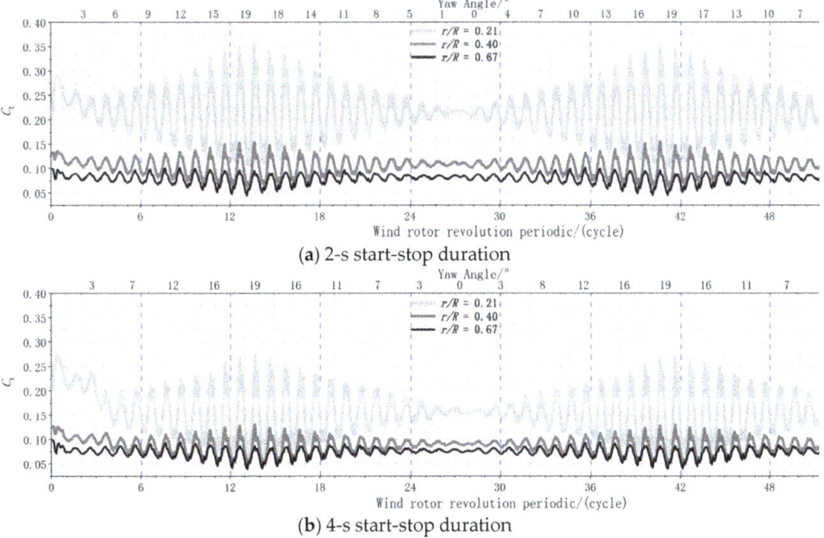

(**a**) 2-s start-stop duration

(**b**) 4-s start-stop duration

Figure 28. Comparison of C_t under different start-stop durations for three different spanwise sections under dynamic yawing.

4. Conclusions

This paper has presented the results of unsteady numerical simulations investigating the static and dynamic aerodynamic performances of a NREL 5-MW HAWT under yawed and yawing conditions. The blades were considered rigid for cases with a prescribed motion, and the tower was not included in the computational domain. The numerical simulation results based on the T-SST turbulent model were consistent with the BEM data. Fluctuating and alternative loads were observed for the yawed case, resulting in the retreating and advancing effect. As the yawing angle decreased, overall performance also decreased according to the cosine law. The results suggest larger variations in power for the dynamic yaw case, compared to yawed cases. The wind turbine suffers from the coupled effects of the rotational axes (yawing axis and rotating axis). For maximum loading of rotor, the combined effects of the yawed condition and yawing resulted in an azimuth around the first periodic, which became larger as the yaw angle increased.

In the simulations of two different yawing start-stop durations (2-s and 4-s), the faster shift of yaw angle (2-s duration) resulted in a larger torque than the 4-s duration, owing to the additional velocity induced by the yawing effect of different sinusoidal frequency and yaw velocity magnitude. Interestingly, the torque of the rotor and blade was much higher in forward yawing compared to backward yawing, caused by reversing the direction of the yaw velocity. The wake deflection occurs in the near wake flow structure and expands more than in the yawed case. The different start-stop durations under dynamic yawing have little influence on the near wake velocity contour. The next step will be to include deformation of the blade in the simulation of multiple axial angular motions.

Author Contributions: S.K. and H.H. contributed to the conception of the study. X.W. and Z.Y. contributed significantly to analysis and manuscript preparation; Z.Y. performed the data analyses and wrote the manuscript; X.W. helped perform the analysis with constructive discussions.

Funding: This research was supported by National Natural Science Foundation of China (51576065), and the Fundamental Research Funds for the Central Universities (2015RCY05, 2018ZD09).

Conflicts of Interest: The authors declare no conflict of interest.

Abbreviations

BEM	Blade Element Momentum method
FVM	Free Vortex Method
VT	Vortex Theory
CFD	Computational Fluid Dynamic
SPIV	Stereo Particle Image Velocimetry
SCADA	Supervisory Control And Data Acquisition
FAST	Fatigue Aerodynamic Structure Turbulence software
FFT	Fast Fourier Transform
URANS	Unsteady Reynolds Averaged Navier-Stokes equations
T-SST	Transient Shear Stress turbulence model
HAWT	Horizontal Axis Wind Turbine
AOA	Angle of Attack

Nomenclature

γ	Yaw Angle
φ	Azimuth Angle
β	Pitch Angle
α_{geom}	Geometric angle of attack
α_{eff}	Effective angle of attack
V_0	Inflow velocity
V_{rel}	Relative velocity
ω	Rotation velocity of wind turbine

T	Time
f	Shedding/Meandering frequency
S_t	Strouhal number
R_e	Reynold number
λ	blade tip speed ratio
C_n	Normal forces coefficient on the local airfoil
C_t	Tangential forces coefficient on the local airfoil
$C_{p\text{-}up}$	Pressure coefficient at the upper side of airfoil
$C_{p\text{-}down}$	Pressure coefficient at the down side of airfoil
C_l	Life forces coefficient
C_d	Drag forces coefficient
C_X	Tangent forces coefficient on the rotational plane
C_Z	Axial forces coefficient on the rotational plane
C_p	Pressure coefficient
C_{po}	Power coefficent
C_t	Thrust coefficent
T	Wind rotor torque
F_t	Wind rotor thrust

References

1. Rahimi, H.; Schepers, J.G.; Shen, W.Z.; García, N.R.; Schneider, M.S.; Micallef, D.; Ferreira, C.J.S.; Jost, E.; Klein, L.; Herráez, I. Evaluation of different methods for determining the angle of attack on wind turbine blades with CFD results under axial inflow conditions. *Renew. Energy* **2018**, *125*, 866–876. [CrossRef]
2. Chehouri, A.; Younes, R.; Ilinca, A.; Perron, J. Review of performance optimization techniques applied to wind turbines. *Appl. Energy* **2015**, *142*, 361–388. [CrossRef]
3. Hand, M.M.; Simms, D.A. *Unsteady Aerodynamics Experiment Phase VI: Wind Tunnel Test Configurations and Available Data Campaigns*; NREL/TP-500-29955; National Renewable Energy Lab: Golden, CO, USA, 2001.
4. Schepers, J.G.; Boorsma, K.; Cho, T. *Analysis of Mexico Wind Tunnel Measurements: Final Report of IEA Task 29, Mexnext (Phase 1)*; Energy Research Centre of the Netherlands: Amsterdam, The Netherlands, 2012.
5. Sant, T. Improving BEM-Based Aerodynamic Models in Wind Turbine Design Codes. Ph.D. Thesis, University of Malta, Msida, Malta, 2007; pp. 45–80.
6. Micallef, D.; Bussel, G.V.; Ferreira, C.S.; Sant, T. An investigation of radial velocities for a horizontal axis wind turbine in axial and yawed flows. *Wind Energy* **2013**, *16*, 529–544. [CrossRef]
7. Vey, S.; Lang, H.M.; Nayeri, C.N.; Paschereit, C.O.; Pechlivanoglou, G.; Weinzierl, G. Utility scale wind turbine yaw from a flow visualization view. In Proceedings of the ASME Turbo Expo 2015: Turbine Technical Conference and Exposition, Montreal, QC, Cananda, 15 June 2015; V009T46A021.
8. Dai, J.; Yang, X.; Hu, W.; Wen, L.; Tan, Y. Effect investigation of yaw on wind turbine performance based on SCADA data. *Energy* **2018**, *149*, 684–696. [CrossRef]
9. Song, D.; Yang, J.; Fan, X.; Liu, Y.; Liu, A.; Chen, G.; Joo, Y.H. Maximum power extraction for wind turbines through a novel yaw control solution using predicted wind directions. *Energy Convers. Manag.* **2018**, *157*, 587–599. [CrossRef]
10. Song, D.; Fan, X.; Yang, J.; Liu, A.; Chen, S.; Joo, Y.H. Power extraction efficiency optimization of horizontal-axis wind turbines through optimizing control parameters of yaw control systems using an intelligent method. *Appl. Energy* **2018**, *224*, 267–279. [CrossRef]
11. Saenz-Aguirre, A.; Zulueta, E.; Fernandez-Gamiz, U.; Lozano, J.; Lopez-Guede, J.M. Artificial Neural Network Based Reinforcement Learning for Wind Turbine Yaw Control. *Energies* **2019**, *12*, 436. [CrossRef]
12. Munters, W.; Meyers, J. Dynamic Strategies for Yaw and Induction Control of Wind Farms Based on Large-Eddy Simulation and Optimization. *Energies* **2018**, *11*, 177. [CrossRef]
13. Ryu, K.; Kang, S.H. Prediction of Aerodynamic Loads for NREL Phase VI Wind Turbine Blade in Yawed Condition. *Int. J. Aeronaut. Space Sci.* **2016**, *17*, 157–166. [CrossRef]
14. Jiménez, Á.; Crespo, A.; Migoya, E. Application of a LES technique to characterize the wake deflection of a wind turbine in yaw. *Wind Energy* **2010**, *13*, 559–572. [CrossRef]

15. Jensen, N.O. *A Note on Wind Generator Interaction*; Risø National Laboratory: Roskilde, Denmark, 1983; ISBN 87-550-0971-9.
16. Gebraad, P.M.O.; Teeuwisse, F.W.; van Wingerden, J.W.; Fleming, P.A.; Ruben, S.D.; Marden, J.R.; Pao, L.Y. Wind plant power optimization through yaw control using a parametric model for wake effects-a CFD simulation study. *Wind Energy* **2014**, *19*, 95–114. [CrossRef]
17. Bastankhah, M.; Porté-Agel, F. Experimental and theoretical study of wind turbine wakes in yawed conditions. *J. Fluid Mech.* **2016**, *806*, 506–541. [CrossRef]
18. Qian, G.W.; Ishihara, T.A. New Analytical Wake Model for Yawed Wind Turbines. *Energies* **2018**, *11*, 665. [CrossRef]
19. Micallef, D.; Sant, T. Chapter 2: A review of wind turbine yaw aerodynamics. In *Wind Turbines-Design, Control and Applications*; InTech: Rijeka, Croatia, 2016; pp. 27–53.
20. Castellani, F.; Astolfi, D.; Natili, F.; Marti, F. The Yawing Behavior of Horizontal-Axis Wind Turbines: A Numerical and Experimental Analysis. *Machines* **2019**, *7*, 15. [CrossRef]
21. Qiu, Y.X.; Wang, X.D.; Kang, S.; Zhao, M.; Liang, J.Y. Predictions of unsteady HAWT aerodynamics in yawing and pitching using the free vortex method. *Renew. Energy* **2014**, *70*, 93–106. [CrossRef]
22. Micallef, D.; Ferreira, C.S.; Sant, T.; Bussel, G.V. Experimental and numerical investigation of tip vortex generation and evolution on horizontal axis wind turbines. *Wind Energy* **2016**, *19*, 1485–1501. [CrossRef]
23. Peters, D.A.; Haquang, N. Dynamic inflow for practical applications. *J. Am. Helicopter Soc.* **1988**, *33*, 64–68. [CrossRef]
24. Gaonkar, G.H.; Peters, D. Review of dynamic inflow modeling for rotorcraft flight dynamics. In *27th Structures, Structural Dynamics and Materials Conference*; VERTICA: Massachusetts, MA, USA, 1988; Volume 12, pp. 213–242.
25. Suzuki, A. Application of Dynamic Inflow Theory to Wind Turbine Rotors. Ph.D. Thesis, The University of Utah, Salt Lake City, UT, USA, 2000.
26. Glauert, H. Airplane propellers. In *Aerodynamic Theory*; Springer: Berlin, Germany, 1935; pp. 169–360.
27. Du, Z.; Selig, M. *A 3D Stall-Delay Model for Horizontal Axis Wind Turbine Performance Predictions*; AIAA-98-0021; ASME: Reno, NV, USA, 1998; pp. 9–19.
28. Buhl, M.L. A new empirical relationship between thrust coefficient and induction factor for the turbulent windmill state. *Natl. Renew. Energy Lab.* **2005**, *47*, 777–780.
29. Yan, S.; Shi, S.P.; Chen, X.M.; Wang, X.D.; Mao, L.Z.; Liu, X.J. Numerical simulations of flow interactions between steep hill terrain and large scale wind turbine. *Energy* **2018**, *151*, 740–747. [CrossRef]
30. Tongchitpakdee, C.; Benjanirat, S.; Sankar, L. Numerical Simulation of the Aerodynamics of Horizontal Axis Wind Turbines Under Yawed Flow Conditions. *J. Sol. Energy Eng.* **2005**, *127*, 464–474. [CrossRef]
31. Yu, D.O.; You, J.Y.; Kwon, O.J. Numerical investigation of unsteady aerodynamics of a Horizontal-axis wind turbine under yawed flow conditions. *Wind Energy* **2013**, *16*, 711–727. [CrossRef]
32. Schulz, C.; Letzgus, P.; Lutz, T.; Kramer, E. CFD study on the impact of yawed inflow on loads, power and near wake of a generic wind turbine. *Wind Energy* **2017**, *20*, 253–268. [CrossRef]
33. Jeong, M.S.; Kim, S.W.; Lee, I.; Yoo, S.J.; Park, K.C. The impact of yaw error on aeroelastic characteristics of a horizontal axis wind turbine blade. *Renew. Energy* **2013**, *60*, 256–268. [CrossRef]
34. Dai, L.; Zhou, Q.; Zhang, Y.W.; Yao, S.G.; Kang, S.; Wang, X.D. Analysis of wind turbine blades aeroelastic performance under yaw conditions. *J. Wind Eng. Ind. Aerodyn.* **2017**, *171*, 273–287. [CrossRef]
35. Leble, V.; Barakos, G. 10-MW wind turbine performance under pitching and yawing motion. *J. Sol. Energy Eng.* **2017**, *139*, 041003. [CrossRef]
36. Wen, B.R.; Tian, X.L.; Dong, X.J.; Peng, Z.K.; Zhang, W.M.; Wei, K.X. A numerical study on the angle of attack to the blade of a horizontal-axis offshore floating wind turbine under static and dynamic yawed conditions. *Energy* **2019**, *168*, 1138–1156. [CrossRef]
37. Jonkman, J.; Butterfield, S.; Musial, W.; Scott, G. *Definition of a 5-MW Reference Wind Turbine for Offshore System Development*; National Renewable Energy Lab. (NREL): Golden, CO, USA, 2009.
38. Tran, T.T.; Kim, D.H. The platform pitching motion of floating offshore wind turbine: A preliminary unsteady aerodynamic analysis. *J. Wind Eng. Ind. Aerodyn.* **2015**, *142*, 65–81. [CrossRef]

39. Zhu, C.; Wang, T.; Zhong, W. Combined Effect of Rotational Augmentation and Dynamic Stall on a Horizontal Axis Wind Turbine. *Energies* **2019**, *12*, 1434. [CrossRef]
40. Wen, B.R.; Dong, X.; Tian, X.; Peng, Z.; Zhang, W.; Wei, K. The power performance of an offshore floating wind turbine in platform pitching motion. *Energy* **2018**, *154*, 508–521. [CrossRef]

Article

Robust LFC Strategy for Wind Integrated Time-Delay Power System Using EID Compensation

Fang Liu *, Kailiang Zhang and Runmin Zou

School of Automation, Central South University, Changsha 410083, China
* Correspondence: csuliufang@csu.edu.cn; Tel.: +86-731-88876750

Received: 5 June 2019; Accepted: 7 August 2019; Published: 21 August 2019

Abstract: This paper presents an active disturbance rejection control (ADRC) technique for load frequency control of a wind integrated power system when communication delays are considered. To improve the stability of frequency control, equivalent input disturbances (EID) compensation is used to eliminate the influence of the load variation. In wind integrated power systems, two area controllers are designed to guarantee the stability of the overall closed-loop system. First, a simplified frequency response model of the wind integrated time-delay power system was established. Then the state-space model of the closed-loop system was built by employing state observers. The system stability conditions and controller parameters can be solved by some linear matrix inequalities (LMIs) forms. Finally, the case studies were tested using MATLAB/SIMULINK software and the simulation results show its robustness and effectiveness to maintain power-system stability.

Keywords: load frequency control (LFC); equivalent input disturbance (EID); active disturbance rejection control (ADRC); wind; linear matrix inequalities (LMI)

---

1. Introduction

Load frequency control (LFC) plays a key role when it comes to measuring the power supply quality of a power system. Ensuring that the frequency is controlled at a fixed value or with small changes in its vicinity is a basic requirement of LFC [1]. In order to maintain the system frequency, some control techniques for power systems have been adopted in LFC such as an adaptive fuzzy logic approach [2], neutral network [3], and robust H_∞ control [4]. Different from conventional energy sources, wind energy has the intrinsic intermittence and fluctuation, which will inevitably bring serious influence on the frequency regulation of a power system [5]. Due to the intermittence of wind power, the large-scale wind power grid operation will affect the stability and balance of power systems [6]. Recently, the LFC problem with wind power sources has attracted much attention. With more and more wind power integrated in power systems, the LFC issue of power systems has become more difficult than before. Therefore, designing an advanced LFC strategy for the wind power generations is of significant value to ensure the stable operation of power systems under the stochastic disturbances of wind power and the random load variation. For multi-area power systems in the presence of wind turbines, a LFC design using the model predictive control (MPC) technique is proposed [7]. In [8], a linear active disturbance rejection control method was applied to power systems with high penetration of wind power. Under the condition of wind speed fluctuation, the linear active disturbance rejection technique has a more prominent control effect than the traditional control method in the doubly-fed wind turbines, which reduces the adjustment time and overshoot [9]. To solve the nonlinearities in the LFC issue of the interconnected power systems, the hybrid neuro-fuzzy scheme was applied in [10]. A low-frequency damping control strategy of a doubly-fed induction generator based on transient energy function analysis of oscillation was proposed in [11].

With the reform of power marketization, the scale of a power system gradually expands, and LFC needs to carry out wide-area information exchange or experience a large amount of data in non-dedicated communication network, which inevitably brings the problem of time delay [12]. A sliding mode control and a robust predictive control strategy for power systems with time-delay and uncertainties of parameter are presented in [13,14] respectively. The authors of [15,16] studied the impulsive control of a nonlinear dynamic system. Considering the time-varying delays, two impulsive control algorithms were designed for the islanded micro-grids in [17]. The delay correlation stability of a LFC scheme is studied by means of the Lyapunov-theory and linear matrix inequality (LMIs) techniques in [18]. The authors of [19] studied the LFC for power systems with communication delays via an event-triggered control method. In [20], a new criterion for the delay-related stability was proposed when the network multi-area LFC system was subjected to an unknown time variant exogenous load disturbance. Time delay not only reduces the control effect of the original LFC, but also makes the controller malfunction, causing the instability of the system and damaging the safe operation of the power grid. Therefore, designing a robust LFC strategy which can perfectly compensate for the influence of time delay becomes an increasingly valuable solution of the wind integrated power system [21].

In this paper, the influence of wind power integration on load frequency of a power system was studied, and the influence of communication delay on the whole system was also considered. An active disturbance rejection control (ADRC) based on equivalent input disturbances (EID) compensation for load frequency control was proposed for a wide integrated power system when communication delays were considered, applied to a two-area power system to dampen its low frequency oscillation. The disturbance information was obtained through the full-order state observer, and the disturbance estimator was designed to compensate for the disturbance. Thus, the disturbance rejection performance of the whole control system was improved.

The remaining sections of this paper are structured as follows: In Section 2, simplified wind turbine models for frequency studies are introduced. In Section 3, ADRC design strategies based on EID are discussed. Some case studies are introduced in Section 4, and the conclusions are drawn in Section 5.

2. System Modelling

Figure 1 shows a two-area interconnected wind integrated power system with two conventional generator units in each area, one aggregated wind turbine model and two controllers based on EID. In the following analysis, the basic parameter description will be listed. The notation Δ indicates the deviation from the normal state.

Currently, the variable speed wind turbine (VSWT) is the most popular type of modern wind turbine. There is a more detailed description of VSWT in [22]. Figure 2 shows a simplified frequency response model of a wind turbine based on doubly-fed induction generator (DFIG). The structure of this model can be described by the following equations [22]:

$$i_{qr} = -\left(\frac{1}{T_1}\right)i_{qr} + \left(\frac{X_2}{T_1}\right)V_{qr} \tag{1}$$

$$\dot{w} = -\left(\frac{X_3}{M_t}\right)i_{qr} + \left(\frac{1}{M_t}\right)T_m \tag{2}$$

$$p_e = wX_3i_{qr} \tag{3}$$

by linearizing, Equation (3) can be rewritten as:

$$p_e = w_{opt}X_3i_{qr} \tag{4}$$

$$T_e = i_{qs} = -\frac{L_m}{L_{ss}}i_{qr} \tag{5}$$

where w_{opt} is the operating point of the rotational speed, T_e is the electromagnetic torque, T_m is the mechanical power change, ω is the rotational speed, P_e is the active power of the wind turbine, i_{qr} is the q-axis component of the rotor current, V_{qr} is the q-axis component of the rotor voltage and M_t is the equivalent inertia constant of the wind turbine.

The main parameters in Figure 2 can be observed from Table 1.

$$L_0 = L_{rr} + \frac{L_m^2}{L_{ss}}, \quad L_{ss} = L_s + L_m, \quad L_{rr} = L_{rs} + L_m$$

where L_m is the magnetizing inductance, R_r and R_s are the rotor and stator resistances, respectively.

L_r and L_s are the rotor and stator leakage inductances, respectively, L_{rr} and L_{ss} are the rotor and stator self-inductances, respectively, w_s is the synchronous speed.

Since each subsystem is connected by power flow through a tie line, a LFC system of each area of the two-area power system should not ignore the control of the interchange power and local frequency with the other control area. Therefore, we take the tie-line power signal into account in the dynamic LFC system model and describes a frequency model for any area i of N power system control areas with an aggregated generator unit in each area [11].

Table 1. Parameters for Figure 1.

X_2	X_3	T_1
$\frac{1}{R_r}$	$\frac{L_m}{L_{ss}}$	$\frac{L_0}{w_s R_s}$

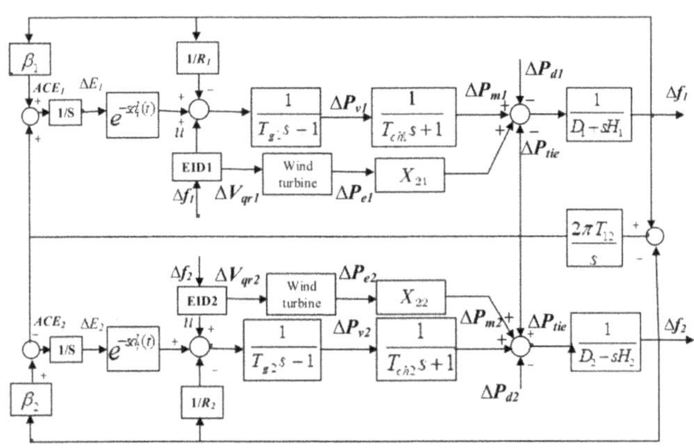

Figure 1. A two-area wind integrated power system for equivalent input disturbances (EID) based load frequency control (LFC).

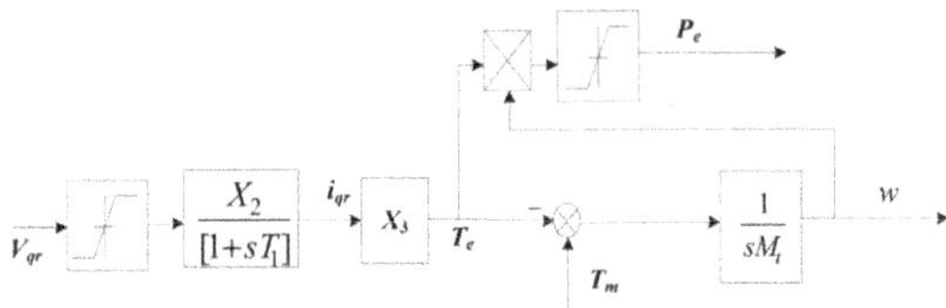

Figure 2. Simplified frequency response model of wind turbine base on doubly-fed induction generator (DFIG).

The LFC dynamic model of one area of the interconnected power system, shown in Figure 3, can be described as follows:

$$
\begin{cases}
\Delta \dot{f}_i = -\frac{D_i}{M_i}\Delta f_i + \frac{1}{M_i}\Delta P_{mi} - \frac{1}{M_i}\Delta P_{tie,i} + \frac{1}{M_i}\Delta P_e - \frac{1}{M_i}\Delta P_{di} \\
\Delta \dot{P}_{mi} = -\frac{1}{T_{chi}}\Delta P_{mi} + \frac{1}{T_{chi}}\Delta P_{vi} \\
\Delta \dot{P}_{vi} = -\frac{1}{R_i T_{gi}}\Delta f_i - \frac{1}{T_{gi}}\Delta P_{vi} - \frac{\Delta E(t-d)}{T_{gi}} + \frac{1}{T_{gi}}\Delta P_{ref} \\
\Delta \dot{P}_{tie,i} = 2\pi\left(\sum\limits_{j=1,j\neq i}^{N} T_{ij}\Delta f_i - \sum\limits_{j=1,j\neq i}^{N} T_{ij}\Delta f_j \right) \\
\Delta \dot{E}_i = ACE_i
\end{cases}
\tag{6}
$$

Additionally, ΔE_i is the area control error (ACE) integral control.

$$
\Delta E_i = \int_0^t ACE(s)ds
\tag{7}
$$

For a multi-area LFC system, the ACE is essentially composed of a regional frequency deviation and a power deviation of the line, and its calculation formula is as follows:

$$
ACE_i = \Delta P_{tie,i} + \beta_i \Delta f_i
\tag{8}
$$

where β_i is the frequency deviation coefficient of the control area, Δf_i is the frequency deviation of area i, $\Delta P_{tie,i}$ is the tie-line power change of area i.

ΔV_i is the control area interface:

$$
\Delta V_i = \sum\limits_{i=1,j\neq i}^{N} T_{ij}\Delta f_j
\tag{9}
$$

where Δf is the frequency deviation, ΔP_m is the generator mechanical power deviation, ΔP_v is the turbine value position deviation, ΔP_d is the load deviation, M and D denote inertia moment and damping coefficient of generator, respectively, T_g, T_{ch} and R denote the governor's time constant, turbine's time constant and speed drop, respectively.

Furthermore, the above equations can be combined in the following state-space model:

$$
\begin{cases}
\dot{x}_i(t) = A_i x_i(t) + A_{di} x_i(t-d) + B_i u_i(t) + B_{wi} w_i(t) \\
y_i(t) = C_i x_i(t)
\end{cases}
\tag{10}
$$

where

$$
x_i(t) = \begin{bmatrix} \Delta f_i & \Delta P_{mi} & \Delta P_{vi} & \Delta E_i & \Delta P_{tie,i} & \Delta i_{qr} & \Delta W \end{bmatrix}^T
$$

$$
A_i = \begin{bmatrix}
-\frac{D_i}{M_i} & \frac{1}{M_i} & 0 & 0 & \frac{-1}{M_i} & \frac{X_3 W_{opt}}{M_i} & 0 \\
0 & \frac{-1}{T_{chi}} & \frac{1}{T_{chi}} & 0 & 0 & 0 & 0 \\
-\frac{1}{R_i T_{gi}} & 0 & \frac{-1}{T_{gi}} & 0 & 0 & 0 & 0 \\
\beta_i & 0 & 0 & 0 & 1 & 0 & 0 \\
2\pi\sum\limits_{j=1,j\neq i}^{N} T_{ij} & 0 & 0 & 0 & 0 & 0 & 0 \\
0 & 0 & 0 & 0 & 0 & \frac{-1}{T_1} & 0 \\
0 & 0 & 0 & 0 & 0 & \frac{-X_3}{M_t} & 0
\end{bmatrix},
$$

$$
A_{di} = \begin{bmatrix}
0 & 0 & 0 & 0 & 0 & 0 & 0 \\
0 & 0 & 0 & 0 & 0 & 0 & 0 \\
0 & 0 & 0 & \frac{-1}{T_{gi}} & 0 & 0 & 0 \\
0 & 0 & 0 & 0 & 0 & 0 & 0 \\
0 & 0 & 0 & 0 & 0 & 0 & 0 \\
0 & 0 & 0 & 0 & 0 & 0 & 0 \\
0 & 0 & 0 & 0 & 0 & 0 & 0
\end{bmatrix}, B_i = \begin{bmatrix}
0 & 0 \\
0 & 0 \\
\frac{1}{T_{gi}} & 0 \\
0 & 0 \\
0 & 0 \\
0 & \frac{X_2}{T_1} \\
0 & 0
\end{bmatrix},
$$

$$B_{wi} = \begin{bmatrix} \frac{-1}{M_i} & 0 & 0 & 0 & 0 & 0 & 0 \\ 0 & 0 & 0 & 0 & 0 & 0 & \frac{1}{M_t} \\ 0 & 0 & 0 & 0 & -2\pi & 0 & 0 \end{bmatrix}^T$$

$$C_i = \begin{bmatrix} 1 & 0 & 0 & 0 & 0 & 0 & 0 \end{bmatrix}, w_i = \begin{bmatrix} \Delta P_{di} & \Delta T_m & \Delta V_i \end{bmatrix}^T$$

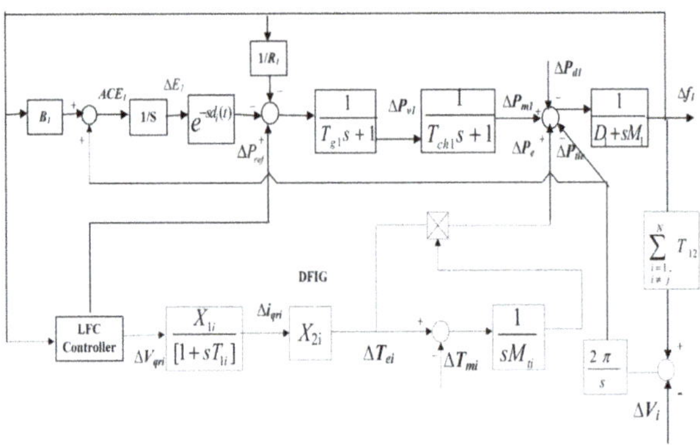

Figure 3. Dynamic model of one area of the interconnected environment.

3. ADRC Design Based on EID Compensation

This section is devoted to explaining the proposed EID-based ADRC design for the LFC. The original EID method cannot be used directly for systems with time delays, so we improved the state observer and extended the method to the time-delay system [23]. EID was originally proposed by She [24,25] and further developed by Liu [26–28]. The method is based on the idea that the effect of actual disturbance, $w(t)$, on the output of a plant (Figure 4a) can be replaced by the disturbance, $w_e(t)$, on the control input channel (Figure 4b). In Figure 4, $w(t)$ and $w_e(t)$ produced exactly the same outputs. Thus, the disturbance $w_e(t)$ is defined as EID.

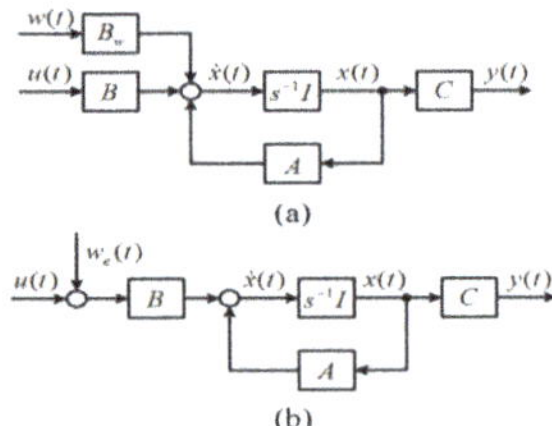

Figure 4. The concept of equivalent input disturbance, (**a**) original plant; (**b**) plant with EID.

Rewriting the plant (Equation (10)) as a plant with EID, we have

$$\begin{cases} \dot{x}\{t = Ax\{t + A_d x\{t - d + B[u[t] + w_e[t]] \\ y\{t = Cx\{t \end{cases} \tag{11}$$

Then, the system (Equation (11)) can be used to design ADRC. We constructed an EID-based closed-loop control system as in Figure 5. The system has five parts: the internal model, the state feedback controller, the disturbance estimator, the modified state observer and the control plant.

3.1. Configuration of the EID-Based Time-Delay System

In Figure 5, a new EID-based control system structure was established to achieve disturbance suppression of the wind integrated time-delay power system.

The following internal model

$$\dot{x}_R(t) = A_M x_M(t) + B_M \left[\Delta f_{ref}[t] - y[t] \right] \tag{12}$$

is still used to ensure accurate tracking of the reference input. When Δf_{ref} is given, A_M and B_M can be directly determined.

A full-order time-delay observer is used to estimate the EID and reconstruct the state of the controlled object, we write the state-space representation of the observer as

$$\begin{cases} \dot{\widetilde{x}}\{t = \Phi \widetilde{x}\{t + A_d \widetilde{x}\{t - d + \Psi[y[t] - C\widetilde{x}[t]] + \Gamma u_f\{t \\ \widetilde{y}\{t = T^{-1}\widetilde{x}\{t \end{cases} \tag{13}$$

where $\widetilde{x}(t)$ is the reconstruction state of $x(t)$. The gain of this full-order time-delay observer is L.

Then, we design the state-feedback controller as

$$u_f(t) = K_M x_M(t) + K_N \widetilde{x}(t) \tag{14}$$

From the above equation, we yield the disturbance estimation of the EID $\hat{w}(t)$ in Figure 5 as

$$\hat{w}(t) = B^+ T^{-1} \Psi C[x[t] - \widetilde{x}[t]] + u_f(t) - u(t) \tag{15}$$

where $B^+ = \left(B^T B \right)^{-1} B^T$

Since the output contains measurement noises, we used a low-pass filter to select angular frequency bandwidth for disturbance estimation. The state-space equation of the filter is described as

$$\begin{cases} \dot{x}_N\{t = A_N x_N\{t + B_N \hat{w}\{t \\ \widetilde{w}\{t = C_N x_N\{t \end{cases} \tag{16}$$

where $x_N(t)$ is the state of filter, $\widetilde{w}(t)$ is the filter disturbance estimation.

Thus, the new control law of the closed-loop control system is

$$u(t) = u_f(t) - \widetilde{w}(t) \tag{17}$$

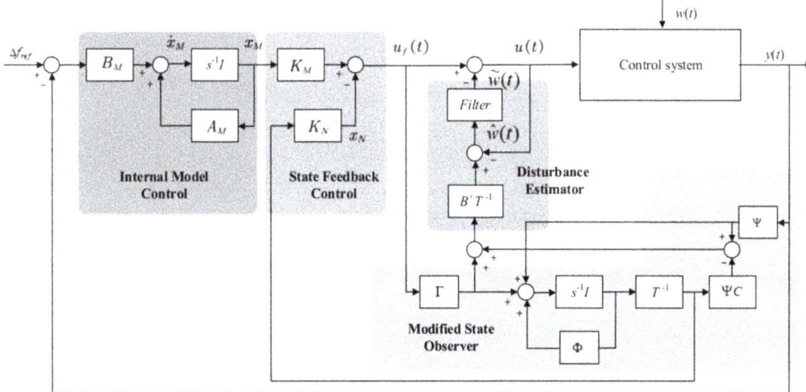

Figure 5. Configuration of the EID-based closed-loop control system.

3.2. Optimal Design of Controller Parameters

Time delay will influence the stability of the system. On account of inherent time delay, the characteristic equation of the system becomes infinitely dimensional. Thus, we propose the parameter of a controller design based on the LMI in this section.

Let $\Delta f_{ref}(t) = 0$, $w(t) = 0$. Then, the time-delay model (Equation (10)) is

$$\begin{cases} \dot{x}(t) = Ax(t) + A_d x(t-d) + Bu(t) \\ y(t) = Cx(t) \end{cases} \tag{18}$$

As shown in Figure 4, there are four states, $\widetilde{x}(t)$, $\Delta x(t)$, $x_N(t)$ and $x_M(t)$. Define

$$\Delta x = x(t) - \widetilde{x}(t) \tag{19}$$

and describe the closed-loop system as

$$\varphi(t) = [\ \widetilde{x}^T(t) \quad \Delta x^T(t) \quad x_N{}^T(t) \quad x_M{}^T(t)\ ]^T \tag{20}$$

Substituting Equation (19) into (13) yields

$$\dot{\widetilde{x}}(t) = A\widetilde{x}(t) + LC\Delta x(t) + Bu_f(t) + A_d\widetilde{x}(t-d) \tag{21}$$

Combining Equations (13), (17)–(19), yields

$$\Delta\dot{x}(t) = [A - LC]\Delta x(t) - BC_N x_N(t) + A_d\Delta x(t-d) \tag{22}$$

Combining Equations (15) and (17), the filter is described as

$$\dot{x}_N(t) = B_N B^+ LC\Delta x(t) + (A_N + B_N C_N)x_N(t) \tag{23}$$

Substituting Equation (19) into (12), the internal model is obtained as:

$$\dot{x}_N(t) = B_N B^+ LC\Delta x(t) + (A_N + B_N C_N)x_N(t) \tag{24}$$

From Equations (21)–(24), the state-space representation of the control-loop system reconstructed according to EID in Figure 4, is as follows:

$$\dot{\varphi}(t) = \overline{A}\varphi(t) + \overline{B}u_f(t) + \overline{A}_d\varphi(t-d) \tag{25}$$

where

$$\overline{A} = \begin{bmatrix} A & LC & 0 & 0 \\ 0 & A-LC & -BC_N & 0 \\ 0 & B_N B^+ LC & A_N + B_N C_N & 0 \\ -B_M C & -B_M C & 0 & A_M \end{bmatrix},$$

$$\overline{A}_d = \begin{bmatrix} A_d & 0 & 0 & 0 \\ 0 & A_d & 0 & 0 \\ 0 & 0 & 0 & 0 \\ 0 & 0 & 0 & 0 \end{bmatrix}, \overline{B} = [\ B^T \quad 0 \quad 0 \quad 0\ ]^T$$

The state-feedback controller law is:

$$u_f = \overline{K}\varphi(t) \tag{26}$$

where

$$\overline{K} = [\ K_N \quad 0 \quad 0 \quad K_M\ ]$$

Substituting Equation (26) into (25) yields the general form of time-delay system:

$$\dot{\varphi}(t) = \hat{A}\varphi(t) + \overline{A}_d\varphi(t - d) \tag{27}$$

where

$$\hat{A} = \begin{bmatrix} A + BK_N & LC & 0 & BK_M \\ 0 & A - LC & -BC_N & 0 \\ 0 & B_NB^+LC & A_N + B_NC_N & 0 \\ -B_MC & -B_MC & 0 & A_M \end{bmatrix}$$

3.3. Stability Analysis

Using the following Lemma 1, the time-delay power system stability analysis was carried out.

Lemma 1. *If there is a positive definite matrix Q and P, which makes the following LMI feasible, then the time-delay system (27) is asymptotically stable* [29,30].

$$\begin{bmatrix} P\hat{A} + \hat{A}^TP + Q & P\overline{A}_d \\ \overline{A}_d{}^TP & -Q \end{bmatrix} < 0 \tag{28}$$

Based on Lemma 1, the controller gain and a sufficient condition for power system stability are obtained as follows.

Theorem 1. *If there is a positive definite matrix X_1, X_{11}, X_{22}, X_3, X_4, Y_1, Y_2, Y_3 and Y_4, and suitable dimension matrices Z_1, Z_2, and Z_3, the following LMI is feasible. For a given positive parameter α and γ, the time-delay system (25) is asymptotically stable under the control law (26).*

$$\begin{bmatrix} \Phi & \Psi & X \\ \Psi^T & -Y & 0 \\ X^T & 0 & -Y \end{bmatrix} < 0 \tag{29}$$

where

$$\Phi = \begin{bmatrix} \Phi_{11} & Z_2C & 0 & \Phi_{14} \\ C^TZ_2{}^T & \Phi_{22} & \Phi_{23} & -X_2C^TB_M{}^T \\ 0 & \Phi_{23}{}^T & \Phi_{33} & 0 \\ \Phi_{14}{}^T & -B_MCX_2{}^T & 0 & \Phi_{44} \end{bmatrix},$$

$$\Psi = \begin{bmatrix} \alpha A_dY_1 & 0 & 0 & 0 \\ 0 & A_dY_2 & 0 & 0 \\ 0 & 0 & 0 & 0 \\ 0 & 0 & 0 & 0 \end{bmatrix},$$

$$X = \begin{bmatrix} \alpha X_1 & 0 & 0 & 0 \\ 0 & X_2 & 0 & 0 \\ 0 & 0 & X_3 & 0 \\ 0 & 0 & 0 & \gamma X_4 \end{bmatrix}, Y = \begin{bmatrix} Y_1 & 0 & 0 & 0 \\ 0 & Y_2 & 0 & 0 \\ 0 & 0 & Y_3 & 0 \\ 0 & 0 & 0 & Y_4 \end{bmatrix}$$

$$\Phi_{11} = \alpha AX_1 + \alpha X_1A^T + \alpha BZ_1 + \alpha Z_1{}^TB^T,$$

$$\Phi_{14} = \gamma BZ_3 - \alpha X_1C^TB_M{}^T,$$

$$\Phi_{22} = AX_2 + X_2A^T - Z_2C - C^TZ_2{}^T,$$

$$\Phi_{23} = -BC_NX_3 + C^TZ_2{}^TB^+{}^TB_N{}^T,$$

$$\Phi_{33} = (A_N + B_NC_N)X_3 + X_3(A_N + B_NC_N)^T,$$

$$\Phi_{44} = \gamma A_MX_4 + \gamma X_4A_M{}^T,$$

and the singular-value decomposition of X_2 is

$$X_2 = V diag\{X_{11}, X_{22}\} V^T$$

The gain of the state feedback controller and the observer is

$$K_N = Z_1 X_1^{-1}, K_M = Z_3 X_4^{-1}, L = Z_2 U S X_{11}^{-1} S^{-1} U^T \tag{30}$$

Additionally, U and V can be obtained from

$$C = U[S, 0] V^T \tag{31}$$

where Equation (31) is a singular value decomposition expression of matrix C.

A detailed proof of Theorem 1 is given in [23], so we omitted the process of proof in this paper.

4. Case Studies

In this section, the proposed ADRC design based on EID compensation is evaluated using MATLAB/SIMULINK software. The basic parameter description of the model is listed in Appendix A [31]. Furthermore, the different profiles of load variation are considered to test the performance of ADRC allocated to each area.

Firstly, the effectiveness of the proposed ADRC method is verified. We study the case that both areas are without ADRC to witness the impact of time delays on two areas of power system and the other case that both areas are equipped with ADRC to confirm the effectiveness of this method. Two-area time-delay power systems with wind farm are considered. It is worth noting that the communication delay is set to be 0.2 s [32,33]. When the hysteresis power system is subjected to random load disturbance as shown in Figure 6a, from Figure 6b we can see that the frequency deviation is large and oscillating, the frequency fluctuation is beyond $[-1\ 1]$ Hz after 2.2 s since the fundamental frequency of a power network is 50 Hz, and even diverges. However, the frequency fluctuations can be damped in the range of $[-0.5\ 0.5] \times 10^{-3}$ pu as shown in Figure 6c by ADRC based on EID compensation.

Next, we verify the superiority of the proposed method. The dynamic responses of wind farm time-delay power system equipped with PID controller and the proposed EID-based ADRC are shown in Figures 7 and 8. Figure 7 shows the dynamic response of frequency in Area 1 and Figure 8 shows the dynamic response of frequency in Area 2. By PID controller, the frequency is varied between $[-1\ 1] \times 10^{-3}$ pu. However, compared with PID controller, the ADRC method based on EID compensation proposed in this paper has higher stability and faster speed. As shown in Figure 7c, Δf is controlled within $[-2\ 2.5] \times 10^{-6}$ pu. Therefore, EID-based ADRC method has better robustness than the PID method.

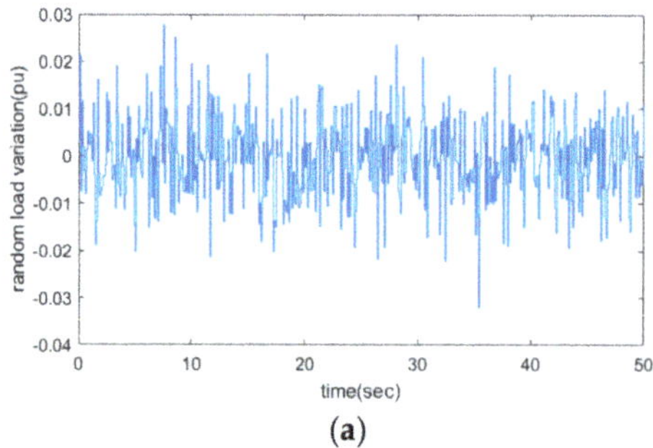

(a)

Figure 6. *Cont.*

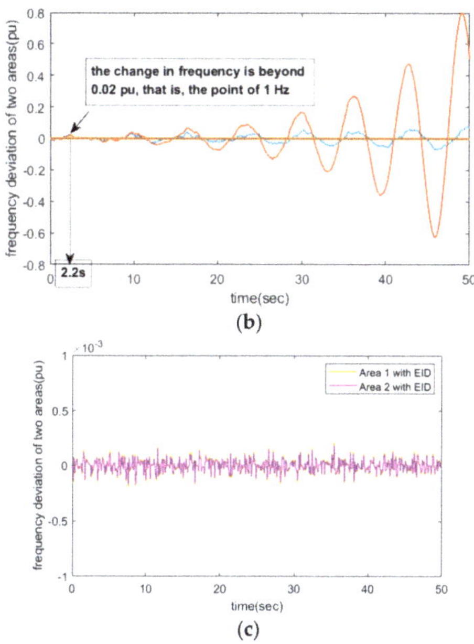

Figure 6. Dynamic response of two areas, (**a**) random load variation; (**b**) frequency response of two areas without ADRC; (**c**) frequency response of two areas with ADRC.

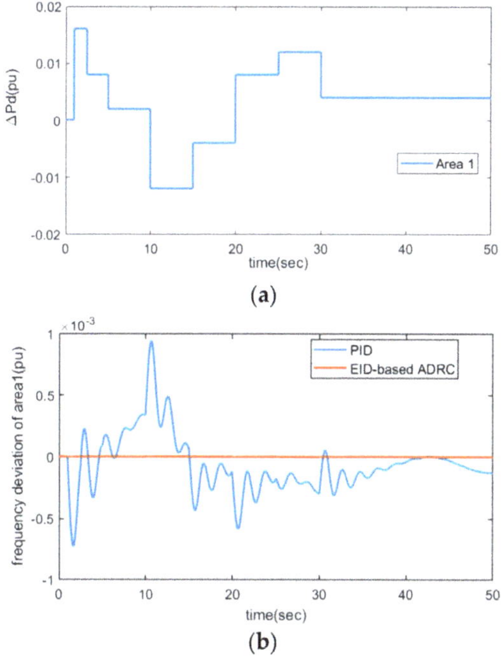

Figure 7. *Cont.*

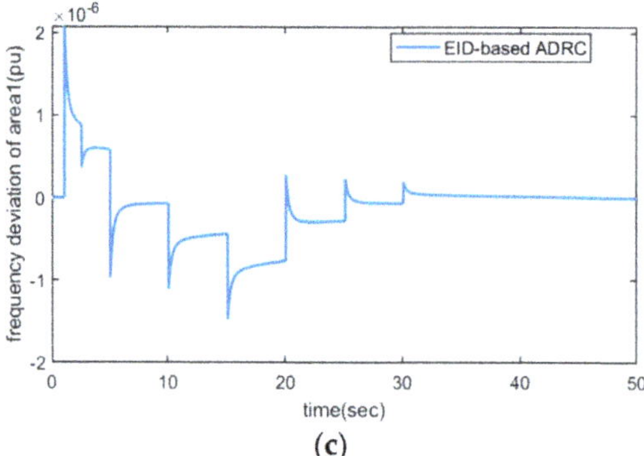

Figure 7. Dynamic response of Area 1, (**a**) load variation; (**b**) frequency deviations of Area 1 with different controller; (**c**) enlarged view under EID.

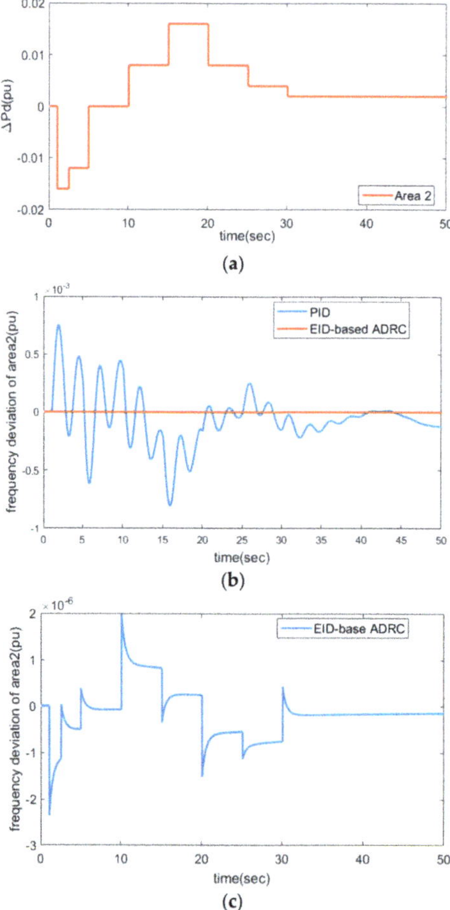

Figure 8. Dynamic response of Area 2, (**a**) load variation of Area 2; (**b**) frequency deviations of Area 2 with different controller; (**c**) enlarged view under EID.

When the system is disturbed by different random load as shown in Figure 9a, the dynamic performances are shown in Figure 9b–d. It can be found that the proposed strategy based on EID can enhance the frequency stability of each area and the control performance is better the PID control strategy.

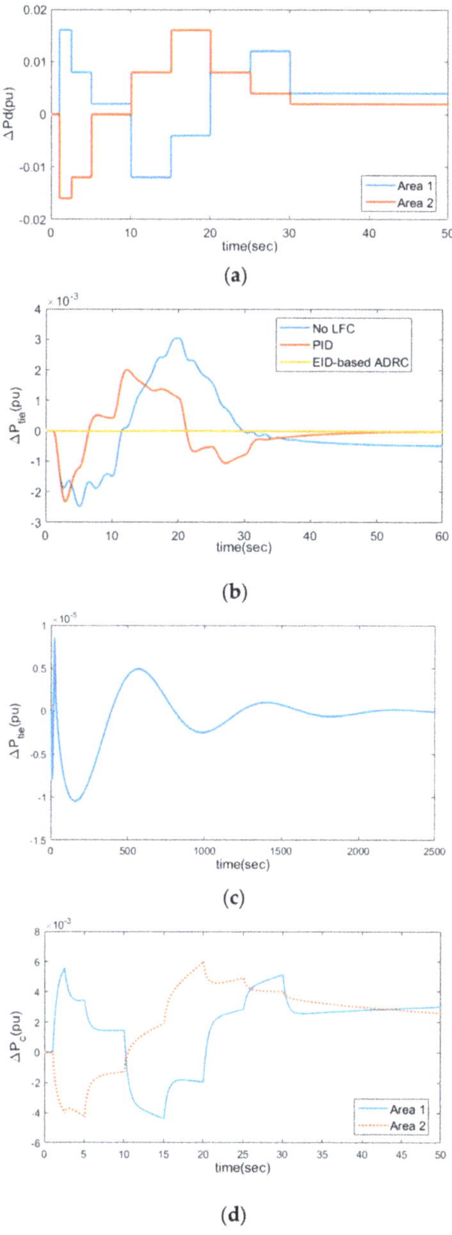

Figure 9. Tie-line power response of a time-delay power system with wind farm under random load, (**a**) load variation; (**b**) tie-line power deviation; (**c**) enlarged view under EID; (**d**) system control signals of EID-LFC under random load disturbance.

Figure 10 shows the simulation results with wind farm participation or without wind farm participation. It has been shown that the control system with the wind farm participation is more stable compared to the system without wind farm participation under the proposed control method.

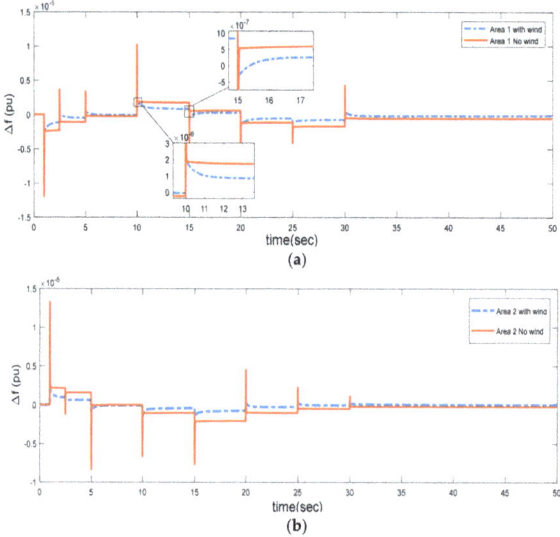

Figure 10. Dynamic response of EID with wind farm participation and EID without wind farm participation (**a**) response of Area 1; (**b**) response of Area 2.

5. Conclusions

In this article, directed at the influence of large-scale wind power integration on the security and stability of power system, and considering the impact of communication delay on this system, an ADRC method with EID compensation was applied to maintain the frequency stability of a time-delay power system with a wind farm. The LFC system model was established, and simulation results validated the effectiveness and superiority of the proposed control method. The disturbance in real time can be estimated and compensated with the control strategy based on EID. Finally, by comparing with traditional LFC methods, the simulation results show that the proposed ADRC method has a significantly higher performance at solving frequency instability under various types of load variations. The EID-based ADRC strategy can quickly and effectively suppress the influence of external disturbances on the frequency of power system.

Author Contributions: All the authors made contributions to the concept and design of the article; F.L. is the main author of this work. R.Z. provided good advice and technical guidance for the manuscript; F.L., K.Z. and R.Z. reviewed and polished the manuscript.

Funding: This work was supported by the Natural Science Foundation of China (NSFC) under Grant 61673398, in part by Natural Science Foundation of Hunan Province of China under Grant 2018JJ2529, and in part by the Huxiang Youth Talent Program of Hunan Province under Grant 2017RS3006.

Conflicts of Interest: The authors declare no conflict of interest.

Appendix A

The following is the parameters of two control area model. In addition, the coupling coefficient between these two areas is 0.1986 pu/rad.

Table A1. Parameters of two control areas.

Parameter	T_{ch} (s)	T_g (s)	R	D	β	M (s)
Area 1	0.3	0.1	0.05	1.0	21.0	10
Area 2	0.4	0.17	0.05	1.5	21.5	12

Table A2. Wind Turbine Parameters And Operating Point.

R_r (pu)	R_s (pu)	X_{lr} (pu)	X_{ls} (pu)	X_m (pu)	M_t (pu)
0.3	0.1	0.05	1.0	21.0	10

References

1. Sasaki, T.; Enomoto, K. Dynamic Analysis of Generation Control Performance Standards. *IEEE Power Eng. Rev.* **2002**, *22*, 54. [CrossRef]
2. Yousef, H. Adaptive fuzzy logic load frequency control of multi-area power system. *Int. J. Electr. Power Energy Syst.* **2015**, *68*, 384–395. [CrossRef]
3. Nag, S.; Philip, N. Application of Neural Networks to Automatic Load Frequency Control. In Proceedings of the International Conference on Swarm, Evolutionary, and Memetic Computing, Calcutta, India, 31 January–2 February 2014; Springer: Cham, Switzerland; Berlin/Heidelberg, Germany, 2013.
4. Chuang, N. Robust H ∞ load-frequency control in interconnected power systems. *IET Control. Theory Appl.* **2016**, *10*, 67–75. [CrossRef]
5. Kumar, A. Impact Study of Dfig Based Wind Power Penetration on Lfc of a Multiarea Power System. In Proceedings of the IEEE India Conference, New Delhi, India, 17–20 December 2015.
6. Rather, Z.H.; Chen, Z.; Thøgersen, P.; Lund, P. Dynamic Reactive Power Compensation of Large-Scale Wind Integrated Power System. *IEEE Trans. Power Syst.* **2015**, *30*, 2516–2526. [CrossRef]
7. Kunya, A.B.; Argin, M. Model Predictive Load Frequency Control of Multi-Area Interconnected Power System. In Proceedings of the 2018 IEEE Texas Power and Energy Conference, College Station, TX, USA, 8–9 February 2018; pp. 1–6.
8. Wang, S.B.; Jiang, Q.Y.; Liu, Z.Y.; Gao, Y.J. Wide-area Damping Control Considering Multiple Delays of Feedback Signals. *Autom. Electr. Power Syst.* **2008**, *32*, 18–22.
9. Yu, X.; Tomsovic, K. Application of Linear Matrix Inequalities for Load Frequency Control with Communication Delays. *IEEE Trans. Power Syst.* **2004**, *19*, 1508–1515. [CrossRef]
10. Bevrani, H.; Hiyama, T. Robust decentralized PI based LFC design for time delay power systems. *Energy Convers. Manag.* **2008**, *49*, 193–204. [CrossRef]
11. Bevrani, H. *Robust Power System Frequency Control*; Springer Science and Business Media LLC: Berlin/Heidelberg, Germany, 2014.
12. Sönmez, S.; Ayasun, S. Stability Region in the Parameter Space of PI Controller for a Single-Area Load Frequency Control System with Time Delay. *IEEE Trans. Power Syst.* **2016**, *31*, 1–2. [CrossRef]
13. Yongjuan, L.; Yang, M.; Yang, Y.; Limin, W. The study of sliding mode load frequency control for single area time delay power system. In Proceedings of the 27th Chinese Control and Decision Conference (CCDC), Qingdao, China, 23–25 May 2015.
14. Ojaghi, P.; Rahmani, M. LMI-Based Robust Predictive Load Frequency Control for Power Systems with Communication Delays. *IEEE Trans. Power Syst.* **2017**, *32*, 4091–4100. [CrossRef]
15. Li, X.; Wu, J. Sufficient stability conditions of nonlinear differential systems under impulsive control with state-dependent delay. *IEEE Trans. Autom. Control* **2017**, *63*, 306–311. [CrossRef]
16. Yang, H.; Wang, X.; Zhong, S.; Shu, L. Synchronization of nonlinear complex dynamical systems via delayed impulsive distributed control. *Appl. Math. Comput.* **2018**, *320*, 75–85. [CrossRef]
17. Lu, X.; Chen, N.; Wang, Y.; Qu, L.; Lai, J. Distributed impulsive control for islanded micro-grids with variable communication delays. *LET Control Theory Appl.* **2016**, *10*, 1732–1739. [CrossRef]
18. Jiang, L.; Yao, W.; Wu, Q.H.; Wen, J.Y.; Cheng, S.J. Delay-Dependent Stability for Load Frequency Control with Constant and Time-Varying Delays. *IEEE Trans. Power Syst.* **2012**, *27*, 932–941. [CrossRef]

19. Wen, S.; Yu, X.; Zeng, Z.; Wang, J. Event-triggering load frequency control for multi-area power systems with communication delays. *IEEE Trans. Ind. Electron.* **2015**, *63*, 1308–1317. [CrossRef]

20. Ramakrishnan, K.; Ray, G. Stability Criteria for Nonlinearly Perturbed Load Frequency Systems with Time-Delay. *IEEE J. Emerg. Sel. Top. Circuits Syst.* **2015**, *5*, 1–10. [CrossRef]

21. Zhao, S.; Gao, Z. Modified active disturbance rejection control for time-delay systems. *ISA Trans.* **2014**, *53*, 882–888. [CrossRef]

22. Mohamed, T.H.; Morel, J.; Bevrani, H.; Hiyama, T. Model predictive based load frequency control_design concerning wind turbines. *Int. J. Electr. Power Energy Syst.* **2012**, *43*, 859–867. [CrossRef]

23. Liu, F.; Xu, Z.; Li, Y.; Sidorov, D. Active disturbance rejection control based on EID compensation for LFC with communication delays. *IFAC J. Syst. Control* **2018**, *6*, 25–32. [CrossRef]

24. She, J.H.; Fang, M.; Ohyama, Y.; Hashimoto, H.; Wu, M. Improving Disturbance-Rejection Performance Based on an Equivalent-Input-Disturbance Approach. *IEEE Trans. Ind. Electron.* **2008**, *55*, 380–389. [CrossRef]

25. She, J.H.; Xin, X.; Yamaura, T. Analysis and Design of Control System with Equivalent-Input-Disturbance Estimation. In Proceedings of the 2006 IEEE International Conference on Control Applications, Munich, Germany, 4–6 October 2006; pp. 1463–1469.

26. Liu, R.J.; Wu, M.; Liu, G.P.; She, J.; Thomas, C. Active Disturbance Rejection Control Based on an Improved Equivalent-Input-Disturbance Approach. *IEEE/ASME Trans. Mechatron.* **2013**, *18*, 1410–1413. [CrossRef]

27. Liu, R.J.; Liu, G.P.; Wu, M.; Nie, Z.Y. Disturbance rejection for time-delay systems based on the equivalent-input-disturbance approach. *J. Frankl. Inst.* **2014**, *c1*, 3364–3377. [CrossRef]

28. Liu, F.; Li, Y.; Cao, Y.; She, J.; Wu, M. A Two-Layer Active Disturbance Rejection Controller Design for Load Frequency Control of Interconnected Power System. *IEEE Trans. Power Syst.* **2016**, *31*, 3320–3321. [CrossRef]

29. Gu, K.; Kharitonov, V.L.; Chen, J. *Stability of Time-Delay Systems*; Birkhauser: Boston, MA, USA, 2003.

30. Xia, Y.; Fu, M.; Shi, P. *Analysis and Synthesis of Dynamical Systems with Time-Delays*; Springer: Berlin/Heidelberg, Germany, 2009.

31. Pourmousavi, S.A.; Nehrir, M.H. Introducing Dynamic Demand Response in the LFC Model. *IEEE Trans. Power Syst.* **2014**, *29*, 1562–1572. [CrossRef]

32. Liu, J.; Yao, Q.; Liu, Y.; Yang, H. Wind Farm Primary Frequency Control Strategy Based on Wind & Thermal Power Joint Control. *Zhongguo Dianji Gongcheng Xuebao Proc. Chin. Soc. Electr. Eng.* **2017**, *37*, 3462–3469.

33. Aho, J.; Pao, L.; Fleming, P. An Active Power Control System for Wind Turbines Capable of Primary and Secondary Frequency Control for Supporting Grid Reliability. In Proceedings of the 51st AIAA Aerospace Sciences Meeting including the New Horizons Forum and Aerospace Exposition, Grapevine, TX, USA, 7–10 January 2013.

Article

Comparative Analysis of Identification Methods for Mechanical Dynamics of Large-Scale Wind Turbine

Jingchun Chu [1], Ling Yuan [1], Yang Hu [2,*], Chenyang Pan [2] and Lei Pan [1]

[1]   Guodian United Power Technology Company Limited, Beijing 102209, China
[2]   School of Control and Computer Engineering, North China Electric Power University, Beijing 102206, China
*   Correspondence: hooyoung@ncepu.edu.cn

Received: 30 July 2019; Accepted: 2 September 2019; Published: 5 September 2019

Abstract: With increasing size and flexibility of modern grid-connected wind turbines, advanced control algorithms are urgently needed, especially for multi-degree-of-freedom control of blade pitches and sizable rotor. However, complex dynamics of wind turbines are difficult to be modeled in a simplified state-space form for advanced control design considering stability. In this paper, grey-box parameter identification of critical mechanical models is systematically studied without excitation experiment, and applicabilities of different methods are compared from views of control design. Firstly, through mechanism analysis, the Hammerstein structure is adopted for mechanical-side modeling of wind turbines. Under closed-loop control across the whole wind speed range, structural identifiability of the drive-train model is analyzed in qualitation. Then, mutual information calculation among identified variables is used to quantitatively reveal the relationship between identification accuracy and variables' relevance. Then, the methods such as subspace identification, recursive least square identification and optimal identification are compared for a two-mass model and tower model. At last, through the high-fidelity simulation demo of a 2 MW wind turbine in the GH Bladed software, multivariable datasets are produced for studying. The results show that the Hammerstein structure is effective for simplify the modeling process where closed-loop identification of a two-mass model without excitation experiment is feasible. Meanwhile, it is found that variables' relevance has obvious influence on identification accuracy where mutual information is a good indicator. Higher mutual information often yields better accuracy. Additionally, three identification methods have diverse performance levels, showing their application potentials for different control design algorithms. In contrast, grey-box optimal parameter identification is the most promising for advanced control design considering stability, although its simplified representation of complex mechanical dynamics needs additional dynamic compensation which will be studied in future.

Keywords: wind turbine; dynamic modeling; grey-box parameter identification; subspace identification; recursive least squares; optimal identification

---

1. Introduction

Upsizing capacity of wind turbines to megawatt-class can increase wind energy capture and has great potential to reduce the LCOE (levelized cost of energy) per kilowatt hour for grid-connected wind power [1]. Yet, larger wind turbine causes higher fatigue load, greater rotational inertia and more challenging control difficulty [2,3]. Then, advanced control of the modern VSVP (variable-speed variable-pitch) wind turbine becomes very important, to fully utilize the multiple-degree-of-freedom control potentiality of blade pitches or sizable rotor while considering the multi-objectives such as maximizing conversion efficiency and alleviating fatigue load [4]. However, advanced control design usually depends heavily on a state-space model of the physical system [5,6]. Nowadays, it is still widely studied by industry and academia.

For wind turbine modeling, there are mainly three routes in the literature, shown in Table 1.

Table 1. Summary of different identification methods.

Methods	Model Forms	Applications	References
Data-driven input-output modeling (black-box)	Machine-learning-based modeling, such as neural network and deep learning neural network.	Dynamic modeling, anomaly identification	[7–10]
	Standard-model-set-based modeling, such as ARX, ARMAX, BJ and OE with LS or PEM criterion	Dynamic modeling, control design	[11–13]
	Subspace identification	Dynamic modeling	[14–16]
Mechanism-oriented Modeling (white-box)	Complex mechanism model	High-fidelity simulation	[17–19]
	Simplified mechanism model	Control verification of theoretic algorithms	[20–22]
Combination-based modeling (grey-box)	RLS parameter identification	Dynamic modeling	[23]
	Optimization-based parameter identification	Dynamic modeling	[24,25]

Data-driven modeling of input–output characteristics is a common way, including machine-learning [7–10], standard-model-set approximation [11–13] and subspace identification [14–16]. In [7–10], machine learning algorithms, such as artificial neural network (ANN), support vector machine (SVM), random forest and deep neural network were useful for black-box modeling. Yet, trial-and-error often existed to select neural network structure and to tune parameters. If optimization is used, computation burden is non-negligible. In [11–13], model structures such as ARX (auto-regressive), ARMAX (auto-regressive moving average), BJ (Box-Jenkins) and OE (Output-Error) were adopted for identification with LS (least square) or PEM (prediction-error method) criterion, where PRBS (pseudo-random binary excitation signal) was often used as input excitation. To get numerical solution, it usually has requirements about forms and amplitudes of excitation signals, open-loop or closed-loop structure and sampling period, etc. In [14–16], subspace identification via MOESP (multivariable output error state space) and PBSIDopt (prediction-based subspace identification) were studied for wind turbine modeling. A state-space model could be obtained via subspace identification, but the reconstructed states did not have physical meanings. Besides, this method suffered great influence from excitation signals, control system structure and sampling period, etc. In general, the above black-box identification methods only focused on the external input–output characteristics of system and bypass the internal operation mechanism. Thus, these black-box models had poor interpretability, unsuitable for control design with stability.

From the first-principle, high-fidelity models of wind turbines were obtained undoubtedly for digital design and simulation [17–19]. Yet, the model structures were too complex to be used for control design. Relatively, control design models were often simplified in continuous or discrete state-space form, just reflecting leading dynamics of the system [20–22]. For an actual wind turbine with high-order dynamics, only leading dynamics represented by simplified models are concerned for control design while how to accurately identify their parameters is a still key problem.

Utilizing a simplified mechanism model of a wind turbine, only the unknown parameters need to be identified where grey-box parameter identification is useful to combine mechanism-oriented modeling and data-driven modeling. In [23], a normal five-order model of DFIG (double-fed induction generator) in d–q axis was adopted and an RLS (recursive least square) algorithm was used to identify model parameters. Through transforming DFIG model into ARX structure, a LS identification problem was built. Then, excitation signals, sampling period and control structure were also concerned. In contrast, optimal parameter identification provided a different way, where an optimization problem is formed to search parameters with optimal objective. In [24], Cava et.al. brought in an evolutionary multi-objective optimization problem to identify parameters of symbolic model under preselected nonlinear structure, where only tower and rotor speed dynamics versus wind speed, pitch angle and aerodynamic torque were studied. In [25], grey-box parameter identification was applied on five-order DFIG model and one-mass drive-train model via particle swarm optimization (PSO). Due to the model structure preselected, only optimal parameters estimation was needed. It was more insensitive to sampling period and more accommodative to closed-loop and process noises. Identification of complexity and difficulty were also greatly reduced. Additionally, value ranges of parameters based on their physical meaning could be set as constraints for optimal search and then to guarantee reasonableness of identified models. It was very important for advanced control design to be based on state-space models considering multi-objectives such as robust H_∞ control or mixed H_2/H_∞ control for maximum power tracking and fatigue load alleviation of the wind turbine [2–4]. Furthermore, intuitiveness of the control design model was helpful not only for steady-state performance but also for transient performance.

In summary, grey-box parameter identification is more advantageous in the three routes to get state-space models, so it is adopted in this paper. Different types of methods will be compared to show their diverse performances and features for advanced control design.

In spite of what control algorithms are used, critical equipment such as aerodynamic system, tower system and drive-train system play the dominant roles. With consideration of multi- objectives for control design such as dispatched-power-point-tracking and fatigue load alleviation, the tower model along fore–aft direction and the two-mass model of drive-train are generally used due to their appropriate representations to mechanical dynamics. However, identification research of them has not been conducted, especially for parameter identification of the two-mass model under closed-loop structure without excitation experiment. In this paper, it will be carefully discussed. As a result, the main contributions of the paper are as follows:

- Structural identifiability analysis of the two-mass model under general closed-loop control conditions across the whole wind speed range is presented and is useful to judge feasibility of parameter identification in the closed-loop.
- Influence of identified data to identification performance is discussed based on mutual information analysis of identified variables and is helpful to select great identified datasets.
- Grey-box identifications for mechanical dynamics of a wind turbine, including the two-mass model of drive-train and the two-order damping model of tower-top, are studied where subspace identification, RLS and optimal identification are compared under wind scenarios with different turbulence intensities.
- Identification performances and features of different methods are analyzed from views of control design, providing guiding opinions for further improvement.

The rest of this paper is organized as follows. Section 2 introduces basic knowledge of the VSVP wind turbine including rationality analysis of the Hammerstein structure and selected models. Section 3 analyzes structural identifiability of models in closed-loop and nonlinear correlations among identified variables. The executable identification procedure is proposed in Section 4. Simulation and comparative analysis are shown in Section 5. Section 6 concludes the paper.

2. Basic Knowledge of The Modern VSVP Wind Turbine

In this section, rationality of the Hammerstein structure is analyzed to simplify the identification process. Meanwhile, the selected mechanism models are introduced for parameter identification.

2.1. Rationality of Hammerstein Structure

In [14], wind turbine dynamics were approximated by the connection of a static nonlinear aerodynamic mapping and a linear time invariant mechanical subsystem—the so-called Hammerstein structure. In this paper, this structure is also adopted, shown in Figure 1. From Figure 1, the modern

VSVP wind turbine usually has mechanical-side, including an aerodynamic system, drive-train system and tower system, and electrical-side, including an electrical system. The electrical-side has faster response than the mechanical-side. In this paper, only the mechanical-side identification is studied under the Hammerstein structure.

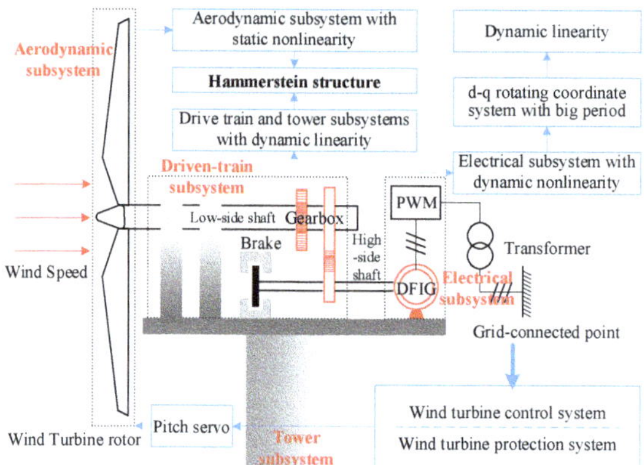

Figure 1. Modern variable-speed variable-pitch (VSVP) wind turbine under the Hammerstein structure.

During wind energy capture, the aerodynamic system contributes the main nonlinearity. Response to varying wind speeds, the three flexible blades and their vibration modes yield both low and high frequency dynamics. However, due to the low-pass filtering effects of the wind turbine rotor, the variables such as rotor thrust, torque and speed often manifest themselves. Their spectrum characteristics mainly concentrate in the low-frequency part. Then, the aerodynamic system can be represented by its static nonlinearity where aerodynamic thrust and torque are used for modeling.

For drive-train with gearbox, the two-mass model is capable enough to represent the required leading dynamics for control design. For the tower system, the motion in side–side direction, caused by drive-train and generator dynamics, is often neglected. Only the motion in fore–aft direction is concerned and can be modeled by a mass-spring-damper. The two-mass model and the mass-spring-damper model are tightly coupled with aerodynamic torque and thrust, respectively. Under the Hammerstein structure, aerodynamic torque and thrust become known nonlinear inputs and then drive-train and tower systems are simplified into linear dynamic models.

The electrical subsystem mainly consists of an asynchronous generator and PWM (pulse width modulation) inverter where the generator produces the main nonlinearity. Taking DFIG for example, although its nonlinearity in the five-order model is greatly weakened under d–q rotating-coordinate system, structural nonlinearity between generator rotor speed and current still exists. However, in a sampling period of generator rotor speed, it can be seen to be fixed, relative to the fast change of the current. Then, the five-order model becomes a linear time invariant one. Considering the rapid response ability of the electrical system, it is often simplified as a first-order inertial process.

In summary, the Hammerstein structure is reasonable for wind turbine modeling and helpful to reduce modeling complexity. Considering the different time-scales, only parameter identification of mechanical-side including drive-train and tower systems will be studied in this paper.

2.2. Selected Subsystem Models

Under the Hammerstein structure, models of the mechanical-side with required leading dynamics for advanced control design are selected and introduced.

2.2.1. Aerodynamic Subsystem

In theory, captured wind power of a wind turbine rotor is defined by:

$$P = \frac{1}{2}\rho\pi R^2 C_P V^3 \tag{1}$$

where ρ is air density; R is rotor radius; V is upstream wind speed; C_P is the dimensionless power coefficient, adjusted by rotor speed ω_r and pitch angle β. The nonlinear relationship among them is:

$$C_P = f(V, \omega_r, \beta) = f(\lambda, \beta) \tag{2}$$

where ω_r is turbine rotor speed; $\lambda = R\omega_r/V$ is tip-speed-ratio. Usually, C_P can be used as fitted three-dimensional surface or look-up table. Torque of turbine rotor is defined by:

$$T_r = \frac{1}{2}\rho\pi R^3 C_T(\lambda, \beta) V^2 = \frac{1}{2}\rho\pi R^3 \frac{C_P(\lambda, \beta)}{\lambda} V^2 = \frac{1}{2}\rho\pi R^5 \frac{C_P(\lambda, \beta)}{\lambda^3}\omega_r^2 = K_r\omega_r^2 \tag{3}$$

where $C_T(\lambda, \beta)$ is torque coefficient; K_r is torque gain. When optimal λ and C_P are adopted, the optimal K_r can be generated. Aerodynamic thrust to tower is defined by:

$$F_t = \frac{1}{2}\rho\pi R^2 C_F(\lambda, \beta) V^2 \tag{4}$$

where C_F is thrust coefficient.

Equations (1)–(4) represent static characteristics, different from that based on BEM (blade element momentum) theory and AEC (aero-elastic code) [26]. Front inflows are often low-pass filtered even facing rapidly changing inflow angles and wake effects. For a larger size megawatt wind turbine, low-pass filtering effect is more significant in yielding relatively steady characteristic.

2.2.2. Drive-Train Subsystem

Drive-train mainly consists of a low-speed shaft, gearbox and high-speed shaft. The torsional stiffness of them can be considered optionally. Total torsional flexibility in the low-speed side is caused by turbine rotor, tower top, rotor hub, low-speed stage of gearbox, yaw bearing roll and low-speed shaft, etc. In the high-speed side, total torsional flexibility involves high-speed stage of gearbox, high-speed shaft and generator rotor, etc. Equivalent inertias of both sides depend on the drive-train structure, which can be measured via the inertial measurement unit. Usually, equivalent inertia of the low-speed side is ten times more than that of the high-speed side while gearbox inertia is less than that of the high-speed side. Thus, gearbox inertia is often merged into the high-speed side. Then, the main inertia sources are simplified into two parts, yielding the two-mass model. Considering torsional flexibility of the low-speed shaft while taking the high-speed shaft to be rigid, the two-mass model is shown in Figure 2. Torque of the turbine rotor is an input from the aerodynamic subsystem. It drives the turbine rotor to rotate. Due to the existence of torsional flexibility of the low-speed shaft, the transmitted torque of the low-speed shaft is different from torque of the turbine rotor, yielding a torsional angle by different rotation speeds and angle displacements of two-ends. Torque of the generator rotor is a reaction input to balance torque of the turbine rotor. Because the high-speed shaft is taken to be rigid, two-ends of the high-speed shaft have the same torque, rotation speed and angle displacement. The mathematical description of the two-mass model is represented by:

$$\begin{cases} J_r\dot{\omega}_r = T_r - T_{ls} \\ T_{ls} = A_{stif}^{ls}(\delta_r - \delta_{ls}) + B_{damp}^{ls}(\omega_r - \omega_{ls}) \\ J_g\dot{\omega}_g = T_{hs} - T_g \end{cases} \tag{5}$$

where J_r and J_g are equivalent inertias of the low-speed and high-speed sides; T_{ls} and T_{hs} are mechanical torques of the low-speed and high-speed shafts in the gear-box; T_g is generator reaction torque; $A_{stif}{}^{ls}$ and $B_{damp}{}^{ls}$ are equivalent stiffness and damping coefficients of the low-speed shaft; ω_{ls} and ω_g are speed of the low-speed shaft and generator rotor; δ_r, δ_{ls}, δ_g and δ_{hs} are angle displacements of the turbine rotor, low-speed shaft, generator rotor and high-speed shaft, respectively. Note that $d(\delta_r)/dt = \omega_r$, $d(\delta_g)/dt = \omega_g$ and $\delta_{hs} = \delta_g$, $\omega_{hs} = \omega_g$, $N_{gear} = T_{ls}/T_{hs} = \delta_{hs}/\delta_{ls} = \omega_{hs}/\omega_{ls}$. N_{gear} is the gearbox ratio. Then, the equivalent of Equation (5) is:

$$\begin{cases} T_r = J_r\dot{\omega}_r + T_{shaf} \\ T_{shaf} = A_{stif}\left(\delta_r - \dfrac{\delta_g}{N_{gear}}\right) + B_{damp}\left(\omega_r - \dfrac{\omega_g}{N_{gear}}\right) \\ -T_g = J_g\dot{\omega}_g - \dfrac{T_{shaf}}{N_{gear}} \end{cases} \tag{6}$$

where $T_{shaf} = T_{ls}$, $A_{stif} = A_{stif}{}^{ls}$ and $B_{damp} = B_{damp}{}^{ls}$; T_{shaf} is internal shaft torque; J_r and J_g can be measured by inertial measurement unit. Torsional flexibility, represented by A_{stif} and B_{damp}, needs to be identified. Effective wind speed can be usually estimated via a LIDAR (Light Detection and Range) system in nacelle or soft sensing methods. Giyanani et al. [27] studied the estimation of effective wind speed using LIDAR data. Assume that C_P, C_T and C_F are known from design parameters. Then, aerodynamic torque T_r can be calculated.

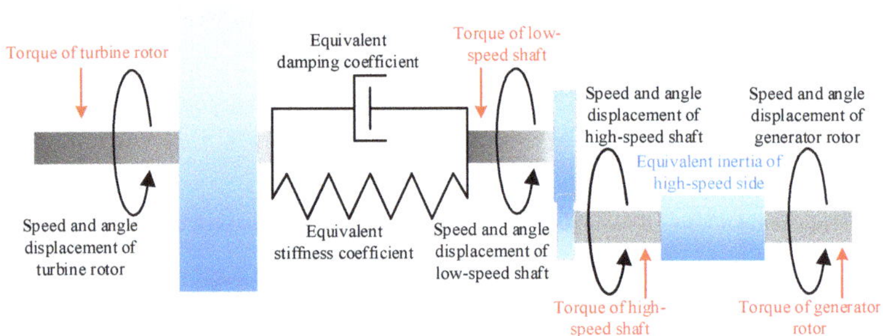

Figure 2. Two-mass model.

In addition to the two-mass model, drive-train can also be modeled as one-mass model or three-mass model. If low-speed and high-speed shafts are both deemed to be rigid, the one-mass model can be obtained. If torsional flexibility is considered, the three-mass model can be obtained. Because the mechanical load is paid more and more attention, the one-mass model is too simplified. The two-mass model is sufficient to represent drive-train dynamic while the three-mass model appears to be complex and unnecessary [26].

2.2.3. Tower Subsystem

Tower dynamics mainly include two aspects, bending motions of fore–aft direction and side–side direction. The former is often caused by wind thrust on turbine rotor. Because the turbine rotor, linked to the low-speed shaft of drive-train, is mounted in the nacelle on a flexible tower top, effects of wind on the rotor can be transmitted to the tower top. Additionally, the latter is caused by coupling effects of drive-train and generator dynamics, etc.

Serving for control design, dominant tower dynamics are considered. In this case, only the tower's first bending mode of fore–aft direction is modeled by an equivalent mass-spring-damper system, where tower torsion deformation, yawing effects and higher bending modes are neglected. As a result, a sufficiently simplified tower model can be obtained as follows:

$$M_t\ddot{d} + D_t\dot{d} + K_t d = F_t \tag{7}$$

where M_t, D_t and K_t are mass, damping and spring coefficients; d is fore–aft motion displacement.

3. Identifiability Analysis under Closed-Loop Condition

3.1. Control Strategy of Modern VSVP Wind Turbine

Below rated wind speed, OTC (optimal torque control) strategy is adopted by many industrial wind turbines and seen as a variable-speed controller. Generator torque demand is given by Equation (3) using optimal K_r. Measuring rotor speed, generator torque demand can be derived to control wind turbine operating around optimal operation points.

Above rated wind speed, generator rotor speed is limited below the maximum or rated value, which can be taken as set point. To track it, a variable-pitch controller is activated to regulate pitch angle. Due to the nonlinear dynamics of a wind turbine, in practice, a gain scheduling PI (proportional-integral) controller is generally used, changing with quasi-steady wind speeds.

Both the two control strategies and their switching mechanism are shown in Figure 3.

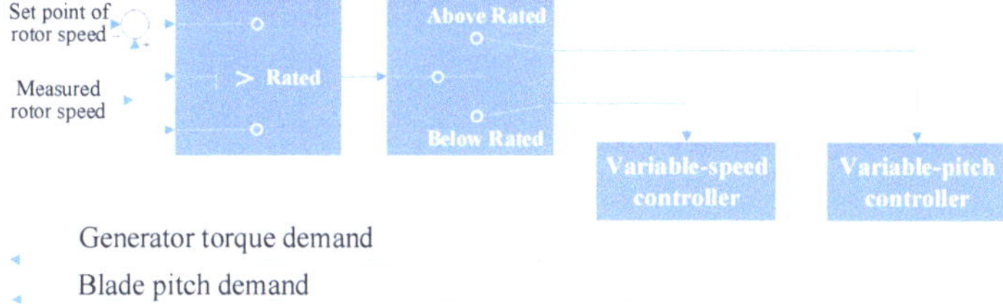

Figure 3. VSVP control loops of modern wind turbine.

3.2. Structural Identifiability Analysis of Closed-Loop Control System

The modern wind turbine works in closed-loop which cannot be cut off during operation. As a result, closed-loop identification is necessary. There are mainly two ways for this—direct and indirect methods. Direct identification uses input–output data of controlled object even under closed-loop condition. The indirect method identifies the augmented closed-loop system and deduces the model of the controlled object based on the known controller structure and parameters. When the input–output data of the forward channel is measurable and disturbance signal exists in the feedback channel, the direct method should be preferentially considered as it is more convenient than indirect method.

A typical closed loop is shown in Figure 4. $P(z^{-1})$ is transfer function in forward channel; $C(z^{-1})$ is that in feedback channel; $R(k)$ is set point signal; $E(k)$ is deviation signal between set point and output; $U(k)$ is controller output with noise; $Z(k)$ is process output with noise; $v(k)$ and $w(k)$ are noise signals. A closed-loop system is identifiable if any one of the following conditions is satisfied [28]:

1. If feedback channel is linear and invariant while no disturbance signal exists and set-point is constant, the identifiable condition is that cancellation between zeros and poles of closed-loop transfer function does not happen, caused by model structure of the feedback channel. Meanwhile, $n_p \geq n_b$, $n_q \geq n_a - d_t$ where n_p and n_q are denominator and numerator order of $C(z^{-1})$; n_a and n_b are those of $P(z^{-1})$; d_t is time delay between output and input of the forward channel.
2. If continuous excitation signal with enough order exists on the feedback channel and is irrelevant with the noise on the forward channel, a closed-loop system is structurally identifiable.
3. If controller is time-varying or nonlinear, a closed-loop is structurally identifiable.

4. If the controller switches among several regulation laws, closed-loop is structurally identifiable. For a multi-variable control system, it requires $l \geq 1 + r/m$, where l is number of feedback controllers; r and m are input and output dimensions of closed-loop.

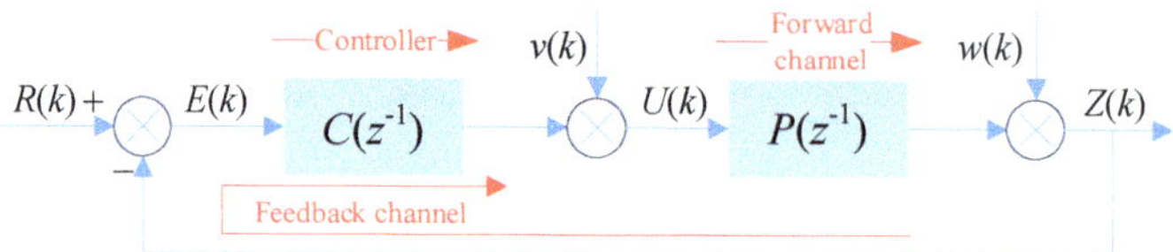

Figure 4. Typical closed-loop control system.

All the above conditions provide a decision support to determine whether the system can be successfully identified from a view of structure. This is a fundamental step before identification.

For a VSVP wind turbine, closed-loops below and above rated wind speed are shown in Figure 5. In Figure 5, the variable-speed controller is nonlinear and variant with rotor speed. The variable-pitch controller is usually a gain scheduling proportional-integral controller which is linear and variant. No continuous excitation signals exist on feedback channels. Both of them are single-variable control systems. When wind speed varies around the rated value, the two control strategies with their regulation laws also switch. Thus, condition 3 can be fulfilled and condition 4 can be partially fulfilled. As a result, the closed-loop of a modern VSVP wind turbine is structurally identifiable. This suggests that system or parameter identification can be executed for the controlled objects on forward channel under a closed-loop condition using direct or indirect methods.

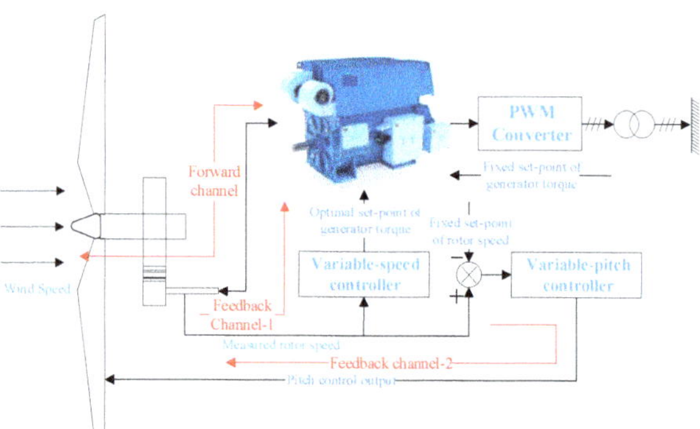

Figure 5. Closed control loops below and above rated wind speed.

3.3. Correlation Analysis of Identified Data

Structural identifiability analysis is qualitative, judging whether the closed-loop identification can succeed. This suggests that enough dynamic information may be contained in the data samples of system variables. In execution, appropriate data samples need to be selected according to their correlations from a relative point. Usually, correlations among system variables are nonlinear. Linear correlation analysis methods such as Person coefficient, linear regression and path analysis, are unsuitable while MI (mutual information) [29] becomes an efficient tool. It is defined by probability density of data without requirements to data distribution and both linear and nonlinear correlation can be analyzed. Firstly, define information entropy $H(X)$ of a time series X as:

$$H(X) = -\int p(x) \log p(x) \mathrm{d}x \tag{8}$$

where $H(X)$ is also called Shannon entropy [28]; $p(X)$ is probability density function of X. Then, the numerical form of uncertainty degree of X can be used to represent its information content. Greater entropy value means stronger information uncertainty. On this basis, MI between X and Y is:

$$I(X,Y) = \iint p_{XY}(x,y) \log \frac{p_{XY}(x,y)}{p_X(x)p_Y(y)} dxdy \tag{9}$$

where $p_{XY}(x, y)$ is joint probability density function. If X and Y are independent, $I(X, Y) = 0$; if X and Y are highly dependent, $I(X, Y)$ becomes greater.

In order to keep proper correlations among variables, MI values can be calculated for quantitative analysis. In this paper, $p_{XY}(x, y)$ are calculated by kernel density estimation [30]. As a result, appropriate data can be determined for identification.

4. Execution of Identification

4.1. Data Acquisition and Preprocessing

For parameter identification, the sampling period is mainly concerned during data acquisition. It depends on maximum or cut-off frequency of the identified object. However, before identification, the frequencies are difficult to be determined, which can be estimated according to the power spectral densities of signals. Of course, trial and error can be used. To avoid missing important information, a smaller sampling period should be preferred to try.

For the data acquired from field, inappropriate low or high frequency components may exist, which affects the parameter identification effect. Low-frequency component mainly refers to slowly changing drift or trend characteristics. High-frequency component refers to interference noise. In this case, band-pass filter design is a feasible solution.

Additionally, if identification results are sensitive to the starting point of time, zero initialization of sampled data may be necessary but optional. Subtracting the mean value of several initial points by the acquired signal can realize zero initialization.

4.2. Optimization Criterion

The two-mass model of drive-train can be seen as a two-input two-output system. Inputs are aerodynamic torque and generator torque. Outputs are generator rotor speed and internal shaft speed. The tower model in fore–aft direction can be seen as a one-input one-output system. The input is aerodynamic thrust. The output is displacement of fore–aft direction. Then, parameter identification of multiple-input multiple-output system appears. Adopting the weighted loss function, optimization criterion in the LS form is defined by:

$$O_{\min} = \sum_{j=1}^{N_{Out}} \sum_{i=1}^{N} \alpha_j \left[y_j(i) - \widetilde{y}_j(i) \right]^2 \tag{10}$$

where N_{Out} is number of outputs; N is number of sampled data points; T_s is sampling period; α_j is weighting coefficient of each output; y is measured output; $\widetilde{y}$ is estimated output. To evaluate the fitting degree between estimated output and measured output, the fitting percent is used

$$r_{FitPercent} = \frac{\|\widetilde{y} - y\|}{\|y - \overline{y}\|} \times 100\% \tag{11}$$

where $\overline{y}$ represents average value. It is actually the normalized root mean squared error.

4.3. Brief Introduction of Identification Algorithms

Facing the advanced control design of the wind turbine in future, only the identification methods applicable for the state-space model are adopted. Among the black-box identification methods, subspace identification is very representative with great performance and convenient executability under state-space structure. Then, for grey-box identification methods using selected mechanism state-space models, there are mainly two categories: RLS and optimal parameter identification. As a result, the three methods are adopted for comparison in this paper.

4.3.1. Subspace Identification

From a statistical perspective, CVA (canonical variate analysis) method was early proposed by Larimore et al. [31]. Based on a geometric concept, Verhaegen et al. [32] proposed MOESP method. Using ARX estimation, SSARX (space state autoregressive exogenous) was given by Jansson et al. [33]. In the simulation, all three methods can be used as candidate methods. For parameter identification of the two-mass model or tower model, the subspace identification method with less MSE (mean squared error) and better fitting percent in Equation (11) can be selected to compare with the other classes of methods. They are applicable for a multi-input multi-output system using Equation (10) as objective.

4.3.2. Grey-Box Parameter Identification via Recursive Least Squares Algorithm

RLS algorithm is used for online parameter estimation of single-input single-output or multi-input single-output system. For the two-mass model (Equation (6)), approximate discretization is used where $\dot{\omega}_r \approx (\omega_r(k+1) - \omega_r(k))/T$, $\dot{\omega}_g \approx \big(\omega_g(k+1) - \omega_g(k)\big)/T$ (T is sampling period). Taking J_r, J_g, A_{stif} and B_{damp} as identified parameters, the discrete model is:

$$\begin{bmatrix} T_r(k) \\ T_g(k) \end{bmatrix} = \begin{bmatrix} \frac{\omega_r(k+1)-\omega_r(k)}{T} & 0 & \delta_r(k) - \frac{\delta_g(k)}{N_{\text{gear}}} & \omega_r(k) - \frac{\omega_g(k)}{N_{\text{gear}}} \\ 0 & \frac{\omega_g(k+1)-\omega_g(k)}{T} & \frac{1}{N_{\text{gear}}}\Big(\delta_r(k) - \frac{\delta_g(k)}{N_{\text{gear}}}\Big) & \frac{1}{N_{\text{gear}}}\Big(\omega_r(k) - \frac{\omega_g(k)}{N_{\text{gear}}}\Big) \end{bmatrix} \begin{bmatrix} J_r \\ J_g \\ A_{\text{stif}} \\ B_{\text{damp}} \end{bmatrix} \tag{12}$$

Taking M_t, D_t and K_t as identified parameters, discrete model of the tower model (Equation (7)) is

$$F_t(k) = \begin{bmatrix} \frac{d(k+1)-2d(k)+d(k-1)}{T^2} \\ \frac{d(k+1)-d(k)}{T} \\ d(k) \end{bmatrix}^T \begin{bmatrix} M_t \\ D_t \\ K_t \end{bmatrix} \tag{13}$$

Based on the discrete model, the infinite-history recursive estimation algorithms [34] via forgetting factor are used for parameter identification.

4.3.3. Grey-Box Parameter Identification via Optimization Algorithm

Using Equations (6) and (7), discrete grey-box models can be established via zero-order holder. Then, optimization algorithms are adopted to optimize model parameters, using Equation (10) as loss function. The methods such as Gauss-Newton, Levenberg-Marquardt and trust-region- reflective are candidate optimization algorithms. Then, dominant dynamics of drive-train and tower systems can be approximated. Optimal parameter identification procedure mainly includes:

Step 1: Set sampling period, acquire data samples and execute data preprocessing.
Step 2: Set weighting coefficients of optimization criterion.
Step 3: Estimate and set initial domains of identified parameters.
Step 4: Identify unknown parameters using optimization algorithms.
Step 5: Test identified dynamics of each input–output channel. If huge deviation happens, adjust certain steps and repeat Step 4. If required performance is fulfilled, the procedure is over.

5. Simulation

Based on the benchmark simulation demo for a 2 MW wind turbine in GH Bladed software, no excitation signals are applied on the controller or the output. Meanwhile, wind speeds with different turbulence intensities are produced to stimulate the benchmark demo where operation data under different wind scenarios can be acquired. The defaulted controllers in the benchmark demo are used as shown in Figures 3 and 5. The controllers can arbitrarily switch with the varying wind speed.

5.1. Parameters Setting

GH Bladed is a high-fidelity software for wind turbine simulation. In this software, a benchmark simulation demo of a 2 MW wind turbine with gear-box and DFIG is adopted. It uses the controllers and control strategies shown in Figures 3 and 5. The main parameters are listed as follows: rated capacity (2 MW), radius (40 m), hub height (61.5 m), gear-box ratio (83.33), cut-in wind speed (4 m/s), rated wind speed (10m/s), cut-out wind speed (25 m/s), generator inertia (60 kg·m^2), stiffness coefficient (1.6×10^8 Nm/rad), damping coefficient (2.5×10^5 Nm·s/rad), variable speed controller (OTC), variable pitch controller (gain scheduling PI controller). Concretely, for this benchmark 2 MW wind turbine model, the horizontal-axis three blades are controlled integratively. For each blade, the blade length is 38.75 m where thickness to chord ratio, Reynolds number, pitching moment center and deployment angle are set as 21%, 2×10^6, 25% and 0°. More information of the turbine blades such as blade structure and aerofoil parameters are shown in Figure A1, Tables A1 and A2 in the 'Appendix A' part.

5.2. Scenarios Setting and Simulation

Under the closed-loop condition, four types of wind scenarios with different turbulence intensities are produced to stimulate the benchmark demo of a 2 MW wind turbine along the whole wind speed range. For each wind scenario, operation data are acquired for identification.

Wind speeds excite wind turbine dynamics. To evaluate volatility of wind speed, turbulence intensity is used, calculated as follows:

$$I_{\text{tur}} = \frac{\sigma_{\text{std}}}{V_{\text{mean}}} \tag{14}$$

$$\sigma_{\text{std}} = \sqrt{\frac{1}{N_{\text{WS}} - 1} \sum_{i=1}^{N_{\text{WS}}} (V_i - V_{\text{mean}})^2} \tag{15}$$

where V_{mean} is mean wind speed in time window T_{WS} with sampling period T_{sp}; $N_{\text{WS}} = T_{\text{WS}}/T_{\text{SP}}$ is number of sampling points; σ_{std} is standard deviation; I_{tur} is turbulence intensity; V_i is sampled values. According to IEC 61400-1 [35], three grades of turbulence intensity such as A(0.16), B(0.14) and C(0.12) are given. In this section, using wind-generation function of GH Bladed [19], wind scenarios with different mean values and turbulence intensities are generated, shown in Figure 6. Equations (14) and (15) were used to calculate mean values and turbulence intensities with different time windows, shown in Table 2. Different wind scenarios' settings have some differences but the whole wind speed range and the whole intensity range of turbulence can both be tested. Finally, the identified datasets with different operation information caused by different wind scenarios and switched controllers can be obtained for further grey-box parameter identification and validation.

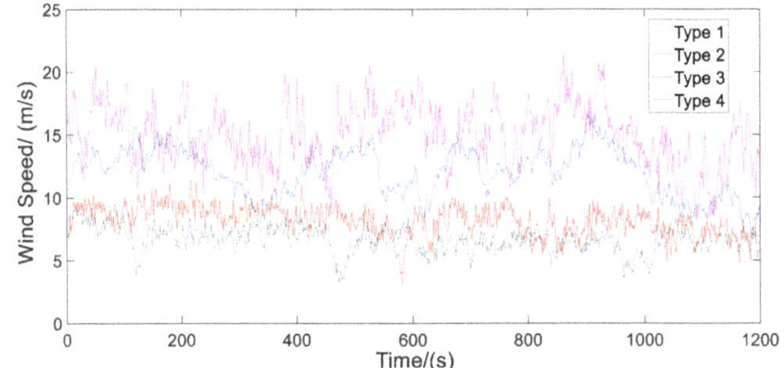

Figure 6. Wind scenarios.

Table 2. Characteristic parameters of wind scenarios.

Scenario Types	Mean Wind Speed (m/s)/Turbulence Intensity		
	0–10 min	10–20 min	0–20 min
Type 1	7.02/0.15	6.30/0.13	6.66/0.15
Type 2	8.62/0.13	7.72/0.14	8.17/0.15
Type 3	12.31/0.13	11.68/0.18	12.00/0.16
Type 4	15.23/0.15	14.76/0.17	15.00/0.16

Using wind speeds in Figure 6 as inputs of a 2 MW wind turbine model in GH Bladed, operation data can be acquired. Using the benchmark demo of a 2 MW wind turbine in the GH Bladed software, the virtual sensors and their measuring points are shown in Figure 7. For the identification of the two-mass model, the measuring points include 'Nominal wind speed at hub position', 'Rotor speed', 'Rotor azimuth angle', 'Generator speed', 'Generator azimuthal position', 'Generator torque', 'Aerodynamic torque' and 'Low speed shaft torque'. For the identification of the tower model, the measuring points include 'Nominal wind speed at hub position', 'Tower F_x–Tower station height = 60 m', 'Nacelle x-deflection', 'Nacelle x-velocity' and 'Nacelle x-acceleration'.

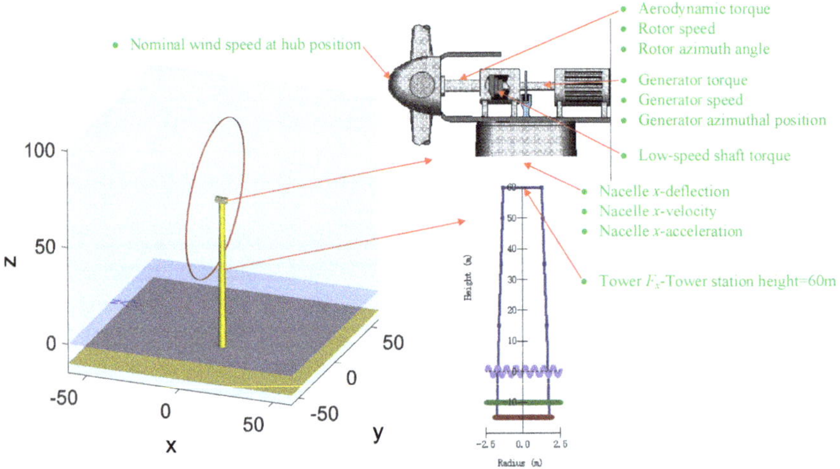

Figure 7. Diagram of wind turbine structure and sensor locations.

Then, MI values of two-mass model and tower model were calculated under different wind scenarios, shown in Tables 3 and 4. In Table 3, divided by the maximum value of each row, normalized MI values of each row can be obtained.

Table 3. Mutual information (MI) values of two-mass model under different wind scenarios.

Scenarios	MI Values of Two-Mass Model									
	T_r–T_g	T_r–ω_r	T_r–ω_g	T_r–T_{shaf}	T_g–ω_r	T_g–ω_g	T_g–T_{shaf}	ω_r–ω_g	ω_r–T_{shaf}	ω_g–T_{shaf}
Type 1	0.4856	0.4555	0.4552	0.5442	3.0705	3.1005	2.5336	5.5664	2.2144	2.2122
Type2	0.6511	0.5651	0.5643	0.7068	2.2504	2.2601	2.5607	5.2716	1.9310	1.9292
Type 3	1.3548	0.2447	0.2417	1.4188	0.2759	0.2751	3.2777	3.2968	0.2760	0.2746
Type 4	0.1381	0.0479	0.0467	0.2241	0.0590	0.0601	0.6571	2.2888	0.0656	0.0658

Scenarios	Normalized MI Values of Two-Mass Model									
	T_r–T_g	T_r–ω_r	T_r–ω_g	T_r–T_{shaf}	T_g–ω_r	T_g–ω_g	T_g–T_{shaf}	ω_r–ω_g	ω_r–T_{shaf}	ω_g–T_{shaf}
Type 1	0.0872	0.0818	0.0818	0.0978	0.5516	0.5570	0.4552	1	0.3978	0.3974
Type2	0.1235	0.1072	0.1070	0.1341	0.4269	0.4287	0.4858	1	0.3663	0.3660
Type 3	0.4109	0.0742	0.0733	0.4304	0.0837	0.0834	0.9942	1	0.0837	0.0833
Type 4	0.0603	0.0209	0.0204	0.0979	0.0258	0.0263	0.2871	1	0.0287	0.0287

Table 4. MI values of tower model under different wind scenarios.

Scenarios		Type 1	Type 2	Type 3	Type 4
MI values	F_t–z	1.3200	1.3080	1.1340	1.0938

According to MI values in Table 3, ω_r–ω_g has the highest correlation. Correlation of T_g–T_{shaf} is relatively high, reflecting the similar dynamics of T_g and T_{shaf}. Correlation of T_g–ω_r is similar to T_g–ω_g. Correlations of ω_r–T_{shaf} and ω_g–T_{shaf} are similar. Correlation of T_r–ω_r is similar to that of T_r–ω_g. Obviously, due to different control strategies and controllers being used below or above rated wind speed, MI values are different under different wind scenarios. From Figure 5, the two-mass model locates at the forward channel of a closed loop. Scenarios Type 1 and Type 2 work below rated wind speed where OTC controller acts. Scenarios Type 3 and Type 4 work above rated wind speed where variable-pitch controller acts. Depending on the acquired data, different data sets yield different MI values. Then, through analysis, connection of control loops and information contained in the acquired data can be preliminarily understood according to control strategies in Figure 5.

The tower model (Equation (7)) is an open-loop system. As a result, in Table 4, correlations between F_t and z are very similar under different wind scenarios.

From the view of basic principle, subspace identification is different from grey-box identification. For subspace identification, estimating the Kalman state vector from the input–output data is the first step and using the Kalman state vector to reconstruct the state space model is the second step. Among them, how to estimate the Kalman state vector from the input–output data is a critical step for subspace identification. For grey-box identification including RLS identification and optimal identification, estimating the model parameters from input–output data based on the selected state-space model structure is the first step and using the obtained state-space model to estimate the Kalman state vector is the second step. The above three identification methods have different basic principles.

Firstly, due to the identified two-mass model including no-self-balancing channel, which is unstable, exploration of different identification methods for direct identification of the two-mass model without excitation experiment under closed-loop condition is mainly studied as follows.

For two-mass model (Equation (6)), define $x = [\omega_r, \omega_g, \Delta\delta]^T$, $u = [T_r, T_g]^T$, $y = [\omega_g, T_{shaf}]^T$. Then, the state-space equation and transfer function matrix can be obtained, shown as following

$$M_{\text{Two-mass}} = \begin{bmatrix} \dfrac{b_1 s^2 + b_2 s + b_3}{s(s^2 + a_1 s + a_2)} & \dfrac{-(b_4 s + b_5)}{s(s^2 + a_1 s + a_2)} \\ \dfrac{b_6 s + b_7}{s^2 + a_1 s + a_2} & \dfrac{b_8 s + b_9}{s^2 + a_1 s + a_2} \end{bmatrix} \tag{16}$$

where $a_1 = B_{damp}/J_r + B_{damp}/[J_g\ (N_{gear})^2]$, $a_2 = A_{stif}/J_r + A_{stif}/[J_g\ (N_{gear})^2]$; $b_1 = 1/J_r$, $b_2 = B_{damp}/[J_r J_g\ (N_{gear})^2]$, $b_3 = A_{stif}/[J_r J_g\ (N_{gear})^2]$; $b_4 = N_{gear}\ b_2$, $b_5 = N_{gear}\ b_3$; $b_6 = B_{damp}\ b_1$, $b_7 = A_{stif}\ b_1$; $b_8 = J_r\ b_4$, $b_9 = J_r\ b_5$. Obviously, for the input–output channels from T_r to ω_g and T_g to ω_g, transfer functions with no-self-balancing ability are obtained. For the input–output channels from T_r to T_{shaf} and T_g to T_{shaf}, transfer functions with self-balancing ability are obtained. In execution, no-self-balancing object is difficult to be identified.

For the two-mass model, the measured data are obtained from operation of the two megawatts wind turbine in GH Bladed using wind speed inputs in Figure 6. Under the Hammerstein structure, inputs are turbine rotor torque, T_r, and generator reaction torque, T_g, for two-mass model. Outputs are generator rotor speed, ω_g, and internal shaft torque, T_{shaf}. Then, measured data can be used for identification. Utilizing the identified state-space model with T_r and T_g as inputs, the estimated generator rotor speed and internal shaft torque can be acquired. Herein, all the measured data for identification are produced by the wind turbine model in GH Bladed with high-order nonlinear characteristics. As a result, comparisons of the three identification methods under different wind scenarios are shown in Table 5 and Figures 8–11.

Table 5. Comparison of identification methods for two-mass model.

Scenarios	Methods	$n_x, J_r, J_g, A_{stif}, B_{damp}$	MSE	Fit-Percent	Stability
Type 1	Subspace (MOESP)	$n_x = 10$	7.964×10^7	-175.6%; 96.3%	$-0.090 \pm 0.951i$; $-0.284 \pm 0.934i$; $-0.010 \pm 0.627i$; $-0.866 \pm 0.365i$; 0.995; -0.437; Instability.
	RLS	3.544×10^6; 1; 2.388×10^5; 1.010×10^7	4.214×10^7 3.073×10^{11}	$-2.785 \times 10^5\%$ -130.064%	-1.458×10^3; -2.363×10^{-2}; 0; Instability.
	Optimization	1.381×10^5; 31.8; 2.101×10^8; 1.655×10^5	8.932×10^8	$-1.37 \times 10^5\%$; 87.67%	$-0.974 \pm 49.719i$; -0. Stability.
Type 2	Subspace (MOESP)	$n_x = 10$	8.883×10^7	53.76%; 89.39%	$0.784 \pm 0.521i$; $0.259 \pm 0.841i$; $-0.286 \pm 0.824i$; $-0.712 \pm 0.592i$; -0.570; 0.994; Instability.
	RLS	5.776×10^6; 1; 1.028×10^5; -2.706×10^6	7.466×10^6; 5.547×10^{10}	-206.69%; -165.14%	390.057; 0; 0.038; Instability.
	Optimization	2.063×10^5; 19.43; 2.062×10^8; 1.702×10^5	7.774×10^8	$-1.69 \times 10^4\%$; 72.67%	$-1.043 \pm 50.266i$; -0; Stability.
Type 3	Subspace (MOESP)	$n_x = 10$	3.502×10^7	90.11%; 89.53%	$-0.878 \pm 0.384i$; $-0.244 \pm 0.939i$; $0.435 \pm 0.796i$; $0.112 \pm 0.974i$; 0.997; 0.680; Instability.
	RLS	6.366×10^6; 1; 6.920×10^4; -4.531×10^6	3.187×10^6; 2.443×10^{10}	$-1.527 \times 10^4\%$ -176.42%	653.17; 0; 0.0152; Instability.
	Optimization	2.197×10^5; 20.06; 2.153×10^8; 1.901×10^5	4.559×10^8	$-1.768 \times 10^4\%$; 63.28%	$-1.115 \pm 50.241i$; -0; Stability.
Type 4	Subspace (MOESP)	$n_x = 10$	3.733×10^8	-449%; 70.33%	$-0.789 \pm 0.405i$; $-0.400 \pm 0.791i$; $0.084 \pm 0.722i$; $0.922 \pm 0.362i$; -0.845; 0.979; Instability.
	RLS	4.686×10^7; 1; 2.767×10^5; 1.620×10^7	4.650×10^7; 3.395×10^{11}	$-4.955 \times 10^5\%$; -794.72%	-2.336×10^3; 0; -0.017; Instability.
	Optimization	2.578×10^5; 15.89; 1.954×10^8; 2.087×10^5	2.742×10^9	$-2.734 \times 10^5\%$; 33.22%	$-1.351 \pm 50.274i$; -0; Stability.

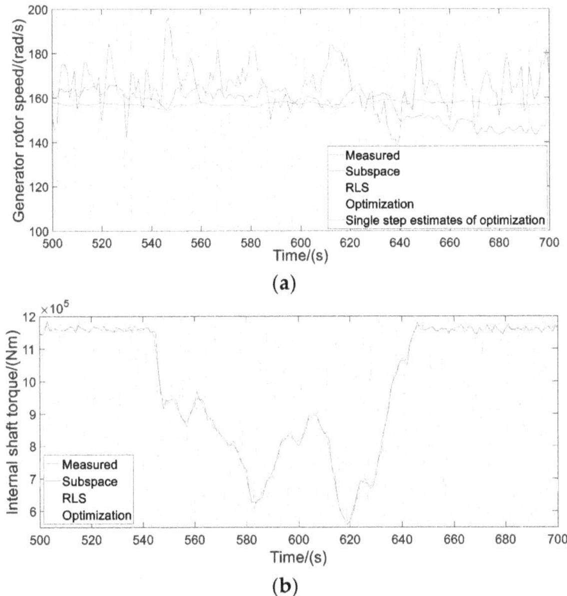

(a)

(b)

Figure 8. (**a**) Comparison of generator rotor speed under scenario Type 1; (**b**) Comparison of internal shaft torque under scenario Type 1.

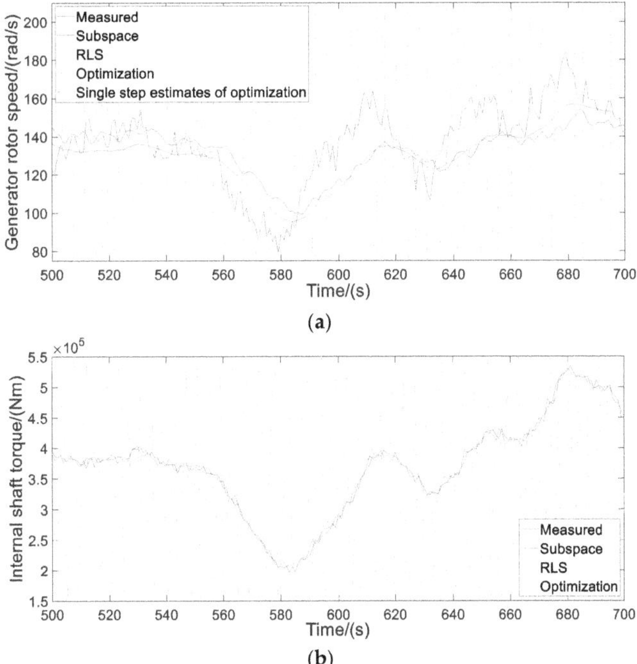

(a)

(b)

Figure 9. (**a**) Comparison of generator rotor speed under scenario Type 2; (**b**) Comparison of internal shaft torque under scenario Type 2.

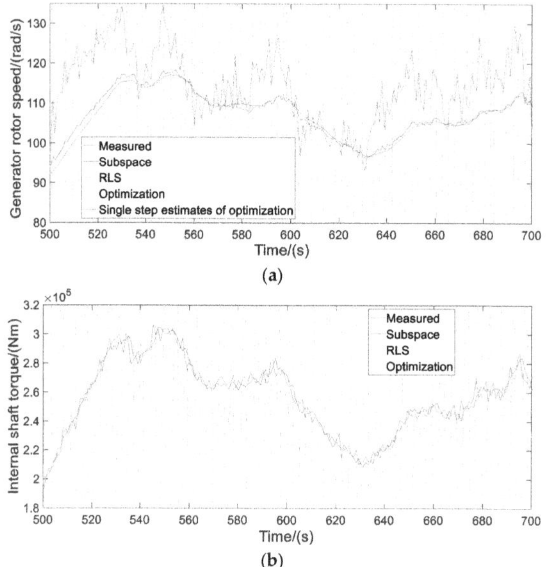

Figure 10. (a) Comparison of generator rotor speed under scenario Type 3; (b) Comparison of internal shaft torque under scenario Type 3.

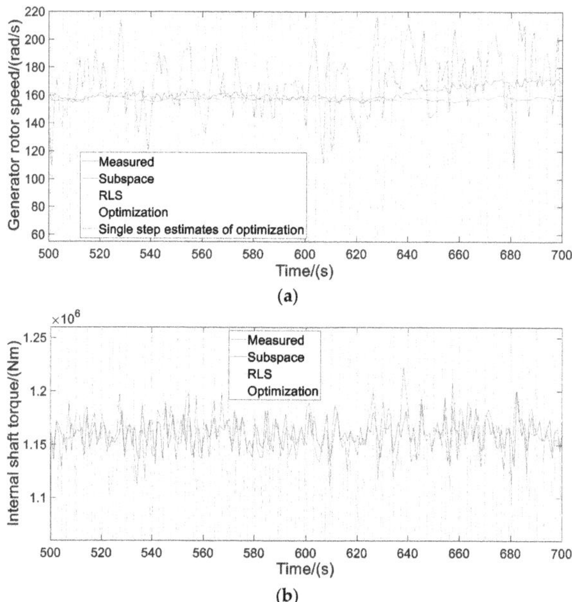

Figure 11. (a) Comparison of generator rotor speed under scenario Type 4; (b) Comparison of internal shaft torque under scenario Type 4.

Subspace identification—a data-driven black-box identification method—is mainly realized via robust numerical solution with QR decomposition or singular value decomposition. All the subspace identification methods assume that input signal and process noise signal are independent with each other. It suggests that the system does not have a feedback loop and all poles strictly fall in the left-half-plane, and thus subspace identification can have unbiased estimation for open-loop system. Under closed-loop condition, input signal becomes dependent with process noise, so the closed-loop system identification via subspace identification is difficult. If directly using subspace identification for the closed-loop condition, a biased estimate or even error estimate may happen. Then, eliminating the dependence between input signal and process noise becomes a basic step when using subspace identification for a closed-loop system. In this case, if sufficient excitation can be designed and injected into the closed-loop system and higher signal-to-noise ratio can be yielded, better results can be obtained by subspace identification for the closed-loop system. Its identification accuracy highly depends on the design of the test signal. However, in this paper, system identification under closed-loop condition without excitation signal is mainly studied for the wind turbine model in GH Bladed with high-order nonlinear characteristics. In theory, the identified system will remain biased or with error if no-excitation signal is used under closed-loop condition. From Table 5 and Figures 8–11, these methods all have limited identification performance. Through comparison, subspace identification via MOESP has the best performance, showing its good adaptability to closed-loops. Even for the mixed input–output channels, where T_g–ω_g and T_r–ω_g represent no-self-balancing channels and T_g–T_{shaf} and T_r–T_{shaf} represent self-balancing channels, subspace identification has good fit-percent using weighted loss function. However, reconstructed states of the discrete high-order system have no-physical-meanings and are disadvantage for control design to improve dynamic performance of the system. Besides, the identified system cannot guarantee its autonomous stability. As a result, subspace identification only yields a good black-box result from the view of fit-percent based on its high-order discrete model structure, whereas the identified system is still unbiased both in theory and in practice.

RLS identification—a black-box or grey-box identification method—is a generalization to LS identification and is mainly solved by numerical solution algorithm. Herein, the drive-train model structure has been selected, so a grey-box problem is formed. RLS identification can have unbiased estimation for an open-loop system without feedback loop where the process noise signal and input signal are independent. Yet, the drive-train model is a no-self-balancing and instable model where closed-loop identification is necessary. Under closed-loop condition, the system needs to be fully excited where a suitable excitation signal should be designed and injected into the system to reduce or even eliminate the dependence between process noise signal and input signal. In this case, the RLS algorithm may have a better identification result. Herein, the simplest identification condition is provided for RLS identification where no-excitation-signal is injected into the closed-loop of drive-train and the data yielded by the closed-loop system are directly acquired for identification. In theory, only unbiased or error estimate may be obtained. RLS identification is just applicable for single output system. Based on Equation (12), T_r is used to estimate parameters. Fit-percent and estimation error of the identification results for the two-mass model performs the worst, as shown in Table 5 and Figures 8–11. Besides, for the identified parameters, negative values often appear. These parameters does not always keep consistent with their physical meanings and cannot guarantee autonomous stability of system. Using the estimated parameters, the simulated outputs deviate a lot to the measured outputs and error estimate happens in practice. As a result, for the direct identification of the two-mass model using operation data under the closed-loop condition, RLS identification is difficult to have proper results. It reflects that RLS method hugely relies on the acquired data and excitation experiment.

Grey-box identification utilizes the physical structure and input–output data to establish a system model where a priori physical knowledge and model uncertainty caused by process noise are both considered. It has obvious advantages for capturing the natural behavior of the actual system. Usually, the selected model structure needs include the noise terms to describe model uncertainty. Yet, many unknown noise signals are difficult to be estimated and improper estimation may yield bad results. Herein, the two-mass model is selected as the identified model structure whereas its noise terms are not added, so the approaching ability of the identified two-mass model may limited to the high-order nonlinearity. Then, based on the two-mass model structure, optimization search algorithm is used to find the parameters with the best estimation performance. Shown in Table 5 and Figures 8–11, grey-box optimization identification has moderate performance. Although it is applicable for multi-output system, it performs poorly for the mixed channels. Because estimation bias always exists and even diverges for the no-self balancing channels, the fit-percent and estimation error have poor performance. To display the performance, single-step estimates of grey-box optimization identification are also shown in Figure 8a, Figure 9a, Figure 10a, and Figure 11a. It can be found that the single-step estimates are convergent and show their trend tracking abilities to the measured data. For the self-balancing channels, only limited estimation performance is obtained due to the simplified model structure without noise terms and weighted identification with the no-self-balancing channels. Even so, grey-box identification has obvious advantages such as better adaptability to identified data for direct identification under closed-loop condition and is a more convenient identification procedure. Besides, it has the most important advantage that the estimated parameters have physical meaning and they can guarantee autonomous stability of the identified system. As a result, even if the grey-box identification may have limited accuracy, it is still very useful for control design. Especially, through choosing better identified data, a better result of grey-box identification can be yielded. For wind scenarios with higher MI values, better identification performance can be obtained. In Type 1, Type 2 and Type 3, T_r–ω_g, T_g–ω_g, T_r–T_{shaf} and T_g–T_{shaf} have higher MI values than that of Type 4, so the identification results of Type 1, Type 2 and Type 3 have less MSE and better fit-percent.

In summary, for parameter identification of the two-mass model, subspace identification gets the best fit-percent. It is suitable for the predictive control design while unsuitable for many advanced optimal control design methods, such as H_∞ control and linear quadratic regulator control, etc. RLS identification is proved to be invalid for the two-mass model under the simple identification condition in this paper. Optimization identification has moderate performance. It is suitable for all kinds of control design algorithms. However, dynamics based on the simplified model structure need to be compensated. Besides, a novel method is needed to realize accurate identification for the no-self-balancing channel. Comparison analyses of above identification results for the two-mass model are briefly summarized and shown in Table 6.

Table 6. Comparison analyses of identification results for two-mass model.

Identification Methods	Identification Condition Herein	Identification Results	Reason Analysis to Identification Results
Subspace (MOESP)	Direct identification under closed-loop condition without excitation signal	Valid: best fit-percent; instability	Black-box high-order model; closed-loop condition without excitation; no-self-balancing channel
RLS		Invalid: worst fit-percent; instability	Grey-box low-order model without process noise term; closed-loop condition without excitation; no-self-balancing channel
Optimization		Valid: moderate fit-percent; stability	Grey-box low-order model without process noise term; no-self-balancing channel

Then, open-loop identification of the tower model is studied as follows. It can directly show the identification performance of different methods.

For tower model (Equation (7)), define $x = [d, \dot{d}]^{\text{T}}$, $u = F_t$, $y = d$. Then, state-space equation and transfer function of this two-order spring-damping model can be obtained, shown as follows:

$$M_{\text{Tower}} = \frac{1}{s(M_t s + D_t) + K_t} \tag{17}$$

For the input–output channel from F_t to d, it has self-balancing ability.

Under open-loop condition, performance of the three identification methods are strongly dependent on that whether the identified data contain fully excited dynamic information. Herein, only a very simple identification condition is set without excitation signal, so the identification result only depends on the acquired operation data.

For tower model, comparisons of three identification methods are shown in Table 7 and Figures 12–15. RLS identification and grey-box optimization identification both performs better than the subspace identification. It suggests the great advantage of priori model structure information to accurately approach the high-order nonlinear dynamics of tower system. Meanwhile, the arbitrarily acquired data may contain insufficient excited dynamic information which affect the estimation of Kalman state vector form the input–output data, so subspace identification performs the worst. It also suggests that subspace identification is a data-driven black-box identification method and its performance strongly depends on the acquired data and excitation signal.

Table 7. Comparison of identification methods for tower model.

Scenarios	Methods	n_x, M_t, D_t, K_t	MSE	Fit-Percent	Stability
Type 1	Subspace (CVA)	$n_x = 7$	2.64×10^4	55.53%	$0.698 \pm 0.633i; -0.573 \pm 0.620i;$ $-0.071 \pm 0.596i; 0.191;$ Instability.
	RLS	$4.54 \times 10^4; -1.68 \times 10^4;$ $1.09 \times 10^6.$	8.70×10^{-5}	74.55%	$0.185 \pm 4.896i;$ Instability.
	Optimization	$1.16 \times 10^4; 1 \times 10^3;$ $1.099 \times 10^6.$	1.38×10^{-4}	67.93%	$-0.129 \pm 9.728i;$ Stability.
Type 2	Subspace (CVA)	$n_x = 3$	2.72×10^{-4}	55.66%	$-0.318 \pm 0.589i; 0.535;$ Instability.
	RLS	$2.88 \times 10^4; -8.93 \times 10^3;$ $1.13 \times 10^6.$	1.04×10^{-4}	72.61%	$0.155 \pm 6.25i;$ Instability.
	Optimization	$1.155 \times 10^4; 1 \times 10^3;$ 1.151×10^6	1.49×10^{-4}	67.33%	$-0.130 \pm 9.98i;$ Stability.
Type 3	Subspace (CVA)	$n_x = 7$	2.20×10^{-4}	48.09%	$-0.549 \pm 0.575i; -0.907 \pm 0.203i;$ $0.707 \pm 0.482i; 0.921;$ Instability.
	RLS	$4.14 \times 10^4; -7.06 \times 10^3;$ 1.18×10^6	9.24×10^{-5}	66.52%	$0.0852 \pm 5.343i;$ Instability.
	Optimization	$1.239 \times 10^4; 1 \times 10^3;$ 1.257×10^6	1.63×10^{-4}	55.53%	$-0.121 \pm 10.072i;$ Stability.
Type 4	Subspace (CVA)	$n_x = 3$	6.72×10^{-4}	38.35%	$0.623; -0.150 \pm 0.347i;$ Instability.
	RLS	$1.00 \times 10^4; 2.12 \times 10^3;$ 1.07×10^6	1.79×10^{-4}	68.35%	$-0.106 \pm 10.358i;$ Stability.
	Optimization	$1.116 \times 10^4; 1 \times 10^3;$ 1.127×10^6	3.07×10^{-4}	58.48%	$-0.134 \pm 10.049i;$ Stability.

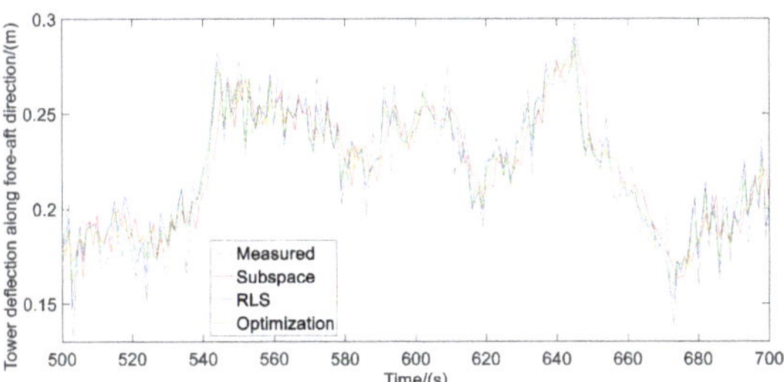

Figure 12. Comparison of tower deflection under scenario Type 1.

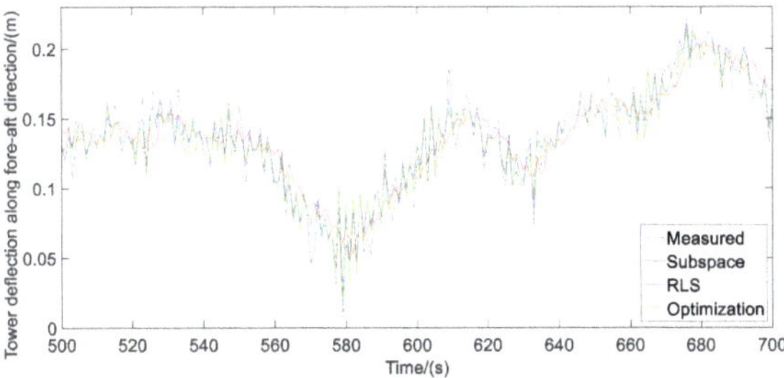

Figure 13. Comparison of tower deflection under scenario Type 2.

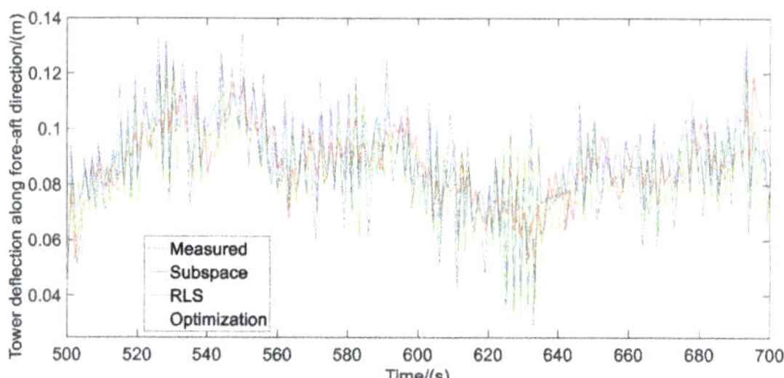

Figure 14. Comparison of tower deflection under scenario Type 3.

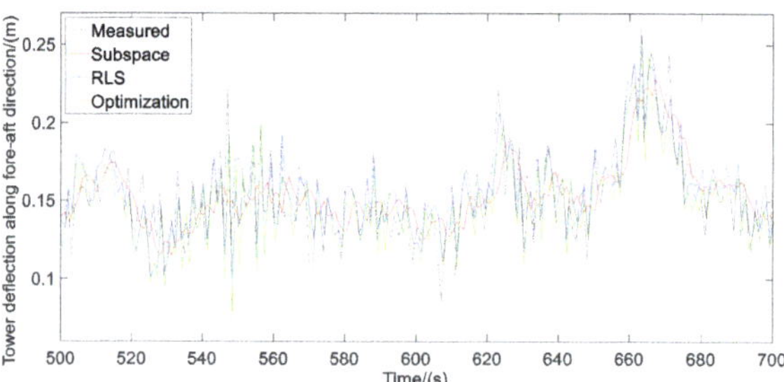

Figure 15. Comparison of tower deflection under scenario Type 4.

Especially, RLS achieves the best fit-percent, while parameters incongruent with its physical meaning may be identified and they cannot guarantee autonomous stability of the system. Subspace identification via CVA has the worst fit-percent, which just uses input–output data and reconstructs the states with no-physical meanings. It also cannot guarantee autonomous stability of the identified system. In contrast, optimization identification has moderate fit-percent, but it is sensitive to the set of initial domains of parameters. Its advantage is that the identified parameters have clear physical meanings and they can guarantee autonomous stability of the identified system. However, the simplified model structure limits its representational ability to the nonlinear tower dynamics. For the subsequent applications, subspace identification is suitable to test identifying feasibility of acquired data in the early stage. RLS is suitable to determine the rough range of identified parameters in the middle stage. Then, optimization identification is suitable for the final stage to obtain parameters with physical meaning while guaranteeing autonomous stability of the open-loop tower system. Of course, if simplified mechanism model is adopted and the fit-percent does not perform well, reasonable dynamic compensation can be added. Besides, because the two-order spring-damping model of the tower system lies in an open-loop, similar MI values are obtained under different wind scenarios. As a result, identification performance under different wind scenarios are very similar, too. Comparison analyses of the above identification results for the tower model are briefly summarized and shown in Table 8.

Table 8. Comparison analyses of identification results for tower model.

Identification Methods	Identification Condition Herein	Identification Results	Reason Analysis to Identification Results
Subspace (CVA)	Direct identification under open-loop condition without excitation signal	Valid: worst fit-percent; instability	Black-box high-order model; insufficient excitation of operation data
RLS		Valid: best fit-percent; instability	Grey-box low-order model without process noise term
Optimization		Valid: moderate fit-percent; stability	Grey-box low-order model without process noise term

The main purpose of this paper is to explore the identification performance of three representative methods under a simple condition with minimal complexity. At last, application potential and features of these identification methods are summarized and shown in Figure 16. Overall, using the minimal complexity for identification, grey-box optimization identification has the best adaptability to obtain a valid state-space model with physical meaning and guaranteed stability. It is very helpful for subsequent control design to improve both the stable and dynamic performance of the system. The work herein will be part of a hybrid modeling method in future where the biased estimation of the identified state-space model will be compensated by a non-parametric machine learning algorithm. As a result, the hybrid modeling method can efficiently balance modeling complexity and accuracy where the state-space model with high-precision approaching ability to high-order nonlinearity can be obtained.

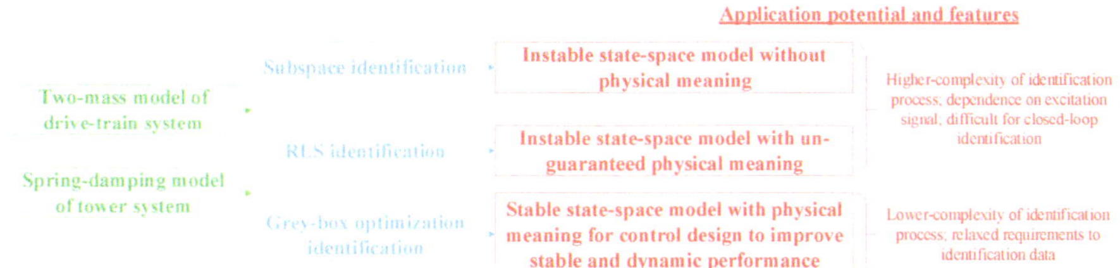

Figure 16. Application potential and features of different identification methods.

6. Conclusions

In this paper, identification of wind turbine mechanical dynamics is studied under non-excitation condition. Identification performance under different wind scenarios is tested using three types of methods. For VSVP wind turbine, the drive-train subsystem is structurally identifiable under closed-loop condition and direct identification is feasible for the two-mass model. Through MI calculation, nonlinear correlations among identified variables can not only validate whether the linkage of these variables are consistent with the control loop but also reveal the relationship between identified data and identification performance. It can be found that higher correlations among identified variables can yield better identification performance. In contrast, state-space model from optimal identification can reflect the physical meaning of parameters and natural stability of the identified system which is important for advanced control algorithms. In summary, grey-box optimal identification shows its feasibility to identify complex wind turbine dynamics and its great potential in advanced control design. Additionally, the limitation of the simplified mechanism model to represent complex and practical dynamics should be paid attention. In future, dynamic compensation to the identified simple mechanism model based on machine-learning will be studied. It may balance modeling complexity and difficulty and would be attractive to the application of digital-twin modeling of wind turbines.

Author Contributions: The individual contributions of the authors are provided as follows: conceptualization, J.C., L.Y. and Y.H.; methodology, Y.H., C.P. and L.P.; validation, Y.H.; writing—review and editing, Y.H.; visualization, Y.H. and C.P.; supervision, L.P.; funding acquisition, J.C. and L.Y.

Funding: This research was funded by 'the research on Intelligent Control Technology of Wind Turbine (Guodian United Power Technology Company Limited), grant number 17001', 'Hebei Provincial Key Research and Development Program, grant number 18214316D'.

Conflicts of Interest: The authors declare no conflict of interest.

Abbreviations

AEC	aero-elastic code
ANN	artificial neural network
ARMAX	auto-regressive moving average
ARX	auto-regressive
BEM	blade element momentum
BJ	Box-Jenkins
CVA	canonical variate analysis
DFIG	double-fed induction generator
LCOE	levelized cost of energy
LIDAR	light detection and range
LS	least square
MI	mutual information
MOESP	multivariable output error state space
MSE	mean squared error
OE	output-error
OTC	optimal torque control
PBSIDopt	prediction-based subspace identification
PEM	prediction-error method
PI	proportional-integral
PRBS	pseudo-random binary excitation signals
PSO	particle swarm optimization
PWM	pulse width modulation
RLS	recursive least square
SSARX	space state autoregressive exogenous
SVM	support vector machine
VSVP	variable-speed variable-pitch

Appendix A

Detail information of blade is shown as following.

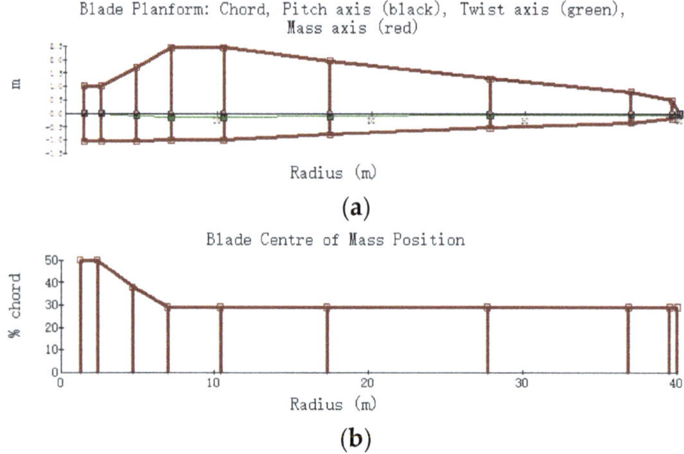

Figure A1. Blade information: (**a**) blade geometry; (**b**) mass and stiffness of blade.

Table A1. Design parameters of blades.

Distance from root (m)	0	3.44444	5.74074	9.18519	16.0741	26.4074	35.5926	38.75
Centre of mass (%)	50	38	29	29	29	29	29	29
Mass per unit length (kg/m)	1084.77	277.356	234.212	209.558	172.577	103.546	55.4713	24.6539
Flapwise stiffness (Nm/rad)	7.472×10^9	1.408×10^9	8.341×10^8	5.561×10^8	2.058×10^8	2.954×10^7	2.259×10^6	3127.98
Edgewise stiffness (Nm/rad)	7.472×10^9	2.085×10^9	1.425×10^9	1.286×10^9	5.648×10^8	1.216×10^8	2.433×10^7	8167.51

Table A2. Aerofoil parameters of blades.

Angle of incidence (deg)	−180	−141.27	−100.65	−19	0	21.61	115.85	150.54	180
Lift coefficient	−0.088	0.716	0.094	−0.354	0.449	0.806	−0.539	−0.674	−0.088
Drag coefficient	0.036	0.772	1.167	0.191	0.007	0.288	1.094	0.466	0.036
Pitch coefficient	−0.041	0.362	0.313	0.042	−0.079	−0.097	−0.363	−0.301	−0.041

References

1. Kumar, Y.; Ringenberg, J.; Depuru, S.S.; Devabhaktuni, V.K.; Lee, J.W.; Nikolaidis, E.; Andersen, B.; Afjeh, A. Wind energy: Trends and enabling technologies. *Renew. Sust. Energ. Rev.* **2016**, *53*, 209–224. [CrossRef]
2. Eduardo, J.N.M.; Alex, M.A.; da Silva, N.; Nadege, S.B. A review on wind turbine control and its associated methods. *J. Clean. Prod.* **2018**, *174*, 945–953. [CrossRef]
3. Njiri, J.G.; Dirk, S. State-of-the-art in wind turbine control: Trends and challenges. *Renew. Sust. Energ. Rev.* **2016**, *60*, 377–393. [CrossRef]
4. Yuan, Y.; Tang, J. On advanced control methods toward power capture and load mitigation in wind turbines. *Engineering* **2017**, *3*, 494–503. [CrossRef]
5. Menezes, E.J.N.; Araújo, A.M.; Rohatgi, J.S.; Foyo, P.M.G. Active load control of large wind turbines using state-space methods and disturbance accommodating control. *Energy* **2018**, *150*, 310–319. [CrossRef]
6. Gao, R.; Gao, Z. Pitch control for wind turbine systems using optimization, estimation and compensation. *Renew. Energy* **2016**, *91*, 501–515. [CrossRef]
7. Kelouwani, S.; Agbossou, K. Nonlinear model identification of wind turbine with a neural network. *IEEE Trans. Energy Conver.* **2004**, *19*, 607–612. [CrossRef]
8. Sun, P.; Li, J.; Wang, C.; Lei, X. A generalized model for wind turbine anomaly identification based on SCADA data. *Appl. Energ.* **2016**, *168*, 550–567. [CrossRef]
9. Prajapat, G.P.; Senroy, N.; Kar, I.N. Wind turbine structural modeling consideration for dynamic studies of DFIG based system. *IEEE T Sustain. Energ.* **2017**, *8*, 1463–1472. [CrossRef]
10. Wang, L.; Zhang, Z.; Long, H.; Xu, X.; Liu, R. Wind turbine gearbox failure identification with deep neural networks. *IEEE T. Ind. Inf.* **2017**, *13*, 1360–1368. [CrossRef]
11. Carcangiu, C.E.; Balaguer, I.F.; Kanev, S.; Rossetti, M. Closed-loop system identification of Alstom 3MW wind turbine. In *Conference Proceedings of the Society for Experimental Mechanics Series*; Springer: New York, USA, 2011; Volume 4, pp. 121–128.
12. Iribas, M.; Landau, I.D. Identification in closed loop operation of models for collective pitch robust controller design. *Wind Energy* **2013**, *16*, 383–399. [CrossRef]
13. Ye, H.Y. *Control Technology of Wind Turbines*, 3rd ed.; China Machine Press: Beijing, China, 2015; ISBN 9787111500179.
14. Van der Veen, G.J.; van Wingerden, J.W.; Fleming, P.A.; Scholbrock, A.K.; Verhaegen, M. Global data-driven modeling of wind turbines in the presence of turbulence. *Control Eng. Pract.* **2013**, *21*, 441–454. [CrossRef]
15. Loh, C.H.; Loh, K.J.; Yang, Y.S.; Hsiung, W.Y.; Huang, Y.T. Vibration-based system identification of wind turbine system. *Struct. Control Hlth.* **2016**, *24*, 1–19. [CrossRef]
16. Aarden, P. System Identification of the 2B6 Wind Turbine-a Regularized PBSIDopt Approach. Master's Thesis, Delft University of Technology, Delft, The Netherlands, 2017.
17. Jonkman, J.M.; Buhl, M.L., Jr. *FAST User's Guide*; Technical Report: NREL/TP-500-38230; National Renewable Energy Laboratory: Golden, CA, USA, 2005.
18. Hsu, M.C.; Akkerman, I.; Bazilevs, Y. Finite element simulation of wind turbine aerodynamics: Validation study using NREL Phase VI experiment. *Wind Energy* **2014**, *17*, 461–481. [CrossRef]

19. DNV GL-Energy. *Theory Manual Bladed*; Garrad Hassan & Partners Ltd: Bristol, UK, 2014.

20. Bianchi, F.D.; De Battista, H.; Mantz, R.J. *Wind Turbine Control Systems: Principles, Modelling and Gain Scheduling Design*; Springer: London, UK, 2007; ISBN 1-84628-492-9.

21. Munteanu, I.; Bratcu, A.I.; Cutululis, N.A.; Ceangă, E. *Optimal Control of Wind Energy Systems-Towards Global Approach*; Springer: London, UK; ISBN 978-1-84800-079-7.

22. Ebadollahi, S.; Saki, S. Wind turbine torque oscillation reduction using soft switching multiple model predictive control based on the gap metric and Kalman filter estimator. *IEEE T. Ind. Electron.* **2018**, *65*, 3890–3898. [CrossRef]

23. Belmokhtar, K.; Ibrahim, H.; Merabet, A. Online parameter identification for a DFIG driven wind turbine generator based on recursive least squares algorithm. In Proceedings of the 2015 IEEE 28th Canadian Conference on Electrical and Computer Engineering (CCECE), Halifax, NS, Canada, 3–6 May 2015.

24. Cava, W.L.; Danai, K.; Spector, L.; Fleming, P.; Wright, A.; Lackner, M. Automatic identification of wind turbine models using evolutionary multi-objective optimization. *Renew. Energ.* **2016**, *87*, 892–902. [CrossRef]

25. Pan, X.; Ju, P.; Wu, F.; Jin, Y. Hierarchical parameter estimation of DFIG and drive train system in a wind turbine generator. *Front. Mech. Eng.* **2017**, *12*, 367–376. [CrossRef]

26. Ackermann, T. *Wind Power in Power System*, 2nd ed.; John Wiley & Sons: West Sussex, UK, 2012; ISBN 9780470974162.

27. Giyanani, A.; Bierbooms, W.A.A.M.; van Bussel, G.J.W. Estimation of rotor effective wind speeds using autoregressive models on Lidar data. *J. Phys. Conf. Ser.* **2016**, *753*, 1–14. [CrossRef]

28. Yang, C.; Zhang, C. *System Identification and Adaptive Control*; Chongqing University Press: Chongqing, China, 2003; ISBN 9787562428176.

29. Shannon, C.E. A mathematical theory of communication. *ACM Sigmobile Mob. Comput. Commun. Rev.* **2001**, *5*, 3–55. [CrossRef]

30. Rosenblatt, M. Remarks on some nonparametric estimates of a density function. *Annals Math. Stat.* **1956**, *27*, 832–837. [CrossRef]

31. Larimore, W.E. Canonical variate analysis in identification, filtering and adaptive control. In Proceedings of the 29th IEEE Conference on Decision and Control, Honolulu, HI, USA, 5–7 December 1990.

32. Verhaegen, M. Identification of the deterministic part of MIMO state space models. *Automatica* **1994**, *30*, 61–74. [CrossRef]

33. Jansson, M. Subspace identification and ARX modeling. In *Proceedings of the 13th IFAC Symposium on System Identification*; IFAC: Rotterdam, The Netherlands, 2003; Volume 36, Issue 16, pp. 1585–1590.

34. Ljung, L. *System Identification: Theory for the User*; Prentice-Hall: Upper Saddle River, NJ, USA, 1986; ISBN 0138816409.

35. IEC. *Wind Turbines-Design Requirements*; Technical Report No. 61400; International Electro-Technical Commission: Geneva, Switzerland, 2005.

MDPI
St. Alban-Anlage 66
4052 Basel
Switzerland
Tel. +41 61 683 77 34
Fax +41 61 302 89 18
www.mdpi.com

Energies Editorial Office
E-mail: energies@mdpi.com
www.mdpi.com/journal/energies